Applied Data Analysis and Modeling for Energy Engineers and Scientists

T. Agami Reddy

Applied Data Analysis and Modeling for Energy Engineers and Scientists

Springer

T. Agami Reddy
The Design School and School of Sustainability
Arizona State University
PO Box 871605, Tempe, AZ 85287-1605
USA
reddyta@asu.edu

ISBN 978-1-4419-9612-1 e-ISBN 978-1-4419-9613-8
DOI 10.1007/978-1-4419-9613-8
Springer New York Dordrecht Heidelberg London

Library of Congress Control Number: 2011931864

Printed on acid-free paper

Springer is part of Springer Science+Business Media (www.springer.com)

Thou must bear the sorrow that thou claimst to heal;
The day-bringer must walk in darkest night.
He who would save the world must share its pain.
If he knows not grief, how shall he find grief's cure?

Savitri—Sri Aurobindo

In loving memory of my father and grandmother

Preface

A Third Need in Engineering Education

At its inception, engineering education was predominantly *process oriented*, while engineering practice tended to be predominantly *system oriented*[1]. While it was invaluable to have a strong fundamental knowledge of the processes, educators realized the need to have courses where this knowledge translated into an ability to design systems; therefore, most universities, starting in the 1970s, mandated that seniors take at least one design/capstone course. However, a third aspect is acquiring increasing importance: *the need to analyze, interpret and model data*. Such a skill set is proving to be crucial in all scientific activities, none so as much as in engineering and the physical sciences. How can data collected from a piece of equipment be used to assess the claims of the manufacturers? How can performance data either from a natural system or a man-made system be respectively used to maintain it more sustainably or to operate it more efficiently? Such needs are driven by the fact that system performance data is easily available in our present-day digital age where sensor and data acquisition systems have become reliable, cheap and part of the system design itself. This applies both to experimental data (gathered from experiments performed according to some predetermined strategy) and to observational data (where one can neither intrude on system functioning nor have the ability to control the experiment, such as in astronomy). Techniques for data analysis also differ depending on the size of the data; smaller data sets may require the use of "prior" knowledge of how the system is expected to behave or how similar systems have been known to behave in the past.

Let us consider a specific instance of observational data: once a system is designed and built, how to evaluate its condition in terms of design intent and, if possible, operate it in an "optimal" manner under variable operating conditions (say, based on cost, or on minimal environmental impact such as carbon footprint, or any appropriate pre-specified objective). Thus, data analysis and data driven modeling methods as applied to this instance can be meant to achieve certain practical ends—for example:

(a) verifying stated claims of manufacturer;
(b) product improvement or product characterization from performance data of prototype;
(c) health monitoring of a system, i.e., how does one use quantitative approaches to reach sound decisions on the state or "health" of the system based on its monitored data?
(d) controlling a system, i.e., how best to operate and control it on a day-to-day basis?
(e) identifying measures to improve system performance, and assess impact of these measures;
(f) verification of the performance of implemented measures, i.e., are the remedial measures implemented impacting system performance as intended?

[1] Stoecker, W.F., 1989. *Design of Thermal Systems*, 3rd Edition, McGraw-Hill, New York

Intent

Data analysis and modeling is not an end in itself; it is a well-proven and often indispensable aid for subsequent decision-making such as allowing realistic assessment and predictions to be made concerning verifying expected behavior, the current operational state of the system and/or the impact of any intended structural or operational changes. It has its roots in statistics, probability, regression, mathematics (linear algebra, differential equations, numerical methods,...), modeling and decision making. Engineering and science graduates are somewhat comfortable with mathematics while they do not usually get any exposure to decision analysis at all. Statistics, probability and regression analysis are usually squeezed into a sophomore term resulting in them remaining "a shadowy mathematical nightmare, and ... a weakness forever"[2] even to academically good graduates. Further, many of these concepts, tools and procedures are taught as disparate courses not only in physical sciences and engineering but in life sciences, statistics and econometric departments. This has led to many in the physical sciences and engineering communities having a pervasive "mental block" or apprehensiveness or lack of appreciation of this discipline altogether. Though these analysis skills can be learnt over several years by some (while some never learn it well enough to be comfortable even after several years of practice), what is needed is a textbook which provides:

1. A review of classical statistics and probability concepts,
2. A basic and unified perspective of the various techniques of data based mathematical modeling and analysis,
3. an understanding of the "process" along with the tools,
4. a proper combination of classical methods with the more recent machine learning and automated tools which the wide spread use of computers has spawned, and
5. well-conceived examples and problems involving real-world data that would illustrate these concepts within the purview of specific areas of application.

Such a text is likely to dispel the current sense of unease and provide readers with the necessary measure of practical understanding and confidence in being able to interpret their numbers rather than merely generating them. This would also have the added benefit of advancing the current state of knowledge and practice in that the professional and research community would better appreciate, absorb and even contribute to the numerous research publications in this area.

Approach and Scope

Forward models needed for system simulation and design have been addressed in numerous textbooks and have been well-inculcated into the undergraduate engineering and science curriculum for several decades. It is the issue of data-driven methods, which I feel is inadequately reinforced in undergraduate and first-year graduate curricula, and hence the basic rationale for this book. Further, this book is not meant to be a monograph or a compilation of information on papers i.e., not a literature review. It is meant to serve as a textbook for senior undergraduate or first-year graduate students or for continuing education professional courses, as well as a self-study reference book for working professionals with adequate background.

[2] Keller, D.K., 2006. *The Tao of Statistics*, Saga Publications, London, U.K

Applied statistics and data based analysis methods find applications in various engineering, business, medical, and physical, natural and social sciences. Though the basic concepts are the same, the diversity in these disciplines results in rather different focus and differing emphasis of the analysis methods. This diversity may be in the process itself, in the type and quantity of data, and in the intended purpose of the analysis. For example, many engineering systems have low "epistemic" uncertainty or uncertainty associated with the process itself, and, also allow easy gathering of adequate performance data. Such models are typically characterized by strong relationships between variables which can be formulated in mechanistic terms and accurate models consequently identified. This is in stark contrast to such fields as economics and social sciences where even qualitative causal behavior is often speculative, and the quantity and uncertainty in data rather poor. In fact, even different types of engineered and natural systems require widely different analysis tools. For example, electrical and specific mechanical engineering disciplines (ex. involving rotary equipment) largely rely on frequency domain analysis methods, while time-domain methods are more suitable for most thermal and environmental systems. This consideration has led me to limit the scope of the analysis techniques described in this book to thermal, energy-related, environmental and industrial systems.

There are those students for whom a mathematical treatment and justification helps in better comprehension of the underlying concepts. However, my personal experience has been that the great majority of engineers do not fall in this category, and hence a more pragmatic approach is adopted. I am not particularly concerned with proofs, deductions and statistical rigor which tend to overwhelm the average engineering student. The intent is, rather, to impart a broad conceptual and theoretical understanding as well as a solid working familiarity (by means of case studies) of the various facets of data-driven modeling and analysis as applied to thermal and environmental systems. On the other hand, this is not a cookbook nor meant to be a reference book listing various models of the numerous equipment and systems which comprise thermal systems, but rather stresses underlying scientific, engineering, statistical and analysis concepts. It should not be considered as a substitute for specialized books nor should their importance be trivialized. A good general professional needs to be familiar, if not proficient, with a number of different analysis tools and how they "map" with each other, so that he can select the most appropriate tools for the occasion. Though nothing can replace hands-on experience in design and data analysis, being familiar with the appropriate theoretical concepts would not only shorten modeling and analysis time but also enable better engineering analysis to be performed. Further, those who have gone through this book will gain the required basic understanding to tackle the more advanced topics dealt with in the literature at large, and hence, elevate the profession as a whole. This book has been written with a certain amount of zeal in the hope that this will give this field some impetus and lead to its gradual emergence as an identifiable and important discipline (just as that enjoyed by a course on modeling, simulation and design of systems) and would ultimately be a required senior-level course or first-year graduate course in most engineering and science curricula.

This book has been intentionally structured so that the same topics (namely, statistics, parameter estimation and data collection) are treated first from a "basic" level, primarily by reviewing the essentials, and then from an "intermediate" level. This would allow the book to have broader appeal, and allow a gentler absorption of the needed material by certain students and practicing professionals. As pointed out by Asimov[3], the Greeks demonstrated that abstraction

[3] Asimov, I., 1966. *Understanding Physics: Light Magnetism and Electricity*, Walker Publications.

(or simplification) in physics allowed a simple and generalized mathematical structure to be formulated which led to greater understanding than would otherwise, along with the ability to subsequently restore some of the real-world complicating factors which were ignored earlier. Most textbooks implicitly follow this premise by presenting "simplistic" illustrative examples and problems. I strongly believe that a book on data analysis should also expose the student to the "messiness" present in real-world data. To that end, examples and problems which deal with case studies involving actual (either raw or marginally cleaned) data have been included. The hope is that this would provide the student with the necessary training and confidence to tackle real-world analysis situations.

Assumed Background of Reader

This is a book written for two sets of audiences: a basic treatment meant for the general engineering and science senior as well as the general practicing engineer on one hand, and the general graduate student and the more advanced professional entering the fields of thermal and environmental sciences. The exponential expansion of scientific and engineering knowledge as well as its cross-fertilization with allied emerging fields such as computer science, nanotechnology and bio-engineering have created the need for a major reevaluation of the thermal science undergraduate and graduate engineering curricula. The relatively few professional and free electives academic slots available to students requires that traditional subject matter be combined into fewer classes whereby the associated loss in depth and rigor is compensated for by a better understanding of the connections among different topics within a given discipline as well as between traditional and newer ones.

It is presumed that the reader has the necessary academic background (at the undergraduate level) of traditional topics such as physics, mathematics (linear algebra and calculus), fluids, thermodynamics and heat transfer, as well as some exposure to experimental methods, probability, statistics and regression analysis (taught in lab courses at the freshman or sophomore level). Further, it is assumed that the reader has some basic familiarity with important energy and environmental issues facing society today. However, special effort has been made to provide pertinent review of such material so as to make this into a sufficiently self-contained book.

Most students and professionals are familiar with the uses and capabilities of the ubiquitous spreadsheet program. Though many of the problems can be solved with the existing (or add-ons) capabilities of such spreadsheet programs, it is urged that the instructor or reader select an appropriate statistical program to do the statistical computing work because of the added sophistication which it provides. This book does not delve into how to use these programs, rather, the focus of this book is *education-based* intended to provide knowledge and skill sets necessary for value, judgment and confidence on how to use them, as against training-based whose focus would be to teach facts and specialized software.

Acknowledgements

Numerous talented and dedicated colleagues contributed in various ways over the several years of my professional career; some by direct association, others indirectly through their textbooks and papers-both of which were immensely edifying and stimulating to me personally. The list of acknowledgements of such meritorious individuals would be very long indeed, and so I have limited myself to those who have either provided direct valuable suggestions on the overview and scope of this book, or have generously given their time in reviewing certain chapters of this book. In the former category, I would like to gratefully mention Drs. David Claridge, Jeff Gordon, Gregor Henze John Mitchell and Robert Sonderegger, while in the latter, Drs. James Braun, Patrick Gurian, John House, Ari Rabl and Balaji Rajagopalan. I am also appreciative of interactions with several exceptional graduate students, and would like to especially thank the following whose work has been adopted in case study examples in this book: Klaus Andersen, Song Deng, Jason Fierko, Wei Jiang, Itzhak Maor, Steven Snyder and Jian Sun. Writing a book is a tedious and long process; the encouragement and understanding of my wife, Shobha, and our children, Agaja and Satyajit, were sources of strength and motivation.

Tempe, AZ, December 2010 T. Agami Reddy

Contents

Mathematical Models and Data Analysis

1

This chapter starts by introducing the benefits of applied data analysis and modeling methods through a case study example pertinent to energy use in buildings. Next, it reviews fundamental notions of mathematical models, illustrates them in terms of sensor response, and differentiates between forward or simulation models and inverse models. Subsequently, various issues pertinent to data analysis and associated uncertainty are described, and the different analysis tools which fall within its purview are discussed. Basic concepts relating to white-box, black-box and grey-box models are then presented. An attempt is made to identify the different types of problems one faces with forward modeling as distinct from inverse modeling and analysis. Notions germane to the disciplines of decision analysis, data mining and intelligent data analysis are also covered. Finally, the various topics covered in each chapter of this book are described.

1.1 Introduction

Applied data analysis and modeling of system performance is historically older than simulation modeling. The ancients, starting as far back as 7000 years ago, observed the movements of the sun, moon and stars in order to predict their behavior and initiate certain tasks such as planting crops or readying for winter. Theirs was a necessity impelled by survival; surprisingly, still relevant today. The threat of climate change and its dire consequences are being studied by scientists using in essence similar types of analysis tools—tools that involve measured data to refine and calibrate their models, extrapolating and evaluating the effect of different scenarios and mitigation measures. These tools fall under the general purview of data analysis and modeling methods, and it would be expedient to illustrate their potential and usefulness with a case study application which the reader can relate to more practically.

One of the current major societal problems facing mankind is the issue of energy, not only due to the gradual depletion of fossil fuels but also due to the adverse climatic and health effects which their burning creates. In 2005, total worldwide energy consumption was about 500 *Exajoules* ($=500\times10^{18}$ J), which is equivalent to about 16 TW ($=16\times10^{12}$ W). The annual growth rate was about 2%, which, at this rate, suggests a doubling time of 35 years. The United States (U.S.) accounts for 23% of the world-wide energy use (with only 5% of the world's population!), while the building sector alone (residential plus commercial buildings) in the U.S. consumes about 40% of the total energy use, close to 70% of the electricity generated, and is responsible for 49% of the SOx and 35% of the CO_2 emitted. Improvement in energy efficiency in all sectors of the economy has been rightly identified as a major and pressing need, and aggressive programs and measures are being implemented worldwide. It has been estimated that industrial countries are likely to see 25–35% in energy efficiency gains over the next 20 years, and more than 40% in developing countries (Jochem 2000). Hence, energy efficiency improvement in buildings is a logical choice for priority action. This can be achieved both by encouraging low energy building designs, but also by operating existing buildings more energy efficiently. In the 2003 Buildings Energy Consumption Survey (CBECS) study by U.S. Department of Energy (USDOE), over 85% of the building stock (excluding malls) was built before 1990. Further, according to USDOE 2008 Building Energy Data book, the U.S. spends $ 785 billion (6.1% of GDP) on new construction and $ 483 billion (3.3% of GDP) on improvements and repairs of existing buildings. A study of 60 commercial buildings in the U.S. found that half of them had control problems and about 40% had problems with the heating and cooling equipment (PECI 1997). This seems to be the norm. Enhanced commissioning processes in commercial/institutional buildings which do not compromise occupant comfort are being aggressively developed which have been shown to reduce energy costs by over 20% and in several cases over 50% (Claridge and Liu 2001). Further, existing techniques and technologies in energy efficiency retrofitting can reduce home energy use by up to 40% per home and lower associated greenhouse gas emissions by up to 160

T. Agami Reddy, *Applied Data Analysis and Modeling for Energy Engineers and Scientists*,
DOI 10.1007/978-1-4419-9613-8_1,

million metric tons annually by the year 2020. Identifying energy conservation opportunities, verifying by monitoring whether anticipated benefits are in fact realized when such measures are implemented, optimal operating of buildings; all these tasks require skills in data analysis and modeling.

Building energy *simulation models* (or forward models) are mechanistic (i.e., based on a mathematical formulation of the physical behavior) and deterministic (i.e. where there is no randomness in the inputs or outputs)[1]. They require as inputs the hourly climatic data of the selected location, the layout, orientation and physical description of the building (such as wall material, thickness, glazing type and fraction, type of shading overhangs,…), the type of mechanical and electrical systems available inside the building in terms of air distribution strategy, performance specifications of primary equipment (chillers, boilers,…), and the hourly operating and occupant schedules of the building. The simulation predicts hourly energy use during the entire year from which monthly total energy use and peak use along with utility rates provide an estimate of the operating cost of the building. The primary benefit of such a forward simulation model is that it is based on sound engineering principles usually taught in colleges and universities, and consequently has gained widespread acceptance by the design and professional community. Major public domain simulation codes (for example, Energy Plus 2009) have been developed with hundreds of man-years invested in their development by very competent professionals. This modeling approach is generally useful for design purposes where different design options are to be evaluated before the actual system is built.

Data analysis and modeling methods, on the other hand, are used when performance data of the system is available, and one uses this data for certain specific purposes, such as predicting or controlling the behavior of the system under different operating conditions, or for identifying energy conservation opportunities, or for verifying the effect of energy conservation measures and commissioning practices once implemented, or even to verify that the system is performing as intended (called condition monitoring). Consider the case of an existing building whose energy consumption is known (either utility bill data or monitored data). Some of the relevant questions which a building professional may apply data analysis methods are:

(a) *Commissioning tests*: How can one evaluate whether a component or a system is installed and commissioned properly?
(b) *Comparison with design intent*: How does the consumption compare with design predictions? In case of discrepancies, are they due to anomalous weather, to unintended building operation, to improper operation or to other causes?
(c) *Demand Side Management* (DSM): How would the consumption reduce if certain operational changes are made, such as lowering thermostat settings, ventilation rates or indoor lighting levels?
(d) *Operation and maintenance* (O&M): How much energy could be saved by retrofits to building shell, changes to air handler operation from constant air volume to variable air volume operation, or due to changes in the various control settings, or due to replacing the old chiller with a new and more energy efficient one?
(e) *Monitoring and verification* (M&V): If the retrofits are implemented to the system, can one verify that the savings are due to the retrofit, and not to other causes, e.g. the weather or changes in building occupancy?
(f) *Automated fault detection, diagnosis and evaluation* (AFDDE): How can one automatically detect faults in heating, ventilating, air-conditioning and refrigerating (HVAC&R) equipment which reduce operating life and/or increase energy use? What are the financial implications of this degradation? Should this fault be rectified immediately or at a later time? What specific measures need to be taken?
(g) *Optimal operation*: How can one characterize HVAC&R equipment (such as chillers, boilers, fans, pumps,…) in their installed state and optimize the control and operation of the entire system?

All the above questions are better addressed by data analysis methods. The forward approach could also be used, by say, (i) going back to the blueprints of the building and of the HVAC system, and repeating the analysis performed at the design stage while using actual building schedules and operating modes, and (ii) performing a calibration or tuning of the simulation model (i.e., varying the inputs in some fashion) since actual performance is unlikely to match observed performance. This process is, however, tedious and much effort has been invested by the building professional community in this regard with only limited success (Reddy 2006). A critical limitation of the calibrated simulation approach is that the data being used to tune the forward simulation model must meet certain criteria, and even then, all the numerous inputs required by the forward simulation model cannot be mathematically identified (this is referred to as an over-parameterized problem). Though awkward, labor intensive and not entirely satisfactory in its current state of development, the calibrated building energy simulation model is still an attractive option and has its place in the toolkit of data analysis methods (discussed at length in Sect. 11.2). The fundamental difficulty is that there is no general and widely-used model or software for dealing with data driven applications as they apply to building energy, though specialized software have been developed which allow certain types of narrow analysis to be performed. In fact, given the wide diversity in applications of data driven models, it is unlikely that any one metho-

[1] These terms will be described more fully in Sect. 1.2.3.

dology or software program will ever suffice. This leads to the basic premise of this book that there exists a crucial need for building energy professionals to be familiar and competent with data analysis methods and tools so that they could select the one which best meets their purpose with the end result that buildings will be operated and managed in a much more energy efficient manner than currently.

Building design simulation tools have played a significant role in lowering energy use in buildings. These are necessary tools and their importance should not be understated. Historically, most of the business revenue in Architectural Engineering and HVAC&R firms was generated from design/build contracts which required extensive use of simulation programs. Hence, the professional community is fairly well knowledgeable in this area, and several universities teach classes geared towards the use of simulation programs. However, there is an increasing market potential in building energy services as evidenced by the number of firms which offer services in this area. The acquisition of the required understanding, skills and tools relevant to this aspect is different from those required during the building design phase. There are other market forces which are also at play. The recent interest in "green" and "sustainable" has resulted in a plethora of products and practices aggressively marketed by numerous companies. Often, the claims that this product can save much more energy that another, and that that device is more environmentally friendly than others, are unfortunately, unfounded under closer scrutiny. Such types of unbiased evaluations and independent verification are imperative, otherwise the whole "green" movement may degrade into mere "greenwashing" and a feel-good attitude as against partially overcoming a dire societal challenge. A sound understanding of applied data analysis is imperative for this purpose and future science and engineering graduates have an important role to play. Thus, the *raison d'etre* of this book is to provide a general introduction and a broad foundation to the mathematical, statistical and modeling aspects of data analysis methods.

1.2 Mathematical Models

1.2.1 Types of Data

Data[2] can be classified in different ways. One classification scheme is as follows (Weiss and Hassett 1982):

- *categorical data (also called nominal or qualitative)* refers to data that has non-numerical qualities or attributes, such as belonging to one of several categories; for example, male/female, type of engineering major, fail/pass, satisfactory/not satisfactory,…;
- *ordinal* data, i.e., data that has some order or rank, such as a building envelope which is leaky, medium or tight, or a day which is hot, mild or cold;
- *metric* data, i.e., data obtained from measurements of such quantities as time, weight and height. Further, there are two different kinds of metric data: (i) data measured on an interval scale which has an arbitrary zero point (such as the Celsius scale); and (ii) data measured on a ratio scale which has a zero point that cannot be arbitrarily changed (such as mass or volume).
- *count* data, i.e., data on the number of individuals or items falling into certain classes or categories.

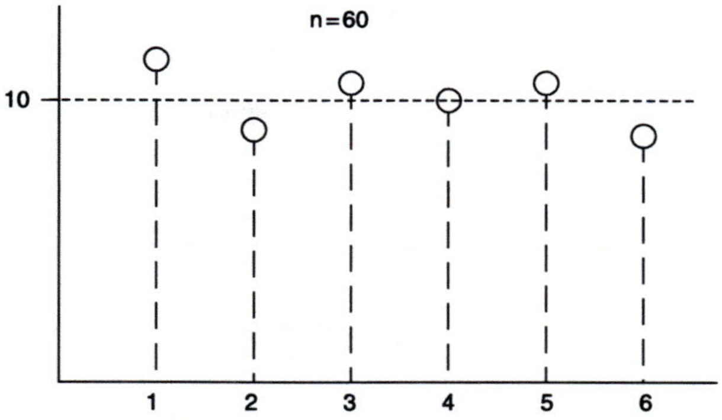

Fig. 1.1 The rolling of a dice is an example of discrete data where the data can only assume whole numbers. If the dice is fair, one would expect that out of 60 throws, numbers 1 through 6 would appear an equal number of times. However, in reality one may get small variations about the expected values as shown in the figure

A common type of classification relevant to metric data is to separate data into:

- *discrete* data which can take on only a finite or countable number of values (most qualitative, ordinal and count data fall in this category). An example is the data one would expect by rolling a dice 60 times (Fig. 1.1);
- *continuous* data which may take on any value in an interval (most metric data is continuous, and hence, is not countable). For example, the daily average outdoor dry-bulb temperature in Philadelphia, PA over a year (Fig. 1.2).

For data analysis purposes, it is important to view data based on their dimensionality, i.e., the number of axes needed to graphically present the data. A univariate data set consists of observations based on a single variable, bivariate those based on two variables, and multivariate those based on more than two variables.

The source or origin of the data can be one of the following:

(a) *Population* is the collection or set of all individuals (or items, or characteristics) representing the same quantity

[2] Several authors make a strict distinction between "data" which is plural and "datum" which is singular and implies a single data point. No such distinction is made throughout this book, and the word "data" is used to imply either.

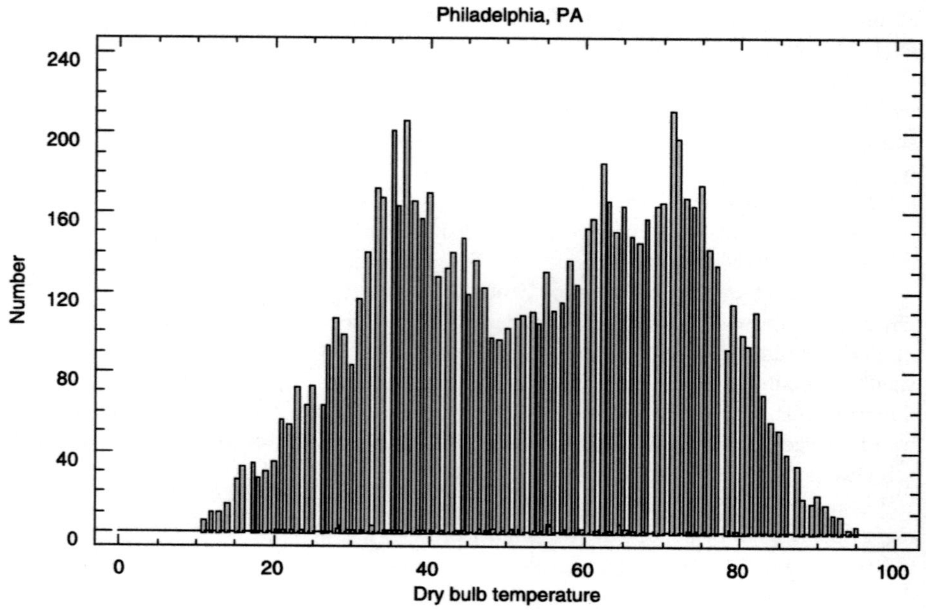

Fig. 1.2 Continuous data separated into a large number of bins (in this case, 300) resulted in the above histogram of the hourly outdoor dry-bulb temperature (in °F) in Philadelphia, PA over a year. A smoother distribution would have resulted if a smaller number of bins had been selected

with a connotation of completeness, i.e., the entire group of items being studied whether they be the freshmen student body of a university, instrument readings of a test quantity, or points on a curve.

(b) *Sample* is a portion or limited number of items from a population from which information or readings are collected. There are again two types of samples:
   - *Single-sample* is a single reading or succession of readings taken at the same time or under different times but under identical conditions;
   - *Multi-sample* is a repeated measurement of a fixed quantity using altered test conditions, such as different observers or different instruments or both.

   Many experiments may appear to be multi-sample data but are actually single-sample data. For example, if the same instrument is used for data collection during different times, the data should be regarded as single-sample not multi-sample.

(c) *Two-stage experiments* are successive staged experiments where the chance results of the first stage determines the conditions under which the next stage will be carried out. For example, when checking the quality of a lot of mass-produced articles, it is frequently possible to decrease the average sample size by carrying out the inspection in two stages. One may first take a small sample and accept the lot if all articles in the sample are satisfactory; otherwise a large second sample is inspected.

Finally, one needs to distinguish between: (i) a *duplicate* which is a separate specimen taken from the same source as the first specimen, and tested at the same time and in the same manner, and (ii) *replicate* which is the same specimen tested again at a different time. Thus, while duplication allows one to test samples till they are destroyed (such a tensile testing of an iron specimen), replicate testing stops short of doing permanent damage to the samples.

One can differentiate between different types of multi-sample data. Consider the case of solar thermal collector testing (as described in Pr. 5.6 of Chap. 5). In essence, the collector is subjected to different inlet fluid temperature levels under different values of incident solar radiation and ambient air temperatures using an experimental facility with instrumentation of pre-specified accuracy levels. The test results are processed according to certain performance models and the data plotted against collector efficiency versus reduced temperature level. The test protocol would involve performing replicate tests under similar reduced temperature levels, and this is one type of multi-sample data. Another type of multi-sample data would be the case when the same collector is tested at different test facilities nation-wide. The results of such a "round-robin" test are shown in Fig. 1.3 where one detects variations around the trend line given by the performance model which can be attributed to differences in both instrumentation and in slight differences in the test procedures from one facility to another.

1.2.2 What is a System Model?

A system is the object under study which could be as simple or as complex as one may wish to consider. It is any ordered, inter-related set of things, and their attributes. A model is a construct which allows one to represent the real-life system

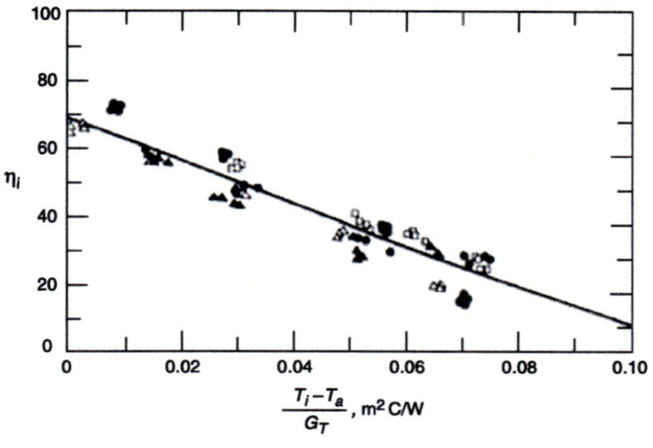

Fig. 1.3 Example of multi-sample data in the framework of a "round-robin" experiment of testing the same solar thermal collector in six different test facilities (*shown by different symbols*) following the same testing methodology. The test data is used to determine and plot the collector efficiency versus the reduced temperature along with uncertainty bands (see Pr. 5.6 for nomenclature). (Streed et al. 1979)

so that it can be used to predict the future behavior of the system under various "what-if" scenarios. The construct could be a scaled down physical version of the actual system (widely followed historically in engineering) or a mental construct, which is what is addressed in this book. The development of a model is not the ultimate objective, in other words, it is not an end by itself. It is a means to an end, the end being a credible means to make decisions which could involve system-specific issues (such as gaining insights about influential drivers and system dynamics, or predicting system behavior, or determining optimal control conditions) as well as those involving a broader context (such as operation management, deciding on policy measures and planning,...).

1.2.3 Types of Models

One differentiates between different types of models:

(i) *intuitive models* (or qualitative or descriptive models) are those where the system's behavior is summarized in non-analytical forms because only general qualitative trends of the system are known. Such a model which relies on quantitative or ordinal data is an aid to thought or to communication. Sociological or anthropological behaviors are typical examples;
(ii) *empirical models* which use metric or count data are those where the properties of the system can be summarized in a graph, a table or a curve fit to observation points. Such models presume knowledge of the fundamental quantitative trends but lack accurate understanding. Econometric models are typical examples; and
(iii) *mechanistic models* (or structural models) which use metric or count data are based on mathematical relationships used to describe physical laws such as Newton's laws, the laws of thermodynamics, etc... Such models can be used for prediction (system design) or for proper system operation and control (data analysis). Further such models can be separated into two sub-groups:
   - *exact* structural models where the model equation is thought to apply rigorously, i.e., the relationship between and variables and parameters in the model is exact, or as close to exact as current state of scientific understanding permits, and
   - *inexact* structural models where the model equation applies only approximately, either because the process is not fully known or because one chose to simplify the exact model so as to make it more usable. A typical example is the dose-response model which characterizes the relation between the amount of toxic agent imbibed by an individual and the incidence of adverse health effect.

Further, one can envision two different types of systems: *open systems* in which either energy and/or matter flows into and out of the system, and *closed systems* in which neither energy nor matter is exchanged to the environment.

A system *model* is a description of the system. Empirical and mechanistic models are made up of three components:

(i) *input variables* (also referred to as regressor, forcing, exciting, exogenous or independent variables in the engineering, statistical and econometric literature) which act on the system. Note that there are two types of such variables: controllable by the experimenter, and uncontrollable or extraneous variables, such as climatic variables;
(ii) *system structure and parameters/properties* which provide the necessary physical description of the systems in terms of physical and material constants; for example, thermal mass, overall heat transfer coefficients, mechanical properties of the elements; and
(iii) *output variables* (also called response, state, endogenous or dependent variables) which describe system response to the input variables.

A structural model of a system is a mathematical relationship between one or several input variables and parameters and one or several output variables. Its primary purpose is to allow better physical understanding of the phenomenon or process or alternatively, to allow accurate prediction of system reaction. This is useful for several purposes, for example, preventing adverse phenomenon from occurring, for proper system design (or optimization) or to improve system performance by evaluating other modifications to the system. A satisfactory mathematical model is subject to two contradictory requirements (Edwards and Penney 1996): it must be sufficiently detailed to represent the phenomenon it is attempting to explain or capture, yet it must be sufficiently simple to make the mathematical analysis practical. This requires judgment and experience of the modeler backed by experimentation and validation[3].

Examples of Simple Models:

(a) Pressure drop Δp of a fluid flowing at velocity v through a pipe of hydraulic diameter D_h and length L:

$$\Delta p = f \frac{L}{D_h} \rho \frac{v^2}{2} \tag{1.1}$$

where f is the friction factor, and ρ is the density of the fluid. For a given system, v can be viewed as the independent or input variable, while the pressure drop is the state variable. The factors f, L and D_h are the system or model parameters and ρ is a property of the fluid. Note that the friction factor f is itself a function of the velocity, thus making the problem a bit more complex.

(b) Rate of heat transfer from a fluid to a surrounding solid:

$$\dot{Q} = UA(T_f - T_o) \tag{1.2}$$

where the parameter *UA* is the overall heat conductance, and T_f and T_o are the mean fluid and solid temperatures (which are the input variables).

(c) Rate of heat added to a flowing fluid:

$$\dot{Q} = \dot{m}\, c_p (T_{out} - T_{in}) \tag{1.3}$$

where $\dot{m}$ is the fluid mass flow rate, c_p is its specific heat at constant pressure, and T_{out} and T_{in} are the exit and inlet fluid temperatures. It is left to the reader to identify the input variables, state variables and the model parameters.

(d) Lumped model of the water temperature T_s in a storage tank with an immersed heating element and losing heat to the environment is given by the first order ordinary differential equation (ODE):

$$Mc_p \frac{dT_s}{dt} = P - UA(T_s - T_i) \tag{1.4}$$

where Mc_p is the thermal heat capacitance of the tank (water plus tank material),
T_i the environment temperature, and P is the auxiliary power (or heat rate) supplied to the tank. It is left to the reader to identify the input variables, state variables and the model parameters.

[3] *Validation* is defined as the process of bringing the user's confidence about the model to an acceptable level either by comparing its performance to other more accepted models or by experimentation.

Table 1.1 Ways of classifying mathematical models

Different classification methods	
1	Distributed vs lumped parameter
2	Dynamic vs static or steady-state
3	Deterministic vs stochastic
4	Continuous vs discrete
5	Linear vs non-linear in the functional model
6	Linear vs non-linear in the model parameters
7	Time invariant vs time variant
8	Homogeneous vs non-homogeneous
9	Simulation vs performance models
10	Physics based (white box) vs data based (black box) and mix of both (grey box)

1.2.4 Classification of Mathematical Models

Predicting the behavior of a system requires a mathematical representation of the system components. The process of deciding on the level of detail appropriate for the problem at hand is called *abstraction* (Cha et al. 2000). This process has to be undertaken with care; (i) over-simplification may result in loss of important system behavior predictability, while (ii) an overly-detailed model may result in undue data and computational resources as well as time spent in understanding the model assumptions and results generated. There are different ways by which mathematical models can be classified. Some of these are shown in Table 1.1 and described below (adapted from Eisen 1988).

(i) Distributed vs Lumped Parameter In a *distributed parameter system*, the elements of the system are continuously distributed along the system geometry so that the variables they influence must be treated as differing not only in time but also in space, i.e., from point to point. Partial differential or difference equations are usually needed. Recall that a partial differential equation (PDE) is a differential equation between partial derivatives of an unknown function against at least two independent variables. One distinguishes between two general cases:

- the independent variables are space variables only
- the independent variables are both space and time variables.

Though partial derivatives of multivariable functions are ordinary derivatives with respect to one variable (the other being kept constant), the study of PDEs is not an easy extension of the theory for ordinary differential equations (ODEs). The solution of PDEs requires fundamentally different approaches. Recall that ODEs are solved by first finding general solutions and then using subsidiary conditions to determine arbitrary constants. However, such arbitrary constants in general solutions of ODEs are replaced by arbitrary functions in PDE, and determination of these arbitrary functions using subsidiary conditions is usually impossible. In other

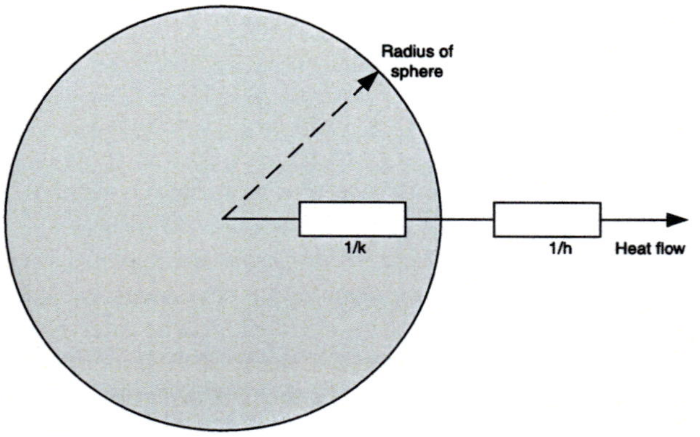

Fig. 1.4 Cooling of a solid sphere in air can be modeled as a lumped model provided the Biot number Bi < 0.1. This number is proportional to the ratio of the heat conductive resistance (1/k) inside the sphere to the convective resistance (1/h) from the outer envelope of the sphere to the air

words, general solutions of ODEs are of limited use in solving PDEs. In general, the solution of the PDEs and subsidiary conditions (called initial or boundary conditions) needs to be determined simultaneously. Hence, it is wise to try to simplify the PDE model as far as possible when dealing with data analysis problems.

In a *lumped parameter system*, the elements are small enough (or the objective of the analysis is such that simplification is warranted) so that each such element can be treated as if it were concentrated (i.e., lumped) at one particular spatial point in the system. The position of the point can change with time but not in space. Such systems usually are adequately modeled by ODE or difference equations. A heated billet as it cools in air could be analyzed as either a distributed system or a lumped parameter system depending on whether the Biot number (Bi) is greater than or less than 0.1 (see Fig. 1.4). The Biot number is proportional to the ratio of the internal to the external heat flow resistances of the sphere, and a small Biot number would imply that the resistance to heat flow attributed to internal body temperature gradient is small enough that it can be neglected without biasing the analysis. Thus, a small body with high thermal conductivity and low convection coefficient can be adequately modeled as a lumped system.

Another example of lumped model representation is the 1-D heat flow through the wall of a building (Fig. 1.5a) using the analogy between heat flow and electricity flow. The internal and external convective film heat transfer coefficients are represented by h_i and h_o respectively, while k,ρ and c_p are the thermal conductivity, density and specific heat of the wall material respectively. In the lower limit, the wall can be discretized into one lumped layer of capacitance C with two resistors as shown by the electric network of Fig. 1.5b (referred to as 2R1C network). In the upper limit, the network can be represented by "n" nodes (see Fig. 1.5c). The 2R1C simplification does lead to some errors, which under certain circumstances is outweighed by the convenience it provides while yielding acceptable results.

(ii) Dynamic vs Steady-State *Dynamic models* are defined as those which allow transient system or equipment behavior

Fig. 1.5 Thermal networks to model heat flow through a homogeneous plane wall of surface area A and wall thickness Δx. **a** Schematic of the wall with the indoor and outdoor temperatures and convective heat flow coefficients, **b** Lumped model with two resistances and one capacitance (2R1C model), **c** Higher n[th] order model with n layers of equal thickness ($\Delta x/n$). While all capacitances are assumed equal, only the (n−2) internal resistances (excluding the two end resistances) are equal

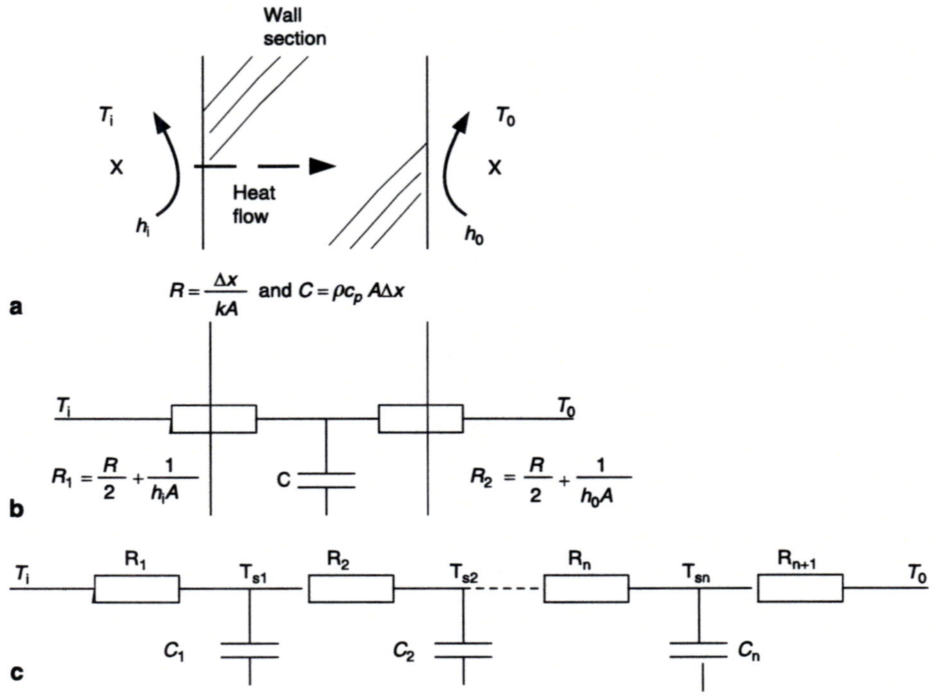

to be captured with explicit recognition of the time varying behavior of both output and input variables. The *steady-state* or static or zeroeth model is one which assumes no time variation in its input variables (and hence, no change in the output variable as well). One can also distinguish an intermediate type, referred to as *quasi-static* models. Cases arise when the input variables (such as incident solar radiation on a solar hot water panel) are constantly changing at a short time scale (say, at the minute scale) while it is adequate to predict thermal output at say hourly intervals. The dynamic behavior is poorly predicted by the solar collector model at such high frequency time scales, and so the input variables can be "time-averaged" so as to make them constant during a specific hourly interval. This is akin to introducing a "low pass filter" for the inputs. Thus, the use of quasi-static models allows one to predict the system output(s) in discrete time variant steps or intervals during a given day with the system inputs averaged (or summed) over each of the time intervals fed into the model. These models could be either zeroeth order or low order ODE.

Dynamic models are usually represented by PDEs or, by ODEs when spatially lumped with respect to time. One could solve them directly, and the simple cases are illustrated in Sect. 1.2.5. Since solving these equations gets harder as the order of the model increases, it is often more convenient to recast the differential equations in a time-series formulation using response functions or transfer functions which are time-lagged values of the input variable(s) only, or of both the inputs and the response respectively. This formulation is discussed in Chap. 9. The *steady-state* or static or zeroeth model is one which assumes no time variation in its inputs or outputs. Its time series formulation results in simple algebraic equations with no time-lagged values of the input variable(s) appearing in the function.

(iii) Deterministic vs Stochastic A *deterministic system* is one whose response to specified inputs under specified conditions is completely predictable (to within a certain accuracy of course) from physical laws. Thus, the response is precisely reproducible time and again. A *stochastic system* is one where the specific output can be predicted to within an uncertainty range only, which could be due to two reasons: (i) that the inputs themselves are random and vary unpredictably within a specified range of values (such as the electric power output of a wind turbine subject to gusting winds), and/or (ii) because the models are not accurate (for example, the dose-response of individuals when subject to asbestos inhalation). Concepts from probability theory are required to make predictions about the response.

The majority of observed data has some stochasticity in them either due to measurement/miscellaneous errors or due to the nature of the process itself. If the random element is so small that it is negligible as compared to the "noise" in the system, then the process or system can be treated in a purely deterministic framework. The orbits of the planets though well described by Kepler's laws have some small disturbances due to other secondary effects, but Newton was able to treat them as deterministic. On the other hand, Brownian motion is purely random, and has to be treated by stochastic methods.

(iv) Continuous vs Discrete A *continuous system* is one in which all the essential variables are continuous in nature and the time that the system operates is some interval (or intervals) of the real numbers. Usually such systems need differential equations to describe them. A *discrete system* is one in which all essential variables are discrete and the time that the system operates is a finite subset of the real numbers. This system can be described by difference equations.

In most applications in engineering, the system or process being studied is fundamentally continuous. However, the continuous output signal from a system is usually converted into a discrete signal by sampling. Alternatively, the continuous system can be replaced by its discrete analog which, of course, has a discrete signal. Hence, analysis of discrete data is usually more relevant in data analysis applications.

(v) Linear vs Non-linear A system is said to be *linear* if and only if, it has the following property: if an input $x_1(t)$ produces an output $y_1(t)$, and if an input $x_2(t)$ produces an output $y_2(t)$, then an input $[c_1\, x_1(t)+c_2\, x_2(t)]$ produces an output $[c_1\, y_1(t) + c_2\, y_2(t)]$ for all pairs of inputs $x_1(t)$ and $x_2(t)$ and all pairs of real number constants a_1 and a_2. This concept is illustrated in Fig. 1.6. An equivalent concept is the *principle of superposition* which states that the response of a linear system due to several inputs acting simultaneously is equal to the sum of the responses of each input acting alone. This is an extremely important concept since it allows the response of a complex system to be determined more simply by decomposing the input driving function into simpler terms, solving the equation for each term separately, and then summing the individual responses to obtain the desired aggregated response.

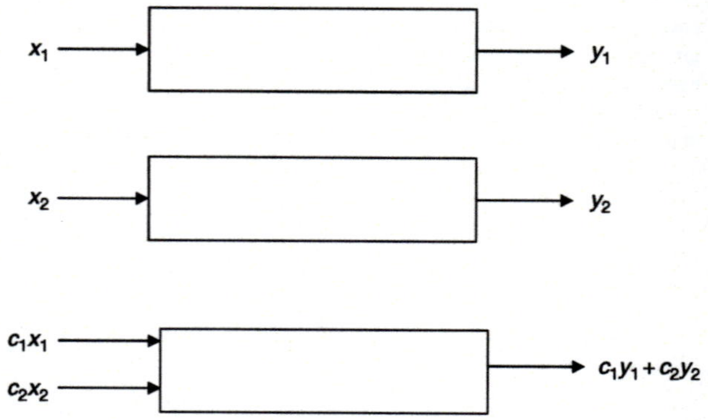

Fig. 1.6 Principle of superposition of a linear system

An important distinction needs to be made between a linear model and a model which is linear in its parameters. For example,

- $y = ax_1 + bx_2$ is linear in both model and parameters a and b,
- $y = a \sin x_1 + bx_2$ is a non-linear model but is linear in its parameters, and
- $y = a \exp(bx_1)$ is non-linear in both model and parameters.

In all fields, linear differential or difference equations are by far more widely used than non-linear equations. Even if the models are non-linear, every attempt is made, due to the subsequent convenience it provides, to make them linear either by suitable transformation (such as logarithmic transform) or by piece-wise linearization, i.e., linear approximation over a smaller range of variation. The advantages of linear systems over non-linear systems are many:

- linear systems are simpler to analyze,
- general theories are available to analyze them,
- they do not have singular solutions (simpler engineering problems rarely have them anyway),
- well-established methods are available, such as the state space approach, for analyzing even relatively complex set of equations. The practical advantage with this type of time domain transformation is that large systems of higher-order ODEs can be transformed into a first order system of simultaneous equations which, in turn, can be solved rather easily by numerical methods using standard computer programs.

(vi) Time Invariant vs Time Variant A system is *time-invariant* or stationary if neither the form of the equations characterizing the system nor the model parameters vary with time under either different or constant inputs; otherwise the system is *time-variant* or non-stationary. In some cases, when the model structure is poor and/or when the data are very noisy, time variant models are used requiring either on-line or off-line updating depending on the frequency of the input forcing functions and how quickly the system responds. Examples of such instances abound in electrical engineering applications. Usually, one tends to encounter time invariant models in less complex thermal and environmental engineering applications.

(vii) Homogeneous vs Non-homogeneous If there are no external inputs and the system behavior is determined entirely by its initial conditions, then the system is called *homogeneous* or unforced or autonomous; otherwise it is called *non-homogeneous* or forced. Consider the general form of a nth order time-invariant or stationary linear ODE:

$$Ay^{(n)} + By^{(n-1)} + \ldots + My'' + Ny' + Oy = P(x) \quad (1.5)$$

where y', y'' and $y^{(n)}$ are the first, second and n^{th} derivatives of y with respect to x, and A, B, … M, N and O are constants. The function P(x) frequently corresponds to some external influence on the system, and is a function of the independent variable. Often, the independent variable is the time variable t. This is intentional since time comes into play when the dynamic behavior of most physical systems is modeled. However, the variable t can be assigned any other physical quantity as appropriate.

To completely specify the problem, i.e., to obtain a unique solution y(x), one needs to specify two additional factors: (i) the interval of x over which a solution is desired, and (ii) a set of n initial conditions. If these conditions are such that y(x) and its first $(n-1)$ derivatives are specified for $x=0$, then the problem is called an *initial value problem*. Thus, one distinguishes between:

(a) the *homogeneous form* where $P(x)=0$, i.e., there is no external driving force. The solution of the differential equation:

$$Ay^{(n)} + By^{(n-1)} + \ldots + My'' + Ny' + Oy = 0 \quad (1.6)$$

yields the *free response* of the system. The homogeneous solution is a general solution whose arbitrary constants are then evaluated using the initial (or boundary) conditions, thus making it unique to the situation.

(b) the *non-homogeneous form* where $P(x) \neq 0$ and Eq. 1.5 applies. The *forced response* of the system is associated with the case when all the initial conditions are identically zero, i.e., $y(0), y'(0), \ldots y^{(n-1)}$ are all zero. Thus, the implication is that the forced response is only dependent on the external forcing function P(x). The *total response* of the linear time-invariant ODE is the sum of the free response and the forced response (thanks to the superposition principle). When system control is being studied, slightly different terms are often used to specify total dynamic system response: (a) the *steady-state response* is that part of the total response which does not approach zero as time approaches infinity, and (b) the *transient response* is that part of the total response which approaches zero as time approaches infinity.

(viii) Simulation Versus Performance Based The distinguishing trait between simulation and performance models is the basis on which the model structure is framed (this categorization is quite important). Simulation models are used to predict system performance during the design phase when no actual system exists and alternatives are being evaluated. A performance based model relies on measured performance data of the actual system to provide insights into model structure and to estimate its parameters. A widely accepted classification involves the following:

Table 1.2 Description of different types of models

Model type	Time variation of system inputs/outputs	Model complexity	Physical understanding	Type of equation
Simulation model	Dynamic Quasi-static	White box Detailed mechanistic	High	PDEs ODEs
Performance model	Quasi-static Steady-state	Gray box Semi-empirical Lumped	Medium	ODEs Algebraic
Performance model	Static or steady-state	Black box Empirical	Low	Algebraic

ODE ordinary differential equations, *PDE* partial differential equations

(a) *White-box models* (also called detailed mechanistic models, reference models or small-time step models) are based on the laws of physics and permit accurate and microscopic modeling of the various fluid flow, heat and mass transfer phenomenon which occur within the equipment or system. These are used for simulation purposes. Usually, temporal and spatial variations are considered, and these models are expressed by PDEs or ODEs. As shown in Table 1.2, a high level of physical understanding is necessary to develop these models, complemented with some expertise in numerical analysis in order to solve these equations. Consequently, these have found their niche in simulation studies which require dynamic and transient operating conditions to be accurately captured.

(b) *Black-box models* (or empirical or curve-fit or data-driven models) are based on little or no physical behavior of the system and rely on the available data to identify the model structure. These belong to one type of performance models which are suitable for predicting future behavior under a similar set of operating conditions to those used in developing the model. However, they provide little or no insights into better understanding of the process or phenomenon dictating system behavior. Statistical methods play a big role in dealing with uncertainties during model identification and model prediction. Historically, these types of models were the first ones developed for engineering systems based on concepts from numerical methods. They are still used when the system is too complex to be modeled physically, or when a "quick-and-dirty" analysis is needed. They are used in both simulation studies (where they are often used to model specific sub-systems or individual equipment of a larger system) and as performance models.

(c) *Gray-box models* fall in-between the two above categories and are best suited for performance models. A small number of possible model structures loosely based on the physics of the underlying phenomena and simplified in terms of time and/or space are posited, and then, the available data is used to identify the best model, and to determine the model parameters. The resulting models are usually lumped models based on first-order ODE or algebraic equations. They are primarily meant to gain better physical understanding of the system behavior and its interacting parts; they can also provide adequate prediction accuracy. The identification of these models which combine *phenomenological plausibility with mathematical simplicity* generally requires both good understanding of the physical phenomenon or of the systems/equipment being modeled, and a competence in statistical methods. These models are a major focus of this book, and they appear in several chapters.

Several authors, for example (Sprent 1998) also use terms such as (i) *data driven models* to imply those which are suggested by the data at hand and commensurate with knowledge about system behavior; this is somewhat akin to our definition of black-box models, and (ii) *model driven approaches* as those which assume a pre-specified model and the data is used to determine the model parameters; this is synonymous with grey-box models as defined here. However, this book makes no such distinction and uses the term "data driven models" interchangeably with performance models so as not to overly obfuscate the reader.

1.2.5 Models for Sensor Response

Let us illustrate steady-state and dynamic system responses using the example of measurement sensors. As stated above, one can categorize models into dynamic or static based on the time-variation of the system inputs and outputs.

Steady-state models (also called zeroeth order models) are the simplest model one can use. As stated earlier, they apply when input variables (and hence, the output variables) are maintained constant. A zeroeth order model for the dynamic performance of measuring systems is used (i) when the variation in the quantity to be measured is very slow as compared to how quickly the instrument responds, or (ii) as a standard of comparison for other more sophisticated models. For a zero-order instrument, the output is directly proportional to the input, such that (Doebelin 1995):

$$a_0 q_o = b_0 q_i \tag{1.7a}$$

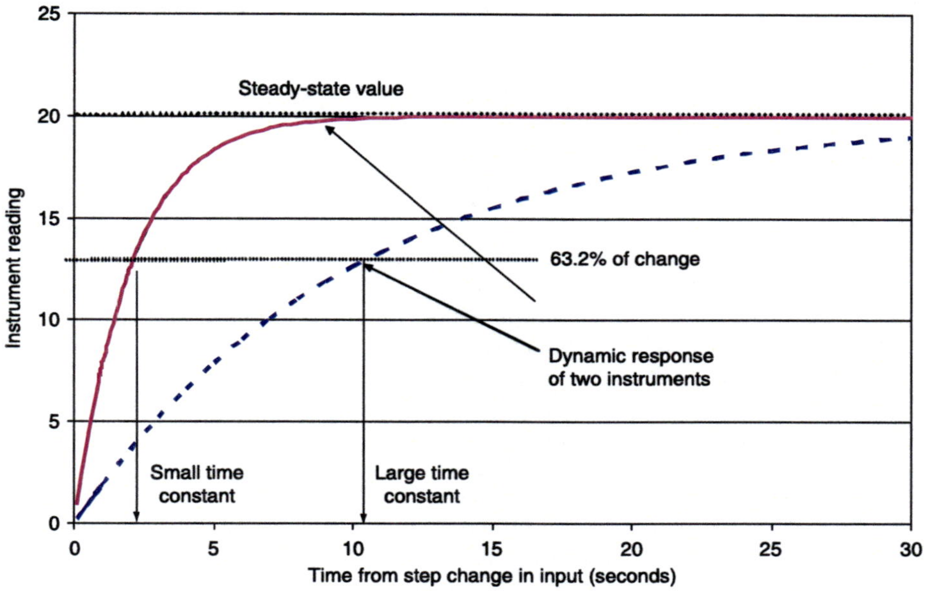

Fig. 1.7 Step-responses of two first-order instruments with different response times with assumed numerical values of time (x-axis) and instrument reading (y-axis). The response is characterized by the time constant which is the time for the instrument reading to reach 63.2% of the steady-state value

or

$$q_o = Kq_i \tag{1.7b}$$

where a_0 and b_0 are the system parameters, assumed time invariant, q_o and q_i are the output and the input quantities respectively, and $K=b_0/a_0$ is called the *static sensitivity* of the instrument.

Hence, only K is required to completely specify the response of the instrument. Thus, the zeroeth order instrument is an ideal instrument; no matter how rapidly the measured variable changes, the output signal faithfully and instantaneously reproduces the input.

The next step in complexity used to represent measuring system response is the *first-order model:*

$$a_1\frac{dq_0}{dt} + a_0q_o = b_0q_i \tag{1.8a}$$

or

$$\tau\frac{dq_0}{dt} + q_o = Kq_i \tag{1.8b}$$

where τ is the *time constant* of the instrument $= a_1/a_0$, and K is the static sensitivity of the instrument which is identical to the value defined for the zeroeth model. Thus, two numerical parameters are used to completely specify a first-order instrument.

The solution to Eq. 1.8b for a step change in input is:

$$q_o(t) = K.q_{is}(1 - e^{-t/\tau}) \tag{1.9}$$

where q_{is} is the value of the input quantity after the step change.

After a step change in the input, the steady-state value of the output will be K times the input q_{is} (just as in the zero-order instrument). This is shown as a dotted horizontal line in Fig. 1.7 with a numerical value of 20. The *time constant* characterizes the speed of response; the smaller its value the faster its response, and vice versa, to any kind of input. Figure 1.7 illustrates the dynamic response and the associated time constants for two instruments when subject to a step change in the input. Numerically, the time constant represents the time taken for the response to reach 63.2% of its final change, or to reach a value within 36.8% of the final value. This is easily seen from Eq. 1.9, by setting $t=\tau$, in which case $\frac{q_o(t)}{K.q_{is}} = (1 - e^{-1}) = 0.632$. Another useful measure of response speed for any instrument is the 5% *settling time,* i.e., the time for the output signal to get to within 5% of the final value. For any first-order instrument, it is equal to 3 times the time constant.

1.2.6 Block Diagrams

Information flow or *block diagram*[4] is a standard shorthand manner of schematically representing the inputs and output quantities of an element or a system as well as the computational sequence of variables. It is a concept widely used during system simulation since a block implies that its output

[4] Block diagrams should not be confused with *material flow diagrams* which for a given system configuration are unique. On the other hand, there can be numerous ways of assembling block diagrams depending on how the problem is framed.

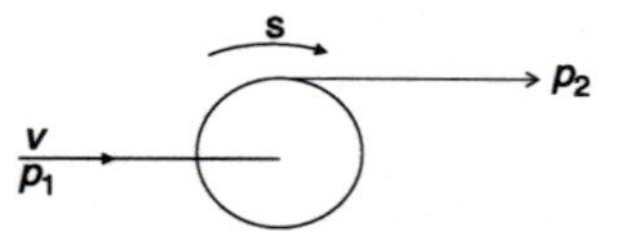

Fig. 1.8 Schematic of a centrifugal pump rotating at speed s (say, in rpm) which pumps a water flow rate v from lower pressure p_1 to higher pressure p_2

can be calculated provided the inputs are known. They are very useful for setting up the set of model equations to solve in order to simulate or analyze systems or components. As illustrated in Fig. 1.8, a centrifugal pump could be represented as one of many possible block diagrams (as shown in Fig. 1.9) depending on which parameters are of interest. If the model equation is cast in a form such that the outlet pressure p_2 is the response variable and the inlet pressure p_1 and the fluid flow volumetric rate v are the forcing variables, then the associated block diagram is that shown in Fig. 1.9a. Another type of block diagram is shown in Fig. 1.9b where flow rate v is the response variable. The arrows indicate the direction of unilateral information or signal flow. Thus, such diagrams depict the manner in which the simulation models of the various components of a system need to be formulated.

In general, a system or process is subject to one or more inputs (or stimulus or excitation or forcing functions) to which it responds by producing one or more outputs (or system response). If the observer is unable to act on the system, i.e., change some or any of the inputs, so as to produce a desired output, the system is not amenable to control. If however, the inputs can be varied, then control is feasible. Thus, a control system is defined as an arrangement of physical components connected or related in such a manner as to command, direct, or regulate itself or another system (Stubberud et al. 1994).

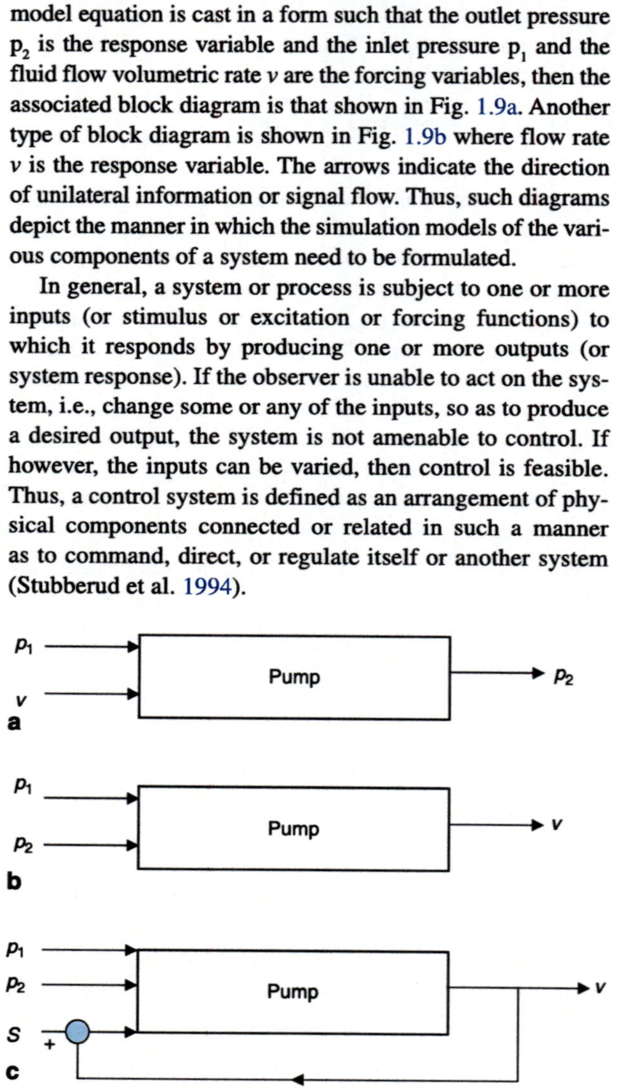

Fig. 1.9 Different block diagrams for modeling a pump depending on how the problem is formulated

One needs to distinguish between open and closed loops, and block diagrams provide a convenient way of doing so.

(a) *Open loop control system* is one in which the control action is independent of the output (see Fig. 1.10a). If the behavior of an open loop system is not completely understood or if unexpected disturbances act on it, then there may be considerable and unpredictable variations in the output. Two important features are: (i) their ability to perform accurately is determined by their *calibration*, i.e., by how accurately one is able to establish the input-output relationship; and (ii) they are generally not unstable. A practical example is an automatic toaster which is simply controlled by a timer.

(b) *Closed loop control system*, also referred to as a feedback control system, is one in which the control action is somehow dependent on the output (see Fig. 1.10b). If the value of the response y(t) is too low or too high, then the control action modifies the manipulated variable (shown as u(t)) appropriately. Such systems are designed to cope with lack of exact knowledge of system behavior, inaccurate component models and unexpected disturbances. Thus, increased accuracy is achieved by reducing the sensitivity of the ratio of output to input to variations in system characteristics (i.e., increased *bandwidth* defined as the range of variation in the inputs over which the system will respond satisfactorily) or due to random perturbations of the system by the environment. They have a serious disadvantage though: they can inadvertently develop unstable oscillations; this issue is an important one by itself, and is treated extensively in control textbooks.

Using the same example of a centrifugal pump but going one step further would lead us to the control of the pump. For example, if the inlet pressure p_1 is specified, and the pump needs to be operated or controlled (i.e., say by varying its rotational speed s) under variable outlet pressure p_2 so as to maintain a constant fluid flow rate v, then some sort of control mechanism or feedback is often used (shown in Fig. 1.9c). The small circle at the intersection of the signal s and the feedback represents a *summing point* which denotes the algebraic operation being carried out. For example, if the feedback signal is summed with the signal s, a "+" sign is placed just outside the summing point. Such graphical representations are called *signal flow diagrams,* and are used in process or system control which requires inverse modeling and parameter estimation.

1.3 Types of Problems in Mathematical Modeling

1.3.1 Background

Let us start with explaining the difference between parameters and variables in a model. A deterministic model is a mathematical relationship, derived from physical consi-

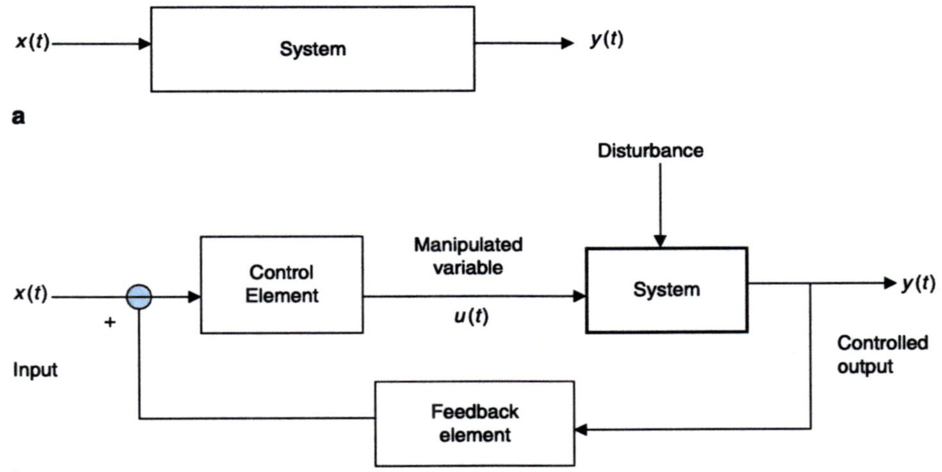

Fig. 1.10 Open and closed loop systems for a controlled output y(t). **a** Open loop. **b** Closed loop

derations, between variables and parameters. The quantities in a model which can be measured independently during an experiment are the "variables" which can be either input or output variables (as described earlier). To formulate the relationship among variables, one usually introduces "constants" which denote inherent properties of nature or of the engineering system called *parameters*. Sometimes, the distinction between both is ambiguous and depends on the context, i.e. the objective of the study and the manner in which the experiment is performed. For example, in Eq. 1.1, pipe length has been taken to be a fixed system parameter since the intention was to study the pressure drop against fluid velocity. However, if the objective is to determine the effect of pipe length on pressure drop for a fixed velocity, the length would then be viewed as the independent variable.

Consider the dynamic model of a component or system represented by the block diagram in Fig. 1.11. For simplicity, let us assume a linear model with no lagged terms in the forcing variables. Then, the model can be represented in matrix form as:

$$\mathbf{Y}_t = \mathbf{A}\mathbf{Y}_{t-1} + \mathbf{B}\mathbf{U}_t + \mathbf{C}\mathbf{W}_t \quad \text{with} \quad \mathbf{Y}_1 = \mathbf{d} \tag{1.10}$$

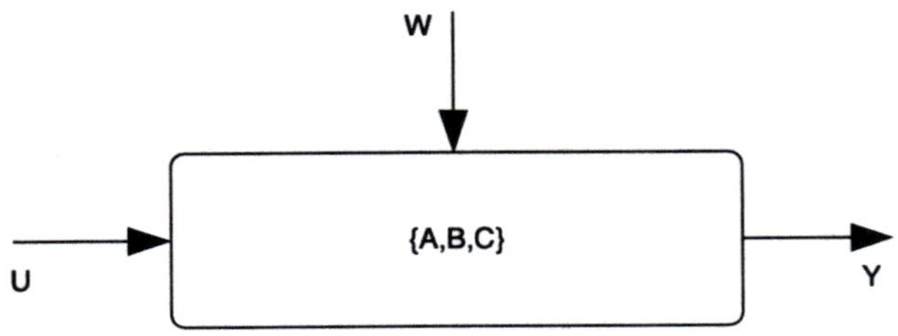

Fig. 1.11 Block diagram of a simple component with parameter vectors $\{\mathbf{A}, \mathbf{B}, \mathbf{C}\}$. Vectors **U** and **W** are the controllable/observable and the uncontrollable/disturbing inputs respectively while **Y** is the state variable or system response

where the output or state variable at time t is $\mathbf{Y}_t$. The forcing (or input or exogenous) variables are of two types: vector **U** denoting observable and controllable input variables, and vector **W** indicating uncontrollable input variables or disturbing inputs. The parameter vectors of the model are $\{\mathbf{A}, \mathbf{B}, \mathbf{C}\}$ while **d** represents the initial condition vector.

As shown in Fig. 1.12, one can differentiate between two broad types of problems; the forward (or well-defined or well-specified or direct) problem and the inverse (or ill-defined or identifiability) problem. The latter can, in turn, be divided into over-constrained (or over-specified or under-parameterized) and under-constrained (or under-specified or over-parameterized) problems which lead to calibration and model selection[5] type of problems respectively. Both of these rely on parameter estimation methods using either calibrated white box models or grey-box or black-box model forms regressed to data. These types of problems and their interactions are discussed at length in Chaps. 10 and 11, while a brief introduction is provided below.

1.3.2 Forward Problems

Such problems are framed as one where:

$$\text{Given } \{\mathbf{U},\mathbf{W}\} \text{ and } \{\mathbf{B},\mathbf{C},\mathbf{d}\}, \text{ determine } \mathbf{Y} \tag{1.11}$$

[5] The term "system identification" is extensively used in numerous texts related to inverse problems (especially in electrical engineering) to denote model structure identification and/or estimating the model parameters. Different authors use it differently, and since two distinct aspects are involved, this does seem to create some confusion. Hence for clarity, this book tries to retain this distinction by explicitly using the terms "model selection" for the process of identifying the functional form or model structure, and "parameter estimation" for the process of identifying the parameters in the functional model.

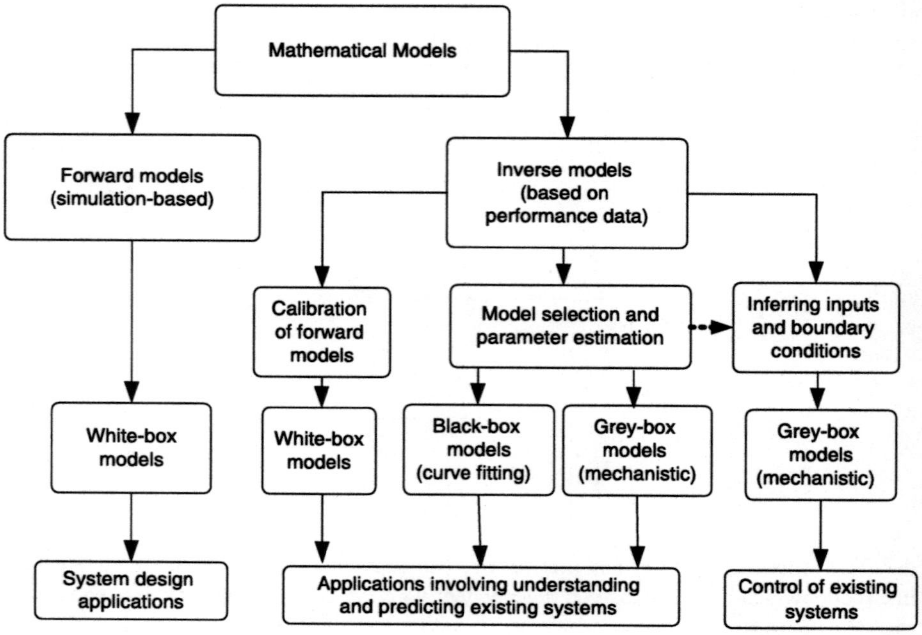

Fig. 1.12 Different types of mathematical models used in forward and inverse approaches. The *dotted line* indicates that control problems often need model selection and parameter estimation as a first step

The objective is to predict the response or state variables of a specified model with known structure and known parameters when subject to specified input or forcing variables (Fig. 1.12). This is also referred to as the "well-defined problem" since it has a unique solution if formulated properly. This is the type of models which is implicitly studied in classical mathematics and also in system simulation design courses. For example, consider a simple steady-state problem wherein the operating point of a pump and piping network are represented by black-box models of the pressure drop (Δp) and volumetric flow rate (V) such as shown in Fig. 1.13:

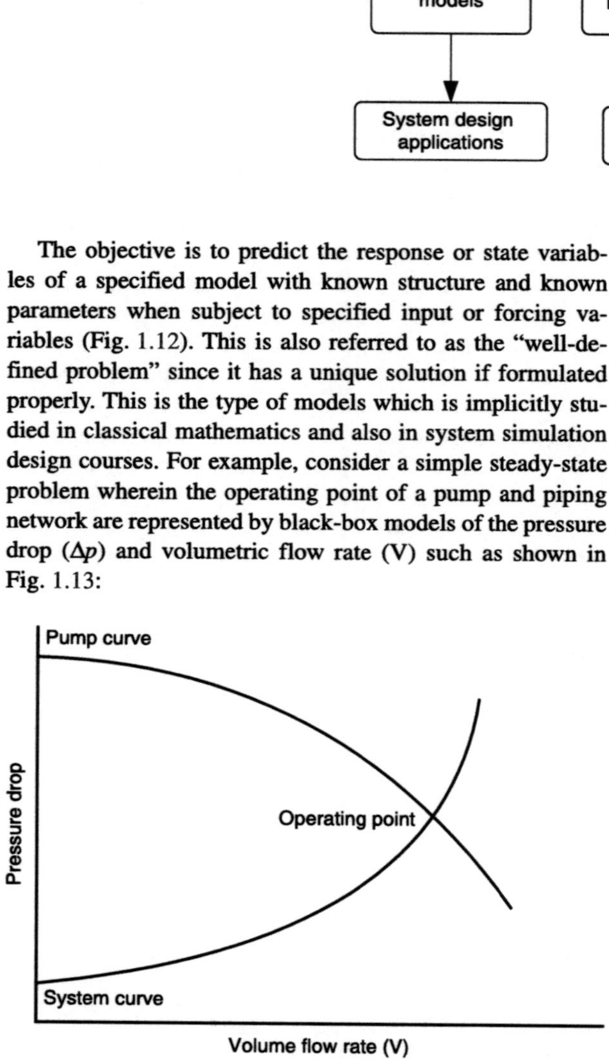

Fig. 1.13 Example of a forward problem where solving two simultaneous equations, one representing the pump curve and the other the system curve, yields the operating point

$$\begin{aligned} \Delta p &= a_1 + b_1 \cdot V + c_1 \cdot V^2 \quad \text{for the pump} \\ \Delta p &= a_2 + b_2 \cdot V + c_2 \cdot V^2 \quad \text{for the pipe network} \end{aligned} \tag{1.12}$$

Solving the two equations simultaneously yields the performance conditions of the operating point, i.e., pressure drop and flow rate (Δp_0,V_0). Note that the numerical values of the model parameters $\{a_i,b_i,c_i\}$ are known, and that (Δp) and V are the two variables, while the two equations provide the two constraints. This simple example has obvious extensions to the solution of differential equations where spatial and temporal response is sought.

In order to ensure accuracy of prediction, the models have tended to become increasingly complex especially with the advent of powerful and inexpensive computing power. The divide and conquer mind-set is prevalent in this approach, often with detailed mathematical equations based on scientific laws used to model micro-elements of the complete system. This approach presumes detailed knowledge of not only the various natural phenomena affecting system behavior but also of the magnitude of various interactions (for example, heat and mass transfer coefficients, friction coefficients, etc.). The main advantage of this approach is that the system need not be physically built in order to predict its behavior. Thus, this approach is ideal in the preliminary design and analysis stage and is most often employed as such. Note that incorporating superfluous variables and needless modeling details does increase computing time and complexity in the numerical resolution. However, if done correctly, it does not compromise the accuracy of the solution obtained.

1.3.3 Inverse Problems

It is rather difficult to succinctly define *inverse problems* since they apply to different classes of problems with applications in diverse areas, each with their own terminology and viewpoints (it is no wonder that it suffers from the "blind men and the elephant" syndrome). Generally speaking, inverse problems are those which involve identification of model structure (system identification) and/or estimates of model parameters (further discussed in Sect. 1.6 and Chaps. 10 and 11) *where the system under study already exists, and one uses measured or observed system behavior to aid in the model building and/or refinement*. Different model forms may capture the data trend; this is why some argue that inverse problems are generally "ill-defined" or "ill-posed".

In terms of mathematical classification[6], there are three types of inverse models all of which require some sort of identification or estimation (Fig. 1.12):

(a) *calibrated forward models* where ones uses a mechanistic model originally developed for the purpose of system simulation, and modifies or "tunes" the numerous model parameters so that model predictions match observed system behavior as closely as possible. Often, only a sub-set or limited number of measurements of system states and forcing function values are available, resulting in a highly over-parameterized problem with more than one possible solution (discussed in Sect. 11.2). Such inverse problems can be framed as:

$$\text{given } \{\mathbf{Y}'', \mathbf{U}'', \mathbf{W}'', \mathbf{d}''\}, \text{ determine } \{\mathbf{A}'', \mathbf{B}'', \mathbf{C}''\} \tag{1.13a}$$

where the $''$ notation is used to represent limited measurements or reduced parameter set;

(b) *model selection and parameter estimation* (using either grey-box or black-box models) where a suite of plausible model structures are formulated from basic scientific and engineering principles involving known influential and physically-relevant regressors, and performing experiments (or identifying system performance data) which allows these competing models to be evaluated and the "best" model identified. If a grey-box model is used, i.e., one which has physical meaning (such as the overall heat loss coefficient, time constant,…), it can then serve to improve our mechanistic understanding of the phenomenon or system behavior, and provide guidance as to ways by which the system behavior can be altered in a pre-specified manner. Different models and parameter estimation techniques need to be adopted depending on whether: (i) the intent is to subsequently predict system behavior *within* the temporal and/or spatial range of input variables—in such cases, simple and well-known methods such as curve fitting may suffice (see Fig. 1.14); (ii) the intent is to subsequently predict system behavior *outside* the temporal and/or spatial range of input variables—in such cases, physically based models are generally required, and this is influenced by the subsequent application of the model. Such problems (also referred to as system identification problems) are examples of under-parameterized problems and can be framed as:

$$\text{given } \{\mathbf{Y}, \mathbf{U}, \mathbf{W}, \mathbf{d}\}, \text{ determine } \{\mathbf{A}, \mathbf{B}, \mathbf{C}\} \tag{1.13b}$$

(c) *models for system control and diagnostics* so as to identify inputs necessary to produce a pre-specified system response, and for inferring boundary or initial conditions. Such problems are framed as:

$$\text{given } \{\mathbf{Y}''\} \text{ and } \{\mathbf{A}, \mathbf{B}, \mathbf{C}\}, \text{ determine } \{\mathbf{U}\ \mathbf{W}, \mathbf{d}\} \tag{1.13c}$$

where $\mathbf{Y}''$ is meant to denote that only limited measurements may be available for the state variable. Such problems require context-specific approximate numerical or analytical solutions for linear and non-linear problems and often involve model selection and parameter estimation as well. The ill-conditioning i.e., the solution is extremely sensitive to the data (see Sect. 10.2) is often due to the repetitive nature of the data collected while the system is under normal operation. There is a rich and diverse body of knowledge on such inverse methods and numerous texts books, monographs and research papers are available on this subject. Chapter 11 address these problems at more length.

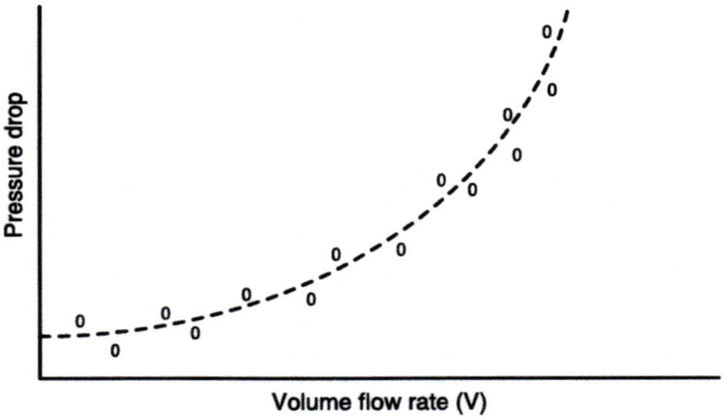

Fig. 1.14 Example of a parameter estimation problem where the model parameters of a presumed function of pressure drop versus volume flow rate are identified from discrete experimental data points

Example 1.3.1: *Simulation of a chiller.*
This example will serve to illustrate a simple application of calibrated simulation, but first, let us discuss the forward

[6] Several authors define inverse methods as applicable uniquely to case (c), and simply use the terms calibrated simulation and system identification for the two other cases.

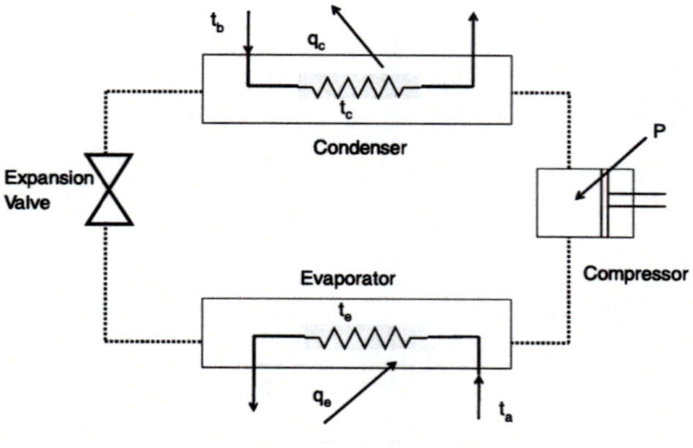

Fig. 1.15 Schematic of the cooling plant for Example 1.3.1

problem. Consider an example of simulating a chilled water cooling plant consisting of the condenser, compressor and evaporator, as shown in Fig. 1.15[7]. We shall use rather simple black-box models for this example for easier comprehension of the underlying concepts. The steady-state cooling capacity q_e (in kWt) and the compressor electric power draw P (in kWe) are function of the refrigerant evaporator temperature t_e and the refrigerant condenser temperature t_c in °C, and are supplied by the equipment manufacturer:

$$\begin{aligned} q_e = {} & 239.5 + 10.073t_e - 0.109t_e^2 - 3.41t_c \\ & - 0.00250t_c^2 - 0.2030t_et_c + 0.00820t_e^2t_c \\ & + 0.0013t_et_c^2 - 0.000080005t_e^2t_c^2 \end{aligned} \tag{1.14}$$

and

$$\begin{aligned} P = {} & -2.634 - 0.3081t_e - 0.00301t_e^2 + 1.066t_c \\ & - 0.00528t_c^2 - 0.0011t_et_c - 0.000306t_e^2t_c \\ & + 0.000567t_et_c^2 + 0.0000031t_e^2t_c^2 \end{aligned} \tag{1.15}$$

Further data has been provided:

- water flow rates through the evaporator: m_e=6.8 kg/s and in the condenser m_c=7.6 kg/s
- thermal conductances of the evaporator: UA_e=30.6 kW/K and condenser UA_c=26.5 kW/K
- and the inlet water temperature to the evaporator t_a=10°C and that to the condenser t_b=25°C

Another equation needs to be introduced for the heat rejected at the condenser q_c (in kWt). This is simply given by a heat balance of the system (i.e., from the first law of thermodynamics) as:

$$q_c = q_e + P \tag{1.16}$$

The forward problem would entail determining the unknown values of Y={t_e,t_c,q_e,P,q_c}. Since there are five unknowns, five equations are needed. In addition to the three equations above, two additional ones are needed. These are the heat balances on the refrigerant side (assuming to be changing phase, and hence, is at a constant temperature) and the coolant water side of both the evaporator and the condenser:

$$q_e = m_e c_p (t_a - t_e)\left[1 - \exp\left(-\frac{UA_e}{m_e c_p}\right)\right] \tag{1.17}$$

and

$$q_c = m_c c_p (t_c - t_b)\left[1 - \exp\left(-\frac{UA_c}{m_c c_p}\right)\right] \tag{1.18}$$

where c_p is the specific heat of water=4.186 kJ/kg K.

Solving the five equations results in:

$$t_e = 2.84°\text{C},\ t_c = 43.05°\text{C},\ q_e = 134.39\text{ kW}$$

and

$$P = 28.34\text{ kW}$$

To summarize, the performance of the various equipment and their interaction have been represented by mathematical equations which allow a single solution set to be determined. This is the case of the well-defined forward problem adopted in system simulation and design studies. Let us discuss how the same system is also amenable to an inverse model approach. Consider the case when a cooling plant similar to that assumed above exists, and the facility manager wishes to instrument the various components in order to: (i) verify that the system is performing adequately, and (ii) vary some of the operating variables so that the power consumed by the compressor is reduced. In such a case, the numerical model coefficients given in Eqs. 1.14 and 1.15 will be unavailable, and so will be the UA values, since either he is unable to find the manufacturer-provided models or the equipment has degraded somewhat that the original models are no longer accurate. The model calibration will involve determining these values from experiment data gathered by appropriately sub-metering the evaporator, condenser and compressor on both the refrigerant and the water coolant side. How best to make these measurements, how accurate should the instrumentation be, what should be the sampling frequency, for how long should one monitor,… are all issues which fall within the purview of design of field monitoring. Uncertainty in the measurements as well as the fact that the assumed models are approximations of reality will introduce model predictions errors and so the verification of the actual system against measured performance will have to consider such aspects properly.

The above example was a simple one with explicit algebraic equations for each component with no feedback loops. Detailed simulation programs are much more complex (with hundreds of variables, complex boundary conditions,…) involving ODEs or PDEs; one example is computational fluid dynamic (CFD) models for indoor air quality studies. Calibrat-

[7] From Stoecker (1989) by permission of McGraw-Hill.

ing such models is extremely difficult given the lack of proper instrumentation which can provide detailed spatial and temporal measurement fields, the inability to conveniently compartmentalize the problem so that inputs and outputs of sub-blocks could be framed and calibrated individually as done in the cooling plant example above. Thus, in view of such limitations in the data, developing a simpler system model consistent with the data available while retaining the underlying mechanistic considerations as far as possible is a more appealing approach; albeit a challenging one—such an approach is shown under the "model selection" branch in Fig. 1.12.

Example 1.3.2: Dose-response models. An example of how inverse models differ from a straightforward curve fit is given below (the same example is treated at much more depth in Sects. 10.4.4 and 11.3.4). Consider the case of models of risk to humans when exposed to toxins (or biological poisons) which are extremely deadly even in small doses. *Dose* is the total mass of toxin which the human body ingests. *Response* is the measurable physiological change in the body produced by the toxin which can have many manifestations; but let us focuses on human cells becoming cancerous. Since different humans (and test animals) react differently to the same dose, the response is often interpreted as a probability of cancer being induced, which can be framed as a risk. Further, tests on lab test animals are usually done at relatively high levels while policy makers would want to know the human response under lower levels of dose. Not only does one have the issue of translating lab specimen results to human response, but also one needs to be able to extrapolate the model to low doses. The manner one chooses to extrapolate the dose-response curve downwards is dependent on either the assumption one makes regarding the basic process itself or how one chooses to err (which has policy-making implications). For example, erring too conservatively in terms of risk would overstate the risk and prompt implementation of more precautionary measures, which some critics would fault as unjustified and improper use of limited resources.

Figure 1.16 illustrates three methods of extrapolating dose-response curves down to low doses (Heinsohn and Cimbala 2003). The dots represent observed laboratory tests performed at high doses. Three types of models are fit to the data and all of them agree at high doses. However, they deviate substantially at low doses because the models are functionally different. While model I is a nonlinear model applicable to highly toxic agents, curve II is generally taken to apply to contaminants that are quite harmless as low doses (i.e., the body is able to metabolize the toxin at low doses). Curve III is an intermediate one between the other two curves. The above models are somewhat empirical (or black-box) and are useful as performance models. However, they provide little understanding of the basic process itself. Models based on simplified but phenomenological considerations of how biological cells become cancerous have also been developed and these are described in Sect. 11.3.

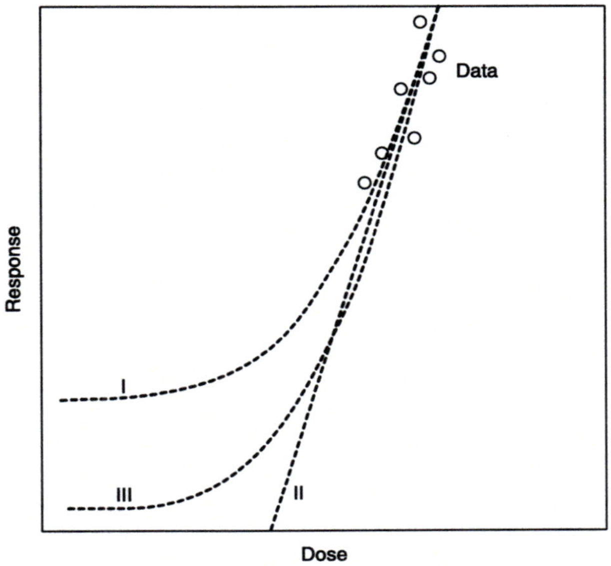

Fig. 1.16 Three different inverse models depending on toxin type for extrapolating dose-response observations at high doses to the response at low doses. (From Heinsohn and Cimbala (2003) by permission of CRC Press)

There are several aspects to this problem relevant to inverse modeling: (i) can the observed data of dose versus response provide some insights into the process which induces cancer in biological cells? (ii) How valid are these results extrapolated down to low doses? (iii) Since laboratory tests are performed on animal subjects, how valid are these results when extrapolated to humans? There are no simple answers to these queries (until the basic process itself is completely understood). Probability is bound to play an important role to the nature of the process, and hence, the adoption of various agencies (such as the U.S. Environmental Protection Agency) of probabilistic methods towards risk assessment and modeling.

1.4 What is Data Analysis?

In view of the diversity of fields to which data analysis is applied, an all-encompassing definition would have to be general. One good definition is: "*an evaluation of collected observations so as to extract information useful for a specific purpose*". The evaluation relies on different mathematical and statistical tools depending on the intent of the investigation. In the area of science, the systematic organization of observational data, such as the orbital movement of the planets, provided a means for Newton to develop his laws of motion. Observational data from deep space allow scientists to develop/refine/verify theories and hypotheses about the structure, relationships, origins, and presence of certain phenomena (such as black holes) in the cosmos. At the other end of the spectrum, data analysis can also be viewed as simply: *"the process of systematically applying statistical and logical techniques to describe, summarize, and compare data"*. From the perspective of an engineer/scientist, data analysis is a process which when applied to system performance data, collected either intrusively or non-intrusively, allows certain conclusions about the state of the system to be drawn, and thereby, to initiate followup actions.

Studying a problem through the use of statistical data analysis usually involves four basic steps (Arsham 2008):

(a) Defining the Problem: The context of the problem and the exact definition of the problem being studied need to be framed. This allows one to design both the data collection system and the subsequent analysis procedures to be followed.

(b) Collecting the Data: In the past (say, 50 years back), collecting the data was the most difficult part, and was often the bottleneck of data analysis. Nowadays, one is overwhelmed by the large amounts of data resulting from the great strides in sensor and data collection technology; and data cleaning, handling, summarizing have become major issues. Paradoxically, the design of data collection systems has been marginalized *"by an apparent belief that extensive computation can make up for any deficiencies in the design of data collection"*. Gathering data without a clear definition of the problem often results in failure or limited success. Data can be collected from existing sources or obtained through observation and experimental studies designed to obtain new data. In an experimental study, the variable of interest is identified. Then, one or more factors in the study are controlled so that data can be obtained about how the factors influence the variables. In observational studies, no attempt is made to control or influence the variables of interest either intentionally or due to the inability to do so (two examples are surveys and astronomical data).

(c) Analyzing the Data: There are various statistical and analysis approaches and tools which one can bring to bear depending on the type and complexity of the problem and the type, quality and completeness of the data available. Section 1.6 describes several categories of problems encountered in data analysis. Probability is an important aspect of data analysis since it provides a mechanism for measuring, expressing, and analyzing the uncertainties associated with collected data and mathematical models used. This, in turn, impacts the confidence in our analysis results: uncertainty in future system performance predictions, confidence level in our confirmatory conclusions, uncertainty in the validity of the action proposed,.... The majority of the topics addressed in this book pertain to this category.

(d) Reporting the Results: The final step in any data analysis effort involves preparing a report. This is the written document that logically describes all the pertinent stages of the work, presents the data collected, discusses the analysis results, states the conclusions reached, and recommends further action specific to the issues of the problem identified at the onset. The final report and any technical papers resulting from it are the only documents which survive over time and are invaluable to other professionals. Unfortunately, the task of reporting is often cursory and not given its due importance.

Recently, the term "intelligent" data analysis has been used which has a different connotation from traditional ones (Berthold and Hand 2003). This term is used not in the sense that it involves added intelligence of the user or analyst in applying traditional tools, but that the statistical tools themselves have some measure of intelligence built into them. A simple example is when a regression model has to be identified from data. The tool evaluates hundreds of built-in functions and presents to the user a prioritized list of models according to their goodness-of-fit. The recent evolution of computer-intensive methods (such as bootstrapping and Monte Carlo methods) along with soft computing algorithms (such as artificial neural networks, genetic algorithms,...) enhance the capability of traditional statistics, model estimation, and data analysis methods. These added capabilities of enhanced computational power of modern-day computers and the sophisticated manner in which the software programs are written allow "intelligent" data analysis to be performed.

1.5 Types of Uncertainty in Data

If the same results are obtained when an experiment is repeated under the same conditions, one says that the experiment is *deterministic*. It is this deterministic nature of science that allows theories or models to be formulated and permits the use of scientific theory for prediction (Hodges and Lehman 1970). However, all observational or experimental data invariably have a certain amount of inherent noise or randomness which introduces a certain degree of uncertainty in the results or conclusions. Due to instrument or measurement technique, or improper understanding of all influential factors, or the inability to measure some of the driving parameters, random and/or bias types of errors usually infect the deterministic data. However, there are also experiments whose results vary due to the very nature of the experiment; for example gambling outcomes (throwing of dice, card games,...). These are called random experiments. Without uncertainty or randomness, there would have been little need for statistics. Probability theory and inferential statistics have been developed to deal with random experiments and the same approach has also been adapted to deterministic experimental data analysis. Both inferential statistics and stochastic model building have to deal with the random nature of observational or experimental data, and thus, require knowledge of probability.

There are several types of uncertainty in data, and all of them have to do with the inability to "determine the true state of affairs of a system" (Haimes 1998). A succinct classification involves the following sources of uncertainties:

(a) purely stochastic variability (or *aleatory uncertainty*) where the ambiguity in outcome is inherent in the nature of the process, and no amount of additional measurements can reduce the inherent randomness. Common examples involve coin tossing, or card games. These processes are inherently random (either on a temporal or spatial basis), and whose outcome, while uncertain, can be anticipated on a statistical basis;

(b) *epistemic uncertainty* or ignorance or lack of complete knowledge of the process which result in certain influential variables not being considered (and, thus, not measured);

(c) *inaccurate measurement* of numerical data due to instrument or sampling errors;

(d) *cognitive vagueness* involving human linguistic description. For example, people use words like tall/short or very important/not important which cannot be quantified exactly. This type of uncertainty is generally associated with qualitative and ordinal data where subjective elements come into play.

The traditional approach is to use probability theory along with statistical techniques to address (a), (b), and (c) types of uncertainties. The variability due to sources (b) and (c) can be diminished by taking additional measurements, by using more accurate instrumentation, by better experimental design and acquiring better insight into specific behavior with which to develop more accurate models. Several authors apply the term "uncertainty" to only these two sources. Finally, source (d) can be modeled using probability approaches though some authors argue that it would be more convenient to use fuzzy logic to model this vagueness in speech.

1.6 Types of Applied Data Analysis and Modeling Methods

Such methods can be separated into the following groups depending on the intent of the analysis:

(a) *Exploratory data analysis and descriptive statistics*, which entails performing "numerical detective work" on the data and developing methods for screening, organizing, summarizing and detecting basic trends in the data (such as graphs, and tables) which would help in information gathering and knowledge generation. Historically, formal statisticians have shied away from exploratory data analysis considering it to be either too simple to warrant serious discussion or too ad hoc in nature to be able to expound logical steps (McNeil 1977). This area had to await the pioneering work by John Tukey and others to obtain a formal structure. This area is not specifically addressed in this book, and the interested reader can refer to Hoagin et al. (1983) or Tukey (1988) for an excellent perspective.

(b) Model building and *point estimation* which involves (i) taking measurements of the various parameters (or regressor variables) affecting the output (or response variables) of a device or a phenomenon, (ii) identifying a causal quantitative correlation between them by regression, and (iii) using it to make predictions about system behavior under future operating conditions. There is a rich literature in this area with great diversity of techniques and level of sophistication.

(c) *Inferential problems* are those which involve making uncertainty inferences or calculating uncertainty or confidence intervals of population estimates from selected samples. They also apply to regression, i.e., uncertainty in model parameters, and in model predictions. When a regression model is identified from data, the data cannot be considered to include the entire "population" data, i.e., all the observations one could possibly conceive. Hence, model parameters and model predictions suffer from uncertainty which needs to be quantified. This takes the form of assigning uncertainty bands around the estimates. Those methods which allow tighter predictions are deemed more "efficient", and hence more desirable.

(d) *Design of experiments* is the process of prescribing the exact manner in which samples for testing need to be selected, and the conditions and sequence under which the testing needs to be performed such that the relationship or model between a response variable and a set of regressor variables can be identified in a robust and accurate manner.

(e) *Classification and clustering problems:* Classification problems are those where one would like to develop a model to statistically distinguish or "discriminate" differences between two or more groups when one knows beforehand that such groupings exist in the data set provided, and, to subsequently assign, allocate or classify a future unclassified observation into a specific group with the smallest probability of error. Clustering, on the other hand, is a more difficult problem, involving situations when the number of clusters or groups is not known beforehand, and the intent is to allocate a set of observation sets into groups which are similar or "close" to one another with respect to certain attribute(s) or characteristic(s).

(f) *Time series analysis and signal processing*. Time series analysis involves the use of a set of tools that include traditional model building techniques as well as those involving the sequential behavior of the data and its noise. They involve the analysis, interpretation and manipulation of time series signals in either time domain or frequency domain. Signal processing is one specific, but important, sub-domain of time series analysis dealing with sound, images, biological signals such as ECG, radar signals, and many others. Vibration analysis of rotating machinery is another example where signal processing tools can be used.

(g) *Inverse modeling* (introduced earlier in Sect. 1.3.3) is an approach to data analysis methods which includes three classes: statistical calibration of mechanistic models, model selection and parameter estimation, and inferring forcing functions and boundary/initial conditions. It combines the basic physics of the process with statistical methods so as to achieve a better understanding of the system dynamics, and thereby use it to predict system performance either within or outside the temporal and/or spatial range used to develop the model. The discipline of inverse modeling has acquired a very important niche not only in the fields of engineering and science but in other disciplines as well (such as biology, medicine,…).

(h) *Risk analysis and decision making*: Analysis is often a precursor to decision-making in the real world. Along with engineering analysis there are other aspects such as making simplifying assumptions, extrapolations into the future, financial ambiguity,… that come into play while making decisions. Decision theory is the study of methods for arriving at "rational" decisions under uncertainty. The decisions themselves may or may not prove to be correct in the long term, but the process provides a structure for the overall methodology by which undesirable events are framed as risks, the chain of events simplified and modeled, trade-offs between competing alternatives assessed, and the risk attitude of the decision-maker captured (Clemen and Reilly 2001). The value of collecting additional information to reduce the risk, capturing heuristic knowledge or combining subjective preferences into the mathematical structure are additional aspects of such problems. As stated earlier, inverse models can be used to make predictions about system behavior. These have inherent uncertainties (which may be large or small depending on the problem at hand), and adopting a certain inverse model over potential competing ones involves the consideration of risk analysis and decision making tools.

1.7 Example of a Data Collection and Analysis System

Data can be separated into experimental or observational depending on whether the system operation can be modified by the observer or not. Consider a system where the initial phase of designing and installing the monitoring system is complete. Figure 1.17 is a flowchart depicting various stages in the collection, analysis and interpretation of data collected from an engineering thermal[8] system while in operation. The various elements involved are:

(a) a measurement system consisting of various sensors of pre-specified types and accuracy. The proper location, commissioning and maintenance of these sensors are important aspects of this element;
(b) data sampling element whereby the output of the various sensors are read at a pre-determined frequency. The low cost of automated data collection has led to increasingly higher sampling rates. Typical frequencies for thermal systems are in the range of 1 s–1 min;
(c) clean raw data for spikes, gross errors, mis-recordings, and missing or dead channels, average (or sum) the data samples and, if necessary, store them in a dynamic fashion (i.e., online) in a central electronic database with an electronic time stamp;
(d) average raw data and store in a database; typical periods are in the range of 1–30 min. One can also include some finer checks for data quality by flagging data when they exceed physically stipulated ranges. This process need not be done online but could be initiated automatically and periodically, say, every day. It is this data set which is queried as necessary for subsequent analysis;
(e) The above steps in the data collection process are performed on a routine basis. This data can be used to advantage, provided one can frame the issues relevant to the client and determine which of these can be satisfied. Examples of such routine uses are assessing overall time-averaged system efficiencies and preparing weekly performance reports, as well as for subtler action such as supervisory control and automated fault detection;
(f) Occasionally the owner would like to evaluate major changes such as equipment change out or addition of new equipment, or would like to improve overall system performance or reliability not knowing exactly how to achieve this. Alternatively, one may wish to evaluate system performance under an exceptionally hot spell of several days. This is when specialized consultants are brought in to make recommendations to the owner. Historically, such analysis were done based on the pro-

[8] Electrical systems have different considerations since they mostly use very high frequency sampling rates.

Fig. 1.17 Flowchart depicting various stages in data analysis and decision making as applied to continuous monitoring of thermal systems

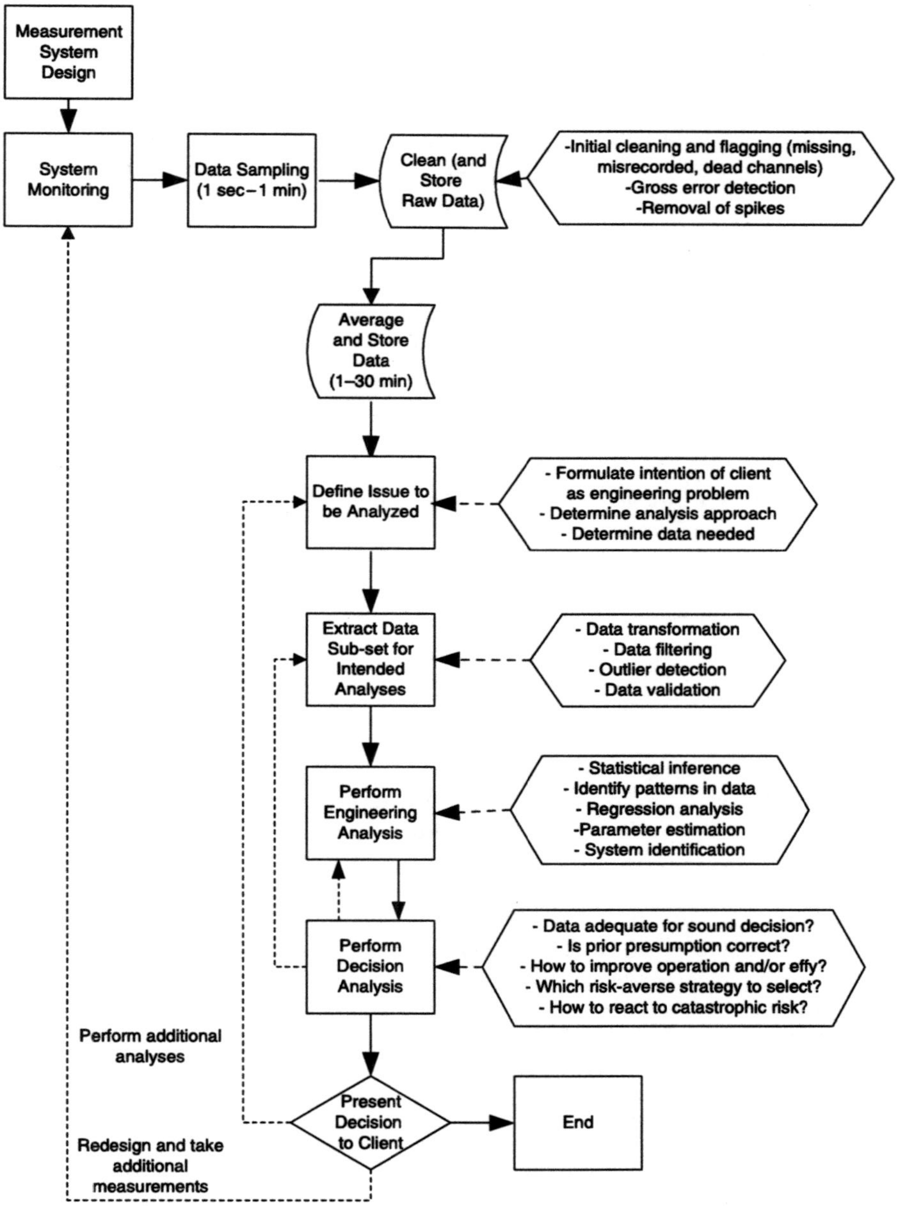

fessional expertise of the consultant with minimal or no measurements of the actual system. However, both financial institutions who would lend the money for implementing these changes or the upper management of the company owning the system are insisting on a more transparent engineering analysis based on actual data. Hence, the preliminary steps involving relevant data extraction and a more careful data proofing and validation are essential;

(g) Extracted data are then subject to certain engineering analyses which can be collectively referred to as data-driven modeling and analysis. These involve statistical inference, identifying patterns in the data, regression analysis, parameter estimation, performance extrapolation, classification or clustering, deterministic modeling,…

(h) Performing a decision analyses, in our context, involves using the results of the engineering analyses and adding an additional layer of analyses that includes modeling uncertainties (involving among other issues a sensitivity analysis), modeling stakeholder preferences and structuring decisions. Several iterations may be necessary between this element and the ones involving engineering analysis and data extraction;

(i) the various choices suggested by the decision analysis are presented to the owner or decision-maker so that a final course of action may be determined. Sometimes, it may be necessary to perform additional analyses or even modify or enhance the capabilities of the measurement system in order to satisfy client needs.

1.8 Decision Analysis and Data Mining

The primary objective of this book is to address element (g) and to some extent element (h) described in the previous section. However, data analysis is not performed just for its own sake; its usefulness lies in the support it provides to such objectives as gaining insight about system behavior which was previously unknown, characterizing current system performance against a baseline, deciding whether retrofits and suggested operational changes to the system are warranted or not, quantifying the uncertainty in predicting future behavior of the present system, suggesting robust/cost effective/risk averse ways to operate an existing system, avoiding catastrophic system failure, etc…

There are two disciplines with overlapping/complementary aims to that of data analysis and modeling which are discussed briefly so as to provide a broad contextual basis to the reader. The first deals with *decision analysis* stated under element (h) above whose objective is to provide both an overall paradigm and a set of tools with which decision makers can construct and analyze a model of a decision situation (Clemen and Reilly 2001). Thus, though it does not give specific answers to problems faced by a person, decision analysis provides a structure, guidance and analytical tools on how to logically and systematically tackle a problem, model uncertainty in different ways, and hopefully arrive at rational decisions in tune with the personal preferences of the individual who has to live with the choice(s) made. While it is applicable to problems without uncertainty but with multiple outcomes, its strength lies in being able to analyze complex multiple outcome problems that are inherently uncertain or stochastic compounded with the utility functions or risk preferences of the decision-maker. There are different sources of uncertainty in a decision process but the one pertinent to data modeling and analysis in the context of this book is that associated with fairly well behaved and well understood engineering systems with relatively low uncertainty in their performance data. This is the reason why historically, engineering students were not subjected to a class in decision analysis. However, many engineering systems are operated wherein the attitudes and behavior of people operating these systems assume importance; in such cases, there is a need to adapt many of the decision analysis tools and concepts with traditional data analysis and modeling techniques. This issue is addressed in Chap. 12.

The second discipline is *data mining* which is defined as the science of extracting useful information from large/enormous data sets. Though it is based on a range of techniques, from the very simple to the sophisticated (involving such methods as clustering techniques, artificial neural networks, genetic algorithms,…), it has the distinguishing feature that it is concerned with shifting through large/enormous amounts of data with no clear aim in mind except to discern hidden information, discover patterns and trends, or summarize data behavior (Dunham 2003). Thus, not only does its distinctiveness lie in the data management problems associated with storing and retrieving large amounts of data from perhaps multiple datasets, but also in it being much more exploratory and less formalized in nature than is statistics and model building where one analyzes a relatively small data set with some specific objective in mind. Data mining has borrowed concepts from several fields such as multivariate statistics and Bayesian theory, as well as less formalized ones such as machine learning, artificial intelligence, pattern recognition, and data management so as to bound its own area of study and define the specific elements and tools involved. It is the result of the digital age where enormous digital databases abound from the mundane (supermarket transactions, credit cards records, telephone calls, internet postings,…) to the very scientific (astronomical data, medical images,…). Thus, the purview of data mining is to explore such data bases in order to find patterns or characteristics (called data discovery) or even in response to some very general research question not provided by any previous mechanistic understanding of the social or engineering system, so that some action can be taken resulting in a benefit or value to the owner. Data mining techniques are not discussed in this book except for those data analysis and modeling issues which are common to both disciplines.

1.9 Structure of Book

The overall structure of the book is depicted in Table 1.3 along with a simple suggestion as to how this book could be used for two courses if necessary. This chapter (Chap. 1) has provided a general introduction of mathematical models, and discussed the different types of problems and analysis tools available for data driven modeling and analysis. Chapter 2 reviews basic probability concepts (both classical and Bayesian), and covers various important probability distributions with emphasis as to their practical usefulness. Chapter 3 reviews rather basic material involving data collection, and preliminary tests within the purview of data validation. It also presents various statistical measures and graphical plots used to describe and scrutinize the data, data errors and their propagation. Chapter 4 covers statistical inference such as hypotheses testing, and ANOVA, as well as non-parametric tests and sampling and re-sampling methods. A brief treatment of Bayesian inference is also provided. Parameter estimation using ordinary least squares (OLS) involving

Table 1.3 Analysis methods covered in this book and suggested curriculum for two courses

Chapter	Topic	First course	Second course
1	Introduction: Mathematical models and data-driven methods	X	X
2	Probability and statistics, important probability distributions	X	
3	Exploratory data analysis and descriptive statistics	X	
4	Inferential statistics, non-parametric tests and sampling	X	
5	OLS regression, residual analysis, point and interval estimation	X	
6	Design of experiments	X	
7	Traditional optimization methods and dynamic programming		X
8	Classification and clustering analysis		X
9	Time series analysis, ARIMA, process monitoring and control		X
10	Parameter estimation methods		X
11	Inverse methods (calibration, system identification, control)		X
12	Decision-making and risk analysis		X

single and multi-linear regression is treated in Chap. 5. Residual analysis, detection of leverage and influential points are also discussed. The material from all these four chapters (Chaps. 2–5) is generally covered in undergraduate statistics and probability classes, and is meant as review or refresher material (especially useful to the general practitioner). Numerous practically-framed examples and problems along with real-world case study examples using actual monitored data are assembled pertinent to energy and environmental issues and equipment (such as solar collectors, pumps, fans, heat exchangers, chillers…). Chapter 6 covers basic classical concepts of experimental design methods, and discusses factorial and response surface methods which allow extending hypothesis testing to multiple variables as well as identifying sound performance models.

Chapter 7 covers traditional optimization methods including dynamic optimization methods. Chapter 8 discusses the basic concepts and some of the analysis methods which allow classification and clustering tasks to be performed. Chapter 9 introduces several methods to smooth time series data analyze time series data in the time domain and to develop forecasting models using both the OLS modeling approach and the ARMA class of models. An overview is also provided of control chart techniques extensively used for process control and condition monitoring. Chapter 10 discusses subtler aspects of parameter estimation such as maximum likelihood estimation, recursive and weighted least squares, robust-fitting techniques, dealing with collinear regressors and error in x models. Computer intensive methods such as bootstrapping are also covered. Chapter 11 presents an overview of the types of problems which fall under inverse modeling: control problems which include inferring inputs and boundary conditions, calibration of white box models and complex linked models requiring computer programs, and system identification using black-box (such as neural networks) and grey-box models (state-space formulation). Illustrative examples are provided in each of these cases. Finally, Chap. 12 covers basic notions relevant to and involving the disciplines of risk analysis and decision-making, and reinforces these by way of examples. It also describes various facets such as framing undesirable events as risks, simplifying and modeling chain of events, assessing trade-offs between competing alternatives, and capturing the risk attitude of the decision-maker. The value of collecting additional information to reduce the risk is also addressed.

Problems

Pr. 1.1 Identify which of the following functions are linear models, which are linear in their parameters (a, b, c) and which are both:

(a) $y = a + bx + cx^2$

(b) $y = a + \frac{b}{x} + \frac{c}{x^2}$

(c) $y = a + b(x-1) + c(x-1)^2$

(d) $y = (a_0 + b_0 x_1 + c_0 x_1^2) + (a_1 + b_1 x_1 + c_1 x_1^2) x_2$

(e) $y = a + b.\sin(c + x)$

(f) $y = a + b\ \sin(cx)$

(g) $y = a + bx^c$

(h) $y = a + bx^{1.5}$

(i) $y = a + b\ e^x$

Pr. 1.2 Recast Eq. 1.1 such that it expresses the fluid volume flow rate (rather than velocity) in terms of pressure drop and other quantities. Draw a block diagram to represent the case when a feedback control is used to control the flow rate from measured pressure drop.

Pr. 1.3 Consider Eq. 1.4 which is a lumped model of a fully-mixed hot water storage tank. Assume initial temperature is

$T_{s,initial}$ = 60°C while the ambient temperature is constant at 20°C.

(i) Deduce the expression for the time constant of the tank in terms of model parameters.
(ii) Compute its numerical value when Mc_p=9.0 MJ/°C and UA=0.833 kW/°C.
(iii) What will be the storage tank temperature after 6 h under cool-down.
(iv) How long will the tank temperature take to drop to 40°C.
(v) Derive the solution for the transient response of the storage tank under electric power input P.
(vi) If P=50 kW, calculate and plot the response when the tank is initially at 30°C (akin to Fig. 1.7).

Pr. 1.4 The first order model of a measurement system is given by Eq. 1.8. Its solution for a step change in the variable being measured results in Eq. 1.9 which is plotted in Fig. 1.7. Derive an analogous model and plot the behavior for a steady sinusoidal variation in the input quantity:

$q_i(t)=A_i \sin(wt)$ where A_i is the amplitude and w the frequency.

Pr. 1.5 Consider Fig. 1.4 where a heated sphere is being cooled. The analysis simplifies considerably if the sphere can be modeled as a lumped one. This can be done if the Biot number $Bi \equiv \frac{hL_e}{k} < 0.1$. Assume that the external heat transfer coefficient is 10 W/m²°C and that the radius of the sphere is 15 cm. The equivalent length of the sphere is $L_e = \frac{\text{Volume}}{\text{Surface area}}$. Determine whether the lumped model assumption is appropriate for spheres made of the following materials:

(a) Steel with thermal conductivity k = 34 W/m K
(b) Copper with thermal conductivity k = 340 W/m K.
(c) Wood with thermal conductivity k = 0.15 W/m K

Pr. 1.6 The thermal network representation of a homogeneous plane is illustrated in Fig. 1.5. Draw the 3R2 C network representation and derive expressions for the three resistors and the two capacitors in terms of the two air film coefficients and the wall properties (Hint: follow the approach illustrated in Fig. 1.5 for the 2R1C network).

Pr. 1.7 *Two pumps in parallel problem viewed from the forward and the inverse perspectives*

Consider Fig. 1.18 which will be analyzed in both the forward and data driven approaches.

(a) *Forward problem*[9]: Two pumps with parallel networks deliver F=0.01 m³/s of water from a reservoir to the destination. The pressure drops in Pascals (Pa) of each network are given by: $\Delta p_1 = (2.1) \cdot 10^{10} \cdot F_1^2$ and $\Delta p_2 = (3.6) \cdot 10^{10} \cdot F_2^2$ where F_1 and F_2 are the flow rates through each branch in m³/s. Assume that pumps and their motor assemblies have the same efficiency. Let P_1 and P_2 be the electric power in Watts (W) consumed by the two pump-motor assemblies.
   (i) Sketch the block diagram for this system with total electric power as the output variable,
   (ii) Frame the total power P as the objective function which needs to be minimized against total delivered water F,
   (iii) Solve the problem for F_1 and P_1 and P.
(b) *Inverse problem*: Now consider the same system in the inverse framework where one would instrument the existing system such that operational measurements of P for different F_1 and F_2 are available.
   (i) Frame the function appropriately using insights into the functional form provided by the forward model.
   (ii) The simplifying assumption of constant efficiency of the pumps is unrealistic. How would the above function need to be reformulated if efficiency can be taken to be a quadratic polynomial (or black-box model) of flow rate as shown below for the first piping branch (with a similar expression applying for the second branch):

$$\eta_1 = a_1 + b_1 \cdot F_1 + c_1 \cdot F_1^2$$

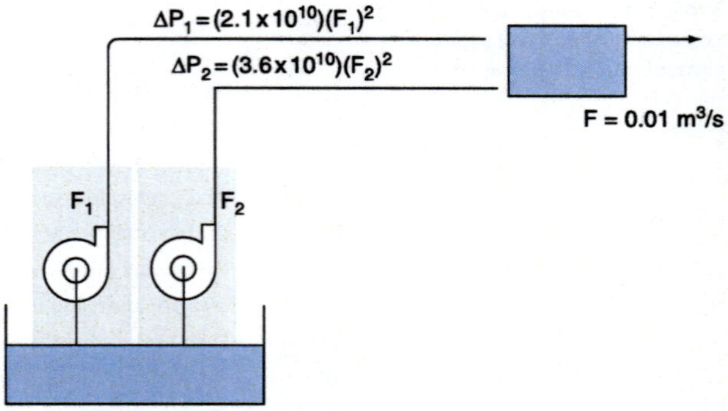

Fig. 1.18 Pumping system with two pumps in parallel

Pr. 1.8 *Lake contamination problem viewed from the forward and the inverse perspectives*

A lake of volume V is fed by an incoming stream with volumetric flow rate Q_s and contaminated with concentration C_s[10] (Fig. 1.19). The outfall of another source (say, the sewage from a factory) also discharges a flow Q_w of the same

[9] From Stoecker (1989) by permission of McGraw-Hill.

[10] From Masters and Ela (2008) by permission of Pearson Education.

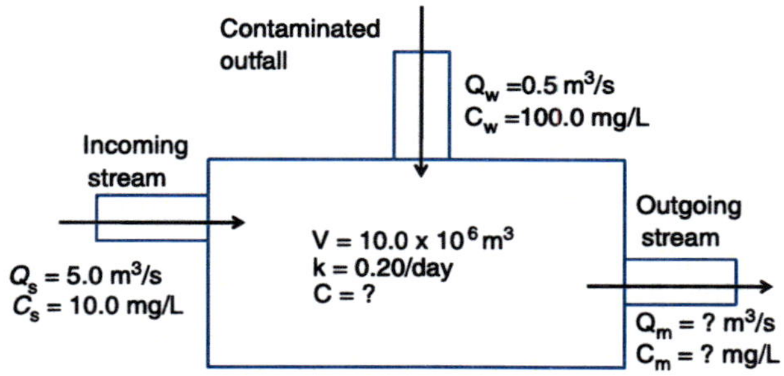

Fig. 1.19 Perspective of the forward problem for the lake contamination situation

pollutant with concentration C_w. The wastes in the stream and sewage have a decay coefficient k.

(a) Let us consider the forward model approach. In order to simplify the problem, the lake will be considered to be a fully mixed compartment and evaporation and seepage losses to the lake bottom will be neglected. In such a case, the concentration of the outflow is equal to that in the lake, i.e., C_m=C. Then, the steady-state concentration in the lake can be determined quite simply: Input rate = Output rate + decay rate
where Input rate = $Q_sC_s+Q_wC_w$, Output rate = Q_mC_m = $(Q_s+Q_w)C_w$, and decay rate =kCV. This results in:

$$C = \frac{Q_sC_s + Q_wC_w}{Q_s + Q_w + kV}$$

Verify the above derived expression, and also check that C = 3.5 mg/L when the numerical values for the various quantities given in Fig. 1.19 are used.

(b) Now consider the inverse control problem when an actual situation can be generally represented by the model treated above. One can envision several scenarios; let us consider a simple one. Flora and fauna downstream of the lake have been found to be adversely affected, and an environmental agency would like to investigate this situation by installing appropriate instrumentation. The agency believes that the factory is polluting the lake, which the factory owner, on the other hand, disputes. Since it is rather difficult to get a good reading of spatial averaged concentrations in the lake, the experimental procedure involves measuring the cross-sectionally averaged concentrations and volumetric flow rates of the incoming, outgoing and outfall streams.
   (i) Using the above model, describe the agency's thought process whereby they would conclude that indeed the factory is the major cause of the pollution.
   (ii) Identify arguments that the factory owner can raise to rebut the agency's findings.

Pr. 1.9 The problem addressed above assumed that only one source of contaminant outfall was present. Rework the problem assuming two sources of outfall with different volumetric flows and concentration levels.

References

Arsham, *http://home.ubalt.edu/ntsbarsh/stat-data/Topics.htm*, downloaded August 2008

Berthold, M. and D.J. Hand (eds.) 2003. *Intelligent Data Analysis*, 2nd Edition, Springer, Berlin.

Cha, P.D., J.J. Rosenberg and C.L. Dym, 2000. *Fundamentals of Modeling and Analyzing Engineering Systems*, 2nd Ed., Cambridge University Press, Cambridge, UK.

Claridge, D.E. and M. Liu, 2001. *HVAC System Commissioning*, Chap. 7.1 *Handbook of Heating, Ventilation and Air Conditioning*, J.F. Kreider (editor), CRC Press, Boca Raton, FL.

Clemen, R.T. and T. Reilly, 2001. *Making Hard Decisions with Decision Tools*, Brooks Cole, Duxbury, Pacific Grove, CA

Energy Plus, 2009. Energy Plus Building Energy Simulation software, developed by the National Renewable Energy Laboratory (NREL) for the U.S. Department of Energy, under the Building Technologies program, Washington DC, USA. *http://www.nrel.gov/buildings/energy_analysis.html#energyplus*.

Edwards, C.H. and D.E. Penney, 1996. *Differential Equations and Boundary Value Problems*, Prentice Hall, Englewood Cliffs, NJ

Eisen, M., 1988. *Mathematical Methods and Models in the Biological Sciences*, Prentice Hall, Englewood Cliffs, NJ.

Doebelin, E.O., 1995. *Engineering Experimentation: Planning, Execution and Reporting*, McGraw-Hill, New York

Dunham, M., 2003. *Data Mining: Introductory and Advanced Topics*, Pearson Education Inc.

Haimes, Y.Y., 1998. *Risk Modeling, Assessment and Management*, John Wiley and Sons, New York.

Heinsohn, R.J. and J.M.Cimbala, 2003, *Indoor Air Quality Engineering*, Marcel Dekker, New York, NY

Hoagin, D.C., F. Moesteller and J.W. Tukey, 1983. *Understanding Robust and Exploratory Analysis*, John Wiley and Sons, New York.

Hodges, J.L. and E.L. Lehman, 1970. *Basic Concepts of Probability and Statistics*, 2nd Edition Holden Day

Jochem, E. 2000. In *Energy End-Use Efficiency in World Energy Assessment*, J. Goldberg, ed., pp. 73–217, United Nations Development Project, New York.

Masters, G.M. and W.P. Ela, 2008. *Introduction to Environmental Engineering and Science*, 3rd Ed. Prentice Hall, Englewood Cliffs, NJ

McNeil, D.R. 1977. *Interactive Data Analysis*, John Wiley and Sons, New York.

PECI, 1997. *Model Commissioning Plan and Guide Commissioning Specifications*, version 2.05, U.S.DOE/PECI, Portland, OR, February.

Reddy, T.A., 2006. Literature review on calibration of building energy simulation programs: Uses, problems, procedures, uncertainty and tools, *ASHRAE Transactions*, 112(1), January

Sprent, P., 1998. *Data Driven Statistical Methods*, Chapman and Hall, London.

Stoecker, W.F., 1989. *Design of Thermal Systems*, 3rd Edition, McGraw-Hill, New York.

Streed, E.R., J.E. Hill, W.C. Thomas, A.G. Dawson and B.D. Wood, 1979. Results and Analysis of a Round Robin Test Program for Liquid-Heating Flat-Plate Solar Collectors, *Solar Energy*, 22, p.235.

Stubberud,A., I. Williams, and J. DiStefano, 1994. *Outline of Feedback and Control Systems*, Schaum Series, McGraw-Hill.

Tukey, J.W., 1988. *The Collected Works of John W. Tukey*, W. Cleveland (Editor), Wadsworth and Brookes/Cole Advanced Books and Software, Pacific Grove, CA

Weiss, N. and M. Hassett, 1982. *Introductory Statistics*, Addison-Wesley. NJ.

2 Probability Concepts and Probability Distributions

This chapter reviews basic notions of probability (or "stochastic variability") which is the formal study of the laws of chance, i.e., where the ambiguity in outcome is inherent in the nature of the process itself. Both the primary views of probability, namely the frequentist (or classical) and the Bayesian, are covered, and some of the important probability distributions are presented. Finally, an effort is made to explain how probability is different from statistics, and to present different views of probability concepts such as absolute, relative and subjective probabilities.

2.1 Introduction

2.1.1 Outcomes and Simple Events

A *random variable* is a numerical description of the outcome of an experiment whose value depends on chance, i.e., whose outcome is not entirely predictable. Tossing a dice is a random experiment. There are two types of random variables:

(i) *discrete* random variable is one that can take on only a finite or countable number of values,

(ii) *continuous* random variable is one that may take on any value in an interval.

The following basic notions relevant to the study of probability apply primarily to *discrete* random variables.

- *Outcome* is the result of a single trial of a random experiment. It cannot be decomposed into anything simpler. For example, getting a {2} when a dice is rolled.
- *Sample space* (some refer to it as "universe") is the set of all possible outcomes of a single trial. For the rolling of a dice, the sample space is $S = \{1, 2, 3, 4, 5, 6\}$.
- *Event* is the combined outcomes (or a collection) of one or more random experiments defined in a specific manner. For example, getting a pre-selected number (say, 4) from adding the outcomes of two dices would constitute a simple event: $A = \{4\}$.
- *Complement of a event* is the set of outcomes in the sample not contained in A. $\bar{A} = \{2, 3, 5, 6, 7, 8, 9, 10, 11, 12\}$ is the complement of the event stated above.

2.1.2 Classical Concept of Probability

Random data by its very nature is indeterminate. So how can a scientific theory attempt to deal with indeterminacy? Probability theory does just that, and is based on the fact that though the result of any particular result of an experiment cannot be predicted, a long sequence of performances taken together reveals a stability that can serve as the basis for fairly precise predictions.

Consider the case when an experiment was carried out a number of times and the anticipated event E occurred in some of them. *Relative frequency* is the ratio denoting the fraction of events when success has occurred. It is usually estimated empirically *after* the event from the following proportion:

$$p(E) = \frac{\text{number of times E occured}}{\text{number of times the experiment was carried out}} \tag{2.1}$$

For certain simpler events, one can determine this proportion without actually carrying out the experiment; this is referred to as "*wise before the event*". For example, the relative frequency of getting heads (selected as a "success" event) when tossing a fair coin is 0.5 In any case, this *apriori* proportion is interpreted as the long run relative frequency, and is referred to as *probability*. This is the classical, or frequentist or traditionalist definition, and has some theoretical basis. This interpretation arises from the *strong law of large numbers* (a well-known result in probability theory) which states that the average of a sequence of independent random variables having the same distribution will converge to the mean of that distribution. If a dice is rolled, the probability of getting a pre-selected number between 1 and 6 (say, 4) will vary from event to event, but on an average will tend to be close to 1/6.

2.1.3 Bayesian Viewpoint of Probability

The classical or traditional probability concepts are associated with the frequentist view of probability, i.e., interpreting

T. Agami Reddy, *Applied Data Analysis and Modeling for Energy Engineers and Scientists*,
DOI 10.1007/978-1-4419-9613-8_2,

probability as the long run frequency. This has a nice intuitive interpretation, hence its appeal. However, people have argued that most processes are unique events and do not occur repeatedly, thereby questioning the validity of the frequentist or objective probability viewpoint. Even when one may have some basic preliminary idea of the probability associated with a certain event, the frequentist view excludes such subjective insights in the determination of probability. The Bayesian approach, however, recognizes such issues by allowing one to update assessments of probability that *integrate prior knowledge with observed events*, thereby allowing better conclusions to be reached. Both the classical and the Bayesian approaches converge to the same results as increasingly more data (or information) is gathered. It is when the data sets are small that the additional benefit of the Bayesian approach becomes advantageous. Thus, the Bayesian view is not an approach which is at odds with the frequentist approach, but rather adds (or allows the addition of) refinement to it. This can be a great benefit in many types of analysis, and therein lies its appeal. The Bayes' theorem and its application to discrete and continuous probability variables are discussed in Sect. 2.5, while Sect. 4.6 (of Chap. 4) presents its application to estimation and hypothesis problems.

2.2 Classical Probability

2.2.1 Permutations and Combinations

The very first concept needed for the study of probability is a sound knowledge of *combinatorial mathematics* which is concerned with developing rules for situations involving permutations and combinations.

(a) Permutation P(n, k) is the number of ways that k objects can be selected from n objects *with the order being important*. It is given by:

$$P(n,k) = \frac{n!}{(n-k)!} \tag{2.2a}$$

A special case is the number of permutations of n objects taken n at a time:

$$P(n,n) = n! = n(n-1)(n-2)...(2)(1) \tag{2.2b}$$

(b) Combinations C(n, k) is the number of ways that k objects can be selected from n objects *with the order not being important*. It is given by:

$$C(n,k) = \frac{n!}{(n-k)!k!} \equiv \binom{n}{k} \tag{2.3}$$

Note that the same equation also defines the *binomial* coefficients since the expansion of $(a+b)^n$ according to the Binomial theorem is

$$(a+b)^n = \sum_{k=0}^{n} \binom{n}{k} a^{n-k} b^k. \tag{2.4}$$

Example 2.2.1: (a) Calculate the number of ways in which three people from a group of seven people can be seated in a row.
This is a case of permutation since the order is important. The number of possible ways is:

$$P(7,3) = \frac{7!}{(7-3)!} = \frac{(7)\cdot(6)\cdot(5)}{1} = 2110$$

(b) Calculate the number of combinations in which three people can be selected from a group of seven.

Here the order is not important and the combination formula can be used. Thus:

$$C(7,3) = \frac{7!}{(7-3)!3!} = \frac{(7)\cdot(6)\cdot(5)}{(3)\cdot(2)} = 35$$ ■

Another type of combinatorial problem is the *factorial problem* to be discussed in Chap. 6 while dealing with design of experiments. Consider a specific example involving *equipment scheduling* at a physical plant of a large campus which includes primemovers (diesel engines or turbines which produce electricity), boilers and chillers (vapor compression and absorption machines). Such equipment need a certain amount of time to come online and so operators typically keep some of them "idling" so that they can start supplying electricity/heating/cooling at a moment's notice. Their operating states can be designated by a binary variable; say "1" for on-status and "0" for off-status. Extensions of this concept include cases where, instead of two states, one could have m states. An example of 3 states is when say two identical boilers are to be scheduled. One could have three states altogether: (i) when both are off (0–0), (ii) when both are on (1–1), and (iii) when only one is on (1–0). Since the boilers are identical, state (iii) is identical to 0–1. In case, the two boilers are of different size, there would be four possible states. The number of combinations possible for "n" such equipment where each one can assume "m" states is given by m^n. Some simple cases for scheduling four different types of energy equipment in a physical plant are shown in Table 2.1.

2.2.2 Compound Events and Probability Trees

A compound or joint or composite event is one which arises from operations involving two or more events. The use of Venn's diagram is a very convenient manner of illustrating and understanding compound events and their probabilities (see Fig. 2.1).

Table 2.1 Number of combinations for equipment scheduling in a large facility

	Status (0- off, 1- on)				Number of Combinations
	Primemovers	Boilers	Chillers-Vapor compression	Chillers-Absorption	
One of each	0–1	0–1	0–1	0–1	$2^4=16$
Two of each-assumed identical	0–0, 0–1, 1–1	0–0, 0–1, 1–1	0–0, 0–1, 1–1	0–0, 0–1, 1–1	$3^4=81$
Two of each-non-identical except for boilers	0–0, 0–1, 1–0, 1–1	0–0, 0–1, 1–0	0–0, 0–1, 1–0, 1–1	0–0, 0–1, 1–0, 1–1	$4^3\times3^1=192$

- The universe of outcomes or sample space is denoted by a rectangle, while the probability of a particular event (say, event A) is denoted by a region (see Fig. 2.1a);
- *union* of two events A and B (see Fig. 2.1b) is represented by the set of outcomes in *either* A or B or both, and is denoted by $A \cup B$ (where the symbol $\cup$ is conveniently remembered as "u" of "union"). An example is the number of cards in a pack which are either hearts or spades (26 nos.);
- *intersection* of two events A and B is represented by the set of outcomes in *both* A and B simultaneously, and is denoted by $A \cap B$. It is represented by the hatched area in Fig. 2.1b. An example is the number of red cards which are jacks (2 nos.);
- *mutually exclusive events* or disjoint events are those which have no outcomes in common (Fig. 2.1c). An example is the number of red cards with spades seven (nil);

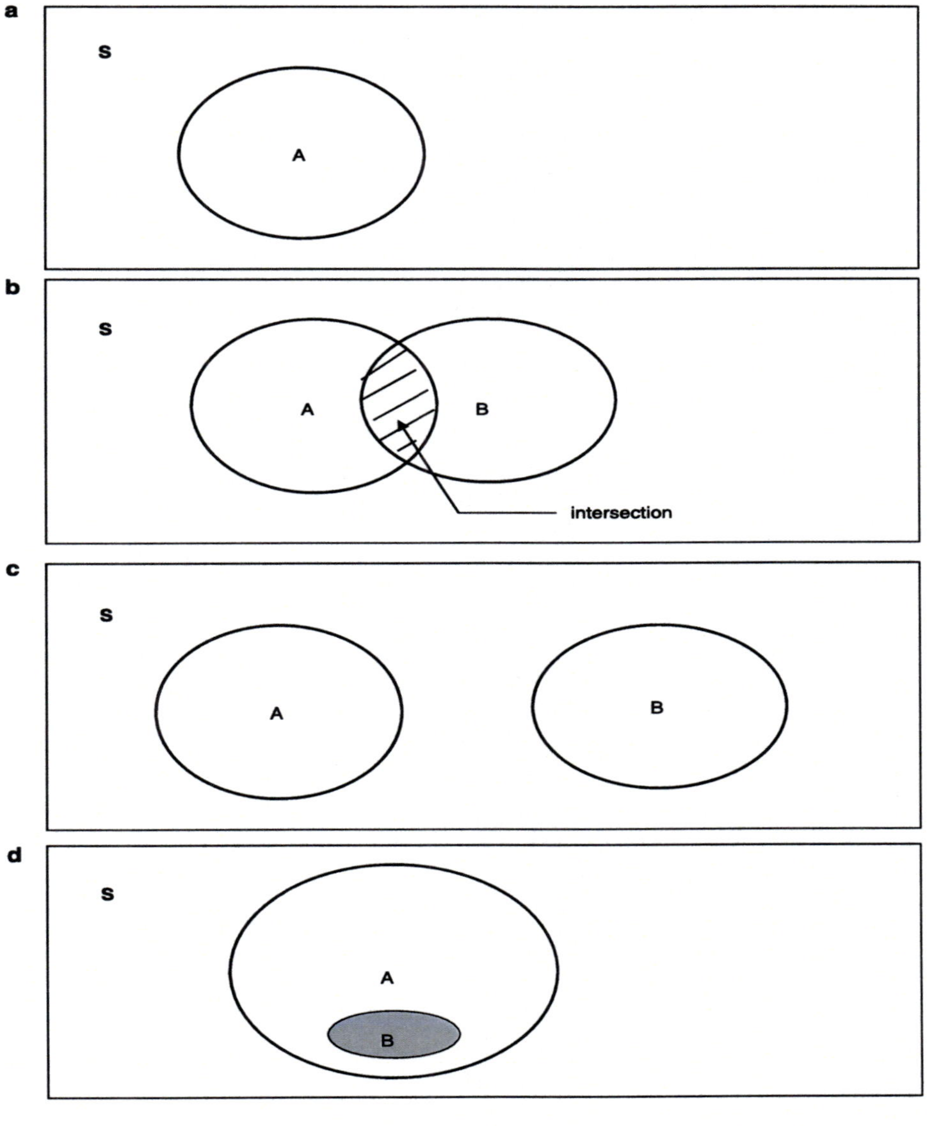

Fig. 2.1 Venn diagrams for a few simple cases. **a** Event A is denoted as a region in space S. Probability of event A is represented by the area inside the circle to that inside the rectangle. **b** Events A and B are intersecting, i.e., have a common overlapping area (shown hatched). **c** Events A and B are mutually exclusive or are disjoint events. **d** Event B is a subset of event A

- event B is inclusive in event A when all outcomes of B are contained in those of A, i.e., B is *a sub-set* of A (Fig. 2.1d). An example is the number of cards less than six (event B) which are red cards (event A).

2.2.3 Axioms of Probability

Let the sample space S consist of two events A and B with probabilities p(A) and p(B) respectively. Then:

(i) probability of any event, say A, cannot be negative. This is expressed as:

$$p(A) \geq 0 \tag{2.5}$$

(ii) probabilities of all events must be unity (i.e., normalized):

$$p(S) \equiv p(A) + p(B) = 1 \tag{2.6}$$

(iii) probabilities of mutually exclusive events add up:

$$p(A \cup B) = p(A) + p(B) \quad \text{if A and B are mutually exclusive} \tag{2.7}$$

If a dice is rolled, the outcomes are mutually exclusive. If event A is the occurrence of 2 and event B that of 3, then p(A or B)=1/6+1/6=1/3. Mutually exclusive events and independent events are not to be confused. While the former is a property of the events themselves, the latter is a property that arises from the event probabilities and their intersections (this is elaborated further below).

Some other inferred relations are:

(iv) probability of the complement of event A:

$$p(\bar{A}) = 1 - p(A) \tag{2.8}$$

(v) probability for *either* A or B (when they are not mutually exclusive) to occur is equal to:

$$p(A \cup B) = p(A) + p(B) - p(A \cap B) \tag{2.9}$$

This is intuitively obvious from the Venn diagram (see Fig. 2.1b) since the hatched area (representing $p(A \cap B)$) gets counted twice in the sum and, so needs to be deducted once. This equation can also be deduced from the axioms of probability. Note that if events A and B are mutually exclusive, then Eq. 2.9 reduces to Eq. 2.7.

2.2.4 Joint, Marginal and Conditional Probabilities

(a) *Joint probability* of two independent events represents the case when both events occur together, i.e. $p(A \text{ and } B) = p(A \cap B)$. It is equal to:

$$p(A \cap B) = p(A) \cdot p(B) \quad \text{if A and B are independent} \tag{2.10}$$

These are called *product models*. Consider a dice tossing experiment. If event A is the occurrence of an even number, then p(A)=1/2. If event B is that the number is less than or equal to 4, then p(B)=2/3. The probability that both events occur when a dice is rolled is p(A and B)=1/2×2/3=1/3. This is consistent with our intuition since events {2,4} would satisfy both the events.

(b) *Marginal probability* of an event A refers to the probability of A in a joint probability setting. For example, consider a space containing two events, A and B. Since S can be taken to be the sum of event space *B* and its complement $\bar{B}$, the probability of A can be expressed in terms of the sum of the disjoint parts of B:

$$p(A) = p(A \cap B) + p(A \cap \bar{B}) \tag{2.11}$$

This notion can be extended to the case of more than two joint events.

Example 2.2.2: Consider an experiment involving drawing two cards from a deck with replacement. Let event A = {first card is a red one} and event B = {card is between 2 and 8 inclusive}. How Eq. 2.11 applies to this situation is easily shown.

Possible events A: hearts (13 cards) plus diamonds (13 cards)

Possible events B: 4 suites of 2, 3, 4, 5, 6, 7, 8.

Also, $p(A \cap B) = \frac{1}{2} \cdot \frac{(7) \cdot (4)}{52} = \frac{14}{52}$ and

$$p(A \cap \bar{B}) = \frac{1}{2} \cdot \frac{(13-7) \cdot (4)}{52} = \frac{12}{52}$$

Consequently, from Eq. 2.11: $p(A) = \frac{14}{52} + \frac{12}{52} = \frac{1}{2}$.

This result of $p(A)=1/2$ is obvious in this simple experiment, and could have been deduced intuitively. However, intuition may mislead in more complex cases, and hence, the usefulness of this approach. ■

(c) *Conditional probability*: There are several situations involving compound outcomes that are sequential or successive in nature. The chance result of the first stage determines the conditions under which the next stage occurs. Such events, called *two-stage* (or multi-stage) events, involve step-by-step outcomes which can be represented as a *probability tree*. This allows better visualization of how the probabilities progress from one stage to the next. If A and B are events, then the probability that event B occurs given that A has already occurred is given by:

$$p(B/A) = \frac{p(A \cap B)}{p(A)} \tag{2.12}$$

A special but important case is when p(B/A)=p(B). In this case, B is said to be *independent* of A because the fact that event A has occurred does not affect the probability of B occurring. Thus, two events A and B are mutually exclusive if *p(B/A)=p(B)*. In this case, one gets back Eq. 2.10.

An example of a conditional probability event is the drawing of a spade from a pack of cards from which a first card was already drawn. If it is known that the first card *was not* a spade, then the probability of drawing a spade the second time is 12/51 = 4/17. On the other hand, if the first card drawn *was* a spade, then the probability of getting a spade on the second draw is 11/51.

Example 2.2.3: A single fair dice is rolled. Let event A= {even outcome} and event B={outcome is divisible by 3}.

(a) List the various events in the sample space: {1 2 3 4 5 6}
(b) List the outcomes in A and find p(A): {2 4 6}, p(A)=1/2
(c) List the outcomes of B and find p(B): {3 6}, p(B)=1/3
(d) List the outcomes in $A \cap B$ and find $p(A \cap B)$: {6}, $p(A \cap B)=1/6$
(e) Are the events A and B independent? Yes, since Eq. 2.10 holds ■

Example 2.2.4: Two defective bulbs have been mixed with 10 good ones. Let event A= {first bulb is good}, and event B={second bulb is good}.

(a) If two bulbs are chosen at random with replacement, what is the probability that both are good?
p(A)=8/10 and *p(B)*=8/10. Then:

$$p(A \cap B) = \frac{8}{10} \cdot \frac{8}{10} = \frac{64}{100} = 0.64$$

(b) What is the probability that two bulbs drawn in sequence (i.e., not replaced) are good where the status of the bulb can be checked after the first draw?
From Eq. 2.12, p(both bulbs drawn are good):

$$p(A \cap B) = p(A) \cdot p(B/A) = \frac{8}{10} \cdot \frac{7}{9} = \frac{28}{45} = 0.622$$

■

Example 2.2.5: Two events A and B have the following probabilities: $p(A) = 0.3$, $p(B) = 0.4$ and $p(\bar{A} \cap B) = 0.28$.

(a) Determine whether the events A and B are independent or not?
From Eq. 2.8, $P(\bar{A}) = 1 - p(A) = 0.7$. Next, one will verify whether Eq. 2.10 holds or not. In this case, one needs to verify whether: $p(\bar{A} \cap B) = p(\bar{A}) \cdot p(B) \cdot$ or whether 0.28 is equal to (0.7×0.4). Since this is correct, one can state that events A and B are independent.
(b) Find $p(A \cup B)$
From Eqs. 2.9 and 2.10:

$$\begin{aligned} p(A \cup B) &= p(A) + p(B) - p(A \cap B) \\ &= p(A) + p(B) - p(A) \cdot p(B) \\ &= 0.3 + 0.4 - (0.3)(0.4) = 0.58 \end{aligned}$$

■

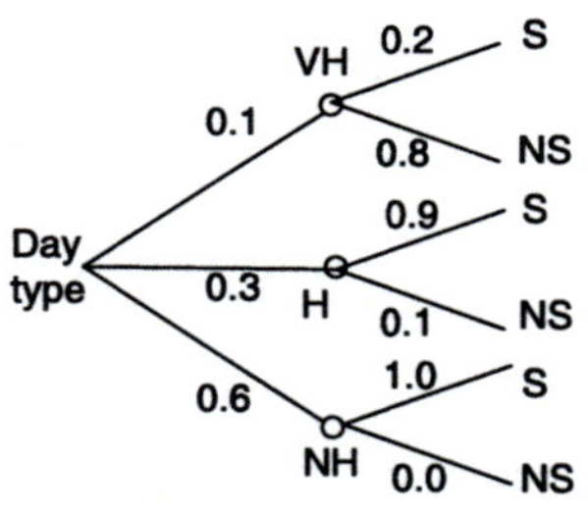

Fig. 2.2 The forward probability tree for the residential air-conditioner when two outcomes are possible (*S* satisfactory or *NS* not satisfactory) for each of three day-types (*VH* very hot, *H* hot and *NH* not hot)

Example 2.2.6: *Generating a probability tree for a residential air-conditioning (AC) system.*

Assume that the AC is slightly under-sized for the house it serves. There are two possible outcomes (S- satisfactory and NS- not satisfactory) depending on whether the AC is able to maintain the desired indoor temperature. The outcomes depend on the outdoor temperature, and for simplicity, its annual variability is grouped into three categories: very hot (VH), hot (H) and not hot (NH). The probabilities for outcomes S and NS to occur in each of the three day-type categories are shown in the probability tree diagram (Fig. 2.2) while the joint probabilities computed following Eq. 2.10 are assembled in Table 2.2.

Note that the relative probabilities of the three branches in both the first stage as well as in each of the two branches of each outcome add to unity (for example, in the Very Hot, the S and NS outcomes add to 1.0, and so on). Further, note that the joint probabilities shown in the table also have to sum to unity (it is advisable to perform such verification checks). The probability of the indoor conditions being satisfactory is determined as: p(S)=0.02+0.27+0.6=0.89 while p(NS)= 0.08+0.03+0=0.11. It is wise to verify that p(S)+p(NS)=1.0. ■

Example 2.2.7: Consider a problem where there are two boxes with marbles as specified:
Box 1: 1 red and 1 white and Box 2: 4 red and 1 green

A box is chosen at random and a marble drawn from it. What is the probability of getting a red marble?

One is tempted to say that since there are 4 red marbles in total out of 6 marbles, the probability is 2/3. However, this is incorrect, and the proper analysis approach requires that one frame this problem as a two-stage experiment. The first stage is the selection of the box, and the second the drawing

Table 2.2 Joint probabilities of various outcomes

$p(VH \cap S) = 0.1 \times 0.2 = 0.02$
$p(VH \cap NS) = 0.1 \times 0.8 = 0.08$
$p(H \cap S) = 0.3 \times 0.9 = 0.27$
$p(H \cap NS) = 0.3 \times 0.1 = 0.03$
$p(NH \cap S) = 0.6 \times 1.0 = 0.6$
$p(NH \cap NS) = 0.6 \times 0 = 0$

Table 2.3 Probabilities of various outcomes

$p(A \cap R) = 1/2 \times 1/2 = 1/4$	$p(B \cap R) = 1/2 \times 3/4 = 3/8$
$p(A \cap W) = 1/2 \times 1/2 = 1/4$	$p(B \cap W) = 1/2 \times 0 = 0$
$p(A \cap G) = 1/2 \times 0 = 0$	$p(B \cap G) = 1/2 \times 1/4 = 1/8$

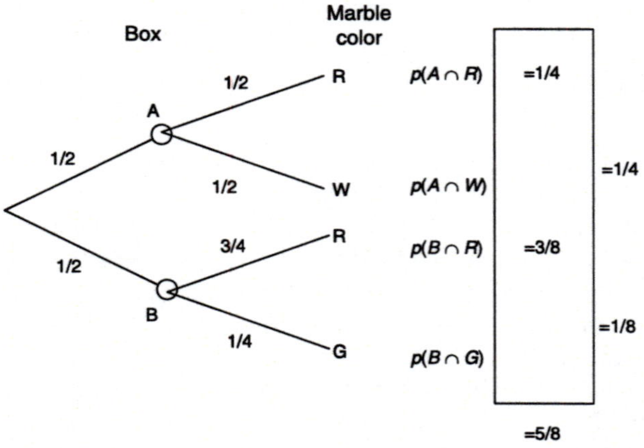

Fig. 2.3 The first stage of the forward probability tree diagram involves selecting a box (either A or B) while the second stage involves drawing a marble which can be red (R), white (W) or green (G) in color. The total probability of drawing a red marble is 5/8

of the marble. Let event A (or event B) denote choosing Box 1 (or Box 2). Let R, W and G represent red, white and green marbles. The resulting probabilities are shown in Table 2.3.

Thus, the probability of getting a red marble=1/4+3/8=5/8. The above example is depicted in Fig. 2.3 where the reader can visually note how the probabilities propagate through the probability tree. This is called the "*forward tree*" to differentiate it from the "reverse" tree discussed in Sect. 2.5.

The above example illustrates how a two-stage experiment has to be approached. First, one selects a box which by itself does not tell us whether the marble is red (since one has yet to pick a marble). Only after a box is selected, can one use the prior probabilities regarding the color of the marbles inside the box in question to determine the probability of picking a red marble. These prior probabilities can be viewed as conditional probabilities; i.e., for example, $p(A \cap R) = p(R/A) \cdot p(A)$ ■

2.3 Probability Distribution Functions

2.3.1 Density Functions

The notions of discrete and continuous random variables were introduced in Sect. 2.1.1. The distribution of a random variable represents the probability of it taking its various possible values. For example, if the y-axis in Fig. 1.1 of the dice rolling experiment were to be changed into a relative frequency (=1/6), the resulting histogram would graphically represent the corresponding *probability density function (PDF)* (Fig. 2.4a). Thus, the probability of getting a 2 in the rolling of a dice is 1/6th. Since, this is a discrete random variable, the function takes on specific values at discrete points of the x-axis (which represents the outcomes). The same type of y-axis normalization done to the data shown in Fig. 1.2 would result in the PDF for the case of continuous random data. This is shown in Fig. 2.5a for the random variable taken to be the hourly outdoor dry bulb temperature over the year at Phila-

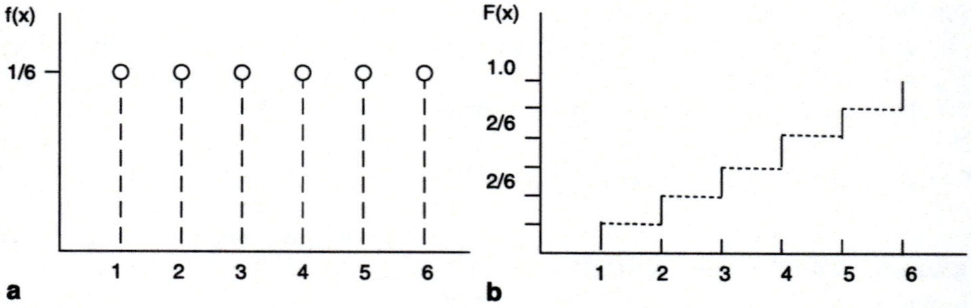

Fig. 2.4 Probability functions for a discrete random variable involving the outcome of rolling a dice. **a** Probability density function. **b** Cumulative distribution function

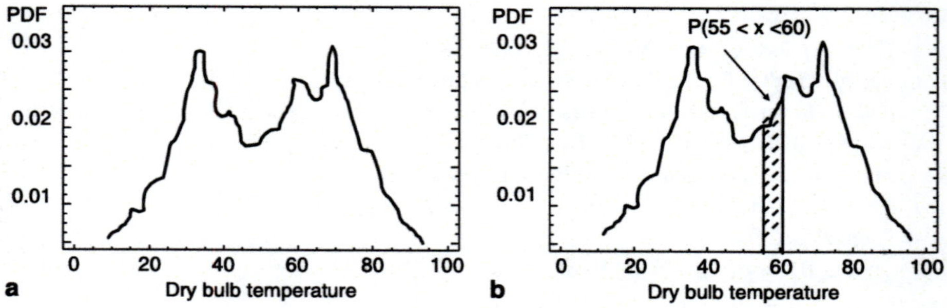

Fig. 2.5 Probability density function and its association with probability for a continuous random variable involving the outcomes of hourly outdoor temperatures at Philadelphia, PA during a year. The probability that the temperature will be between 55° and 60°F is given by the shaded area. **a** Density function. **b** Probability interpreted as an area

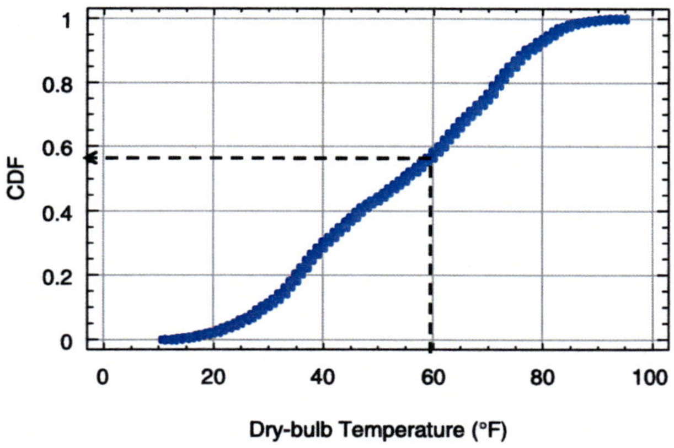

Fig. 2.6 The cumulative distribution function (CDF) for the PDF shown in Fig. 2.5. Such a plot allows one to easily determine the probability that the temperature is less than 60°F

delphia, PA. Notice that this is the envelope of the histogram of Fig. 1.2. Since the variable is continuous, it is implausible to try to determine the probability of, say temperature outcome of 57.5°F. One would be interested in the probability of outcomes within a range, say 55°–60°F. The probability can then be determined as the area under the PDF as shown in Fig. 2.5b. It is for such continuous random variables that the *cumulative distribution function (CDF)* is useful. It is simply the cumulative area under the curve starting from the lowest value of the random variable to the current value (Fig. 2.6). The vertical scale directly gives the probability (or, in this case, the fractional time) that X is less than or greater than a certain value. Thus, the probability $(x \le 60)$ is about 0.58. The concept of CDF also applies to discrete variables as illustrated in Fig. 2.4b for the dice rolling example.

To restate, depending on whether the random variable is discrete or continuous, one gets discrete or continuous probability distributions. Though most experimentally gathered data is discrete, the underlying probability theory is based on the data being continuous. Replacing the integration sign by the summation sign in the equations that follow allows extending the following definitions to discrete distributions. Let f(x) be the probability distribution function associated with a random variable X. This is a function which provides the probability that a discrete random variable X takes on some specific value x among its various possible values. The axioms of probability (Eqs. 2.5 and 2.6) for the discrete case are expressed for the case of continuous random variables as:

- PDF cannot be negative:

$$f(x) \ge 0 \qquad -\infty < x < \infty \tag{2.13}$$

- Probability of the sum of all outcomes must be unity

$$\int_{-\infty}^{\infty} f(x)dx = 1 \tag{2.14}$$

The cumulative distribution function (CDF) or F(a) represents the area under f(x) enclosed in the range $-\infty < x < a$:

$$F(a) = p\{X \le a\} = \int_{-\infty}^{a} f(x)dx \tag{2.15}$$

The inverse relationship between f(x) and F(a), provided a derivative exists, is:

$$f(x) = \frac{dF(x)}{dx} \tag{2.16}$$

This leads to the probability of an outcome $a \le X \le b$ given by:

$$\begin{aligned} p\{a \le X \le b\} &= \int_{a}^{b} f(x)dx \\ &= \int_{-\infty}^{b} f(x)dx - \int_{-\infty}^{a} f(x)dx \\ &= F(b) - F(a) \end{aligned} \tag{2.17}$$

Notice that the CDF for discrete variables will be a step function (as in Fig. 2.4b) since the PDF is defined at discrete values only. Also, the CDF for continuous variables is a function which increases monotonically with increasing x. For example, the probability of the outdoor temperature being between 55° and 60°F is given by $p\{55 \le X \le 60\} = F(b) - F(a) = 0.58 - 0.50 = 0.08$ (see Fig. 2.6).

The concept of probability distribution functions can be extended to the treatment of simultaneous outcomes of multiple random variables. For example, one would like to study how temperature of quenching of a particular item made of steel affects its hardness. Let X and Y be the two random variables. The probability that they occur together can be represented by a function $f(x, y)$ for any pair of values (x, y) within the range of variability of the random variables X and Y. This function is referred to as the *joint probability density function* of X and Y which has to satisfy the following properties for continuous variables:

$$f(x, y) \ge 0 \quad \text{for all} \quad (x, y) \tag{2.18}$$

$$\int_{-\infty}^{\infty}\int_{-\infty}^{\infty} f(x, y)dxdy = 1 \tag{2.19}$$

$$p[(X, Y) \in A] = \iint_{A} f(x, y)dxdy \tag{2.20}$$

where A is any region in the xy plane.

If X and Y are two *independent* random variables, their joint PDF will be the product of their marginal ones:

$$f(x,y) = f(x) \cdot f(y) \tag{2.21}$$

Note that this is the continuous variable counterpart of Eq. 2.10 which gives the joint probability of two discrete events.

The *marginal distribution* of X given two jointly distributed random variables X and Y is simply the probability distribution of X ignoring that of Y. This is determined for X as:

$$g(x) = \int_{-\infty}^{\infty} f(x,y)dy \tag{2.22}$$

Finally, the *conditional probability distribution* of X given that X=x for two jointly distributed random variables X and Y is:

$$f(y/x) = \frac{f(x,y)}{g(x)} \qquad g(x) > 0 \tag{2.23}$$

Example 2.3.1: Determine the value of c so that each of the following functions can serve as probability distributions of the discrete random variable X:

(a) $f(x) = c(x^2 + 4) \quad for\ x = 0, 1, 2, 3$

(b) $f(x) = ax^2 \qquad for\ -1 < x < 2$

(a) One uses the discrete version of Eq. 2.14, i.e., $\sum_{i=0}^{3} f(x_i) = 1$ leads to 4c+5c+8c+13c=1 from which c=1/30
(b) One uses Eq. 2.14 modified for the limiting range in x: $\int_{-1}^{2} ax^2 dx = 1$ from which $\left[\frac{ax^3}{3}\right]_{-1}^{2} = 1$ resulting in $a = 1/3$. ■

Example 2.3.2: The operating life in weeks of a high efficiency air filter in an industrial plant is a random variable X having the PDF:

$$f(x) = \frac{20}{(x+100)^3} \qquad for\ x > 0$$

Find the probability that the filter will have an operating life of:
(a) at least 20 weeks
(b) anywhere between 80 and 120 weeks

First, determine the expression for the CDF from Eq. 2.14. Since the operating life would decrease with time, one needs to be careful about the limits of integration applicable to this case. Thus,

$$\text{CDF} = \int_{x}^{0} \frac{20}{(x'+100)^3} dx' = \left[-\frac{10}{(x'+100)^2}\right]_{x}^{0}$$

(a) with x=20, the probability that the life is at least 20 weeks:

$$p(20 < X < \infty) = \left[-\frac{10}{(x+100)^2}\right]_{20}^{\infty} = 0.000694$$

(b) for this case, the limits of integration are simply modified as follows:

$$p(80 < X < 120) = \left[-\frac{10}{(x+100)^2}\right]_{80}^{120} = 0.000102$$ ■

Example 2.3.3: Consider two random variables X and Y with the following joint density function:

$$f(x,y) = \frac{2}{5}(2x + 3y)$$
$$for\ 0 \le x \le 1, 0 \le y \le 1$$

(a) Verify whether the normalization criterion is satisfied. This is easily verified from Eq. 2.19:

$$\int_{-\infty}^{\infty}\int_{-\infty}^{\infty} f(x,y)dxdy = \int_{0}^{1}\int_{0}^{1} \frac{2}{5}(2x+3y)dxdy$$
$$= \int_{0}^{1} \left[\frac{2x^2}{5} + \frac{6xy}{5}\right]_{x=0}^{x=1} dy$$
$$= \int_{0}^{1} \left(\frac{2}{5} + \frac{6y}{5}\right) dy = \frac{2}{5} + \frac{3}{5} = 1$$

(b) Determine the joint probability in the region $(0 < x < 1/2,\ 1/4 < y < 1/2)$. In this case, one uses Eq. 2.20 as follows:

$$p(0 < X < 1/2, 1/4 < Y < 1/2) = \int_{1/4}^{1/2}\int_{0}^{1/2} \frac{2}{5}(2x+3y)dxdy = \frac{13}{160}$$

(c) Determine the marginal distribution g(x). From Eq. 2.22:

$$g(x) = \int_{0}^{1} \frac{2}{5}(2x+3y)dy = \left[\frac{4xy}{5} + \frac{6y^2}{10}\right]_{y=0}^{y=1} = \frac{4x+3}{5}$$ ■

Table 2.4 Computing marginal probabilities from a probability table

Age (Y)	Income (X)			Marginal probability of Y
	>$ 40,000	40,000–90,000	<90,000	
Under 25	0.15	0.09	0.05	0.29
Between 25–40	0.10	0.16	0.12.	0.38
Above 40	0.08	0.20	0.05	0.33
Marginal probability of X	0.33	0.45	0.22	Should sum to 1.00 both ways

Example 2.3.4: The percentage data of annual income versus age has been gathered from a large population living in a certain region— see Table 2.4. Let X be the income and Y the age. The marginal probability of X for each class is simply the sum of the probabilities under each column and that of Y the sum of those for each row. Thus, $p(X \geq 40,000) = 0.15 + 0.10 + 0.08 = 0.33$, and so on. Also, verify that the sum of the marginal probabilities of X and Y sum to 1.00 (so as to satisfy the normalization condition). ■

2.3.2 Expectation and Moments

This section deals with ways by which one can summarize the characteristics of a probability function using a few important measures. Commonly, the mean or the expected value E[X] is used as a measure of the central tendency of the distribution, and the variance var[X] as a measure of dispersion of the distribution about its mean. These are very similar to the notions of arithmetic mean and variance of a set of data. As before, the equations which apply to continuous random variables are shown below; in case of discrete variables, the integrals have to be replaced with summations.

- *expected value of the first moment or mean*

$$E[X] \equiv \mu = \int_{-\infty}^{\infty} xf(x)dx. \tag{2.24}$$

The mean is exactly analogous to the physical concept of center of gravity of a mass distribution. This is the reason why PDF are also referred to as the "mass distribution function". The concept of symmetry of a PDF is an important one implying that the distribution is symmetric about the mean. A distribution is symmetric if: $p(\mu - x) = p(\mu + x)$ for every value of x.

- *variance*

$$\mathrm{var}[X] \equiv \sigma^2 = E[(X-\mu)^2] = \int_{-\infty}^{\infty} (x-\mu)^2 f(x)dx \tag{2.25a}$$

Alternatively, it can be shown that for any discrete distribution:

$$\mathrm{var}[X] = E[X^2] - \mu^2 \tag{2.25b}$$

Notice the appearance of the expected value of the second moment $E[X^2]$ in the above equation. The variance is analogous to the physical concept of the moment of inertia of a mass distribution about its center of gravity.

In order to express the variance which is a measure of dispersion in the same units as the random variable itself, the square root of the variance, namely the *standard deviation* σ is used. Finally, errors have to be viewed, or evaluated, in terms of the magnitude of the random variable. Thus, the *relative error* is often of more importance than the actual error. This has led to the widespread use of a dimensionless quantity called the *Coefficient of Variation* (CV) defined as the percentage ratio of the standard deviation to the mean:

$$CV = 100 \cdot \left(\frac{\sigma}{\mu}\right) \tag{2.26}$$

2.3.3 Function of Random Variables

The above definitions can be extended to the case when the random variable X is a function of several random variables; for example:

$$X = a_0 + a_1 X_1 + a_2 X_2 ... \tag{2.27}$$

where the a_i coefficients are constants and X_i are random variables.

Some important relations regarding the mean:

$$\begin{aligned} &E[a_0] = a_0 \\ &E[a_1 X_1] = a_1 E[X_1] \\ &E[a_0 + a_1 X_1 + a_2 X_2] = a_0 + a_1 E[X_1] + a_2 E[X_2] \end{aligned} \tag{2.28}$$

Similarly there are a few important relations that apply to the variance:

$$\begin{aligned} &\mathrm{var}[a_0] = 0 \\ &\mathrm{var}[a_1 X_1] = a_1^2 \mathrm{var}[X_1] \end{aligned} \tag{2.29}$$

Again, if the two random variables are independent,

$$\mathrm{var}[a_0 + a_1 X_1 + a_2 X_2] = a_1^2 \mathrm{var}[X_1] + a_2^2 \mathrm{var}[X_2] \tag{2.30}$$

The notion of *covariance* of two random variables is an important one since it is a measure of the tendency of two random variables to vary together. The covariance is defined as:

$$\mathrm{cov}[X_1, X_2] = E[(X_1 - \mu_1) \cdot (X_2 - \mu_2)] \tag{2.31}$$

where μ_1 and μ_2 are the mean values of the random variables X_1 and X_2 respectively. Thus, for the case of two random

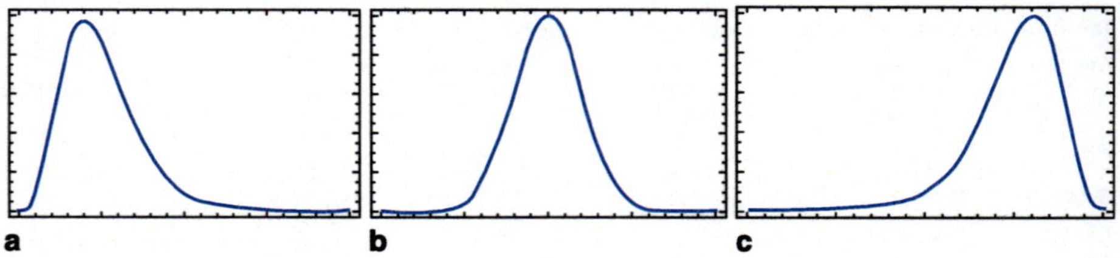

Fig. 2.7 Skewed and symmetric distributions. **a** Skewed to the right. **b** Symmetric. **c** Skewed to the left

variables which are not independent, Eq. 2.30 needs to be modified into:

$$\begin{aligned}\text{var}[a_0 + a_1 X_1 + a_2 X_2] = a_1^2 \text{var}[X_1] + a_2^2 \text{var}[X_2] \\ + 2a_1 a_2.\text{cov}[X_1, X_2]\end{aligned} \tag{2.32}$$

Moments higher than the second are sometimes used. For example, the third moment yields the *skewness* which is a measure of the symmetry of the PDF. Figure 2.7 shows three distributions: one skewed to the right, a symmetric distribution, and one skewed to the left. The fourth moment yields the *coefficient of kurtosis* which is a measure of the peakiness of the PDF.

Two commonly encountered terms are the median and the mode. The value of the random variable at which the PDF has a peak is the *mode*, while the *median* divides the PDF into two equal parts (each part representing a probability of 0.5).

Finally, distributions can also be described by the number of "humps" they display. Figure 2.8 depicts the case of unimodal and bi-modal distributions, while Fig. 2.5 is the case of a distribution with three humps.

Example 2.3.5: Let X be a random variable representing the number of students who fail a class. Its PDF is given in Table 2.5.
The discrete event form of Eqs. 2.24 and 2.25 is used to compute the mean and the variance:

$$\begin{aligned}\mu &= (0)(0.51) + (1)(0.38) + (2)(0.10) + (3)(0.01) \\ &= 0.61\end{aligned}$$

Further:

$$\begin{aligned}E(X^2) &= (0)(0.51) + (1^2)(0.38) + (2^2)(0.10) + (3^2)(0.01) \\ &= 0.87\end{aligned}$$

Hence:

$$\sigma^2 = 0.87 - (0.61)^2 = 0.4979$$ ■

Table 2.5 PDF of number of students failing a class

X	0	1	2	3
f(x)	0.51	0.38	0.10	0.01

Example 2.3.6: Consider Example 2.3.2 where a PDF of X is defined. Let g(x) be a function of this PDF such that $g(x)=4x+3$.
One wishes to determine the expected value of g(X). From Eq. 2.24,

$$E[f(x)] = \int_{\infty}^{0} \frac{20x}{(x+100)^3} dx = 0.1$$

Then from Eq. 2.28

$$E[g(X)] = 3 + 4.E[f(X)] = 3.4$$ ■

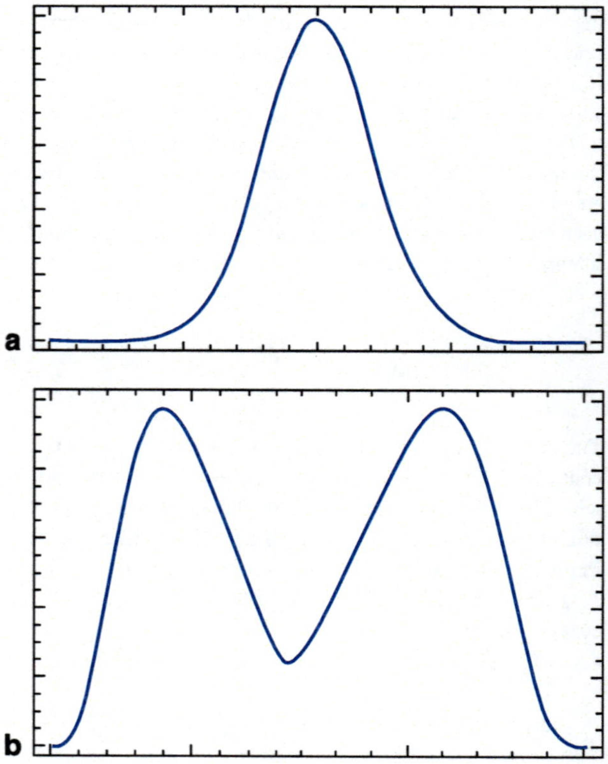

Fig. 2.8 Unimodal and bi-modal distributions. **a** Unimodal, **b** Bi-modal

2.4 Important Probability Distributions

2.4.1 Background

Data arising from an occurrence or phenomenon or descriptive of a class or a group can be viewed as a distribution of a random variable with a PDF associated with it. A majority of data sets encountered in practice can be described by one (or two) among a relatively few PDFs. The ability to characterize data in this manner provides distinct advantages to the analysts in terms of: understanding the basic dynamics of the phenomenon, in prediction and confidence interval specification, in classification, and in hypothesis testing (discussed in Chap. 4). Such insights eventually allow better decision making or sounder structural model identification since they provide a means of quantifying the random uncertainties inherent in the data. Surprisingly, most of the commonly encountered or important distributions have a common genealogy, shown in Fig. 2.9 which is a useful mnemonic for the reader.

2.4.2 Distributions for Discrete Variables

(a) Bernouilli Process. Consider an experiment involving repeated trials where only two complementary outcomes are possible which can be labeled either as a "success" or a "failure". Such a process is called a Bernouilli process: (i) if the successive trials are independent, and (ii) if the probability of success p remains constant from one trial to the next. Note that the number of partitions or combinations of n outcomes into two groups with x in one group and (n-x) in the other is equal to

$$C(n,x) = \binom{n}{x}$$

(b) Binomial Distribution. The number of successes in n Bernouilli trials is called a binomial random variable. Its PDF is called a Binomial distribution (so named because of its association with the terms of the binomial expansion). It is a unimodal distribution which gives the probability of x successes in n independent trials, if the probability of success in any one trial is p. Note that the outcomes must be Bernouilli trials. This distribution is given by:

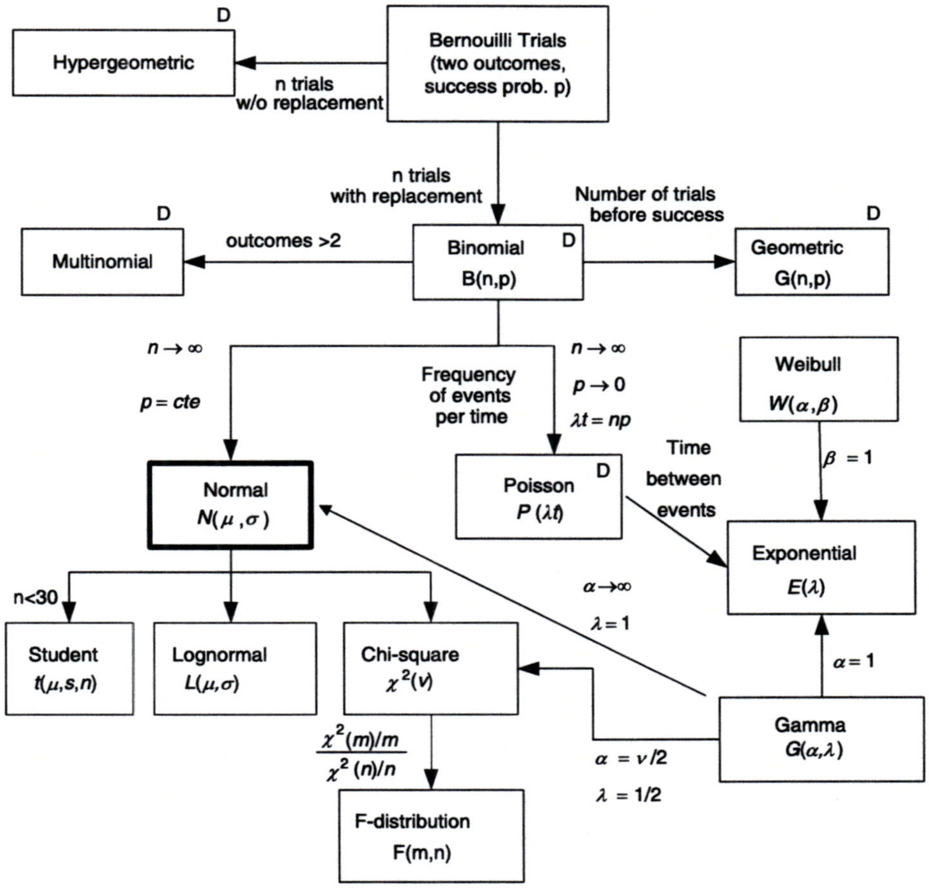

Fig. 2.9 Genealogy between different important probability distribution functions. Those that are discrete functions are represented by "D" while the rest are continuous functions. (Adapted with modification from R. E. Lave, Jr. of Stanford University)

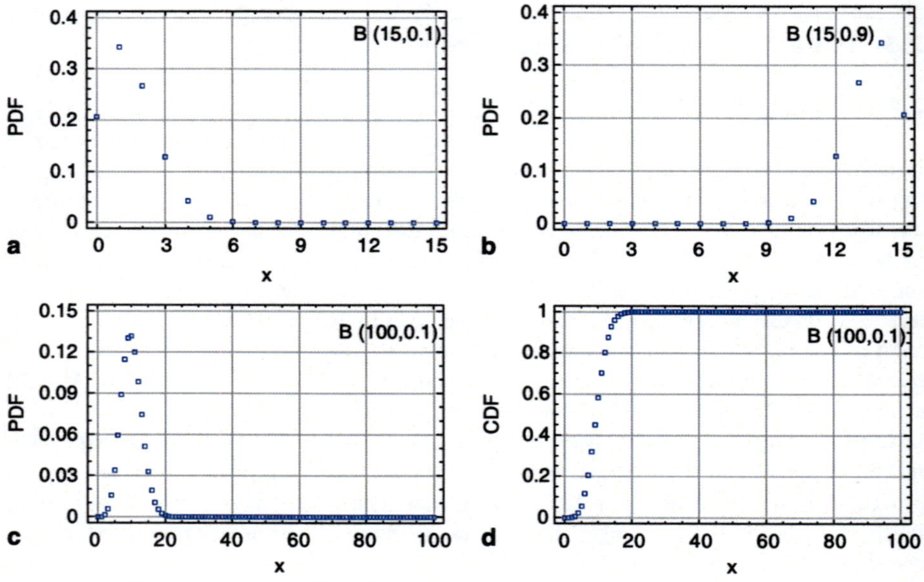

Fig. 2.10 Plots for the Binomial distribution illustrating the effect of probability of success p with X being the probability of the number of "successes" in a total number of n trials. Note how the skewness in the PDF is affected by p (frames a and b), and how the number of trials affects the shape of the PDF (frame a and c). Instead of vertical bars at discrete values of X as is often done for discrete distributions such as the Binomial, the distributions are shown as contour points so as to be consistent with how continuous distributions are represented. **a** n=15 and p=0.1, **b** n=15 and p=0.9, **c** n=100 and p=0.1, **d** n=100 and p=0.1

$$B(x; n, p) = \binom{n}{x} p^x (1-p)^{n-x} \qquad (2.33a)$$

with mean: $\mu = (n.p)$ and variance

$$\sigma^2 = np(1-p) \qquad (2.33b)$$

When n is small, it is easy to compute Binomial probabilities using Eq. 2.33a. For large values of n, it is more convenient to refer to tables which apply not to the PDF but to the corresponding cumulative distribution functions, referred to here as *Binomial probability sums,* defined as:

$$B(r; n, p) = \sum_{x=0}^{r} B(x; n, p) \qquad (2.33c)$$

There can be numerous combinations of n, p and r, which is a drawback to such tabular determinations. Table A1 in Appendix A illustrates the concept only for n=15 and n=20 and for different values of p and r. Figure 2.10 illustrates how the skewness of the Binomial distribution is affected by p, and by the total number of trials n.

Example 2.4.1: Let k be the number of heads in n=4 independent tosses of a coin. Then the mean of the distribution=(4)·(1/2)=2, and the variance σ^2=(4)·(1/2)·(1−1/2)=1. From Eq. 2.33a, the probability of two successes in four tosses =

$$B(2; 4, 0.5) = \binom{4}{2}\left(\frac{1}{2}\right)^2\left(1-\frac{1}{2}\right)^{4-2}$$
$$= \frac{4\times 3}{2}\times\frac{1}{4}\times\frac{1}{4} = \frac{3}{8}$$

■

Example 2.4.2: The probability that a patient recovers from a type of cancer is 0.6. If 15 people are known to have contracted this disease, then one can determine probabilities of various types of cases using Table A1. Let X be the number of people who survive.

(a) The probability that at least 5 survive is:

$$p(X \geq 5) = 1 - p(X < 5) = 1 - \sum_{x=0}^{4} B(x; 15, 0.6)$$
$$= 1 - 0.0094 = 0.9906$$

(b) The probability that there will be 5 to 8 survivors is:

$$P(5 \leq X \leq 8) = \sum_{x=0}^{8} B(x; 15, 0.6) - \sum_{x=0}^{4} B(x; 15, 0.6)$$
$$= 0.3902 - 0.0094 = 0.3808$$

(c) The probability that exactly 5 survive:

$$p(X = 5) = \sum_{x=0}^{5} B(x; 15, 0.6) - \sum_{x=0}^{4} B(x; 15, 0.6)$$
$$= 0.0338 - 0.0094 = 0.0244$$

■

(c) Geometric Distribution. Rather than considering the number of successful outcomes, there are several physical instances where one would like to ascertain the time interval for a certain probability event to occur the first time (which could very well destroy the physical system). This probability (p) is given by the geometric distribution which can be derived from the Binomial distribution. Consider N to be the random variable representing the number of trials until the event does occur. Note that if an event occurs the first

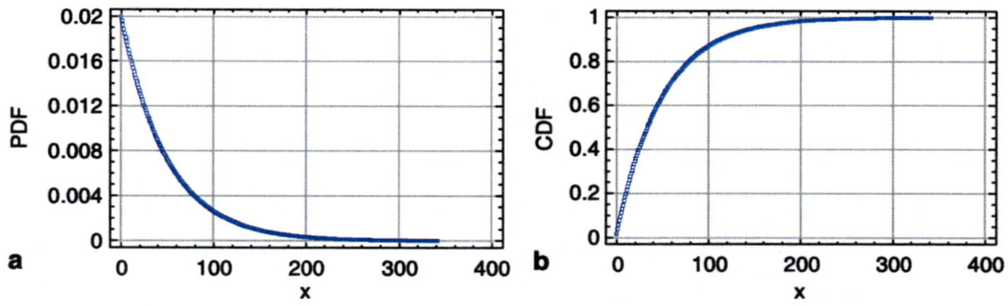

Fig. 2.11 Geometric distribution G(x;0.02) for Example 2.4.3 where the random variable is the number of trials until the event occurs, namely the "50 year design wind" at the coastal location in question. **a** PDF. **b** CDF

time during the nth trial then it did not occur during the previous (n−1) trials. Then, the geometric distribution is given by:

$$G(n;p) = p \cdot (1-p)^{n-1} \quad n = 1, 2, 3, \ldots \tag{2.34a}$$

An extension of the above concept relates to the time between two successive occurrences of the same event, called the recurrence time. Since the events are assumed independent, the mean recurrence time denoted by random variable T between two consecutive events is simply the expected value of the Bernouilli distribution:

$$\begin{aligned}\bar{T} = E(T) &= \sum_{t=1}^{\infty} t \cdot p(1-p)^{t-1} \\ &= p[1 + 2(1-p) + 3(1-p)^2] \approx \frac{1}{p}\end{aligned} \tag{2.34b}$$

Example 2.4.3:[1] *Using geometric PDF for 50 year design wind problems*

The design code for buildings in a certain coastal region specifies the 50-year wind as the "design wind", i.e., a wind velocity with a return period of 50 years, or one which may be expected to occur once every 50 years. What are the probabilities that:

(a) the design wind is encountered in any given year. From Eq. 2.34b, $p = \frac{1}{\bar{T}} = \frac{1}{50} = 0.02$

(b) the design wind is encountered during the fifth year of a newly constructed building (from Eq. 2.34a): $G(5; 0.02) = (0.02).(1 - 0.02)^4 = 0.018$

(c) the design wind is encountered within the first 5 years:

$$\begin{aligned}G(n \le 5; p) &= \sum_{t=1}^{5} (0.02).(1-0.02)^{t-1} = 0.02 \\ &\quad + 0.0196 + 0.0192 + 0.0188 + 0.0184 = 0.096\end{aligned}$$

Figure 2.11 depicts the PDF and the CDF for the geometric function corresponding to this example. ■

(d) Hypergeometric Distribution. The Binomial distribution applies in the case of independent trials or when sampling from a batch of items is done with replacement. Another type of dependence arises when sampling is done *without* replacement. This case occurs frequently in areas such as acceptance sampling, electronic testing and quality assurance where the item is destroyed during the process of testing. If n items are to be selected without replacement from a set of N items which contain k items that pass a success criterion, the PDF of the number X of successful items is given by the hypergeometric distribution:

$$H(x; N, n, k) = \frac{C(k,x) \cdot C(N-k, n-x)}{C(N,n)} = \frac{\binom{k}{x}\binom{N-k}{n-x}}{\binom{N}{n}} \tag{2.35a}$$

$x = 0, 1, 2, 3 \ldots$

$$\text{with mean } \mu = \frac{nk}{N} \text{ and} \\ \text{variance } \sigma^2 = \frac{N-n}{N-1} \cdot n \cdot \frac{k}{N} \cdot \left(1 - \frac{k}{N}\right) \tag{2.35b}$$

Note that C(k, x) is the number of ways x items can be chosen from the k "successful" set, while C(N-k, n-x) is the number of ways that the remainder (n-x) items can be chosen from the "unsuccessful" set of (N-k) items. Their product divided by the total number of combinations of selecting equally likely samples of size n from N items is represented by Eq. 2.35a.

Example 2.4.4: Lots of 10 computers each are called acceptable if they contain no fewer than 2 defectives. The procedure for sampling the lot is to select 5 computers at random and test for defectives. What is the probability that exactly one defective is found in the sample if there are 2 defectives in the entire lot?

Using the hypergeometric distribution given by Eq. 2.35a with n=5, N=10, k=2 and x=1:

[1] From Ang and Tang (2007) by permission of John Wiley and Sons.

$$H(1;10,5,2) = \frac{\binom{2}{1}\binom{10-2}{5-1}}{\binom{10}{5}} = 0.444 \quad \blacksquare$$

(e) Multinomial Distribution. A logical extension to Bernouilli experiments where the result is a two-way outcome, either success/good or failure/defective, is the multinomial experiment where k possible outcomes are possible. An example of k=5 is when the grade of a student is either A, B, C, D or F. The issue here is to find the number of combinations of n items which can be partitioned into k independent groups (a student can only get a single grade for the same class) with x_1 being in the first group, x_2 in the second,... This is represented by:

$$\binom{n}{x_1, x_2, \ldots x_k} = \frac{n!}{x_1!x_2!\ldots x_k!} \tag{2.36a}$$

with the conditions that $(x_1 + x_2 + \ldots + x_k) = n$ and that all partitions are mutually exclusive and occur with equal probability from one trial to the next. It is intuitively obvious that when n is large and k is small, the hypergeometric distribution will tend to closely approximate the Binomial.

Just like Bernouilli trials lead to the Binomial distribution, the multinomial experiment leads to the multinomial distribution which gives the probability distribution of k random variables $x_1, x_2, \ldots x_k$ in n independent trials occurring with probabilities $p_1, p_2, \ldots p_k$:

$$f(x_1, x_2, \ldots x_k) = \binom{n}{x_1, x_2, ..x_k} p_1^{x_1} \cdot p_2^{x_2} \ldots p_k^{x_k} \tag{2.36b}$$

with $\sum_{i=1}^{k} x_i = n$ and $\sum_{i=1}^{k} p_i = 1$

Example 2.4.5: Consider an examination given to 10 students. The instructor, based on previous years' experience, expects the distribution given in Table 2.6.
On grading the exam, he finds that 5 students got an A, 3 got a B and 2 got a C, and no one got either D or F. What is the probability that such an event could have occurred purely by chance?

This answer is directly provided by Eq. 2.36b which yields the corresponding probability of the above event taking place:

Table 2.6 PDF of student grades for a class

X	A	B	C	D	F
p(X)	0.2	0.3	0.3	0.1	0.1

$$f(A,B,C,D,F) = \binom{10}{5,3,2,0,0} (0.2^5) \cdot (0.3^3) \cdot (0.3^2) \cdot (0.1^0) \cdot (0.1^0) \approx 0.00196$$

This is very low, and hence this occurrence is unlikely to have occurred purely by chance. ■

(f) Poisson Distribution. Poisson experiments are those that involve the number of outcomes of a random variable X which occur per unit time (or space); in other words, as describing the occurrence of isolated events in a continuum. A Poisson experiment is characterized by: (i) independent outcomes (also referred to as memoryless), (ii) probability that a single outcome will occur during a very short time is proportional to the length of the time interval, and (iii) probability that more than one outcome occurs during a very short time is negligible. These conditions lead to the Poisson distribution which is the limit of the Binomial distribution when $n \to \infty$ and $p \to 0$ in such a way that the product $(n.p) = \lambda t$ remains constant. It is given by:

$$p(x;\lambda t) = \frac{(\lambda t)^x \exp(-\lambda t)}{x!} \quad x = 0, 1, 2, 3 \ldots \tag{2.37a}$$

where λ is called the "*mean occurrence rate*", i.e., the average number of occurrences of the event per unit time (or space) interval t. A special feature of this distribution is that its mean or average number of outcomes μ per time t and its variance σ^2 are such that

$$\mu(X) = \sigma^2(X) = \lambda t = n \cdot p \tag{2.37b}$$

Akin to the Binomial distribution, tables for certain combinations of the two parameters allow the cumulative Poisson distribution to be read off directly (see Table A2) with the latter being defined as:

$$P(r;\lambda t) = \sum_{x=0}^{r} P(x;\lambda t) \tag{2.37c}$$

Applications of the Poisson distribution are widespread: the number of faults in a length of cable, number of suspended particles in a volume of gas, number of cars in a fixed length of roadway or number of cars passing a point in a fixed time interval (traffic flow), counts of a-particles in radio-active decay, number of arrivals in an interval of time (queuing theory), the number of noticeable surface defects found by quality inspectors on a new automobile,...

Example 2.4.6: During a laboratory experiment, the average number of radioactive particles passing through a counter in 1 millisecond is 4. What is the probability that 6 particles enter the counter in any given millisecond?

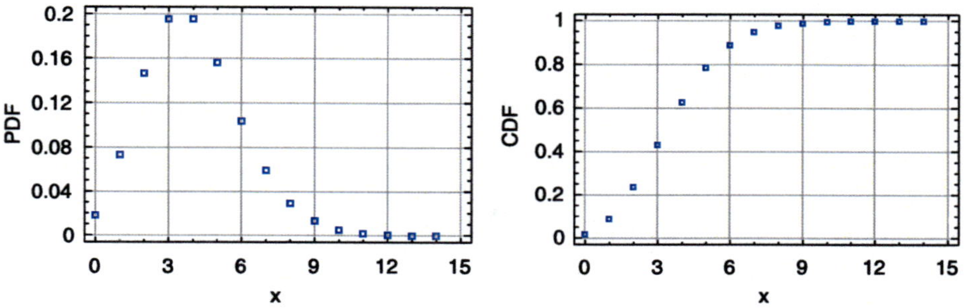

Fig. 2.12 Poisson distribution for the number of storms per year where $\lambda t=4$

Using the Poisson distribution function (Eq. 2.37a) with x=6 and λt=4:

$$P(6;4) = \frac{4^6.e^{-4}}{6!} = 0.1042 \quad \blacksquare$$

Example 2.4.7: The average number of planes landing at an airport each hour is 10 while the maximum number it can handle is 15. What is the probability that on a given hour some planes will have to be put on a holding pattern?
In this case, Eq. 2.37c is used. From Table A2, with $\lambda t = 10$

$$P(X > 15) = 1 - P(X \leq 15) = 1 - \sum_{x=0}^{15} P(x;10)$$
$$= 1 - 0.9513 = 0.0487 \quad \blacksquare$$

Example 2.4.8: *Using Poisson PDF for assessing storm frequency*
Historical records at Phoenix, AZ indicate that on an average there are 4 dust storms per year. Assuming a Poisson distribution, compute the probabilities of the following events using Eq. 2.37a:
(a) that there would not be any storms at all during a year:

$$p(X=0) = \frac{(4)^0 \cdot e^{-4}}{0!} = 0.018$$

(b) the probability that there will be four storms during a year:

$$p(X=4) = \frac{(4)^4 \cdot e^{-4}}{4!} = 0.195$$

Note that though the average is four, the probability of actually encountering four storms in a year is less than 20%. Figure 2.12 represents the PDF and CDF for different number of X values for this example. ■

2.4.3 Distributions for Continuous Variables

(a) Gaussian Distribution. The Gaussian distribution or normal error function is the best known of all continuous distributions. It is a special case of the Binomial distribution with the same values of mean and variance but applicable when *n* is sufficiently large (n>30). It is a two-parameter distribution given by:

$$N(x;\mu,\sigma) = \frac{1}{\sigma(2\pi)^{1/2}} \exp\left[-\frac{(x-\mu)}{\sigma}\right]^2 \qquad (2.38a)$$

where μ and σ are the mean and standard deviation respectively of the random variable X. Its name stems from an erroneous earlier perception that it was the natural pattern followed by distributions and that any deviation from it required investigation. Nevertheless, it has numerous applications in practice and is the most important of all distributions studied in statistics. Further, it is the parent distribution for several important continuous distributions as can be seen from Fig. 2.9. It is used to model events which occur by chance such as variation of dimensions of mass-produced items during manufacturing, experimental errors, variability in measurable biological characteristics such as people's height or weight,... Of great practical import is that normal distributions apply in situations where the random variable is the result of a sum of several other variable quantities acting independently on the system.

The shape of the normal distribution is unimodal and symmetrical about the mean, and has its maximum value at $x=\mu$ with points of inflexion at $x = \mu \pm \sigma$. Figure 2.13 illustrates its shape for two different cases of μ and σ. Further, the normal distribution given by Eq. 2.38a provides a convenient approximation for computing binomial probabilities for large number of values (which is tedious), provided $[n \cdot p \cdot (1-p)] > 10$.

In problems where the normal distribution is used, it is more convenient to standardize the random variable into a new random variable $z \equiv \frac{x-\mu}{\sigma}$ with mean zero and variance of unity. This results in the *standard normal curve* or *z-curve*:

$$N(z;0,1) = \frac{1}{\sqrt{2\pi}} \exp(-z^2/2). \qquad (2.38b)$$

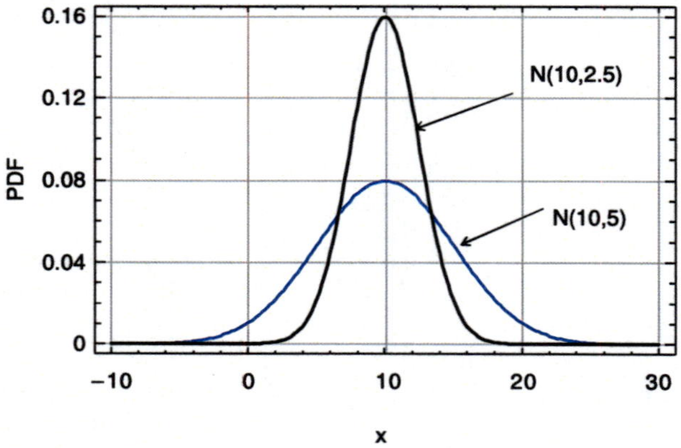

Fig. 2.13 Normal or Gaussian distributions with same mean of 10 but different standard deviations. The distribution flattens out as the standard deviation increases

In actual problems, the standard normal distribution is used to determine the probability of the variate having a value within a certain interval, say z between z_1 and z_2. Then Eq. 2.38a can be modified into:

$$N(z_1 \leq z \leq z_2) = \int_{z_1}^{z_2} \frac{1}{\sqrt{2\pi}} \exp(-z^2/2)dz \qquad (2.38c)$$

The shaded area in Table A3 permits evaluating the above integral, i.e., determining the associated probability assuming $z_1=-\infty$. Note that for z=0, the probability given by the shaded area is equal to 0.5. Since not all texts adopt the same format in which to present these tables, the user is urged to use caution in interpreting the values shown in such tables.

Example 2.4.9: *Graphical interpretation of probability using the standard normal table*

Resistors made by a certain manufacturer have a nominal value of 100 ohms but their actual values are normally distributed with a mean of $\mu=100.6$ ohms and standard deviation $\sigma=3$ ohms. Find the percentage of resistors that will have values:

(i) higher than the nominal rating. The standard normal variable $z(x=100)=(100-100.6)/3=-0.2$. From Table A3, this corresponds to a probability of $(1-0.4207)=0.5793$ or 57.93%.

(ii) within 3 ohms of the nominal rating (i.e., between 97 and 103 ohms).The lower limit $z_1=(97-100.6)/3=-1.2$, and the tabulated probability from Table A3 is $p(z=-1.2)=0.1151$ (as illustrated in Fig. 2.14a). The upper limit is: $z_2=(103-100.6)/3=0.8$. However, care should be taken in properly reading the corresponding value from Table A3 which only gives probability values of $z<0$. One first determines the probability about the negative value symmetric about 0, i.e., $p(z=-0.8)=0.2119$ (shown in Fig. 2.14b). Since the total area under the curve is 1.0, $p(z=0.8)=1.0-0.2119=0.7881$. Finally, the required probability $p(-1.2<z<0.8)=(0.7881-0.1151)=0.6730$ or 67.3%. ■

Inspection of Table A3 allows the following statements which are important in statistics:

The interval $\mu \pm \sigma$ contains approximately [1−2(0.1587)] = 0.683 or 68.3% of the observations,

The interval $\mu \pm 2\sigma$ contains approximately 95.4% of the observations,

The interval $\mu \pm 3\sigma$ contains approximately 99.7% of the observations,

Another manner of using the standard normal table is for the "backward" problem. Instead of being specified the z value and having to deduce the probability, such a problem arises when the probability is specified and the z value is to be deduced.

Example 2.4.10: Reinforced and pre-stressed concrete structures are designed so that the compressive stresses are carried mostly by the concrete itself. For this and other reasons the main criterion by which the quality of concrete is assessed is its compressive strength. Specifications for concrete used in civil engineering jobs may require specimens of specified size and shape (usually cubes) to be cast and tested on site. One can assume the normal distribution to apply. If the mean and standard deviation of this distribution are μ and σ, the civil engineer wishes to determine the "statistical minimum strength" x specified as the strength below which

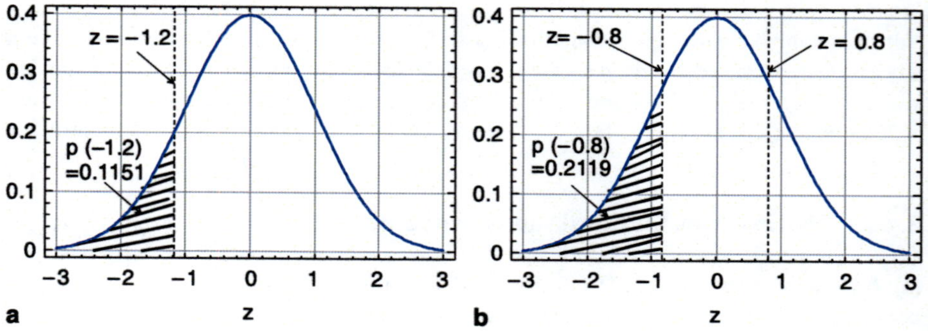

Fig. 2.14 Figures meant to illustrate that the shaded areas are the physical representations of the tabulated standardized probability values in Table A3. **a** Lower limit. **b** Upper limit

only say 5% of the cubes are expected to fail. One searches Table A3 and determines the value of z for which the probability is 0.05, i.e., p(z=−1.645)=0.05. Hence, one infers that this would correspond to $x=\mu-1.645\sigma$.

(b) Student t Distribution. One important application of the normal distribution is that it allows making statistical inferences about population means from random samples (see Sect. 4.2). In case the random samples are small (n<30), then the t-student distribution, rather than the normal distribution, should be used. If one assumes that the sampled population is approximately normally distributed, then the random variable $t=\frac{x-\mu}{s\sqrt{n}}$ has the Student t-distribution $t(\mu, s, \nu)$ where s is the sample standard deviation and ν is the degrees of freedom=(n−1). Thus, the number of degrees of freedom (d.f.) equals the number of data points minus the number of constraints or restrictions placed on the data. Table A4 (which is set up differently from the standard normal table) provides numerical values of the t-distribution for different degrees of freedom at different confidence levels. How to use these tables will be discussed in Sect. 4.2. Unlike the z curve, one has a family of t-distributions for different values of ν. Qualitatively, the t-distributions are similar to the standard normal distribution in that they are symmetric about a zero mean, while they are but slightly wider than the corresponding normal distribution as indicated in Fig. 2.15. However, in terms of probability values represented by areas under the curves as in Example 2.4.9, the differences between the normal and the student-t distributions are large enough to warrant retaining this distinction.

(c) Lognormal Distribution. This distribution is appropriate for non-negative outcomes which are the product of a number of quantities. In such cases, the data are skewed and the symmetrical normal distribution is no longer appropriate. If a variate X is such that log(X) is normally distributed, then

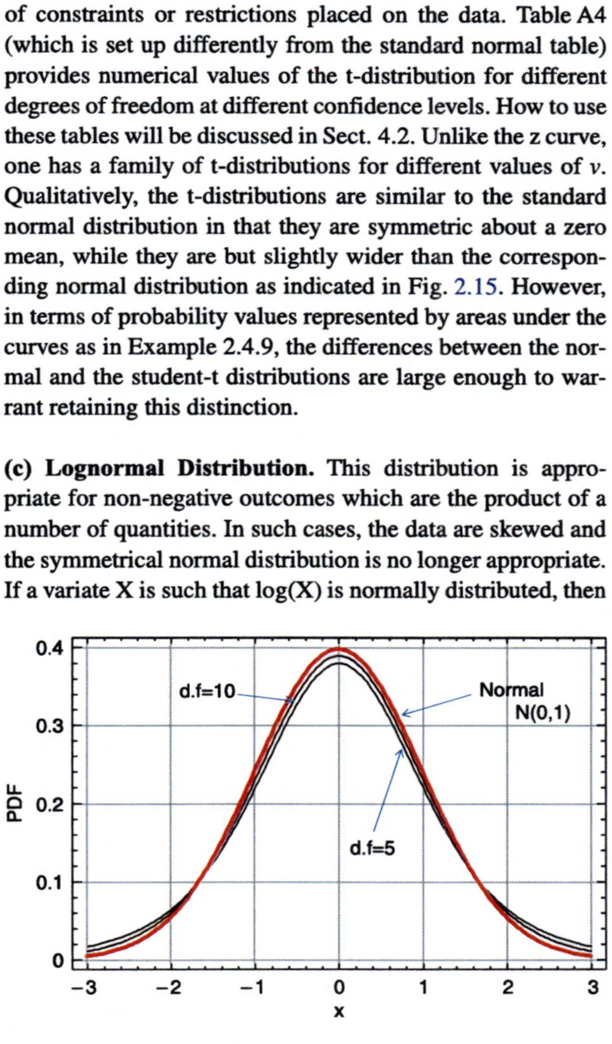

Fig. 2.15 Comparison of the normal (or Gaussian) z curve to two Student-t curves with different degrees of freedom (d.f.). As the d.f. increase, the PDF for the Student-t distribution flattens out and deviates increasingly from the normal distribution

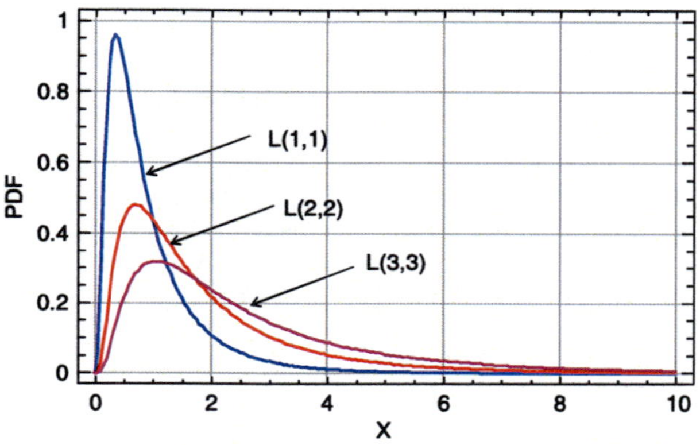

Fig. 2.16 Lognormal distributions for different mean and standard deviation values

the distribution of X is said to be lognormal. With X ranging from $-\infty$ to $+\infty$, log(X) would range from 0 to $+\infty$. Not only does the lognormal model accommodate skewness, but it also captures the non-negative nature of many variables which occur in practice. It is characterized by two parameters, the mean and variance (μ, σ), as follows:

$$L(x;\mu,\sigma) = \frac{1}{\sigma.x(\sqrt{2\pi})}\exp\left[-\frac{(\ln x-\mu)^2}{2\sigma^2}\right] \quad \text{when } x \ge 0$$

$$= 0 \quad \text{elsewhere} \tag{2.39}$$

The lognormal curves are a family of skewed curves as illustrated in Fig. 2.16. Lognormal failure laws apply when the degradation in lifetime is proportional to the previous amount of degradation. Typical applications in civil engineering involve flood frequency, in mechanical engineering with crack growth and mechanical wear, and in environmental engineering with pollutants produced by chemical plants and threshold values for drug dosage.

Example 2.4.11: *Using lognormal distributions for pollutant concentrations*

Concentration of pollutants produced by chemical plants is known to resemble lognormal distributions and is used to evaluate issues regarding compliance of government regulations. The concentration of a certain pollutant, in parts per million (ppm), is assumed lognormal with parameters $\mu=4.6$ and $\sigma=1.5$. What is the probability that the concentration exceeds 10 ppm?

One can use Eq. 2.39, or simpler still, use the z tables (Table A3) by suitable transformations of the random variable.

$$L(X>10) = N[\ln(10), 4.6, 1.5] = N\left[\frac{\ln(10)-4.6}{1.5}\right]$$

$$= N(-1.531) = 0.0630$$

■

(d) Gamma Distribution. There are several processes where distributions other than the normal distribution are warranted. A distribution which is useful since it is versatile in the shapes it can generate is the gamma distribution (also called the *Erlang distribution*). It is a good candidate for modeling random phenomena which can only be positive and are unimodal. The gamma distribution is derived from the gamma function for positive values of α, which one may recall from mathematics, is defined by the integral:

$$\Gamma_x(\alpha) = \int_0^x x^{\alpha-1} \cdot e^{-x} dx \tag{2.40a}$$

Recall that for non-negative integers k:

$$\Gamma(k+1) = k! \tag{2.40b}$$

The continuous random variable X has a gamma distribution with positive parameters a and λ if its density function is given by:

$$G(x;\alpha,\lambda) = \begin{cases} \lambda^{\alpha} e^{-\lambda x} \dfrac{x^{\alpha-1}}{(\alpha-1)!} & x > 0 \\ = 0 & \text{elsewhere} \end{cases} \tag{2.40c}$$

The mean and variance of the gamma distribution are:

$$\mu = \alpha/\lambda \quad \text{and} \quad \sigma^2 = \alpha/\lambda^2 \tag{2.40d}$$

Variation of the parameter α (called the shape factor) and λ (called the scale parameter) allows a wide variety of shapes to be generated (see Fig. 2.17). From Fig. 2.9, one notes that the Gamma distribution is the parent distribution of many other distributions discussed below. If $\alpha \to \infty$ and $\lambda = 1$, the gamma distribution approaches the normal (see Fig. 2.9). When $\alpha = 1$, one gets the exponential distribution. When $\alpha = \upsilon/2$ and $\lambda = 1/2$, one gets the chi-square distribution (discussed below).

(e) Exponential Distribution. A special case of the gamma distribution for $\alpha = 1$ is the exponential distribution. It is the continuous distribution analogue to the geometric distribution which applied to the discrete case. It is used to model the interval between two occurrences, e.g. the distance between consecutive faults in a cable, or the time between chance failures of a component (such as a fuse) or a system, or the time between consecutive emissions of α-particles, or the time between successive arrivals at a service facility. Its PDF is given by

$$E(x;\lambda) = \begin{cases} \lambda \cdot e^{-\lambda x} & \text{if } x > 0 \\ = 0 & \text{otherwise} \end{cases} \tag{2.41a}$$

where λ is the mean value per unit time or distance. The mean and variance of the exponential distribution are:

$$\mu = 1/\lambda \quad \text{and} \quad \sigma^2 = 1/\lambda^2 \tag{2.41b}$$

The distribution is represented by a family of curves for different values of λ (see Fig. 2.18). Exponential failure laws apply to products whose current age does not have much effect on their remaining lifetimes. Hence, this distribution is said to be "memoryless". Notice the relationship between the exponential and the Poisson distributions. While the latter represents the *number* of failures per unit time, the exponential represents the *time* between successive failures. Its CDF is given by:

$$CDF[E(a,\lambda)] = \int_0^a \lambda . e^{-\lambda x} dx = 1 - e^{-\lambda a} \tag{2.41c}$$

Example 2.4.12: Temporary disruptions to the power grid can occur due to random events such as lightning, transformer failures, forest fires,.. The Poisson distribution has been known to be a good function to model such failures. If these occur, on average, say, once every 2.5 years, then λ =1/2.5=0.40 per year.

(a) What is the probability that there will be at least one disruption next year?

$$CDF[E(X \le 1;\lambda)] = 1 - e^{-0.4(1)} = 1 - 0.6703 = 0.3297$$

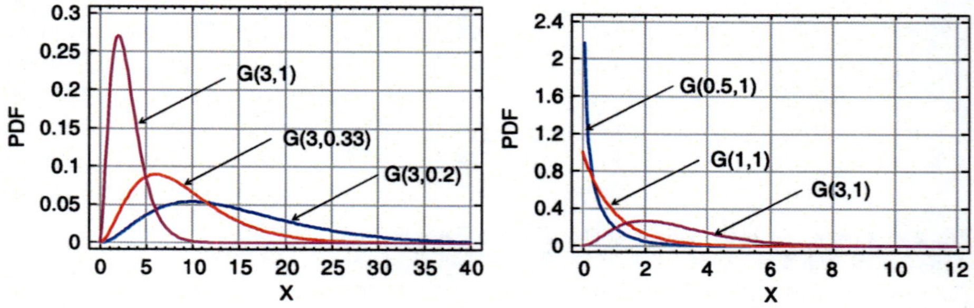

Fig. 2.17 Gamma distributions for different combinations of the shape parameter α and the scale parameter $\beta = 1/\lambda$

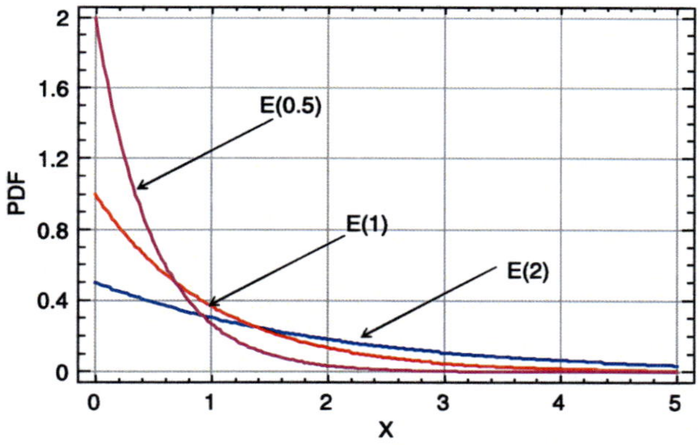

Fig. 2.18 Exponential distributions for three different values of the parameter λ

(b) What is the probability that there will be no more than two disruptions next year?

This is the complement of at least two disruptions.

$$\begin{aligned}\text{Probability} &= 1 - CDF[E(X \leq 2; \lambda)] \\ &= 1 - [1 - e^{-0.4(2)}] = 0.4493\end{aligned}$$ ■

(f) Weibull Distribution. Another versatile and widely used distribution is the Weibull distribution which is used in applications involving reliability and life testing; for example, to model the time of failure or life of a component. The continuous random variable X has a Weibull distribution with parameters α and β (shape and scale factors respectively) if its density function is given by:

$$W(x;\alpha,\beta) = \frac{\alpha}{\beta^{\alpha}} \cdot x^{\alpha-1} \cdot \exp[-(x/\beta)^{\alpha}] \quad \text{for } x > 0 \qquad (2.42a)$$
$$= 0 \quad \text{elsewhere}$$

with mean

$$\mu = \beta \cdot \Gamma\left(1 + \frac{1}{\alpha}\right) \qquad (2.42b)$$

Figure 2.19 shows the versatility of this distribution for different sets of α and β values. Also shown is the special case of W(1,1) which is the exponential distribution. For $\beta > 1$, the curves become close to bell-shaped and somewhat resemble the normal distribution. The Weibull distribution has been found to be very appropriate to model reliability of a system i.e., the failure time of the weakest component of a system (bearing, pipe joint failure,…).

Example 2.4.13: ***Modeling wind distributions using the Weibull distribution***

The Weibull distribution is also widely used to model the hourly variability of wind velocity in numerous locations worldwide. The mean wind speed and its distribution on an annual basis, which are affected by local climate conditions, terrain and height of the tower, are important in order to determine annual power output from a wind turbine of a certain design whose efficiency changes with wind speed. It has been found that the shape factor α varies between 1 and 3 (when $\alpha = 2$, the distribution is called the *Rayleigh distribution*). The probability distribution shown in Fig. 2.20 has a mean wind speed of 7 m/s. Determine:

(a) the numerical value of the parameter β assuming the shape factor $\alpha = 2$

One calculates the gamma function $\Gamma(1 + \frac{1}{2}) = 0.8862$ from which $\beta = \frac{\mu}{0.8862} = 7.9$

(b) using the PDF given by Eq. 2.42, it is left to the reader to compute the probability of the wind speed being equal to 10 m/s (and verify the solution against the figure which indicates a value of 0.064). ■

(g) Chi-square Distribution. A third special case of the gamma distribution is when $\alpha = v/2$ and $\lambda = 1/2$ where v is a positive integer, and is called the degrees of freedom. This distribution called the chi-square (χ^2) distribution plays an important role in inferential statistics where it is used as a test of significance for hypothesis testing and analysis of variance type of problems. Just like the t-statistic, there is a family of distributions for different values of v (Fig. 2.21). Note that the distribution cannot assume negative values, and that it is positively skewed. Table A5 assembles critical values of the Chi-square distribution for different values of the degrees of freedom parameter v and for different signifi-

Fig. 2.19 Weibull distributions for different values of the two parameters α and β (the shape and scale factors respectively)

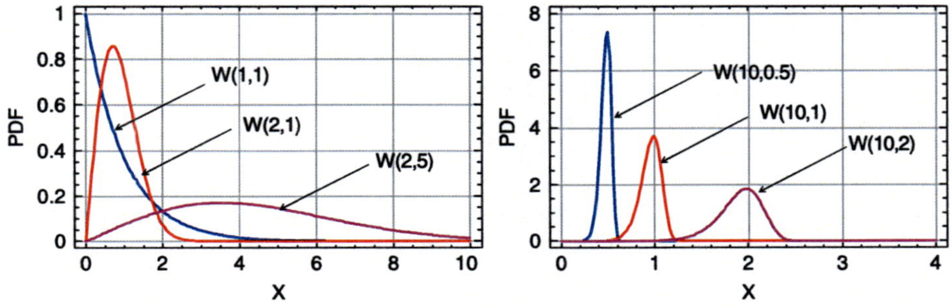

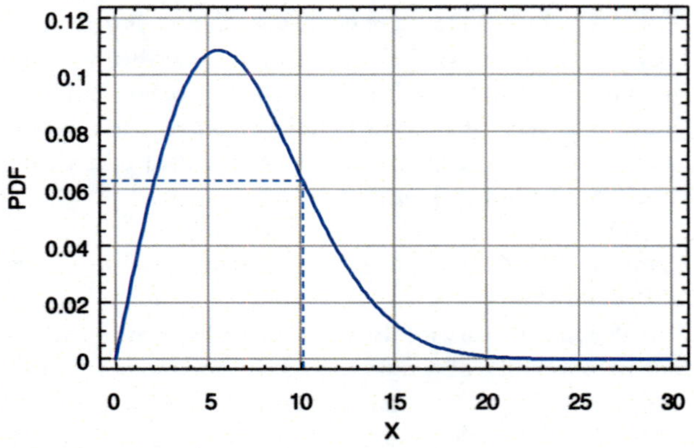

Fig. 2.20 PDF of the Weibull distribution W(2, 7.9)

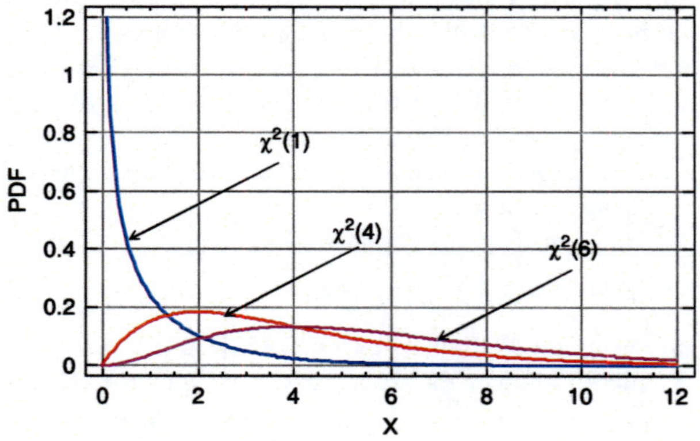

Fig. 2.21 Chi-square distributions for different values of the variable υ denoting the degrees of freedom

cance levels. The usefulness of these tables will be discussed in Sect. 4.2.

The PDF of the chi-square distribution is:

$$\chi^2(x;\nu) = \frac{1}{2^{\nu/2}\Gamma(\upsilon/2)} \cdot x^{\nu/2-1} \cdot e^{-x/2} \quad x > 0 \qquad (2.43a)$$
$$= 0 \qquad \text{elsewhere}$$

while the mean and variance values are :

$$\mu = \nu \quad \text{and} \quad \sigma^2 = 2\nu \qquad (2.43b)$$

(h) F-Distribution. While the t-distribution allows comparison between two sample means, the F distribution allows comparison between two or more sample variances. It is defined as the ratio of two independent chi-square random variables, each divided by its degrees of freedom. The F distribution is also represented by a family of plots (see Fig. 2.22) where each plot is specific to a set of numbers representing the degrees of freedom of the two random variables (v_1, v_2). Table A6 assembles critical values of the F-distri-

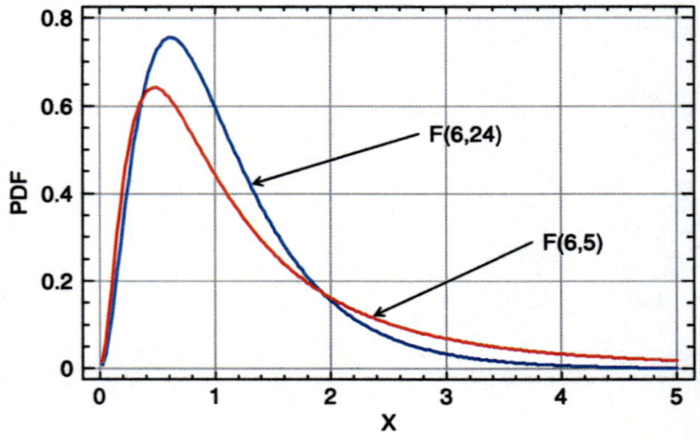

Fig. 2.22 Typical F distributions for two different combinations of the random variables (υ_1 and υ_2)

butions for different combinations of these two parameters, and its use will be discussed in Sect. 4.2.

(i) Uniform Distribution. The uniform probability distribution is the simplest of all PDFs and applies to both continuous and discrete data whose outcomes are all equally likely, i.e. have equal probabilities. Flipping a coin for heads/tails or rolling a dice for getting numbers between 1 and 6 are examples which come readily to mind. The probability density function for the discrete case where X can assume values $x_1, x_2, \ldots x_k$ is given by:

$$U(x;k) = \frac{1}{k} \qquad (2.44a)$$

$$\text{with mean } \mu = \frac{\sum_{i=1}^{k} x_i}{k} \quad \text{and} \qquad (2.44b)$$
$$\text{variance } \sigma^2 = \frac{\sum_{i=1}^{k} (x_i - \mu)^2}{k}$$

For random variables that are continuous over an interval (c,d) as shown in Fig. 2.23, the PDF is given by:

$$U(x) = \frac{1}{d-c} \quad \text{when } c < x < d \qquad (2.44c)$$
$$= 0 \qquad \text{otherwise}$$

The mean and variance of the uniform distribution (using notation shown in Fig. 2.23) are given by:

Fig. 2.23 The uniform distribution assumed continuous over the interval [c, d]

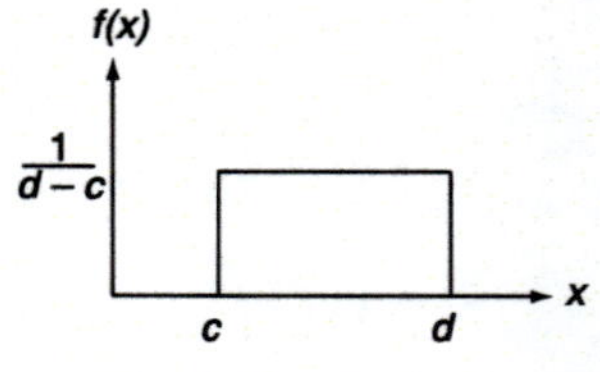

$$\mu = \frac{c+d}{2} \quad \text{and} \quad \sigma^2 = \frac{(d-c)^2}{12} \qquad (2.44d)$$

The probability of random variable X being between say x_1 and x_2 is:

$$U(x_1 \leq X \leq x_2) = \frac{x_2 - x_1}{d - c} \qquad (2.44e)$$

Example 2.4.14: A random variable X has a uniform distribution with c=−5 and d=10 (see Fig. 2.23). Determine:
(a) On an average, what proportion will have a negative value? (Answer: 1/3)
(b) On an average, what proportion will fall between −2 and 2? (Answer: 4/15) ■

(j) Beta Distribution. A very versatile distribution is the Beta distribution which is appropriate for discrete random variables between 0 and 1 such as representing proportions. It is a two parameter model which is given by:

$$Beta(x; p, q) = \frac{(p+q+1)!}{(p-1)!(q-1)!} x^{p-1}(1-x)^{q-1} \qquad (2.45a)$$

Depending on the values of p and q, one can model a wide variety of curves from u shaped ones to skewed distributions (see Fig. 2.24). The distributions are symmetrical when p and q are equal, with the curves becoming peakier as the numerical values of the two parameters increase. Skewed distributions are obtained when the parameters are unequal.

The mean of the Beta distribution $\mu = \frac{p}{p+q}$ and

$$\text{variance } \sigma^2 = \frac{pq}{(p+q)^2(p+q+1)} \qquad (2.45b)$$

This distribution originates from the Binomial distribution, and one can detect the obvious similarity of a two-outcome affair with specified probabilities. The usefulness of this distribution will become apparent in Sect. 2.5.3 dealing with the Bayesian approach to probability problems.

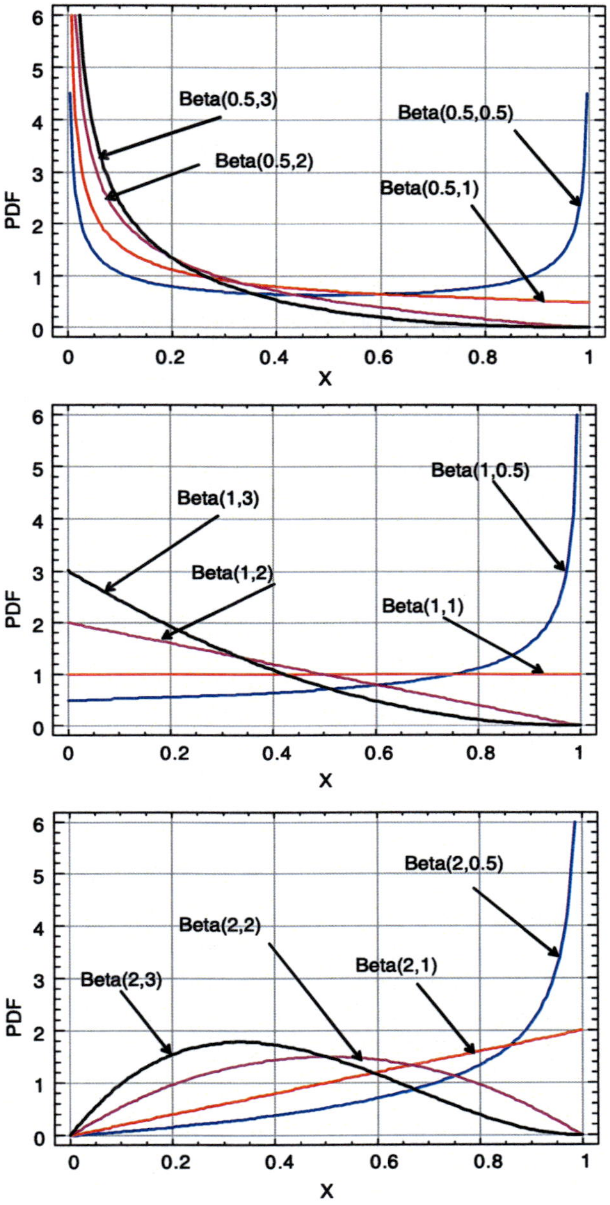

Fig. 2.24 Various shapes assumed by the Beta distribution depending on the values of the two model parameters

2.5 Bayesian Probability

2.5.1 Bayes' Theorem

It was stated in Sect. 2.1.4 that the Bayesian viewpoint can enhance the usefulness of the classical frequentist notion of probability[2]. Its strength lies in the fact that it provides a framework to include prior information in a two-stage (or multi-stage) experiment. If one substitutes the term p(A) in Eq. 2.12 by that given by Eq. 2.11, one gets :

$$p(B/A) = \frac{p(A \cap B)}{p(A \cap B) + p(A \cap \bar{B})} \qquad (2.46)$$

Also, one can re-arrange Eq. 2.12 into: $p(A \cap B) = p(A) \cdot p(B/A)$ or $= p(B) \cdot p(A/B)$. This allows expressing

[2] There are several texts which deal with Bayesian statistics; for example, Bolstad (2004).

Eq. 2.46 into the following expression referred to as the *law of total probability or Bayes' theorem:*

$$p(B/A) = \frac{p(A/B) \cdot p(B)}{p(A/B) \cdot p(B) + p(A/\bar{B}) \cdot p(\bar{B})} \quad (2.47)$$

Bayes theorem, superficially, appears to be simply a restatement of the conditional probability equation given by Eq. 2.12. The question is why is this reformulation so insightful or advantageous? First, the probability is now re-expressed in terms of its disjoint parts $\{B, \bar{B}\}$, and second the probabilities have been "flipped", i.e., p(B/A) is now expressed in terms of p(A/B). Consider the two events A and B. If event A is observed while event B is not, this expression allows one to infer the "flip" probability, i.e. probability of occurrence of B from that of the observed event A. In Bayesian terminology, Eq. 2.47 can be written as:

$$\text{Posterior probability of event B given event A} = \frac{(\text{Likelihood of A given B}) \cdot (\text{Prior probability of B})}{\text{Prior probability of A}} \quad (2.48)$$

Thus, the probability p(B) is called the *prior probability* (or unconditional probability) since it represents opinion *before* any data was collected, while p(B/A) is said to be the *posterior probability* which is reflective of the opinion revised in light of new data. The likelihood is identical to the conditional probability of A given B i.e., p(A/B).

Equation 2.47 applies to the case when only one of two events is possible. It can be extended to the case of more than two events which partition the space S. Consider the case where one has *n* events, $B_1 \ldots B_n$ which are disjoint and make up the entire sample space. Figure 2.25 shows a sample space of 4 events. Then, the law of total probability states that the probability of an event A is the sum of its disjoint parts:

$$p(A) = \sum_{j=1}^{n} p(A \cap B_j) = \sum_{j=1}^{n} p(A/B_j) \cdot p(B_j) \quad (2.49)$$

$$\text{Then } \underbrace{p(B_i/A)}_{\text{posterior probability}} = \frac{p(A \cap B_i)}{p(A)} = \frac{p(A/B_i) \cdot p(B_i)}{\sum_{j=1}^{n} \underbrace{p(A/B_j)}_{\text{likelihood}} \cdot \underbrace{p(B_j)}_{\text{prior}}} \quad (2.50)$$

which is known as *Bayes' theorem for multiple events*. As before, the marginal or prior probabilities $p(B_i)$ for $i = 1, \ldots, n$ are assumed to be known in advance, and the intention is to update or revise our "belief" on the basis of the observed evidence of event A having occurred. This is captured by the probability $p(B_i/A)$ for $i = 1, \ldots, n$ called the posterior probability or the weight one can attach to each event B_i after event A is known to have occurred.

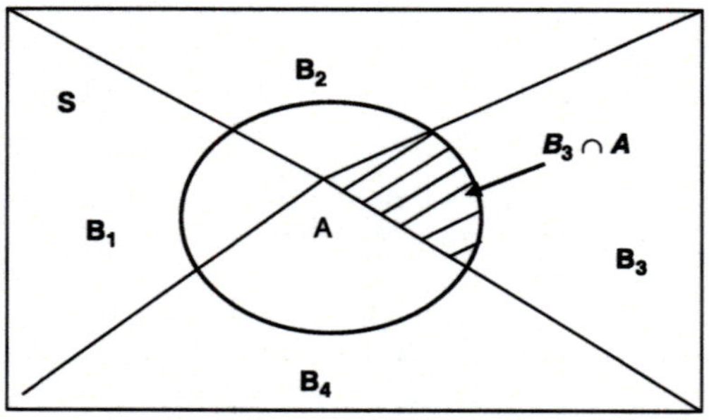

Fig. 2.25 Bayes theorem for multiple events depicted on a Venn diagram. In this case, the sample space is assumed to be partitioned into four discrete events $B_1 \ldots B_4$. If an observable event A has already occurred, the conditional probability of B_3 : $p(B_3/A) = \frac{p(B_3 \cap A)}{p(A)}$. This is the ratio of the hatched area to the total area inside the ellipse

Example 2.5.1: Consider the two-stage experiment of Example 2.2.7. Assume that the experiment has been performed and that a red marble has been obtained. One can use the information known beforehand i.e., the prior probabilities R, W and G to determine from which box the marble came from. Note that the probability of the red marble having come from box A represented by p(A/R) is now the conditional probability of the "flip" problem. This is called

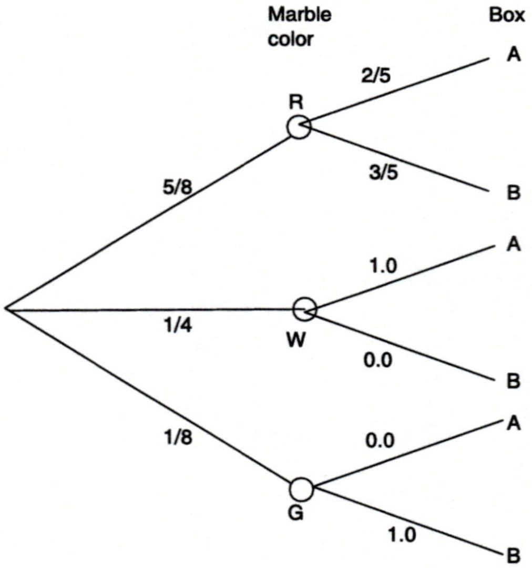

Fig. 2.26 The probabilities of the reverse tree diagram at each stage are indicated. If a red marble (R) is picked, the probabilities that it came from either Box A or Box B are 2/5 and 3/5 respectively

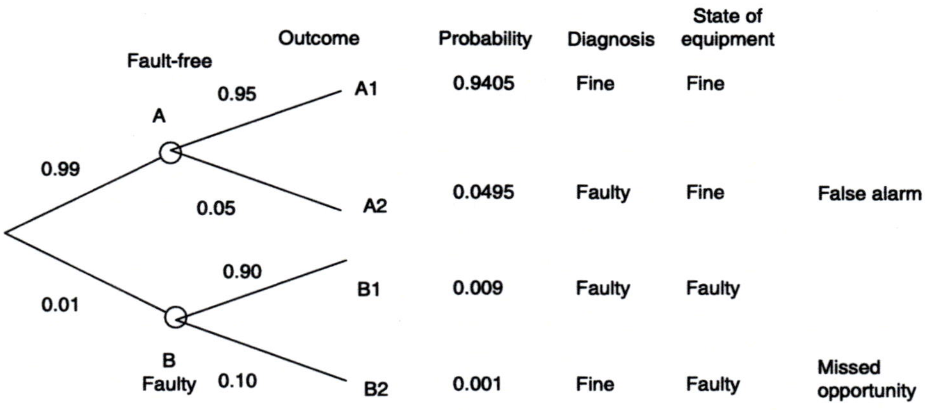

Fig. 2.27 The forward tree diagram showing the four events which may result when monitoring the performance of a piece of equipment

the *posterior probabilities* of event A with R having occurred, i.e., they are relevant after the experiment has been performed. Thus, from the law of total probability (Eq. 2.47):

$$p(B/R) = \frac{\frac{1}{2}\cdot\frac{3}{4}}{\frac{1}{2}\cdot\frac{1}{2}+\frac{1}{2}\cdot\frac{3}{4}} = \frac{3}{5}$$

and

$$p(A/R) = \frac{\frac{1}{2}\cdot\frac{1}{2}}{\frac{1}{2}\cdot\frac{1}{2}+\frac{1}{2}\cdot\frac{3}{4}} = \frac{2}{5}$$

The reverse probability tree for this experiment is shown in Fig. 2.26. The reader is urged to compare this with the forward tree diagram of Example 2.2.7. The probabilities of 1.0 for both W and G outcomes imply that there is no uncertainty at all in predicting where the marble came from. This is obvious since only Box A contains W, and only Box B contains G. However, for the red marble, one cannot be sure of its origin, and this is where a probability measure has to be determined. ■

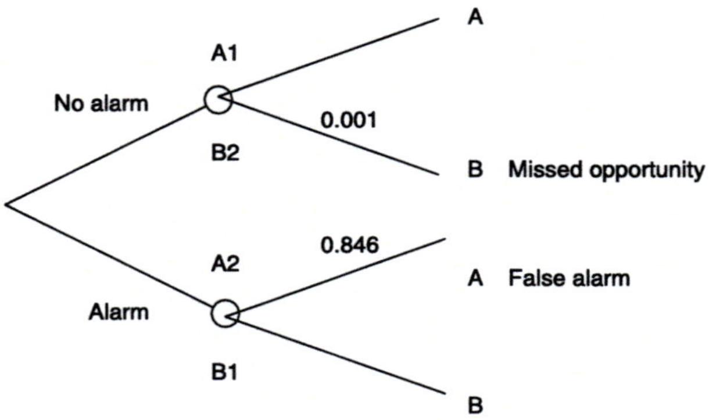

Fig. 2.28 Reverse tree diagram depicting two possibilities. If an alarm sounds, it could be either an erroneous one (outcome A from A2) or a valid one (B from B1). Further, if no alarm sounds, there is still the possibility of missed opportunity (outcome B from B2). The probability that it is a false alarm is 0.846 which is too high to be acceptable in practice. How to decrease this is discussed in the text

Example 2.5.2: *Forward and* reverse *probability trees for fault detection of equipment*

A large piece of equipment is being continuously monitored by an add-on fault detection system developed by another vendor in order to detect faulty operation. The vendor of the fault detection system states that their product correctly identifies faulty operation when indeed it is faulty (this is referred to as *sensitivity*) 90% of the time. This implies that there is a probability p=0.10 of a false negative occurring (i.e., a missed opportunity of signaling a fault). Also, the vendor quoted that the correct status prediction rate or *specificity* of the detection system (i.e., system identified as healthy when indeed it is so) is 0.95, implying that the false positive or false alarm rate is 0.05. Finally, historic data seem to indicate that the large piece of equipment tends to develop faults only 1% of the time.

Figure 2.27 shows how this problem can be systematically represented by a forward tree diagram. State A is the fault-free state and state B is represented by the faulty state. Further, each of these states can have two outcomes as shown. While outcomes A1 and B1 represent correctly identified fault-free and faulty operation, the other two outcomes are errors arising from an imperfect fault detection system. Outcome A2 is the "*false negative*" (or false alarm or error type II which will be discussed at length in Sect. 4.2 of Chap. 4), while outcome B2 is the *false positive* rate (or missed opportunity or error type I). The figure clearly illustrates that the probabilities of A and B occurring along with the conditional probabilities p(A1/A)=0.95 and p(B1/B)=0.90, result in the probabilities of each the four states as shown in the figure.

The reverse tree situation, shown in Fig. 2.28, corresponds to the following situation. A fault has been signaled. What is the probability that this is a false alarm? Using Eq. 2.47:

$$p(A/A2) = \frac{(0.99).(0.05)}{(0.99).(0.05) + (0.01).(0.90)} = \frac{0.0495}{0.0495 + 0.009} = 0.846$$

This is very high for practical situations and could well result in the operator disabling the fault detection system altogether. One way of reducing this false alarm rate, and thereby enhance robustness, is to increase the sensitivity of the detection device from its current 90% to something higher by altering the detection threshold. This would result in a higher missed opportunity rate, which one has to accept for the price of reduced false alarms. For example, the current missed opportunity rate is:

$$p(B/B1) = \frac{(0.01)\cdot(0.10)}{(0.01)\cdot(0.10) + (0.99)\cdot(0.95)} = \frac{0.001}{0.001 + 0.9405} = 0.001$$

This is probably lower than what is needed, and so the above suggested remedy is one which can be considered. Note that as the piece of machinery degrades, the percent of time when faults are likely to develop will increase from the current 1% to something higher. This will have the effect of lowering the false alarm rate (left to the reader to convince himself why). ∎

Bayesian statistics provide the formal manner by which prior opinion expressed as probabilities can be revised in the light of new information (from additional data collected) to yield posterior probabilities. When combined with the relative consequences or costs of being right or wrong, it allows one to address decision-making problems as pointed out in the example above (and discussed at more length in Sect. 12.2.9). It has had some success in engineering (as well as in social sciences) where subjective judgment, often referred to as intuition or experience gained in the field, is relied upon heavily.

The Bayes' theorem is a consequence of the probability laws and is accepted by all statisticians. It is the interpretation of probability which is controversial. Both approaches differ in how probability is defined:

- classical viewpoint: long run relative frequency of an event
- Bayesian viewpoint: degree of belief held by a person about some hypothesis, event or uncertain quantity (Phillips 1973).

Advocates of the classical approach argue that human judgment is fallible while dealing with complex situations, and this was the reason why formal statistical procedures were developed in the first place. Introducing the vagueness of human judgment as done in Bayesian statistics would dilute the "purity" of the entire mathematical approach. Advocates of the Bayesian approach, on the other hand, argue that the "personalist" definition of probability should not be interpreted as the "subjective" view. Granted that the prior probability is subjective and varies from one individual to the other, but with additional data collection all these views get progressively closer. Thus, with enough data, the initial divergent opinions would become indistinguishable. Hence, they argue, the Bayesian method brings consistency to informal thinking when complemented with collected data, and should, thus, be viewed as a mathematically valid approach.

2.5.2 Application to Discrete Probability Variables

The following example illustrates how the Bayesian approach can be applied to discrete data.

Example 2.5.3:[3] *Using the Bayesian approach to enhance value of concrete piles testing*
Concrete piles driven in the ground are used to provide bearing strength to the foundation of a structure (building, bridge,…). Hundreds of such piles could be used in large construction projects. These piles could develop defects such as cracks or voids in the concrete which would lower compressive strength. Tests are performed by engineers on piles selected at random during the concrete pour process in order to assess overall foundation strength. Let the random discrete variable be the proportion of defective piles out of the entire lot which is taken to assume five discrete values as shown in the first column of Table 2.7. Consider the case where the prior experience of an engineer as to the proportion of defective piles from similar sites is given in the second column of the table below.

Before any testing is done, the *expected value* of the probability of finding one pile to be defective is: p = (0.20)(0.30) + (0.4)(0.40) + (0.6)(0.15) + (0.8)(0.10) + (1.0)

Table 2.7 Illustration of how a prior PDF is revised with new data

Proportion of defectives (x)	Probability of being defective				
	Prior PDF of defectives	After one pile tested is found defective	After two piles tested are found defective		Limiting case of infinite defectives
0.2	0.30	0.136	0.049	…	0.0
0.4	0.40	0.364	0.262		0.0
0.6	0.15	0.204	0.221		0.0
0.8	0.10	0.182	0.262		0.0
1.0	0.05	0.114	0.205	…	1.0
Expected probability of defective pile	0.44	0.55	0.66		1.0

[3] From Ang and Tang (2007) by permission of John Wiley and Sons.

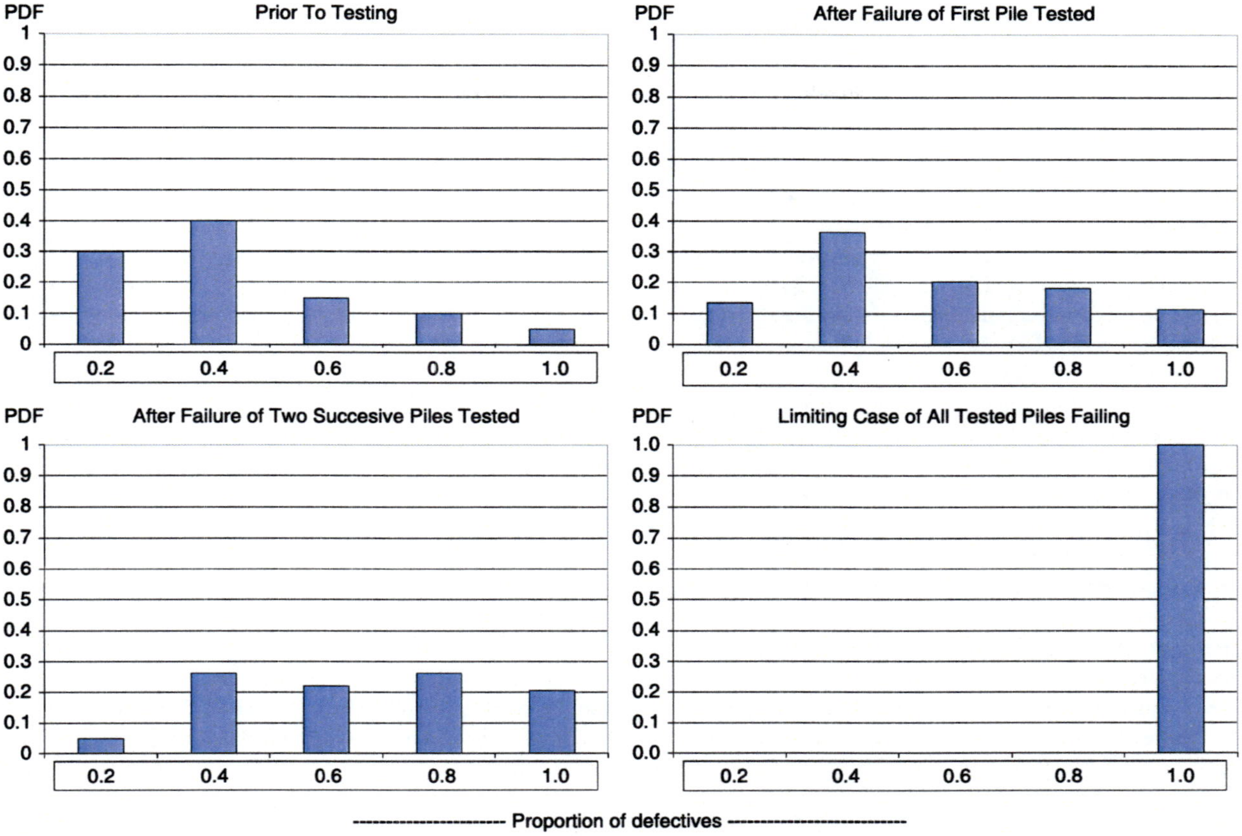

Fig. 2.29 Illustration of how the prior discrete PDF is affected by data collection following Bayes' theorem

(0.05)=0.44 (as shown in the last row under the second column). This is the prior probability.

Suppose the first pile tested is found to be defective. How should the engineer revise his prior probability of the proportion of piles likely to be defective? This is given by Bayes' theorem (Eq. 2.50). For proportion x=0.2, the posterior probability is:

$$\begin{aligned} p(x=0.2) &= \frac{(0.2)(0.3)}{(0.2)(0.3)+(0.4)(0.4)+(0.6)(0.15)+(0.8)(0.10)+(1.0)(0.05)} \\ &= \frac{0.06}{0.44} \\ &= 0.136 \end{aligned}$$

This is the value which appears in the first row under the third column. Similarly the posterior probabilities for different values of x can be determined as well as the expected value E (x=1) which is 0.55. Hence, a single inspection has led to the engineer revising his prior opinion upward from 0.44 to 0.55. Had he drawn a conclusion on just this single test without using his prior judgment, he would have concluded that all the piles were defective; clearly, an over-statement. The engineer would probably get a second pile tested, and if it also turns out to be defective, the associated probabilities are shown in the fourth column of Table 2.7. For example, for x=0.2:

$$p(x=0.2) = \frac{(0.2)(0.136)}{(0.2)(0.136)+(0.4)(0.364)+(0.6)(0.204)+(0.8)(0.182)+(1.0)(0.114)} = 0.049$$

Table 2.8 Prior pdf of defective proportion

X	0.1	0.2
f(x)	0.6	0.4

The expected value in this case increases to 0.66. In the limit, if each successive pile tested turns out to be defective, one gets back the classical distribution, listed in the last column of the table. The progression of the PDF from the prior to the infinite case is illustrated in Fig. 2.29. Note that as more piles tested turn out to be defective, the evidence from the data gradually overwhelms the prior judgment of the engineer. However, it is only when collecting data is so expensive or time consuming that decisions have to be made from limited data that the power of the Bayesian approach becomes evident. Of course, if one engineer's prior judgment is worse than that of another engineer, then his conclusion from the same data would be poorer than the other engineer. It is this type of subjective disparity which antagonists of the Bayesian approach are uncomfortable with. On the other hand, proponents of the Bayesian approach would argue that experience (even if intangible) gained in the field is a critical asset in engineering applications and that discarding this type of knowledge entirely is naïve, and a severe handicap. ■

There are instances when no previous knowledge or information is available about the behavior of the random variable; this is sometime referred to as *prior of pure ignorance*. It can be shown that this assumption of the prior leads to results identical to those of the traditional probability approach (see Examples 2.5.5 and 2.5.6).

Example 2.5.4:[4] Consider a machine whose prior pdf of the proportion x of defectives is given by Table 2.8.
If a random sample of size 2 is selected, and one defective is found, the Bayes estimate of the proportion of defectives produced by the machine is determined as follows.

Let y be the number of defectives in the sample. The probability that the random sample of size 2 yields one defective is given by the Binomial distribution since this is a two-outcome situation:

$$f(y/x) = B(y; n, x) = \binom{2}{y} x^y (1-x)^{2-y}; \; y = 0, 1, 2$$

If x=0.1, then

$$f(1/0.1) = B(1; 2, 0.1) = \binom{2}{1} (0.1)^1 (0.9)^{2-1}$$
$$= 0.18$$

Similarly, for $x = 0.2, \; f(1/0.2) = 0.32$.

[4] From Walpole et al. (2007) by permission of Pearson Education.

Thus, the total probability of finding one defective in a sample size of 2 is:

$$\begin{aligned} f(y=1) &= (0.18)(0.6) + (0.32)(0.40) \\ &= (0.108) + (0.128) \\ &= 0.236 \end{aligned}$$

The posterior probability f(x/y=1) is then given:

- for x=0.1: 0.108/0.236=0.458
- for x=0.2: 0.128/0.236=0.542

Finally, the Bayes' estimate of the proportion of defectives x is:

$$x = (0.1)(0.458) + (0.2)(0.542) = 0.1542$$

which is quite different from the value of 0.5 given by the classical method. ■

2.5.3 Application to Continuous Probability Variables

The Bayes' theorem can also be extended to the case of continuous random variables (Ang and Tang 2007). Let x be the random variable with a prior PDF denoted by p(x). Though any appropriate distribution can be chosen, the Beta distribution (given by Eq. 2.45) is particularly convenient[5], and is widely used to characterize prior PDF. Another commonly used prior is the uniform distribution called a *diffuse prior*.

For consistency with convention, a slightly different nomenclature than that of Eq. 2.50 is adopted. Assuming the Beta distribution, Eq. 2.45a can be rewritten to yield the prior:

$$p(x) \propto x^a (1-x)^b \tag{2.51}$$

Recall that higher the values of the exponents a and b, the peakier the distribution indicative of the prior distribution being relatively well defined.

Let L(x) represent the conditional probability or likelihood function of observing y "successes" out of n observations. Then, the posterior probability is given by:

$$f(x/y) \propto L(x) \cdot p(x) \tag{2.52}$$

In the context of Fig. 2.25, the *likelihood* of the unobservable events $B_1 \ldots B_n$ is the conditional probability that A has occurred given B_i for $i=1, \ldots, n$, or by $p(A/B_i)$. The likelihood function can be gleaned from probability considerations in many cases. Consider Example 2.5.3 involving testing the foundation piles of buildings. The Binomial distribution gives the probability of x failures in n independent Bernoulli

[5] Because of the corresponding mathematical simplicity which it provides as well as the ability to capture a wide variety of PDF shapes

trials, provided the trials are independent and the probability of failure in any one trial is p. This applies to the case when one holds p constant and studyies the behavior of the pdf of defectives x. If instead, one holds x constant and lets p(x) vary over its possible values, one gets the *likelihood function*. Suppose n piles are tested and y piles are found to be defective or sub-par. In this case, the likelihood function is written as follows for the Binomial PDF:

$$L(x) = \binom{n}{y} x^y (1-x)^{n-y} \quad 0 \leq x \leq 1 \tag{2.53}$$

Notice that the Beta distribution is the same form as the likelihood function. Consequently, the posterior distribution given by Eq. 2.53 assumes the form:

$$f(x/y) = k \cdot x^{a+y}(1-x)^{b+n-y} \tag{2.54}$$

where k is independent of x and is a normalization constant. Note that (1/k) is the denominator term in Eq. 2.54 and is essentially a constant introduced to satisfy the probability law that the area under the PDF is unity. What is interesting is that the information contained in the prior has the net result of "artificially" augmenting the number of observations taken. While the classical approach would use the likelihood function with exponents y and (n−y) (see Eq. 2.51), these are inflated to (a+y) and (b+n−y) in Eq. 2.54 for the posterior distribution. This is akin to having taken more observations, and supports the previous statement that the Bayesian approach is particularly advantageous when the number of observations is low.

Three examples illustrating the use of Eq. 2.54 are given below.

Example 2.5.5: Repeat Example 2.5.4 assuming that no information is known about the prior. In this case, assume a uniform distribution.

The prior pdf can be found from the Binomial distribution:

$$f(y/x) = B(1;2,x) = \binom{2}{1} x^1(1-x)^{2-1}$$
$$= 2x(1-x)$$

The total probability of one defective is now given by:

$$f(y=1) = \int_0^1 2x(1-x)dx = \frac{1}{3}$$

The posterior probability is then found by dividing the above two expressions (Eq. 2.54):

$$f(x/y=1) = \frac{2x(1-x)}{1/3} = 6x(1-x)$$

Finally, the Bayes' estimate of the proportion of defectives x is:

$$x = 6\int_0^1 x^2(1-x)dx = 0.5$$

which can be compared to the value of 0.5 given by the classical method. ■

Example 2.5.6: Let us consider the same situation as that treated in Example 2.5.3. However, the proportion of defectives x is now a continuous random variable for which no prior distribution can be assigned. This implies that the engineer has no prior information, and in such cases, a uniform distribution is assumed:

$$p(x) = 1.0 \quad \text{for} \quad 0 \leq x \leq 1$$

The likelihood function for the case of the single tested pile turning out to be defective is x, i.e. L(x)=x. The posterior distribution is then:

$$f(x/y) = k \cdot x(1.0)$$

The normalizing constant

$$k = \left[\int_0^1 x dx\right]^{-1} = 2$$

Hence, the posterior probability distribution is:

$$f(x/y) = 2x \quad \text{for} \quad 0 \leq x \leq 1$$

The Bayesian estimate of the proportion of defectives is:

$$p = E(x/y) = \int_0^1 x \cdot 2x dx = 0.667$$ ■

Example 2.5.7:[6] *Enhancing historical records of wind velocity using the Bayesian approach*

Buildings are designed to withstand a maximum wind speed which depends on the location. The probability *x* that the wind speed will not exceed 120 km/h more than once in 5 years is to be determined. Past records of wind speeds of a nearby location indicated that the following beta distribution would be an acceptable prior for the probability distribution (Eq. 2.45):

$$p(x) = 20x^3(1-x) \quad \text{for} \quad 0 \leq x \leq 1$$

In this case, the likelihood that the annual maximum wind speed will exceed 120 km/h in 1 out of 5 years is given by:

$$L(x) = \binom{5}{4} x^4(1-x) = 5x^4(1-x)$$

[6] From Ang and Tang (2007) by permission of John Wiley and Sons.

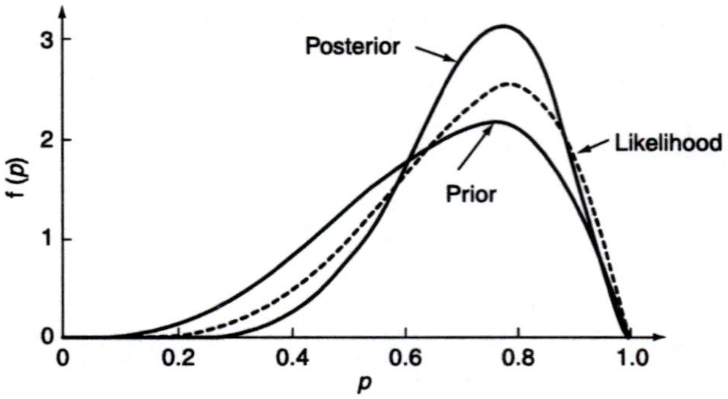

Fig. 2.30 Probability distributions of the prior, likelihood function and the posterior. (From Ang and Tang 2007 by permission of John Wiley and Sons)

Hence, the posterior probability is deduced following Eq. 2.54:

$$\begin{aligned} f(x/y) &= k \cdot [5x^4(1-x)] \cdot [20x^3(1-x)] \\ &= 100k \cdot x^7 \cdot (1-x)^2 \end{aligned}$$

where the constant k can be found from the normalization criterion:

$$k = \left[\int_0^1 100x^7(1-x)^2 dx \right]^{-1} = 3.6$$

Finally, the posterior PDF is given by

$$f(x/y) = 360x^7(1-x)^2 \quad \text{for} \quad 0 \le x \le 1$$

Plots of the prior, likelihood and the posterior functions are shown in Fig. 2.30. Notice how the posterior distribution has become more peaked reflective of the fact that the single test data has provided the analyst with more information than that contained in either the prior or the likelihood function. ■

2.6 Probability Concepts and Statistics

The distinction between probability and statistics is often not clear cut, and sometimes, the terminology adds to the confusion[7]. In its simplest sense, probability generally allows one to predict the behavior of the system "before" the event under the stipulated assumptions, while statistics refers to a body of knowledge whose application allows one to make sense out of the data collected. Thus, probability concepts provide the theoretical underpinnings of those aspects of statistical analysis which involve random behavior or noise in the actual data being analyzed. Recall that in Sect. 1.5, a distinction had been made between four types of uncertainty or unexpected variability in the data. The first was due to the stochastic or inherently random nature of the process itself which no amount of experiment, even if done perfectly, can overcome. The study of probability theory is mainly mathematical, and applies to this type, i.e., to situations/processes/systems whose random nature is known to be of a certain type or can be modeled so that its behavior (i.e., certain events being produced by the system) can be predicted in the form of probability distributions. Thus, probability deals with the idealized behavior of a system under a *known type of randomness*. Unfortunately, most natural or engineered systems do not fit neatly into any one of these groups, and so when performance data is available of a system, the objective may be:

(i) to try to understand the overall nature of the system from its measured performance, i.e., to explain what caused the system to behave in the manner it did, and
(ii) to try to make inferences about the general behavior of the system from a limited amount of data.

Consequently, some authors have suggested that probability be viewed as a "deductive" science where the conclusion is drawn without any uncertainty, while statistics is an "inductive" science where only an imperfect conclusion can be reached, with the added problem that this conclusion hinges on the types of assumptions one makes about the random nature of the underlying drivers! Here is a simple example to illustrate the difference. Consider the flipping of a coin supposed to be fair. The probability of getting "heads" is ½. If, however, "heads" come up 8 times out of the last 10 trials, what is the probability the coin is not fair? Statistics allows an answer to this type of enquiry, while probability is the approach for the "forward" type of questioning.

The previous sections in this chapter presented basic notions of classical probability and how the Bayesian viewpoint is appropriate for certain types of problems. Both these viewpoints are still associated with the concept of probability as the relative frequency of an occurrence. At a broader context, one should distinguish between three kinds of probabilities:

(i) *Objective or absolute probability* which is the classical one where it is interpreted as the "long run frequency". This is the same for everyone (provided the calculation is done correctly!). It is an informed guess of an event which in its simplest form is a constant; for example, historical records yield the probability of flood occurring this year or of the infant mortality rate in the U.S.

Table 2.9 assembles probability estimates for the occurrence of natural disasters with 10 and 1000 fatalities per event (indicative of the severity level) during different time spans (1, 10 and 20 years). Note that floods and tornados have relatively small return times for small

[7] For example, "statistical mechanics" in physics has nothing to do with statistics at all but is a type of problem studied under probability.

Table 2.9 Estimates of absolute probabilities for different natural disasters in the United States. (Adapted from Barton and Nishenko 2008)

Exposure Times	10 fatalities per event				1000 fatalities per event			
Disaster	1 year	10 years	20 years	Return time (yrs)	1 year	10 years	20 years	Return time (yrs)
Earthquakes	0.11	0.67	0.89	9	0.01	0.14	0.26	67
Hurricanes	0.39	0.99	>0.99	2	0.06	0.46	0.71	16
Floods	0.86	>0.99	>0.99	0.5	0.004	0.04	0.08	250
Tornadoes	0.96	>0.99	>0.99	0.3	0.006	0.06	0.11	167

Table 2.10 Leading causes of death in the United States, 1992. (Adapted from Kolluru et al. 1996)

Cause	Annual deaths (×1000)	Percent %
Cardiovascular or heart disease	720	33
Cancer (malignant neoplasms)	521	24
Cerebrovascular diseases (strokes)	144	7
Pulmonary disease (bronchitis, asthma..)	91	4
Pneumonia and influenza	76	3
Diabetes mellitus	50	2
Nonmotor vehicle accidents	48	2
Motor vehicle accidents	42	2
HIV/AIDS	34	1.6
Suicides	30	1.4
Homicides	27	1.2
All other causes	394	18
Total annual deaths (rounded)	*21,77*	*100*

events while earthquakes and hurricanes have relatively short times for large events. Such probability considerations can be determined at a finer geographical scale, and these play a key role in the development of codes and standards for designing large infrastructures (such as dams) as well as small systems (such as residential buildings).

(ii) *Relative probability* where the chance of occurrence of one event is stated in terms of another. This is a way of comparing the effect of different types of adverse events happening on a system or on a population when the absolute probabilities are difficult to quantify. For example, the relative risk for lung cancer is (approximately) 10 if a person has smoked before, compared to a nonsmoker. This means that he is 10 times more likely to get lung cancer than a nonsmoker. Table 2.10 shows leading causes of death in the United States in the year 1992. Here the observed values of the individual number of deaths due to various causes are used to determine a relative risk expressed as % in the last column. Thus, heart disease which accounts for 33% of the total deaths is more than 16 times more likely than motor vehicle deaths. However, as a note of caution, these are values aggregated across the whole population, and need to be interpreted accordingly. State and government analysts separate such relative risks by age groups, gender and race for public policy-making purposes.

(iii) *Subjective probability* which differs from one person to another is an informed or best guess about an event which can change as our knowledge of the event increases. Subjective probabilities are those where the objective view of probability has been modified to treat two types of events: (i) when the occurrence is unique and is unlikely to repeat itself, or (ii) when an event has occurred but one is unsure of the final outcome. In such cases, one still has to assign some measure of likelihood of the event occurring, and use this in our analysis. Thus, a subjective interpretation is adopted with the probability representing a degree of belief of the outcome selected as having actually occurred. There are no "correct answers", simply a measure reflective of our subjective judgment. A good example of such subjective probability is one involving forecasting the probability of whether the impacts on gross world product of a 3°C global climate change by 2090 would be large or not. A survey was conducted involving twenty leading researchers working on global warming issues but with different technical backgrounds, such as scientists, engineers, economists, ecologists, and politicians who were asked to assign a probability estimate (along with 10% and 90% confidence intervals). Though this was not a scientific study as such since the whole area of expert opinion elicitation is still not fully mature, there was nevertheless a protocol in how the questioning was performed, which led to the results shown in Fig. 2.31. The median, 10% and 90% confidence intervals predicted by different respondents show great scatter, with the ecologists estimating impacts to be 20–30 times higher (the two right most bars in the figure), while the economists on average predicted the chance of large consequences to have only a 0.4% loss in gross world product. An engineer or a scientist may be uncomfortable with such subjective probabilities, but there are certain types of problems where this is the best one can do with current knowledge. Thus, formal analysis methods have to accommodate such information, and it is here that Bayesian techniques can play a key role.

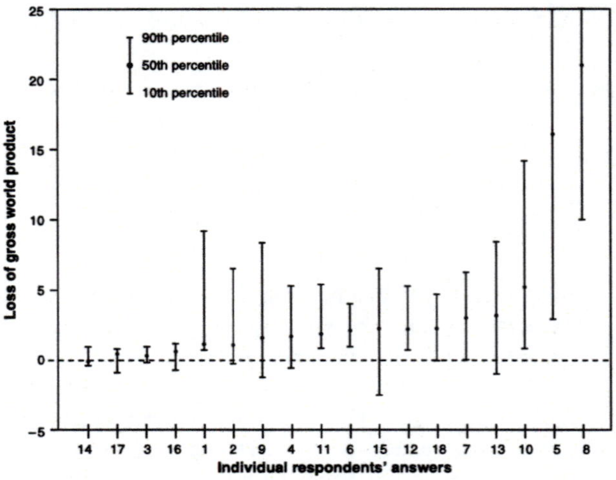

Fig. 2.31 Example illustrating large differences in subjective probability. A group of prominent economists, ecologists and natural scientists were polled so as to get their estimates of the loss of gross world product due to doubling of atmosphereic carbon dioxide (which is likely to occur by the end of the twenty-first century when mean global temperatures increase by 3°C). The two ecologists predicted the highest adverse impact while the lowest four individuals were economists. (From Nordhaus 1994)

Problems

Pr. 2.1 An experiment consists of tossing two dice.
(a) List all events in the sample space
(b) What is the probability that both outcomes will have the same number showing up both times?
(c) What is the probability that the sum of both numbers equals 10?

Pr. 2.2 Expand Eq. 2.9 valid for two outcomes to three outcomes: $p(A \cup B \cup C) = ...$

Pr. 2.3 A solar company has an inspection system for batches of photovoltaic (PV) modules purchased from different vendors. A batch typically contains 20 modules, while the inspection system involves taking a random sample of 5 modules and testing all of them. Suppose there are 2 faulty modules in the batch of 20.
(a) What is the probability that for a given sample, there will be one faulty module?
(b) What is the probability that both faulty modules will be discovered by inspection?

Pr. 2.4 A county office determined that of the 1000 homes in their area, 600 were older than 20 years (event A), that 500 were constructed of wood (event B), and that 400 had central air conditioning (AC) (event C). Further, it is found that events A and B occur in 300 homes, that all three events occur in 150 homes, and that no event occurs in 225 homes. If a single house is picked, determine the following probabilities:
(a) that it is older than 20 years and has central AC?
(b) that it is older than 20 years and does not have central AC?
(c) that it is older than 20 years and is not made of wood?
(d) that it has central air and is made of wood?

Pr. 2.5 A university researcher has submitted three research proposals to three different agencies. Let E_1, E_2 and E_3 be the events that the first, second and third bids are successful with probabilities: $p(E_1)=0.15$, $p(E_2)=0.20$, $p(E_3)=0.10$. Assuming independence, find the following probabilities:
(a) that all three bids are successful
(b) that at least two bids are successful
(c) that at least one bid is successful

Pr. 2.6 Consider two electronic components A and B with probability rates of failure of $p(A)=0.1$ and $p(B)=0.25$. What is the failure probability of a system which involves connecting the two components in (a) series and (b) parallel.

Pr. 2.7[8] A particular automatic sprinkler system for a high-rise apartment has two different types of activation devices for each sprinkler head. Reliability of such devices is a measure of the probability of success, i.e., that the device will activate when called upon to do so. Type A and Type B devices have reliability values of 0.90 and 0.85 respectively. In case, a fire does start, calculate:
(a) the probability that the sprinkler head will be activated (i.e., at least one of the devices works),
(b) the probability that the sprinkler will not be activated at all, and
(c) the probability that both activation devices will work properly.

Pr. 2.8 Consider the two system schematics shown in Fig. 2.32. At least one pump must operate when one chiller is operational, and both pumps must operate when both chillers are on. Assume that both chillers have identical reliabilities of 0.90 and that both pumps have identical reliabilities of 0.95.
(a) Without any computation, make an educated guess as to which system would be more reliable overall when (i) one chiller operates, and (ii) when both chillers operate.
(b) Compute the overall system reliability for each of the configurations separately under cases (i) and (ii) defined above.

[8] From McClave and Benson (1988) by permission of Pearson Education.

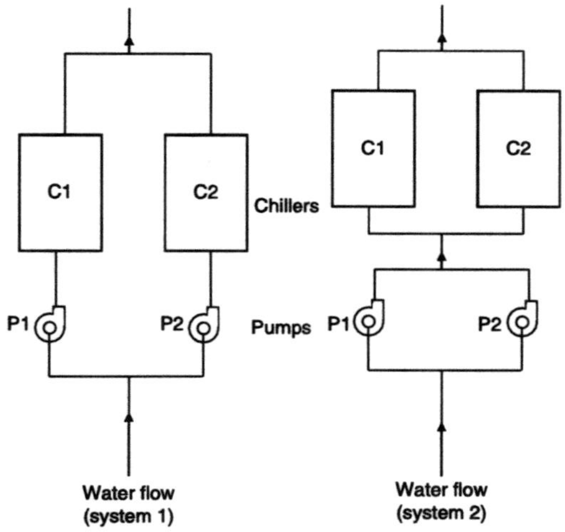

Fig. 2.32 Two possible system configurations

Pr. 2.9 Consider the following CDF:

$$F(x) = 1 - \exp(-2x) \quad \text{for } x > 0$$
$$= 0 \quad x \leq 0$$

(a) Construct and plot the cumulative distribution function
(b) What is the probability of $x<2$
(c) What is the probability of $3<x<5$

Pr. 2.10 The joint density for the random variables (X,Y) is given by:

$$f(x,y) = 10xy^2 \quad 0 < x < y < 1$$
$$= 0 \quad \text{elsewhere}$$

(a) Verify that Eq. 2.19 applies
(b) Find the marginal distributions of X and Y
(c) Compute the probability of $0 < x < 1/2$, $1/4 < y < 1/2$

Pr. 2.11[9] Let X be the number of times a certain numerical control machine will malfunction on any given day. Let Y be the number of times a technician is called on an emergency call. Their joint probability distribution is:

f(x,y)	X	1	2	3
Y	1	0.05	0.05	0.1
	2	0.05	0.1	0.35
	3	0	0.2	0.1

(a) Determine the marginal distributions of X and of Y
(b) Determine the probability $p(x<2, y>2)$

[9] From Walpole et al. (2007) by permission of Pearson Education.

Pr. 2.12 Consider the data given in Example 2.2.6 for the case of a residential air conditioner. You will use the same data to calculate the flip problem using Bayes' law.
(a) During a certain day, it was found that the air-conditioner was operating satisfactorily. Calculate the probabilities that this was a "NH= not hot" day.
(b) Draw the reverse tree diagram for this case.

Pr. 2.13 Consider a medical test for a disease. The test has a probability of 0.95 of correctly or positively detecting an infected person (this is the *sensitivity*), while it has a probability of 0.90 of correctly identifying a healthy person (this is called the *specificity*). In the population, only 3% of the people have the disease.
(a) What is the probability that a person testing positive is actually infected?
(b) What is the probability that a person testing negative is actually infected?

Pr. 2.14 A large industrial firm purchases several new computers at the end of each year, the number depending on the frequency of repairs in the previous year. Suppose that the number of computers X purchased each year has the following PDF:

X	0	1	2	3
f(x)	0.2	0.3	0.2	0.1

If the cost of the desired model will remain fixed at $ 1500 throughout this year and a discount of $ $50x^2$ is credited towards any purchase, how much can this firm expect to spend on new computers at the end of this year?

Pr. 2.15 Suppose that the probabilities of the number of power failure in a certain locality are given as:

X	0	1	2	3
f(x)	0.4	0.3	0.2	0.1

Find the mean and variance of the random variable X.

Pr. 2.16 An electric firm manufacturers a 100 W light bulb, which is supposed to have a mean life of 900 and a standard deviation of 50 h. Assume that the distribution is symmetric about the mean. Determine what percentage of the bulbs fails to last even 700 h if the distribution is found to follow: (i) a normal distribution, (ii) a lognormal, (iii) a Poisson, and (iii) a uniform distribution.

Pr. 2.17 Sulfur dioxide concentrations in air samples taken in a certain region have been found to be well represented by a lognormal distribution with parameters $\mu=2.1$ and $\sigma=0.8$.

(a) What proportion of air samples have concentrations between 3 and 6?
(b) What proportion do not exceed 10?
(c) What interval (a,b) is such that 95% of all air samples have concentration values in this interval, with 2.5% have values less than a, and 2.5% have values exceeding b?

Pr. 2.18 The average rate of water usage (in thousands of gallons per hour) by a certain community can be modeled by a lognormal distribution with parameters μ=4 and σ=1.5. What is the probability that that the demand will:
(a) be 40,000 gallons of water per hour
(b) exceed 60,000 gallons of water per hour

Pr. 2.19 Suppose the number of drivers who travel between two locations during a designated time period is a Poisson distribution with parameter λ=30. In the long run, in what proportion of time periods will the number of drivers:
(a) Be at most 20?
(b) Exceed 25?
(c) Be between 10 and 20.

Pr. 2.20 Suppose the time, in hours, taken to repair a home furnace can be modeled as a gamma distribution with parameters a=2 and λ=1/2. What is the probability that the next service call will require:
(a) at most 1 h to repair the furnace?
(b) at least 2 h to repair the furnace?

Pr. 2.21[10] In a certain city, the daily consumption of electric power, in millions of kilowatts-hours (kWh), is a random variable X having a gamma distribution with mean=6 and variance=12.
(a) Find the values of the parameters α and λ
(b) Find the probability that on any given day the daily power consumption will exceed 12 million kWh.

Pr. 2.22 The life in years of a certain type of electrical switches has an exponential distribution with an average life in years of λ=5. If 100 of these switches are installed in different systems,
(a) what is the probability that at most 20 will fail during the first year?
(b) How many are likely to have failed at the end of 3 years?

Pr. 2.23 *Probability models for global horizontal solar radiation.*
Probability models for predicting solar radiation at the surface of the earth was the subject of several studies in the last several decades. Consider the daily values of global (beam plus diffuse) radiation on a horizontal surface at a specified location. Because of the variability of the atmospheric conditions at any given location, this quantity can be viewed as a random variable. Further, global radiation has an underlying annual pattern due to the orbital rotation of the earth around the sun. A widely adopted technique to filter out this deterministic trend is:
(i) to select the random variable not as the daily radiation itself but as the daily clearness index K defined as the ratio of the daily global radiation on the earth's surface for the location in question to that outside the atmosphere for the same latitude and day of the year, and
(ii) to truncate the year into 12 monthly time scales since the random variable K for a location changes appreciably on a seasonal basis.

Gordon and Reddy (1988) proposed an expression for the PDF of the random variable $X = (K/\bar{K})$ where $\bar{K}$ is the monthly mean value of the daily values of K during a month. The following empirical model was shown to be of general validity in that it applied to several locations (temperate and tropical) and months of the year with the same variance in K:

$$p(X; A, n) = AX^n[1 - (X/X_{\max})] \tag{2.55}$$

where A, n and X_{max} are model parameters which have been determined from the normalization of p(X), knowledge of $\bar{K}$ (i.e., $\bar{X} = 1$) and knowledge of the variance of X or $\sigma^2(X)$. Derive the following expressions for the three model parameters:

$$n = -2.5 + 0.5[9 + (8/\sigma^2(X))]^{1/2}$$
$$X_{\max} = (n+3)/(n+1)$$
$$A = (n+1)(n+2)/X_{\max}^{n+1}$$

Note that because of the manner of normalization, the random variable selected can assume values greater than unity. Figure 2.33 shows the proposed distribution for a number of different variance values.

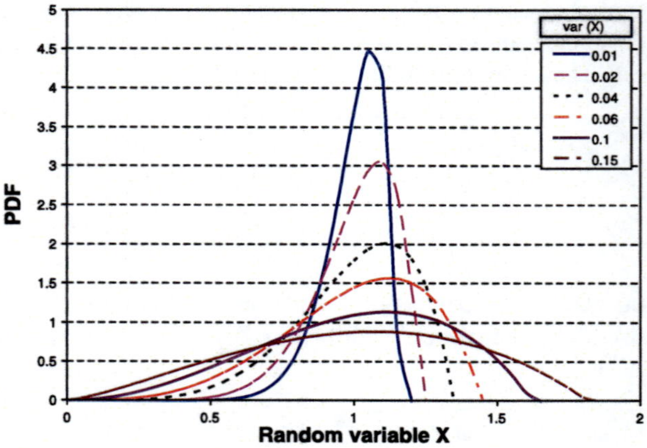

Fig. 2.33 Probability distributions of solar radiation given by Eq. 2.55 as proposed by Gordon and Reddy (1988)

[10] From Walpole et al. (2007) by permission of Pearson Education.

Pr. 2.24 *Cumulative distribution and utilizability functions for horizontal solar radiation*

Use the equations described below to derive the CDF and the utilizability functions for the Gordon-Reddy probability distribution function described in Pr. 2.23.

Probability distribution functions for solar radiation (as in Pr. 2.23 above) and also for ambient air temperatures are useful to respectively predict the long-term behavior of solar collectors and the monthly/annual heating energy use of small buildings. For example, the annual/monthly space heating load Q_{Load} (in MJ) is given by:

$$Q_{Load} = (UA)_{Bldg} \cdot DD \cdot (86,400\frac{s}{day}) \cdot (10^{-6}\frac{MJ}{J})$$

where $(UA)_{Bldg}$ is the building overall energy loss/gain per unit temperature difference in W/°C; and DD is the degree-day given by:

$$DD = \sum_{d=1}^{N} (18.3 - \bar{T}_d)^+ \quad \text{in °C} - \text{day}$$

where $\bar{T}_d$ is the daily average ambient air temperature and N=365 if annual time scales are considered. The "+" sign indicates that only positive values within the brackets should contribute to the sum, while negative values should be set to zero. Physically this implies that only when the ambient air is lower than the design indoor temperature, which is historically taken as 18.3°C, would there be a need for heating the building. It is clear that the DD is the sum of the differences between each day's mean temperature and the design temperature of the conditioned space. It can be derived from knowledge of the PDF of the daily ambient temperature values at the location in question.

A similar approach has also been developed for predicting the long-term energy collected by a solar collector either at the monthly or annual time scale involving the concept of *solar utilizability* (Reddy 1987). Consider Fig. 2.34a which shows the PDF function P(X) for the normalized variable X described in Pr. 2.23, and bounded by X_{min} and X_{max}. The area of the shaded portion corresponds to a specific value X' of the CDF (see Fig. 2.34b):

$$F(X') = probability(X \leq X') = \int_{X_{min}}^{X'} P(X)dX \qquad (2.56a)$$

The long-term energy delivered by a solar thermal collector is proportional to the amount of solar energy above a certain critical threshold X_C, and this is depicted as a shaded area in Fig. 2.34b). This area is called the solar utilizability, and is functionally described by:

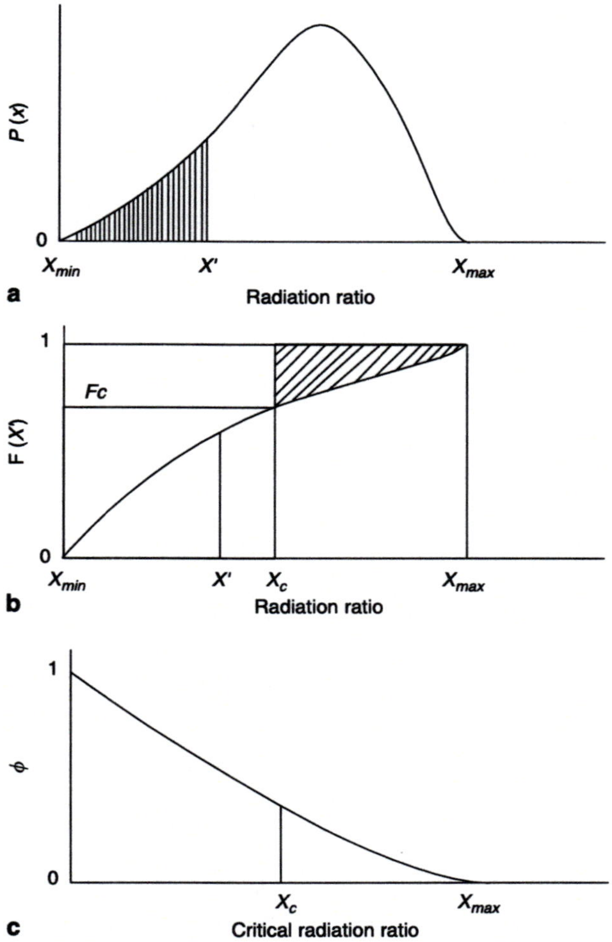

Fig. 2.34 Relation between different distributions. **a** Probability density curve (shaded area represents the cumulative distribution value *F*(*X'*)). **b** Cumulative distribution function (shaded area represents utilizability fraction ar *Xc*). **c** Utilizability curve. (From Reddy 1987)

$$\phi(X_C) = \int_{F_C}^{1} (X' - X_C)dF = \int_{X_C}^{X_{max}} [1 - F(X')dX' \qquad (2.56b)$$

The value of the utilizability function for such a critical radiation level X_C is shown in Fig. 2.34c.

Pr. 2.25 *Generating cumulative distribution curves and utilizability curves from measured data.*

The previous two problems involved probability distributions of solar radiation and ambient temperature, and how these could be used to derive functions for quantities of interest such as the solar utilizability or the Degree-Days. If monitored data is available, there is no need to delve into such considerations of probability distributions, and one can calculate these functions numerically.

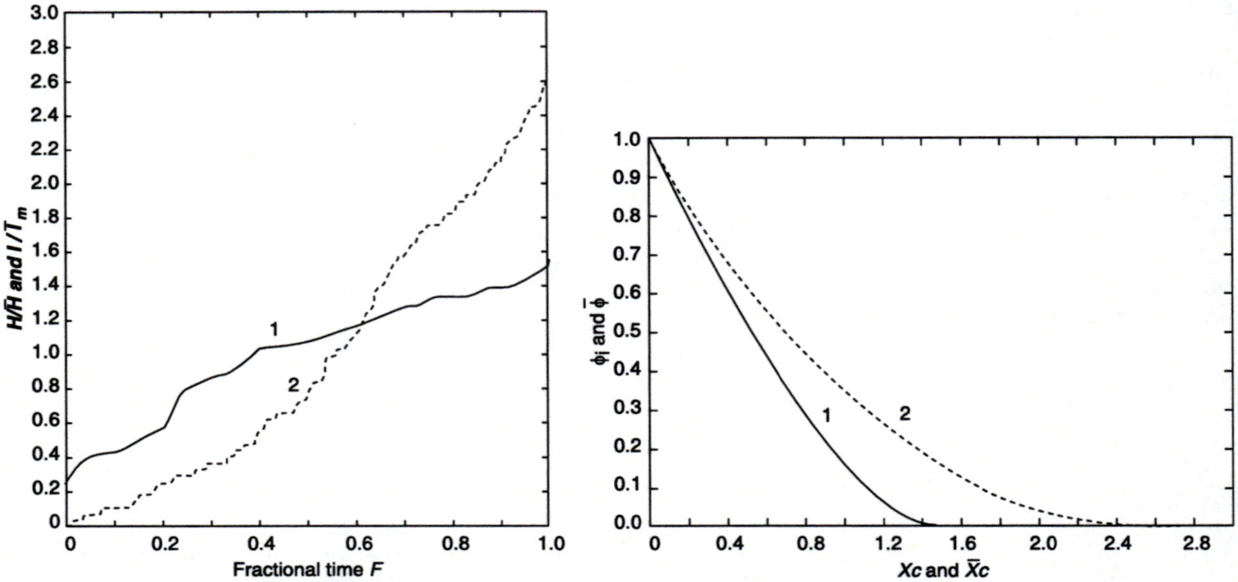

a Cumulative distribution curves of daily (Curve 1) and hourly (Curve 2) radiation values at Quezon City, Manila

b Utilizability curves of daily (Curve 1) and hourly (Curve 2) radiation values at Quezon City, Manila

Fig. 2.35 Distribution for Quezon City, Manila during October 1980. (From Reddy 1987)

Consider Table P2.25 (in Appendix B) which assembles the global solar radiation on a horizontal surface at Quezon City, Manila during October 1980 (taken from Reddy 1987). You are asked to numerically generate the CDF and the utilizability functions (Eq. 2.56a, b) and compare your results to Fig. 2.35. The symbols I and H denote hourly and daily radiation values respectively. Note that instead of normalizing the radiation values by the extra-terrestrial solar radiation (as done in Pr. 2.23), here the corresponding average values $\bar{I}$ (for individual hours of the day) and $\bar{H}$ have been used.

References

Ang, A.H.S. and W.H. Tang, 2007. *Probability Concepts in Engineering*, 2nd Ed., John Wiley and Sons, USA

Barton, C. and S. Nishenko, 2008. *Natural Disasters: Forecasting Economic and Life Losses*, U.S. Geological Survey, Marine and Coastal Geology Program.

Bolstad, W.M., 2004. *Introduction to Bayesian Statistics*, Wiley and Sons, Hoboken, NJ.

Gordon, J. M., and T.A. Reddy, 1988. Time series analysis of daily horizontal solar radiation, *Solar Energy*, 41(3), pp.215-226

Kolluru, R.V., S.M. Bartell, R.M. Pitblado, and R.S. Stricoff, R.S., 1996. *Risk Assessment and Management Handbook*, McGraw-Hill, New York.

McClave, J.T. and P.G. Benson, 1988. *Statistics for Business and Economics*, 4th Ed., Dellen and Macmillan, London.

Nordhaus, W.D., 1994. Expert opinion on climate change, *American Scientist*, 82: pp.45-51.

Phillips, L.D., 1973. *Bayesian Statistics for Social Scientists*, Thomas Nelson and Sons, London, UK

Reddy, T.A., 1987. *The Design and Sizing of Active Solar Thermal Systems*, Oxford University Press, Oxford, U.K.

Walpole, R.E., R.H. Myers, S.L. Myers, and K. Ye, 2007, *Probability and Statistics for Engineers and Scientists*, 8th Ed., Prentice Hall, Upper Saddle River, NJ.

3 Data Collection and Preliminary Data Analysis

This chapter starts by presenting some basic notions and characteristics of different types of data collection systems and types of sensors. Next, simple ways of validating and assessing the accuracy of the data collected are addressed. Subsequently, salient statistical measures to describe univariate and multivariate data are presented along with how to use them during basic exploratory data and graphical analyses. The two types of measurement uncertainty (bias and random) are discussed and the concept of confidence intervals is introduced and its usefulness illustrated. Finally, three different ways of determining uncertainty in a data reduction equation by propagating individual variable uncertainty are presented; namely, the analytical, numerical and Monte Carlo methods.

3.1 Generalized Measurement System

There are several books, for example Doebelin (1995) or Holman and Gajda (1984) which provide a very good overview of the general principles of measurement devices and also address the details of specific measuring devices (the common ones being those that measure physical quantities such as temperature, fluid flow, heat flux, velocity, force, torque, pressure, voltage, current, power,…). This section will be limited to presenting only those general principles which would aid the analyst in better analyzing his data. The generalized measurement system can be divided into three parts (Fig. 3.1):

(i) *detector-transducer* stage, which detects the value of the physical quantity to be measured and transduces or transforms it into another form, i.e., performs either a mechanical or an electrical transformation to convert the signal into a more easily measured and usable form (either digital or analog);

(ii) *intermediate* stage, which modifies the direct signal by amplification, filtering, or other means so that an output within a desirable range is achieved. If there is a known correction (or calibration) for the sensor, it is done at this stage; and

(iii) *output or terminating* stage, which acts to indicate, record, or control the variable being measured. The output could be digital or analog.

Ideally, the output or terminating stage should indicate only the quantity to be measured. Unfortunately, there are various spurious inputs which could contaminate the desired measurement and introduce errors. Doebelin (1995) groups the inputs that may cause contamination into two basic types: modifying and interfering (Fig. 3.2).

(i) *Interfering inputs* introduce an error component to the output of the detector-transducer stage in a rather direct manner, just as does the desired input quantity. For example, if the quantity being measured is temperature of a solar collector's absorber plate, improper shielding of the thermocouple would result in an erroneous reading due to the solar radiation striking the thermocouple junction. As shown in Fig. 3.3, the calibration of the instrument is no longer a constant but is affected by the time at which the measurement is made, and since this may differ from one day to the next, the net result is improper calibration. Thus, solar radiation would be thought of as an interfering input.

(ii) *Modifying inputs* have a more subtle effect, introducing errors by modifying the input/output relation between the desired value and the output measurement (Fig. 3.2). An example of this occurrence is when the oil used to lubricate the various intermeshing mechanisms of a system has deteriorated, and the resulting change in viscosity can lead to the input/output relation getting altered in some manner.

One needs also to distinguish between the *analog* and the *digital* nature of the sensor output or signal (Doebelin 1995). For analog signals, the precise value of the quantity (voltage, temperature,…) carrying the information is important. Digital signals, however, are basically binary in nature (on/off), and variation in numerical values is associated with changes in the logical state (true/false). Consider a digital electronic system where any voltage in the range of +2 to +5 V produces the on-state, while signals of 0 to +1.0 V correspond to the

T. Agami Reddy, *Applied Data Analysis and Modeling for Energy Engineers and Scientists*,
DOI 10.1007/978-1-4419-9613-8_3, © Springer Science+Business Media, LLC 2011

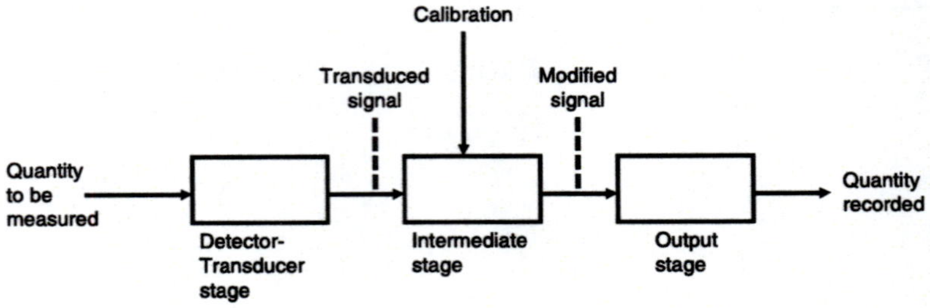

Fig. 3.1 Schematic of the generalized measurement system

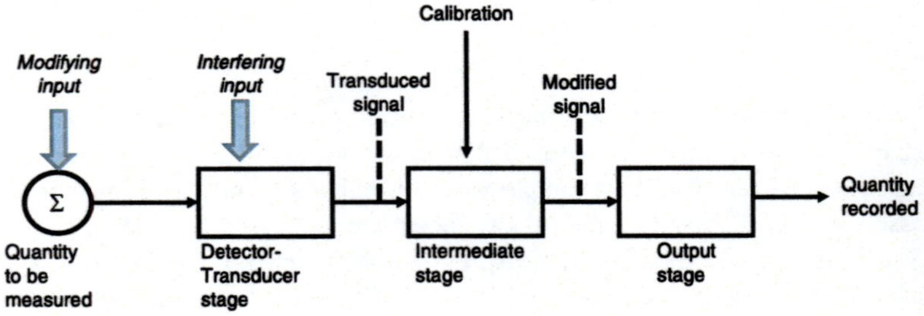

Fig. 3.2 Different types of inputs and noise in a measurement system

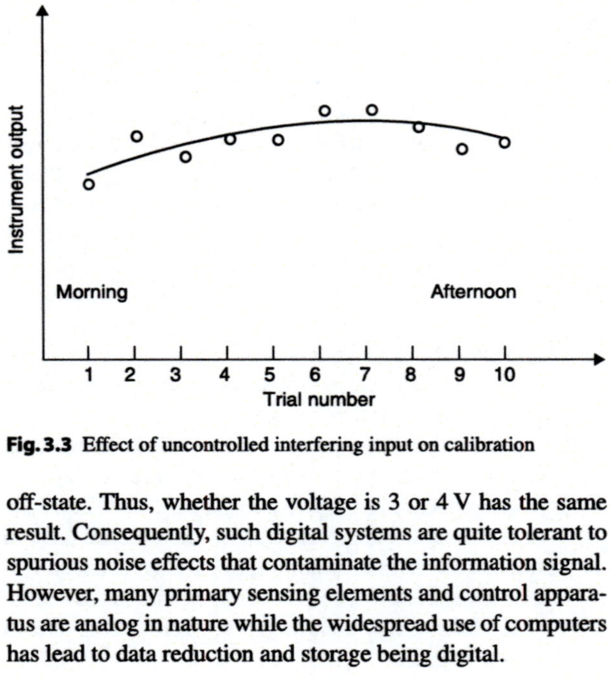

Fig. 3.3 Effect of uncontrolled interfering input on calibration

off-state. Thus, whether the voltage is 3 or 4 V has the same result. Consequently, such digital systems are quite tolerant to spurious noise effects that contaminate the information signal. However, many primary sensing elements and control apparatus are analog in nature while the widespread use of computers has lead to data reduction and storage being digital.

3.2 Performance Characteristics of Sensors and Sensing Systems

There are several terms that are frequently used in connection with sensors and data recording systems. These have to do with their performance characteristics, both static and dynamic, and these will be briefly discussed below.

3.2.1 Sensors

(a) ***Accuracy*** is the ability of an instrument to indicate the true value of the measured quantity. As shown in Fig. 3.4a, the accuracy of an instrument indicates the deviation between one, or an average of several, reading(s) from a known input or accepted reference value. The spread in the target holes in the figure is attributed to random effects.

(b) ***Precision*** is the closeness of agreement among repeated measurements of the same physical quantity. The precision of an instrument indicates its ability to reproduce a certain reading with a given accuracy. Figure 3.4b illustrates the case of precise marksmanship but which is inaccurate due to the bias.

(c) ***Span*** (also called dynamic range) of an instrument is the range of variation (minimum to maximum) of the physical quantity which the instrument can measure.

(d) ***Resolution or least count*** is the smallest incremental value of a measured quantity that can be reliably measured and reported by an instrument. Typically, this is half the smallest scale division of an analog instrument, or the least significant bit of an analog to digital system. In case of instruments with non-uniform scale, the resolution will vary with the magnitude of the output signal being measured. When resolution is measured at the origin of the calibration curve, it is called the threshold of the instrument (see Fig. 3.5). Thus, the *threshold* is the smallest detectable value of the measured quantity while resolution is the smallest perceptible change over its operable range.

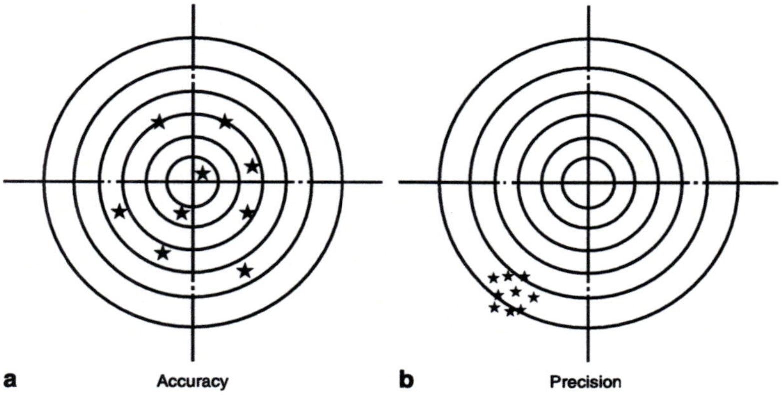

Fig. 3.4 Concepts of accuracy and precision illustrated in terms of shooting at a target

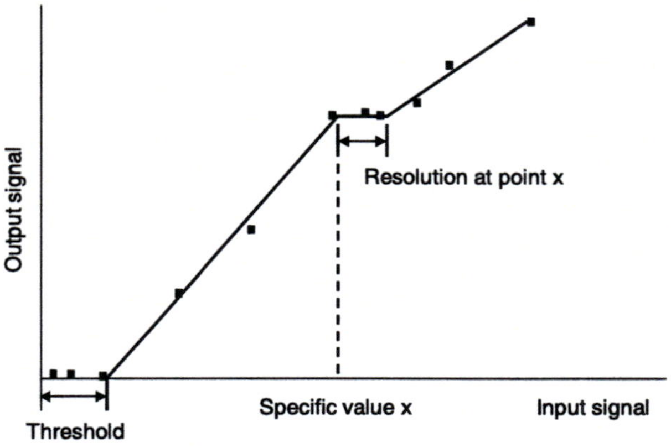

Fig. 3.5 Concepts of threshold and resolution

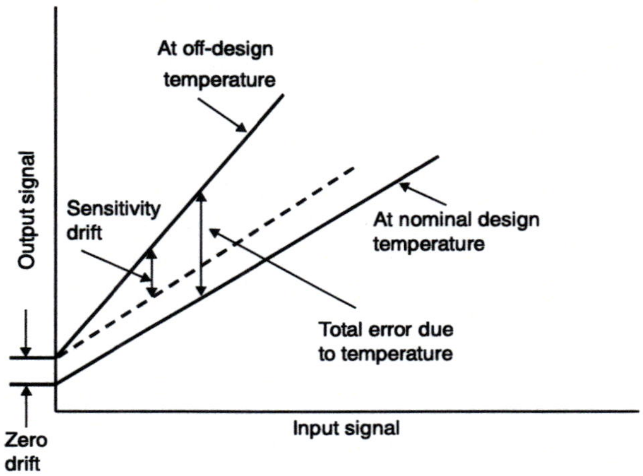

Fig. 3.6 Zero drift and sensitivity drift

(e) ***Sensitivity*** of an instrument is the ratio of the linear movement of the pointer on an analog instrument to the change in the measured variable causing this motion. For example, a 1 mV recorder with a 25 cm scale-length, would have a sensitivity of 25 cm/mV if the measurements were linear over the scale. It is thus representative of the slope of the input/output curve if assumed linear. All things being equal, instruments with larger sensitivity are preferable, but this would generally lead to the range of such an instrument to be smaller. Figure 3.6 shows a linear relationship between the output and the input. Spurious inputs due to the modifying and interfering inputs can cause a *zero drift* and a *sensitivity drift* from the nominal design curve. Some "smart" transducers have inbuilt corrections for such effects which can be done on a continuous basis. Note finally, that sensitivity should not be confused with accuracy which is entirely another characteristic.

(f) ***Hysteresis*** (also called dead space or dead band) is the difference in readings depending on whether the value of the measured quantity is approached from above or below (see Fig. 3.7). This is often the result of mechanical friction, magnetic effects, elastic deformation, or thermal effects. Another cause could be when the experimenter does not allow enough time between measurements to reestablish steady-state conditions. The band can vary over the range of variation of the variables, as shown in the figure.

(g) ***Calibration*** is the checking of the instrument output against a known standard, and then correcting for bias. The standards can be either a primary standard (say, at the National Institute of Standards and Technology), or a secondary standard with a higher accuracy than the instrument to be calibrated, or a known input source (say, checking a flowmeter against direct weighing of the fluid). Doebelin (1995) suggests that, as a rule of thumb, the primary standard should be about 10 times more accurate than the instrument being calibrated.
Figure 3.8 gives a table and a graph of the results of calibrating a pressure measuring device. The data points denoted by circles have been obtained during the calibration process when the pressure values have been incrementally increased while the data points denoted

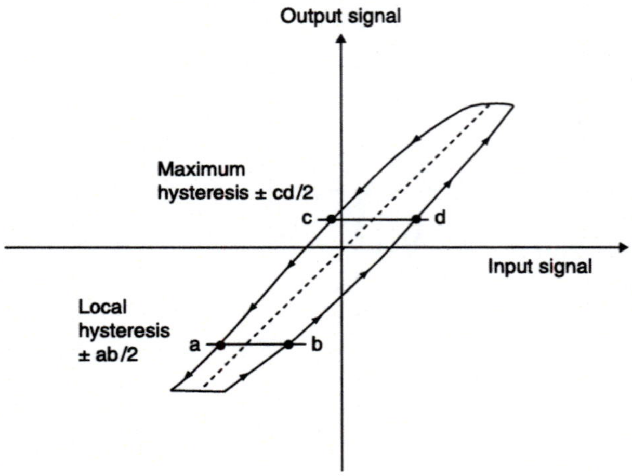

Fig. 3.7 Illustrative plot of a hysteresis band of a sensor showing local and maximum values

by triangles are those obtained when the magnitude of the pressure has been incrementally decreased. The difference in both sets of points is due to the hysteresis of the instrument. Further the "true" value and the instrument value may have a bias (or systematic error) and an uncertainty (or random error) as shown in Fig. 3.8. A linear relationship is fit to the data points to yield the calibration curve. Note that the fitted line need not necessarily be linear though practically instruments are designed to have such a linearity because of the associated convenience of usage and interpretation. When a calibration is completed, it is used to convert an instrument reading of an unknown quantity into a best estimate of the true value. Thus, the calibration curve corrects for bias and puts numerical limits (say ±2 standard deviations) on the random errors of the observations.

The above terms basically describe the *static response*, i.e., when the physical quantity being measured does not change with time. Section 1.2.5 also introduced certain simple models for static and dynamic response of sensors. Usually the physical quantity will change with time, and so the *dynamic response* of the sensor or instrument has to be considered. In such cases, new ways of specifying accuracy are required.

(h) ***Rise time*** is the delay in the sensor output response when the physical quantity being measured undergoes a step change (see Fig. 3.9).

(i) ***Time constant*** of the sensor is defined as the time taken for the sensor output to attain a value of 63.2% of the difference between the final steady-state value and the initial steady-state value when the physical quantity being measured undergoes a step change. Though the concept of time constant is strictly applicable to linear systems only (see Sect. 1.2.5), the term is commonly used to all types of sensors and data recording systems.

(j) ***Distortion*** is a very general term that is used to describe the variation of the output signal from the sensor from its true form characterized by the variation of the physical quantity being measured. Depending on the sensor, the distortion could result either in poor frequency response or poor phase-shift response (Fig. 3.10). For pure electrical measurements, electronic devices are used to keep distortion to a minimum.

3.2.2 Types and Categories of Measurements

Measurements are categorized as either primary measurements or derived measurements.

(i) A *primary measurement* is one that is obtained directly from the measurement sensor. This can be temperature, pressure, velocity, etc. The basic criterion is that a primary measurement is of a single item from a specific measurement device.

(ii) A *derived measurement* is one that is calculated using one or more measurements. This calculation can occur at the sensor level (an energy meter uses flow and temperature difference to report an energy rate), by a data logger, or can occur during data processing. Derived measurements can use both primary and other derived measurements.

Further, measurements can also be categorized by type:

(i) *Stationary data* does not change with time. Examples of stationary data include the mass of water in a tank, the area of a room, the length of piping or the volume of a building. Therefore, whenever the measurement is replicated, the result should be the same, independently of time, within the bounds of measurement uncertainty.

(ii) *Time dependent data* varies with time. Examples of time dependent data include the pollutant concentration in a water stream, temperature of a space, the chilled water flow to a building, and the electrical power use of a facility. A time-dependent reading taken now could be different than a reading taken in the next five minutes, the next day, or the next year. Time dependent data can be recorded either as time-series or cross-sectional:
   - Time-series data consist of a multiplicity of data taken at a single point or location over fixed intervals of time, thus retaining the time sequence nature.
   - Cross-sectional data are data taken at single or multiple points at a single instant in time with time not being a variable in the process.

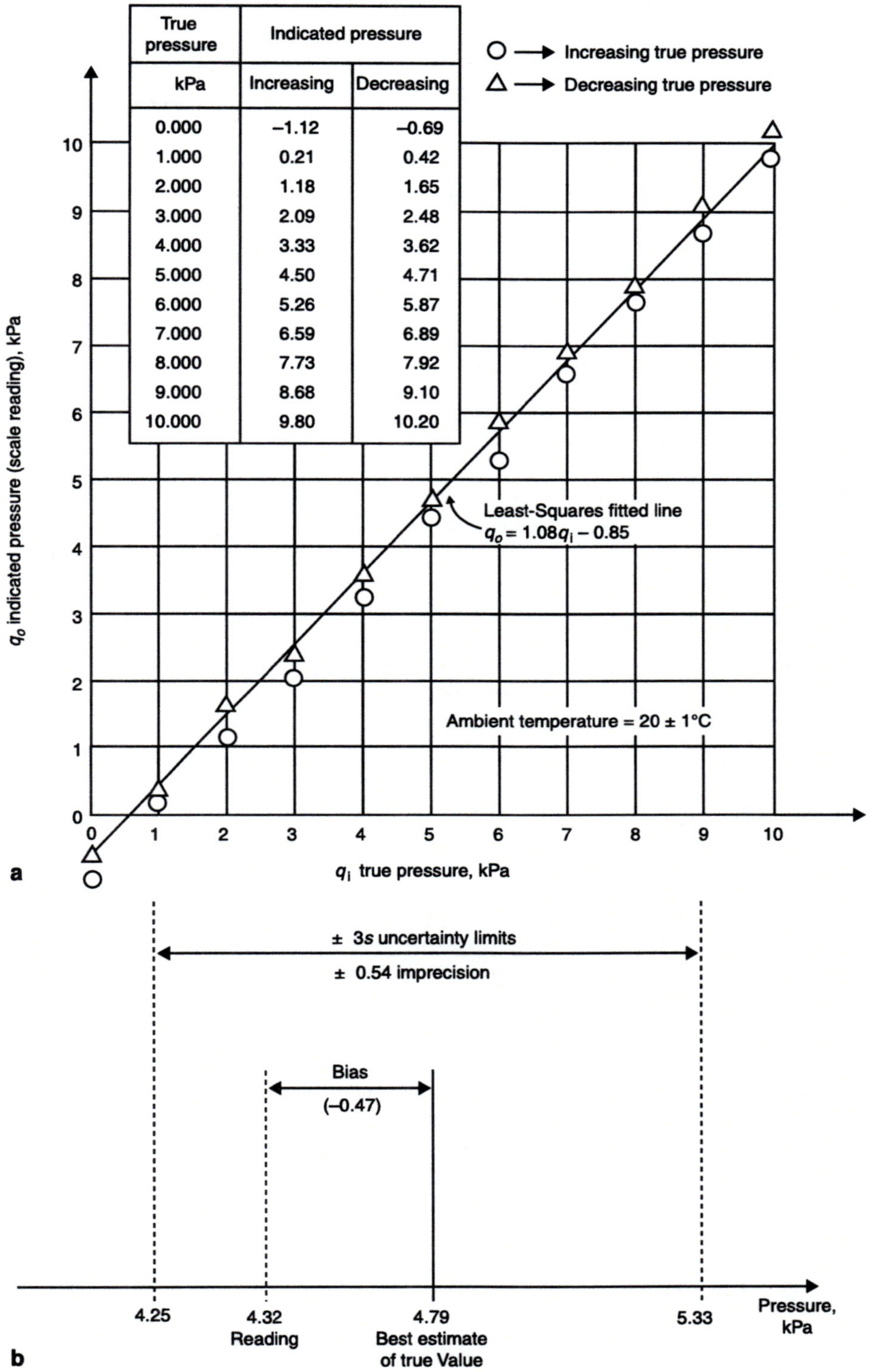

True pressure	Indicated pressure	
kPa	Increasing	Decreasing
0.000	−1.12	−0.69
1.000	0.21	0.42
2.000	1.18	1.65
3.000	2.09	2.48
4.000	3.33	3.62
5.000	4.50	4.71
6.000	5.26	5.87
7.000	6.59	6.89
8.000	7.73	7.92
9.000	8.68	9.10
10.000	9.80	10.20

Fig. 3.8 Static calibration to define bias and random variation or uncertainty. Note that s is the standard deviation of the differences between measurement and the least squares model. (From Doebelin (1995) by permission of McGraw-Hill)

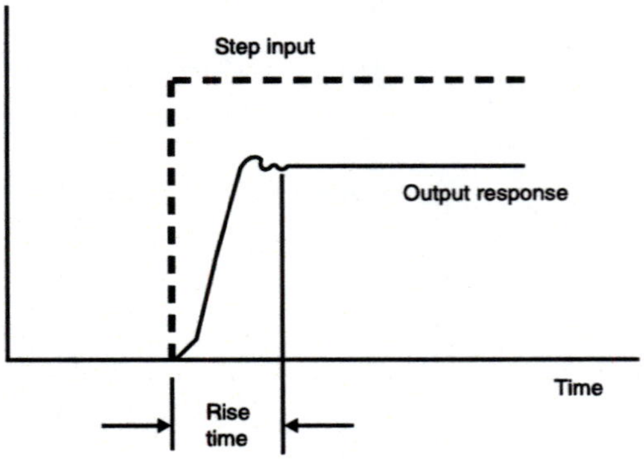

Fig. 3.9 Concept of rise time of the output response to a step input

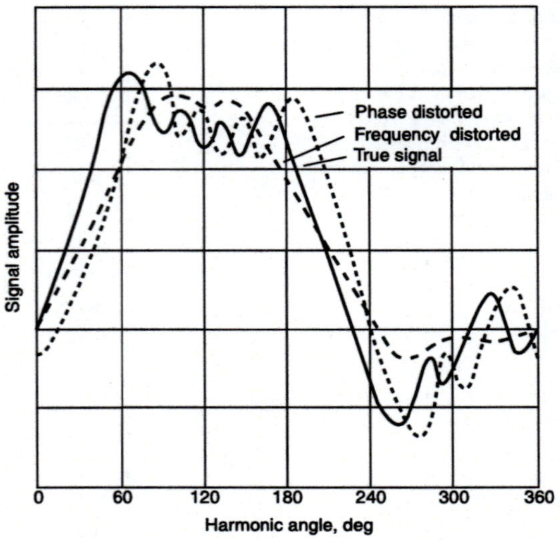

Fig. 3.10 Effects of frequency response and phase-shift response on complex waveforms. (From Holman and Gajda (1984) by permission of McGraw-Hill)

3.2.3 Data Recording Systems

The above concepts also apply to data recording or logging systems, where, however, additional ones need also be introduced:

(a) ***Recording interval*** is the time period or intervals at which data is recorded (a typical range for a thermal systems could be 1–15 min)
(b) ***Scan rate*** is the frequency with which the recording system samples individual measurements; this is often much smaller than the recording interval (with electronic loggers, a typical value could be one sample per second)
(c) ***Scan interval*** is the minimum interval between separate scans of the complete set of measurements which includes several sensors (a typical value could be 10–15 s)
(d) ***Non-process data trigger***. Care must be taken that averaging of the physical quantities that are subsequently recorded does not include non-process data (i.e., temperature data when the flow in a pipe is stopped but the sensor keeps recording the temperature of the fluid at rest). Often data acquisition systems use a threshold trigger to initiate acceptance of individual samples in the final averaged value or monitor the status of an appropriate piece of equipment (for example, whether a pump is operational or not).

3.3 Data Validation and Preparation

The aspects of data collection, cleaning, validation and transformation are crucial. However, these aspects are summarily treated in most books, partly because their treatment involves adopting circumstance specific methods, and also because it is (alas) considered neither of much academic interest nor a field worthy of scientific/statistical endeavor. This process is not so much a bag of tricks, but rather a process of critical assessment, exploration, testing and evaluation which comes by some amount of experience.

Data reduction involves the distillation of raw data into a form that can be subsequently analyzed. It may involve averaging multiple measurements, quantifying necessary conditions (e.g., steady state), comparing with physical limits or expected ranges, and rejecting outlying measurements. Data validation or proofing data for consistency is a process for detecting and removing *gross* or "egregious" errors in the monitored data. It is extremely important to do this proofing or data quality checking at the very beginning, even before any sort of data analysis is attempted. Few such data points could completely overwhelm even the most sophisticated analysis procedures one could adopt. Note that statistical screening (discussed later) is more appropriate for detecting outliers and not for detecting gross errors. There are several types of data proofing, as described below (ASHRAE 2005).

3.3.1 Limit Checks

Fortunately, many of the measurements made in the context of engineering systems have identifiable limits. Limits are useful in a number of experimental phases such as establishing a basis for appropriate instrumentation and measurement techniques, rejecting individual experimental observations, and bounding/bracketing measurements. Measurements can often be compared with one or more of the following limits: physical, expected and theoretical.

(a) *Physical Limits*. Appropriate physical limits should be identified in the planning phases of an experiment so that they can be used to check the reasonableness of both raw and post-processed data. Under no circumstance can experimental observations exceed physical limits. For example, in psychrometrics:
- dry bulb temperature ≥ wet bulb temperature ≥ dew point temperature
- 0 ≤ relative humidity ≤ 100%

Examples in refrigeration systems is that refrigerant saturated condensing temperature should always be greater than the outdoor air dry bulb for air-cooled condensers. Another example in solar radiation measurement is that global radiation on a surface should be greater than the beam radiation incident on the same surface.

Experimental observations or processed data that exceed physical limits should be flagged and closely scrutinized to determine the cause and extent of their deviation from the limits. The reason for data being persistently beyond physical limits is usually instrumentation bias or errors in data analysis routines/methods. Data that infrequently exceed physical limits may be caused by noise or other related problems. Resolving problems associated with observations that sporadically exceed physical limits is often difficult. However, if they occur, the experimental equipment and data analysis routines should be inspected and repaired. In situations where data occasionally exceed physical limits, it is often justifiable to purge such observations from the dataset prior to undertaking any statistical analysis or testing of hypotheses.

(b) *Expected Limits*. In addition to identifying physical limits, expected upper and lower bounds should be identified for each measured variable. During the planning phase of an experiment, determining expected ranges for measured variables facilitates the selection of instrumentation and measurement techniques. Prior to taking data, it is important to ensure that the measurement instruments have been calibrated and are functional over the range of expected operation. An example is that the relative humidity in conditioned office spaces should be in the range between 30–65%. During the execution phase of experiments, the identified bounds serve as the basis for flagging potentially suspicious data. If individual observations exceed the upper or lower range of expected values, those points should be flagged and closely scrutinized to establish their validity and reliability. Another suspicious behavior is constant values when varying values are expected. Typically, this is caused by an incorrect lower or upper bound in the data reporting system so that limit values are being reported instead of actual values.

(c) *Theoretical Limits*. These limits may be related to physical properties of substances (e.g., fluid freezing point), thermodynamic limits of a subsystem or system (e.g., Carnot efficiency for a vapor compression cycle), or thermodynamic definitions (e.g., heat exchanger effectiveness between zero and one). During the execution phase of experiments, theoretical limits can be used to bound measurements. If individual observations exceed theoretical values, those points should be flagged and closely scrutinized to establish their validity and reliability.

3.3.2 Independent Checks Involving Mass and Energy Balances

In a number of cases, independent checks can be used to establish viability of data once the limit checks have been performed. Examples of independent checks include comparison of measured (or calculated) values with those of other investigators (reported in the published literature) and *intra-experiment comparisons* (based on component conservation principles) which involve collecting data and applying appropriate conservation principles as part of the validation procedures. The most commonly applied conservation principles used for independent checks include mass and energy balances on components, subsystems and systems. All independent checks should agree to within the range of expected uncertainty of the quantities being compared. An example of heat balance conservation check as applied to vapor compression chillers is that the chiller cooling capacity and the compressor power should add up to the heat being rejected at the condenser.

Another sound practice is to design some amount of *redundancy* into the experimental design. This allows consistency and conservation checks to be performed. A simple example of consistency check is during the measurement of say the pressure differences between indoors and outdoors of a two-story residence. One could measure the pressure difference between the first floor and the outdoors and the second floor and the outdoors, and deduce the difference in pressure between both floors as the difference between both measurements. Redundant consistency checking would involve also measuring the first floor and second floor pressure difference and verifying whether the three measurements are consistent or not. Of course such checks would increase the cost of the instrumentation, and their need would depend on the specific circumstance.

3.3.3 Outlier Rejection by Visual Means

This phase is undertaken after limit checks and independent checks have been completed. Unless there is a definite reason for suspecting that a particular observation is invalid, indiscriminate outlier rejection is not advised. The sensible

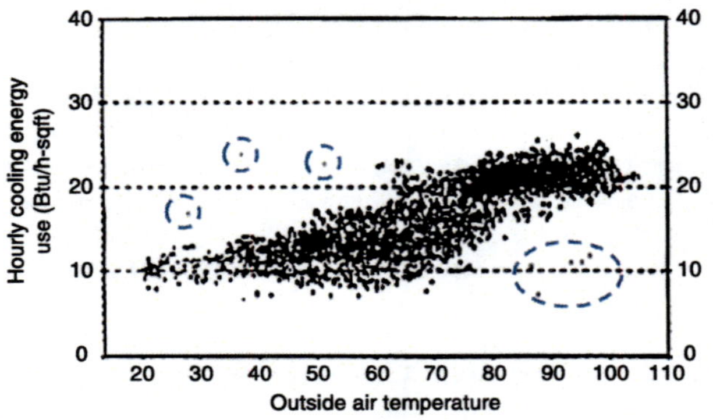

Fig. 3.11 Scatter plot of the hourly chilled water consumption in a commercial building. Some of the obvious outlier points are circled. (From Abbas and Haberl 1994 by permission of Haberl)

approach is to select a reasonable rejection criterion, which may depend on the specific circumstance, *and* couple this with a visual inspection and a computational diagnostics of the data. A commonly used rejection criterion in case the distribution is normal is to eliminate data points which are outside the (3 × standard deviation) range (see Fig. 3.8). Some analysts advocate doing the analytical screening first; rather, it is suggested here that the graphical screening be done first since it also reveals the underlying distribution of the data.

When dealing with correlated bivariate data, relational scatter plots (such as x–y scatter plots) are especially useful since they also allow outliers to be detected with relative ease by visual scrutiny. The hourly chilled water energy use in a commercial building is plotted against outside dry-bulb temperature in Fig. 3.11. One can clearly detect several of the points which fall away from the cloud of data points and which could be weeded out. Further, in cases when the physical process is such that its behavior is known at a limit (for example, both variables should be zero together), one could visually extrapolate the curve and determine whether this is more or less true. Outlier rejection based on statistical considerations is treated in Sect. 3.6.6.

3.3.4 Handling Missing Data

Data is said to be missing, as against bad data during outlier detection, when the channel goes "dead" indicating either a zero value or a very small value which is constant over time when the physics of the process would strongly indicate otherwise. Missing data are bound to occur in most monitoring systems, and can arise from a variety of reasons. First, one should spend some time trying to ascertain the extent of the missing data and whether it occurs preferentially, i.e., whether it is non-random. For example, certain sensors (such as humidity sensors, flow meters or pollutant concentration) develop faults more frequently than others, and the data set becomes biased. This non-random nature of missing data is more problematic than the case of data missing at random.

There are several approaches to handling missing data. It is urged that the data be examined first before proceeding to rehabilitate it. These approaches are briefly described below:

(a) *Use observations with complete data only*: This is the simplest and most obvious, and is adopted in most analysis. Many of the software programs allow such cases to be handled. Instead of coding missing values as zero, analysts often use a default value such as –99 to indicate a missing value. This approach is best suited when the missing data fraction is small enough not to cause the analysis to become biased.

(b) *Reject variables*: In case only one or a few channels indicate high levels of missing data, the judicious approach is to drop these variables from the analysis itself. If these variables are known to be very influential, then more data needs to be collected with the measurement system modified to avoid such future occurrences.

(c) *Adopt an imputation method*: This approach, also called *data rehabilitation*, involves estimating the missing values based on one of the following methods:
   (i) substituting the missing values by a constant value is easy to implement but suffers from the drawback that it would introduce biases, i.e., it may distort the probability distribution of the variable, its variance and its correlation with other variables;
   (ii) substituting the missing values by the mean of the missing variable deduced from the valid data. It suffers from the same distortion as (i) above, but would perhaps add a little more realism to the analysis;
   (iii) univariate interpolation where missing data from a specific variable are predicted using time series methods. One can use numerical methods used to interpolate between tabular data as is common in many engineering applications (see any appropriate textbook on numerical methods; for example, Ayyub and McCuen 1996). One method is that of *undetermined coefficients* where a nth order polynomial (usually second or third order suffice) is used as the interpolation function whose numerical values are obtained by solving n simultaneous equations. The *Gregory-Newton method* results in identifying a similar polynomial function without requiring a set of simultaneous equations to be solved. Another common interpolation method is the *Lagrange polynomials method* (applicable to data taken at unequal intervals). One could also use trigonometric functions with time as the regressor variable provided the data exhibits such periodic

Table 3.1 Saturation water pressure with temperature

T (C)	P (kPa)
50	12.335
54	15.002
58	18.147 ← Point to be estimated
62	21.84
66	26.15

variations (say, the diurnal variation of outdoor dry-bulb temperature). This approach works well when the missing data gaps are short and the process is sort of stable;

(iv) regression methods which use a regression model between the variable whose data is missing and other variables with complete data. Such regression models can be simple regression models or could be multivariate models depending on the specific circumstance. Many of the regression methods (including *splines* which are accurate especially for cases where data exhibits large sudden changes and which are described in Sect. 5.7.2) can be applied. However, the analyst should be cognizant of the fact that such a method of rehabilitation always poses the danger of introducing, sometimes subtle, biases in the final analysis results. The process of rehabilitation may have unintentionally given a structure or an interdependence which may not have existed in the phenomena or process.

Example 3.3.1 *Example of interpolation.*
Consider Table 3.1 showing the saturated water vapor pressure (P) against temperature (T). Let us assume that the mid-point (T=58°C) is missing (see Fig. 3.12). The use of different interpolation methods to determine this point using

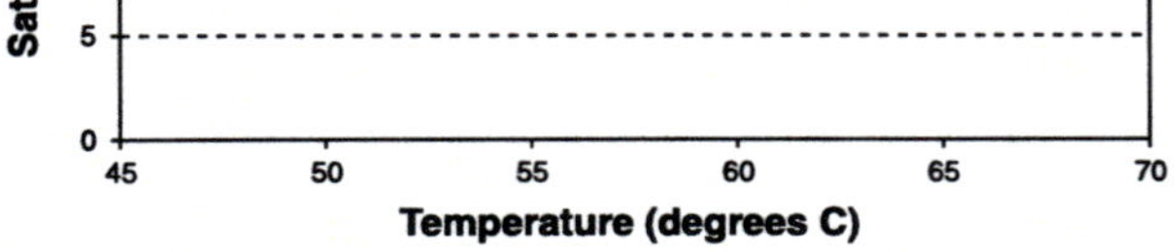

Fig. 3.12 Simple linear interpolation to estimate value of missing point

two data points on either side of the missing point is illustrated below.

(a) Simple linear interpolation: Since the x-axis data are at equal intervals, once would estimate P(58°C)=(15.002+21.84)/2=18.421 which is 1.5% too high.

(b) Method of undetermined coefficients using third order model: In this case, a more flexible functional form of the type: $P=a+b \cdot T+c \cdot T^2+d \cdot T^3$ is assumed, and using data from the four sets of points, the following four simultaneous equations need to be solved for the four coefficients:

$$12.335 = a + b \cdot (50) + c \cdot (50)^2 + d \cdot (50)^3$$

$$15.002 = a + b \cdot (54) + c \cdot (54)^2 + d \cdot (54)^3$$

$$21.840 = a + b \cdot (62) + c \cdot (62)^2 + d \cdot (62)^3$$

$$26.150 = a + b \cdot (66) + c \cdot (66)^2 + d \cdot (66)^3$$

Once the polynomial function is known, it can be used to predict the value of P at T=58°C.

(c) Gregory-Newton method takes the form:

$$y = a_1 + a_2(x - x_1) + a_3(x - x_1)(x - x_2) + \ldots$$

Substituting each set of data point in turn results in

$$a_1 = y_1, a_2 = \frac{y_2 - a_1}{x_2 - x_1},$$
$$a_3 = \frac{(y_3 - a_1) - a_2 \cdot (x_3 - x_1)}{(x_3 - x_1) \cdot (x_3 - x_2)}, \ldots$$

and so on
It is left to the reader to use these formulae and estimate the value of P at T=58°C ■

3.4 Descriptive Measures for Sample Data

3.4.1 Summary Statistical Measures

Descriptive summary measures of sample data are meant to characterize salient statistical features of the data for easier reporting, understanding, comparison and evaluation. The following are some of the important ones:

(a) **Mean** (or arithmetic mean or average) of a set or sample of n numbers is:

$$x_{mean} \equiv \bar{x} = \frac{1}{n}\sum_{i=1}^{n} x_i \quad (3.1)$$

where n=sample size, and x_i=individual reading

(b) **Weighted mean** of a set of n numbers is:

$$\bar{x} = \frac{\sum_{i=1}^{n} x_i w_i}{\sum_{i=1}^{n} w_i} \quad (3.2)$$

where w_i is the weight for group i.

(c) **Geometric mean** is more appropriate when studying phenomenon that exhibit exponential behavior (like population growth, biological processes,…). This is defined as the nth root of the product of n data points:

$$x_{geometric} = [x_1.x_2....x_n]^{1/n} \quad (3.3)$$

(d) **Mode** is the value of the variate which occurs most frequently. When the variate is discrete, the mean may turn out to have a value that cannot actually be taken by the variate. In case of continuous variates, the mode is the value where the frequency density is highest. For example, a survey of the number of occupants in a car during the rush hour could yield a mean value of 1.6 which is not physically possible. In such cases, using a value of 2 (i.e., the mode) is more appropriate.

(e) **Median** is the middle value of the variates, i.e., half the numbers have numerical values below the median and half above. The mean is unduly influenced by extreme observations, and in such cases the median is a more robust indicator of the central tendency of the data. In case of an even number of observations, the mean of the middle two numbers is taken to be the median.

(f) **Range** is the difference between the largest and the smallest observation values.

(g) **Percentiles** are used to separate the data into bins. Let p be a number between 0 and 1. Then, the (100p)th percentile (also called pth *quantile*), represents the data value where 100p% of the data values are lower. Thus, 90% of the data will be below the 90th percentile, and the median is represented by the 50th percentile.

(h) **Inter-quartile range** (IQR) cuts out the more extreme values in a distribution. It is the range which covers the middle 50% of the observations and is the difference between the lower quartile (i.e., the 25th percentile) and the upper quartile (i.e., the 75th percentile). In a similar manner, *deciles* divide the distribution into tenths, and *percentiles* into hundreths.

(i) **Deviation** of a number x_i in a set of n numbers is a measure of dispersion of the data from the mean, and is given by:

$$d_i = (x_i - \bar{x}) \quad (3.4)$$

(j) The **mean deviation** of a set of n numbers is the mean of the absolute deviations:

$$\bar{d} = \frac{1}{n}\sum_{i=1}^{n} |d_i| \quad (3.5)$$

(k) The **variance** or the mean square error (MSE) of a set of n numbers is:

$$s_x^2 = \frac{1}{n-1}\sum_{i=1}^{n}(x_i - \bar{x})^2 = \frac{s_{xx}}{n-1} \quad (3.6a)$$

where

$$s_{xx} = \text{sum of squares} = \sum_{i=1}^{n}(x_i - \bar{x})^2 \quad (3.6b)$$

(l) The **standard deviation** of a set of n numbers

$$S_x = \left[\frac{s_{xx}}{n-1}\right]^{1/2} \quad (3.7)$$

The more variation there is in the data set, the bigger the standard deviation. This is a measure of the actual absolute error. For large samples (say, $n > 100$), one can replace $(n-1)$ by n in the above equation with acceptable error.

(m) **Coefficient of variation** is a measure of the relative error, and is often more appropriate than the standard deviation. It is defined as the ratio of the standard deviation and the mean:

$$CV = s_x / \bar{x} \quad (3.8)$$

This measure is also used in other disciplines: the reciprocal of the "*signal to noise ratio*" is widely used in electrical engineering, and also as a measure of "*risk*" in financial decision making.

(n) **Trimmed mean**. The sample mean may be very sensitive to outliers, and, hence, may bias the analysis results. The sample median is more robust since it is impervious to outliers. However, non-parametric tests which use the median are less efficient than parametric tests in general. Hence, a compromise is to use the trimmed mean value which is less sensitive to outliers than the mean but is more sensitive than the median. One selects a trimming percentage 100r% with the recommendation that $0 < r < 0.25$. Suppose one has a data set with $n = 20$. Selecting $r = 0.1$ implies that the trimming percentage is 10% (i.e., two observations). Then, two of the largest values and two of the smallest values of the data set are rejected prior to subsequent analysis. Thus, a specified percentage of the extreme values can be removed.

Example 3.4.1 *Exploratory data analysis of utility bill data*
The annual degree-day number (DD) is a statistic specific to the climate of the city or location which captures the annual variation of the ambient dry-bulb temperature usually above a pre-specified value such as 65°F or 18.3°C (see Pr. 2.24 for description). Gas and electric utilities have been using the DD method to obtain a first order estimate of the gas and electric

Table 3.2 Values of the heat loss coefficient for 90 homes (Example 3.4.1)

2.97	4.00	5.20	5.56	5.94	5.98	6.35	6.62	6.72	6.78
6.80	6.85	6.94	7.15	7.16	7.23	7.29	7.62	7.62	7.69
7.73	7.87	7.93	8.00	8.26	8.29	8.37	8.47	8.54	8.58
8.61	8.67	8.69	8.81	9.07	9.27	9.37	9.43	9.52	9.58
9.60	9.76	9.82	9.83	9.83	9.84	9.96	10.04	10.21	10.28
10.28	10.30	10.35	10.36	10.40	10.49	10.50	10.64	10.95	11.09
11.12	11.21	11.29	11.43	11.62	11.70	11.70	12.16	12.19	12.28
12.31	12.62	12.69	12.71	12.91	12.92	13.11	13.38	13.42	13.43
13.47	13.60	13.96	14.24	14.35	15.12	15.24	16.06	16.90	18.26

use of residences in their service territory. The annual heating consumption Q of a residence can be predicted as:

$$Q = U \times A \times DD$$

where U is the overall heat loss coefficient of the residence (includes heat conduction as well as air infiltration,...) and A is the house floor area.

Based on gas bills, a certain electric company calculated the U value of 90 homes in their service territory in an effort to determine which homes were "leaky", and hence are good candidates for weather stripping so as to reduce their energy use. These values (in units which need not concern us here) are given in Table 3.2.

An exploratory data analysis would involve generating the types of pertinent summary statistics or descriptive measures given in Table 3.3. Note that no value is given for "Mode" since there are several possible values in the table. What can one say about the variability in the data? If all homes whose U values are greater than twice the mean value are targeted for further action, how many such homes are there? Such questions and answers are left to the reader to explore. ■

Table 3.3 Summary statistics for values of the heat loss coefficient (Example 3.4.1)

Count	90
Average	10.0384
Median	9.835
Mode	
Geometric mean	9.60826
5% Trimmed mean	9.98444
Variance	8.22537
Standard deviation	2.86799
Coeff. of variation	28.5701%
Minimum	2.97
Maximum	18.26
Range	15.29
Lower quartile	7.93
Upper quartile	12.16
Interquartile range	4.23

3.4.2 Covariance and Pearson Correlation Coefficient

Though a scatter plot of bivariate numerical data gives a good visual indication of how strongly variables x and y vary together, a quantitative measure is needed. This is provided by the covariance which represents the strength of the *linear* relationship between the two variables:

$$\mathrm{cov}(xy) = \frac{1}{n-1} \cdot \sum_{i=1}^{n} (x_i - \bar{x}) \cdot (y_i - \bar{y}) \tag{3.9}$$

where $\bar{x}$ and $\bar{y}$ are the mean values of variables x and y.

To remove the effect of magnitude in the variation of x and y, the *Pearson correlation coefficient* r is probably more meaningful than the covariance since it standardizes the coefficients x and y by their standard deviations:

$$r = \frac{\mathrm{cov}(xy)}{s_x s_y} \tag{3.10}$$

where s_x and s_y are the standard deviations of x and y.

Hence the absolute value of r is less than or equal to unity. r=1 implies that all the points lie on a straight line, while r=0 implies no linear correlation between x and y. It is pertinent to point out that for linear models $r^2=R^2$ (the well known coefficient of determination used in regression and discussed in Sect. 5.3.2), the use of lower case and upper case to denote the same quantity being a historic dichotomy. Figure 3.13 illustrates how the different data scatter affect the magnitude and sign of r. Note that a few extreme points may exert undue influence on r especially when data sets are small. As a general thumb rule[1], for applications involving engineering data where the random uncertainties are low:

$$\begin{array}{ll} \mathrm{abs}(r) > 0.9 & \rightarrow \text{strong linear correlation} \\ 0.7 < \mathrm{abs}(r) < 0.9 & \rightarrow \text{moderate} \\ 0.7 > \mathrm{abs}(r) & \rightarrow \text{weak} \end{array} \tag{3.11}$$

[1] A more statically sound procedure is described in Sect. 4.2.7 which allows one to ascertain whether observed correlation coefficients are significant or not.

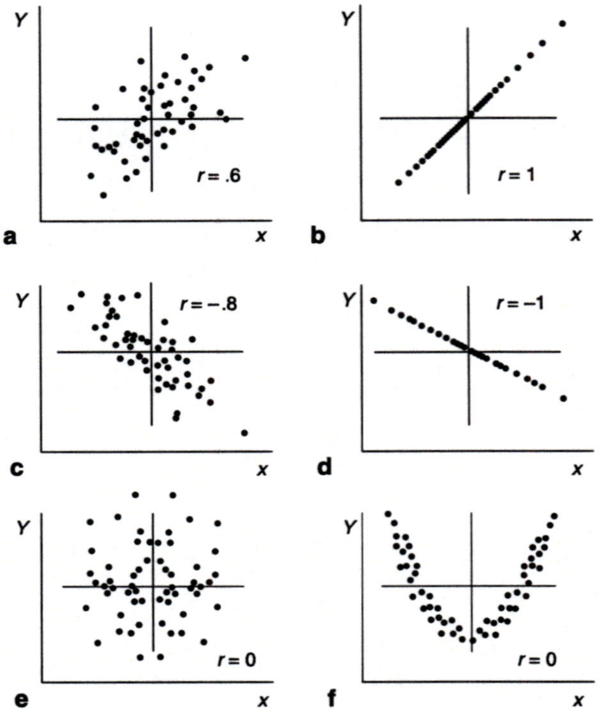

Fig. 3.13 Illustration of various plots with different correlation strengths. (From Wonnacutt and Wonnacutt (1985) by permission of John Wiley and Sons)

It is very important to note that inferring non-association of two variables x and y from inspection of their correlation coefficient is misleading since it only indicates *linear relationship*. Hence, a poor correlation does not mean that no relationship exist between them (for example, a second order relation may exist between x and y; see Fig. 3.13f). Note also that correlation analysis does not indicate whether the relationship is causal, i.e. one cannot assume causality just because a correlation exists. Finally, keep in mind that the correlation analysis does not provide an equation for predicting the value of a variable—this is done under model building (see Chap. 5).

Example 3.4.2 The following observations are taken of the extension of a spring under different loads (Table 3.4).

The standard deviations of load and extension are 3.7417 and 18.2978 respectively, while the correlation coefficient=0.9979. This indicates a very strong positive correlation between the two variables as one should expect. ■

3.4.3 Data Transformations

Once the above validation checks have been completed, the raw data can be transformed to one on which subsequent engineering analyses can be performed. Examples include converting into appropriate units, taking ratios, transforming variables, … Sometimes, normalization methods may be required which are described below:

Table 3.4 Extension of a spring with applied load

Load (Newtons)	2	4	6	8	10	12
Extension (mm)	10.4	19.6	29.9	42.2	49.2	58.5

(a) **Decimal scaling** moves the decimal point but still preserves most of the original data. The specific observations of a given variable may be divided by 10^x where x is the minimum value so that all the observations are scaled between −1 and 1. For example, say the largest value is 289 and the smallest value is −150, then since x=3, all observations are divided by 1000 so as to lie between [0.289 and −0.150].

(b) **Min-max scaling** allows for better distribution of observations over the range of variation than does decimal scaling. It does this by redistributing the values to lie between [−1 and 1]. Hence, each observation is normalized as follows:

$$z_i = \frac{x_i - x_{\min}}{x_{\max} - x_{\min}} \tag{3.12}$$

where x_{max} and x_{min} are the maximum and minimum numerical values respectively of the x variable. Note that though this transformation may look very appealing, the scaling relies largely on the minimum and maximum values, which are generally not very robust and may be error prone.

(c) **Standard deviation scaling** is widely used for distance measures (such as in multivariate statistical analysis) but transforms data into a form unrecognizable from the original data. Here, each observation is normalized as follows:

$$z_i = \frac{x_i - \bar{x}}{s_x} \tag{3.13}$$

where $\bar{x}$ and s_x are the mean and standard deviation respectively of the x variable.

3.5 Plotting Data

Graphs serve two purposes. During exploration of the data, they provide a better means of assimilating broad qualitative trend behavior of the data than can be provided by tabular data. Second, they provide an excellent manner of communicating to the reader what the author wishes to state or illustrate (recall the adage "a picture is worth a thousand words").

Hence, they can serve as mediums to communicate *information*, not just to explore data trends (an excellent reference is Tufte 2001). However, it is important to be clear as to the intended message or purpose of the graph, and also tailor it as to be suitable for the intended audience's background and understanding. A pretty graph may be visually appealing but may obfuscate rather than clarify or highlight the necessary aspects being communicated. For example, unless one is experienced, it is difficult to read numerical values off of 3-D graphs. Thus, graphs should present data clearly and accurately without hiding or distorting the underlying intent. Table 3.5 provides a succinct summary of graph formats appropriate for different applications.

Graphical methods are recommended after the numerical screening phase is complete since they can point out unflagged data errors. Historically, the strength of a graphical analysis was to visually point out to the analyst relationships (linear or non-linear) between two or more variables in instances when a sound physical understanding is lacking, thereby aiding in the selection of the appropriate regression model. Present day graphical visualization tools allow much more than this simple objective, some of which will become apparent below. There are a very large number of graphical ways of presenting data, and it is impossible to cover them all. Only a small representative and commonly used plots will be presented below, while operating manuals of several high-end graphical software programs describe complex, and sometimes esoteric, plots which can be generated by their software.

Table 3.5 Type and function of graph message determines format. (Downloaded from http://www.eia.doe.gov/neic/graphs/introduc.htm)

Type of message	Function	Typical format
Component	Shows relative size of various parts of a whole	– Pie chart (for 1 or 2 important components) – Bar chart – Dot chart – Line chart
Relative amounts	Ranks items according to size, impact, degree, etc.	– Bar chart – Line chart – Dot chart
Time series	Shows variation over time	– Bar chart (for few intervals) – Line chart
Frequency	Shows frequency of distribution among certain intervals	– Histogram – Line chart – Box-and-Whisker
Correlation	Shows how changes in one set of data is related to another set of data	– Paired bar – Line chart – Scatter diagram

3.5.1 Static Graphical Plots

Graphical representations of data are the backbone of exploratory data analysis. They are usually limited to one-, two- and three-dimensional data. In the last few decades, there has been a dramatic increase in the types of graphical displays largely due to the seminal contributions of Tukey (1988), Cleveland (1985) and Tufte (1990, 2001). A particular graph is selected based on its ability to emphasize certain characteristics or behavior of one-dimensional data, or to indicate relations between two- and three-dimension data. A simple manner of separating these characteristics is to view them as being:

(i) cross-sectional (i.e., the sequence in which the data has been collected is not retained),
(ii) time series data,
(iii) hybrid or combined, and
(iv) relational (i.e., emphasizing the joint variation of two or more variables).

An emphasis on visualizing data to be analyzed has resulted in statistical software programs becoming increasingly convenient to use and powerful towards this end. Any data analysis effort involving univariate and bivariate data should start by looking at basic plots (higher dimension data require more elaborate plots discussed later).

(a) *for univariate data*:
Commonly used graphics for cross-sectional representation are mean and standard deviation, steam-and-leaf, histograms, box-whisker-mean plots, distribution plots, bar charts, pie charts, area charts, quantile plots. Mean and standard deviation plots summarize the data distribution using the two most basic measures; however, this manner is of limited use (and even misleading) when the distribution is skewed. For univariate data, plotting of *histograms* is very useful since they provide insight into the underlying parent distribution of data dispersion, and can flag outliers as well. There are no hard and fast rules of how to select the number of bins (N_{bins}) or classes in case of continuous data, probably because there is no theoretical basis. Generally, the larger the number of observations n, the more classes can be used, though as a guide it should be between 5 and 20. Devore and Fornum (2005) suggest:

$$\text{Number of bins or classes} = N_{bins} = (n)^{1/2} \quad (3.14)$$

which would suggest that if $n=100$, $N_{bins}=10$

Doebelin (1995) proposes another equation:

$$N_{bins} = 1.87.(\text{n} - 1)^{0.4} \quad (3.15)$$

which would suggest that if $n=100$, $N_{bins}=12$.

The box and whisker plots also summarize the distribution, but at different percentiles (see Fig. 3.14). The lower and upper box values (or *hinges*) correspond to the 25th and 75th percentiles (i.e., the interquartile range (IQR) defined

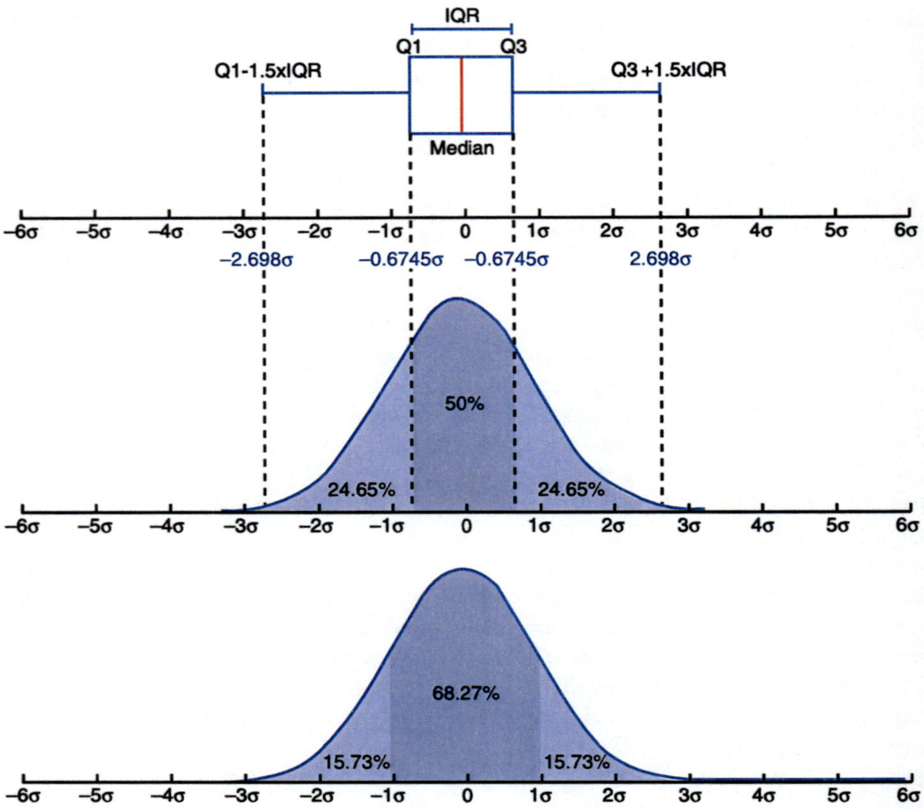

Fig. 3.14 Box and whisker plot and its association with a normal distribution. The box represents the 50th percentile range while the whiskers extend 1.5 times the inter-quartile range (IQR) on either side. (From Wikipedia website)

in Sect. 3.4.1) while the whiskers extend to 1.5 times the IQR on either side. These allow outliers to be detected. Any observation farther than (3.0×IQR) from the closest quartile is taken to be an *extreme outlier*, while if farther than (1.5×IQR), it is considered to be a *mild outlier*.

Though plotting a box-and-whisker plot or a plot of the distribution itself can suggest the shape of the underlying distribution, a better visual manner of ascertaining whether a presumed underlying parent distribution applies to the data being analyzed is to plot a *quantile plot* (also called the probability plot). The observations are plotted against the parent distribution (which could be any of the standard probability distributions presented in Sect. 2.4), and if the points fall on a straight line, this suggests that the assumed distribution is plausible. The example below illustrates this concept.

Example 3.5.1 An instructor wishes to ascertain whether the time taken by his students to complete the final exam follows a normal or Gaussian distribution. The values in minutes shown in Table 3.6 have been recorded.

The quantile plot for this data assuming the parent distribution to be Gaussian is shown in Fig. 3.15. The pattern is obviously nonlinear, so a Gaussian distribution is implausible for this data. The apparent break appearing in the data on the right side of the graph is indicative of data that contains outliers (caused by five students taking much longer to complete the exam). ■

Example 3.5.2 Consider the same data set as for Example 3.4.1. The following plots have been generated (Fig. 3.16):

(b) Box and whisker plot
(c) Histogram of data (assuming 9 bins)
(d) Normal probability plots
(e) Run chart

It is left to the reader to identify and briefly state his observations regarding this data set. Note that the run chart is meant to retain the time series nature of the data while the other graphics do not. The manner in which the run chart has been generated is meaningless since the data seems to have been entered into the spreadsheet in the wrong sequence, with data entered column-wise instead of row-wise. The run chart, had the data been entered correctly, would have

Table 3.6 Values of time taken (in minutes) for 20 students to complete an exam

37.0	37.5	38.1	40.0	40.2	40.8	41.0
42.0	43.1	43.9	44.1	44.6	45.0	46.1
47.0	62.0	64.3	68.8	70.1	74.5	

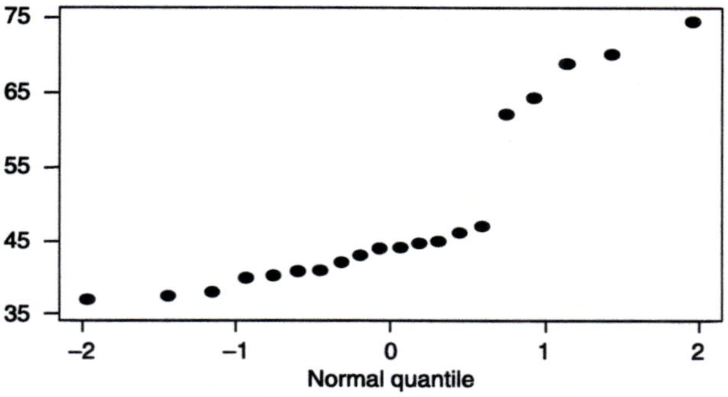

Fig. 3.15 Quantile plot of data in Table 3.15 assuming a Gaussian normal distribution

resulted in a monotonically increasing curve and been more meaningful. ■

(b) *for bi-variate and multi-variate data*

There are numerous graphical representations which fall in this category and only an overview of the more common plots will be provided here. Multivariate stationary data of worldwide percentages of total primary energy sources can be represented by the widely used **pie chart** (Fig. 3.17a) which allows the relative aggregate amounts of the variables to be clearly visualized. The same information can also be plotted as a bar chart (Fig. 3.17b) which is not quite as revealing.

More elaborate **Bar charts** (such as those shown in (Fig. 3.18) allow numerical values of more than one variable to be plotted such that their absolute and relative amounts are clearly highlighted. The plots depict differences between the electricity sales during each of the four different quarters of the year over 6 years. Such plots can be drawn as compounded plots to allow better visual inter-comparisons (Fig. 3.18a). Column charts or stacked charts (Fig. 3.18b, c) show the same information as that in Fig. 3.18a but are stacked one above another instead of showing the numerical values side-by-side. One plot shows the stacked values normalized such that the sum adds to 100%, while another stacks them so as to retain their numerical values. Finally, the same information can be plotted as an area chart wherein both the time series trend and the relative magnitudes are clearly highlighted.

Time series plots or relational plots or **scatter plots** (such as x–y plots) between two variables are the most widely used types of graphical displays. Scatter plots allows visual determination of the trend line between two variables and the extent to which the data scatter around the trend line (Fig. 3.19).

Another important issue is that the manner of selecting the range of the variables can be misleading to the eye. The same data is plotted in Fig. 3.20 on two different scales, but one would erroneously conclude that there is more data scatter around the trend line for (b) than for (a). This is referred to as the *lie factor* defined as the ratio of the apparent size of effect in the graph and the actual size of effect in the data (Tufte 2001). The data at hand and the intent of the analy-

Fig. 3.16 Various exploratory plots for data in Table 3.2

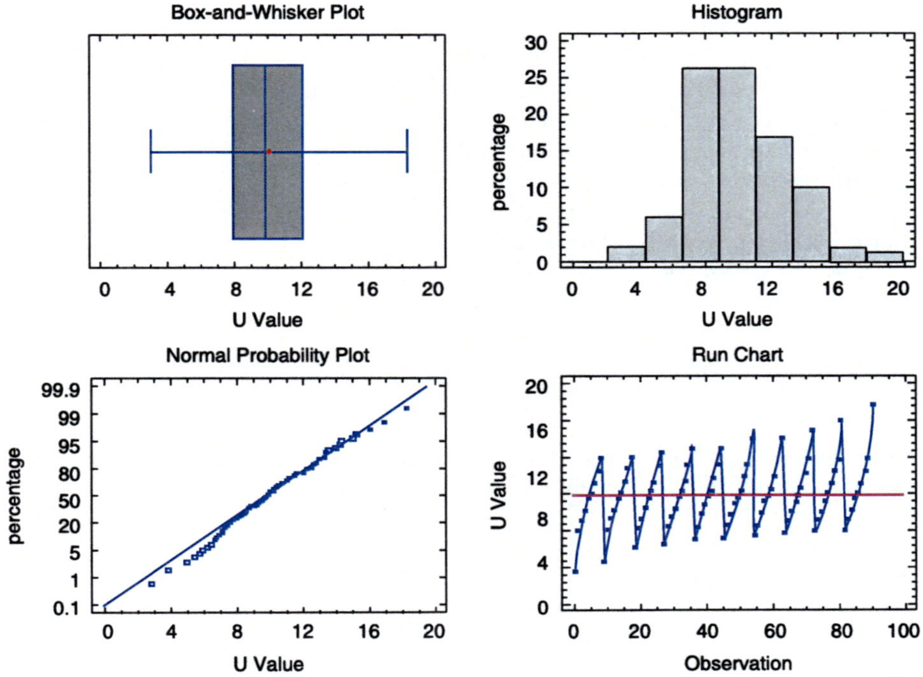

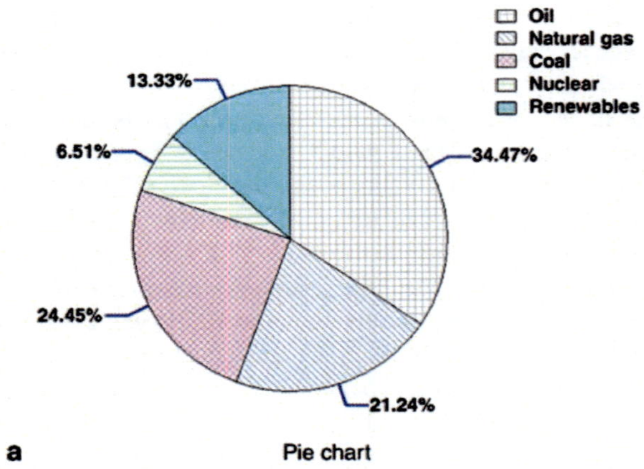

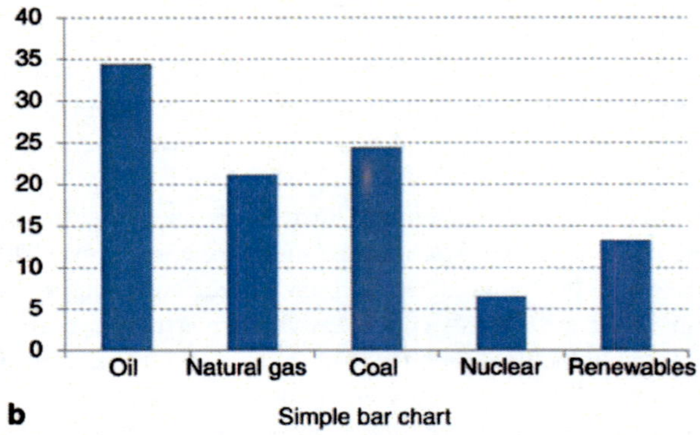

Fig. 3.17 Two different ways of plotting stationary data. Data corresponds to worldwide percentages of total primary energy supply in 2003. (From IEA, World Energy Outlook, IEA, Paris, France, 2004)

sis should dictate the scale of the two axes, but it is difficult in practice to determine this heuristically[2]. It is in such instances that statistical measures can be used to provide an indication of the magnitude of the graphical scales.

Dot plots are simply one dimensional plots where each dot is an observation on an univariate scale. The 2-D version of such plots is the well-known x–y scatter plot. An additional variable representative of a magnitude can be included by increasing the size of the plot to reflect this magnitude. Figure 3.21 shows such a representation for the commute patterns in major U.S. cities in 2008.

Combination charts can take numerous forms, but in essence, are those where two different basic ways of representing data are combined together. One example is Fig. 3.22 where the histogram depicts actual data spread, the distribution of which can be visually evaluated against the standard normal curve.

For purposes of data checking, x–y plots are perhaps most appropriate as discussed in Sect. 3.3.3. The x–y scatter plot (Fig. 3.11) of hourly cooling energy use of a large institutional building versus outdoor temperature allowed outliers to be detected. The same data could be summarized by combined **box and whisker plots** (first suggested by Tukey 1988) as shown in Fig. 3.23. Here the x-axis range is subdivided into discrete bins (in this case, 5°F bins), showing the median values (joined by a continuous line) along with the 25th percentiles on either side of the mean (shown boxed) and the 10th and 90th percentiles indicated by the vertical whiskers from the box, and the values less than the 10th percentile and those greater than the 90th percentile are shown as individual pluses (+).[3] Such a representation is clearly a useful tool for data quality checking, for detecting underlying patterns in data at different sub-ranges of the independent variable, and also for ascertaining the shape of the data spread around this pattern.

(c) *for higher dimension data:*
Some of the common plots are multiple trend lines, contour plots, component matrix plots, and three-dimension charts. In case the functional relationship between the independent and dependent variables changes due to known causes, it is advisable to plot these in different frames. For example, hourly energy use in a commercial building is known to change with time of day but the functional relationship is quite different dependent on the season (time of year). **Component-effect plots** are multiple plots between the variables for cold, mild and hot periods of the year combined with box and whisker type of presentation. They provide more clarity in underlying trends and scatter as illustrated in Fig. 3.24 where the time of year is broken up into three temperature bins.

Three dimensional (or 3-D) plots are being increasingly used from the past few decades. They allow plotting variation of a variable when it is influenced by two independent factors (Fig. 3.25). They also allow trends to be gauged and are visually appealing but the numerical values of the variables are difficult to read.

Another benefit of such 3-D plots is their ability to aid in the identification of oversights. For example, energy use data collected from a large commercial building could be improperly time-stamped; such as, overlooking daylight savings shift or misalignment of 24-hour holiday profiles (Fig. 3.26). One negative drawback associated with these graphs is the difficulty in viewing exact details such as the specific hour or specific day on which a misalignment occurs. Some analysts complain that 3-D surface plots obscure data that is behind "hills" or in "valleys". Clever use of color or dotted lines have been suggested to make it easier to interpret such graphs.

[2] Generally, it is wise, at least at the onset, to adopt scales starting from zero, view the resulting graphs and make adjustments to the scales as appropriate.

[3] Note that the whisker end points are different than those described earlier in Sect. 3.5.1. Different textbooks and papers adopt slightly different selection criteria.

Fig. 3.18 Different types of bar plots to illustrate year by year variation (over 6 years) in quarterly electricity sales (in GigaWatt-hours) for a certain city

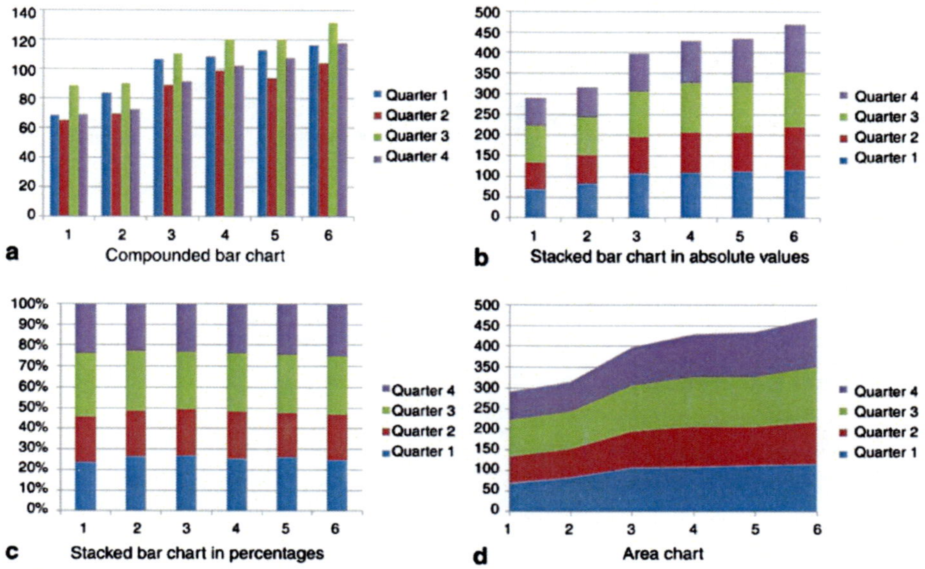

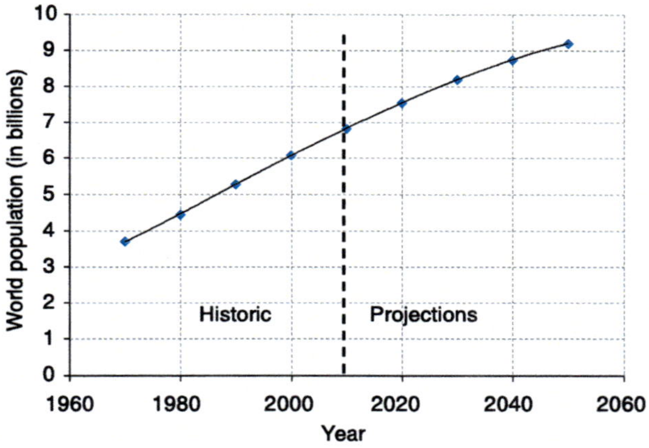

Fig. 3.19 Scatter plot (or x-y plot) with trend line through the observations. In this case, a second order quadratic regression model has been selected as the trend line

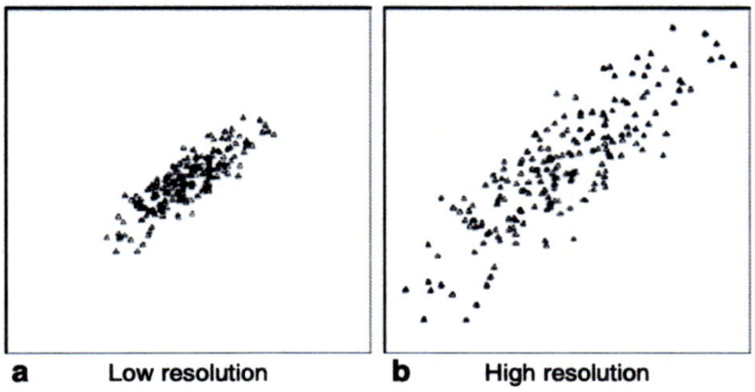

Fig. 3.20 Figure to illustrate how the effect of resolution can mislead visually. The same data is plotted in the two plots but one would erroneously conclude that there is more data scatter around the trend line for **(b)** than for **(a)**.

Often values of physical variables need to be plotted against two physical variables. One example is the well-know *psychrometric chart* which allows one to determine (for a given elevation) the various properties of air-water mixtures (such as relative humidity, specific volume, enthalpy, wet bulb temperature) when the mixture is specified by its dry-bulb temperature and the humidity ratio. In such cases, a series of lines are drawn for each variable at selected numerical values. A similar and useful representation is a **contour plot** which is a plot of iso-lines of the dependent variable at different preselected magnitudes drawn over the range of variation of the two independent variables. An example is provided by Fig. 3.27 where the total power of a condenser loop of a cooling system is the sum of the pump power and the cooling tower fan.

Another visually appealing plot is the *sun-path diagram* which allows one to determine the position of the sun in the sky (defined by the solar altitude and the solar azimuth angles) at different times of the day and the year for a location of latitude 40° N (Fig. 3.28). Such a representation has also been used to determine periods of the year when shading occurs from neighboring obstructions. Such considerations are important while siting solar systems or designing buildings.

Figure 3.29 called **carpet plots** (or scatter plot matrix) is another useful representation of visualizing multivariate data. Here the various permutations of the variables are shown as individual scatter plots. The idea, though not novel, has merit because of the way the graphs are organized and presented. The graphs are arranged in rows and columns such that each row or column has all the graphs relating a certain variable to all the others; thus, the variables have shared axes. Though

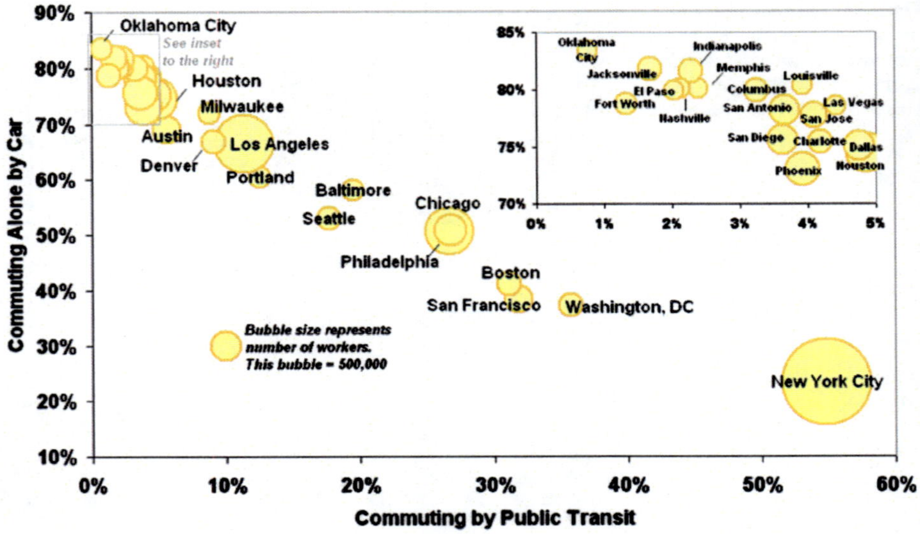

Fig. 3.21 Commute patterns in major U.S. cities in 2008 shown as enhanced dot plots with the size of the dot representing the number of commuters. (From Wikipedia website)

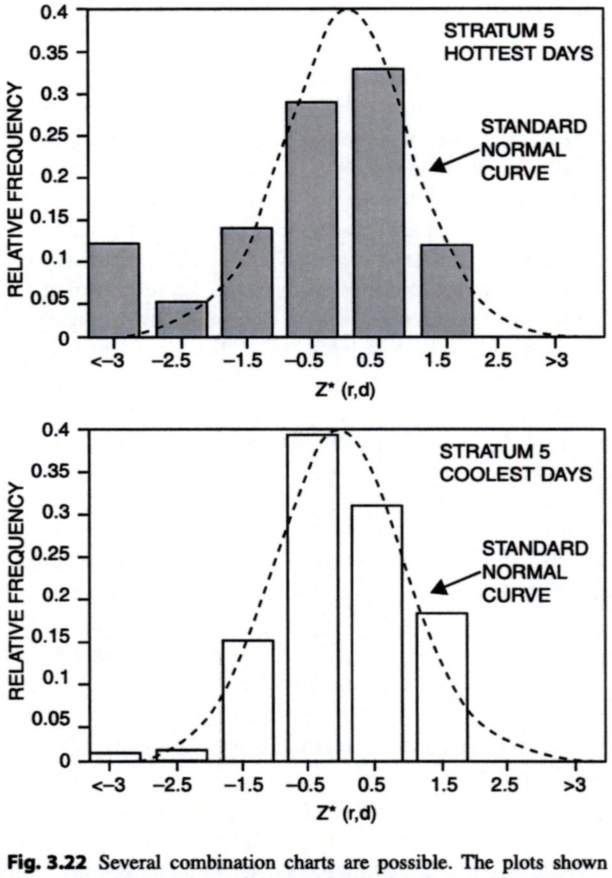

Fig. 3.22 Several combination charts are possible. The plots shown allows visual comparison of the standardized (subtracted by the mean and divided by the standard deviation) hourly whole-house electricity use in a large number of residences against the standard normal distribution. (From Reddy 1990)

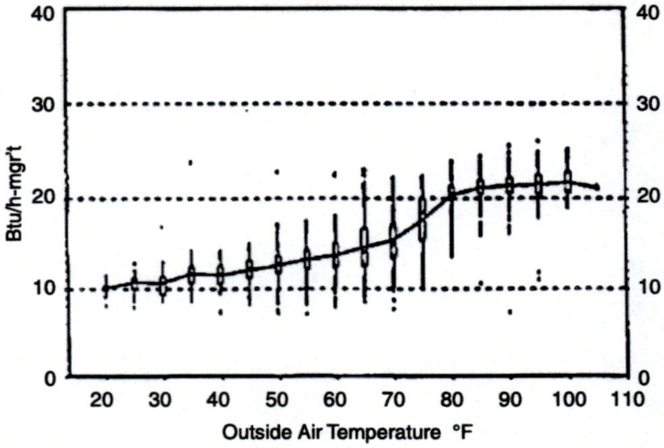

Fig. 3.23 Scatter plot combined with box-whisker-mean (BWM) plot of the same data as shown in Fig. 3.11. (From Haberl and Abbas (1998) by permission of Haberl)

there are twice as many graphs as needed minimally (since each graph has another one with the axis interchanged), the redundancy is sometimes useful to the analyst in better detecting underlying trends.

3.5.2 High-Interaction Graphical Methods

The above types of plots can be generated by relatively low end data analysis software programs. More specialized software programs called data visualization software are available which provide much greater insights into data trends, outliers and local behavior, especially when large amounts of data are being considered. Animation has also been used to advantage in understanding system behavior from monitored

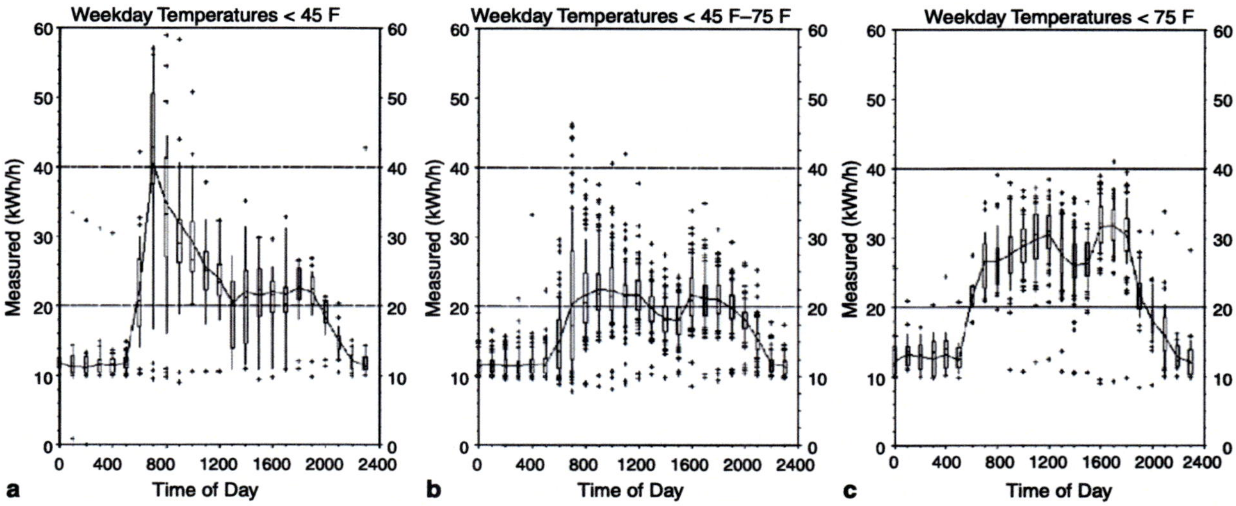

Fig. 3.24 Example of a combined box-whisker-component plot depicting how hourly energy use varies with hour of day during a year for different outdoor temperature bins for a large commercial building. (From ASHRAE 2002 © American Society of Heating, Refrigerating and Air-conditioning Engineers, Inc., www.ashrae.org)

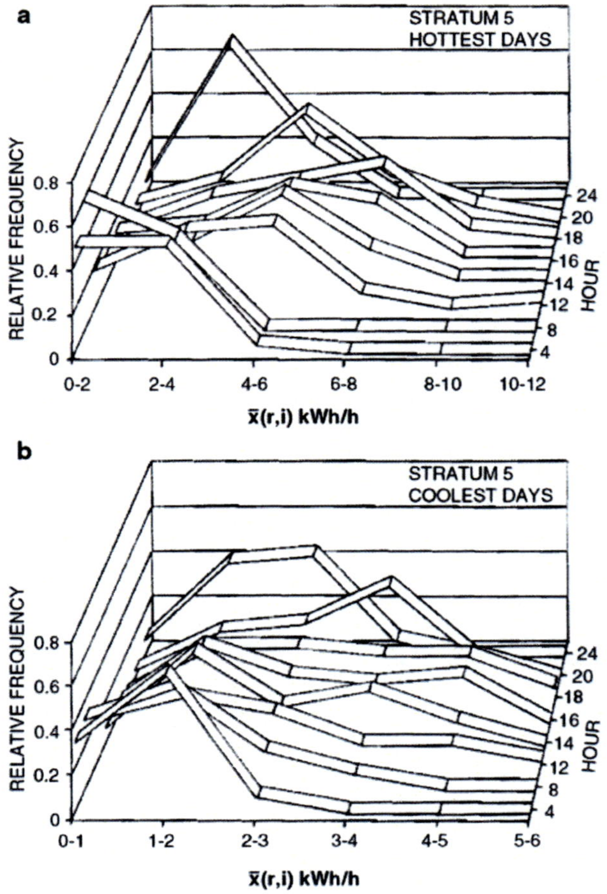

Fig. 3.25 Three dimensional surface charts of mean hourly whole-house electricity during different hours of the day across a large number of residences. (From Reddy 1990)

data since time sequence can be retained due to, say, seasonal differences. Animated scatter plots of the x and y variables as well as animated contour plots, with color superimposed, which can provide better visual diagnostics have also been developed.

More sophisticated software is available which, however, requires higher user skill. Glaser and Ubbelohde (2001) describe novel high performance visualization techniques for reviewing time dependent data common to building energy simulation program output. Some of these techniques include: (i) brushing and linking where the user can investigate the behavior during a few days of the year, (ii) tessellating a 2-D chart into multiple smaller 2-D charts giving a 4-D view of the data such that a single value of a representative sensor can be evenly divided into smaller spatial plots arranged by time of day, (iii) magic lenses which can zoom into a certain portion of the room, and (iv) magic brushes. These techniques enable rapid inspection of trends and singularities which cannot be gleaned from conventional viewing methods.

3.5.3 Graphical Treatment of Outliers

No matter how carefully an experiment is designed and performed, there always exists the possibility of serious errors. These errors could be due to momentary instrument malfunction (say dirt sticking onto a paddle-wheel of a flow meter), power surges (which may cause data logging errors), or the engineering system deviating from its intended operation due to random disturbances. Usually, it is difficult to pin-

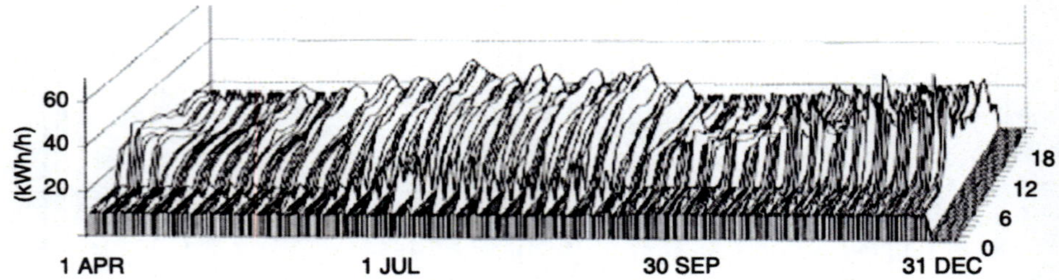

Fig. 3.26 Example of a three-dimensional plots of measured hourly electricity use in a commercial building over nine months. (From ASHRAE 2002 © American Society of Heating, Refrigerating and Air-conditioning Engineers, Inc., www.ashrae.org)

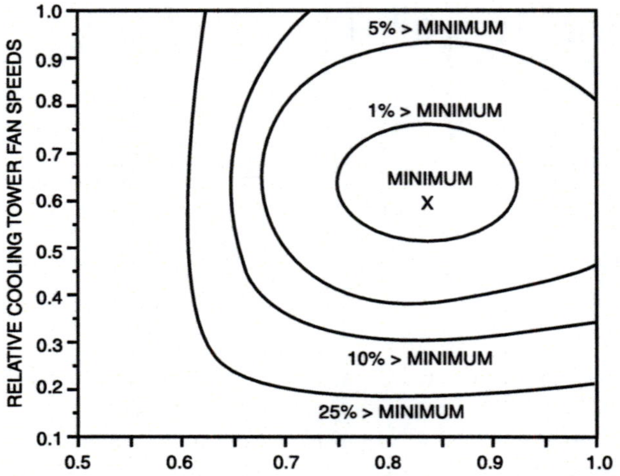

Fig. 3.27 Contour plot characterizing the sensitivity of total power consumption (condenser water pump power plus tower fan power) to condenser water-loop controls for a single chiller load, ambient wet-bulb temperature and chilled water supply temperature. (From Braun et al. (1989) © American Society of Heating, Refrigerating and Air-conditioning Engineers, Inc., www.ashrae.org)

point the cause of the anomalies. The experimenter is often not fully sure whether the outlier is anomalous, or whether it is a valid or legitimate data point which does not conform to what the experimenter "thinks" it should. In such cases, throwing out a data point may amount to data "tampering" or fudging of results. Usually, data which exhibit such anomalous tendency are a minority. Even then, if the data analyst retains these questionable observations, they can bias the results of the entire analysis since they exert an undue influence and can dominate a computed relationship between two variables.

Let us consider the case of outliers during regression for the univariate case. Data points are said to be outliers when their model residuals are large relative to the other points. Instead of blindly using a statistical criterion, a better way is to visually look at the data, and distinguish between end points and center points. For example, point A of Fig. 3.30 is quite obviously an outlier, and if the rejection criterion orders its removal, one should proceed to do so. On the other hand, point B which is near the end of the data domain, may not be

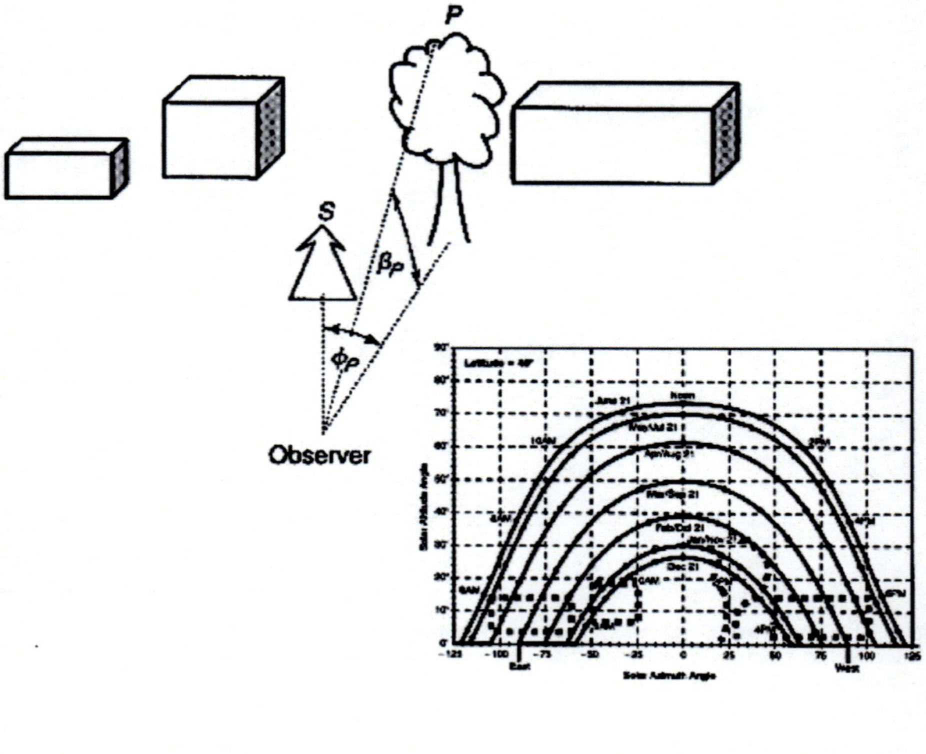

Fig. 3.28 Figure illustrating an overlay plot for shading calculations. The sun-path diagram is generated by computing the solar declination and azimuth angles for a given latitude (for 40° N) during different times of the day and times of the year. The "obstructions" from trees and objects are drawn over the graph to yield important information of potential shading on the collector. (From Kreider et al. 2009 by permission of CRC Press)

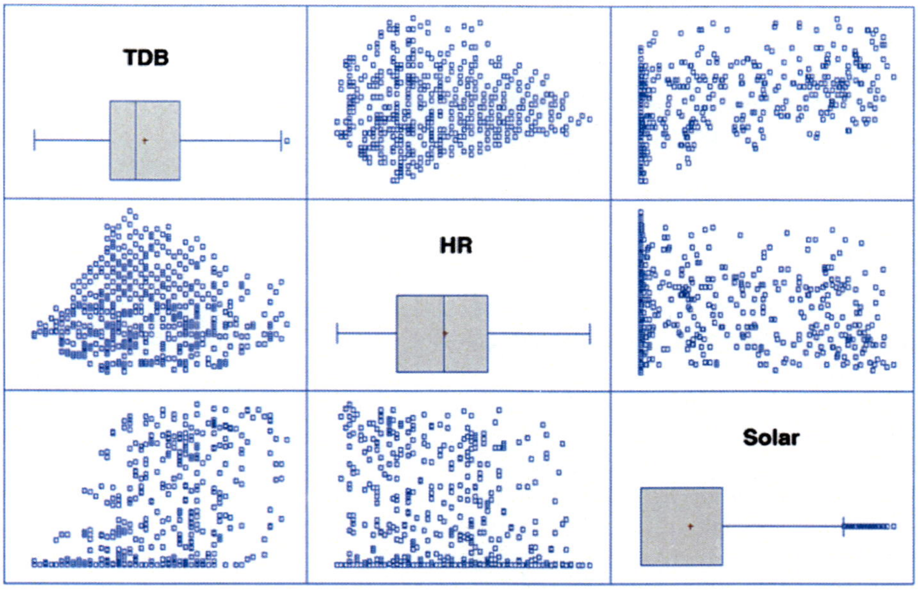

Fig. 3.29 Scatter plot matrix or carpet plots for multivariable graphical data analysis. The data corresponds to hourly climatic data for Phoenix, AZ for January 1990. The bottom left hand corner frame indicates how solar radiation in Btu/hr-ft² (x-axis) varies with dry-bulb temperature (in °F) and is a flipped and rotated image of that at the top right hand corner. The HR variable represents humidity ratio (in lbm/lba). Points which fall distinctively outside the general scatter can be flagged as outliers

a bad point at all, but merely the beginning of a new portion of the curve (say, the onset of turbulence in an experiment involving laminar flow). Similarly, even point C may be valid and important. Hence, the only way to remove this ambiguity is to take more observations at the lower end. Thus, a modification of the statistical rejection criterion is that one should do so *only if the points to be rejected are center points*.

Several advanced books present formal analytical treatment of outliers which allow diagnosing whether the regressor data set is ill-conditioned or not, as well as identifying and rejecting, if needed, the necessary outliers that cause ill-conditioning (for example, Belsley et al. 1980). Consider Fig. 3.31a. The outlier point will have little or no influence on the regression parameters identified, and in fact retaining it would be beneficial since it would lead to a reduction in model parameter variance. The behavior shown in Fig. 3.31b is more troublesome because the estimated slope is almost wholly determined by the extreme point. In fact, one may view this situation as a data set with only two data points, or one may view the single point as a spurious point and remove it from the analysis. Gathering more data at that range would be advisable, but may not be feasible; this is where the judgment of the analyst or prior information about the underlying trend line are useful. How and the extent to which each of the data points will affect the outcome of the regression line will determine whether that particular point is an influence point or not. This aspect is treated more formally in Sect. 5.6.2.

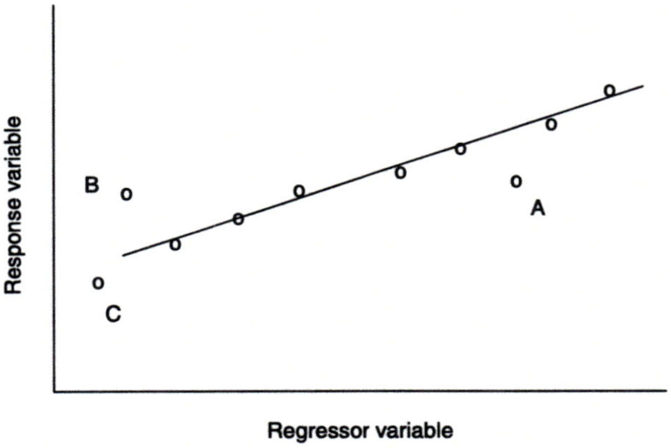

Fig. 3.30 Illustrating different types of outliers. Point A is very probably a doubtful point; point B might be bad but could potentially be a very important point in terms of revealing unexpected behavior; point C is close enough to the general trend and should be retained until more data is collected

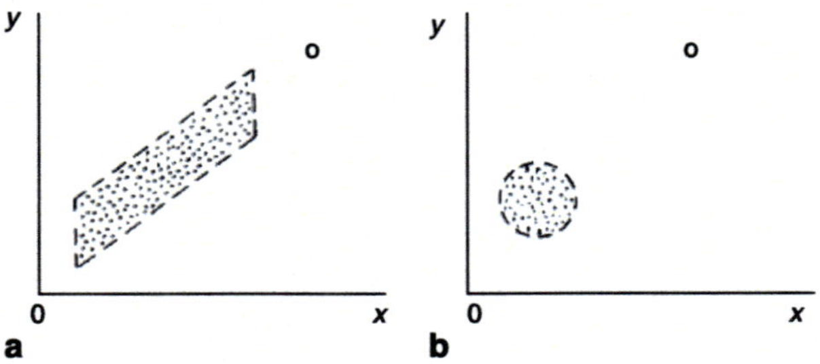

Fig. 3.31 Two other examples of outlier points. While the outlier point in (**a**) is most probably a valid point, it is not clear for the outlier point in (**b**). Either more data has to be collected, failing which it is advisable to delete this data from any subsequent analysis. (From Belsley et al. (1980) by permission of John Wiley and Sons)

3.6 Overall Measurement Uncertainty

The International Organization of Standardization (ISO) and six other organizations have published guides which have established the experimental uncertainty standard (an example is ANSI/ASME 1990). The following material is largely drawn from Guideline 2 (ASHRAE 2005) which deals with engineering analysis of experimental data.

3.6.1 Need for Uncertainty Analysis

Any measurement exhibits some difference between the measured value and the true value and, therefore, has an associated uncertainty. A statement of measured value without an accompanying uncertainty statement has limited meaning. *Uncertainty* is the interval around the measured value within which the true value is expected to fall with some stated confidence level. "Good data" does not describe data that yields the desired answer. It describes data that yields a result within an acceptable uncertainty interval or, in other words, provides the acceptable degree of confidence in the result.

Measurements made in the field are especially subject to potential errors. In contrast to measurements made under the controlled conditions of a laboratory setting, field measurements are typically made under less predictable circumstances and with less accurate and less expensive instrumentation. Furthermore, field measurements are vulnerable to errors arising from:

(a) Variable measurement conditions so that the method employed may not be the best choice for all conditions;
(b) Limited instrument field calibration, because it is typically more complex and expensive than laboratory calibration;
(c) Simplified data sampling and archiving methods; and
(d) Limitations in the ability to adjust instruments in the field.

With appropriate care, many of these sources of error can be minimized: (i) through the systematic development of a procedure by which an uncertainty statement can be ascribed to the result, and (ii) through the optimization of the measurement system to provide maximum benefit for the least cost. The results of a practitioner who does not consider sources of error are likely to be questioned by others, especially since the engineering community is increasingly becoming sophisticated and mature about the proper reporting of measured data.

3.6.2 Basic Uncertainty Concepts: Random and Bias Errors

The latest ISO standard is described in Coleman and Steele (1999) and involves treating bias and random errors in a certain manner. However, the previous version is slightly more simplified, and gives results which in many practical instances are close enough. It is this which is described below (ANSI/ASME 1990). The bias and random errors are treated as random variables, with however, different confidence level multipliers applied to them as explained below (while the latest ISO standard suggests a combined multiplier).

(a) *Bias or systematic error* (or precision or fixed error) is an unknown error that persists and is usually due to the particular instrument or technique of measurement (see Fig. 3.32). It is analogous to the sensor precision (see Sect. 3.2.1). Statistics is of limited use in this case. The best corrective action is to ascertain the extent of the bias (say, by recalibration of the instruments) and to correct the observations accordingly. Fixed (bias) errors are the constant deviations that are typically the hardest to estimate or document. They include such items as mis-calibration as well as improper sensor placement. Biases are essentially offsets from the true value that are constant over time and do not change when the number of observations is increased. For example, a bias is present if a temperature sensor always reads 1°C higher than the true value from a certified calibration procedure. Note that the magnitude of the bias is unknown for the specific situation; and so measurements cannot be simply corrected.
(b) *Random error* (or inaccuracy error) is an error due to the unpredictable and unknown variations in the experiment that causes readings to take random values on either side of some mean value. Measurements may be *precise or imprecise* depending on how well an instrument can reproduce the subsequent readings of an unchanged input (see Fig. 3.32). Only random errors can be treated by statistical methods. There are two types of random errors: (i) *additive* errors that are independent of the magnitude of the observations, and (ii) *multiplicative* errors which are dependent on the magnitude of the observations (Fig. 3.33). Usually instrument accuracy is stated in terms of percent of full scale, and so uncertainty of a reading is taken to be additive, i.e., irrespective of the magnitude of the reading.

Random errors are differences from one observation to the next due to both sensor noise and extraneous conditions affecting the sensor. The random error changes from one observation to the next, but its mean (average value) over a very large number of observations is taken to approach zero. Random error generally has a well-defined probability distribution that can be used to bound its variability in statistical terms as described in the next two sub-sections when a finite number of observations is made of the same variable.

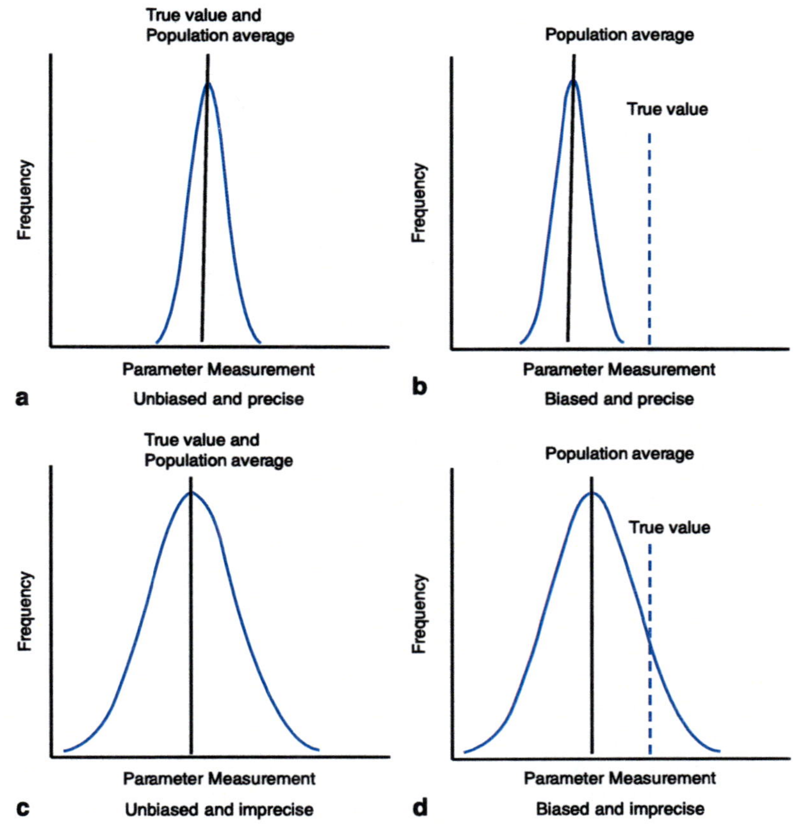

Fig. 3.32 Effect of measurement bias and precision errors

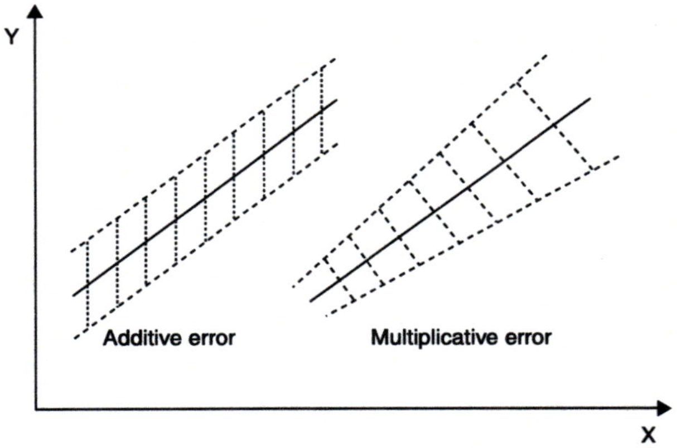

Fig. 3.33 Conceptual figures illustrating how additive and multiplicative errors affect the uncertainty bands around the trend line

3.6.3 Random Uncertainty of a Measured Variable

Based on measurements of a variable X, the true value of X can be specified to lie in the interval ($X_{best} \pm U_x$) where X_{best} is usually the mean value of the measurements taken and U_x is the uncertainty in X that corresponds to the estimate of the effects of combining fixed and random errors.

The uncertainty being reported is specific to a *confidence level*[4]. The confidence level defines the range of values or the *confidence limits* (CL) that can be expected to include the true value with a stated probability. For example, a statement that the 95% CL are 5.1 to 8.2 implies that the true value will be contained between the interval bounded by 5.1 and 8.2 in 19 out of 20 predictions (95% of the time), or that one is 95% confident that the true value lies between 5.1 and 8.2, or that there is a 95% probability that the actual value is contained in the interval {5.1, 8.2}.

An uncertainty statement with a low confidence level is usually of little use. For the example in the previous example, if a confidence level of 40% is used instead of 95%, the interval becomes a tight 7.6 to 7.7. However, only 8 out of 20 predictions will likely lie between 7.6 and 7.7. Conversely, it is useless to seek a 100% CL since then the true value of some quantity would lie between plus and minus infinity.

Multi-sample data (repeated measurements of a fixed quantity using altered test conditions, such as different observers or different instrumentation or both) provides greater reliability and precision than single sample data

[4] Several publications cite uncertainty levels without specifying a corresponding confidence level; such practice should be avoided.

(measurements by one person using a single instrument). For the majority of engineering cases, it is impractical and too costly to perform a true multi-sample experiment. While, strictly speaking, merely taking repeated readings with the same procedure and equipment does not provide multi-sample results, such a procedure is often accepted by the engineering community as a fair approximation of a multi-sample experiment.

Depending upon the sample size of the data (greater or less than about 30 samples), different statistical considerations and equations apply. The issue of estimating confidence levels is further discussed in Chap. 4, but operational equations are presented below. These levels or limits are directly based on the Gaussian and the Student-t distributions presented in Sect. 2.4.3a and b.

(a) *Random Uncertainty in large samples* (n > about 30): The best estimate of a variable x is usually its mean value given by $\bar{x}$. The limits of the confidence interval are determined from the sample standard deviation s_x. The typical procedure is then to assume that the individual data values are scattered about the mean following a certain probability distribution function, within (±z. standard deviation s_x) of the mean. Usually a *normal probability curve* (Gaussian distribution) is assumed to represent the dispersion in experimental data, unless the process is known to follow one of the standard distributions (discussed in Sect. 2.4). For a normal distribution, the standard deviation indicates the following degrees of dispersion of the values about the mean (see Table A3). For z = 1.96, 95% of the data will be within ($\pm 1.96 s_x$) of the mean. Thus, the z multiplier has a direct relationship with the confidence level selected (assuming a known probability distribution). The confidence interval (CL) for the mean of n number of multi-sample random data, i.e., data *which do not have any fixed error* is:

$$\bar{x}_{\min} = \bar{x} - \left(\frac{z.s_x}{\sqrt{n}}\right) \text{ and } \bar{x}_{\max} = \bar{x} + \left(\frac{z.s_x}{\sqrt{n}}\right) \quad (3.16)$$

(b) *Random uncertainty in small samples* (n < about 30). In many circumstances, the analyst will not be able to collect a large number of data points, and may be limited to a data set of less than 30 values (n < 30). Under such conditions, the mean value and the standard deviation are computed as before. The z value applicable for the normal distribution cannot be used for small samples. The new values, called t-values, are tabulated for different degrees of freedom d.f. ($\nu = n - 1$) and for the acceptable degree of confidence (see Table A4[5]). The confidence interval for the mean value of x, when no fixed (bias) errors are present in the measurements, is given by:

$$\bar{x}_{\min} = \bar{x} - \left(\frac{t.s_x}{\sqrt{n}}\right) \text{ and } \bar{x}_{\max} = \bar{x} + \left(\frac{t.s_x}{\sqrt{n}}\right) \quad (3.17)$$

For example, consider the case of d.f. = 10 and two-tailed significance level $\alpha = 0.05$. One finds from Table A4 that t = 2.228 for 95% CL. Note that this increases to t = 2.086 for d.f. = 20 and reaches the z value for 1.96 for d.f. = ∞.

Example 3.6.1 *Estimating confidence intervals*

(a) The length of a field is measured 50 times. The mean is 30 with a standard deviation of 3. Determine the 95% CL.
This is a large sample case, for which the z multiplier is 1.96. Hence, the 95% CL are
$$= 30 \pm \frac{(1.96)\cdot(3)}{(50)^{1/2}} = 30 \pm 0.83 = \{29.17, 30.83\}$$
(b) Only 21 measurements are taken and the same mean and standard deviation as in *(a)* are found. Determine the 95% CL.
This is a small sample case for which the t-value = 2.086 for d.f. = 20. Then, the 95% CL will turn out to be wider:
$$30 \pm \frac{(2.086)\cdot(3)}{(21)^{1/2}} = 30 \pm 1.37 = \{28.63, 31.37\}$$ ■

3.6.4 Bias Uncertainty

Estimating the bias or fixed error at a specified confidence level (say, 95% confidence) is described below. The fixed error B_x for a given value x is assumed to be a single value drawn from some larger distribution of possible fixed errors. The treatment is similar to that of random errors with the major difference that only one value is considered even though several observations may be taken. Lacking further knowledge, a normal distribution is usually assumed. Hence, if a manufacturer specifies that the fixed uncertainty B_x is ±1.0°C with 95% confidence (compared to some standard reference device), then one assumes that the fixed error belongs to a larger distribution (taken to be Gaussian) with a standard deviation S_B = 0.5°C (since the corresponding z-value = 2.0).

3.6.5 Overall Uncertainty

The overall uncertainty of a measured variable x has to combine the random and bias uncertainty estimates. Though several forms of this expression appear in different texts, a convenient working formulation is as follows:

$$U_x = \sqrt{{B_x}^2 + \left(t\frac{s_x}{\sqrt{n}}\right)^2} \quad (3.18)$$

[5] Table A4 applies to critical values for one-tailed distributions, while most of the discussion here applies to the two-tailed case. See Sect. 4.2.2 for the distinction between both.

where:
U_x = overall uncertainty in the value x at a *specified confidence level*
B_x = uncertainty in the bias or fixed component at the specified confidence level
s_x = standard deviation estimates for the random component
n = sample size
t = t-value at the specified confidence level for the appropriate degrees of freedom

Example 3.6.2: For a single measurement, the statistical concept of standard deviation does not apply Nonetheless, one could estimate it from manufacturer's specifications if available. It is desired to estimate the overall uncertainty at 95% confidence level in an individual measurement of water flow rate in a pipe under the following conditions:
(a) full scale meter reading 150 L/s
(b) actual flow reading 125 L/s
(c) random error of instrument is ±6% of full-scale reading at 95% CL
(d) fixed (bias) error of instrument is ±4% of full-scale reading at 95% CL

The solution is rather simple since all stated uncertainties are at 95% CL. It is implicitly assumed that the normal distribution applies. The random error = 150 × 0.06 = ±9 L/s. The fixed error = 150 × 0.04 = ±6 L/s. The overall uncertainty can be estimated from Eq. 3.18 with n = 1:

$$U_x = (6^2 + 9^2)^{1/2} = \pm 10.82\ \text{L/s}$$

The fractional overall uncertainty at 95% CL $= \dfrac{U_x}{x} = \dfrac{10.82}{125} = 0.087 = 8.7\%$ ■

Example 3.6.3: Consider Example 3.6.2. In an effort to reduce the overall uncertainty, 25 readings of the flow are taken instead of only one reading. The resulting uncertainty in this case is determined as follows.

The bias error remains unchanged at ±6 L/s.

The random error decreases by a factor of $\sqrt{n}$ to $9/(25)^{1/2} = \pm 1.8$ L/s

The overall uncertainty is thus: $U_x = (6^2 + 1.8^2)^{1/2} = \pm 6.26$ L/s

The fractional overall uncertainty at 95% confidence level $= \dfrac{U_x}{x} = \dfrac{6.26}{125} = 0.05 = 5.0\%$

Increasing the number of readings from 1 to 25 reduces the relative uncertainty in the flow measurement from ±8.7% to ±5.0%. Because of the large fixed error, further increase in the number of readings would result in only a small reduction in the overall uncertainty. ■

Example 3.6.4: A flow meter manufacturer stipulates a random error of 5% for his meter at 95.5% CL (i.e., at z = 2). Once installed, the engineer estimates that the bias error due to the placement of the meter in the flow circuit is 2% at 95.5% CL. The flow meter takes a reading every minute, but only the mean value of 15 such measurements is recorded once every 15 min. Estimate the overall uncertainty at 99% CL of the mean of the recorded values.

The bias uncertainty can be associated with the normal tables. From Table A3, z = 2.575 has an associated probability of 0.01 which corresponds to the 99% CL. Since 95.5% CL corresponds to z = 2, the bias uncertainty at one standard deviation = 1%.

Since the number of observations is less than 30, the student-t table has to be used for the random uncertainty component. From Table A4, the critical t value for d.f. = 15 − 1 = 14 and significance level of 0.01 is equal to 2.977. Also, the random uncertainty at one standard deviation $= \dfrac{5.0}{2} = 2.5\%$

Hence, the overall uncertainty of the recorded values at 99% CL

$$= U_x = \left\{[(2.575).1]^2 + \left[\frac{(2.977).(2.5)}{(15)^{1/2}}\right]^2\right\}^{1/2}$$
$$= 0.0322 = 3.22\%$$ ■

3.6.6 Chauvenet's Statistical Criterion of Data Rejection

The statistical considerations described above can lead to analytical screening methods which can point out data errors not flagged by graphical methods alone. Though several types of rejection criteria can be formulated, perhaps the best known is the *Chauvenet's criterion*. This criterion, which presumes that the errors are normally distributed and have constant variance, specifies that any reading out of a series of n readings shall be rejected if the magnitude of its deviation d_{max} from the mean value of the series is such that the probability of occurrence of such a deviation exceeds (1/2n). It is given by:

$$\frac{d_{max}}{s_x} = 0.819 + 0.544.\ln(n) - 0.02346.\ln(n^2) \quad (3.19)$$

where s_x is the standard deviation of the series and n is the number of data points. The deviation ratio for different number of readings is given in Table 3.7. For example, if one takes 10 observations, an observation shall be discarded if its deviation from the mean is $d_{max} \geq (1.96)s_x$.

This data rejection should be done only once, and more than one round of elimination using the Chauvenet criterion is not advised. Note that the Chauvenet criterion has inherent assumptions which may not be justified. For example, the

Table 3.7 Table for Chauvenet's criterion of rejecting outliers following Eq. 3.19

Number of readings N	Deviation ratio d_{max}/S_x
2	1.15
3	1.38
4	1.54
5	1.65
6	1.73
7	1.80
10	1.96
15	2.13
20	2.31
25	2.33
30	2.51
50	2.57
100	2.81
300	3.14
500	3.29
1000	3.48

underlying distribution may not be normal, but could have a longer tail. In such a case, one may be throwing out good data. A more scientific manner of dealing with outliers which also yields similar results is to use *weighted regression or robust regression*, where observations farther away from the mean are given less weight than those from the center (see Sect. 5.6 and 9.6.1 respectively).

3.7 Propagation of Errors

In many cases, the variable of interest is not directly measured, but values of several associated variables are measured, which are then combined using a data reduction equation to obtain the value of the desired result. The objective of this section is to present the methodology to estimate overall uncertainty from knowledge of the uncertainties in the individual variables. The random and fixed components, which together constitute the overall uncertainty, have to be estimated separately. The treatment that follows, though limited to random errors, could also apply to bias errors.

3.7.1 Taylor Series Method for Cross-Sectional Data

In general, the standard deviation of a function $y=y(x_1, x_2, \ldots, x_n)$, whose independently measured variables are all given with the same confidence level, is obtained by the first order expansion of the Taylor series:

$$s_y = \sqrt{\left(\sum_{i=1}^{n} \left(\frac{\partial y}{\partial x_i} s_{x,i}\right)^2\right)} \tag{3.20}$$

where:

s_y = function standard deviation

$s_{x,i}$ = standard deviation of the measured quantity x_i

Neglecting terms higher than the first order (as implied by a first order Taylor Series expansion), the propagation equations for some of the basic operations are given below. Let x_1 and x_2 have standard deviations s_1 and s_2. Then:

Addition or subtraction: $y = x_1 \pm x_2$ and

$$s_y = (s_{x1}^2 + s_{x2}^2)^{1/2} \tag{3.21}$$

Multiplication: $y = x_1.x_2$ and

$$s_y = (x_1.x_2).\left[\left(\frac{s_{x1}}{x_1}\right)^2 + \left(\frac{s_{x2}}{x_2}\right)^2\right]^{1/2} \tag{3.22}$$

Division: $y = x_1/x_2$ and

$$s_y = \left(\frac{x_1}{x_2}\right).\left[\left(\frac{s_{x1}}{x_1}\right)^2 + \left(\frac{s_{x2}}{x_2}\right)^2\right]^{1/2} \tag{3.23}$$

For multiplication and division, the fractional error is given by the same expression. If $y = \frac{x_1 x_2}{x_3}$, then the **fractional standard deviation**:

$$\frac{s_y}{y} = \left(\frac{s_{x1}^2}{x_1^2} + \frac{s_{x2}^2}{x_2^2} + \frac{s_{x3}^2}{x_3^2}\right)^{1/2} \tag{3.24}$$

The uncertainty in the result depends on the squares of the uncertainties in the independent variables. This means that if the uncertainty in one variable is larger than the uncertainties in the other variables, then it is the largest uncertainty that dominates. To illustrate, suppose there are three variables with an uncertainty of magnitude 1 and one variable with an uncertainty of magnitude 5. The uncertainty in the result would be $(5^2+1^2+1^2+1^2)^{0.5}=(28)^{0.5}=5.29$. Clearly, the effect of the largest uncertainty dominates the others.

An analysis involving relative magnitude of uncertainties plays an important role during the design of an experiment and the procurement of instrumentation. Very little is gained by trying to reduce the "small" uncertainties since it is the "large" ones that dominate. Any improvement in the overall experimental result must be achieved by improving the instrumentation or experimental technique connected with these relatively large uncertainties. This concept is illustrated in Example 3.7.2 below.

Equation 3.20 applies when the measured variables are uncorrelated. If they are correlated, their interdependence can be quantified by the covariance (defined by Eq. 3.9). If two variables x_1 and x_2 are correlated, then the standard deviation of their sum is given by:

Table 3.8 Error table of the four quantities that define the Reynolds number (Example 3.7.2)

Quantity	Minimum flow	Maximum flow	Random error at full flow (95% CL)	% errors[a] Minimum	Maximum
Velocity m/s (V)	1	20	0.1	10	0.5
Pipe diameter m (d)	0.2	0.2	0	0	0
Density kg/m³ (ρ)	1000	1000	1	0.1	0.1
Viscosity kg/m -s (μ)	1.12×10^{-3}	1.12×10^{-3}	0.45×10^{-3}	0.4	0.4

[a] Note that the last two columns under "% error" are computed from the previous three columns of data

$$s_y = {s_{x1}}^2 + {s_{x2}}^2 + 2.\mathrm{cov}(x_1, x_2).x_1.x_2 \tag{3.25}$$

Another method of dealing with propagation of errors is to adopt a *perturbation approach*. To simplify this computation, a computer routine can be written to perform the task of calculating uncertainties approximately. One method is based on approximating partial derivatives by a central finite-difference approach. If $y=y(x_1, x_2, \ldots x_n)$, then:

$$\frac{\partial y}{\partial x_1} = \frac{y(x_1+\Delta x_1,\ x_2, \ldots) - y(x_1-\Delta x_1,\ x_2, \ldots)}{2.\Delta x_1}$$
$$\frac{\partial y}{\partial x_2} = \frac{y(x_1,\ x_2+\Delta x_2, \ldots) - y(x_1,\ x_2-\Delta x_2 \ldots)}{2.\Delta x_2} \quad \text{etc...} \tag{3.26}$$

No strict rules for the size of the perturbation or step size Δx can be framed since they would depend on the underlying shape of the function. Perturbations in the range of 1–4% of the value are reasonable choices, and one should evaluate the stability of the partial derivative computed numerically by repeating the calculations for a few different step sizes. In cases involving complex experiments with extended debugging phases, one should update the uncertainty analysis whenever a change is made in the data reduction program. Commercial software programs are also available with in-built uncertainty propagation formulae. This procedure is illustrated in Example 3.7.4 below.

Example 3.7.1: *Uncertainty in overall heat transfer coefficient*

The equation of the over-all heat-transfer coefficient U of a heat exchanger consisting of a fluid flowing inside and another fluid flowing outside a steel pipe of negligible thermal resistance is $U=(1/h_1+1/h_2)^{-1}=(h_1 h_2/(h_1+h_2))$ where h_1 and h_2 are the individual coefficients of the two fluids. If $h_1=15$ W/m²°C with a fractional error of 5% at 95% CL and $h_2=20$ W/m²°C with a fractional error of 3%, also at 95% CL, what will be the fractional error in random uncertainty of the U coefficient at 95% CL assuming bias error to be zero?

In order to use the propagation of error equation, the partial derivatives need to be computed. One could proceed to do so analytically using basic calculus. Then:

$$\left(\frac{\partial U}{\delta h_1}\right)_{h_2} = \frac{h_2(h_1+h_2) - h_1 h_2}{(h_1+h_2)^2} = \frac{h_2^2}{(h_1+h_2)^2} \tag{3.27a}$$

and

$$\left(\frac{\partial U}{\delta h_2}\right)_{h_1} = \frac{h_1(h_1+h_2) - h_1 h_2}{(h_1+h_2)^2} = \frac{h_1^2}{(h_1+h_2)^2} \tag{3.27b}$$

The expression for the fractional uncertainty in the overall heat transfer coefficient U is:

$$\frac{S_U}{U} = \sqrt{\frac{{h_2}^2}{(h_1+h_2)^2}\left(\frac{{S_{h1}}^2}{{h_1}^2}\right) + \frac{{h_1}^2}{(h_1+h_2)^2}\left(\frac{{S_{h2}}^2}{{h_2}^2}\right)} \tag{3.28}$$

Plugging numerical values, one gets U=8.571, while the partial derivatives given by Eqs. 3.27 are computed as:

$$\frac{\partial U}{\partial h_1} = 0.3265 \quad \text{and} \quad \frac{\partial U}{\partial h_2} = 0.1837$$

The two terms on the right hand side of Eq. 3.28 provide insight into the relative contributions of h_2 and h_1. These are estimated as 16.84% and 83.16% indicating that the latter is the dominant one.

Finally, $S_U=0.2686$ yielding a fractional error $(S_U/U)=$ 3.1% at 95% CL ■

Example 3.7.2[6]: *Relative error in Reynolds number of flow in a pipe*

Water is flowing in a pipe at a certain measured rate. The temperature of the water is measured and the viscosity and density are then found from tables of water properties. Determine the probable errors of the Reynolds numbers (Re) at the low and high flow conditions given the following information (Table 3.8):

Recall that $\mathrm{Re} = \frac{\rho V d}{\mu}$. From Eq. 3.24, at minimum flow condition, the relative error in Re is:

$$\frac{\Delta \mathrm{Re}}{\mathrm{Re}} = \left[\left(\frac{0.1}{1}\right)^2 + \left(\frac{1}{1000}\right)^2 + \left(\frac{0.45}{1.12}\right)^2\right]^{1/2}$$
$$= (0.1^2 + 0.001^2 + 0.004^2)^{1/2} = 0.1 \text{ or } 10\%$$

[6] Adapted from Schenck (1969) by permission of Mc Graw-Hill.

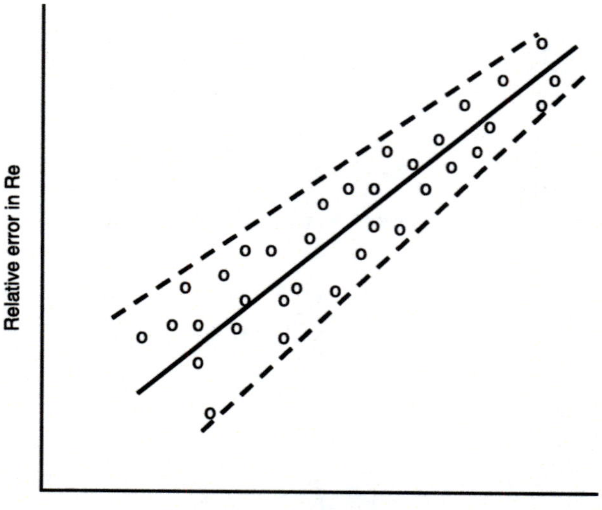

Fig. 3.34 Expected variation in experimental relative error with magnitude of Reynolds number (Example 3.7.2)

(to within 4 decimal points)—note that there is no error in pipe diameter value. At maximum flow condition, the percentage error is:

$$\frac{\Delta \text{Re}}{\text{Re}} = (0.005^2 + 0.001^2 + 0.004^2)^{1/2} = 0.0065 \text{ or } 0.65\%$$

The above example reveals that (i) at low flow conditions the error is 10% which reduces to 0.65% at high flow conditions, and (ii) at low flow conditions the other sources of error are absolutely dwarfed by the 10% error due to flow measurement uncertainty. Thus, the only way to improve the experiment is to improve flow measurement accuracy. If the experiment is run without changes, one can confidently expect the data at the low flow end to show a broad scatter becoming smaller as the velocity is increased. This phenomenon is captured by the confidence intervals shown in Fig. 3.34. ■

Example 3.7.3: *Selecting instrumentation during the experimental design phase*
An experimental program is being considered involving continuous monitoring of a large chiller under field conditions. The objective of the monitoring is to determine the chiller Coefficient of Performance (COP) on an hourly basis. The fractional uncertainty in the COP should not be greater than 5% at 95% CL. The rated full load is 450 tons of cooling (1 ton = 12,000 BTU/h). The chiller is operated under constant chilled water and condenser water flow rates. Only random errors are to be considered.

The COP of a chiller is defined as the ratio of the amount of cooling at the evaporator (Q_{ch}) to the electric power (E) consumed:

$$COP = \frac{Q_{ch}}{E} \quad (3.29)$$

while power E can be measured directly, the amount of cooling Q_{ch} has to be determined by individual measurements of the chilled water volumetric flow rate and the difference between the supply and return chilled water temperatures along with water properties.

$$Q_{ch} = \rho V c \Delta T \quad (3.30)$$

where:
ρ = density of water,
V = chilled water volumetric flow rate, assumed constant during operation (= 1080 gpm),
c = specific heat of water,
ΔT = temperature difference between the entering and leaving chilled water at the evaporator (which changes during operation)

The fractional uncertainty in COP (neglecting the small effect of uncertainties in the density and specific heat) is:

$$\frac{U_{COP}}{COP} = \sqrt{\left(\frac{U_V}{V}\right)^2 + \left(\frac{U_{\Delta T}}{\Delta T}\right)^2 + \left(\frac{U_E}{E}\right)^2} \quad (3.31)$$

Note that since this is a preliminary uncertainty analysis, only random (precision) errors are considered.

1. Let us assume that the maximum flow reading of the selected meter is 1500 gpm and has 4% uncertainty at 95% CL. This leads to an absolute uncertainty of (1500 × 0.04) = 60 gpm. The first term $\frac{U_V}{V}$ is a constant and does not depend on the chiller load since the flow through the evaporator is maintained constant. The rated chiller flow rate is 1080 gpm, Thus

$$\left(\frac{U_V}{V}\right)^2 = \left(\frac{60}{1080}\right)^2 = 0.0031 \text{ and } \frac{U_V}{V} = \pm 0.056.$$

2. Assume that for the power measurement, the instrument error at 95% CL is 4.0, calculated as 1% of the instrument full scale value of 400 kW. The chiller rated capacity is 450 tons of cooling, with an assumed realistic lower bound of 0.8 kW per tons of cooling. The anticipated electric draw at full load of the chiller = 0.8 × 450 = 360 kW. The fractional uncertainty at full load is then:

$$\left(\frac{U_E}{E}\right)^2 = \left(\frac{4.0}{360}\right)^2 = 0.00012 \text{ and } \frac{U_E}{E} = \pm 0.011$$

Thus, the fractional uncertainty in the power is about five times less that of the flow rate.

3. The random (precision) error at 95% CL for the type of commercial grade sensor to be used for temperature measurement is 0.2°F. Consequently, the error in the measurement of temperature difference $\Delta T=(0.2^2+0.2^2)^{1/2}=$ 0.28°F. From manufacturer catalogs, the temperature difference between supply and return chilled water temperatures at full load can be assumed to be 10°F. The fractional uncertainty at full load is then

$$\left(\frac{U_{\Delta T}}{\Delta T}\right)^2 = \left(\frac{0.28}{10}\right)^2 = 0.00078 \text{ and } \frac{U_{\Delta T}}{\Delta T} = \pm 0.078$$

4. Propagation of the above errors yields the fractional uncertainty at 95% CL at full chiller load of the measured COP:

$$\left(\frac{U_{COP}}{COP}\right) = (0.0031 + 0.00012 + 0.00078)^{1/2}$$
$$= 0.063 = 6.3\%$$

It is clear that the fractional uncertainty of the proposed instrumentation is not satisfactory for the intended purpose.

The logical remedy is to select a more accurate flow meter or one with a lower maximum flow reading. ■

Example 3.7.4: *Uncertainty in exponential growth models*
Exponential growth models are used to model several commonly encountered phenomena, from population growth to consumption of resources. The amount of resource consumed over time Q(t) can be modeled as:

$$Q(t) = \int_0^t P_0 e^{rt} dt = \frac{P_0}{r}(e^{rt} - 1) \qquad (3.32a)$$

where P_0=initial consumption rate, and r=exponential rate of growth

The world coal consumption in 1986 was equal to 5.0 billion (short) tons and the estimated recoverable reserves of coal were estimated at 1000 billion tons.

(a) If the growth rate is assumed to be 2.7% per year, how many years will it take for the total coal reserves to be depleted?

Rearranging Eq. 3.32a results in

$$t = \frac{1}{r}\left[\ln\left(1 + \frac{Q.r}{P_0}\right)\right] \qquad (3.32b)$$

Or

$$t = \frac{1}{0.027} \cdot \ln\left[1 + \frac{(1000)(0.027)}{5}\right] = 68.75 \text{ years}$$

(b) Assume that the growth rate r and the recoverable reserves are subject to random uncertainty. If the uncertainties of both quantities are taken to be normal with one standard deviation values of 0.2% (absolute) and 10% (relative) respectively, determine the lower and upper estimates of the years to depletion at the 95% confidence level.

Though the partial derivatives can be derived analytically, the use of Eq. 3.26 will be illustrated so as to compute them numerically. Let us use Eq. 3.32b with a perturbation multiplier of 1% to both the base values of r (=0.027) and of Q (=1000). The pertinent results are assembled in Table 3.9.

Table 3.9 Numerical computation of the partial derivatives of t with Q and r

Multiplier	Assuming Q=1000		Assuming r=0.027	
	r	t (from Eq. 3.32b)	Q	t (from Eq. 3.32b)
0.99	0.02673	69.12924	990	68.43795
1.00	0.027	68.75178	1000	68.75178
1.01	0.02727	68.37917	1010	69.06297

From here:

$$\frac{\partial t}{\partial r} = \frac{(68.37917 - 69.12924)}{(0.02727 - 0.02673)} = -1389 \text{ and}$$
$$\frac{\partial t}{\partial Q} = \frac{(69.06297 - 68.43795)}{(1010 - 990)} = 0.03125$$

Then:

$$s_t = \left[\left(\frac{\partial t}{\partial r} s_r\right)^2 + \left(\frac{\partial t}{\partial Q} s_Q\right)^2\right]^{1/2}$$

$$= \{[-1389)(0.002)]^2 + [(0.03125)(0.1)(1000)]^2\}^{1/2}$$
$$= (2.778^2 + 3.125^2)^{1/2} = 4.181$$

Thus, the lower and upper limits at the 95% CL (with the z=1.96) is

$$= 68.75 \pm (1.96)4.181 = \{60.55, 76.94\} \text{ years}$$

The analyst should repeat the above procedure with, say, a perturbation multiplier of 2% in order to evaluate the stability of the numerically derived partial derivatives. If these differ substantially, it is urged that the function be plotted and scrutinized for irregular behavior around the point of interest. ■

3.7.2 Taylor Series Method for Time Series Data

Uncertainty in time series data differs from that of stationary data in two regards:

(a) the uncertainty in the dependent variable y_t at a given time t depends on the uncertainty at the previous time y_{t-1}, and thus, uncertainty compounds over consecutive time steps, i.e., over time; and

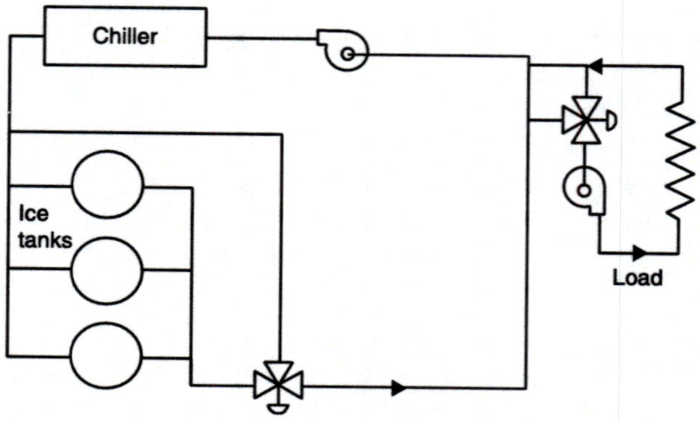

Fig. 3.35 Schematic layout of a cool storage system with the chiller located upstream of the storage tanks for Example 3.7.5. (From Dorgan and Elleson 1994 © American Society of Heating, Refrigerating and Air-conditioning Engineers, Inc., www.ashrae.org)

(b) some or all of the independent variables x may be cross-correlated, i.e., they have a tendency to either increase or decrease in unison.

The effect of both these factors is to increase the uncertainty as compared to stationary data (i.e., data without time-wise behavior). Consider the function shown below:

$$y = f(x_1, x_2, x_3) \quad (3.33)$$

Following Eq. 3.25, the equation for the propagation of random errors for a data reduction function with variables that exhibit cross correlation (case (b) above) is given by:

$$\begin{aligned} U_y^2 = &[(U_{x_1}.SC_{x_1})^2 + (U_{x_2}.SC_{x_2})^2 + (U_{x_3}.SC_{x_3})^2 \\ &+ 2.r_{x_1x_2}.SC_{x_1}.SC_{x2}.U_{x_1}U_{x_2} \\ &+ 2.r_{x_1x_3}.SC_{x_1}.SC_{x_3}.U_{x_1}U_{x3} \\ &+ 2.r_{x_2x_3}.SC_{x_2}.SC_{x_3}.U_{x_2}U_{x_3}] \end{aligned} \quad (3.34)$$

where

U_{xi} is the uncertainty of variable x_i

SC_{xi}=the sensitivity coefficients of y to variable $x_i = \frac{\partial y}{\partial x_i}$, and

$r_{x_ix_j}$=correlation coefficient between variables x_i and x_j.

Example 3.7.5: *Temporal Propagation of Uncertainty in ice storage inventory*

The concept of propagation of errors can be illustrated with time-wise data for an ice storage system. Figure 3.35 is a schematic of a typical cooling system comprising of an upstream chiller charging a series of ice tanks. The flow to these tanks can be modulated by means of a three-way valve when partial charging or discharging is to be achieved. The building loads loop also has its dedicated pump and three-way valve. It is common practice to charge and discharge the tanks uniformly. Thus, they can be considered to be one large consolidated tank for analysis purposes. The inventory of the tank is the cooling capacity available at any given time, and is an important quantity for the system operator to know because it would dictate the operation of the chiller and how much to either charge or discharge the chiller at any given time. Unfortunately, the direct measurement of this state is difficult. Sensors can be embedded inside the tanks, but this measurement is usually unreliable. Hence, it is more common for analysts to use the heat balance method to deduce the state of charge. An energy balance on the tank yields:

$$\frac{dQ}{dt} = q_{in} - q_{loss} \quad (3.35)$$

where

Q=stored energy amount or inventory of the storage system (say, in kWh or Ton-hours)

t=time

q_{in}=rate of energy flow into (or out of) the tank due to the secondary coolant loop during charging (or discharging)

q_{loss}=rate of heat lost by tank to surroundings

The rate of energy flow into or out of the tank can be deduced by measurements from:

$$q_{in} = V\rho c_p(T_{in} - T_{out}) \quad (3.36)$$

where

V=volumetric flow rate of the secondary coolant

ρ=density of the coolant

c_p=specific heat of coolant

T_{out}=exit temperature of coolant from tank

T_{in}=inlet temperature of coolant to tank

The two temperatures and the flow rate can be measured, and thereby q_{in} can be deduced.

The rate of heat loss from the tank to the surroundings can also be calculated as:

$$q_{loss} = UA(T_s - T_{amb}) \quad (3.37)$$

where

UA=effective overall heat loss coefficient of tank

T_{amb}=ambient temperature

T_s=average storage temperature

The UA value can be determined from the physical construction of the tank and the ambient temperature measured.

Combining all three above equations:

$$\frac{dQ}{dt} = V\rho c_p(T_{in} - T_{out}) - UA(T_s - T_{amb}) \quad (3.38)$$

Expressing the time rate change of heat transfer in terms of finite differences results in an expression for stored energy at time (t) with respect to time (t−1):

$$\Delta Q \equiv Q_t - Q_{t-1} = \Delta t.[C.\Delta T - UA.(T_s - T_{amb})] \quad (3.39a)$$

where

Δt=time step at which observations are made (say, 1 h),

Table 3.10 Storage inventory and uncertainty propagation table for Example 3.7.5

Hour Ending (t)	Mode of storage	Storage variables					95% CL Uncertainty in storage capacity			
		Change in storage ΔQ_t (kWh)	Storage capacity Q_t (kWh)	Inlet fluid temp (°C)	Exit fluid temp (°C)	Total flow rate V (L/s)	Change in Uncertainty (Eq. 3.40c)	$U_{Q,t}$	Absolute $U_{Q,t}/Q_{max}$	Relative $U_{Q,t}/Q_t$
8	Idle	0	2967	–	–	–	0.00	0.00	0.000	0.000
9	Idle	0	2967	–	–	–	0.00	0.00	0.000	0.000
10	Discharging	–183	2784	4.9	0.1	9.08	1345.72	36.68	0.012	0.013
11	"	–190	2594	5.5	0.2	8.54	1630.99	54.56	0.018	0.021
12	"	–327	2266	6.9	0.9	12.98	2116.58	71.37	0.024	0.031
13	"	–411	1855	7.8	2.1	17.17	1960.29	83.99	0.028	0.045
14	"	–461	1393	8.3	3.1	21.11	1701.69	93.57	0.032	0.067
15	"	–443	950	8.1	3.4	22.44	1439.15	100.97	0.034	0.106
16	"	–260	689	6.2	1.7	13.76	1223.73	106.86	0.036	0.155
17	"	–165	524	5.3	1.8	11.22	744.32	110.28	0.037	0.210
18	Idle	0	524	–	–	–	0.00	110.28	0.037	0.210
19	"	0	524	–	–	–	0.00	110.28	0.037	0.210
20	"	0	524	–	–	–	0.00	110.28	0.037	0.210
21	"	0	524	–	–	–	0.00	110.28	0.037	0.210
22	"	0	524	–	–	–	0.00	110.28	0.037	0.210
23	Charging	265	847	–3.3	–0.1	19.72	721.59	113.51	0.038	0.134
24	"	265	1112	–3.4	–0.2	19.72	721.59	116.64	0.039	0.105
1	"	265	1377	–3.4	–0.2	19.72	721.59	119.70	0.040	0.087
2	"	265	1642	–3.6	–0.3	19.12	750.61	122.79	0.041	0.075
3	"	265	1907	–3.6	–0.4	19.72	721.59	125.70	0.042	0.066
4	"	265	2172	–3.8	–0.6	19.72	721.59	128.53	0.043	0.059
5	"	265	2437	–4	–0.8	19.72	721.59	131.31	0.044	0.054
6	"	265	2702	–4.4	–1.1	19.12	750.61	134.14	0.045	0.050
7	"	265	2967	–4.8	–1.6	19.72	721.59	136.80	0.046	0.046

ΔT=temperature difference between inlet and outlet fluid streams, and

C=heat capacity rate of the fluid=$V.\rho.c_p$

So as to simplify this example, the small effect of heat losses is neglected (in practice, it is small but not negligible). Then Eq. 3.39a reduces to:

$$\Delta Q \equiv Q_t - Q_{t-1} = \Delta t.C.\Delta T \tag{3.39b}$$

Thus, knowing the state of charge Q_{t-1} where (t−1) could be the start of the operational cycle when the storage is fully charged, one can keep track of the state of charge over the day by repeating the calculation at hourly time steps. Unfortunately, the uncertainty of the inventory compounds because of the time series nature of how the calculations are made. Hence, determining this temporal uncertainty is a critical aspect.

Since the uncertainties in the property values for density and specific heat of commonly used coolants are much smaller than the other terms, the effect of their uncertainty can be neglected. Therefore, the following equation can be used to calculate the random error propagation of time-wise data results for this example.

$$U_{Q,t}^2 - U_{Q,t-1}^2 = \Delta t.[(U_C.\Delta T)^2 + (C.U_{\Delta T})^2 + 2r_{C.\Delta T}.C.\Delta T.U_C U_{\Delta T}] \tag{3.40a}$$

where C is the heat capacity rate of the fluid which changes hourly.

Assuming further that variables C and ΔT are uncorrelated, Eq. 3.40a reduces to:

$$U_{Q,t}^2 - U_{Q,t-1}^2 = \Delta t.[(U_C.\Delta T)^2 + (C.U_{\Delta T})^2] \tag{3.40b}$$

If needed, a similar expression can be used for the fixed error. Finally, the quadratic sum of both uncertainties would yield the total uncertainty.

Table 3.10 assembles hourly results of an example structured similarly to one from a design guide (Dorgan and Elleson 1994). This corresponds to the hour by hour performance of a storage system such as that shown in Fig. 3.35. The storage is fully charged at the end of 7:00 am where the daily cycle is assumed to start. The status of the storage inventory is indicated as either charging/discharging/idle, while the amount of heat flow in or out and the running inventory capacity of the tank are shown in columns 3 and 4. The two

Table 3.11 Magnitude and associated uncertainty of various quantities used in Example 3.7.5

Quantity	Symbol	Value	Random uncertainty at 95% CL
Density of water	ρ	1000 kg/m³	0.0
Specific heat of water	c_p	4.2 kJ/kg°C	0.0
Temperature	T	°C	0.1°C
Flow rate	V	L/s	U_V=6% of full scale reading of 30 L/s=1.8 L/s=6.48 m³/hr
Temperature difference	ΔT	°C	$U_{\Delta T}=(0.1^2+0.1^2)^{1/2}=0.141$

inlet and outlet temperatures and the fluid flow through the tank are also indicated. These are the operational variables of the system. Table 3.11 gives numerical values of the pertinent variables and their uncertainty values which are used to compute the last four columns of the table.

The uncertainty at 95% CL in the fluid flow rate into the storage is:

$$U_C = \rho c_p U_V = (1000).(4.2)(6.48) = 27,216 \text{ kJ/hr-}^{\circ}C$$
$$= 7.56 \text{ kWh/hr-}^{\circ}C$$

Inserting numerical values in Eq. 3.40b and setting the time step as one hour, one gets

$$U_{Q,t}^2 - U_{Q,t-1}^2 = [(7.56)\Delta T]^2 + [C.(0.141)]^2 \quad \text{kWh/hr-}^{\circ}C \tag{3.40c}$$

The uncertainty at the start of the calculation of the storage inventory is taken to be 0% while the maximum storage capacity Q_{max}=2967 kWh. Equation 3.40c is used at each time step, and the time evolution of the uncertainty is shown in the last two columns both as a fraction of the maximum storage capacity (referred to as "absolute", i.e., $[U_{Q,t}/Q_{max}]$) and as a relative uncertainty, i.e., as $[U_{Q,t}/Q_t]$. The variation of both these quantities is depicted graphically in Fig. 3.36. Note that the absolute uncertainty at 95% CL increases to 4.6% during the course of the day, while the relative uncertainty goes up to 21% during the hours of the day when the storage is essentially depleted. Further, note that various simplifying assumptions have been made during the above analysis; a detailed evaluation can be quite complex, and so, whenever possible, simplifications should be made depending on the specific system behavior and the accuracy to which the analysis is being done. ■

3.7.3 Monte Carlo Method

The previous method of ascertaining uncertainty, namely based on the first order Taylor series expansion is widely used; but it has limitations. If uncertainty is large, this method may be inaccurate for non-linear functions since it assumes derivatives based on local functional behavior. Further, an implicit assumption is that errors are normally distributed. Finally, in many cases, deriving partial derivatives of complex analytical functions is a tedious and error-prone affair, and even the numerical approach described and illustrated above is limited to cases of small uncertainties. A more general manner of dealing with uncertainty propagation is to use *Monte Carlo methods*, though these are better suited for more complex situations (and treated at more length in Sects. 11.2.3 and 12.2.7). These methods are numerical methods for solving problems involving random numbers and require considerations of probability. Monte Carlo, in essence, is a process where the individual basic variables or inputs are sampled randomly from their prescribed probability distributions so as to form one repetition (or run or trial). The corresponding numerical solution is one possible outcome of the function. This process of generating runs is repeated a large number of times resulting in a distribution of the functional values which can then be represented as probability distributions, or as histograms, or by summary statistics or by confidence intervals for any percentile threshold chosen. The last option is of great importance in certain types of studies. The accuracy of the results improves with the number of runs in a square root manner. Increasing the number of runs 100 times will approximately reduce the uncertainty by a factor of 10. Thus, the process is computer intensive and requires thousands of runs be performed. However, the entire process is simple and easily implemented on spreadsheet programs (which have inbuilt functions for generating pseudo-random numbers of selected distributions). Specialized software programs are also available.

There is a certain amount of uncertainty associated with the process because Monte Carlo simulation is a numerical method. Several authors propose approximate formulae for determining the number of trials, but a simple method is as follows. Start with a large number of trials (say, 1000), and generate pseudo random numbers with the assumed probability distribution. Since they are pseudo-random, the mean and the distribution (say, the standard deviation) may deviate somewhat from the desired ones (which depend on the accuracy of the algorithm used). Generate a few such sets and pick one which is closest to the desired quantities. Use this set to simulate the corresponding values of the function. This can be repeated a few times till one finds that the mean and standard deviations stabilize around some average values which are taken to be the answer. It is also urged that the analyst evaluate the effect of the results with different number of trials; say, using 3000 trials, and ascertaining that the results of both the 1000 trial and 3000 trials are similar. If they are not, sets with increasingly large number of trials should be used till the results converge.

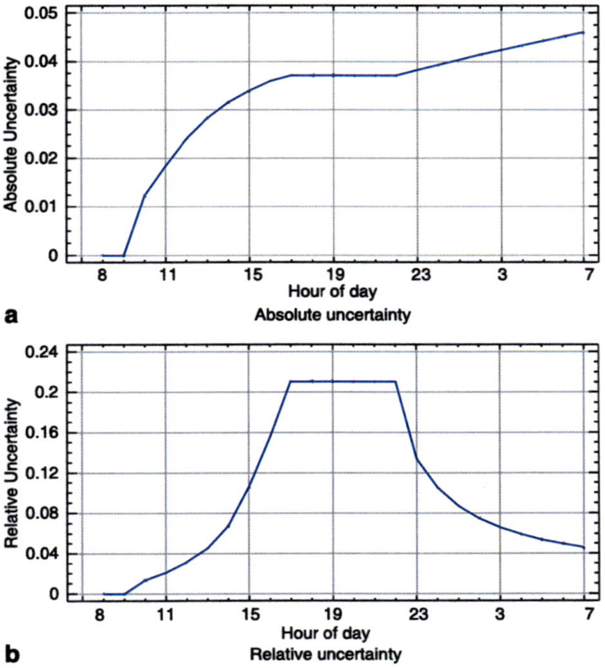

Fig. 3.36 Time variation of the absolute and relative uncertainties at 95% CL of the ice storage inventory for Example 3.7.5

The approach is best understood by means of a simple example.

Example 3.7.6: ***Using Monte Carlo to determine uncertainty in exponential growth models***
Let us solve the problem given in Example 3.7.4 by the Monte Carlo method. The approach involves setting up a spreadsheet table as shown in Table 3.12. Since only two variables (namely Q and r) have uncertainty, one needs only assign two columns to these and a third column to the desired quantity, i.e. time t over which the total coal reserves will be depleted. The first row shows the calculation using the mean values and one sees that the value of t=68.75 as found in part (a) of Example 3.7.4 is obtained (this is done for verifying the cell formula). The analyst then generates random numbers of Q and r with the corresponding mean and standard deviations as specified and shown in the first row of the table.

Monte Carlo methods, being numerical methods, require that a large sample be generated in order to obtain reliable results. In this case, 1000 normal distribution samples were generated, and the first few and last few rows are shown in Table 3.12 Even with 1000 samples, one finds that the sample mean and standard deviation deviate somewhat from the desired ones because of the pseudo-random nature of the random numbers generated by the spreadsheet program. For example, instead of having (1000, 100) for the mean and standard deviation of Q, the 1000 samples have (1005.0, 101.82). On the other hand, the differences for r are negligible. The corresponding mean and standard deviation

Table 3.12 The first few and last few calculations used to determine uncertainty in variable t using the Monte Carlo method (Example 3.7.6)

Run #	Q (1000, 100)	r (0.027, 0.002)	t (years)
1	1000.0000	0.0270	68.7518
2	1050.8152	0.0287	72.2582
3	1171.6544	0.0269	73.6445
4	1098.2454	0.0284	73.2772
5	1047.5003	0.0261	69.0848
6	1058.0283	0.0247	67.7451
7	946.8644	0.0283	68.5256
8	1075.5269	0.0277	71.8072
9	967.9137	0.0278	68.6323
10	1194.7164	0.0262	73.3758
11	747.9499	0.0246	57.2155
12	1099.7061	0.0269	71.5707
13	1074.3923	0.0254	69.1221
14	1000.2640	0.0265	68.2233
15	1071.4876	0.0274	71.3437
983	1004.2355	0.0282	70.1973
984	956.4792	0.0277	68.1372
985	1001.2967	0.0293	71.3534
986	1099.9830	0.0306	75.7549
987	1033.7338	0.0267	69.4667
988	934.5567	0.0279	67.6464
989	1055.7171	0.0282	71.8201
990	1133.6639	0.0278	73.6712
991	997.0123	0.0252	66.5173
992	896.6957	0.0257	63.8175
993	1056.2361	0.0283	71.9108
994	1033.8229	0.0298	72.8905
995	1078.6051	0.0295	73.9569
996	1137.8546	0.0276	73.4855
997	950.8749	0.0263	66.3670
998	1023.7800	0.0264	68.7452
999	950.2093	0.0248	64.5692
1000	849.0252	0.0247	61.0231
mean	*1005.0*	*0.0272*	*68.91*
stdev.	*101.82*	*0.00199*	*3.919*

of t are found to be (68.91, 3.919) compared to the previously estimated values of (68.75, 4.181). This difference is not too large, but the pseudo-random generation of the values for Q is rather poor and ought to be improved. Thus, the analyst should repeat the Monte Carlo simulation a few times with different seeds for the random number generator; this is likely to result in more robust estimates. ■

3.8 Planning a Non-intrusive Field Experiment

Any experiment should be well-planned involving several rational steps (for example, ascertaining that the right sensors and equipment are chosen, that the right data collection protocol and scheme are followed, and that the appropriate data analysis procedures are selected). It is advisable to explicitly adhere to the following steps (ASHRAE 2005):

(a) *Identify experimental goals and acceptable accuracy*
Identify realistic experimental goals (along with some measure of accuracy) that can be achieved within the time and budget available for the experiment.

(b) *Identify variables and relationships*
Identify the entire list of relevant measurable variables that should be examined. If some are inter-dependent, or if some are difficult to measure, find alternative variables.

(c) *Establish measured variables and limits*
For each measured variable, determine its theoretical limits and expected bounds to match the selected instrument limits. Also, determine *instrument limits* – all sensor and measurement instruments have physical limits that restrict their ability to accurately measure quantities of interest.

(d) *Preliminary instrumentation selection*
Selection of the equipment should be based on accuracy, repeatability and features of the instrument increase, as well as cost. Regardless of the instrument chosen, it should have been calibrated within the last twelve months or within an interval required by the manufacturer, whichever is less. The required accuracy of the instrument will depend upon the acceptable level of uncertainty for the experiment.

(e) *Document uncertainty of each measured variable*
Utilizing information gathered from manufacturers or past experience with specific instrumentation, document the uncertainty for each measured variable. This information will then be used in estimating the overall uncertainty of results using propagation of error methods.

(f) *Perform preliminary uncertainty analysis*
An uncertainty analysis of proposed measurement procedures and experimental methodology should be completed before the procedures and methodology are finalized in order to estimate the uncertainty in the final results. The higher the accuracy required of measurements, the higher the accuracy of sensors needed to obtain the raw data. The uncertainty analysis is the basis for selection of a measurement system that provides acceptable uncertainty at least cost. How to perform such a preliminary uncertainty analysis was discussed in Sect. 3.6 and 3.7.

(g) *Final instrument selection and methods*
Based on the results of the preliminary uncertainty analysis, evaluate earlier selection of instrumentation. Revise selection if necessary to achieve the acceptable uncertainty in the experiment results.

(h) *Install instrumentation*
Instrumentation should be installed in accordance with manufacturer's recommendations. Any deviation in the installation from the manufacturer's recommendations should be documented and the effects of the deviation on instrument performance evaluated. A change in instrumentation or location may be required if in-situ uncertainty exceeds acceptable limits determined by the preliminary uncertainty analysis.

(i) *Perform initial data quality verification*
To ensure that the measurements taken are not too uncertain and represent reality, instrument calibration and independent checks of the data are recommended. Independent checks can include sensor validation, energy balances, and material balances (see Sect. 3.3).

(j) *Collect data*
The challenge for data acquisition in any experiment is to collect the required amount of information while avoiding collection of superfluous information. Superfluous information can overwhelm simple measures taken to follow the progress of an experiment and can complicate data analysis and report generation. The relationship between the desired result, either static, periodic stationary or transient, and time is the determining factor for how much information is required. A static, non-changing result requires only the steady-state result and proof that all transients have died out. A periodic stationary result, the simplest dynamic result, requires information for one period and proof that the one selected is one of three consecutive periods with identical results within acceptable uncertainty. Transient or non-repetitive results, whether a single pulse or a continuing, random result, require the most information. Regardless of the result, the dynamic characteristics of the measuring system and the full transient nature of the result must be documented for some relatively short interval of time. Identifying good models requires a certain amount of diversity in the data, i.e., should cover the spatial domain of variation of the independent variables (discussed in Sect. 6.2). Some basic suggestions pertinent to controlled experiments are summarized below which are also pertinent for non-intrusive data collection.

(i) *Range of variability:* The most obvious way in which an experimental plan can be made compact and efficient is to space the variables in a predetermined manner. If a functional relationship between an independent variable X and a dependent variable Y is sought, the most obvious way is to select end points or limits of the test, thus covering the test envelope or domain that encloses the complete family of data. For a model of the type Z=f(X,Y), a plane area or map is formed (see Fig. 3.37). Functions involving more variables are usually broken down to a series of maps. The above discussion relates to controllable regressor variables. Extraneous variables, by their very nature, cannot

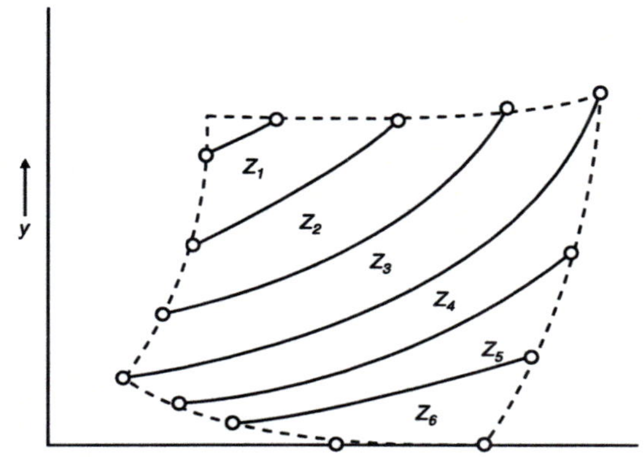

Fig. 3.37 A possible XYZ envelope with Z as the independent variable. The *dashed lines* enclose the total family of points over the feasible domain space

be varied at will. An example is phenomena driven by climatic variables. As an example, the energy use of a building is affected by outdoor dry-bulb temperature, humidity and solar radiation. Since these cannot be varied at will, a proper experimental data collection plan would entail collecting data during different seasons of the year.

(ii) *Grid spacing considerations:* Once the domains or ranges of variation of the variables are defined, the next step is to select the grid spacing. Being able to anticipate the system behavior from theory or from prior publications would lead to a better experimental design. For a relationship between X and Y which is known to be linear, the optimal grid is to space the points at the two extremities. However, if a linear relationship between X and Y is sought for a phenomenon which can be approximated as linear, then it would be best to space the x points evenly.

For non-linear or polynomial functions, an equally spaced test sequence in X is clearly not optimal. Consider the pressure drop through a new fitting as a function of flow. It is known that the relationship is quadratic. Choosing an experiment with equally spaced X values would result in a plot such as that shown in Fig. 3.38a. One would have more observations in the low pressure drop region and less in the higher range. One may argue that an optimal spacing would be to select the velocity values such that the pressure drop readings are more or less spaced (see Fig. 3.38b). Which one of two is better depends on the instrument precision. If the pressure drop instrument has constant relative precision during the entire range of variation of the experiment, then test spacing as shown in Fig. 3.38b is clearly better. But if the fractional uncertainty of the instrument decreases with increasing pressure drop values, then the point spacing sequence shown in Fig. 3.38a is better.

(k) *Accomplish data reduction and analysis*
Data reduction involves the distillation of raw data into a form that is usable for further analysis. Data reduction may involve averaging multiple measurements, quantifying necessary conditions (e.g., steady state), comparing with physical limits or expected ranges, and rejecting outlying measurements.

(l) *Perform final uncertainty analysis*
A detailed final uncertainty analysis is done after the entire experiment has been completed and when the results of the experiments are to be documented or reported. This will take into account unknown field effects and variances in instrument accuracy during the experiment. A final uncertainty analysis involves the following steps: (i) Estimate fixed (bias) error based upon instrumentation calibration results, and (ii) document the random error due to the instrumentation based upon instrumentation calibration results. As pointed out by Coleman and Steele (1999), the fixed errors needed for the detailed uncertainty analysis are usually more difficult to estimate with a high degree of certainty.

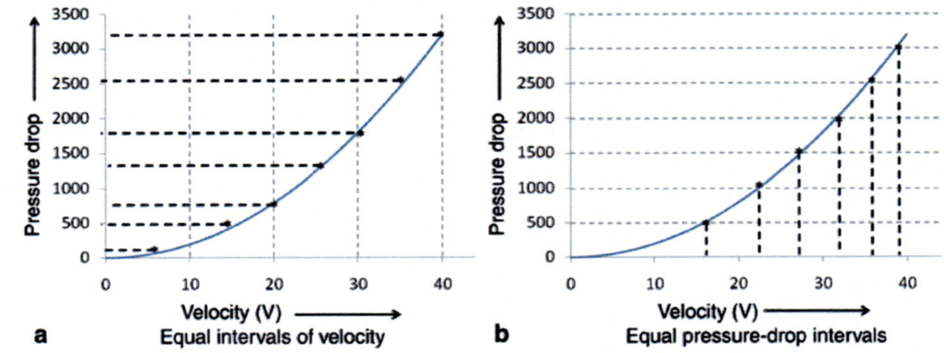

Fig. 3.38 Two different experimental designs for proper identification of the parameter (k) appearing in the model for pressure drop versus velocity of a fluid flowing through a pipe assuming $\Delta P = kV^2$. The grid spacing shown in (**a**) is the more common one based on equal increments in the regressor variable, while that in (**b**) is likely to yield more robust estimation but would require guess-estimating the range of variation for the pressure drop

Minimizing fixed errors can be accomplished by careful calibration with referenced standards.

(m) *Reporting results*
Reporting is the primary means of communicating the results from an experiment. The report should be structured to clearly explain the goals of the experiment and the evidence gathered to achieve the goals. It is assumed that data reduction, data analysis and uncertainty analysis have processed all data to render them understandable by the intended audiences. Different audiences require different reports with various levels of detail and background information. In any case, all reports should include the results of the uncertainty analysis to an identified confidence level (typically 95%). Uncertainty limits can be given as either absolute or relative (in percentages). Graphical and mathematical representations are often used. On graphs, error bars placed vertically and horizontally on representative points are a very clear way to present expected uncertainty. A data analysis section and a conclusion are critical sections, and should be prepared with great care while being succinct and clear.

Problems

Pr. 3.1 Consider the data given in Table 3.2. Determine
(a) the 10% trimmed mean value
(b) which observations can be considered to be "mild" outliers (>1.5×IQR)
(c) which observations can be considered to be "extreme" outliers (>3.0×IQR)
(d) identify outliers using Chauvenet's criterion given by Eq. 3.19
(e) compare the results from (b), (c) and (d).

Pr. 3.2 Consider the data given in Table 3.6. Perform an exploratory data analysis involving computing pertinent statistical summary measures, and generating pertinent graphical plots.

Pr. 3.3 A nuclear power facility produces a vast amount of heat which is usually discharged into the aquatic system. This heat raises the temperature of the aquatic system resulting in a greater concentration of chlorophyll which in turn extends the growing season. To study this effect, water samples were collected monthly at three stations for one year. Station A is located closest to the hot water discharge, and Station C the farthest (Table 3.13).

You are asked to perform the following tasks and annotate with pertinent comments:
(a) flag any outlier points
(b) compute pertinent statistical descriptive measures
(c) generate pertinent graphical plots
(d) compute the correlation coefficients.

Table 3.13 Data table for Problem 3.3

Month	Station A	Station B	Station C
January	9.867	3.723	4.410
February	14.035	8.416	11.100
March	10.700	20.723	4.470
April	13.853	9.168	8.010
May	7.067	4.778	34.080
June	11.670	9.145	8.990
July	7.357	8.463	3.350
August	3.358	4.086	4.500
September	4.210	4.233	6.830
October	3.630	2.320	5.800
November	2.953	3.843	3.480
December	2.640	3.610	3.020

Pr. 3.4 Consider Example 3.7.3 where the uncertainty analysis on chiller COP was done at full load conditions. What about part-load conditions, especially since there is no collected data? One could use data from chiller manufacturer catalogs for a similar type of chiller, or one could assume that part-load operation will affect the inlet minus the outlet chilled water temperatures (ΔT) in a proportional manner, as stated below.
(a) Compute the 95% CL uncertainty in the COP at 70% and 40% full load assuming the evaporator water flow rate to be constant. At part load, the evaporator temperatures difference is reduced proportionately to the chiller load, while the electric power drawn is assumed to increase from a full load value of 0.8 kW/t to 1.0 kW/t at 70% full load and to 1.2 kW/t at 40% full load.
(b) Would the instrumentation be adequate or would it be prudent to consider better instrumentation if the fractional COP uncertainty at 95% CL should be less than 10%.
(c) Note that fixed (bias) errors have been omitted from the analysis, and some of the assumptions in predicting part-load chiller performance can be questioned. A similar exercise with slight variations in some of the assumptions, called a sensitivity study, would be prudent at this stage. How would you conduct such an investigation?

Pr. 3.5 Consider the uncertainty in the heat transfer coefficient illustrated in Example 3.7.1. The example was solved analytically using the Taylor's series approach. You are asked to solve the same example using the Monte Carlo method:
(a) using 500 data points
(b) using 1000 data points
Compare the results from this approach with those in the solved example.

Pr. 3.6 You will repeat Example 3.7.6. Instead of computing the standard deviation, plot the distribution of the time variable t in order to evaluate its shape. Numerically determine the uncertainty bands for the 95% CL.

Pr. 3.7 *Determining cooling coil degradation based on effectiveness*

The thermal performance of a cooling coil can also be characterized by the concept of effectiveness widely used for thermal modeling of traditional heat exchangers. In such coils, a stream of humid air flows across a coil supplied by chilled water and is cooled and dehumidified as a result. In this case, the effectiveness can be determined as:

$$\varepsilon = \frac{\text{actual heat transfer rate}}{\text{maximum possible heat transfer rate}} = \frac{(h_{ai} - h_{ao})}{(h_{ai} - h_{ci})} \tag{3.41}$$

where h_{ai} and h_{ao} are the enthalpies of the air stream at the inlet and outlet respectively, and h_{ci} is the enthalpy of entering chilled water.

The effectiveness is independent of the operating conditions provided the mass flow rates of air and chilled water remain constant. An HVAC engineer would like to determine whether the coil has degraded after it has been in service for a few years. For this purpose he assembles the following coil performance data at identical air and water flow rates corresponding to when originally installed (done during start-up commissioning) and currently (Table 3.14).

Note that the uncertainty in determining the air enthalpies are relatively large due to the uncertainty associated with measuring bulk air stream temperatures and humidities. However, the uncertainty in the enthalpy of the chilled water is only half of that of air.

(a) Asses, at 95% CL, whether the cooling coil has degraded or not. Clearly state any assumptions you make during the evaluation.
(b) What are the relative contributions of the uncertainties in the three enthalpy quantities to the uncertainty in the effectiveness value? Do these differ from the installed period to the time when current tests were performed?

Pr. 3.8[7] Consider a basic indirect heat exchanger where heat rates of the heat exchange associated with the cold and hot sides is given by:

$$\begin{aligned} Q_{actual} &= m_c.c_{pc}.(T_{c,o} - T_{c,i}) \quad \text{(cold side heating)} \\ Q_{actual} &= m_h.c_{ph}.(T_{h,i} - T_{h,o}) \quad \text{(hot side cooling)} \end{aligned} \tag{3.42a}$$

Table 3.14 Data table for Problem 3.7

	Units	When installed	Current	95% Uncertainty
Entering air enthalpy (h_{ai})	Btu/lb	38.7	36.8	5%
Leaving air enthalpy (h_{ao})	Btu/hr	27.2	28.2	5%
Entering water enthalpy (h_{ci})	Btu/hr	23.2	21.5	2.5%

[7] From ASHRAE (2005) © American Society of Heating, Refrigerating and Air-conditioning Engineers, Inc., www.ashrae.org).

Table 3.15 Parameters and uncertainties to be assumed (Pr. 3.8)

Parameter	Nominal value	95% Uncertainty
c_{pc}	1 Btu/lb°F	±5%
m_c	475,800 lb/h	±10%
$T_{c,i}$	34°F	±1°F
$T_{c,o}$	46°F	±1°F
c_{hc}	0.9 Btu/hr°F	±5%
m_h	450,000 lb/h	±10%
$T_{h,i}$	55°F	±1°F
$T_{h,o}$	40°F	±1°F

where m, T and c are the mass flow rate, temperature and specific heat respectively, while the subscripts 0 and i stand for outlet and inlet, and c and h denote cold and hot streams respectively.

The effectiveness of the sensible heat exchanger is given by:

$$\begin{aligned} \varepsilon &= \frac{\text{actual heat transfer rate}}{\text{maximum possible heat transfer rate}} \\ &= \frac{Q_{actual}}{(mc_p)_{\min}(T_{hi} - T_{ci})} \end{aligned} \tag{3.42b}$$

Assuming the values and uncertainties of various parameters shown in the table (Table 3.15):

(i) compute the heat exchanger loads and the uncertainty ranges for the hot and cold sides
(ii) compute uncertainty in the effectiveness determination
(iii) what would you conclude regarding the heat balance checks?

Pr. 3.9 The following table (Table 3.16) (EIA 1999) indicates the total electricity generated by five different types of primary energy sources as well as the total emissions associated by each. Clearly coal and oil generate a lot of emissions or pollutants which are harmful not only to the environment but also to public health. France, on the other hand, has a mix of 21% coal and 79% nuclear.

Table 3.16 Data table for Problem 3.9

US power generation mix and associated pollutants

Fuel	Electricity		Short Tons (=2000 lb/t)		
	kWh (1999)	% Total	SO_2	NO_x	CO_2
Coal	1.77E+12	55.7	1.13E+07	6.55E+06	1.90E+09
Oil	8.69E+10	2.7	6.70E+05	1.23E+05	9.18E+07
Nat. Gas	2.96E+11	9.3	2.00E+03	3.76E+05	1.99E+08
Nuclear	7.25E+11	22.8	0.00E+00	0.00E+00	0.00E+00
Hydro/ Wind	3.00E+11	9.4	0.00E+00	0.00E+00	0.00E+00
Totals	3.18E+12	100.0	1.20E+07	7.05E+06	2.19E+09

Table 3.17 Data table for Problem 3.10

Symbol	Description	Value	95% Uncertainty
HP	Horse power of the end use device	40	5%
Hours	Number of operating hours in the year	6500	10%
η_{old}	Efficiency of the old motor	0.85	4%
η_{new}	Efficiency of the new motor	0.92	2%

(a) Calculate the total and percentage reductions in the three pollutants should the U.S. change its power generation mix to mimic that of France (Hint: First normalize the emissions per kWh for all three pollutants)
(b) The generation mix percentages (coal, oil, natural gas, nuclear and hydro/wind) have an inherent uncertainty of 5% at the 95% CL, while the uncertainties of the three pollutants are 5, 8 and 3% respectively. Assuming normal distributions for all quantities, compute the uncertainty of the reduction values estimated in (a) above.

Pr. 3.10 *Uncertainty in savings from energy conservation retrofits*
There is great interest in implementing retrofit measures meant to conserve energy in individual devices as well as in buildings. These measures have to justified economically, and including uncertainty in the estimated energy savings is an important element of the analysis. Consider the rather simple problem involving replacing an existing electric motor with a more energy efficient one. The annual energy savings E_{save} in kWh/yr are given by:

$$E_{save} = (0.746).(HP).(Hours).\left(\frac{1}{\eta_{old}} - \frac{1}{\eta_{new}}\right) \quad (3.43)$$

with the symbols described in Table 3.17 along with their numerical values.
(i) Determine the absolute and relative uncertainties in E_{save} under these conditions.
(ii) If this uncertainty had to be reduced, which variable will you target for further refinement?
(iii) What is the minimum value of η_{new} under which the lower bound of the 95% CL interval is greater than zero.

Pr. 3.11 *Uncertainty in estimating outdoor air fraction in HVAC systems*
Ducts in heating, ventilating and air-conditioning (HVAC) systems supply conditioned air (SA) to the various spaces in a building, and also exhaust the air from these spaces, called return air (RA). A sketch of an all-air HVAC system is shown in Fig. 3.39. Occupant comfort requires that a certain amount of outdoor air (OA) be brought into the HVAC systems while an equal amount of return air is exhausted to the outdoors. The OA and the RA mix at a point just before the air-handler unit. Outdoor air ducts have dampers installed in order to control the OA since excess OA leads to unnecessary energy wastage. One of the causes for recent complaints from occupants has been identified as inadequate OA, and sensors installed inside the ducts could modulate the dampers accordingly. Flow measurement is always problematic on a continuous basis. Hence, OA flow is inferred from measurements of the air temperature T_R inside the RA stream, of T_O inside the OA stream and T_M inside the mixed air (MA) stream. The supply air is deduced by measuring the fan speed with a tachometer, using a differential pressure gauge to measure static pressure rise, and using manufacturer equation for the fan curve. The random error of the sensors is 0.2°F at 95% CL with negligible bias error.

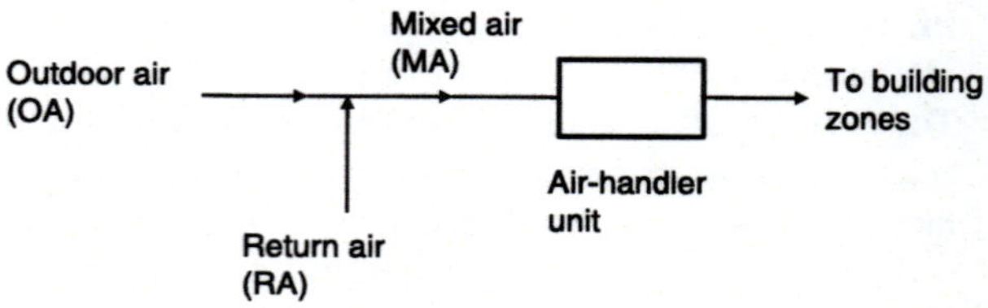

Fig. 3.39 Sketch of an all-air HVAC system supplying conditioned air to indoor rooms of a building

(a) From a sensible heat balance where changes in specific heat with temperature are neglected, derive the following expression for the fraction of outdoor air fraction (ratio of outdoor air and mixed air) $OA_f = (T_R - T_M)/(T_R - T_O)$
(b) Derive the expression for the uncertainty in OA_f and calculate the 95% CL in the OA_f if T_R=70°F, T_O=90°F and T_M=75°F.

Pr. 3.12 *Sensor placement in HVAC ducts with consideration of flow non-uniformity*
Consider the same situation as in Pr. 3.11. Usually, the air ducts have large cross-sections. The problem with inferring outdoor air flow using temperature measurements is the large thermal non-uniformity usually present in these ducts due to both stream separation and turbulence effects. Moreover, temperature (and, hence density) differences between the OA and MA streams result in poor mixing. The following table gives the results of a *traverse* in the mixed air duct with 9 measurements (using an equally spaced grid of 3×3 designated by numbers in bold in Table 3.18). The measurements were replicated four times under the same outdoor conditions. The random error of the sensors is 0.2°F at 95% CL with negligible bias error. Determine:
(a) the worst and best grid locations for placing a single sensor (to be determined based on analyzing the recordings at each of the 9 grid locations and for all four time periods)

Table 3.18 Table showing the temperature readings (in °F) at the nine different sections (S#1–S#9) of the mixed air (MA) duct (Pr. 3.12)

55.6, 54.6, 55.8, 54.2 *S#1*	56.3, 58.5, 57.6, 63.8 *S#2*	53.7, 50.2, 59.0, 49.4 *S#3*
58.0, 62.4, 62.3, 65.8 *S#4*	66.4, 67.8, 68.7, 67.6 *S#5*	61.2, 56.3, 64.7, 58.8 *S#6*
63.5, 65.0, 63.6, 64.8 *S#7*	67.4, 67.4, 66.8, 65.7 *S#8*	63.9, 61.4, 62.4, 60.6 *S#9*

(b) the maximum and minimum errors at 95% CL one could expect in the average temperature across the duct cross-section, if the best grid location for the single sensor was adopted.

Pr. 3.13 *Uncertainty in estimated proportion of exposed subjects using Monte Carlo method*
Dose-response modeling is the process of characterizing the relation between the dose of an administered/exposed agent and the incidence of an adverse health effect. These relationships are subject to large uncertainty because of the paucity of data as well as the fact that they are extrapolated from laboratory animal tests. Haas (2002) suggested the use of an exponential model for mortality rate due to inhalation exposure by humans to anthrax spores (characterized by the number of colony forming units or cfu):

$$p = 1 - \exp(-kd) \tag{3.44}$$

where p is the expected proportion of exposed individuals likely to die, d is the average dose (in cfu) and k is the dose response parameter (in units of 1/cfu). A value of $k=0.26\times10^{-5}$ has been suggested. One would like to determine the shape and magnitude of the uncertainty distribution of d at p=0.5 assuming that the one standard deviation (or uncertainty) of k is 30% of the above value and is normally distributed. Use the Monte Carlo method with 1000 trials to solve this problem. Also, investigate the shape of the error probability distribution, and ascertain the upper and lower 95% CL.

Pr. 3.14 *Uncertainty in the estimation of biological dose over time for an individual*
Consider an occupant inside a building in which an accidental biological agent has been released. The dose (D) is the cumulative amount of the agent to which the human body is subjected, while the response is the measurable physiological change produced by the agent. The widely accepted approach for quantifying dose is to assume functional forms based on first-order kinetics. For biological and radiological agents where the process of harm being done is cumulative, one can use Haber's law (Heinsohn and Cimbala 2003):

$$D(t) = k\int_{t_1}^{t_2} C(t)dt \tag{3.45a}$$

where $C(t)$ is the indoor concentration at a given time t, k is a constant which includes effects such as the occupant breathing rate, the absorption efficiency of the agent or species,… and t_1 and t_2 are the start and end times. This relationship is often used to determine health-related exposure guidelines for toxic substances. For a simple one-zone building, the free response, i.e., the temporal decay is given in terms of the initial concentration $C(t_1)$ by:

$$C(t) = C(t_1).\exp[(-a(t-t_1)] \tag{3.45b}$$

where the model parameter "a" is a function of the volume of the space and the outdoor and supply air flow rates. The above equation is easy to integrate during any time period from t_1 to t_2, thus providing a convenient means of computing total occupant inhaled dose when occupants enter or leave the contaminated zones at arbitrary times. Let a=0.017186 with 11.7% uncertainty while $C(t_1)=7000$ cfu/m³ (cfu–colony forming units). Assume k=1.

(a) Determine the total dose to which the individual is exposed to at the end of 15 min.
(b) Compute the uncertainty of the total dose at 1 min time intervals over 15 min (similar to the approach in Example 3.7.6)
(c) Plot the 95% CL over 15 min at 1 min intervals

Pr. 3.15 *Propagation of optical and tracking errors in solar concentrators*
Solar concentrators are optical devices meant to increase the incident solar radiation flux density (power per unit area) on a receiver. Separating the solar collection component (viz., the reflector) and the receiver can allow heat losses per collection area to be reduced. This would result in higher fluid operating temperatures at the receiver. However, there are several sources of errors which lead to optical losses:

(i) Due to non-specular or diffuse reflection from the reflector, which could be due to improper curvature of the reflector surface during manufacture (shown in Fig. 3.40a) or to progressive dust accumulation over the surface over time as the system operates in the field;
(ii) Due to tracking errors arising from improper tracking mechanisms as a result of improper alignment sensors or non-uniformity in drive mechanisms (usually, the tracking is not continuous; a sensor activates a motor every few minutes which re-aligns the reflector to the solar radiation as it moves in the sky). The result is a spread in the reflected radiation as illustrated in Fig. 3.40b;
(iii) Improper reflector and receiver alignment during the initial mounting of the structure or due to small ground/pedestal settling over time).

The above errors are characterized by root mean square (or rms) random errors (bias errors such as that arising from structural mismatch can often be corrected by one-time or regular corrections), and their combined effect can be deter-

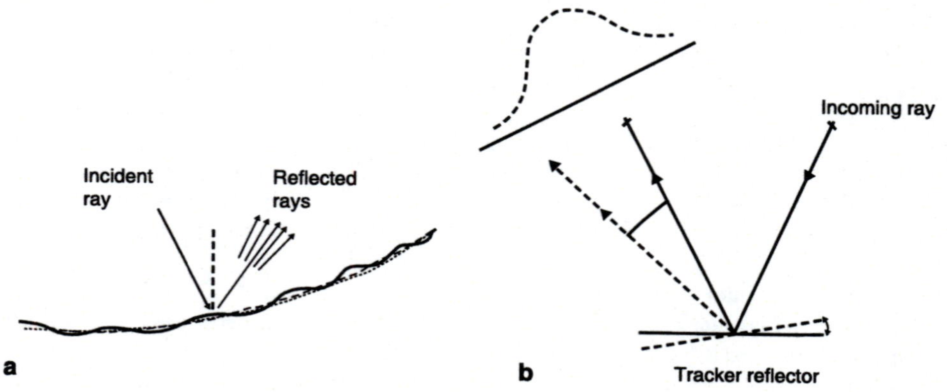

Fig. 3.40 Different types of optical and tracking errors. **a** Micro-roughness in solar concentrator surface leads to a spread in the reflected radiation. The roughness is illustrated as a dotted line for the ideal reflector surface and as a solid line for the actual surface. **b** Tracking errors lead to a spread in incoming solar radiation shown as a normal distribution. Note that a tracker error of σ_{track} results in a reflection error $\sigma_{reflec} = 2.\sigma_{track}$ from Snell's law. Factor of 2 also pertains to other sources based on the error occurring as light both enters and leaves the optical device (see Eq. 3.46)

mined statistically following the basic propagation of errors formula. Note that these errors need not be normally distributed, but such an assumption is often made in practice. Thus, rms values representing the standard deviations of these errors are used for such types of analysis.

The finite angular size of the solar disc results in incident solar rays that are not parallel but subtend an angle of about 33 min or 9.6 mrad.

(a) You will analyze the absolute and relative effects of this source of radiation spread at the receiver considering various other optical errors described above, and using the numerical values shown in Table 3.19.

$$\sigma_{totalspread} = [(\sigma_{solardisk})^2 + (2\sigma_{manuf})^2 + (2\sigma_{dustbuild})^2 + [(2\sigma_{sensor})^2 + (2\sigma_{drive})^2 + (\sigma_{rec-misalign})^2]^{1/2} \quad (3.46)$$

(b) Plot the variation of the total error as a function of the tracker drive non-uniformity error for three discrete values of dust building up (0, 1 and 2 mrad).

Table 3.19 Data table for Problem 3.15

Component	Source of error	RMS error	
		Fixed value	Variation over time
Solar disk	Finite angular size	9.6 mrad	–
Reflector	Curvature manufacture	1.0 mrad	–
	Dust buildup	–	0–2 mrad
Tracker	Sensor mis-alignment	2.0 mrad	–
	Drive non-uniformity	–	0–10 mrad
Receiver	Misalignment	2.0 mrad	–

References

Abbas, M., and J.S. Haberl,1994. "Development of indices for browsing large building energy databases", Proc. Ninth Symp. Improving Building Systems in Hot and Humid Climates, pp. 166–181, Dallas, TX, May.

ANSI/ASME,1990. "*Measurement Uncertainty: Instruments and Apparatus*", ANSI/ASME Standard PTC 19.1–1985, American Society of Mechanical Engineers, New York, NY.

ASHRAE 14, 2002. *Guideline 14–2002: Measurement of Energy and Demand Savings*, American Society of Heating, Refrigerating and Air-Conditioning Engineers, Atlanta.

ASHRAE, 2005. *Guideline 2- 2005: Engineering Analysis of Experimental Data,* American Society of Heating, Refrigerating and Air-Conditioning Engineers, Atlanta, GA.

Ayyub, B.M. and R.H. McCuen,1996. *Numerical Methods for Engineers*, Prentice-Hall, Upper Saddle River, NJ

Belsley, D.A., E. Kuh and R.E. Welsch, 1980, *Regression Diagnostics*, John Wiley & Sons, New York

Braun, J.E., S.A. Klein, J.W. Mitchell and W.A. Beckman,1989. Methodologies for optimal control of chilled water systems without storage, *ASHRAE Trans*., 95(1), American Society of Heating, Refrigerating and Air-Conditioning Engineers, Atlanta, GA.

Cleveland, W.S.,1985. *The Elements of Graphing Data*, Wadsworth and Brooks/Cole, Pacific Grove, California.

Coleman, H.W. and H.G. Steele,1999. *Experimentation and Uncertainty Analysis for Engineers*, 2nd Edition, John Wiley and Sons, New York.

Dorgan, C.E. and J.S. Elleson,1994. Design Guide for Cool Thermal Storage, American Society of Heating, Refrigerating and Air-Conditioning Engineers, Atlanta, GA.

Devore J., and N. Farnum, 2005. *Applied Statistics for Engineers and Scientists*, 2nd Ed., Thomson Brooks/Cole, Australia.

Doebelin, E.O.,1995. *Measurement Systems: Application and Design*, 4th Edition, McGraw-Hill, New York

EIA, 1999. Electric Power Annual 1999, Vol.II, October 2000, DOE/EIA-0348(99)/2, Energy Information Administration, US DOE, Washington, D.C. 20585–065 http://www.eia.doe.gov/eneaf/electricity/epav2/epav2.pdf.

Glaser, D. and S. Ubbelohde, 2001. "Visualization for time dependent building simulation", *7th IBPSA Conference*, pp. 423–429, Rio de Janeiro, Brazil, Aug. 13–15.

Haas, C. N., 2002. "On the risk of mortality to primates exposed to anthrax spores." *Risk Analysis* vol. **22**(2): pp.189–93.

Haberl, J.S. and M. Abbas,1998. Development of graphical indices for viewing building energy data: Part I and Part II, *ASME J. Solar Energy Engg.*, vol. 120, pp. 156–167

Heinsohn, R.J. and J.M. Cimbala, 2003, *Indoor Air Quality Engineering*, Marcel Dekker, New York, NY

Holman, J.P. and W.J. Gajda, 1984. *Experimental Methods for Engineers*, 5th Ed., McGraw-Hill, New York

Kreider, J.K., P.S. Curtiss and A. Rabl, 2009. *Heating and Cooling of Buildings*, 2nd Ed., CRC Press, Boca Raton, FL.

Reddy, T.A., 1990. Statistical analyses of electricity use during the hottest and coolest days of summer for groups of residences with and without air-conditioning. *Energy,* vol. **15**(1): pp. 45–61.

Schenck, H., 1969. *Theories of Engineering Experimentation*, 2nd Edition, McGraw-Hill, New York.

Tufte, E.R., 1990. *Envisioning Information*, Graphic Press, Cheshire, CN.

Tufte, E.R., 2001. *The Visual Display of Quantitative Information*, 2nd Edition, Graphic Press, Cheshire, CN

Tukey, J., 1988. *The Collected Works of John W. Tukey*, W. Cleveland (Editor), Wadsworth and Brookes/Cole Advanced Books and Software, Pacific Grove, CA

Wonnacutt, R.J. and T.H. Wonnacutt,1985. *Introductory Statistics*, 4th Ed., John Wiley & Sons, New York.

Making Statistical Inferences from Samples

4

This chapter covers various concepts and methods dealing with statistical inference, namely point estimation, interval or confidence interval estimation, hypothesis testing and significance testing. These methods are used to infer point and interval estimates about a population from sample data using knowledge of probability and probability distributions. Classical univariate and multivariate techniques as well as non-parametric and Bayesian methods are presented. Further, various types of sampling methods are also described, which is followed by a discussion on estimators and their desirable properties. Finally, resampling methods are treated which, though computer intensive, are conceptually simple, versatile, and allow robust point and interval estimation.

4.1 Introduction

The primary reason for resorting to sampling as against measuring the whole population is to reduce expense, or to make quick decisions (say, in case of a production process), or often, it is impossible to do otherwise. *Random sampling*, the most common form of sampling, involves selecting samples from the population in a random manner which should also be independent. If done correctly, it reduces or eliminates bias while enabling inferences to be made about the population from the sample. Such inferences or estimates, usually involving descriptive measures such as the mean value or the standard deviation, are called *estimators*. These are mathematical expressions to be applied to sample data in order to deduce the *estimate* of the true parameter. For example, Eqs. 3.1 and 3.7 in Chap. 3 are the estimators for deducing the mean and standard deviation of a data set. Unfortunately, certain unavoidable, or even undetected, biases may creep into the supposedly random sample, and this could lead to improper or biased inferences. This issue, as well as a more complete discussion of sampling and sampling design is covered in Sect. 4.7.

Parameter tests on population estimates assume that the sample data are random and independently drawn. It is said that, in the case of finite populations, the sampling fraction should be smaller than about 1/10th the population size. Further, the data of the random variable is assumed to be close to being normally distributed. There is an entire field of inferential statistics based on *nonparametric or distribution-free tests* which can be applied to population data with unknown probability distributions. Though nonparametric tests are unencumbered by fewer restrictive assumptions, are easier to apply and understand, they are less efficient than parametric tests (in that their uncertainty intervals are larger). These are briefly discussed in Sect. 4.5, while Bayesian statistics, whereby one uses prior information to enhance the inference-making process, is addressed in Sect. 4.6.

4.2 Basic Univariate Inferential Statistics

4.2.1 Sampling Distribution and Confidence Limits of the Mean

(a) Sampling distribution of the mean Consider a population from which many random samples are taken. What can one say about the distribution of the sample estimators? Let μ and $\bar{x}$ be the population mean and sample mean respectively, and σ and s_x be the population standard deviation and sample standard deviation respectively. Then, regardless of the shape of the population frequency distribution:

$$\mu = \bar{x} \tag{4.1}$$

and the standard deviation of the population mean (also referred to as SE or *standard error of the mean*)

$$\sigma = \frac{s_x}{(n)^{1/2}} \tag{4.2}$$

where s_x is given by Eq. 3.7 and n is the number of samples selected or picked.

T. Agami Reddy, *Applied Data Analysis and Modeling for Energy Engineers and Scientists*,
DOI 10.1007/978-1-4419-9613-8_4, © Springer Science+Business Media, LLC 2011

In case the population sample is small and sampling is done without replacement, then the above standard deviation has to be modified to

$$\sigma = \frac{s_x}{(n)^{1/2}} \left(\frac{N-n}{N-1} \right)^{1/2} \tag{4.3}$$

where N is the population size. Note that if N>>n, one effectively gets back Eq. 4.2.

The sampling distribution of the mean provides an indication of the confidence, or the degree of certainty, one can place about the accuracy involved in using the sample mean to estimate the population mean. This confidence is interpreted as a probability, and is given by the very important law stated below.

The *Central Limit Theorem* (one of the most important theorems in probability) states that if a random sample of n observations is selected from a *population with any distribution*, then the sampling distribution of $\bar{x}$ will be approximately a Gaussian distribution when n is sufficiently large (n > 30). The larger the sample n, the closer does the sampling distribution approximate the Gaussian (Fig. 4.1)[1]. A consequence of the theorem is that it leads to a simple method of computing approximate probabilities of sums of independent random variables. It explains the remarkable fact that the empirical frequencies of so many natural "populations" exhibit bell-shaped (i.e., a normal) curves. Let x_1, x_2,…x_n be a sequence of independent identically distributed random variables with mean μ and variance σ^2. Then the distribution of the random variable z (Sect. 2.4.3)

$$z = \frac{\bar{x} - \mu}{\sigma/\sqrt{n}} \tag{4.4}$$

tends to be *standard normal* as n tends towards infinity. Note that this theorem is valid for any distribution of x; herein lies its power.

Probabilities for random quantities can be found by determining areas under the standard normal curve as described in Sect. 2.4.3. Suppose one takes a random sample of size n from a population of mean μ and standard deviation σ. Then the random variable z has (i) *approximately* the standard normal distribution if n>30 regardless of the distribution of the population, and (ii) *exactly* the standard normal distribution if the population itself is normally distributed regardless of the sample size (Fig. 4.1).

Note that when sample sizes are small (n<30) and the underlying distribution is unknown, the t-student distribution which has wider uncertainty bands (Sect. 2.4.3), should be used with (n−1) degrees of freedom instead of the Gaussian (Fig. 2.15 and Table A4). Unlike the z-curve, there are several t-curves depending on the degrees of freedom (d.f.). At the limit of infinite d.f.s, the t-curve collapses into the z-curve.

(b) Confidence limits for the mean In the sub-section above, the behavior of many samples, all taken from one population, was considered. Here, *only one large random sample* from a population is selected, and analyzed so as to make an educated guess on properties (or estimators) of the population such as its mean and standard deviation. This process is called *inductive reasoning* or arguing backwards from a set of observations to a reasonable hypothesis. However, the benefit provided by having to select only a sample of the population comes at a price: one has to accept some uncertainty in our estimates. Based on a sample taken from a population:

(a) one can deduce interval bounds of the population mean at a specified confidence level (this aspect is covered in this sub-section), and
(b) one can test whether the sample mean differs from the presumed population mean (this is covered in the next sub-section).

The concept of confidence intervals (CL) was introduced in Sect. 3.6.3 in reference to instrument errors. This concept pertinent to random variables in general is equally applicable to sampling. A 95% CL is commonly interpreted as implying that there is a 95% probability that the actual population estimate will lie within this confidence interval[2]. The range is obtained from the z-curve by finding the value at which the area under the curve (i.e., the probability) is equal to 0.95. From Table A3, the corresponding critical value $z_{c/2}$ is 1.96 (note that the critical value for a two-tailed confidence level, as in this case, is determined as that value of z in Table A3 which corresponds to a probability value of [(1−0.95)/2]=0.025). This implies that the probability is:

$$p\left(-1.96 < \frac{\bar{x} - \mu}{s_x/\sqrt{n}} < 1.96\right) \approx 0.95$$
$$\text{or} \quad \bar{x} - 1.96\frac{s_x}{\sqrt{n}} < \mu < \bar{x} + 1.96\frac{s_x}{\sqrt{n}} \tag{4.5a}$$

Thus the *confidence interval* of

$$\mu = \bar{x} \pm z_{c/2}\frac{s_x}{\sqrt{n}}. \tag{4.5b}$$

This formula is valid for any shape of the population distribution provided, of course, that the sample is large (say, n>30).

[1] That the sum of two Gaussian distributions from a population would be another Gaussian variable (a property called invariant under addition) is intuitive. Why the sum of two non-Gaussian distributions should gradually converge to a Gaussian is less so, and hence the importance of this theorem.

[2] It will be pointed out in Sect. 4.6.2 that this statement can be debated, but this is a common interpretation and somewhat simpler to comprehend than the more accurate one.

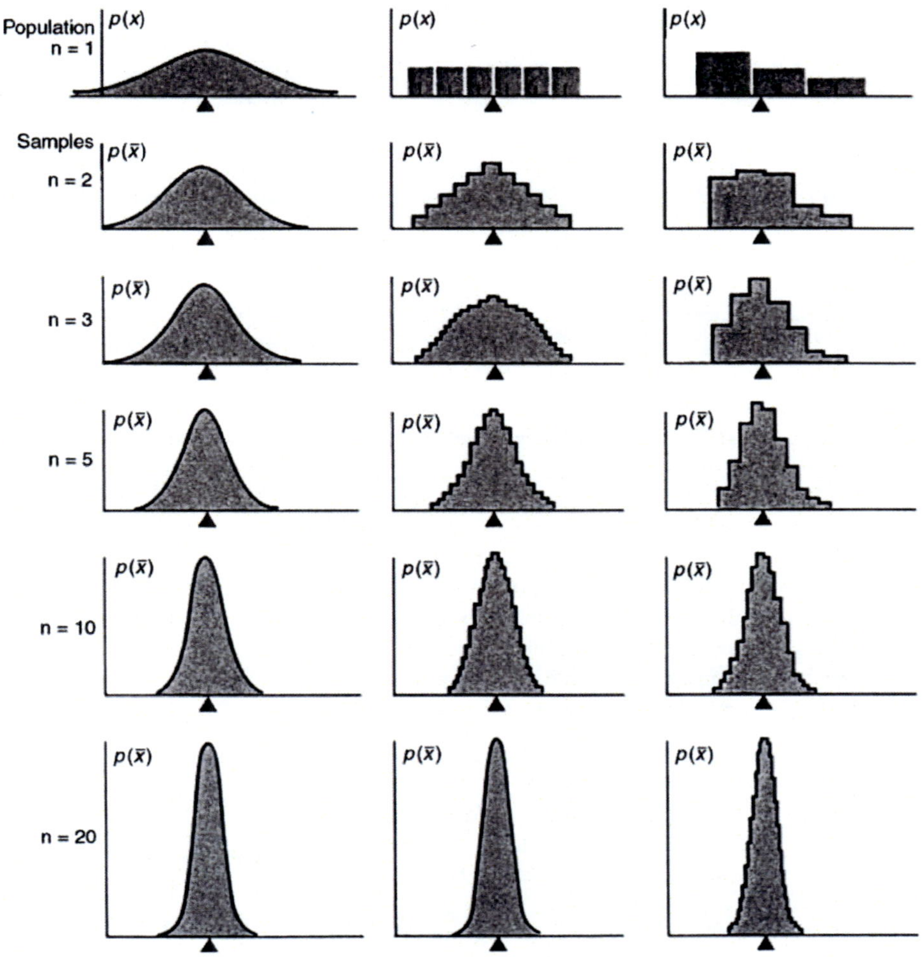

Fig. 4.1 Illustration of the important law of strong numbers. The sampling distribution of $\bar{X}$ contrasted with the parent population distribution for three cases. The first case (*left column of figures*) shows sampling from a normal population. As sample size n increases, the standard error of $\bar{X}$ decreases. The next two cases show that even though the populations are not normal, the sampling distribution still becomes approximately normal as n increases. (From Wonnacutt and Wonnacutt (1985) by permission of John Wiley and Sons)

The half-width of the 95% CL is $(1.96\frac{s_x}{\sqrt{n}})$ and is called the *bound of the error of estimation*. For small samples, instead of random variable z, one uses the student-t variable.

Note that Eq. 4.5 refers to the long-run bounds, i.e., in the long run roughly 95% of the intervals will contain μ. If one is interested in predicting a single x value that has yet to be observed, one uses the following equation (Devore and Farnum 2005):

$$\textit{Prediction interval of } x = \bar{x} \pm t_{c/2} \cdot s_x \left(1 + \frac{1}{n}\right)^{1/2} \tag{4.6}$$

where $t_{c/2}$ is the two-tailed critical value determined from the t-distribution at d.f.=n−1 at the desired confidence level.

It is clear that the prediction intervals are much wider than the confidence intervals because the quantity "1" within the brackets of Eq. 4.6 will generally dominate (1/n). This means that there is a lot more uncertainty in predicting the value of a single observation x than there is in estimating a mean value μ.

Example 4.2.1: *Evaluating manufacturer-quoted lifetime of light bulbs from sample data*
A manufacturer of zenon light bulbs for street lighting claims that the distribution of the lifetimes of his best model has a mean μ=16 years and a standard deviation $s_x=2$ years when the bulbs are lit for 12 h every day. Suppose that a city official wants to check the claim by purchasing a sample of 36 of these bulbs and subjecting them to tests that determine their lifetimes.

(i) Assuming the manufacturer's claim to be true, describe the sampling distribution of the mean lifetime of a sample of 36 bulbs. Even though the shape of the distribution is unknown, the Central Limit Theorem suggests that the normal distribution can be used. Thus

$$\mu=\bar{x} = 16 \text{ and } \sigma = \frac{2}{\sqrt{36}} = 0.33 \text{ years.}$$

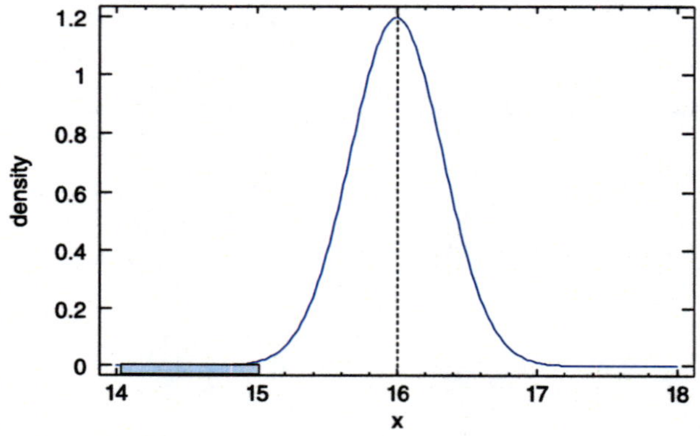

Fig. 4.2 Sampling distribution of $\bar{X}$ for a normal distribution N(16, 0.33). Shaded area represents the probability of the mean life of the bulb being <15 years (Example 4.2.1)

(ii) What is the probability that the sample purchased by the city officials has a mean-lifetime of 15 years or less?
The normal distribution N(16, 0.33) is drawn and the darker shaded area to the left of x=15 as shown in Fig. 4.2 provides the probability of the city official observing a mean life of 15 years or less ($\bar{x} \leq 15$). Next, the standard normal statistic is computed as: $z = \frac{\bar{x} - \mu}{\sigma/\sqrt{n}} = \frac{15 - 16}{2/\sqrt{36}} = -3.0$. This probability or p-value can be read off from Table A3 as $p(\bar{z} \leq -3.0)=0.0013$. Consequently, the probability that the consumer group will observe a sample mean of 15 or less is only 0.13%.

(iii) If the manufacturer's claim is correct, compute the 95% prediction interval of a single bulb from the sample of 36 bulbs. From the t-tables (Table A4), the critical value is $t_c=1.691 \cong 1.7$ for d.f. = 36 − 1 = 35, and CL = 95% corresponding to the one-tailed distribution. Thus, 95% prediction interval of $x=16 \pm (1.70).2.\left(1 + \frac{1}{36}\right)^{1/2} =$ 12.6 to 19.4 years. ■

The above example is one type of problem which can be addressed by one-sample statistical tests. However, the classical hypothesis testing approach is slightly different, and is addressed next.

4.2.2 Hypothesis Test for Single Sample Mean

The previous sub-sections dealt with estimating confidence intervals of certain estimators of the underlying population from a single drawn sample. During hypothesis testing, on the other hand, the intent is to decide which of two competing claims is true. For example, one wishes to support the hypothesis that women live longer than men. Samples from each of the two populations are taken, and a test, called *statistical inference* is performed to prove (or disprove) this claim. Since there is bound to be some uncertainty associated with such a procedure, one can only be confident of the results to a degree that can be stated as a probability. If this probability value is higher than a pre-selected threshold probability, called *significance level of the test*, then one would conclude that women do live longer than men; otherwise, one would have to accept that the test was non-conclusive.

Thus, a test of hypotheses is performed based on information deduced from the sample data involving its mean and its probability distribution, which is assumed to be close to a normal distribution. Once this is gathered, the following steps are performed:

(i) formulate the hypotheses: the null or status quo, and the alternate (which are complementary)
(ii) identify a test statistic that will be used to assess the evidence against the null hypothesis
(iii) determine the probability (or p-value) that the null hypothesis can be true
(iv) compare this value with a threshold probability corresponding to a pre-selected significance level α (say, 0.01 or 0.05)
(v) rule out the null hypothesis only if p-value $\leq \alpha$, and accept the alternate hypothesis.

This procedure can be applied to two sample tests as well, and is addressed in the subsequent sub-sections. The following example illustrates this procedure for *single sample means* where one would like to prove or disprove sample behavior from a previously held notion about the underlying population.

Example 4.2.2: *Evaluating whether a new lamp bulb has longer burning life than traditional ones*
The traditional process of light bulbs manufacture results in bulbs with a mean life of $\mu=1200$ h and a standard deviation $\sigma=300$ h. A new process of manufacture is developed and whether this is superior is to be determined. Such a problem involves using the classical test whereby one proceeds by defining two hypotheses:

(a) The *null hypothesis* which represents the status quo, i.e., that the new process is no better than the previous one (unless the data provides convincing evidence to the contrary). In our example, the null hypothesis is $H_0: \mu=1200$ h,
(b) The *research or alternative hypothesis* (H_a) is the premise that $\mu \neq 1200$ h.

Assume a sample size of n=100 of bulbs manufactured by the new process, and set the significance or error level of the test to be $\alpha = 0.05$ assuming a one-tailed test (since the new bulb manufacturing process should have *a longer life,* not just different from that of the traditional process). The mean life $\bar{x}$ of the sample of 100 bulbs can be assumed to be normally distributed with mean 1200 and standard deviation $\sigma/\sqrt{n} = 300/\sqrt{100} = 30$. From the standard normal table (Table A3), the critical z-value is: $z_{\alpha=0.05} = 1.64$. Recalling that the critical value is defined as: $z_c = \dfrac{\bar{x}_c - \mu_0}{\sigma/\sqrt{n}}$, leads to $\bar{x}_c = 1200 + 1.64 \times 300/(100)^{1/2} = 1249$ or about 1250.

Suppose testing of the 100 tubes yields a value of $\bar{x} = 1260$. As $\bar{x} > \bar{x}_c$, one would reject the null hypothesis at the 0.05 significance (or error) level. This is akin to jury trials where the null hypothesis is taken to be that the accused is innocent, and *the burden of proof during hypothesis testing is on the alternate hypothesis*, i.e., on the prosecutor to show overwhelming evidence of the culpability of the accused. If such overwhelming evidence is absent, the null hypothesis is preferentially favored. ■

There is another way of looking at this testing procedure (Devore and Farnum 2005):

(a) H_0 is true, but one has been exceedingly unlucky and got a very improbable sample with mean $\bar{x}$. In other words, the observed difference turned out to be significant when, in fact, there is no real difference. Thus, the null hypothesis has been rejected erroneously. The innocent man has been falsely convicted;

(b) H_0 is not true after all. Thus, it is no surprise that the observed $\bar{x}$ value was so high, or that the accused is indeed culpable.

The second explanation is likely to be more plausible, but there is always some doubt because statistical decisions inherently contain probabilistic elements. In other words, statistical tests of hypothesis do not always yield conclusions with absolute certainty: they have in-built margins of error just like jury trials are known to hand down wrong verdicts. Hence, two types of errors can be distinguished:

(i) Concluding that the null hypothesis is false, when in fact it is true, is called a *Type I* error, and represents the probability α (i.e., the pre-selected significance level) of erroneously rejecting the null hypothesis. This is also called the "*false negative*" or "false alarm" rate. The upper normal distribution shown in Fig. 4.3 has a mean value of 1200 (equal to the population or claimed mean value) with a standard deviation of 30. The area to the right of the critical value of 1250 represents the probability of Type I error occurring.

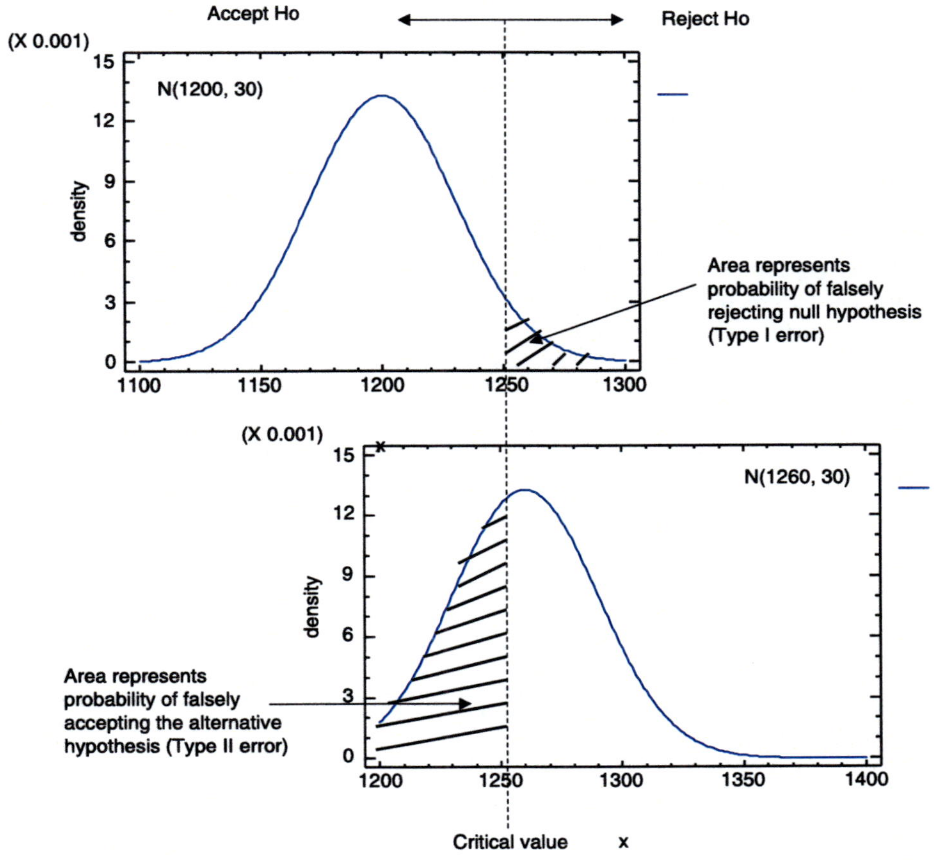

Fig. 4.3 The two kinds of error that occur in a classical test. **a** If H_0 is true, then significance level α =probability of erring (rejecting the true hypothesis H_0). **b** If H_a is true, then β =probability of erring (judging that the false hypothesis H_0 is acceptable). The numerical values correspond to data from Example 4.2.2

(ii) The flip side, i.e. concluding that the null hypothesis is true, when in fact it is false, is called a *Type II* error and represents the probability β of erroneously accepting the alternate hypothesis, also called the "*false positive*" rate. The lower plot of the normal distribution shown in Fig. 4.3 now has a mean of 1260 (the mean value of the sample) with a standard deviation of 30, while the area to the left of the critical value $\bar{x}_c$ indicates the probability β of being in error of Type II.

The two types of error are inversely related as is clear from the vertical line in Fig. 4.3 drawn through both figures. A decrease in probability of one type of error is likely to result in an increase in the probability of the other. Unfortunately, one cannot simultaneously reduce both by selecting a smaller value of α. The analyst would select the significance level depending on the tolerance, or seriousness of the consequences of either type of error specific to the circumstance. Recall that the probability of making a *type I error is called the significance level of the test*. This probability of correctly rejecting the null hypothesis is also referred to as the *statistical power*. The only way of reducing both types of errors is to increase the sample size with the expectation that the standard deviation would decrease and the sample mean would get closer to the population mean.

An important concept needs to be clarified, namely when does one *use one-tailed* as against *two-tailed* tests. In the two-tailed test, one is testing whether the sample is *different* (i.e., smaller or larger) than the stipulated population. In cases where one wishes to test whether the sample is specifically *larger* (or specifically smaller) than the stipulated population, then the one tailed test is used (as in Examples 4.2.1 and 4.2.2). The tests are set up and addressed in like manner, the difference being in how the p-level is finally determined. The shaded areas of the normal distributions shown in Fig. 4.4 illustrate the difference in both types of tests assuming a significance level corresponding to p=0.05 for the two-tailed test and half the probability value (or p=0.025) for the one-tailed test.

One final issue relates to the selection of the test statistic. One needs to distinguish between the following two instances:

(i) if the population variance σ is known and for sample sizes $n>30$, then the z statistic is selected for performing the test along with the standard normal tables (as done for Example 4.2.2 above);

(ii) if the population variance is unknown or if the sample size $n<30$, then the t-statistic is selected (using the sample standard deviation s instead of σ) for performing the test using Student-t tables with the appropriate degree of freedom.

4.2.3 Two Independent Sample and Paired Difference Tests on Means

As opposed to hypothesis tests for a single population mean, there are hypothesis tests that allow one to compare values of two population means from samples taken from each population. Two basic presumptions for the tests (described below) to be valid are that the standard deviations of the populations are reasonably close, and that the populations are approximately normally distributed.

(a) Two independent sample test The test is based on the information (namely, the mean and the standard deviation) obtained from taking two *independent* random samples from the two populations under consideration whose variances are unknown and unequal (but reasonably close). Using the same notation as before for population and sample and using subscripts 1 and 2 to denote the two samples, the random variable

$$z = \frac{(\bar{x}_1 - \bar{x}_2) - (\mu_1 - \mu_2)}{\left(\frac{s_1^2}{n_1} + \frac{s_2^2}{n_2}\right)^{1/2}} \tag{4.7}$$

is said to approximate the standard normal distribution for large samples ($n_1>30$ and $n_2>30$) where s_1 and s_2 are the standard deviations of the two samples. The denominator is called the *standard error* (SE) and is a measure of the total variability of both samples combined (remember that variances of quantities which are independent add in quadrature).

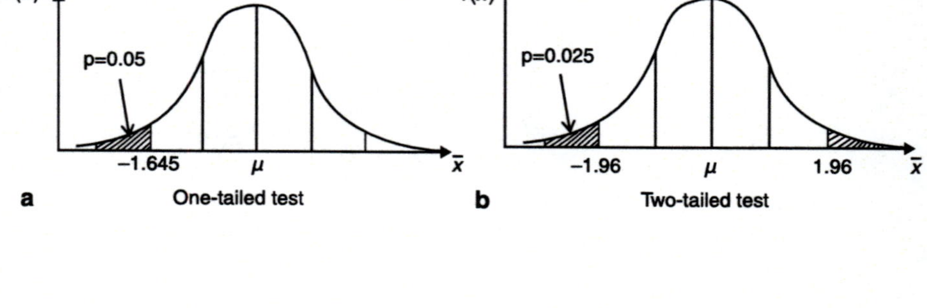

Fig. 4.4 Illustration of critical cutoff values between one tailed and two-tailed tests assuming the normal distribution. The *shaded areas* represent the probability values corresponding to 95% CL or 0.05 significance level or p=0.05. The critical values shown can be determined from Table A3

The confidence intervals of the difference in the population means can be determined as:

$$\mu_1 - \mu_2 = (\bar{x}_1 - \bar{x}_2) \pm z_c \cdot SE(\bar{x}_1, \bar{x}_2)$$
$$\text{where } SE(\bar{x}_1, \bar{x}_2) = \left(\frac{s_1^2}{n_1} + \frac{s_2^2}{n_2}\right)^{1/2} \tag{4.8}$$

where z_c is the critical value at the selected significance level. Thus, the testing of the two samples involves a single random variable combining the properties of both.

For smaller sample sizes, Eq. 4.8 still applies, but the z standardized variable is replaced with the student-t variable. The critical values are found from the student t-tables with degrees of freedom d.f. $= n_1 + n_2 - 2$. If the variances of the population are known, then these should be used instead of the sample variances.

Some textbooks suggest the use of *"pooled variances"* when the samples are small and the variances of both populations are close. Here, instead of using individual standard deviation values s_1 and s_2, a new quantity called the pooled variance s_p is used:

$$s_p^2 = \frac{(n_1 - 1)s_1^2 + (n_2 - 1)s_2^2}{n_1 + n_2 - 2} \quad \text{with d.f.} = n_1 + n_2 - 2 \tag{4.9}$$

Note that the pooled variance is simply the weighted average of the two sample variances. The use of the *pooled variance approach is said to result in tighter confidence intervals,* and hence its appeal. The random variable approximates the t-distribution, and the confidence intervals of the difference in the population means are:

$$\mu_1 - \mu_2 = (\bar{x}_1 - \bar{x}_2) \pm t_c \cdot SE(\bar{x}_1, \bar{x}_2)$$
$$\text{where} \quad SE(\bar{x}_1, \bar{x}_2) = \left[s_p^2\left(\frac{1}{n_1} + \frac{1}{n_2}\right)\right]^{1/2} \tag{4.10}$$

Devore and Farnum (2005) strongly discourage the use of the pooled variance approach as a general rule, and so the better approach, when in doubt, is to use Eq. 4.8 so as to be conservative.

Figure 4.5 illustrates, in a simple conceptual manner, the four characteristic cases which can arise when comparing the means of two populations based on sampled data. Recall that the box and whisker plot is a type of graphical display of the shape of the distribution where the solid line denotes the median, the upper and lower hinges of the box indicate the interquartile range values (25th and 75th percentiles) with the whiskers extending to 1.5 times this range. Case (a) corresponds to the case where the two whisker bands do not overlap, and one could state with confidence that the two population means are very likely to be different at the 95% confidence level. Case (b) also suggests difference between population means, but will a little less certitude. Case (d) illustrates the case where the two whisker bands are practically identical, and so the population means are very likely to be statistically similar. It is when cases as illustrated in frames (b) and (c) occur that the value of statistical tests becomes apparent. As a rough thumb rule, if the 25th percentile for one sample exceeds the median line of the other sample, one could conclude that the mean are likely to be different (Walpole et al. 2007).

Manly (2005) states that the independent random sample test is fairly robust to the assumptions of normality and equal population variance especially when the sample size exceeds 20 or so. The assumption of equal population variances is said not to be an issue if the ratio of the two variances is within 0.4 to 2.5.

Example 4.2.3: *Verifying savings from energy conservation measures in homes*

Certain electric utilities with limited generation capacities fund contractors to weather strip residences in an effort to

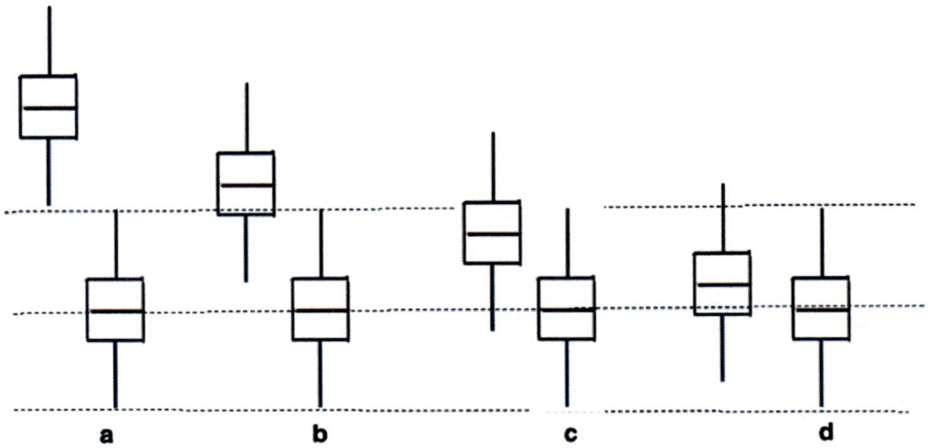

Fig. 4.5 Conceptual illustration of four characteristic cases that may arise during two-sample testing of medians. The box and whisker plots provide some indication as to the variability in the results of the tests. Case (**a**) clearly indicates that the samples are very much different, while the opposite applies to case (**d**). However, it is more difficult to draw conclusions from cases (**b**) and (**c**), and it is in such cases that statistical tests are useful

reduce infiltration losses which lower electricity needs[3]. Suppose an electric utility wishes to determine the cost-effectiveness of their weather-stripping program by comparing the annual electric energy use of 200 similar residences in a given community, half of which were weather-stripped, and the other half were not. Samples collected from both types of residences yield:

Control sample: $\bar{x}_1 = 18{,}750$; $s_1 = 3{,}200$ and $n_1 = 100$.
Weather-stripped sample: $\bar{x}_2 = 15{,}150$; $s_2 = 2{,}700$ and $n_2 = 100$.

The mean difference $(\bar{x}_1 - \bar{x}_2) = 18{,}750 - 15{,}150 = 3{,}600$, i.e., the mean saving in each weather-stripped residence is 19.2% (=3,600/18,750) of the mean baseline or control home. However, there is an uncertainty associated with this mean value since only a sample has been analyzed. This uncertainty is characterized as a bounded range for the mean difference. At the 95% CL, corresponding to a significance level $\alpha = 0.05$ for a one-tailed distribution, $z_c = 1.645$ from Table A3, and from Eq. 4.8:

$$\mu_1 - \mu_2 = (18{,}750 - 15{,}150) \pm 1.645\left(\frac{s_1^2}{100} + \frac{s_2^2}{100}\right)^{1/2}$$

To complete the calculation of the confidence interval, it is assumed, given that the sample sizes are large, that the sample variances are reasonably close to the population variances. Thus, our confidence interval is approximately: $3{,}600 \pm 1.645\left(\frac{3{,}200^2}{100} + \frac{2{,}700^2}{100}\right)^{1/2} = 3{,}600 \pm 689 = (2{,}911$ and $4{,}289)$. These intervals represent the lower and upper values of saved energy at the 95% CL. To conclude, one can state that the savings are positive, i.e., one can be 95% confident that there is an energy benefit in weather-striping the homes. More specifically, the mean saving is 19.2% of the baseline value with an uncertainty of 19.1% (=689/3,600) in the savings at the 95% CL. Thus, the uncertainty in the savings estimate is as large as the estimate itself which casts doubt on the efficacy of the conservation program. Increasing the sample size or resorting to stratified sampling are obvious options and are discussed in Sect. 4.7. Another option is to adopt a less stringent confidence level; 90% CL is commonly adopted. This example reflects a realistic concern in that energy savings in homes from energy conservation measures are often difficult to verify accurately. ■

(b) Paired difference test The previous section dealt with independent samples from two populations with close to normal probability distributions. There are instances when the samples are somewhat correlated, and such *interdependent samples* are called paired samples. This interdependence can also arise when the samples are taken at the same time, and are affected by a time-varying variable which is not explicitly considered in the analysis. Rather than the individual values, the difference is taken as the only random sample since it is likely to exhibit much less variation than those of the two samples. Thus, the confidence intervals calculated from paired data will be narrower than those calculated from two independent samples. Let d_i be the difference between individual readings of two small paired samples ($n<30$), and $\bar{d}$ their mean value. Then, the t-statistic is taken to be:

$$t = \bar{d}/SE \quad \text{where} \quad SE = \left(s_d/\sqrt{n}\right) \tag{4.11a}$$

and the confidence interval around $\bar{d}$ is:

$$\mu_d = \bar{d} \pm t_c\left(s_d/\sqrt{n}\right) \tag{4.11b}$$

Hypothesis testing of means for paired samples is done the same way as that for a single independent mean, and is usually (but not always) superior to an independent sample test. Paired difference tests are used for comparing "before and after" or "with and without" type of experiments done on the same group in turn, say, to assess effect of an action performed. For example, the effect of an additive in gasoline meant to improve gas mileage can be evaluated statistically by considering a set of data representing the difference in the gas mileage of n cars which have each been subjected to tests involving "no additive" and "with additive". Its usefulness is illustrated by the following example which is another type of application for which paired difference tests can be used.

Example 4.2.4: *Comparing energy use of two similar buildings based on utility bills—the wrong way*
Buildings which are designed according to certain performance standards are eligible for recognition as energy-efficient buildings by federal and certification agencies. A recently completed building (B2) was awarded such an honor. The federal inspector, however, denied the request of another owner of an identical building (B1) close by who claimed that the differences in energy use between both buildings were within statistical error. An energy consultant was hired by the owner to prove that B1 is as energy efficient as B2. He chose to compare the monthly mean utility bills over a year between the two commercial buildings based on the data recorded over the same 12 months and listed in Table 4.1. This problem can be addressed using the two sample test method described earlier.

The null hypothesis is that the mean monthly utility charges μ_1 and μ_2 for the two buildings are equal against the alternative hypothesis that they differ. Since the sample sizes are less than 30, the t-statistic has to be used instead of the standard normal z statistic. The pooled variance approach given by Eq. 4.9 is appropriate in this instance. It is computed as:

[3] This is considered more cost effective to utilities in terms of deferred capacity expansion costs than the resulting revenue loss in electricity sales due to such conservation measures.

Table 4.1 Monthly utility bills and the corresponding outdoor temperature for the two buildings being compared-Example 4.2.4

Month	Building B1 Utility cost ($)	Building B2 Utility cost ($)	Difference in Costs (B1−B2)	Outdoor temperature (°C)
1	693	639	54	3.5
2	759	678	81	4.7
3	1005	918	87	9.2
4	1074	999	75	10.4
5	1449	1302	147	17.3
6	1932	1827	105	26
7	2106	2049	57	29.2
8	2073	1971	102	28.6
9	1905	1782	123	25.5
10	1338	1281	57	15.2
11	981	933	48	8.7
12	873	825	48	6.8
Mean	*1,349*	*1,267*	*82*	
Std. Deviation	*530.07*	*516.03*	*32.00*	

$$s_p^2 = \frac{(12-1)\cdot(530.07)^2 + (12-1)\cdot(516.03)^2}{12+12-2}$$
$$= 273{,}630.6$$

while the t-statistic can be deduced from Eq. 4.10 and is given by

$$t = \frac{(1349-1267)-0}{\left[(273,630.6)\left(\frac{1}{12}+\frac{1}{12}\right)\right]^{1/2}}$$
$$= \frac{82}{213.54} = 0.38$$
$$\text{for d.f.} = 12+12-2 = 22$$

The t-value is very small, and will not lead to the rejection of the null hypothesis even at significance level $\alpha=0.02$ (from Table A4, the one-tailed critical value is 1.321 for CL=90% and d.f.=22). Thus, the consultant would report that insufficient statistical evidence exists to state that the two buildings are different in their energy consumption.

Example 4.2.5: ***Comparing energy use of two similar buildings based on utility bills—the right way***

There is, however, a problem with the way the energy consultant performed the test. Close observation of the data as plotted in Fig. 4.6 would lead one not only to suspect that this conclusion is erroneous, but also to observe that the utility bills of the two buildings tend to rise and fall together because of seasonal variations in the outdoor temperature. *Hence the condition that the two samples are independent is violated.* It is in such circumstances that a paired test is relevant. Here, the test is meant to determine whether the monthly mean of the *differences* in utility charges between both buildings ($\bar{x}_D$) is zero or not. The null hypothesis is that this is zero, while the alternate hypothesis is that it is different from zero. Thus:

$$\text{t-statistic} = \frac{\bar{x}_D - 0}{s_D/\sqrt{n_D}} = \frac{82}{32/\sqrt{12}} = 8.88$$
$$\text{with d.f.} = 12-1 = 11$$

where the values of 82 and 32 are found from Table 4.1.

For a significance level of 0.05 and using a one-tailed test, Table A4 suggests a critical value $t_{0.05}=1.796$. Because 8.88 is much higher than this critical value, one can safely reject the null hypothesis. In fact, Bldg 1 is less energy efficient than Bldg 2 even at a significance level of 0.0005 (or CL=99.95%), and the owner of B1 does not have a valid case at all! This illustrates how misleading results can be obtained

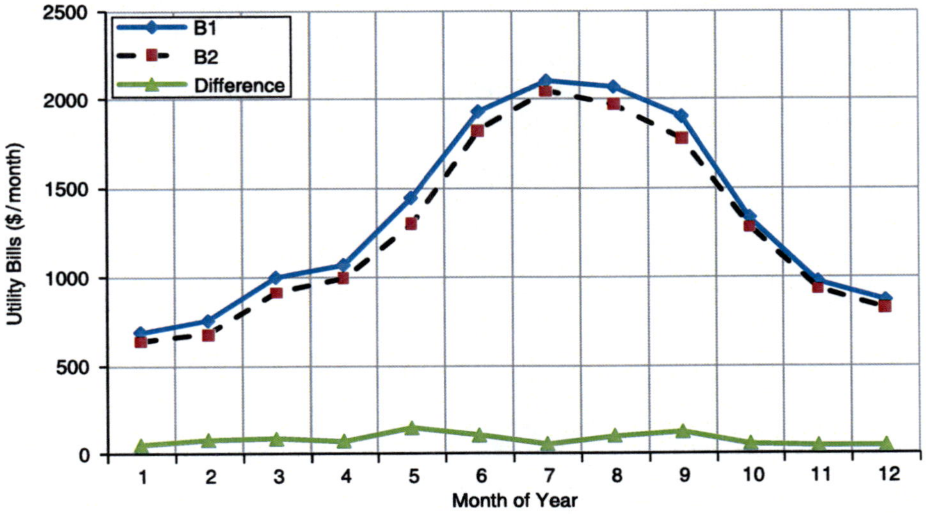

Fig. 4.6 Illustrating variation of the utility bills for the two buildings B1 and B2 (Example 4.2.5)

if inferential tests are misused, or if the analyst ignores the underlying assumptions behind a particular test.

4.2.4 Single and Two Sample Tests for Proportions

There are several cases where surveys are performed in order to determine fractions or proportions of populations who either have preferences of some sort or have a certain type of equipment. For example, the gas company may wish to determine what fraction of their customer base has gas heating as against oil heat or electric heat pumps. The company performs a survey on a random sample from which it would like to extrapolate and ascertain confidence limits on this fraction. It is in such cases which can be interpreted as either a "success" (the customer has gas heat) or a "failure"—in short, a *binomial experiment* (see Sect. 2.4.2b)—that the following test is useful.

(a) Single sample test Let p be the population proportion one wishes to estimate from the sample proportion $\hat{p}$ which can be determined as: $\hat{p} = \dfrac{\text{number of successes in sample}}{\text{total number of trials}} = \dfrac{x}{n}$.
Then, provided the sample is large ($n \geq 30$), proportion $\hat{p}$ is an unbiased estimator of p with approximately normal distribution. Dividing the expression for standard deviation of the Bernoulli trials (Eq. 2.33b) by "n²", yields the standard deviation of the sampling distribution of $\hat{p}$:

$$\left[\hat{p}\left(1-\hat{p}\right)/n\right]^{1/2} \tag{4.12}$$

Thus, the large sample confidence interval for $\hat{p}$ for the two tailed case at a significance level z is given by:

$$\hat{p} \pm z_{\alpha/2}\left[\hat{p}\left(1-\hat{p}\right)/n\right]^{1/2} \tag{4.13}$$

Example 4.2.6: In a random sample of n = 1000 new residences in Scottsdale, AZ, it was found that 630 had swimming pools. Find the 95% confidence interval for the fraction of buildings which have pools.

In this case, n = 1000, while $\hat{p} = \dfrac{630}{1000} = 0.63$. From Table A3, the one-tailed critical value $z_{0.025} = 1.96$, and hence from Eq. 4.13, the two tailed 95% confidence interval for p is:

$$0.63 - 1.96\left[\frac{0.63(1-0.63)}{100}\right]^{1/2} < p <$$

$$0.63 + 1.96\left[\frac{0.63(1-0.63)}{100}\right]^{1/2} \quad \text{or } 0.5354 < p < 0.7246.$$

■

Example 4.2.7: The same equations can also be used to determine sample size in order for p not to exceed a certain range or error e. For instance, one would like to determine from Example 4.6 data, the sample size which will yield an estimate of p within 0.02 or less at 95% CL

Then, recasting Eq. 4.13 results in a sample size:

$$\begin{aligned} n &= \frac{z^2{}_{\alpha/2}\hat{p}(1-\hat{p})}{e^2} \\ &= \frac{(1.96^2)(0.63)(1-0.63)}{(0.02)^2} \\ &= 2239 \end{aligned}$$

■

It must be pointed out that the above example is somewhat misleading since one does not know the value of $\hat{p}$ beforehand. One may have a preliminary idea, in which case, the sample size n would be an approximate estimate and this may have to be revised once some data is collected.

(b) Two sample tests The intent here is to estimate whether statistically significant differences exist between proportions of two populations based on one sample drawn from each population. Assume that the two samples are large and independent. Let $\hat{p}_1$ and $\hat{p}_2$ be the sampling proportions. Then, the sampling distribution of $(\hat{p}_1 - \hat{p}_2)$ is approximately normal with $(\hat{p}_1 - \hat{p}_2)$ being an unbiased estimator of $(p_1 - p_2)$ and the standard deviation given by:

$$\left[\frac{\hat{p}_1(1-\hat{p}_1)}{n_1} + \frac{\hat{p}_2(1-\hat{p}_2)}{n_2}\right]^{1/2} \tag{4.14}$$

The following example illustrates the procedure.

Example 4.2.8: *Hypothesis testing of increased incidence of lung ailments due to radon in homes*
The Environmental Protection Agency (EPA) would like to determine whether the fraction of residents with health problems living in an area known to have high radon concentrations is statistically different from one where levels of radon are negligible. Specifically, it wishes to test the hypothesis at the 95% CL that the fraction of residents with lung ailments in radon prone areas is higher than one with low radon levels. The following data is collected:

High radon level area: $n_1 = 100,\ \hat{p}_1 = 0.38$

Low radon area: $n_2 = 225,\ \hat{p}_2 = 0.22$

Then

null hypothesis $H_0 : (p_1 - p_2) = 0$
alternative hypothesis $H_1 : (p_1 - p_2) \neq 0$

One calculates the random variable

$$z = \frac{(\hat{p}_1 - \hat{p}_2)}{\left[\frac{\hat{p}_1(1-\hat{p}_1)}{n_1} + \frac{\hat{p}_2(1-\hat{p}_2)}{n_2}\right]^{1/2}} = \frac{(0.38 - 0.22)}{\left[\frac{(0.38)(0.62)}{100} + \frac{(0.22)(0.78)}{225}\right]^{1/2}} = 2.865$$

A one-tailed test is appropriate, and from Table A3 the critical value of $z_{0.05} = 1.65$ for the 95% CL. Since the calculated z value > z_c, this would suggest that the null hypothesis can be rejected. Thus, one would conclude that those living in areas of high radon levels have statistically higher lung ailments than those who do not. Further inspection of Table A3 reveals that $z_c = 2.865$ corresponds to a probability value of 0.021 or close to 98% CL. Should the EPA require mandatory testing of all homes at some expense to all homeowners or should some other policy measure be adopted? These types of considerations fall under the purview of decision making discussed in Chap. 12. ■

4.2.5 Single and Two Sample Tests of Variance

Recall that when a sample mean is used to provide an estimate of the population mean μ, it is more informative to give a confidence interval for μ instead of simply stating the value $\bar{x}$. A similar approach can be adopted for estimating the population variance from that of a sample.

(a) Single sample test The confidence intervals for a population variance σ^2 based on sample variance s^2 are to be determined. To construct such confidence intervals, one will use the fact that if a random sample of size n is taken from a *population that is normally distributed* with variance σ^2, then the random variable

$$\chi^2 = \frac{n-1}{\sigma^2} s^2 \tag{4.15}$$

has the chi-square distribution with $\nu = (n-1)$ degrees of freedom (described in Sect. 2.4.3). The advantage of using χ^2 instead of s^2 is similar to the advantage of standardizing a variable to a normal random variable. Such a transformation allows standard tables (such as Table A5) to be used for determining probabilities irrespective of the magnitude of s^2. The basis of these probability tables is again akin to finding the areas under the chi-square curves.

Example 4.2.9: A company which makes boxes wishes to determine whether their automated production line requires major servicing or not. They will base their decision on whether the weight from one box to another is significantly different from a maximum permissible population variance value of $\sigma^2 = 0.12$ kg². A sample of 10 boxes is selected, and their variance is found to be $s^2 = 0.24$ kg². Is this difference significant at the 95% CL?

From Eq. 4.15, the observed chi-square value is $\chi^2 = \frac{10-1}{0.12}(0.24) = 18$. Inspection of Table A5 for $\nu = 9$ degrees of freedom, reveals that for a significance level $\alpha = 0.05$, the critical chi-square value $\chi^2_c = 16.92$ and, for $\alpha = 0.025$, $\chi^2_c = 19.02$. Thus, the result is significant at $\alpha = 0.05$ or 95% CL. However, the result is not significant at the 97.5% CL. Whether to service the automated production line based on these statistical tests involves performing a decision analysis. ■

(b) Two sample tests This instance applies to the case when two independent random samples are taken from *two populations that are normally distributed,* and one needs to determine whether the variances of the two populations are different or not. Such tests find applications prior to conducting t-tests on two means which presumes equal variances. Let σ_1 and σ_2 be the standard deviations of both the populations, and s_1 and s_2 be the sample standard deviations. If $\sigma_1 = \sigma_2$, then the random variable

$$F = \frac{s_1^2}{s_2^2} \tag{4.16}$$

has the F-distribution (described in Sect. 2.4.3) with degrees of freedom (d.f.) = (ν_1, ν_2) where $\nu_1 = (n_1 - 1)$ and $\nu_2 = (n_2 - 1)$. Note that the distributions are different for different combinations of ν_1 and ν_2. The probabilities for F can be determined using areas under the F curves or from tabulated values (Table A6). Note that the F-test applies to independent samples, and, unfortunately, is known to be rather sensitive to the assumption of normality. Hence, some argue against its use altogether for two sample testing (Manly 2005).

Example 4.2.10: *Comparing variability in daily productivity of two workers*
It is generally acknowledged that worker productivity increases if his environment is conditioned so as to meet the stipulated human comfort conditions. One is interested in comparing the mean productivity of two office workers. However, before undertaking that evaluation, one is unsure about the assumption of equal variances in productivity of the workers (i.e., in how consistent the workers are from one day to another). This test can be used to check the validity of this assumption. Suppose the following data has been collected

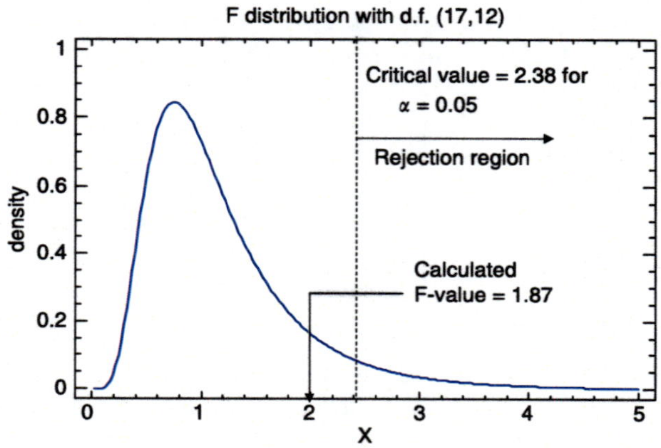

Fig. 4.7 Since the calculated F value is lower than the critical value, one is forced to accept the null hypothesis (Example 4.2.10)

for two workers under the same environment and performing similar tasks. An initial analysis of the data suggests that the normality condition is met for both workers:

Worker A: $n_1=13$ days, mean $\bar{x}_1=26.3$ production units, standard deviation $s_1=8.2$ production units.

Worker B: $n_2=18$ days, mean $\bar{x}_2=19.7$ production units, standard deviation $s_2=6.0$ production units.

The intent here is to compare not the means but the standard deviations. The F-statistic is determined by *always* choosing the larger variance as the numerator. Then $F=(8.2/6.0)^2=1.87$. From Table A6, the critical F value $F_c=2.38$ for $(13-1)=12$ and $(18-1)=17$ degrees of freedom at a significance level $\alpha=0.05$. Thus, as illustrated in Fig. 4.7, one is forced to accept the null hypothesis, and conclude that the data provides not enough evidence to indicate that the population variances of the two workers are statistically different at $\alpha=0.05$. Hence, one can now proceed to use the two-sample t-test with some confidence to determine whether the difference in the means between both workers is statistically significant or not.

4.2.6 Tests for Distributions

The Chi-square (χ^2) statistic applies to discrete data. It is used to statistically test the hypothesis that a set of empirical or sample data does not differ significantly from that which would be expected from some specified theoretical distribution. In other words, it is a goodness-of-fit test to ascertain whether the distribution of proportions of one group differs from another or not. The chi-square statistic is computed as:

$$\chi^2 = \sum_k \frac{(f_{obs} - f_{\exp})^2}{f_{\exp}} \qquad (4.17)$$

where f_{obs} is the observed frequency of each class or interval, f_{exp} is the expected frequency for each class predicted by the theoretical distribution, and k is the number of classes or intervals. If $\chi^2=0$, then the observed and theoretical frequencies agree exactly. If not, the larger the value of χ^2, the greater the discrepancy. Tabulated values of χ^2 are used to determine significance for different values of degrees of freedom $\upsilon=k-1$ (see Table A5). Certain restrictions apply for proper use of this test. The sample size should be greater than 30, and none of the expected frequencies should be less than 5 (Walpole et al. 2007). In other words, a long tail of the probability curve at the lower end is not appropriate. The following example serves to illustrate the process of applying the chi-square test.

Example 4.2.11: *Ascertaining whether non-code compliance infringements in residences is random or not*
A county official was asked to analyze the frequency of cases when home inspectors found new homes built by one specific builder to be non-code compliant, and determine whether the violations were random or not. The following data for 380 homes were collected:

No. of code infringements	0	1	2	3	4
Number of homes	242	94	38	4	2

The underlying random process can be characterized by the Poisson distribution (see Sect. 2.4.2): $P(x) = \dfrac{\lambda^x \exp(-\lambda)}{x!}$.
The null hypothesis, namely that the sample is drawn from a population that is Poisson distributed, is to be tested at the 0.05 significance level.

$$\text{The sample mean } \lambda = \frac{0(242)+1(94)+2(38)+3(4)+4(2)}{380}$$

$= 0.5$ infringements per home

For a Poisson distribution with $\lambda=0.5$, the underlying or expected values are found for different values of x as shown in Table 4.2.

The last three categories have expected frequencies that are less than 5, which do not meet one of the requirements

Table 4.2 Expected number of homes for different number of non-code compliance values if the process is assumed to be a Poisson distribution with sample mean of 0.5

X=number of non-code compliance values	P(x)·n	Expected no
0	(0.6065)·380	230.470
1	(0.3033)·380	115.254
2	(0.0758)·380	28.804
3	(0.0126)·380	4.788
4	(0.0016)·380	0.608
5 or more	(0.0002)·380	0.076
Total	*(1.000)·380*	*380*

for using the test (as stated above). Hence, these will be combined into a new category called "3 or more cases" which will have an expected frequency of 4.7888+0.608+0.076=5.472. The following statistic is calculated first:

$$\chi^2 = \frac{(242-230.470)^2}{230.470} + \frac{(94-115.254)^2}{115.254} + \frac{(38-28.804)^2}{28.804} + \frac{(6-5.472)^2}{5.472} = 7.483$$

Since there are only 4 groups, the degrees of freedom $\upsilon = 4-1=3$, and from Table A5, the critical value at 0.05 significance level is $\chi^2_{critical} = 7.815$. Hence, the null hypothesis cannot be rejected at the 0.05 significance level; this is, however, marginal. ■

Example 4.2.12[4]**:** *Evaluating whether injuries in males and females is independent of circumstance*

Chi-square tests are also widely used as tests of independence using contingency tables. In 1975, more than 59 million Americans suffered injuries. More males (33.6 million) were injured than females (25.6 million). These statistics do not distinguish whether males and females tend to be injured in similar circumstances. A safety survey of n=183 accident reports were selected at random to study this issue in a large city, as summarized in Table 4.3.

The null hypothesis is that the circumstance of an accident (whether at work or at home) is independent of the gender of the victim. It is decided to check this hypothesis at a significance level of $\alpha=0.01$. The degrees of freedom d.f.= $(r-1)\cdot(c-1)$ where r is the number of rows and c the number of categories. Hence, d.f.=$(3-1)\cdot(2-1)=2$. From Table A5, the critical value is $\chi^2_c=9.21$ at $\alpha=0.01$ for d.f.=2.

The expected values for different joint occurrences (male/work, male/home, male/other, female/work, female/home, female/other) are shown in italics in the table and correspond to the case when the occurrences are really independent. Recall from basic probability (Eq. 2.10) that if events A and B are independent, then $p(A \cap B) = p(A).p(B)$ where p indicates the probability. In our case, if being male and being involved in an accident at work were truly independent, then $p(work \cap male) = p(work).p(male)$. Consider the cell corresponding to male/at work. Its expected value = $n\cdot p(work \cap male) = n\cdot p(work)\cdot p(male) = 183\cdot\frac{45}{183}\cdot\frac{107}{183} = \frac{(45)\cdot(107)}{183} = 26.3$ (as shown in the table). Expected values for other joint occurrences shown in the table have been computed in like manner.

Thus, the chi-square statistics is $\chi^2 = \frac{(40-26.3)^2}{26.3} + \frac{(5-18.7)^2}{18.7} + \ldots + \frac{(13-12.9)^2}{12.9} = 24.3.$

Since, $\chi^2_c < 24.3$, the null hypothesis can be safely rejected at a significance level of 0.01. Hence, the gender does have a bearing on the circumstance in which the accidents occur. ■

Table 4.3 Observed and computed (assuming gender independence) number of accidents in different circumstances

	Male		Female		Total
Circumstance	Observed	Expected	Observed	Expected	Observed
At work	40	26.3	5	18.7	45
At home	49	62.6	58	44.4	107
Other	18	18.1	13	12.9	31
Total	*107*		*76*		*183=n*

[4] From Weiss (1987) by © permission of Pearson Education.

4.2.7 Test on the Pearson Correlation Coefficient

Recall that the Pearson correlation coefficient was presented in Sect. 3.4.2 as a means of quantifying the linear relationship between samples of two variables. One can also define a population correlation coefficient ρ for two variables. Section 4.2.1 presented methods by which the uncertainty around the population mean could be ascertained from the sample mean by determining confidence limits. Similarly, one can make inferences about the population correlation coefficient ρ from knowledge of the sample correlation coefficient r. Provided both the variables are normally distributed (called a bivariate normal population), then Fig. 4.8 provides a convenient way of ascertaining the 95% CL of the population correlation coefficient for different sample sizes. Say, r=0.6 for a sample n=10 pairs of observations, then the 95% CL for the population correlation coefficient are $(-0.05 < \rho < 0.87)$, which are very wide. Notice how increasing the sample size shrinks these bounds. For n=100, the intervals are $(0.47 < \rho < 0.71)$.

Table A7 lists the critical values of the sample correlation coefficient r for testing the null hypothesis that the population correlation coefficient is statistically significant (i.e., $\rho \neq 0$) at the 0.05 and 0.01 significance levels for one and two tailed tests. The interpretation of these values is of some importance in many cases, especially when dealing with small data sets. Say, analysis of the 12 monthly bills of a residence revealed a linear correlation of r=0.6 with degree-days at the location. Assume that a one-tailed test applies. The sample correlation suggests the presence of a correlation at a significance level $\alpha=0.05$ (the critical value from Table A7 is $\rho_c=0.497$) while none at $\alpha=0.01$, (for which $\rho_c=0.658$). Whether observed sample correlations are significant or not can be evaluated statistically as illustrated

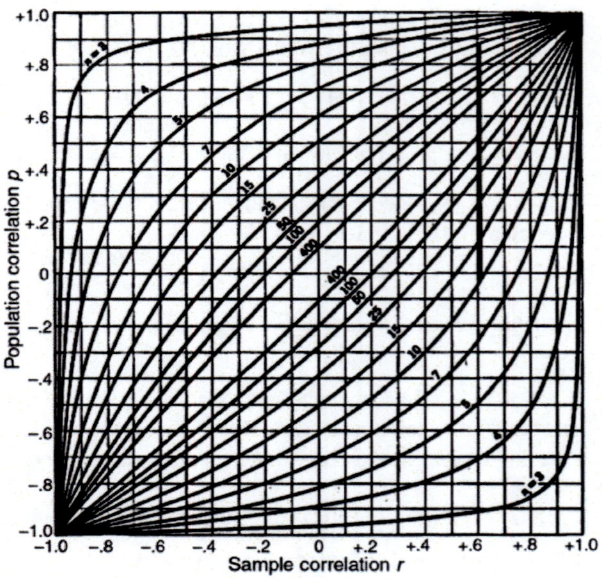

Fig. 4.8 Plot depicting 95% confidence bands for population correlation in a bivariate normal population for various sample sizes n. The *bold vertical line* defines the lower and upper limits of ρ when r=0.6 from a data set of 10 pairs of observations. (From Wonnacutt and Wonnacutt (1985) by permission of John Wiley and Sons)

above. Note that certain simplified suggestions on interpreting values of r in terms of whether they are strong, moderate or weak were given by Eq. 3.11; these are to be used with caution and were meant as thumb-rules only.

4.3 ANOVA Test for Multi-Samples

The statistical methods known as ANOVA (analysis of variance) are a broad set of widely used and powerful techniques meant to identify and measure sources of variation within a data set. This is done by partitioning the total variation in the data into its component parts. Specifically, ANOVA uses variance information from several samples in order to make inferences about the means of the populations from which these samples were drawn (and, hence, the appellation). Recall that z-tests and t-tests described previously are used to test for differences in one random variable (namely, their mean values) between two independent groups. This random experimental variable is called a *factor* in designed experiments and hypothesis testing. It is obvious that several of the cases treated in Sect. 4.2 involve single-factor hypothesis tests. ANOVA is an extension of such tests to multiple factors or experimental variables; even more generally, multiple ANOVA (called MANOVA) analysis can be used to test for multiple factor differences of multiple groups. Thus, ANOVA allows one to test whether the mean values of sampled data taken from different groups are essentially equal or not, i.e., whether the samples emanate from different populations or whether they are from the same population.

This section deals with single factor (or single variable) ANOVA methods since they are a logical lead-in to multivariate techniques (discussed in Sect. 4.4) as well as experimental design methods involving several variables which are discussed at more length in Chap. 6.

4.3.1 Single-Factor ANOVA

The ANOVA procedure uses *just one test* for comparing k sample means, just like that followed by the two-sample test. The following example allows a conceptual understanding of the approach. Say, four random samples have been selected, one from each of four populations. Whether the sample means differ enough to suggest different parent populations can be ascertained from the *within-sample variation* to the variation between the four samples. The more the sample means differ, the larger will be the *between-samples variation*, as shown in Fig. 4.9b, and the less likely is the probability that the samples arise from the same population. The reverse is true if the ratio of between-samples variation to that of the within-samples is small (Fig. 4.9a).

ANOVA methods test the null hypothesis of the form:

$$\begin{aligned} &H_0: \mu_1 = \mu_2 = \ldots = \mu_k \\ &H_a: \text{at least two of the } \mu_i\text{'s are different} \end{aligned} \tag{4.18}$$

Adopting the following notation:

Sample sizes: $n_1, n_2 \ldots, n_k$
Sample means: $\bar{x}_1, \bar{x}_2 \ldots \bar{x}_k$
Sample standard deviations: $s_1, s_2 \ldots s_k$
Total sample size: $n = n_1 + n_2 \ldots + n_k$
Grand average: $\langle \bar{x} \rangle$ = weighted average of all n responses

Then, one defines between-sample variation called "*treatment sum of squares*[5]" (SSTr) as:

$$\text{SSTr} = \sum_{i=1}^{k} n_i (\bar{x}_i - \langle \bar{x} \rangle)^2 \quad \text{with} \quad \text{d.f.} = \text{k} - 1 \tag{4.19}$$

and within-samples variation or "*error sum of squares*" (SSE) as:

$$\text{SSE} = \sum_{i=1}^{k} (n_i - 1) s_i^2 \quad \text{with} \quad \text{d.f.} = \text{n} - \text{k} \tag{4.20}$$

[5] The term "treatment" was originally coined for historic reasons where one was interested in evaluating the effect of treatments or changes in a product development process. It is now used synonymously to mean "classes" from which the samples are drawn.

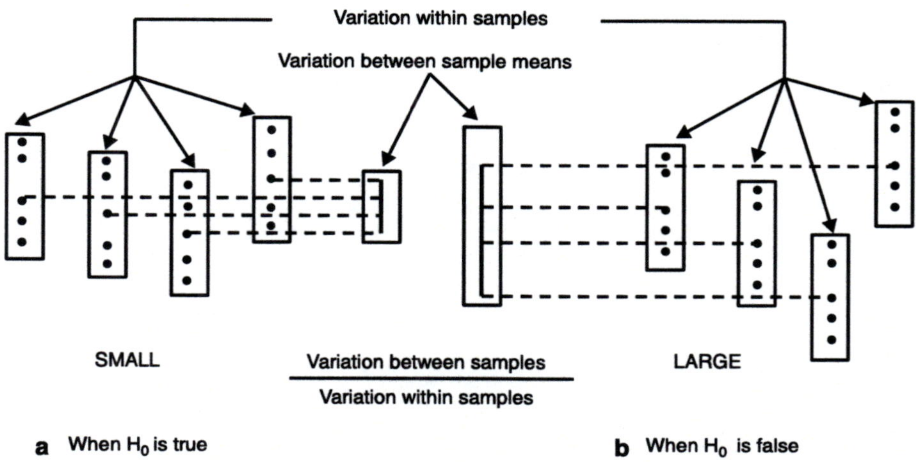

Fig. 4.9 Conceptual explanation of the basis of an ANOVA test

Together these two sources of variation comprise the "*total sum of squares*" (SST):

$$SST = SSTr + SSE = \sum_{i=1}^{k}\sum_{j=1}^{n}\left(\hat{x}_{ij} - \langle\bar{x}\rangle\right)^2 \quad \text{with} \quad \text{d.f.} = n - 1 \tag{4.21}$$

SST is simply the sample variance of the combined set of n data points$= (n_i - 1)s^2$ where s is the standard deviation of all the n data points.

The statistic defined below as the ratio of two variances is said to follow the F-distribution:

$$F = \frac{MSTr}{MSE} \tag{4.22}$$

where MSTr is the mean between-sample variation

$$= SSTr/(k - 1) \tag{4.23}$$

and MSE is the mean total sum of squares

$$= SSE/(n - k) \tag{4.24}$$

Recall that the p-value is the area of the F curve for (k−1, n−k) degrees of freedom to the right of F value. If p-value $\leq \alpha$ (the selected significance level), then the null hypothesis can be rejected. Note that the test is meant to be used for normal populations and equal population variances.

Example 4.3.1:[6] *Comparing mean life of five motor bearings*
A motor manufacturer wishes to evaluate five different motor bearings for motor vibration (which adversely results in reduced life). Each type of bearing is installed on different random samples of six motors. The amount of vibration (in microns) is recorded when each of the 30 motors are running. The data obtained is assembled in Table 4.4.

Table 4.4 Vibration values (in microns) for five brands of bearings tested on six motor samples (Example 4.3.1)

Sample	Brand 1	Brand 2	Brand 3	Brand 4	Brand 5
1	13.1	16.3	13.7	15.7	13.5
2	15.0	15.7	13.9	13.7	13.4
3	14.0	17.2	12.4	14.4	13.2
4	14.4	14.9	13.8	16.0	12.7
5	14.0	14.4	14.9	13.9	13.4
6	11.6	17.2	13.3	14.7	12.3
Mean	*13.68*	*15.95*	*13.67*	*14.73*	*13.08*
Std. dev.	*1.194*	*1.167*	*0.816*	*0.940*	*0.479*

Determine whether the bearing brands have an effect on motor vibration at the $\alpha = 0.05$ significance level. In this example, k=5, and n=30. The one-way ANOVA table is first generated as shown in Table 4.5.

From the F tables (Table A6) and for $\alpha = 0.05$, the critical F value for d.f.=(4,25) is $F_c = 2.76$, which is less than F=8.44 computed from the data. Hence, one is compelled to reject the null hypothesis that all five means are equal, and conclude that type of bearing motor does have a significant effect on motor vibration. In fact, this conclusion can be reached even at the more stringent significance level of $\alpha = 0.001$.

The results of the ANOVA analysis can be conveniently illustrated by generating an *effects plot*, as shown in Fig. 4.10a. This illustrates clearly the relationship between the mean values of the response variable, i.e., vibration level

Table 4.5 ANOVA table for Example 4.3.1

Source	d.f.	Sum of Squares	Mean Square	F-value
Factor	5−1=4	SSTr=30.855	MSTr=7.714	8.44
Error	30−5=25	SSE=22.838	MSE=0.9135	
Total	30−1=29	SST=53.694		

[6] From Devore and Farnum (2005) by © permission of Cengage Learning.

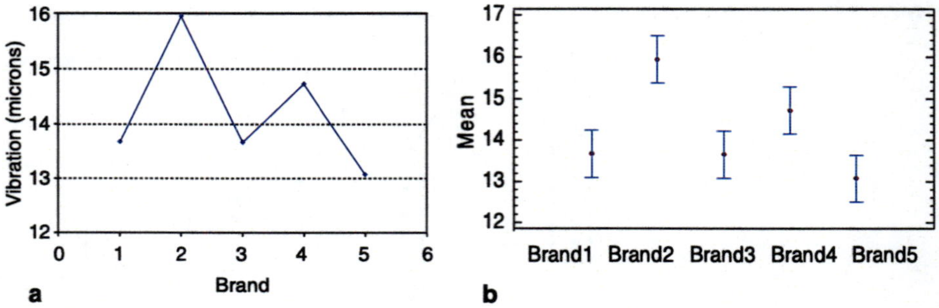

Fig. 4.10 a Effect plot. b Means plot showing the 95% CL intervals around the mean values of the 5 brands (Example 4.3.1)

for the five different motor bearing brands. Brand 5 gives the lowest average vibration, while Brand 2 has the highest. Note that such plots, though providing useful insights, are not generally a substitute for an ANOVA analysis. Another way of plotting the data is a *means plot* (Fig. 4.10b) which includes 95% CL intervals as well as the information provided in Fig. 4.10a. Thus, a sense of the variation within samples can be gleaned. ■

4.3.2 Tukey's Multiple Comparison Test

A limitation with the ANOVA test is that, in case the null hypothesis is rejected, one is unable to determine the exact cause. For example, one poor motor bearing brand could have been the cause of this rejection in the example above even though the four other brands could be essentially similar. Thus, one needs to be able to pinpoint the sample which leads one to conclude that the test was not significant overall. One could, of course, perform paired comparisons of two brands one at a time. In the case of 5 sets, one would then make 10 such tests. Apart from the tediousness of such a procedure, making independent paired comparisons leads to a decrease in sensitivity, i.e., type I errors are magnified. Hence, procedures that allow multiple comparisons to be made simultaneously have been proposed for this purpose (see Manly 2005). One such method is discussed in Sect. 4.4.2.

In this section, the Tukey's significant difference procedure based on paired comparisons is described which is limited to cases of *equal sample sizes*. This procedure allows the simultaneous formation of prespecified confidence intervals for all paired comparisons using the Student t-distribution. Separate tests are conducted to determine whether $\mu_i = \mu_j$ for each pair (i,j) of means in an ANOVA study of k population means. Tukey's procedure is based on comparing the distance (or absolute value) between any two sample means $|\bar{x}_i - \bar{x}_j|$ to a threshold value T that depends on significance level α as well as on the mean square error (MSE) from the ANOVA test. The T value is calculated as:

$$T = q_\alpha \left(\frac{MSE}{n_i} \right)^{1/2} \tag{4.25}$$

where n_i is the size of the sample drawn from each population, q_α values are called the studentized range distribution values and are given in Table A8 for $\alpha = 0.05$ for d.f.= (k, n−k)

If $|\bar{x}_i - \bar{x}_j| > T$, then one concludes that $\mu_i \neq \mu_j$ at the corresponding significance level. Otherwise, one concludes that there is no difference between the two means. Tukey also suggested a convenient visual representation to keep track of the results of all these pairwise tests. The Tukey's procedure and this representation are illustrated in the following example.

Example 4.3.2:[7] Using the same data as that in Example 4.3.1, conduct a multiple comparison procedure to distinguish which of the motor bearing brands are superior to the rest.

Following Tukey's procedure given by Eq. 4.25, the critical distance between sample means at $\alpha = 0.05$ is:

$$T = q_\alpha \left(\frac{MSE}{n_i} \right)^{1/2} = 4.15 \left(\frac{0.913}{6} \right)^{1/2} = 1.62$$

where q_α is found by interpolation from Table A8 based on d.f.=(k, n−k)=(5, 25).

The pairwise distances between the five sample means shown in Table 4.6 can be determined, and appropriate inferences made.

Thus, the distance T between the following pairs is less than 1.62: {1,3;1,4;1,5}, {2,4}, {3,4;3,5}. This information is visually summarized in Fig. 4.11 by arranging the five sample means in ascending order and then drawing rows of bars connecting the pairs *whose distances do not exceed* T=1.62. It is now clear that though brand 5 has the lowest mean value, it is not significantly different from brands 1 and 3. Hence, the final selection of which motor bearing to pick can be made from these three brands only. ■

[7] From Devore and Farnum (2005) by © permission of Cengage Learning.

Table 4.6 Pairwise analysis of the five samples following Tukey's procedure

Samples	Distance	Conclusion[a]
1,2	$\|13.68-15.95\|=2.27$	$\mu_i \neq \mu_j$
1,3	$\|13.68-13.67\|=0.01$	
1,4	$\|13.68-14.73\|=1.05$	
1,5	$\|13.68-13.08\|=0.60$	
2,3	$\|15.95-13.67\|=2.28$	$\mu_i \neq \mu_j$
2,4	$\|15.95-14.73\|=1.22$	
2,5	$\|15.95-13.08\|=2.87$	$\mu_i \neq \mu_j$
3,4	$\|13.67-14.73\|=1.06$	
3,5	$\|13.67-13.08\|=0.59$	
4,5	$\|14.73-13.08\|=1.65$	$\mu_i \neq \mu_j$

[a] Only if distance > critical value of 1.62

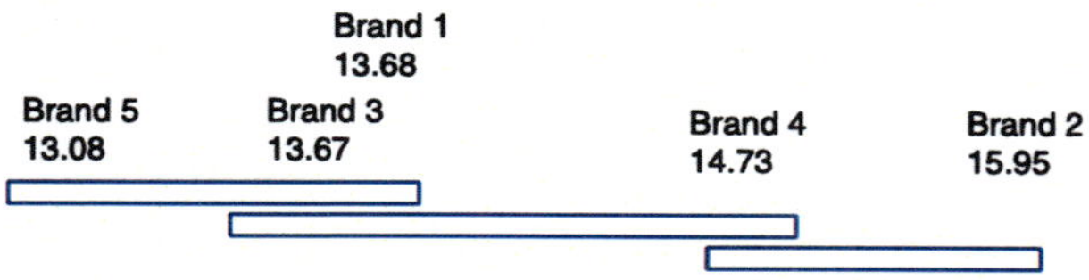

Fig. 4.11 Graphical depiction summarizing the ten pairwise comparisons following Tukey's procedure. Brand 2 is significantly different from Brands 1, 3 and 5, and so is Brand 4 from Brand 5 (Example 4.3.2)

4.4 Tests of Significance of Multivariate Data

4.4.1 Introduction to Multivariate Methods

Multivariate analysis (also called multifactor analysis) is the branch of statistics that deals with statistical inference and model building as applied to multiple measurements made from one or several samples taken from one or several populations. Multivariate methods can be used to make inferences about sample means and variances. Rather than treating each measure separately as done in t-tests and single-factor ANOVA, multivariate inferential methods allow the analyses of multiple measures simultaneously as a system of measurements. This generally results in sounder inferences to be made, a point elaborated below.

The univariate probability distributions presented in Sect. 2.4 can also be extended to bivariate and multivariate distributions. Let x_1 and x_2 be two variables of the same type, say both discrete (the summations in the equations below need to be replaced with integrals for continuous variables). Their joint distribution is given by:

$$f(x_1, x_2) \geq 0 \quad \text{and} \quad \sum_{all(x_1,x_2)} f(x_1, x_2) = 1 \tag{4.26}$$

Consider two sets of multivariate data each consisting of p variables. However, they could be different in size, i.e., the number of observations in each set may be different, say n_1 and n_2. Let $\bar{\mathbf{X}}_1$ and $\bar{\mathbf{X}}_2$ be the sample mean vectors of dimension p. For example,

$$\bar{\mathbf{X}}_1 = \left[\bar{x}_{11}, \bar{x}_{12}, \ldots \bar{x}_{1i}, \ldots \bar{x}_{1p}\right] \tag{4.27}$$

where $\bar{x}_{1i}$ is the sample average over n_1 observations of parameter i for the first set.

Further, let $\mathbf{C}_1$ and $\mathbf{C}_2$ be the sample covariance matrices of size (p×p) for the two sets respectively (the basic concepts of covariance and correlation were presented in Sect. 3.4.2). Then, the sample matrix of variances and covariances for the first data set is given by:

$$\mathbf{C}_1 = \begin{bmatrix} c_{11} & c_{12} & .. & c_{1p} \\ c_{21} & c_{22} & .. & c_{2p} \\ .. & .. & .. & .. \\ c_{p1} & c_{p2} & .. & c_{pp} \end{bmatrix} \tag{4.28}$$

where c_{ii} is the variance for parameter i and c_{ik} the covariance for parameters i and k.

Similarly, the sample correlation matrix where the diagonal elements are equal to unity and other terms scaled appropriately, is given by

$$\mathbf{R}_1 = \begin{bmatrix} 1 & r_{12} & .. & r_{1p} \\ r_{21} & 1 & .. & r_{2p} \\ .. & .. & .. & .. \\ r_{p1} & r_{p2} & .. & 1 \end{bmatrix} \tag{4.29}$$

Both matrices contain the correlations between each pair of variables, and they are symmetric about the diagonal since, say, $c_{12}=c_{21,}$ and so on. This redundancy is simply meant to allow easier reading. These matrices provide a convenient visual representation of the extent to which the different sets of variables are correlated with each other, thereby allowing strongly correlated sets to be easily identified. Note that correlations are not affected by shifting and scaling the data. Thus, standardizing the variables obtained by subtracting each observation by the mean and dividing by the standard deviation will still retain the correlation structure of the original data set while providing certain convenient interpretations of the results.

Underlying assumptions for multivariate tests of significance include the fact that the two samples have close to multivariate normal distributions with equal population covariance matrices. The multivariate normal distribution is a generalization of the univariate normal distribution when $p \geq 2$ where p is the number of dimensions or parameters. Figure 4.12 illustrates how the bivariate normal distribution is distorted in the presence of correlated variables. The contour lines are circles for uncorrelated variables and ellipses for correlated ones.

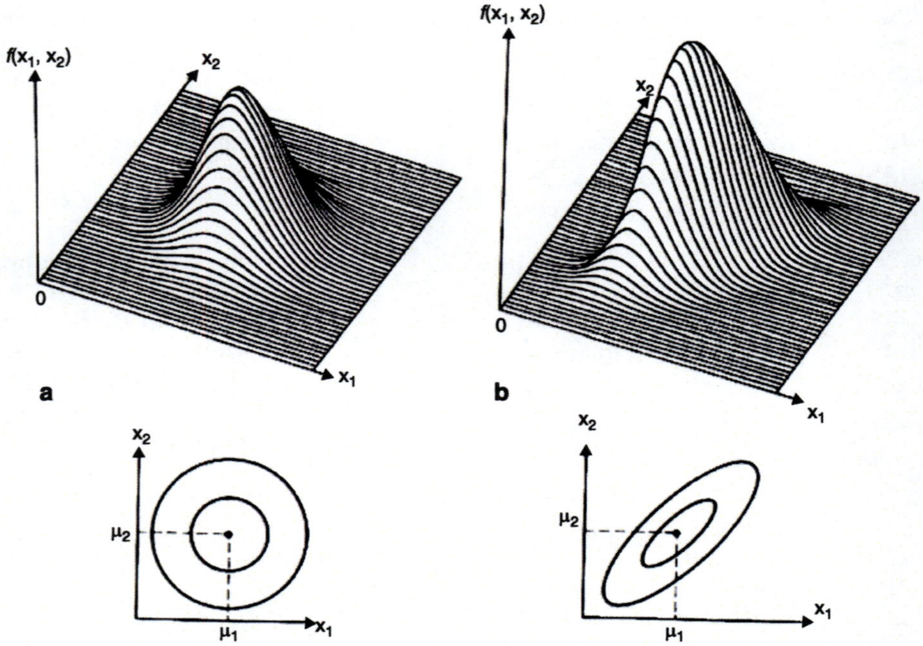

Fig. 4.12 Two bivariate normal distributions and associated 50% and 90% contours assuming equal standard deviations for both variables. However, the *left hand side plots* presume the two variables to be uncorrelated, while those on the *right* have a correlation coefficient of 0.75 which results in elliptical contours. (From Johnson and Wichern (1988) by © permission of Pearson Education)

4.4.2 Hotteling T^2 Test

The simplest extension of univariate statistical tests is the situation when two or more samples are evaluated to determine whether they originate from populations with: (i) different means and (ii) different variances/covariances. One can distinguish between the following types of multivariate inference tests involving more than one parameter (Manly 2005):

(a) comparison of mean values for two samples is best done using the Hotteling T^2-test;
(b) comparison of variation for two samples (several procedures have been proposed; the best known are the Box's M-test, the Levene's test based on T^2-test, and the Van Valen test);
(c) comparison of mean values for several samples (several tests are available; the best known are the Wilks' lambda statistic test, Roy's largest root test, and Pillai's trace statistic test);
(d) comparison of variation for several samples (using the Box's M-test).

Only case (a) will be described here, while the others are treated in texts such as Manly (2005). Consider two samples with sample sizes n_1 and n_2. One wishes to compare differences in p random variables among the two samples. Let $\bar{\mathbf{X}}_1$ and $\bar{\mathbf{X}}_2$ be the mean vectors of the two samples. A pooled estimate of covariance matrix is:

$$\mathbf{C} = \{(n_1 - 1)\mathbf{C}_1 + (n_2 - 1)\mathbf{C}_2\}/(n_1 + n_2 - 2) \tag{4.30}$$

where $\mathbf{C}_1$ and $\mathbf{C}_2$ are the covariance vectors given by Eq. 4.28

Then, the Hotteling's T^2-statistic is defined as:

$$T^2 = \frac{n_1 n_2 \left(\bar{\mathbf{X}}_1 - \bar{\mathbf{X}}_2\right)' \mathbf{C}^{-1} \left(\bar{\mathbf{X}}_1 - \bar{\mathbf{X}}_2\right)}{(n_1 + n_2)} \tag{4.31}$$

A large numerical value of this statistic suggests that the two population mean vectors are different. The null hypothesis test uses the transformed statistic:

$$F = \frac{(n_1 + n_2 - p - 1)T^2}{(n_1 + n_2 - 2)p} \tag{4.32}$$

which follows the F-distribution with the number of p and $(n_1 + n_2 - p - 1)$ degrees of freedom.

Since, the T^2 statistic is quadratic, it can also be written in double sum notation as:

$$T^2 = \left[\frac{n_1 n_2}{(n_1 + n_2)}\right] \sum_{i=1}^{p} \sum_{k=1}^{p} (\bar{x}_{1i} - \bar{x}_{2i}) c_{ik} (\bar{x}_{1k} - \bar{x}_{2k}) \tag{4.33}$$

Example 4.4.1:[8] *Comparing mean values of two samples by pairwise and by Hotteling T^2 procedures*
Consider two samples of 5 parameters (p=5) with paired samples. Sample 1 has 21 observations and sample 2 has 28. The mean and covariance matrices of both these samples have been calculated and shown below:

[8] From Manly (2005) by © permission of CRC Press.

$$\bar{\mathbf{X}}_1 = \begin{bmatrix} 157.381 \\ 241.000 \\ 31.433 \\ 18.500 \\ 20.810 \end{bmatrix}$$

$$\text{and} \quad \mathbf{C}_1 = \begin{bmatrix} 11.048 & 9.100 & 1.557 & 0.870 & 1.286 \\ 9.100 & 17.500 & 1.910 & 1.310 & 0.880 \\ 1.557 & 1.910 & 0.531 & 0.189 & 0.240 \\ 0.870 & 1.310 & 0.189 & 0.176 & 0.133 \\ 1.286 & 0.880 & 0.240 & 0.133 & 0.575 \end{bmatrix}$$

$$\bar{\mathbf{X}}_2 = \begin{bmatrix} 158.429 \\ 241.571 \\ 31.479 \\ 18.446 \\ 20.839 \end{bmatrix}$$

$$\text{and} \quad \mathbf{C}_2 = \begin{bmatrix} 15.069 & 17.190 & 2.243 & 1.746 & 2.931 \\ 17.190 & 32.550 & 3.398 & 2.950 & 4.066 \\ 2.243 & 3.398 & 0.728 & 0.470 & 0.559 \\ 1.746 & 2.950 & 0.470 & 0.434 & 0.506 \\ 2.931 & 4.066 & 0.559 & 0.506 & 1.321 \end{bmatrix}$$

If one performed paired t-tests with each parameter taken one at a time (as described in Sect. 4.2.3), one would compute the pooled variance for the first parameter as:

$$s_1^2 = [(21-1)(11.048) + (28-1)(15.069)]/(21+28-2) = 13.36$$

And the t-statistic as:

$$t = \frac{(157.381 - 158.429)}{\left[13.36\left(\frac{1}{21} + \frac{1}{28}\right)\right]} = -0.99$$

with 47 degrees of freedom. This is not significantly different from zero as one can note from the p-value indicated in Table A4. Table 4.7 assembles similar results for all other parameters. One would conclude that none of the five parameters in both data sets are statistically different.

In order to perform the multivariate test, one first calculates the pooled sample covariance matrix (Eq. 4.30):

Table 4.7 Paired t-tests for each of the five parameters taken one at a time

Parameter	First data set		Second data set		t-value (47 d.f.)	p-value
	Mean	Variance	Mean	Variance		
1	157.38	11.05	158.43	15.07	−0.99	0.327
2	241.00	17.50	241.57	32.55	−0.39	0.698
3	31.43	0.53	31.48	0.73	−0.20	0.842
4	18.50	0.18	18.45	0.43	0.33	0.743
5	20.81	0.58	20.84	1.32	−0.10	0.921

$$\mathbf{C} = \left(\frac{20C_1 + 27C_2}{47}\right)$$

$$= \begin{bmatrix} 13.358 & 13.748 & 1.951 & 1.373 & 2.231 \\ 13.748 & 26.146 & 2.765 & 2.252 & 2.710 \\ 1.951 & 2.765 & 0.645 & 0.350 & 0.423 \\ 1.373 & 2.252 & 0.350 & 0.324 & 0.347 \\ 2.231 & 2.710 & 0.423 & 0.347 & 1.004 \end{bmatrix}$$

where, for example, the first entry is: $(20\times11.048 + 27\times15.069)/47 = 13.358$.

The inverse of the matrix C yields

$$\mathbf{C}^{-1} = \begin{bmatrix} 0.2061 & -0.0694 & -0.2395 & 0.0785 & -0.1969 \\ -0.0694 & 0.1234 & -0.0376 & -0.5517 & 0.0277 \\ -0.2395 & -0.0376 & 4.2219 & -3.2624 & -0.0181 \\ 0.0785 & -0.5517 & -3.2624 & 11.4610 & -1.2720 \\ -0.1969 & 0.0277 & -0.0181 & -1.2720 & 1.8068 \end{bmatrix}$$

Substituting the elements of the above matrix in Eq. 4.33 results in:

$$T^2 = \left\{\frac{(21)(28)}{(21+28)}[(157.381 - 158.429)(0.2061)\right.$$
$$(157.381 - 158.429) - (157.318 - 158.429)$$
$$(0.0694)(241.000 - 241.571) + \cdots + (20.810 - 20.839)$$
$$(1.8068)(20.810 - 20.839) = 2.824$$

which from Eq. 4.32 results in a F-statistic$=\dfrac{(21+28-5-1)(2.824)}{(21+28-2)(5)} = 0.517$ with 5 and 43 d.f.

This is clearly not significant since $F_{critical}=2.4$ (from Table A6), and so there is no evidence to support that the population means of the two groups are statistically different when all five parameters are simultaneously considered. In this case one could have drawn such a conclusion directly from Table 4.7 by looking at the pairwise p-values, but this may not happen always. ■

Other than the elegance provided, there are two distinct advantages of performing a single multivariate test as against a series of univariate tests. The probability of finding a type-I result purely by accident increases as the number of variables increase, and the multivariate test takes *proper account of the correlation between variables.* The above example illustrated the case where no significant differences in population means could be discerned either from univariate tests performed individually or from an overall multivariate test. However, there are instances, when the latter test turns out to be significant as a result of the cumulative effects of all parameters while any one parameter is not significantly different. The converse may also hold, the evidence provided from one significantly different parameters may be swamped by lack of differences between the other

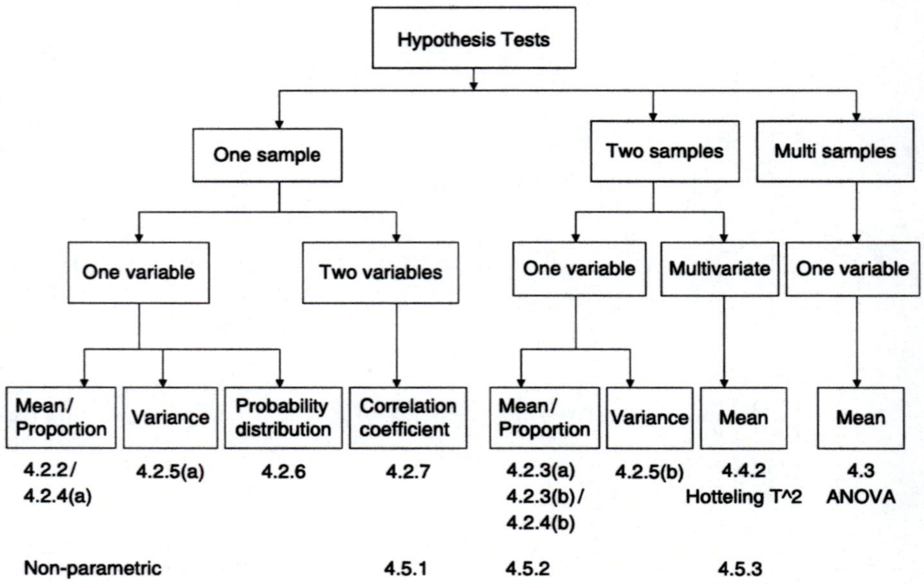

Fig. 4.13 Overview of various types of parametric hypothesis tests treated in this chapter along with section numbers. The lower set of three sections treat non-parametric tests

parameters. Hence, it is advisable to perform tests as illustrated in the above example.

The above sections (Sect. 4.2–4.4) treated several cases of hypothesis testing. An overview of these cases is provided in Fig. 4.13 for greater clarity. The specific sub-section of each of the cases is also indicated. The ANOVA case corresponds to the lower right box, namely testing for differences in the means of a single factor or variable which is sampled from several populations, while the Hotteling T^2-test corresponds to the case when the mean of several variables from two samples are evaluated. As noted above, formal use of statistical methods can become very demanding mathematically and computationally when multivariate and multisamples are considered, and hence the advantage of using numerical based resampling methods (discussed in Sect. 4.8).

4.5 Non-parametric Methods

The parametric tests described above have implicit built-in assumptions regarding the distributions from which the samples are taken. Comparison of populations using the t-test and F-test can yield misleading results when the random variables being measured are not normally distributed and do not have equal variances. It is obvious that fewer the assumptions, broader would be the potential applications of the test. One would like that the significance tests used lead to sound conclusions, or that the risk of coming to wrong conclusions be minimized. Two concepts relate to the latter aspect. The concept of *robustness* of a test is inversely proportional to the sensitivity of the test and to violations of the underlying assumptions. The *power of a test*, on the other hand, is a measure of the extent to which cost of experimentation is reduced.

There are instances when the random variables are not quantifiable measurements but can only be ranked in order of magnitude. For example, a consumer survey respondent may rate one product as better than another but is unable to assign quantitative values to each product. Data involving such "*preferences*" cannot also be subject to the t and F tests. It is under such cases that one has to resort to nonparametric statistics. Rather than use actual numbers, nonparametric tests usually use *relative ranks* by sorting the data by rank (or magnitude), and discarding their specific numerical values. Nonparametric tests are generally less powerful than parametric ones, but on the other hand, are more robust and less sensitive to outlier points (much of the material that follows is drawn from McClave and Benson 1988).

4.5.1 Test on Spearman Rank Correlation Coefficient

The Pearson correlation coefficient (Sect. 3.4.2) was a parametric measure meant to quantify the correlation between two quantifiable variables. The Spearman rank correlation coefficient r_s is exactly similar but applies to relative ranks. The same equation as Eq. 3.10 can be used to compute this measure, with its magnitude and sign interpreted in the same fashion. However, a simpler formula is often used:

$$r_s = 1 - \frac{6\sum d_i^2}{n(n^2-1)} \tag{4.34}$$

Table 4.8 Data table for Example 4.5.1 showing how to conduct the non-parametric correlation test

Faculty	Research grants ($)	Teaching evaluation	Research Rank (u_i)	Teaching Rank (v_i)	Difference d_i	Diff squared d_i^2
1	1480,000	7.05	5	7	−2	4
2	890,000	7.87	1	8	−7	49
3	3360,000	3.90	10	2	8	64
4	2210,000	5.41	8	5	3	9
5	1820,000	9.02	7	9	−2	4
6	1370,000	6.07	4	6	−2	4
7	3180,000	3.20	9	1	8	64
8	930,000	5.25	2	4	−2	4
9	1270,000	9.50	3	10	−7	49
10	1610,000	4.45	6	3	3	9
					Total	*260*

where n is the number of paired measurements, and the difference between the ranks for the ith measurement for ranked variables u and v is $d_i = u_i - v_i$.

Example 4.5.1: *Non-parametric testing of correlation between the sizes of faculty research grants and teaching evaluations*

The provost of a major university wants to determine whether a statistically significant correlation exists between the research grants and teaching evaluation rating of its senior faculty. Data over three years has been collected as assembled in Table 4.8 which also shows the manner in which ranks have been generated and the quantities $d_i = u_i - v_i$ computed.

Using Eq. 4.34 with n=10:

$$r_s = 1 - \frac{6(260)}{10(100-1)} = -0.576$$

Thus, one notes that there exists a negative correlation between the sample data. However, whether this is significant for the population correlation coefficient ρ_s can be ascertained by means of a statistical test:

$H_0 : \rho_s = 0$ (there is no significant correlation)
$H_a : \rho_s \neq 0$ (there is sigificant correlation)

Table A10 in Appendix A gives the absolute cutoff values for different significance levels. For n=10, the critical value for $\alpha = 0.05$ is 0.564, which suggests that the correlation can be deemed to be significant at the 0.05 significance level, but not at the 0.025 level whose critical value is 0.648. ■

4.5.2 Wilcoxon Rank Tests—Two Sample and Paired Tests

Rather than compare specific parameters (such as the mean and the variance), the non-parametric tests evaluate whether the probability distributions of the sampled populations are different or not. The test is nonparametric and no restriction is placed on the distribution other than it needs to be continuous and symmetric.

(a) The **Wilcoxon rank sum test** is meant for independent samples where the individual observations can be ranked by magnitude. The following example illustrates the approach.

Example 4.5.2: *Ascertaining whether oil company researchers and academics differ in their predictions of future atmospheric carbon dioxide levels*

The intent is to compare the predictions in the change of atmospheric carbon dioxide levels between researchers who are employed by oil companies and those who are in academia. The gathered data shown in Table 4.9 in percentage increase in carbon dioxide from the current level over the next 10 years from 6 oil company researchers and seven academics. Perform a statistical test at the 0.05 significance level in order to evaluate the following hypotheses:

(a) Predictions made by oil company researchers differ from those made by academics.
(b) Predictions made by oil company researchers tend to be lower than those made by academics.

Table 4.9 Wilcoxon rank test calculation for paired independent samples (Example 4.5.2)

	Oil Company Researchers		Academics	
	Prediction (%)	Rank	Prediction (%)	Rank
1	3.5	4	4.7	6
2	5.2	7	5.8	9
3	2.5	2	3.6	5
4	5.6	8	6.2	11
5	2.0	1	6.1	10
6	3.0	3	6.3	12
7	–	–	6.5	13
Sum		25		66

(a) First, ranks are assigned as shown for the two groups of individuals combined. Since there are 13 predictions, the ranks run from 1 through 13 as shown in the table. The test statistic is based on the *sum totals* of each group (and hence its name). If they are close, the implication is that there is no evidence that the probability distributions of both groups are different; and vice versa.

Let T_A and T_B be the rank sums of either group. Then

$$T_A + T_B = \frac{n(n+1)}{2} = \frac{13(13+1)}{2} = 91 \quad (4.35)$$

where $n = n_1 + n_2$ with $n_1 = 6$ and $n_2 = 7$. Note that n_1 should be selected as the one with fewer observations. A small value of T_A implies a large value of T_B, and vice versa. Hence, greater the difference between both the rank sums, greater the evidence that the samples come from different populations. Since one is testing whether the predictions by both groups are different or not, the two-tailed significance test is appropriate. Table A11 provides the lower and upper cutoff values for different values of n_1 and n_2 for both the one-tailed and the two-tailed tests. Note that the lower and higher cutoff values are (28, 56) at 0.05 significance level for the two-tailed test. The computed statistics of $T_A = 25$ and $T_B = 66$ are outside the range, the null hypothesis is rejected, and one would conclude that the predictions from the two groups are different.

(b) Here one wishes to test the hypothesis that the predictions by oil company researchers is lower than those made by academics. Then, one uses a one-tailed test whose cutoff values are given in part (b) of Table A11. These cutoff values at 0.05 significance level are (30, 54) but only the lower value of 30 is used in this case. The null hypothesis will be rejected only if $T_A < 30$. Since this is so, the above data suggests that the null hypothesis can be rejected at a significance level of 0.05. ■

(b) **The Wilcoxon signed rank test** is meant for paired tests where samples taken are not independent. This is analogous to the two sample paired difference test treated in Sect. 4.2.3b. As before, one deals with one variable involving paired differences of observations or data. This is illustrated by the following example.

Example 4.5.3: *Evaluating predictive accuracy of two climate change models from expert elicitation*

A policy maker wishes to evaluate the predictive accuracy of two different climate change models for predicting short-term (say, 30 years) carbon dioxide changes in the atmosphere. He consults 10 experts and asks them to grade these models on a scale from 1 to 10, with 10 being extremely accurate. Clearly, this data is not independent since the same expert is asked to make two value judgments about the models being evaluated. Data shown in Table 4.10 is obtained

Table 4.10 Wilcoxon signed rank test calculation for paired non-independent samples (Example 4.5.3)

Expert	Model A	Model B	Difference (A−B)	Absolute difference	Rank
1	6	4	2	2	5
2	8	5	3	3	7.5
3	4	5	−1	1	2
4	9	8	1	1	2
5	4	1	3	3	7.5
6	7	9	−2	2	5
7	6	2	4	4	9
8	5	3	2	2	5
9	6	7	−1	1	2
10	8	2	6	6	10
				Sum of positive ranks T_+	=46
				Sum of negative ranks T_-	=9

(note that these are not ranked values, except for the last column but the grades from 1 to 10 assigned by the experts):

The paired differences are first computed (as shown in the table) from which the ranks are generated based on the absolute differences, and finally the sums of the positive and negative ranks are computed. Note how the ranking has been assigned since there are repeats in the absolute difference values. There are three "1" in the absolute difference column. Hence a mean value of rank "2" has been assigned for all 3 three. Similarly for the three absolute differences of "2", the rank is given as "5", and so on. For the highest absolute difference of "6", the rank is assigned as "10". The values shown in last two rows of the table are also simple to deduce. The values of the difference (A−B) column are either positive or negative. One simply adds up all the rank values corresponding to the cases when (A−B) is positive and also when they are negative. These are found to be 46 and 9 respectively.

The test statistic for the null hypothesis is $T = \min(T_-, T_+)$. In our case, T=9. *The smaller the value of T, the stronger the evidence that the difference between both distributions is important.* The rejection region for T is determined from Table A12. The two-tailed critical value for n=10 at 0.05 significance level is 8. Since the computed value for T is higher, one cannot reject the null hypothesis, and so one would conclude that there is not enough evidence to suggest that one of the models is more accurate than the other at the 0.05 significance level. Note that if a significance level of 0.10 were selected, the null hypothesis would have been rejected.

Looking at the ratings shown in the table, one notices that these seem to be generally higher for model A than model B. In case one wishes to test the hypothesis, at a significance level of 0.05, that researchers deem model B to be less accurate than model A, one would have used T_- as the test statistic and compared it to the critical value of a one-tailed column values of Table A12. Since the critical value is 11 for n=10, which is greater than 9, one would reject the null hypothesis. ■

4.5.3 Kruskall-Wallis—Multiple Samples Test

Recall that the single-factor ANOVA test was described in Sect. 4.3.1 for inferring whether mean values from several samples emanate from the same population or not, with the necessary assumption of normal distributions. The Kruskall-Wallis H test is the nonparametric equivalent. It can also be taken to be the extension or generalization of the rank-sum test to more than two groups. Hence, the test applies to the case when one wishes to compare more than two groups which may not be normally distributed. Again, the evaluation is based on the rank sums where the ranking is made based on samples of all k groups combined. The test is framed as follows:

H_0 : All populations have identical probability distributions
H_a : Probability distributions of at least two populations are different

Let R_1, R_2, R_3 denote as the rank sums of say, three samples. The H-test statistic measures the extent to which the three samples differ with respect to their relative ranks, and is given by:

$$H = \left[\frac{12}{n(n+1)} \sum_{j=1}^{k} \left(\frac{R_j^2}{n_j} \right) \right] - 3(n+1) \qquad (4.36)$$

where k is the number of groups, n_j is the number of observations in the jth sample and n is the total sample size. Thus, if the H statistic is close to zero, one would conclude that all groups have the same mean rank, and vice versa. The distribution of the H statistic is approximated by the chi-square distribution, which is used to make statistical inferences. The following example illustrates the approach.

Example 4.5.4:[9] *Evaluating probability distributions of number of employees in three different occupations using a non-parametric test*

One wishes to compare, at a significance level of 0.05, the number of employees in companies representing each of three different business classifications, namely agriculture, manufacturing and service. Samples from ten companies each were gathered which are shown in Table 4.11. Since the distributions are unlikely to be normal (there are some large numbers), a nonparametric test is appropriate.

[9] From McClave and Benson (1988) by © permission of Pearson Education.

Table 4.11 Data table for Example 4.5.4

	Agriculture		Manufacturing		Service	
	# employees	Rank	# employees	Rank	# employees	Rank
1	10	5	244	25	17	9.5
2	350	27	93	19	249	26
3	4	2	3532	30	38	15
4	26	13	17	9.5	5	3
5	15	8	526	29	101	20
6	106	21	133	22	1	1
7	18	11	14	7	12	6
8	23	12	192	23	233	24
9	62	17	443	28	31	14
10	8	4	69	18	39	16
		R_1= 120		R_2= 210.5		R_3= 134.5

First, the ranks for all samples from the three classes are generated as shown tabulated under the 2nd, 4th and 6th columns. The values of the sums R_j are also computed and shown in the last row. Note that n=30, while n_j=10.The test statistic H is computed first:

$$H = \frac{12}{30(31)} \left[\frac{120^2}{10} + \frac{210.5^2}{10} + \frac{134.5^2}{10} \right] - 3(31)$$
$$= 99.097 - 93 = 6.097$$

The degrees of freedom is the number of groups minus one, or 3−1=2. From the Chi-square tables (Table A5), the critical value at α =0.05 is 5.991. Since the computed H value exceeds this threshold, one would reject the null hypothesis at 95% CL and conclude that at least two of the three probability distributions describing the number of employees in the sectors are different. However, the verdict is marginal since the computed H statistic is close to the critical value. It would be wise to consider the practical implications of the statistical inference test, and perform a decision analysis study. ■

4.6 Bayesian Inferences

4.6.1 Background

The Bayes' theorem and how it can be used for probability related problems has been treated in Sect. 2.5. Its strength lies in the fact that it provides a framework for including prior information in a two-stage (or multi-stage) experiment whereby one could draw stronger conclusions than one could with observational data alone. It is especially advantageous for small data sets, and it was shown that its predictions converge with those of the classical method for two cases: (i) as the data set of observations gets larger; and (ii) if the prior distribution is modeled as a uniform distribution. It was pointed out that advocates of the Bayesian approach view probability as a degree of belief held by a person about an uncertainty issue as compared to the objective view of long run relative frequency held by traditionalists. This section will discuss how the Bayesian approach can also be used to make statistical inferences from samples about an uncertain quantity, and also used for hypothesis testing problems.

4.6.2 Inference About One Uncertain Quantity

Consider the case when the population mean μ is to be estimated (point and interval estimates) from the sample mean $\bar{x}$ with the population assumed to be Gaussian with a known standard deviation σ.This case is given by the sampling distribution of the mean $\bar{x}$ treated in Sect. 4.2.1. The probability P of a two-tailed distribution at significance level α can be expressed as:

$$P\left(\bar{x} - z_{\alpha/2}\cdot\frac{\sigma}{n^{1/2}} < \mu < \bar{x} + z_{\alpha/2}\cdot\frac{\sigma}{n^{1/2}}\right) = 1 - \alpha \tag{4.37}$$

where n is the sample size and z is the value from the standard normal tables. The traditional interpretation is that one can be $(1-\alpha)$ confident that the above interval contains the true population mean. However, the interval itself should not be interpreted as a probability interval for the parameter.

The Bayesian approach uses the same formula but the mean and standard deviation are modified since the posterior distribution is now used which includes the sample data as well as the prior belief. The confidence interval is usually narrower than the traditional one and is referred to as the *credible interval* or the *Bayesian confidence interval*. The interpretation of this credible interval is somewhat different from the traditional confidence interval: there is a $(1-\alpha)$ probability that the population mean falls within the interval. Thus, the traditional approach leads to a probability statement about the interval, while the Bayesian about the population parameter (Phillips 1973).

The relevant procedure to calculate the credible intervals for the case of a Gaussian population and a Gaussian prior is presented without proof below (Wonnacutt and Wonnacutt 1985). Let the prior distribution, assumed normal, be characterized by a mean μ_0 and variance σ_0^2 , while the sample values are $\bar{x}$ and s_x. Selecting a prior distribution is equivalent to having a quasi-sample of size n_0 whose size is given by:

$$n_0 = \frac{{s_x}^2}{\sigma_0^2} \tag{4.38}$$

The posterior mean and standard deviation μ^* and σ^* are then given by:

$$\mu^* = \frac{n_0\mu_0 + n\bar{x}}{n_0 + n} \text{ and } \sigma^* = \frac{s_x}{(n_0+n)^{1/2}} \tag{4.39}$$

Note that the expression for the posterior mean is simply the weighted average of the sample and the prior mean, and is likely to be less biased than the sample mean alone. Similarly, the standard deviation is divided by the total normal sample size and will result in increased precision. However, had a different prior rather than the normal distribution been assumed above, a slightly different interval would have resulted which is another reason why traditional statisticians are uneasy about fully endorsing the Bayesian approach.

Example 4.6.1: *Comparison of classical and Bayesian confidence intervals*

A certain solar PV module is rated at 60 W with a standard deviation of 2 W. Since the rating varies somewhat from one shipment to the next, a sample of 12 modules has been selected from a shipment and tested to yield a mean of 65 W and a standard deviation of 2.8 W. Assuming a Gaussian distribution, determine the 95% confidence intervals by both the traditional and the Bayesian approaches.

(a) Traditional approach:

$$\mu = \bar{x} \pm 1.96\frac{s_x}{n^{1/2}} = 65 \pm 1.96\frac{2.8}{12^{1/2}} = 65 \pm 1.58$$

(b) Bayesian approach. Using Eq. 4.38 to calculate the quasi-sample size inherent in the prior:

$$n_0 = \frac{2.8^2}{2^2} = 1.96 \simeq 2.0$$

i.e., the prior is equivalent to information from an additional 2 modules tested.

Next, Eq. 4.39 is used to determine the posterior mean and standard deviation:

$$\mu^* = \frac{2(60) + 12(65)}{2 + 12} = 64.29$$

$$\text{and} \quad \sigma^* = \frac{2.8}{(2+12)^{1/2}} = 0.748$$

The Bayesian 95% confidence interval is then:

$$\mu = \mu^* \pm 1.96\,\sigma^* = 64.29 \pm 1.96(0.748)$$
$$= 62.29 \pm 1.47$$

Since prior information has been used, the Bayesian interval is likely to be centered better and be more precise (with a narrower interval) than the classical interval.

4.6.3 Hypothesis Testing

Section 4.2 dealt with the traditional approach to hypothesis testing where one frames the problem in terms of two competing claims. The application areas discussed involved testing for single sample mean, testing for two sample and paired differences, testing for single and two sample variances, testing for distributions, and testing on the Pearson correlation coefficient. In all these cases, one proceeds by defining two hypotheses:

(a) The *null hypothesis* which represents the status quo, i.e., that the hypothesis will be accepted unless the data provides convincing evidence of the contrary.

(b) The *research or alternative hypothesis* (H_a) which is the premise that the variation observed in the data sample cannot be ascribed to random variability or chance alone, and that there must be some inherent structural or fundamental cause.

Thus, the traditional or frequentist approach is to divide the sample space into an acceptance region and a rejection region, and posit that the null hypothesis can be rejected only if the probability of the test statistic lying in the rejection region can be ascribed to chance or randomness at the preselected significance level α. Advocates of the Bayesian approach have several objections to this line of thinking (Phillips 1973):

(i) the null hypothesis is rarely of much interest. The precise specification of, say, the population mean is of limited value; rather, ascertaining a range would be more useful;

(ii) the null hypothesis is only one of many possible values of the uncertain variable, and undue importance being placed on this value is unjustified;

(iii) as additional data is collected, the inherent randomness in the collection process would lead to the null hypothesis to be rejected in most cases;

(iv) erroneous inferences from a sample may result if prior knowledge is not considered.

The Bayesian approach to hypothesis testing is not to base the conclusions on a traditional significance level like $p<0.05$. Instead it makes use of the *posterior credible interval* introduced in the previous section. The procedure is summarized below for the instance when one wishes to test the population mean μ of the sample collected against a prior mean value μ_0 (Bolstad 2004).

(a) **One sided hypothesis test:** Let the posterior distribution of the mean value be given by $g(\mu/x_1, \ldots x_n)$. The hypothesis test is set up as:

$$H_0 : \mu \le \mu_0 \text{ versus } H_1 : \mu > \mu_0 \tag{4.40}$$

Let α be the significance level assumed (usually 0.10, 0.05 or 0.01). Then, the posterior probability of the null hypothesis, for the special case when the posterior distribution is Gaussian:

$$P(H_0 : \mu \le \mu_0/x_{1\ldots}x_n) = P\left(z \le \frac{\mu_0 - \mu^*}{\sigma^*}\right) \tag{4.41}$$

where z is the standard normal variable with μ^* and σ^* given by Eq. 4.39. If the probability is less than our selected value of α, the null hypothesis is rejected, and one concludes that $\mu > \mu_0$.

(b) **Two sided hypothesis test:** In this case, one is testing for

$$H_0 : \mu = \mu_0 \text{ versus } H_1 : \mu \neq \mu_0 \tag{4.42}$$

A slightly different approach is warranted since one is dealing with continuous variables for which the probability of them assuming a specific value is nil. Here, one calculates the $(1-\alpha)$ credible interval for μ using our posterior distribution. If μ_0 is outside these intervals, the null hypothesis is rejected; and vice versa.

Example 4.6.2: *Traditional and Bayesian approaches to determining confidence levels*

The life of a certain type of smoke detector battery is specified as having a mean of 32 months and a standard deviation of 0.5 months. The variable can be assumed to have a Gaussian distribution. A building owner decides to test this claim at a significance level of 0.05. He tests a sample of 9 batteries and finds a mean of 31 and a sample standard deviation of 1 month. Note that this is a one-side hypothesis test case.

(a) The traditional approach would entail testing $H_0 : \mu \le 32$ versus $H_1 : \mu > 32$. The Student t value: $t = \dfrac{31-32}{1/\sqrt{9}} = -3.0$. From Table A4, the critical value for d.f.=8 is $t_{0.05} = -1.86$. Thus, he can reject the null hypothesis, and state that the claim of the manufacturer is incorrect.

(b) The Bayesian approach, on the other hand, would require calculating the posterior probability of the null hypothesis. The prior distribution has a mean $\mu_0 = 32$ and variance $\sigma_0^2 = 0.5^2$.

First, use Eq. 4.38, and determine $n_0 = \dfrac{1^2}{0.5^2} = 4$, i.e., the prior information is "equivalent" to increasing the sample size by 4. Next, use Eq. 4.39 to determine the posterior mean and standard deviation:

$$\mu^* = \frac{4(32) + 9(31)}{4+9} = 31.3$$

$$\text{and } \sigma^* = \frac{1.0}{(4+9)^{1/2}} = 0.277.$$

From here: $t = \dfrac{32.0 - 31.3}{0.277} = 2.53$. From the student t table (Table A4) for d.f.=(9+4−1)=12, this corresponds to a confidence level of less than 99% or a probability of less than 0.01. Since this is lower than the selected significance level $\alpha = 0.05$, he can reject the null hypothesis. In this case, both approaches gave the same result, but sometimes one would reach different conclusions especially when sample sizes are small. ■

4.7 Sampling Methods

4.7.1 Types of Sampling Procedures

A sample is a portion or limited number of items from a larger entity called population of which information and characteristic traits are sought. Point and interval estimation as well as notions of inferential statistics covered in the previous sections involved the use of samples drawn from some underlying population. The premise was that finite samples would reduce the expense associated with the estimation; this being viewed as more critical than the associated uncertainty which would consequently creep into the estimation process. It is quite clear that the sample drawn must be representative of the population. However, there are different ways by which one could draw samples; this aspect falls under the purview of *sampling design*. Since these have different implications, they are discussed in this section.

There are three general rules of sampling design:

(i) the more representative the sample, the better the results;
(ii) all else being equal, larger samples yield better results, i.e., the results are more precise;
(iii) larger samples cannot compensate for a poor sampling design plan or a poorly executed plan.

Some of the common sampling methods are described below:

(a) **random sampling** (also called simple random sampling) is the simplest conceptually, and is most widely used. It involves selecting the sample of n elements in such as way that all possible samples of n elements have the same chance of being selected. Two important strategies of random sampling involve:
   (i) *sampling with replacement*, in which the object selected is put back into the population pool and has the possibility to be selected again in subsequent picks, and
   (ii) *sampling without replacement*, where the object picked is not put back into the population pool prior to picking the next item.

   Random sampling without replacement of N objects from a population n could be practically implemented in one of several ways. The most common is to order the objects of the population (say as 1, 2, 3 … n), use a random number generator to generate N numbers from 1 to n without replication, and pick only the objects whose numbers have been generated. This approach is illustrated by means of the following example. A consumer group wishes to select a sample of 5 cars from a lot of 500 cars for crash testing. It assigns integers from 1 to 500 to each and every car on the lot, uses a random number generator to select a set of 5 integers, and thereby select the 5 cars corresponding to the 5 integers picked randomly.

   Dealing with random samples has several advantages: (i) any random sub-sample of a random sample or its complement is also a random sample; (ii) after a random sample has been selected, any random sample from its complement can be added to it to form a larger random sample.

(b) **non-random sampling** occurs when the selection of members from the population is done according to some method or pre-set process which is not random. Often it occurs unintentionally or unwittingly with the experimenter thinking that he is dealing with random samples while he is not. In such cases, bias or skewness is introduced, and one obtains misleading confidence limits which may lead to erroneous inferences depending on the degree of non-randomness in the data set. However, in some cases, the experimenter intentionally selects the samples in a non-random manner and analyzes the data accordingly. This can result in the required conclusions being reached with reduced sample sizes, thereby saving resources. There are different types of nonrandom sampling (ASTM E 1402 1996), and some of the important ones are listed below:
   (i) *stratified sampling* in which the target population is such that it is amenable to partitioning into disjoint subsets or strata based on some criterion. Samples are selected independently from each stratum, possibly of different sizes. This improves efficiency of the sampling process in some instances, and is discussed at more length in Sect. 4.7.4;
   (ii) *cluster sampling* in which strata are first generated (these are synonymous to clusters), then random sampling is done to identify a subset of clusters, and finally all the elements in the picked clusters are selected for analysis;
   (iii) *sequential sampling* is a quality control procedure where a decision on the acceptability of a batch of products is made from tests done on a sample of the batch. Tests are done on a preliminary sample, and depending on the results, either the batch is accepted or further sampling tests are performed. This procedure usually requires, on an average, fewer samples to be tested to meet a pre-stipulated accuracy.

(iv) *composite sampling* where elements from different samples are combined together;

(v) *multistage or nested sampling* which involves selecting a sample in stages. A larger sample is first selected, and then subsequently smaller ones. For example, for testing indoor air quality in a population of office buildings, the design could involve selecting individual buildings during the first stage of sampling, choosing specific floors of the selected buildings in the second stage of sampling, and finally, selecting specific rooms in the floors chosen to be tested during the third and final stage.

(vi) *convenience sampling*, also called opportunity sampling, is a method of choosing samples arbitrarily following the manner in which they are acquired. If the situation is such that a planned experimental design cannot be followed, the analyst has to make do with the samples in the sequence they are acquired. Though impossible to treat rigorously, it is commonly encountered in many practical situations.

4.7.2 Desirable Properties of Estimators

The parameters from a sample are random variables since different sets of samples will result in different values for the parameters. Recall the definition of two seemingly analogous, but distinct, terms: an *estimate* is a specific number, while an *estimator* is a random variable. Since the search for estimators is the crux of the parameter estimation process, certain basic notions and desirable properties of estimators need to be explicitly recognized (a good discussion is provided by Pindyck and Rubinfeld 1981). Many of these concepts are logical extensions of the concepts applicable to errors, and also apply to regression models treated in Chap. 5. For example, consider the case where inferences about the population mean parameter μ are to be made from the sample mean estimator $\bar{x}$.

(a) **Lack of bias:** A very desirable property is for the distribution of the estimator to have the parameter as its mean value (see Fig. 4.14). Then, if the experiment were repeated many times, one would at least be assured that one would be right on an average. In such a case, the bias in $E(\bar{x} - \mu) = 0$, where E represents the expected value.

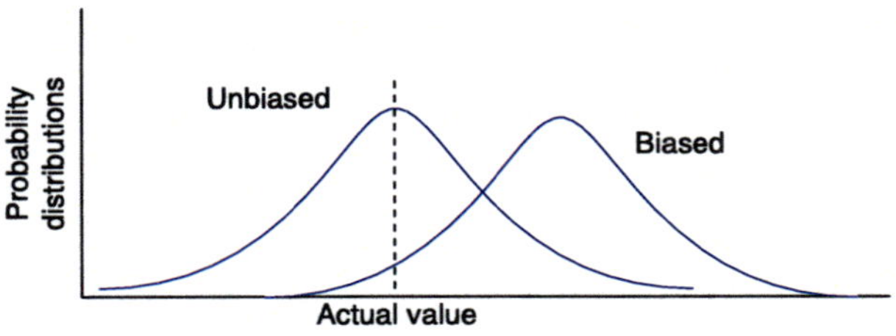

Fig. 4.14 Concept of biased and unbiased estimators

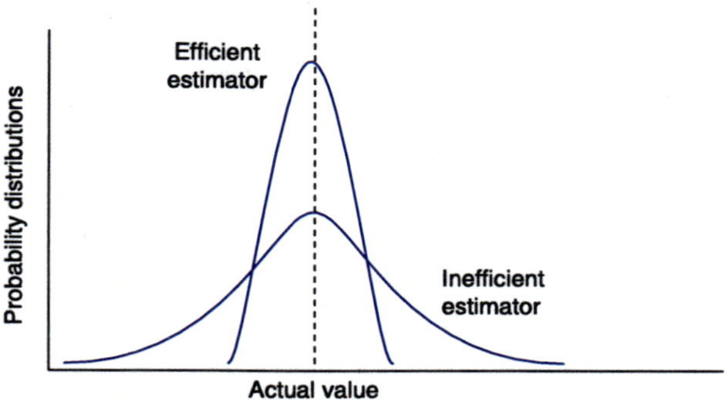

Fig. 4.15 Concept of efficiency of estimators

(b) **Efficiency:** Lack of bias provides no indication regarding the variability. Efficiency is a measure of how small the dispersion can possibly get. The value $\bar{x}$ is said to be an efficient unbiased estimator if, for a given sample size, the variance of $\bar{x}$ is smaller than the variance of any unbiased estimator (see Fig. 4.15) and is the smallest limiting variance that can be achieved. More often a relative order of merit, called the **relative efficiency**, is used which is defined as the ratio of both variances. Efficiency is desirable since the greater the efficiency associated with an estimation process, the stronger the statistical or inferential statements one can make about the estimated parameters.
Consider the following example (Wonnacutt and Wonnacutt 1985). If a population being sampled is symmetric, its center can be estimated without bias by either the sample mean $\bar{x}$ or its median $\tilde{x}$. For some populations $\bar{x}$ is more efficient; for others $\tilde{x}$ is more efficient. In case of a normal parent distribution, the standard error of $\tilde{x}=SE(\tilde{x})=1.25\,\sigma/\sqrt{n}$. Since $SE(\bar{x})=\sigma/\sqrt{n}$, efficiency of $\bar{x}$ relative to

$$\tilde{x} \equiv \frac{\text{var}\,\tilde{x}}{\text{var}\,\bar{x}} = 1.25^2 = 1.56. \quad (4.43)$$

(c) **Mean square error:** There are many circumstances in which one is forced to trade off bias and variance of estimators. When the goal of a model is to maximize the precision of predictions, for example, an estimator with very low variance and some bias may be more desirable than an unbiased estimator with high variance (see Fig. 4.16). One criterion which is useful in this regard is the goal of minimizing mean square error (MSE), defined as:

$$\text{MSE}(\bar{x}) = E(\bar{x} - \mu)^2 = [\text{Bias}(\bar{x})]^2 + \text{var}(\bar{x}) \quad (4.44)$$

where E(x) is the expected value of x.

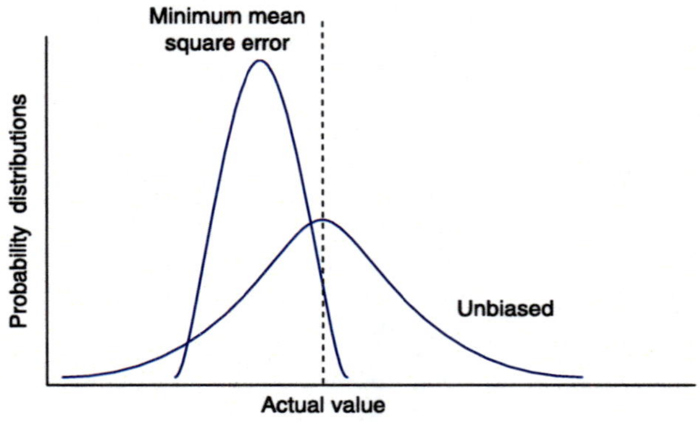

Fig. 4.16 Concept of mean square error which includes bias and efficiency of estimators

Thus, when $\bar{x}$ is unbiased, the mean square error and variance of the estimator $\bar{x}$ are equal. MSE may be regarded as a generalization of the variance concept. This leads to the generalized definition of the relative efficiency of two estimators, whether biased or unbiased: "*efficiency is the ratio of both MSE values.*"

(d) **Consistency:** Consider the properties of estimators as the sample size increases. In such cases, one would like the estimator $\bar{x}$ to converge to the true value, or the probability limit of $\bar{x}$ (*plim* $\bar{x}$) should equal μ as sample size n approaches infinity (see Fig. 4.17). This leads to the criterion of consistency: $\bar{x}$ is a consistent estimator of μ if plim $(\bar{x})=\mu$. In other words, as the sample size grows larger, a consistent estimation would tend to approximate the true parameters, i.e., the mean square error of the estimator approaches zero. Thus, one of the conditions that make an estimator consistent is that *both* its bias and variance approach zero in the limit. However, it does not necessarily follow that an unbiased estimator is a consistent estimator. Although consistency is an abstract concept, it often provides a useful preliminary criterion for sorting out estimators. However, to finally determine the best estimator, the efficiency is a more powerful criterion. As discussed earlier, the sample mean is preferable to the median for estimating the center of a normal population because the former is more efficient though both estimators are clearly consistent and unbiased.

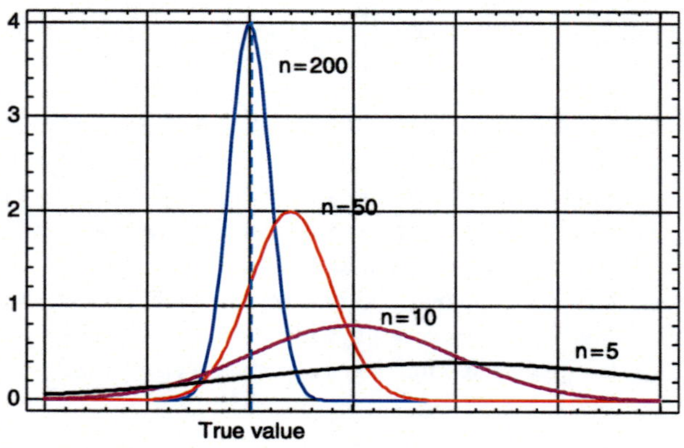

Fig. 4.17 A consistent estimator is one whose distribution becomes gradually peaked as the sample size n is increased

As a general rule, one tends to be more concerned with consistency than with lack of bias as an estimation criterion. A biased yet consistent estimator may not equal the true parameter on average, but will approximate the true parameter as the sample information grows larger. This is more reassuring practically than the alternative of finding a parameter estimate which is unbiased initially, yet will continue to deviate substantially from the true parameter as the sample size gets larger.

4.7.3 Determining Sample Size During Random Surveys

Population census, market surveys, pharmaceutical field trials, etc.. are examples of survey sampling. These can be done in one of two ways which are discussed in this section and in the next. The discussion and equations presented in the previous sub-sections pertain to *random sampling*. Survey sampling frames the problem using certain terms slightly different from those presented above. Here, a major issue is to determine the sample size which can meet a certain pre-stipulated precision at predefined confidence levels.

The estimates from the sample should be close enough to the population characteristic so as to be useful for drawing conclusions and taking subsequent decisions. One generally assumes the underlying probability distribution to be normal. Let RE be the *relative error* (also called the margin of error or bound on error of estimation) of the population mean μ at a confidence level $(1-\alpha)$, which, for a two-tailed distribution, is defined as:

$$RE_{1-\alpha} = z_{\alpha/2}.\frac{\sigma}{\mu} \tag{4.45}$$

where σ is the standard error given by Eq. 4.2.

A measure of variability in the population needs to be introduced, and this is done through the coefficient of variation (CV) defined as:

$$CV = \frac{\text{std.dev.}}{\text{true mean}} = \frac{s_x}{\mu}$$

where s_x is the sample standard deviation. The maximum value of s_x which would allow the confidence level to be met is:

$$s_{x,1-\alpha} = z_{\alpha/2}.CV_{1-\alpha}.\bar{x} \qquad (4.46)$$

One could deduce the required sample size n from the above equation to reach the target $RE_{1-\alpha}$ as follows: First, a simplifying assumption is made by replacing (N−1) by N in Eq. 4.3 which is the expression for the standard error of the mean for small samples. Then

$$\sigma^2 = \frac{s_x^2}{n}\left(\frac{N-n}{N}\right) = \frac{s_x^2}{n} - \frac{s_x^2}{N} \qquad (4.47)$$

Finally, using the definitions of RE and CV stated above, the required sample size is:

$$n = \frac{1}{\frac{\sigma^2}{s_x^2} + \frac{1}{N}} = \frac{1}{\left(\frac{RE_{1-\alpha}}{z_{\alpha/2}.CV_{1-\alpha}}\right)^2 + \frac{1}{N}} \qquad (4.48)$$

This is the functional form normally used in survey sampling in order to determine sample size *provided some prior estimate of the population mean and standard deviation are known.*

Example 4.7.1: *Determination of random sample size needed to verify peak reduction in residences at preset confidence levels*

An electric utility has provided financial incentives to a large number of their customers to replace their existing air-conditioners with high efficiency ones. This rebate program was initiated in an effort to reduce the aggregated electric peak during hot summer afternoons which is dangerously close to the peak generation capacity of the utility. The utility analyst would like to determine the sample size necessary to assess whether the program has reduced the peak as projected such that the relative error RE≤10% at 90% CL. The following information is given:

The total number of customers: N=20000
Estimate of the mean peak saving $\mu = 2$ kW (from engineering calculations)
Estimate of the standard deviation s_x=1 kW (from engineering calculations)

This is a two-tailed distribution problem with 90% CL which corresponds to a one-tailed significance level of $\alpha/2$ =(100−90)/2/100=0.05. Then, from Table A4, $z_{0.05} = 1.65$.

Inserting values of RE=0.1 and $CV = \frac{s_x}{\mu} = \frac{1}{2} = 0.5$ in Eq. 4.48, the required sample size is:

$$n = \frac{1}{\left[\frac{0.1}{(1.65)\cdot(0.5)}\right]^2 + \frac{1}{20{,}000}} = 658.2 \approx 660$$

It would be advisable to perform some sensitivity runs given that many of the assumed quantities are guess-estimates. It is simple to use the above approach to generate figures such as Fig. 4.18 for assessing tradeoff between increased accuracy with sample size and increase in cost of instrumentation, installation and monitoring as sample size is increased.

Note that accepting additional error reduces sample size in a hyperbolic manner. For example, lowering the requirement that RE $\leq \pm 10\%$ to $\leq \pm 15\%$ decreases n from 660 to about 450, while increasing it to $\leq \pm 5\%$ would require a sample size of about 1300 (about twice the initial estimate). On the other hand, there is not much one could do about varying CV since this represents an inherent variability in the population if random sampling is adopted. However, non-random stratified sampling, described next, could be one approach to reduce sample sizes. ■

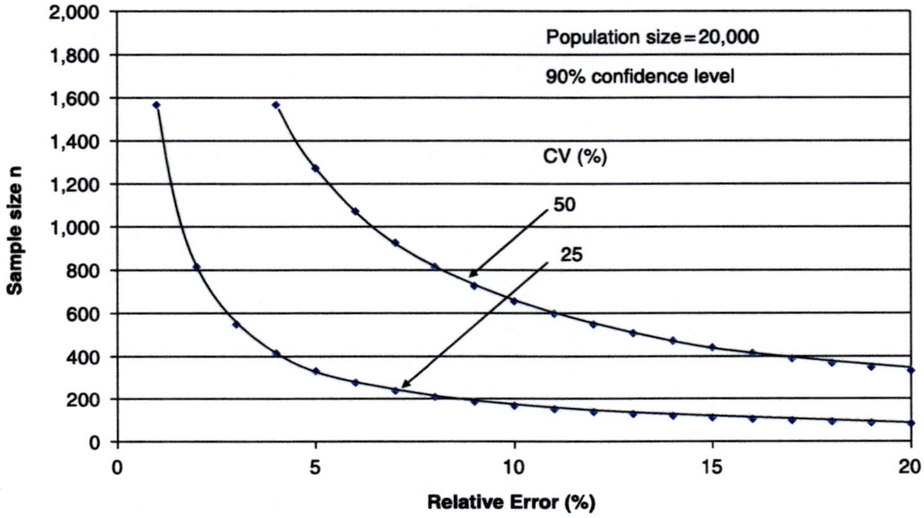

Fig. 4.18 Size of random sample needed to achieve different relative errors of the population mean for two different values of population variability (CV of 25% and 50%, Example 4.7.1)

4.7.4 Stratified Sampling for Variance Reduction

Variance reduction techniques are a special type of sample estimating procedures which rely on the principle that prior knowledge about the structure of the model and the properties of the input can be used to increase the precision of estimates for a fixed sample size, or, conversely decrease the sample size required to obtain a fixed degree of precision. These techniques distort the original problem so that special techniques can be used to obtain the desired estimates at a lower cost.

Variance can be decreased by considering a larger sample size which involves more work. So the effort with which a parameter is estimated can be evaluated as: efficiency= (variance × work)$^{-1}$. This implies that a reduction in variance is not worthwhile if the work needed to achieve it is excessive. A common recourse among social scientists to increase efficiency is to use *stratified sampling*, which counts as a variance reduction technique. In stratified sampling, the distribution function to be sampled is broken up into several pieces, each piece is then sampled separately, and the results are later combined into a single estimate. The specification of the strata to be used is based on prior knowledge about the characteristics of the population to be sampled. Often an order of magnitude variance reduction is achieved by stratified sampling as compared to the standard random sampling approach.

Example 4.7.2:[10] *Example of stratified sampling for variance reduction*
Suppose a home improvement center wishes to estimate the mean annual expenditure of its local residents in the hardware section and the drapery section. It is known that the expenditures by women differ more widely than those by men. Men visit the store more frequently and spend annually approximately $ 50; expenditures of as much as $ 100 or as little as $ 25 per year are found occasionally. Annual expenditures by women can vary from nothing to over $ 500. The variance for expenditures by women is therefore much greater, and the mean expenditure more difficult to estimate.

Assume that 80% of the customers are men and that a sample size of 15 is to be taken. If simple random sampling were employed, one would expect the sample to consist of approximately 12 men (80% of 15) and 3 women. However, assume that a sample that included 5 men and 10 women was selected instead (more women have been preferentially selected because their expenditures are more variable). Suppose the annual expenditures of the members of the sample turned out to be:

Men: 45, 50, 55, 40, 90
Women: 80, 50, 120, 80, 200, 180, 90, 500, 320, 75

It is intuitively clear that such data will lead to a more accurate estimate of the overall average than would the expenditures of 12 men and 3 women.

The appropriate weights must be applied to the original sample data if one wishes to deduce the overall mean. Thus, if M_i and W_i are used to designate the ith sample of men and women, respectively,

$$\begin{aligned}\overline{X} &= \frac{1}{15}\left[\sum_{i=1}^{5}\frac{0.80}{0.33}M_i + \sum_{i=1}^{10}\frac{0.20}{0.67}W_i\right] \\ &= \frac{1}{15}\left[\frac{0.80}{0.33}280 + \frac{0.20}{0.67}1695\right] \approx \$79\end{aligned}$$

where 0.80 and 0.20 are the original weights in the population, and 0.33 and 0.67 the sample weights respectively.

This value is likely to be a more realistic estimate than if the sampling had been done based purely on the percentage of the gender of the customers. The above example is a simple case of stratified sampling where the customer base was first stratified into the two genders, and then these were sampled disproportionately. There are statistical formulae which suggest near-optimal size of selecting stratified samples, for which the interested reader can refer to Devore and Farnum (2005) and other texts.

4.8 Resampling Methods

4.8.1 Basic Concept and Types of Methods

The precision of a population related estimator can be improved by drawing multiple samples from the population, and inferring the confidence limits from these samples rather than determining them from classical analytical estimation formulae based on a single sample only. However, this is infeasible in most cases because of the associated cost and time of assembling multiple samples. The basic rationale behind *resampling methods* is to draw one sample, treat this original sample as a surrogate for the population, and *generate numerous sub-samples by simply resampling the sample itself*. Thus, resampling refers to the use of given data, or a data generating mechanism, to produce new samples from which the required estimator can be deduced numerically. It is obvious that the sample must be unbiased and be reflective of the population (which it will be if the sample is drawn randomly), otherwise the precision of the method is severely compromised.

Efron and Tibshirani (1982) have argued that given the available power of computing, one should move away from the constraints of traditional parametric theory with its over-

[10] From Hines and Montgomery (1990) by © permission of John Wiley and Sons.

reliance on a small set of standard models for which theoretical solutions are available, and substitute computational power for theoretical analysis. This parallels the manner in which numerical methods have in large part replaced closed forms solution techniques in almost all fields of engineering mathematics. Thus, versatile numerical techniques allow one to overcome such problems as the lack of knowledge of the probability distribution of the errors of the variables, and even determine sampling distributions of such quantities as the median or of the inter-quartile range or even the 5th and 95th percentiles for which no traditional tests exist. The methods are conceptually simple, requiring low levels of mathematics, while the needed computing power is easily provided by present-day personal computers. Hence, they are becoming increasingly popular and are used to complement classical/traditional parametric tests.

Resampling methods can be applied to diverse problems (Good 1999): (i) for determining probability in complex situations, (ii) to estimate confidence levels of an estimator during univariate sampling of a population, (iii) hypothesis testing to compare estimators of two samples, (iv) to estimate confidence bounds during regression, and (v) for classification. These problems can all be addressed by classical methods provided one makes certain assumptions regarding probability distributions of the random variables. The analytic solutions can be daunting to those who use these statistical analytic methods rarely, and one can even select the wrong formula by error. Resampling is much more intuitive and provides a way of simulating the physical process without having to deal with the, sometimes obfuscating, statistical constraints of the analytic methods. They are based on a direct extension of ideas from statistical mathematics and have a sound mathematical theory. A big virtue of resampling methods is that they extend classical statistical evaluation to cases which cannot be dealt with mathematically.

The downside to the use of these methods in that they require large computing resources (of the order of 1000 and more samples). This issue is no longer a constraint because of the computing power of modern day personal computers. Resampling methods are also referred to as *computer-intensive methods*, though other techniques discussed in Sect. 10.6 are more often associated with this general appellation. It has been suggested that one should use a parametric test when the samples are large, say number of observations is greater than 40, or when they are small (<5) (Good 1999). The resampling provides protection against violation of parametric assumptions.

The creation of multiple sub-samples from the original sample can be done in several ways and distinguishes one method against the other. The three most common resampling methods are:

(a) *Permutation method* (or randomization method) is one where *all possible subsets* of r items (which is the subsample size) out of the total n items (the sample size) are generated, and used to deduce the population estimator and its confidence levels or its percentiles. This may require some effort in many cases, and so, an equivalent and less intensive deviant of this method is to use only a sample of all possible subsets. The size of the sample is selected based on the accuracy needed, and about 1000 samples are usually adequate.
The use of the permutation method when making inferences about the medians of two populations is illustrated below. The null hypothesis is that the there is no difference between the two populations. First, one samples both populations to create two independent random samples. The difference in the medians between both samples is computed. Next, two subsamples *without replacement* are created from the two samples, and the difference in the medians between both resampled groups recalculated. This is done a large number of times, say 1000 times. The resulting distribution contains the necessary information regarding the statistical confidence in the null hypothesis of the parameter being evaluated. For example, if the difference in the median between the two original samples was lower in 50 of 1000 possible subsamples, then one concludes that the one-tailed probability of the original event was only 0.05. It is clear that such a sampling distribution can be done for any statistic of interest, not just the median. However, the number of randomizations become quickly very large, and so one has to select the number of randomizations with some care.

(b) *The jackknife method* creates subsamples with replacement. The jackknife method, introduced by Quenouille in 1949 and later extended by Tukey in 1958, is a technique of universal applicability that allows confidence intervals to be determined of an estimate calculated from a sample while reducing bias of the estimator. There are several numerical schemes for implementing the jackknife scheme. One version is: (i) to divide the random sample of n observations into g groups of equal size (ii) omit one group at a time and determine what are called pseudo-estimates from the $(g-1)$ groups, (iii) estimate the actual confidence intervals of the parameters. A more widespread method of implementation is to simply create n subsamples with $(n-1)$ data points wherein a single different observation is omitted in each subsample.

(c) *The bootstrap method* (popularized by Efron in 1979) is similar but differs in that no groups are formed but the different sets of data sequences are generated by simply sampling with replacement from the observational data set (Davison and Hinkley 1997). Individual estimators deduced from such samples permit estimates and confidence intervals to be determined. The analyst has to select the number of randomizations while the sample size is selected to be equal to that of the original sample. The method would appear to be circular, i.e., how can one acquire more insight by resampling the same sample? The simple explanation is that "*the population is to the sample as the sample is to the bootstrap sample*". Though the jackknife is a viable method, it has been supplanted by the bootstrap method which has emerged as the most efficient of the resampling methods in that better estimates of standard errors and confidence limits are obtained. Several improvements to the naïve bootstrap have been proposed (such as the bootstrap-t method) especially for long-tailed distributions or for time series data. The bootstrap method will be discussed at more length in Sect. 4.8.3 and Sect. 10.6.

4.8.2 Application to Probability Problems

How resampling methods can be used for solving probability type of problems are illustrated below (Simon 1992). Consider a simple example, where one has six balls labeled 1 to 6. What is the probability that three balls will be picked such that they have 1, 2, 3 in that order if this is done with replacement. The traditional probability equation would yield $(1/6)^3$. The same result can be determined by simulating the 3-ball selection a large number of times. This approach, though tedious, is more intuitive since this is exactly what the traditional probability of the event is meant to represent; namely, the long run frequency. One could repeat this 3-ball selection say a million times, and count the number of times one gets 1, 2, 3 in sequence, and from there infer the needed probability. The procedure rules or the sequence of operations of drawing samples has to be written in computer code, after which the computer does the rest. Much more difficult problems can be simulated in this manner, and its advantages lie in its versatility, its low level of mathematics required, and most importantly, its direct bearing with the intuitive interpretation of probability as the long-run frequency.

4.8.3 Application of Bootstrap to Statistical Inference Problems

The use of the bootstrap method to two types of instances is illustrated: determining confidence intervals and for correlation analysis. At its simplest, the algorithm of the bootstrap method consists of the following steps (Devore and Farnum 2005):

1. Obtain a random sample of size n from the population
2. Generate a random sample of size n with replacement from the original sample in step 1.

Table 4.12 Data table for Example 4.8.1

62	50	53	57	41	53	55	61
59	64	50	53	64	62	50	68
54	55	57	50	55	50	56	55
46	55	53	54	52	47	47	55
57	48	63	57	57	55	53	59
53	52	50	55	60	50	56	58

3. Calculate the statistic of interest for the sample in step 2
4. Repeat steps 2 and 3 a large number of times to form an approximate sampling distribution of the statistic.

It is important to note that bootstrapping requires that sampling be done *with replacement*, and about 1000 samples are required. It is advised that the analyst perform a few evaluations with different number of samples in order to be more confident about his results. The following example illustrates the implementation of the bootstrap method.

Example 4.8.1:[11] *Using the bootstrap method for deducing confidence intervals*
The data in Table 4.12 corresponds to the breakdown voltage (in kV) of an insulating liquid which is indicative of its dielectric strength. Determine the 95% CL.

First, use the large sample confidence interval formula to estimate the 95% CL intervals of the mean. Summary quantities are : sample size n=48, $\sum x_i = 2646$ and $\sum x_i^2 = 144{,}950$ from which $\bar{x} = 54.7$ and standard deviation s=5.23. The 95% CL interval is then:

$$54.7 \pm 1.96\frac{5.23}{\sqrt{48}} = 54.7 \pm 1.5 = (53.2, 56.2)$$

The confidence intervals using the bootstrap method are now recalculated in order to evaluate differences. A histogram of 1000 samples of n=48 each, drawn with replacement, is shown in Fig. 4.19. The 95% confidence intervals correspond to the two-tailed 0.05 significance level. Thus, one

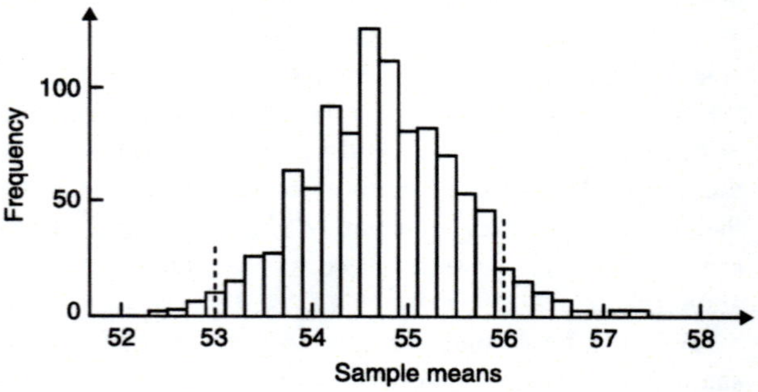

Fig. 4.19 Histogram of bootstrap sample means with 1000 samples (Example 4.8.1)

[11] From Devore and Farnum (2005) by © permission of Cengage Learning.

selects 1000(0.05/2)=25 units from each end of the distribution, i.e., the value of the 25th and that of the 975th largest values which yield (53.2, 56.1) which are very close to the parametric range determined earlier. This example illustrates the fact that bootstrap intervals usually agree with traditional parametric ones when all the assumptions underlying the latter are met. It is when they do not, that the power of the bootstrap stands out. ■

The following example illustrates the versatility of the bootstrap method for determining correlation between two variables, a problem which is recast as comparing two sample means.

Example 4.8.2:[12] *Using the bootstrap method with a nonparametric test to ascertain correlation of two variables*
One wishes to determine whether there exists a correlation between athletic ability and intelligence level. A sample of 10 high school athletes was obtained involving their athletic and I.Q. scores. The data is listed in terms of descending order of athletic scores in the first two columns of Table 4.13.

A *nonparametric* approach is adopted to solve this problem. The athletic scores and the I.Q. scores are rank ordered from 1 to 10 as shown in the last two columns of the table. The two observations (athletic rank, I.Q. rank) are treated together since one would like to determine their joint behavior. The table is split into two groups of five "high" and five "low". An even split of the group is advocated since it uses the available information better and usually leads to better "efficiency". The sum of the observed I.Q. ranks of the five top athletes $=(3+1+7+4+2)=17$. The resampling scheme will involve numerous trials where a subset of 5 numbers is drawn randomly from the set $\{1\ldots10\}$. One then adds these five numbers for each individual trial. If the observed sum across trials is consistently higher than 17, this will indicate that the best athletes will not have earned the observed I.Q. scores purely by chance. The probability can be directly estimated from the proportion of trials whose sum exceeded 17. Figure 4.20 depicts the histogram of the sum of 5 random observations using 100 trials (a rather low number of trials meant for illustration purposes). Note that in only 2% of the trials was the sum 17 or lower. Hence, one can state to within 98% confidence, that there does exist a correlation between athletic ability and I.Q. level. ■

Table 4.13 Data table for Example 4.8.2 along with ranks

Athletic score	I.Q. Score	Athletic rank	I.Q. Rank
97	114	1	3
94	120	2	1
93	107	3	7
90	113	4	4
87	118	5	2
86	101	6	8
86	109	7	6
85	110	8	5
81	100	9	9
76	99	10	10

[12] From Simon (1992) by © permission of Duxbury Press.

Problems

Pr. 4.1 The specification to which solar thermal collectors are being manufactured requires that their lengths be between 8.45–8.65 feet and their width between 1.55–1.60 ft. The modules produced by a certain assembly line have lengths that are normally distributed about a mean of 8.56 ft with standard deviation 0.05 ft, and widths also normally distributed with a mean of 1.58 ft with standard deviation 0.01 ft. For the modules produced by this assembly line, find:

(a) the % that will not be within the specified limits for length; state the implicit assumption in this approach
(b) the % that will not be within the specified limits for width; state the implicit assumption in this approach
(c) the % that will not meet the specifications; state the implicit assumption in this approach.

Pr. 4.2 The pH of a large lake is to be determined for which purpose 9 test specimens were collected and tested to give: $\{6.0, 5.7, 5.8, 6.5, 7.0, 6.3, 5.6, 6.1, 5.0\}$.

(a) Calculate the mean pH for the 9 specimens
(b) Find an unbiased estimate of the standard deviation of the population of all pH samples
(c) Find the 95% CL interval for the mean of this population if it is known from past experience that the pH values have a standard deviation of 0.6. State the implicit assumption in this approach
(d) Find the 95% CL interval for the mean of this population if no previous information about pH value is available. State the implicit assumption in this approach.

Pr. 4.3 Two types of cement brands A and B were evaluated by testing the compressive strength of concrete blocks made from them. The results for 7 test pieces for A and 5 for B are shown in Table 4.14.

(a) Estimate the mean difference between both types of concrete
(b) Estimate the 95% confidence limits of the difference of both types of concrete
(c) Repeat the above using the bootstrap method with 1000 samples and compare results.

Pr. 4.4 *Using classical and Bayesian approaches to verify claimed benefit of gasoline additive*
An inventor claims to have developed a gasoline additive which increases gas mileage of cars. He specifically states

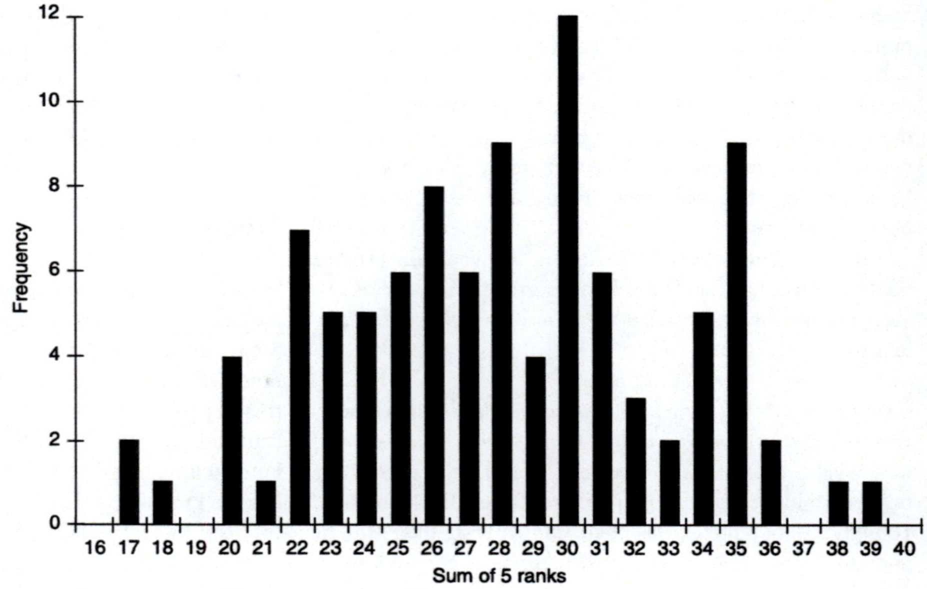

Fig. 4.20 Histogram based on 100 trials of the sum of 5 random ranks from the sample of 10. Note that in only 2% of the trials was the sum equal to 17 or lower (Example 4.8.2)

Table 4.14 Data table for Problem 4.3

Brand A	2.18	3.17	2.46	2.70	2.78	3.35	3.52
Brand B	3.13	3.07	3.92	3.51	2.92		

Table 4.15 Data table for Problem 4.4

Regular	395	420	405	417	399	410
With additive	440	436	447	453	444	426

Table 4.16 Data table for Problem 4.5

4.04	4.38	2.90	4.47	2.73	0.74
4.60	5.05	2.87	1.72	3.08	4.01
5.73	4.04	6.48	3.26	3.25	1.22
5.39	3.48	3.74	6.01	6.08	
2.37	5.25	3.99	3.40	5.15	
5.39	1.80	0.89	3.96	2.73	
4.60	4.93	3.72	2.82	5.87	
5.05	3.83	3.51	3.41	3.77	

that tests using a specific model and make of car resulting in an increase of 50 miles per filling. An independent testing group repeated the tests on six identical cars and obtained the results shown in Table 4.15.

(a) Assuming normal distribution, perform parametric tests at the 95% CL to verify the inventor's claim
(b) Repeat the problem using the bootstrap method with 1000 samples and compare results.

Pr. 4.5[13] *Analyzing distribution of radon concentration in homes*

The Environmental Protection Agency (EPA) determined that an indoor radon concentration in homes of 4 pCi/L was acceptable though there is an increased cancer risk level for humans of 10^{-6}. Indoor radon concentrations in 43 residences were measured randomly, as shown in Table 4.16.

The Binomial distribution is frequently used in risk assessment since only two states or outcomes can exist: either one has cancer or one does not.

(a) Determine whether the normal or the t-distributions better represent the data
(b) Compute the standard deviation
(c) Use the one-tailed test to evaluate whether the mean value is less than the threshold value of 4 pCi/L at the 90% CL
(d) At what confidence level can one state that the true mean is less than 6 pCi/L ?
(e) Compute the range for the 90% confidence intervals.

Pr. 4.6 Table 4.17 assembles radon concentrations in pCi/L for U.S. homes. Clearly it is not a normal distribution. Researchers have suggested using the lognormal distribution or the power law distribution with exponent of 0.25. Evaluate which of these two functions is more appropriate:

(a) graphically using quantile plots
(b) using appropriate statistical tests for distributions.

Pr. 4.7 *Using survey sample to determine proportion of population in favor of off-shore wind farms*

A study is initiated to estimate the proportion of residents in a certain coastal region who do not favor the construction of an off-shore wind farm. The state government decides that

[13] From Kammen and Hassenzahl (1999) by © permission of Princeton University Press.

Table 4.17 Data table for Problem 4.6

Concentration Level (pCi/L)	% homes	Concentration Level (pCi/L)	% homes
0.25	16	2.75	2
0.50	18	3.00	3
0.75	13	3.25	1
1.00	10	3.50	2
1.25	8	3.75	1
1.50	6	4.00	2
1.75	5	4.25	2
2.00	4	4.50	1
2.25	3	4.75	1
2.50	2	5.00	0

if the fraction of those against wind farms at the 95% CL drops to less than 0.50, then the wind farm permit will likely be granted.

(a) A survey of 200 residents at random is taken from which 90 state that they are not in favor. Based on this data, would the permit be granted or not?
(b) If the survey size is increased, what factors could intervene which could result in a reversal of the above course of action?
(c) A major public relation campaign is initiated by the wind farm company in an effort to sway public opinion in their favor. After the campaign, a new sample of 100 residents at random was taken, and now only 30 stated that they were not in favor. Did the fraction of residents change significantly at the 95% CL from the previous fraction?
(d) Would a permit likely to be granted in this case?

Pr. 4.8 *Using ANOVA to evaluate pollutant concentration levels at different times of day*
The transportation department of a major city is concerned with elevated air pollution levels during certain times of the day at some key intersections. Samples of SO_2 in ($\mu g/m^3$) are taken at three locations during three different times of the day as shown in Table 4.18.

(a) Conduct an ANOVA test to determine whether the mean concentrations of SO_2 differ during the three collection periods at $\alpha = 0.05$
(b) Create an effects plot of the data
(c) Use Tukey's multiple comparison procedure to determine which collection periods differ from one another.

Table 4.18 Data table for Problem 4.8

Collection time	Location A	Location B	Location C
7 am	50	80	62
Noon	45	52	48
6 pm	57	74	68

Table 4.19 Data table for Problem 4.9

Test #	1	2	3	4	5	6	7	8
Fan A	55	52	51	59	60	56	54	54
Fan B	46	55	59	50	47	62	53	55

Pr. 4.9 *Using non-parametric tests to identify the better of two fan models*
The facility manager of a large campus wishes to replace the fans in the HVAC system of his buildings. He narrows down the possibilities to two manufacturers and wishes to use Wilcoxon Rank sum at significance level $\alpha = 0.05$ to identify the better fan manufacturer based on the number of hours of operation prior to servicing. Table 4.19 assembles such data (in hundreds of hours) generated by an independent testing agency:

Pr. 4.10 *Parametric test to evaluate relative performance of two PV systems from sample data*
A consumer advocate group wishes to evaluate the performance of two different types of photovoltaic (PV) panels which are very close in terms of rated performance and cost. They convince a builder of new homes to install 2 panels of each brand on two homes in the same locality with care taken that their tilt and orientation towards the sun are identical. The test protocol involves monitoring these two PV panels for 15 weeks and evaluating the performance of the two brands based on their weekly total electrical output. The weekly total electrical output in kWh is listed in Table 4.20. The monitoring equipment used is identical in both locations and has an absolute error of 3 kWh/week at 95% uncertainty level. Evaluate using parametric tests whether the two brands are different at a significance level of $\alpha = 0.05$ with measurement errors being explicitly considered.

Pr. 4.11 *Comparing two instruments using parametric, non-parametric and bootstrap methods*
A pyranometer meant to measure global solar radiation is being cross-compared with a primary reference instrument. Several simultaneous observations in (kW/m^2) were taken with both instruments deployed side by side as shown in Table 4.21. Determine, at a significance level $\alpha = 0.05$, whether the secondary field instrument differs from the primary based on:

(a) Parameteric tests
(b) Non-parametric tests
(c) The bootstrap method with a sample size of 1000.

Pr. 4.12 Repeat Example 4.2.1 using the Bayesian approach assuming:

(a) the sample of 36 items tested have a mean of 15 years and a standard deviation of 2.5 years
(b) the same mean and standard deviation but the sample consists of 9 items only.

Table 4.20 Weekly electrical output in kWh (Problem 4.10)

Week	Brand A	Brand B	Week	Brand A	Brand B	Week	Brand A	Brand B
1	197	189	6	203	187	11	174	170
2	202	199	7	165	160	12	225	218
3	148	142	8	121	115	13	242	232
4	246	248	9	146	138	14	206	213
5	173	176	10	189	173	15	197	193

Table 4.21 Data table for Problem 4.11

Observation	Reference	Secondary	Observation	Reference	Secondary	Observation	Reference	Secondary
1	0.96	0.93	6	0.89	0.91	11	0.84	0.81
2	0.82	0.78	7	0.64	0.62	12	0.59	0.55
3	0.75	0.76	8	0.81	0.77	13	0.94	0.87
4	0.61	0.64	9	0.68	0.63	14	0.91	0.86
5	0.77	0.74	10	0.65	0.62			

Pr. 4.13 An electrical motor company states that one of their product lines of motors has a mean life 8100 h with a standard deviation of 200 h. A wholesale dealer purchases a consignment and tests 10 of the motors. The sample mean and standard deviation are found to be 7800 h with a standard deviation of 100 h. Assume normal distribution. Compute:

(a) The 95 % confidence interval based on the classical approach
(b) The 95 % confidence interval based on the Bayesian approach
(c) The probability that the consignment has a mean value less than 4000 h.

Pr. 4.14[14] The average cost of electricity to residential customers during the three summer months is to be determined. A sample of electric cost in 25 residences is collected as shown in Table 4.22. Assume a normal distribution with standard deviation of 80.

(a) If the prior value is a Gaussian with N(325, 80), find the posterior distribution for the mean μ
(b) Find a 95% Bayesian credible interval for μ
(c) Compare the interval with that from the traditional method
(d) Perform a traditional test for: $H_0 : \mu = 350$ versus $H_1' : \mu \neq 350$ at the 0.05 significance level
(e) Perform a Bayesian test of the hypothesis: $H_0 : \mu \leq 350$ versus $H_1' : \mu > 350$ at the 0.05 significance level.

Table 4.22 Data table for Problem 4.14

514	536	345	440	427
443	386	418	364	483
506	385	410	561	275
306	294	402	350	343
480	334	324	414	296

[14] From Bolstad (2004) by © permission of John Wiley and Sons.

Pr. 4.15 *Comparison of human comfort correlations between Caucasian and Chinese subjects*
Human indoor comfort can be characterized by to the occupants' feeling of well-being in the indoor environment. It depends on several interrelated and complex phenomena involving subjective as well as objective criteria. Research initiated over 50 years back and subsequent chamber studies have helped define acceptable thermal comfort ranges for indoor occupants. Perhaps the most widely used standard is ASHRAE Standard 55-2004 (ASHRAE 2004). The basis of the standard is the thermal sensation scale determined by the votes of the occupants following the scale in Table 4.23.

The individual votes of all the occupants are then averaged to yield the predicted mean vote (PMV). This is one of the two indices relevant to define acceptability of a large population of people exposed to a certain indoor environment. PMV=0 is defined as the neutral state (neither cool nor warm), while positive values indicate that occupants feel warm, and vice versa. The mean scores from the chamber studies are then regressed against the influential environmental parameters so as to yield an empirical correlation which can be used as a means of prediction:

$$PMV = a^* T_{db} + b^* P_v + c^* \tag{4.49}$$

where T_{db} is the indoor dry-bulb temperature (degrees C), P_v is the partial pressure of water vapor (kPa), and the nu-

Table 4.23 ASHRAE thermal sensation classes

+3	+2	+1	0	−1	−2	−3
Hot	Warm	Slightly warm	Neutral	Slightly cool	Cool	Cold

Table 4.24 Regression parameters of the ASHRAE PMV model (Eq. 4.49) for 3 h exposure

Sex	a*	b*	c*
Male	0.212	0.293	−5.949
Female	0.275	0.255	−8.622
Combined	0.243	0.278	−6.802

merical values of the coefficients a*, b* and c* are dependent on such factors as sex, age, hours of exposure, clothing levels, type of activity, …. The values relevant to healthy adults in an office setting for a 3 h exposure period are given in Table 4.24.

In general, the distribution of votes will always show considerable scatter. The second index is the percentage of people dissatisfied (PPD), defined as people voting outside the range of −1 to +1 for a given value of PMV. When the PPD is plotted against the mean vote of a large group characterized by the PMV, one typically finds a distribution such as that shown in Fig. 4.21. This graph shows that even under optimal conditions (i.e., a mean vote of zero), at least 5% are dissatisfied with the thermal comfort. Hence, because of individual differences, it is impossible to specify a thermal environment that will satisfy everyone. A correlation between PPD and PMV has also been suggested:

$$PPD = 100 - 95 \exp\left[-0.03353.PMV^4 + 0.2179.PMV^2)\right] \tag{4.50}$$

Note that the overall approach is consistent with the statistical approach of approximating distributions by the two primary measures, the mean and the standard deviation. However, in this instance, the standard deviation (characterized by PPD) has been empirically found to be related to the mean value, namely PMV (Eq. 4.50).

A research study was conducted in China by Jiang (2001) in order to evaluate whether the above types of correlations, developed using American and European subjects, are applicable to Chinese subjects as well. The environmental chamber test protocol was generally consistent with previous Western studies. The total number of Chinese subjects in the pool was about 200, and several tests were done with smaller batches (about 10–12 subjects per batch evenly split between males and females). Each batch of subjects first spent some time in a pre-conditioning chamber after which they were moved to the main chamber. The environmental conditions (dry-bulb temperature T_{db}, relative humidity RH and air velocity) of the main chamber were controlled such that: $T_{db}(\pm 0.3^0C)$, $RH(\pm 5\%)$ and air velocity<0.15 m/s. The subjects were asked to vote about every ½ hr over 2½ h in accordance with the 7-point thermal sensation scale. However, in this problem we consider only the data relating to averages of the two last votes corresponding to 2 and 2½ h since only then was it found that the voting had stabilized (this feature of the length of exposure is also consistent with American/European tests).

Three separate sets each consisting of 18 tests were performed; one for females only, one for males only, and one for combined[15]. The chamber T_{db},RH and the associated partial pressure of water needed in Eq. 4.49 (which can be determined from psychrometric relations) along with the PMV and PPD measures are tabulated as shown in Table P4.15 (see Appendix B). The conditions under which these were done is better visualized if plotted on a psychrometric chart shown in Fig. 4.22. Based on this data, one would like to determine whether the psychological responses of Chinese people are different from those of American/European people.

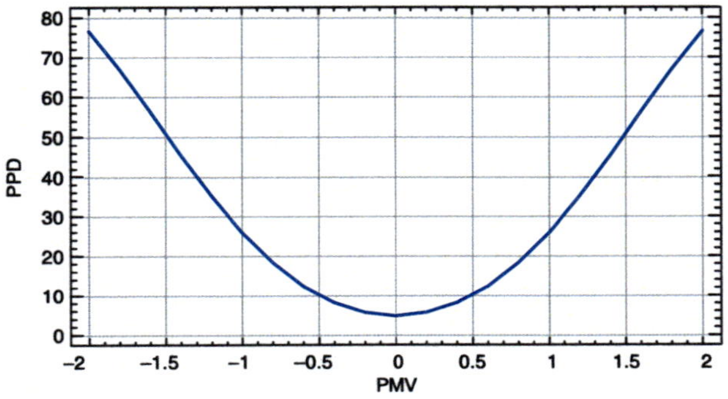

Fig. 4.21 Predicted percentage of dissatisfied (PPD) as function of predicted mean vote (PMV) following Eq. 4.50

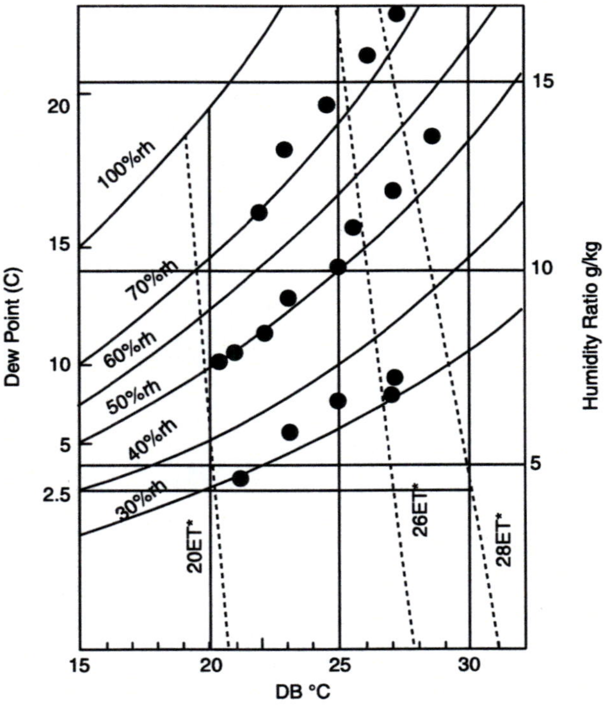

Fig. 4.22 Chamber test conditions plotted on a psychrometric chart for Chinese subjects

[15] This data set was provided by Wei Jiang for which we are grateful.

Hint: One of the data points is suspect. Also use Eqs. 4.49 and 4.50 to generate the values pertinent to Western subjects prior to making comparative evaluations.

(a) Formulate the various different types of tests one would perform stating the intent of each test
(b) Perform some or all of these tests and draw relevant conclusions
(c) Prepare a short report describing your entire analysis.

References

ASHRAE, 2004. *ASHRAE Standard 55-2004: Thermal Environmental Conditions for Human Occupancy* (ANSI Approved), American Society of Heating, Refrigerating and Air-Conditioning Engineers, Atlanta, GA.

ASTM E 1402, 1996. *Standard Terminology Relating to Sampling*, ASTM Standard, pub. January 1997.

Bolstad, W.M.,2004. *Introduction to Bayesian Statistics*, John Wiley and Sons, Hoboken, NJ.

Davison, A.C. and D.V.Hinkley,1997. *Bootstrap Methods and their Applications*, Cambridge University Press, Cambridge

Devore J., and N. Farnum,2005. *Applied Statistics for Engineers and Scientists*, 2nd Ed., Thomson Brooks/Cole, Australia.

Efron, B. and J. Tibshirani,1982. *An Introduction to the The Bootstrap*, Chapman and Hall, New York

Good, P.I.,1999. *Resampling Methods: A Practical Guide to Data Analysis*, Birkhauser, Boston.

Hines, W.W. and D.C. Montgomery,1990. *Probability and Statistics in Engineering and Management Sciences*, 3rd Edition, Wiley Interscience, New York.

Jiang, Wei, 2001. "Experimental Research on Thermal Comfort" (in Chinese), Master of Science thesis, School of Civil Engineering, Tianjin University, China, June.

Johnson, R.A. and D.W. Wichern,1988. *Multivariate Analysis*, Prentice-Hall, Englewood Cliffs, NJ.

Kammen, D.M. and D.M. Hassenzahl,1999. *Should We Risk It*, Princeton Univ. Press, Princeton, NJ

Manly, B.J.F.,2005. *Multivariate Statistical Methods: A Primer*, 3rd Ed.,Chapman & Hall/CRC, Boca Raton, FL

McClave, J.T. and P.G. Benson,1988. *Statistics for Business and Economics*, 4th Ed., Dellen and Macmillan, London

Phillips, L.D.,1973. *Bayesian Statistics for Social Scientists*, Thomas Nelson and Sons, London, UK

Pindyck, R.S. and D.L. Rubinfeld,1981. *Econometric Models and Economic Forecasts*, 2nd Edition, McGraw-Hill, New York, NY.

Simon, J.,1992. *Resampling: The New Statistic*, Duxbury Press, Belmont, CA

Walpole, R.E., R.H. Myers, S.L. Myers, and K. Ye,2007, *Probability and Statistics for Engineers and Scientists*, 8th Ed., Pearson Prentice Hall, Upper Saddle River, NJ.

Weiss, N., 1987. *Introductory Statistics*, Addison-Wesley. NJ.

Wonnacutt, R.J. and T.H. Wonnacutt,1985. *Introductory Statistics*, 4th Ed., John Wiley & Sons, New York.

Estimation of Linear Model Parameters Using Least Squares

5

This chapter deals with methods to estimate parameters of linear parametric models using ordinary least squares (OLS). The univariate case is first reviewed along with equations for the uncertainty in the model estimates as well as in the model predictions. Several goodness-of-fit indices to evaluate the model fit are also discussed, and the assumptions inherent in OLS are highlighted. Next, multiple linear models are treated, and several notions specific to correlated regressors are presented. The insights which residual analysis provides are discussed, and different types of remedial actions to improper model residuals are addressed. Other types of linear models such as splines and models with indicator variables are discussed. Finally, a real-world case study analysis which was meant to verify whether actual field tests supported the claim that a refrigerant additive improved chiller thermal performance is discussed.

5.1 Introduction

The analysis of observational data or data obtained from designed experiments often requires the identification of a statistical model or relationship which captures the underlying structure of the system from which the sample data was drawn. A model is a relation between the variation of one variable (called the *dependent or response variable*) against that of other variables (called *independent or regressor variables*). If observations (or data) are taken of both response and regressor variables under various sets of conditions, one can build a mathematical model from this information which can then be used as a predictive tool under different sets of conditions. How to analyze the relationships among variables and determine a (if not "the") optimal relation, falls under the realm of *regression model building or regression analysis*.

Models, as stated in Sect. 1.1, can be of different forms, with mathematical models being of sole concern in this book. These can divided into:

(i) *parameteric models* which can be a single function (or a set of functions) capturing the variation of the response variable in terms of the regressors. The intent is to identify both the model function and determine the values of the parameters of the model along with some indication of their uncertainty; and
(ii) *nonparametric models* where the relationship between response and regressors is such that a mathematical model in the conventional sense is inadequate. Nonparameteric models are treated in Sect. 9.3 in the framework of time series models and in Sect. 11.3.2 when dealing with artificial neural network models.

The parameters appearing in parametric models can be estimated in a number of ways, of which ordinary least squares (OLS) is the most common and historically the oldest. Other estimation techniques are described in Chap. 10. There is a direct link between how the model parameters are estimated and the underlying joint probability distributions of the variables, which is discussed below and in Chap. 10. In this chapter, only *models linear in the parameters* are addressed which need not necessarily be linear models (see Sect. 1.2.4 for relevant discussion). However, often, the former are loosely referred to as linear parametric models.

5.2 Regression Analysis

5.2.1 Objective of Regression Analysis

The objectives of regression analysis are: (i) to identify the "best" model among several candidates in case the physics of the system does not provide an unique mechanistic relationship, and (ii) to determine the "best" values of the model parameters; with "best" being based on some criterion yet to be defined. Desirable properties of estimators, which are viewed as random variables, have been described in Sect. 4.7.2, and most of these concepts apply to parameter estimation of regression models as well.

T. Agami Reddy, *Applied Data Analysis and Modeling for Energy Engineers and Scientists*,
DOI 10.1007/978-1-4419-9613-8_5, © Springer Science+Business Media, LLC 2011

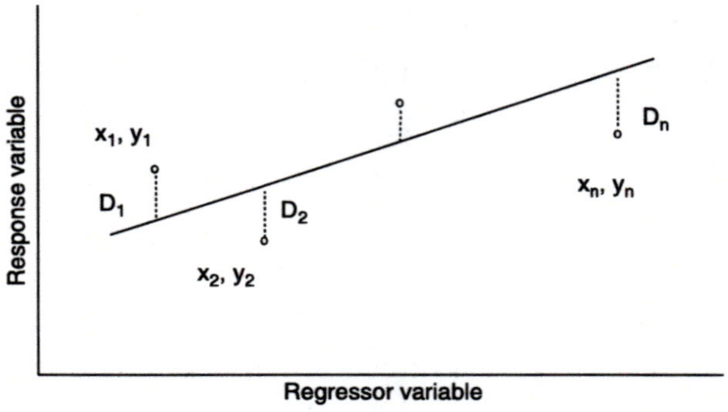

Fig. 5.1 Ordinary Least Squares Regression (OLS) is based on finding the model parameters which minimize the squared sum of the vertical deviations or residuals or $\left(\sum_{i=1}^{n} D_i^2\right)^{1/2}$

5.2.2 Ordinary Least Squares

Once a set of data is available, what is the best model which can be fit to the data. Consider the (x, y) set of n data points shown in Fig. 5.1. The criterion for "best fit" should be objective, intuitively reasonable and relatively easy to implement mathematically. One would like to minimize the deviations of the points from the prospective regression line. The method most often used is the *method of least squares* where, as the name implies, the "best fit" line is interpreted as one which minimizes the sum of the squares of the residuals. Since it is based on minimizing the squared deviations, it is also referred to as the Method of Moments Estimation (MME). The most common and widely used sub-class of least squares is the ordinary least squares (OLS) where, as shown in Fig. 5.1, squared sum of the vertical differences between the line and the observation points are minimized, i.e., $\min\left(D_1^2 + D_2^2 + \cdots + D_n^2\right)$. Another criterion for determining the best fit line could be to minimize the sum of the absolute deviations, i.e., $\min(|D_1| + |D_2| + \cdots |D_n|)$. However, the mathematics to deal with absolute quantities becomes cumbersome and restrictive, and that is why historically, the method of least squares was proposed and developed. Inferential statistics plays an important part in regression model building because identification of the system structure via a regression line from sample data has an obvious parallel to inferring population mean from sample data (discussed in Sect. 4.2.1). The intent of regression is to capture or "explain" via a model the variation in y for different x values. Taking a simple mean value of y (see Fig. 5.2a) leaves a lot of the variation in y unexplained. Once a model is fit, however, the unexplained variation is much reduced as the regression line accounts for some of the variation that is due to x (see Fig. 5.2b, c). Further, the assumption of normally distributed variables, often made in inferential theory, is also presumed for the distribution of the population of y values at each x value (see Fig. 5.3). Here, one notes that when slices of data are made at different values of x, the individual y distributions are *close to normal with equal variance*.

5.3 Simple OLS Regression

5.3.1 Traditional Simple Linear Regression

Let us consider a simple linear model with two parameters, a and b, given by:

$$y = a + b \cdot x \tag{5.1}$$

The parameter 'a' denotes the model intercept, i.e., the value of y at x=0, while the parameter 'b' is the slope of the straight line represented by the simple model (see Fig. 5.4). The objective of the regression analysis is to determine the numerical values of the parameters a and b which result in the model given by Eq. 5.1 able to *best* explain the variation of y about its mean $\bar{y}$ as the numerical value of the regressor variable x changes. Note that the slope parameter b explains the *variation* in y due to that in x. It does not necessarily follow that this parameter accounts for more of the observed *absolute* magnitude in y than does the intercept parameter term a. For any y value, the total deviation can be partitioned into two pieces: explained and unexplained (recall the ANOVA approach presented in Sect. 4.3 which is based on the same conceptual approach). Mathematically,

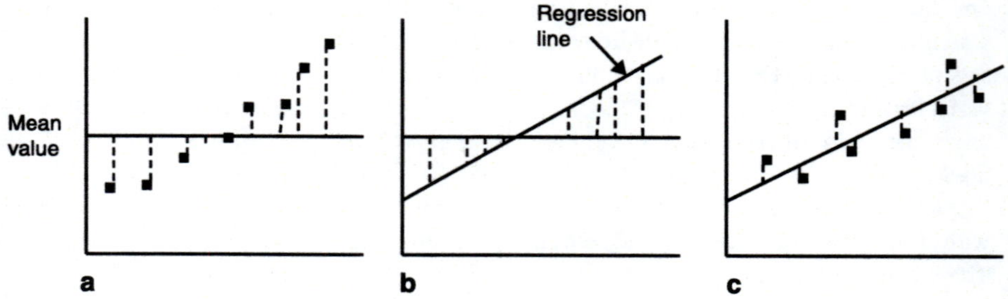

Fig. 5.2 Conceptual illustration of how regression explains or reduces unexplained variation in the response variable. It is important to note that the variation in the response variable is taken in reference to its mean value. **a** total variation (before regression), **b** explained variation (due to regression), **c** residual variation (after regression)

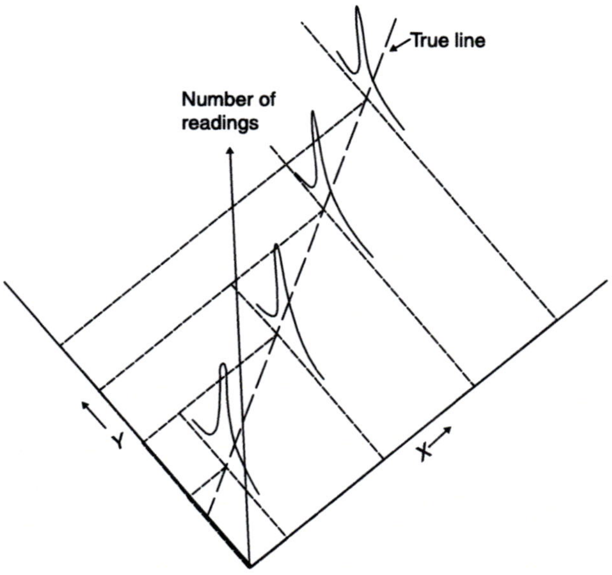

Fig. 5.3 Illustration of normally distributed errors with equal variances at different discrete slices of the regressor variable values. Normally distributed errors is one of the basic assumptions in OLS regression analysis. (From Schenck 1969 by permission of McGraw-Hill)

$$\sum_{i=1}^{n}(y_i-\bar{y})^2=\sum_{i=1}^{n}(y_i-\hat{y}_i)^2+\sum_{i=1}^{n}(\hat{y}_i-\bar{y})^2 \quad (5.2)$$
$$\text{or}\quad SST = SSE + SSR$$

where
y_i is the individual response at observation i,
$\bar{y}$ the mean value of y_i of the n observations,
$\hat{y}_i$ the value of y estimated from the regression model for observation i,
SST = total sum of squares,
SSE = error sum of squares or sum of the residuals which reflects the variation about the regression line (similar to Eq. 4.20 when dealing with ANOVA type of problems), and
SSR = regression sum of squares which reflects the amount of variation in y explained by the model (similar to treatment sum of squares of Eq. 4.19).

These quantities are conceptually illustrated in Fig. 5.4. The sum of squares minimization implies that one wishes to minimize SSE, i.e.

$$\sum_{i=1}^{n}(y_i-\hat{y}_i)^2=\sum_{i=1}^{n}(y_i-a-bx_i)^2=\sum_{i=1}^{n}\varepsilon_i^2 \quad (5.3)$$

where ε is called the model residuals or error.

From basic calculus, the model residuals are minimized when:

$$\frac{\partial\sum\varepsilon^2}{\partial a}=0 \quad \text{and} \quad \frac{\partial\sum\varepsilon^2}{\partial b}=0$$

The above two equations lead to the following equations (called the *normal equations*):

$$na+b\sum x=\sum y \quad \text{and}$$
$$a\sum x+b\sum x^2=\sum xy \quad (5.4)$$

where n is the number of observations. This leads to the following expressions of the most "probable" OLS values of a and b:

$$b=\frac{n\sum x_iy_i-(\sum x_i)(\sum y_i)}{n\sum x_i^2-(\sum x_i)^2}=\frac{S_{xy}}{S_{xx}} \quad (5.5a)$$

$$a=\frac{(\sum y_i)(\sum x_i^2)-(\sum x_iy_i)(\sum x_i)}{n\sum x_i^2-(\sum x_i)^2}=\bar{y}-b\cdot\bar{x} \quad (5.5b)$$

where

$$S_{xy}=\sum_{i=1}^{n}(x_i-\bar{x})(y_i-\bar{y}) \quad \text{and}$$
$$S_{xx}=\sum_{i=1}^{n}(x_i-\bar{x})^2 \quad (5.6)$$

Fig. 5.4 The value of regression in reducing unexplained variation in the response variable as illustrated by using a single observed point. The total variation from the mean of the response variable is partitioned into two portions: one that is explained by the regression model and the other which is the unexplained deviation, also referred to as model residual

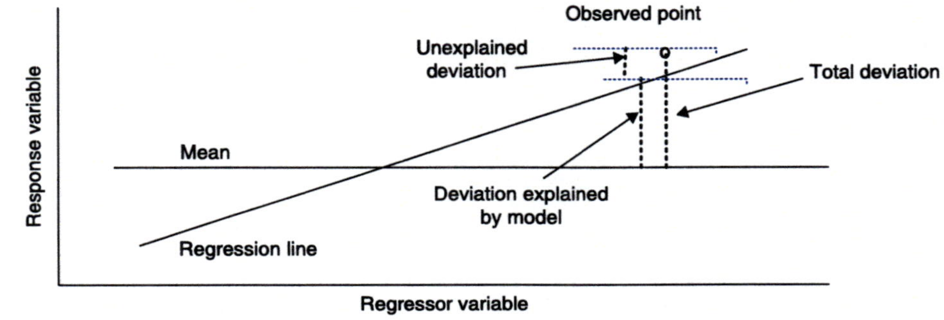

Table 5.1 Data table for Example 5.3.1

Solids reduction x (%)	Chemical oxygen demand, y (%)	Solids reduction x (%)	Chemical oxygen demand, y (%)
3	5	36	34
7	11	37	36
11	21	38	38
15	16	39	37
18	16	39	36
27	28	39	45
29	27	40	39
30	25	41	41
30	35	42	40
31	30	42	44
31	40	43	37
32	32	44	44
33	34	45	46
33	32	46	46
34	34	47	49
36	37	50	51
36	38		

How the least squares regression model reduces the unexplained variation in the response variable is conceptually illustrated in Fig. 5.4.

Example 5.3.1:[1] *Water pollution model between solids reduction and chemical oxygen demand*
In an effort to determine a regression model between tannery waste (expressed as solids reduction) and water pollution (expressed as chemical oxygen demand), sample data (33 observation sets) shown in Table 5.1 were collected. Estimate the parameters of a linear model.

The regression line is estimated by first calculating the following quantities:

$$\sum_{i}^{33} x_i = 1104, \quad \sum_{i}^{33} y_i = 1124,$$

$$\sum_{i}^{33} x_i \cdot y_i = 41{,}355, \quad \sum_{i}^{33} x_i^2 = 41{,}086$$

Subsequently Eqs. 5.5a and b are used to compute:

$$b = \frac{(33)(41{,}355) - (1104)(1124)}{(33)(41{,}086) - (1104)^2} = 0.9036$$

$$a = \frac{1124 - (0.903643)(1104)}{33} = 3.8296$$

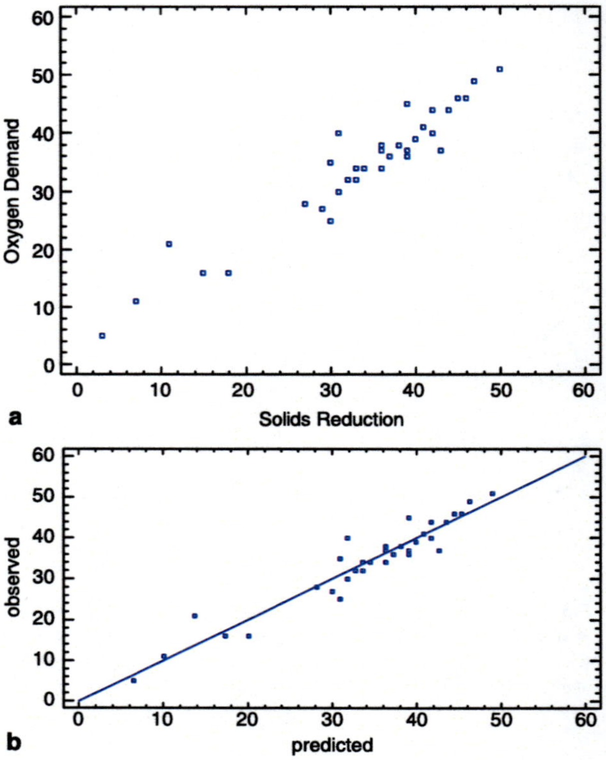

Fig. 5.5 **a** Scatter plot of data **b** Plot of observed versus OLS model predicted values of the y variable

Thus, the estimated regression line is:

$$\hat{y} = 3.8296 + 0.9036 \cdot x$$

The above data is plotted as a scatter plot in Fig. 5.5a. How well the regression model performs compared to the measurements is conveniently assessed from the observed vs predicted plot such as Fig. 5.5b. Tighter scatter of the data points around the regression line indicates more accurate model fit.

The regression line can be used for prediction purposes. The value of y at, say, x=50 is simply:

$$\hat{y} = 3.8296 + (0.9036)(50) = 49$$ ■

5.3.2 Model Evaluation

(a) The most widely used measure of model adequacy or goodness-of-fit is the *coefficient of determination* R^2 where $0 \leq R^2 \leq 1$:

$$R^2 = \frac{\text{explained variation of y}}{\text{total variation of y}} = \frac{SSR}{SST} \tag{5.7a}$$

[1] From Walpole et al. (1998) by © permission of Pearson Education.

For a perfect fit $R^2=1$, while $R^2=0$ indicates that either the model is useless or that no relationship exists. For a univariate linear model, R^2 is identical to the square of the Pearson correlation coefficient r (see Sect. 3.4.2). R^2 is a misleading statistic if models with different number of regressor variables are to be compared. The reason for this is that R^2 does not account for the number of degrees of freedom, it cannot but increase as additional variables are included in the model even if these variables have very little explicative power.

(b) A more desirable goodness-of-fit measure is *the corrected or adjusted* $\bar{R}^2$, computed as

$$\bar{R}^2 = 1 - (1 - R^2)\frac{n-1}{n-k} \qquad (5.7b)$$

where n is the total number of observation sets, and k is the number of model parameters (for a simple linear model, k=2).

Since $\bar{R}^2$ concerns itself with variances and not variation, this eliminates the incentive to include additional variables in a model which have little or no explicative power. Thus, $\bar{R}^2$ is the right measure to use during identification of a parsimonious[2] model when multiple regressors are in contention. However, it should not be used to decide whether an intercept is to be added or not. For the intercept model, $\bar{R}^2$ is the proportion of variability measured by the sum of squares *about the mean* which is explained by the regression. Hence, for example, $\bar{R}^2 = 0.92$ would imply that 92% of the variation in the dependent variable about its mean value is explained by the model.

(c) Another widely used estimate of the magnitude of the absolute error of the model is the *root mean square error* (RMSE), defined as follows:

$$\mathrm{RMSE} = \left(\frac{\mathrm{SSE}}{\mathrm{n-k}}\right)^{1/2} \qquad (5.8a)$$

where SSE is the sum of square error defined as

$$SSE = \sum (y_i - \hat{y}_i)^2 = \sum (y_i - a - b \cdot x_i)^2. \qquad (5.8b)$$

The RMSE is an absolute measure and its range is $0 \le \mathrm{RMSE} \le \infty$. Its units are the same as those of the y variable. It is also referred to as "*standard error of the estimate*".

A normalized measure is often more appropriate: the *coefficient of variation* of the RMSE (or CVRMSE or simply CV), defined as:

$$CV = \frac{RMSE}{\bar{y}} \qquad (5.8c)$$

Hence, a CV value of say 12% implies that the root mean value of the unexplained variation in the dependent variable y is 12% of the mean value of y.

[2] Parsimony in the context of regression model building is a term denoting the most succinct model, i.e., one without any statistically superfluous regressors.

Note that the CV defined thus is based on absolute errors. Hence, it tends to place less emphasis on deviations between model predictions and observations which occur at lower numerical values of y than at the high end. Consequently, the measure may inadequately represent the goodness of fit of the model over the entire range of variation under certain circumstances. An alternative definition of CV based on *relative mean deviations* is:

$$CV^* = \left\{\frac{1}{(n-k)}\sum_{i=1}^{n}\left[\frac{(y_i - \hat{y}_i)}{y_i}\right]^2\right\}^{1/2} \qquad (5.8d)$$

If CV and CV* indices differ appreciably for a particular model, this would suggest that the model may be inadequate at the extreme range of variation of the response variable. Specifically, if CV* > CV, this would indicate that the model deviates more at the lower range, and vice versa.

(d) The *mean bias error* (MBE) is defined as the mean difference between the actual data values and model predicted values:

$$\mathrm{MBE} = \frac{\sum_{i=1}^{n}(y_i - \hat{y}_i)}{n-k} \qquad (5.9a)$$

Note that when a model is identified by OLS, the model MBE of the original set of regressor variables used to identify the model should be zero (to within round-off errors of the computer). Only when, say, the model identified from a first set of observations is used to predict the value of the response variable under a second set of conditions will MBE be different than zero. Under such circumstances, the MBE is also called the mean *simulation or prediction error*. A normalized MBE (or NMBE) is often used, and is defined as the MBE given by Eq. 5.9a divided by the mean value $\bar{y}$:

$$\mathrm{NMBE} = \frac{MBE}{\bar{y}} \qquad (5.9b)$$

Competing models can be evaluated based on the CV and the NMBE values; i.e., those that have low CV and NMBE values. Under certain circumstances, one model may be preferable to another in terms of one index but not the other. The analysts is then perplexed as to which index to pick as the primary one. In such cases, the specific intent of how the model is going to be subsequently applied should be considered which may suggest the model selection criterion.

While fitting regression models, there is the possibility of "overfitting", i.e., the model fits part of the noise in the data along with the system behavior. In such cases, the model is likely to have poor predictive ability which often the analyst is unaware of. A statistical index is defined later (Eq. 5.42) which can be used to screen against this possibility. A better way to minimize this effect is to randomly partition the data set into two (say, in proportion of 80/20), use the 80% portion of the data to develop the model, calculate the *internal predictive indices* CV and NMBE (following Eqs. 5.8c and 5.9b), use the 20% portion of the data and predict the y values using the already identified model, and finally calculate the *external or simulation* indices CV and NMBE. The competing models can then be compared, and a selection made, based on both the internal and external predictive indices. The simulation indices will generally be poorer than the internal predictive indices; larger discrepancies are suggestive of greater over-fitting, and vice versa. This method of model evaluation which can avoid model over-fitting is referred to as *holdout sample cross-validation* or simply cross-validation. Note, however, that though the same equations are used to compute the CV and NMBE indices, the degrees of freedom (df) are different. While df=n-k for computing the internal predictive errors where n is the number of observations used for model building, df=m for computing the external indices where m is the number of observations in the cross-validation set.

(e) The *mean absolute deviation* (MAD) is defined as the mean *absolute* difference between the actual data values and model predicted values:

$$\text{MAD} = \frac{\sum_{i=1}^{n} |y_i - \hat{y}_i|}{n - k} \tag{5.10}$$

Example 5.3.2: Using the data from Example 5.3.1 repeat the exercise using your spreadsheet program. Calculate, R^2, RMSE and CV values.

From Eq. 5.2, SSE=323.3 and SSR=3713.88. From this SST=SSE+SSR=4037.2.

Then from Eq. 5.7a, R^2=92.0%, while from Eq. 5.8a, RMSE=3.2295, from which CV=0.095=9.5%. ■

5.3.3 Inferences on Regression Coefficients and Model Significance

Even after the overall regression model is found, one must guard against the fact that there may not be a significant relationship between the response and the regressor variables, in which case the entire identification process becomes suspect. The *F-statistic*, which tests for significance of the overall regression model, is defined as:

$$\begin{aligned} F &= \frac{\text{variance explained by the regression}}{\text{variance not explained by the regression}} \\ &= \frac{SSR}{SSE} \cdot \frac{n-k}{k-1} \end{aligned} \tag{5.11}$$

Thus, the smaller the value of F, the poorer the regression model. It will be noted that the F-statistic is directly related to R^2 as follows:

$$F = \frac{R^2}{(1-R^2)} \cdot \frac{n-k}{k-1} \tag{5.12}$$

Hence, the F-statistic can alternatively be viewed as being a *measure to test the R^2 significance itself.* In the case of univariate regression, the F-test is really the same as a t-test for the significance of the slope coefficient. In the general case, the F-test allows one to test the joint hypothesis of whether *all* coefficients of the regressor variables are equal to zero or not.

Example 5.3.3: Calculate the F-statistic for the model identified in Example 5.3.1. What can you conclude about the significance of the fitted model? From Eq. 5.11,

$$F = \left(\frac{3713.88}{323.3}\right) \cdot \left(\frac{33-2}{2-1}\right) = 356$$

which clearly indicates that the overall regression fit is significant. The reader can verify that Eq. 5.12 also yields an identical value of F. ■

Note that the values of coefficients a and b based on the given sample of n observations are only estimates of the true model parameters α and β. If the experiment is repeated over and over again, the estimates of a and b are likely to vary from one set of experimental observations to another. OLS estimation assumes that the model residual ε is a random variable with zero mean. Further, it is assumed that the residuals ε_i at specific values of x are randomly distributed, which is akin to saying that the distributions shown in Fig. 5.3 at specific values of x are normal and have equal variance.

After getting an overall picture of the regression model, it is useful to study the significance of each individual regressor on the overall statistical fit in the presence of all other regressors. The *student t-statistic* is widely used for this purpose and is applied to each regression parameter:

For the slope parameter:

$$t = \frac{b-1}{s_b} \tag{5.13a}$$

where the estimated standard deviation of parameter "b" is $s_b = RMSE/\sqrt{S_{xx}}$.

For the intercept parameter:

$$t = \frac{a-0}{s_a} \tag{5.13b}$$

where the estimated standard deviation of parameter "a" is

$$s_a = RMSE.\left(\frac{\sum_i^n x_i^2}{n.S_{xx}}\right)^{1/2}$$

where b and a are the estimated slope and intercept coefficients, β and α the hypothesized true values, and RMSE is given by Eq. 5.8a. Estimated standard deviations of the coefficients b and a, given by Eqs. 5.13a and b, are usually referred to as *standard errors of the coefficients*. Basically, the t-test as applied to regression model building is a formal statistical test to determine how significantly different an individual coefficient is from zero in the presence of the remaining coefficients. Stated simply, it enables an answer to the following question: would the fit become poorer if the regressor variable in question is not used in the model at all?

The confidence intervals, assuming the model residuals to be normally distributed, are given by:

For the slope:

$$b - \frac{t_{\alpha/2} \cdot RMSE}{\sqrt{S_{xx}}} < \beta < b + \frac{t_{\alpha/2} \cdot RMSE}{\sqrt{S_{xx}}} \quad (5.14a)$$

For the intercept:

$$a - \frac{t_{\alpha/2} \cdot RMSE \cdot \sqrt{\sum_i^n x_i^2}}{\sqrt{n \cdot S_{xx}}} < \alpha < a + \frac{t_{\alpha/2} \cdot RMSE \cdot \sqrt{\sum_i^n x_i^2}}{\sqrt{n \cdot S_{xx}}} \quad (5.14b)$$

where $t_{\alpha/2}$ is the value of the t distribution with df=(n−2) and S_{xx} is defined by Eq. 5.6.

Example 5.3.4: In Example 5.3.1, the estimated value of b=0.9036. Test the hypothesis that $\beta = 1.0$ as against the alternative that < 1.0.

H_0: $\beta = 1.0$
H_1: $\beta < 1.0$

From Eq. 5.6a, $S_{xx} = 4152.1$. Using Eq. 5.13a, with *RMSE*=3.2295

$$t = \frac{0.9036 - 1.0}{3.2295/\sqrt{4152.18}} = -1.92$$

with n−2=31 degrees of freedom.

From Table A.4, the one-sided critical t-value for 95% CL=1.697. Since the computed t-value is greater than the critical value, one can reject the null hypothesis and conclude that there is strong evidence to support $\beta < 1$ at the 95% confidence level. ■

Example 5.3.5: In Example 5.3.1, the estimated value of a=3.8296. Test the hypothesis that $\alpha = 0$ as against the alternative that $\alpha \neq 0$ at the 95% confidence level.

H_0: $\alpha = 0$
H_1: $\alpha \neq 0$

Using Eq. 5.13b,

$$t = \frac{3.8296 - 0}{3.2295/\sqrt{41{,}086/(33)(4152.18)}} = 2.17$$

with n−2=31 degrees of freedom.

Again, one can reject the null hypothesis, and conclude that $\alpha \neq 0$ at 95% CL. ■

Example 5.3.6: Find the 95% confidence interval for the slope term of the linear model identified in Example 5.3.1.

Assuming a two-tailed test, $t_{0.05/2} = 2.045$ for 31 degrees of freedom. Therefore, the 95% confidence interval for β given by Eq. 5.14a is:

$$0.9036 - \frac{(2.045)(3.2295)}{(4152.18)^{1/2}} < \beta < 0.9036 + \frac{(2.045)(3.2295)}{(4152.18)^{1/2}}$$
$$0.8011 < \beta < 1.0061$$ ■

Example 5.3.7: Find the 95% confidence interval for the intercept term of the linear model identified in Example 5.3.1.

Again, assuming a two-tailed test, and using Eq. 5.14b, the 95% confidence interval for α is:

$$3.8296 - \frac{(2.045)(3.2295)\sqrt{41{,}086}}{[(33)(4152.18)]^{1/2}} < \alpha < 3.8296 + \frac{(2.045)(3.2295)\sqrt{41{,}086}}{[(33)(4152.18)]^{1/2}}$$
$$0.2131 < \alpha < 7.4461$$ ■

5.3.4 Model Prediction Uncertainty

A regression equation can be used to predict future values of y *provided* the x value is within the domain of the original data from which the model was identified. One differentiates between the two types of predictions (similar to the confidence limits of the mean treated in Sect. 4.2.1.b):

(a) **mean response** or *standard error of regression* where one would like to predict the mean value of y for a large number of repeated x_0 values. The mean value is directly deduced from the regression equation while the variance is:

$$\sigma^2(\hat{y}_0) = MSE \cdot \left[\frac{1}{n} + \frac{(x_0 - \bar{x})^2}{S_{xx}}\right] \quad (5.15)$$

Note that the first term within the brackets, namely (MSE/n) is the standard error of the mean (see Eq. 4.2) while the other term is a result of the standard error of the slope coefficient. The latter has the effect of widening the uncertainty bands at either end of the range of variation of x.

(b) **individual or specific response** or *standard error of prediction* where one would like to predict the specific value of y for a specific value x_0. This error is larger than the error in the mean response by an amount equal to the RMSE. Thus,

$$\sigma^2(\hat{y}_0) = MSE \cdot \left[1 + \frac{1}{n} + \frac{(x_0 - \bar{x})^2}{S_{xx}}\right] \quad (5.16)$$

Finally, the 95% CL for the individual response at level x_0 is:

$$y_0 = \hat{y}_0 \pm t_{0.05/2} \cdot \sigma(\hat{y}_0) \quad (5.17)$$

where $t_{0.05/2}$ is the value of the t-student distribution at a significance level of 0.05 for a two-tailed error distribution. It is obvious that the *prediction intervals* for individual responses are wider than those of the mean response called *confidence levels* (see Fig. 5.6). Note that Eqs. 5.16 and 5.17 strictly apply when the errors are normally distributed.

Some texts state that the data set should be at least five to eight times larger than the number of model parameters to be identified. In case of *short data sets,* OLS may not yield robust estimates of model uncertainty and resampling methods are advocated (see Sect. 10.6.2).

Example 5.3.8: Calculate the 95% confidence limits (CL) for predicting the mean response for x=20.

First, the regression model is used to calculate $\hat{y}_0$ at x_0=20:

$$\hat{y}_0 = 3.8296 + (0.9036)(20) = 21.9025$$

Using Eq. 5.15,

$$\sigma(\hat{y}_0) = (3.2295)\left[\frac{1}{33} + \frac{(20 - 33.4545)^2}{4152.18}\right]^{1/2} = 0.87793$$

Further, from Table A.4, $t_{0.05/2}$=2.04 for d.f. = 33–2 = 31. Using Eq. 5.15 yields the confidence interval for the mean response

$$21.9025 - (2.04)(0.87793) < \mu(\hat{y}_{20}) < 21.9025 + (2.04)(0.87793)$$

or,

$$20.112 < \mu(\hat{y}_{20}) < 23.693 \text{ at 95\% CL.}$$ ■

Example 5.3.9: Calculate the 95% prediction limits (PL) for predicting the individual response for x=20.

Using Eq. 5.16,

$$\sigma(\hat{y}_0) = (3.2295)\left[1 + \frac{1}{33} + \frac{(20 - 33.4545)^2}{4152.18}\right]^{1/2} = 3.3467$$

Further, $t_{0.05/2}$=2.04. Using Eq. 5.17 yields

$$21.9025 - (2.04)(3.3467) < \hat{y}_{20} < 21.9025 + (2.04)(3.3467)$$

or

$$15.075 < \hat{y}_{20} < 28.730.$$ ■

5.4 Multiple OLS Regression

Regression models can be classified as:

(i) *single variate or multivariate,* depending on whether only one or several regressor variables are being considered;

(ii) *single equation or multi-equation* depending on whether only one or several response variables are being considered; and

(iii) *linear or non-linear,* depending on whether the model is linear or non-linear in its function. Note that the distinction is with respect to the parameters (and not its variables). Thus, a regression equation such as $y = a + b \cdot x + c \cdot x^2$ is said to be linear in its parameters {a, b, c} though it is non-linear in the regressor variable

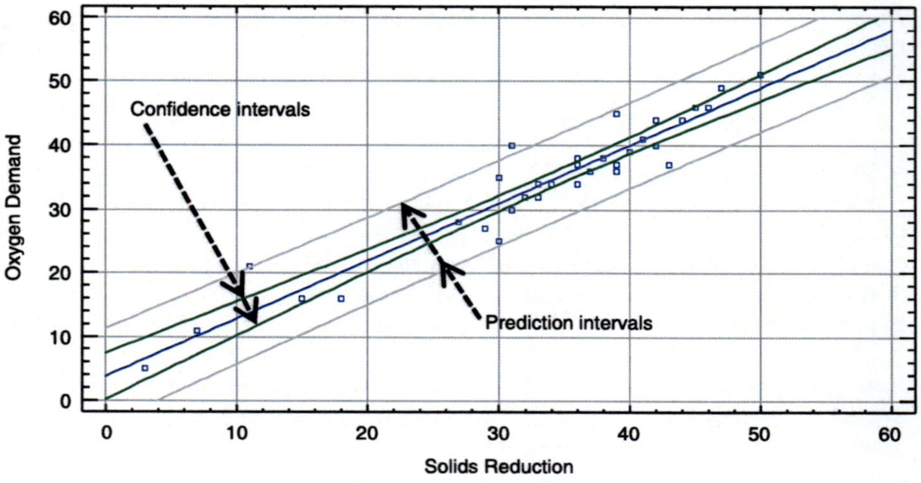

Fig. 5.6 95% confidence intervals and 95% prediction intervals about the regression line

x (see Sect. 1.2.3 for a discussion on classification of mathematical models).

Certain simple single variate equation models are shown in Fig. 5.7. Frame (a) depicts simple linear models (one with a positive slope and another with a negative slope), while (b) and (c) are higher order polynomial models which, though non-linear in the function, are models linear in their parameters. The other figures depict non-linear models. Because of the relative ease in linear model building, data analysts often formulate a linear model even if the relationship of the data is not strictly linear. If a function such as that shown in frame (d) is globally non-linear, and if the domain of the experiment is limited say to the right knee of the curve (bounded by points c and d), then a linear function in this region could be postulated. Models tend to be preferentially framed as linear ones largely due to the simplicity in the subsequent analysis and the prevalence of solution methods based on matrix algebra.

5.4.1 Higher Order Linear Models: Polynomial, Multivariate

When more than one regressor variable is known to influence the response variable, a multivariate model will explain more of the variation and provide better predictions than a single

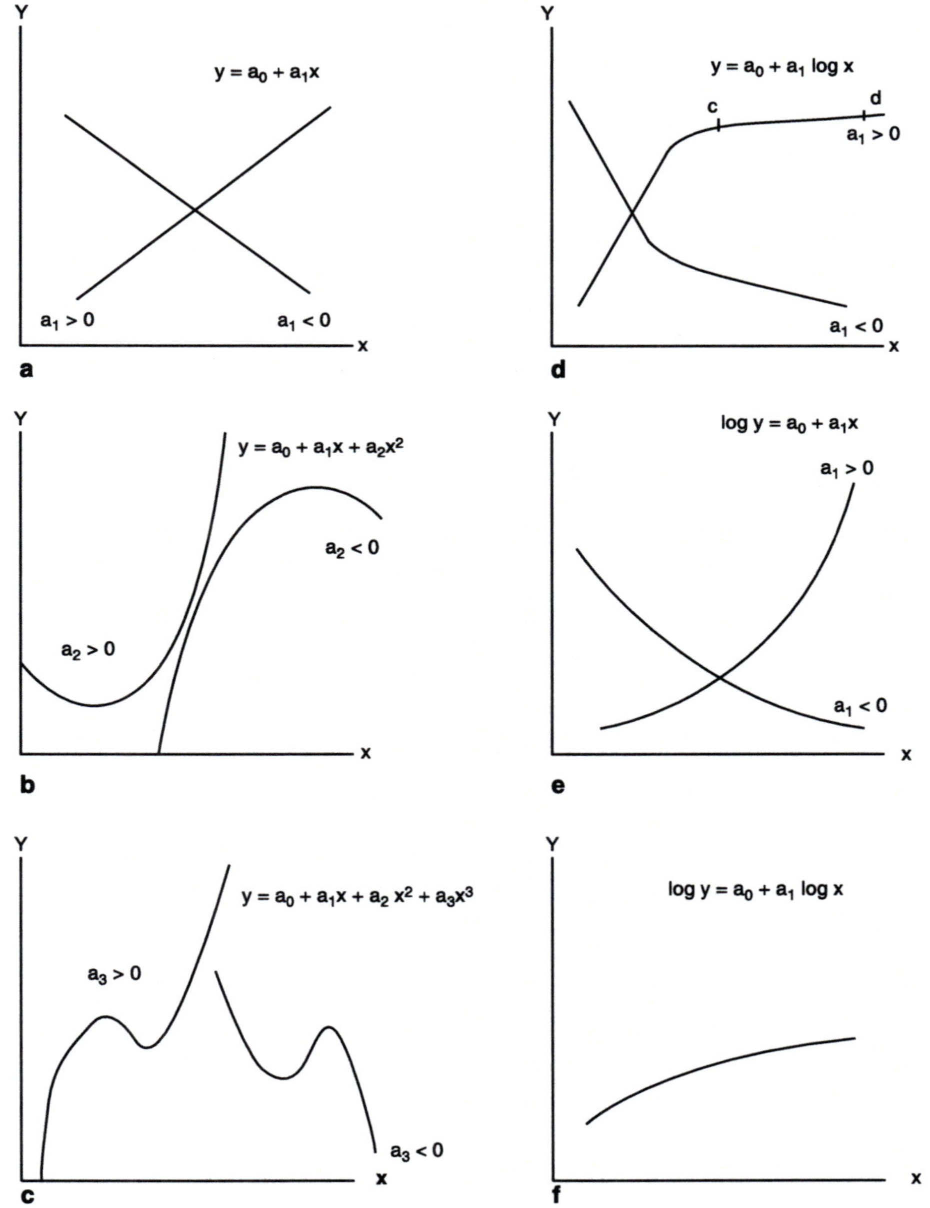

Fig. 5.7 General shape of regression curves. (From Shannon 1975 by © permission of Pearson Education)

variate model. The parameters of such a model need to be identified using multiple regression techniques. This section will discuss certain important issues regarding multivariate, single-equation models linear in the parameters. For now, the treatment is limited to regressors which are *uncorrelated or independent.* Consider a data set of n readings that include k regressor variables. The corresponding form, called the *additive multiple linear regression* model, is:

$$y = \beta_0 + \beta_1 x_1 + \beta_2 x_2 + \cdots + \beta_k x_k + \varepsilon \quad (5.18a)$$

where ε is the error or unexplained variation in y. Note the lack of any interaction terms, and hence the term "additive". The simple interpretation of the model parameters is that β_i measures the unit influence of x_i on y (i.e., denotes the slope $\frac{dy}{dx_i}$). Note that this is strictly true only when the variables are really independent or uncorrelated, which, often, they are not.

The same model formulation is equally valid for a k-th degree polynomial regression model which is a special case of Eq. 5.18a with $x_1=x$, $x_2=x^2$ …

$$y = \beta_0 + \beta_1 x + \beta_2 x^2 + \cdots + \beta_k x^k + \varepsilon \quad (5.19)$$

Let x_{ij} denote the ith observation of parameter j. Then Eq. 5.18a can be re-written as

$$y_i = \beta_0 + \beta_1 x_{i1} + \beta_2 x_{i2} + \cdots + \beta_k x_{ik} + \varepsilon_i \quad (5.18b)$$

Often, it is most convenient to consider the "normal" transformation where the regressor variables are expressed as a difference from the mean (the reason why this form is important will be discussed in Sect. 6.3 while dealing with experimental design methods). Specifically, Eq. 5.18a transforms into

$$\begin{aligned} y =& \beta_0' + \beta_1(x_1 - \bar{x}_1) + \beta_2(x_2 - \bar{x}_2) \\ &+ \cdots + \beta_k(x_k - \bar{x}_k) + \varepsilon \end{aligned} \quad (5.18c)$$

An important special case is the quadratic regression model when k=2. The straight line is now replaced by parabolic curves depending on the value of β (i.e., either positive or negative). Multivariate model development utilizes some of the same techniques as discussed in the single variable case. The first step is to identify all variables that can influence the response as predictor variables. It is the analyst's responsibility to identify these potential predictor variables based on his or her knowledge of the physical system. It is then possible to plot the response against all possible predictor variables in an effort to identify any obvious trends. The greatest single disadvantage to this approach is the sheer labor involved when the number of possible predictor variables is high.

A situation that arises in multivariate regression is the concept of variable synergy, or commonly called *interaction between variables* (this is a consideration in other problems; for example, when dealing with design of experiments). This occurs when two or more variables interact and impact system response to a degree greater than when the variables operate independently. In such a case, the *first-order linear model with two interacting regressor variables* takes the form:

$$y = \beta_0 + \beta_1 x_1 + \beta_2 x_2 + \beta_3 x_1 \cdot x_2 + \varepsilon \quad (5.20)$$

How the interaction parameter affects the shape of the family of curves is illustrated in Fig. 5.8. The origin of this model function is easy to derive. The lines for different values of regressor x_1 are essentially parallel, and so the slope terms for both models are equal. Let the model with the first regressor be: $y = a' + bx_1$, while the intercept be given by: $a' = f(x_2) = a + cx_2$. Combining both equations results in: $y = a + bx_1 + cx_2$. This corresponds to Fig. 5.8a. For the interaction case, both the slope and the intercept terms are function of x_2. Hence, representing $a' = a + bx_1$ and $b' = c + dx_1$, then:

$$y = a + bx_1 + (c + dx_1)x_2 = a + bx_1 + cx_2 + dx_1x_2$$

which is identical in structure to Eq. 5.20.

Simple linear functions have been assumed above. It is straightforward to derive expressions for higher order models by analogy. For example, the *second-order (or quadratic) model without interacting variables* is:

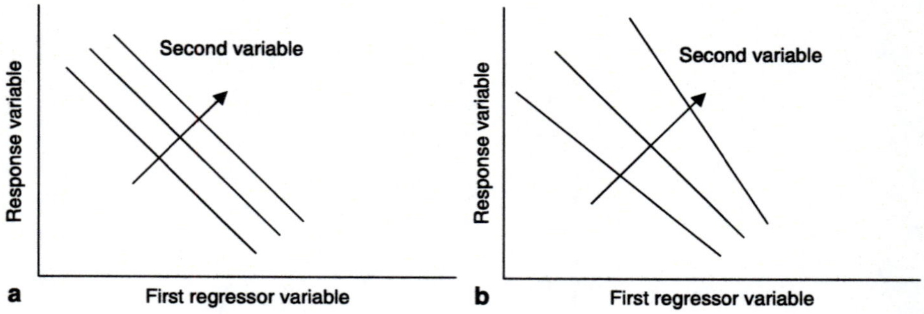

Fig. 5.8 Plots illustrating the effect of interaction among the regressor variables. **a** Non-interacting. **b** Interacting

$$y = \beta_0 + \beta_1 x_1 + \beta_2 x_2 + \beta_3 x_1^2 + \beta_4 x_2^2 + \varepsilon \quad (5.21)$$

For a second order model *with interacting terms,* the corresponding expression can be derived as follows:

Consider the linear polynomial model with one regressor:

$$y = b_0 + b_1 x_1 + b_2 {x_1}^2 \quad (5.22)$$

If the parameters $\{b_0, b_1, b_2\}$ can themselves be expressed as second-order polynomials of another regressor x_2, the full model which has nine regression parameters is:

$$\begin{aligned} y &= b_{00} + b_{10}x_1 + b_{01}x_2 + b_{11}x_1x_2 \\ &+ b_{20}x_1^2 + b_{02}x_2^2 + b_{21}x_1^2x_2 \\ &+ b_{12}x_1x_2^2 + b_{22}x_1^2x_2^2. \end{aligned} \quad (5.23)$$

The most general additive model, which imposes little structure to the relationship is given by:

$$y = \beta_0 + f_1(x_1) + f_2(x_2) + \cdots + f_k(x_k) + \varepsilon \quad (5.24)$$

where the form of $f_i(x_i)$ are unspecified.

Note that synergistic behavior can result in two or more variables working together to "overpower" another variable's prediction capability. As a result, it is necessary to always check the importance (the relative value of either the t- or F-values) of each individual predictor variable while performing multivariate regression. Those variables with t- or F-values that are insignificant should be omitted from the model and the remaining predictors used to estimate the model parameters. The stepwise regression method described in Sect. 5.7.4 is based on this approach.

5.4.2 Matrix Formulation

When dealing with multiple regression, it is advantageous to resort to matrix algebra because of the compactness and ease of manipulation it offers without loss in clarity. Though the solution is conveniently provided by a computer, a basic understanding of matrix formulation is nonetheless useful. In matrix notation (with **y'** denoting the transpose of **y**), the linear model given by Eq. 5.18 can be expressed as follows (with the matrix dimension shown in subscripted brackets):

$$Y_{(n,1)} = X_{(n,p)}\beta_{(p,1)} + \varepsilon_{(n,1)} \quad (5.25)$$

where p is the number of parameters in the model = k + 1 (for a linear model), n is the number of observations and

$$Y' = [y_1\ y_2 \ldots y_n], \quad \beta' = [\beta_0\ \beta_1 \ldots \beta_k], \quad \varepsilon' = [\varepsilon_1\ \varepsilon_2 \ldots \varepsilon_n] \quad (5.26a)$$

and

$$X = \begin{bmatrix} 1 & x_{11} & \cdots & x_{1k} \\ 1 & x_{21} & \cdots & \cdots \\ \cdots & \cdots & \cdots & \cdots \\ 1 & x_{n1} & \cdots & x_{nk} \end{bmatrix}. \quad (5.26b)$$

The descriptive measures applicable for a single variable can be extended to multivariables of order p (= k + 1), and written in compact matrix notation.

5.4.3 OLS Parameter Identification

The approach involving minimization of SSE for the univariate case (Sect. 5.3.1) can be generalized to multivariate linear regression. Here, the parameter set β is to be identified such that the sum of squares function L is minimized:

$$L = \sum_{i=1}^{n} \varepsilon_i^2 = \varepsilon'\varepsilon = (Y - X\beta)'(Y - X\beta) \quad (5.27)$$

or,

$$\frac{\partial L}{\partial \beta} = -2X'Y + 2X'X\beta = 0 \quad (5.28)$$

which leads to the system of normal equations

$$X'Xb = X'Y. \quad (5.29)$$

From here,

$$b = (X'X)^{-1}X'Y \quad (5.30)$$

provided matrix X is not singular and where b is the least square estimator matrix of β.

Note that X'X is called the *variance-covariance matrix* of the estimated regression coefficients. It is a symmetrical matrix with the main diagonal elements being the sum of squares of the elements in the columns of X (i.e., the variances) and the off-diagonal elements being the sum of the cross-products (i.e., the covariances). Specifically,

$$\mathbf{x}'\mathbf{x} = \begin{bmatrix} n & \sum_{i=1}^{n} x_{i1} & \cdots & \sum_{i=1}^{n} x_{ik} \\ \sum_{i=1}^{n} x_{i1} & \sum_{i=1}^{n} x_{i1}^2 & \cdots & \sum_{i=1}^{n} x_{i1} \cdot x_{ik} \\ \cdots & \cdots & \cdots & \cdots \\ \sum_{i=1}^{n} x_{ik} & \sum_{i=1}^{n} x_{ik} \cdot x_{i1} & \cdots & \sum_{i=1}^{n} x_{ik}^2 \end{bmatrix}. \quad (5.31)$$

Under OLS regression, **b** is an unbiased estimator of β with the variance-covariance matrix var(b) given by:

$$\text{var}(b) = \sigma^2 (X'X)^{-1} \tag{5.32}$$

where σ^2 is the mean square error of the model error terms

$$= (\text{sum of square errors})/(n-p) \tag{5.33}$$

An unbiased estimator of σ^2 is s^2, where

$$s^2 = \frac{\varepsilon'\varepsilon}{n-p} = \frac{y'y - b'x'y}{n-p} = \frac{SSE}{n-p} \tag{5.34}$$

For predictions within the range of variation of the original data, the mean and individual response values are normally distributed with the variance given by the following:

(a) For the *mean* response at a specific set of x_0 values, called the *confidence level,* under OLS

$$\text{var}(\hat{y}_0) = s^2 \left[X_0(X'X)^{-1}X_0'\right] \tag{5.35}$$

(b) The variance of an *individual* prediction, called the *prediction level,* is

$$\text{var}(\hat{y}_0) = s^2 \left[\mathbf{1} + X_0(X'X)^{-1}X_0'\right] \tag{5.36}$$

where **1** is a column vector of unity.
Confidence limits at a significance level α are:

$$y_0 \pm t(n-k, \alpha/2) \cdot \text{var}^{1/2}(\hat{y}_0) \tag{5.37}$$

Example 5.4.1: *Part load performance of fans (and pumps)*
Part-load performance curves do not follow the idealized fan laws due to various irreversible losses. For example, decreasing the flow rate by half of the rated flow does not result in a 1/8th decrease in its rated power consumption. Hence, actual tests are performed for such equipment under different levels of loading. The performance tests of the flow rate and the power consumed are then normalized by the rated or 100% load conditions called part load ratio (PLR) and fractional full-load power (FFLP) respectively. Polynomial models can then be fit between these two quantities. Data assembled in Table 5.2 were obtained from laboratory tests on a variable speed drive (VSD) control which is a very energy efficient control option.

(a) What is the matrix X in this case if a second order polynomial model is to be identified of the form $y = \beta_0 + \beta_1 x_1 + \beta_2 x_1^2$?

Table 5.2 Data table for Example 5.4.1

PLR	0.2	0.3	0.4	0.5	0.6	0.7	0.8	0.9	1.0
FFLP	0.05	0.11	0.19	0.28	0.39	0.51	0.68	0.84	1.00

(b) Using the data given in the table, identify the model and report relevant statistics on both parameters and overall model fit.

(c) Compute the confidence bands and the prediction bands at 0.05 significance level for the response at values of PLR=0.2 and 1.00 (i.e., the extreme points).

Solution

(a) The independent variable matrix X given by Eq. 5.26b is:

$$X = \begin{bmatrix} 1 & 0.2 & 0.05 \\ 1 & 0.3 & 0.11 \\ 1 & 0.4 & 0.19 \\ 1 & 0.5 & 0.28 \\ 1 & 0.6 & 0.39 \\ 1 & 0.7 & 0.51 \\ 1 & 0.8 & 0.68 \\ 1 & 0.9 & 0.84 \\ 1 & 1 & 1 \end{bmatrix}$$

(b) The results of the regression are shown below:

Parameter	Estimate	Standard error	t-statistic	P-value
CONSTANT	−0.0204762	−0.0173104	−1.18288	0.2816
PLR	0.179221	0.0643413	2.78547	0.0318
PLR^2	0.850649	0.0526868	16.1454	0.0000

Analysis of Variance

Source	Sum of squares	Df	Mean square	F-ratio	P-value
Model	0.886287	2	0.443144	5183.10	0.0000
Residual	0.000512987	6	0.0000854978		
Total (Corr.)	0.8868	8			

Goodness-of-fit R^2=99.9%, Adjusted R^2=99.9%, RMSE=0.009246

Mean absolute error (MAD)=0.00584. The equation of the fitted model is (with appropriate rounding)

$$\text{FFLP} = -0.0205 + 0.1792^*\text{PLR} + 0.8506^*\text{PLR}^2$$

Since the P-value in the ANOVA table is less than 0.05, there is a statistically significant relationship between FFLP and PLR at the 95% confidence level. However, the p-value of the constant term is large, and a model without an intercept term is probably more appropriate; thus, such an analysis ought to be performed, and its results evaluated. The values shown are those provided by the software package. There are too many significant decimals, and so the analyst should round these off appropriately while reporting the results (as shown above).

(c) The 95% confidence and the prediction intervals are shown in Fig. 5.9. Because the fit is excellent, these are very

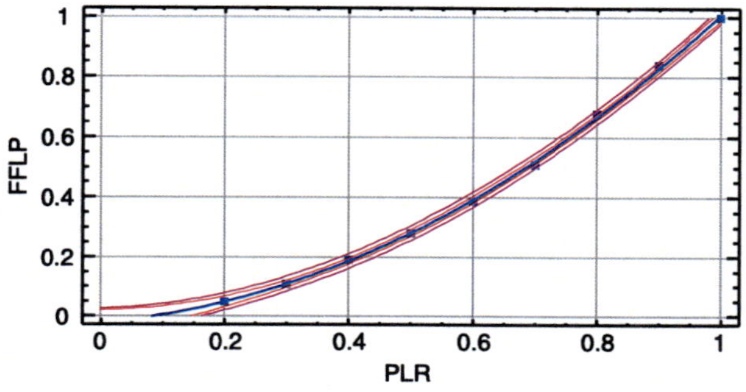

Fig. 5.9 Plot of fitted model with 95% CL and 95% PL bands

narrow and close to each other. The predicted values as well as the 95% CL and PL for the two data points are given in the table below. Note that the uncertainty range is relatively much larger at the lower value than at the higher range.

	Predicted	95% Prediction Limits		95% Confidence Limits	
x	y	Lower	Upper	Lower	Upper
0.2	0.0493939	0.0202378	0.0785501	0.0310045	0.0677834
1.0	1.00939	0.980238	1.03855	0.991005	1.02778

■

Example 5.4.2: Table 5.3 gives the solubility of oxygen in water in (mg/L) at 1 atm pressure for different temperatures and different chloride concentrations in (mg/L).

(a) Plot the data and formulate two different models to be evaluated

(b) Evaluate both models and identify the better one. Give justification for your choice

(c) Report pertinent statistics for model parameters as well as overall model fit

(a) The above data is plotted in Fig. 5.10a. One notes that the series of plots are slightly non-linear but parallel suggesting a higher order model without interaction terms. Hence, first order and second order polynomial models without interaction are logical models to investigate.

Table 5.3 Solubility of oxygen in water (mg/L) with temperature and chloride concentration

Temperature (°C)	Chloride concentration in water (mg/L)			
	0	5,000	10,000	15,000
0	14.62	13.73	12.89	12.10
5	12.77	12.02	11.32	10.66
10	11.29	10.66	10.06	9.49
15	10.08	9.54	9.03	8.54
20	9.09	8.62	8.17	7.75
25	8.26	7.85	7.46	7.08
30	7.56	7.19	6.85	6.51

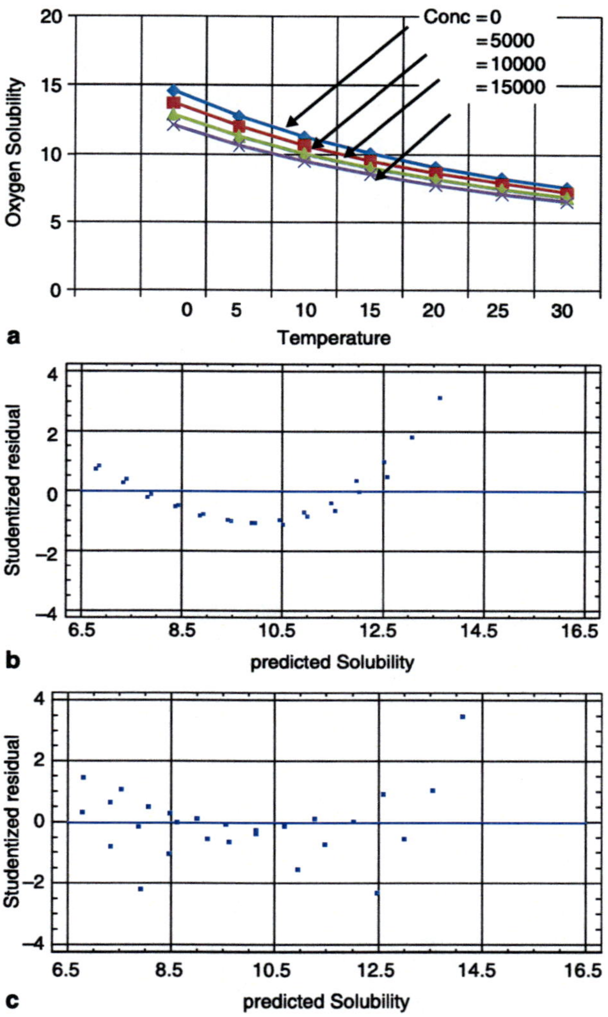

Fig. 5.10 **a** Plot of data. **b** Residual pattern for the first order model. **c** Residual pattern for the second order model

(b1) Analysis results of the first order model without interaction term:

R^2=96.83%, Adjusted R^2=96.57%, RMSE=0.41318

Parameter	Estimate	Standard error	t-statistic	P-value
CONSTANT	13.6111	0.175471	77.5686	0.0000
Chloride Concentration	−0.000109857	0.000013968	−7.86489	0.0000
Temperature	−0.206786	0.00780837	−26.4826	0.0000

Analysis of Variance

Source	Sum of squares	Df	Mean square	F-ratio	P-value
Model	130.289	2	65.1445	381.59	0.0000
Residual	4.26795	25	0.170718		
Total (Corr.)	134.557	27			

The equation of the fitted model is:

Solubility = 13.6111 − 0.000109857 * Chloride Concentration − 0.206786 * Temperature

The model has excellent R^2 with all coefficients being statistically significant, but the model residuals are very ill-behaved since a distinct pattern can be seen (Fig. 5.10b). This issue of how model residuals can provide diagnostic insights into model building will be explored in detail in Sect. 5.6.

(b2) Analysis results for the second order model without interaction term:

The OLS regression results in R^2=99.26%, Adjusted R^2=99.13%, RMSE=0.20864, Mean absolute error=0.14367. This model is distinctly better with higher R^2 and lower RMSE. Except for one term (the square of the concentration), all parameters are statistically significant. The residual pattern is less distinct, but the residuals are still patterned (Fig. 5.10c). It would be advisable to investigate other functional forms, probably non-linear or based on some mechanistic insights.

Parameter	Estimate	Standard error	t-statistic	P-value
CONSTANT	14.1183	0.112448	125.554	0.0000
Temperature	−0.325	0.0142164	−22.8609	0.0000
Chloride concentration	−0.000118643	0.0000246866	−4.80596	0.0001
Temperature^2	0.00394048	0.000455289	8.65489	0.0000
Chloride concentration^2	5.85714E-10	1.57717E-9	0.371371	0.7138

Analysis of Variance

Source	Sum of squares	Df	Mean square	F-ratio	P-value
Model	133.556	4	33.3889	767.02	0.0000
Residual	1.0012	23	0.0435305		
Total (Corr.)	134.557	27			

■

5.4.4 Partial Correlation Coefficients

The simple correlation coefficient between two variables has already been introduced previously (Sect. 3.4.2). Consider the multivariate linear regression (MLR) model given by Eq. 5.18. If the regressors are uncorrelated, then the simple correlation coefficients provide a direct indication of the influence of the individual regressors on the response variable. Since regressors are often "somewhat" correlated, the concept of the simple correlation coefficient can be modified to handle such interactions. This leads to the concept of *partial correlation coefficients*. Assume a MLR model with only two regressors: x_1 and x_2. The procedure to compute the partial correlation coefficient r_{yx_1} between y and x_1 will make the concept clear:

Step 1: Regress y vs x_2 so as to identify a prediction model for $\hat{y}$

Step 2: Regress x_1 vs x_2 so as to identify a prediction model for $\hat{x}_1$

Step 3: Compute new variables (in essence, the model residuals): $y^* = y - \hat{y}$ and $x_1^* = x_1 - \hat{x}_1$

Step 4: The partial correlation r_{yx_1} between y and x_1 is the simple correlation coefficient between y^* and $x_1{}^*$

Note that the above procedure allows the linear influence of x_2 to be removed from both y and x_1, thereby enabling the partial correlation coefficient to describe only the effect of x_2 on y which is not accounted for by the other variables in the model. This concept plays a major role in the process of stepwise model identification described in Sect. 5.7.4.

5.4.5 Beta Coefficients and Elasticity

Beta coefficients β^* are occasionally used to make statements about the relative importance of the regressor variables in a multiple regression model (Pindyck and Rubinfeld 1981). These coefficients are the parameters of a linear regression model with each variable normalized by subtracting its mean and dividing by its standard deviation:

$$\frac{y-\bar{y}}{\sigma_y} = \beta_1^* \frac{x_1-\bar{x}_1}{\sigma_{x1}} + \beta_2^* \frac{x_2-\bar{x}_2}{\sigma_{x2}} + \cdots \varepsilon \quad (5.38)$$

or

$$y^* = \beta_1^* x_1{}^* + \beta_2^* x_2{}^* + \cdots \varepsilon$$

The β^* matrix can be directly deduced from the original slope parameter "b" of the un-normalized MLR model as:

$$\beta^* = b \cdot \frac{\sigma_x}{\sigma_y} \quad (5.39)$$

For example, the beta coefficient β^*=0.7 can be interpreted to mean that one standard deviation in the regressor variable leads to a 0.7 standard deviation in the dependent variable. For a two-variable model, β^* is the simple correlation between the two variables. The rescaling associated with the normalized regression makes it possible to compare the individual values of β^* directly, i.e., the relative importance of the different regressors can be directly evaluated against each other, provided the regressors are uncorrelated with each other. A variable with a high β^* coefficient should account for more of the variance in the response variable (variance is not to be confused with contribution). The square of the β^* weights are indicative of the relative effects of the respective variables on the variation of the response variable.

The beta coefficients indicate or represent the marginal effect of the standardized regressors on the standardized response variable. Often, one is interested in deducing the effect of a fractional (or percentage) change of a regressor j on the dependent variable. This is provided by the *elasticity*

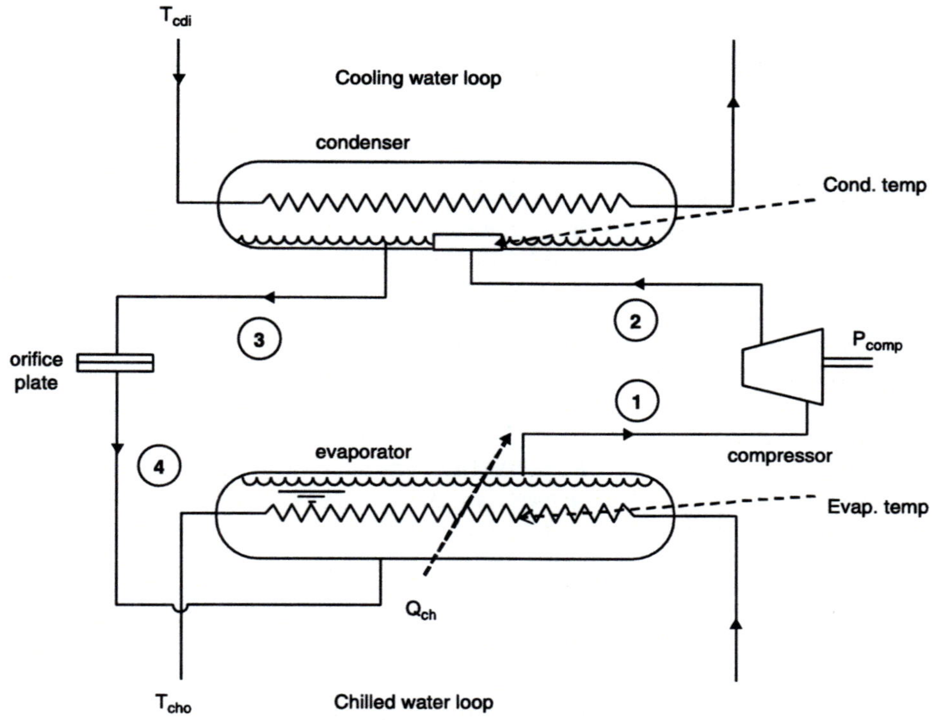

Fig. 5.11 Sketch of a flooded-type centrifugal chiller with two water loops showing the various regressors often used to develop the performance model for COP

of y with respect to say x_j which is usually evaluated at their mean values as:

$$E_j = b_j \cdot \frac{\bar{x}_j}{\bar{y}} \approx \frac{\partial y}{\bar{y}} / \frac{\partial x_j}{\bar{x}_j} \quad (5.40)$$

Elasticities can take both positive or negative values. Large values of elasticity imply that the regressor variable is very responsive to changes in the regressor variables. For non-linear functions, elasticities can also be calculated at the point of interest rather than at the mean point. The interpretation of elasticities is straightforward. If $E_j = 1.5$, this implies that a 1% increase in the mean of the regressor variable will result in a 1.5% increase in y.

Example 5.4.3: *Beta coefficients for ascertaining importance of driving variables for chiller thermal performance*

The thermal performance of a centrifugal chiller is characterized by the Coefficient of Performance (COP) which is the dimensionless ratio of the cooling thermal capacity (Q_{ch}) and the compressor electric power (P_{comp}) in consistent units. A commonly used performance model for the COP is one with three regressors, namely the cooling load Q_{ch}, the condenser inlet temperature T_{cdi} and chiller leaving temperature T_{cho} (see Fig. 5.11). The condenser and evaporator temperatures shown are those of the refrigerant as it changes phase.

A data set of 107 performance points from an actual chiller was obtained whose summary statistics are shown in the table below. An OLS regression yielded a model with $R^2=90.1\%$ whose slope coefficients b_j are also shown in Table 5.4 along with the beta coefficients and the elasticity computed from Eqs. 5.39 and 5.40 respectively. One would conclude looking at the elasticity values that T_{cdi} has the most influence on COP followed by Q_{ch}, while that of T_{cho} is very small. A 1% increase in Q_{ch} increases COP by 0.431% while a 1% increase in T_{cdi} would decrease COP by 0.603%. The beta coefficients, on the other hand, take into account the range of variation of the variables. For example, the load variable Q_{ch} can change from 20 to 100% while T_{cdi} usually changes only by 15°C or so. Thus, beta coefficients express the change in the COP of 0.839 in terms of one standard deviation change in Q_{ch} (i.e., a load change of 88.1 kW) while a comparable one standard deviation change in T_{cdi} (of 4.28°C) would result in a *decrease* of 0.496 in COP. ■

Table 5.4 Associated statistics of the four variables, results of the OLS regression and beta coefficients

	Response	Regressors		
	COP	Q_{ch} (kW)	T_{cdi} (°C)	T_{cho} (°C)
Mean	3.66	205.8	23.66	7.37
St. dev	0.806	88.09	4.283	2.298
Min	2.37	86	16.01	3.98
Max	4.98	361.4	29.95	10.94
Slope coeff. b		0.0077	−0.0933	0.0354
beta_coeff. (Eq. 5.39)		0.839	−0.496	0.101
Elasticity (Eq. 5.40)		0.431	−0.603	0.071

5.5 Assumptions and Sources of Error During OLS Parameter Estimation

5.5.1 Assumptions

The ordinary least square (OLS) regression method:

(i) enables simple or multiple linear regression models to be identified from data, which can then be used for future prediction of the response variable along with its uncertainty bands, and

(ii) allows statistical statements to be made about the estimated model parameters.

No statistical assumptions are used to obtain the OLS estimators for the model coefficients. When nothing is known regarding measurement errors, OLS is often the best choice for estimating the parameters. However, in order to make statistical statements about these estimators and the model predictions, it is necessary to acquire information regarding the measurement errors. Ideally, one would like the error terms ε_i to be normally distributed, without serial correlation, with mean zero and constant variance. The implications of each of these four assumptions, as well as a few additional ones, will be briefly addressed below since some of these violations may lead to biased coefficient estimates as well as distorted estimates of the standard errors, confidence intervals, and statistical tests.

(a) *Errors should have zero mean:* If this is not true, the OLS estimator of the intercept will be biased. The impact of this assumption not being correct is generally viewed as the least critical among the various assumptions. Mathematically, this implies that expected value $E(\varepsilon_i)=0$.

(b) *Errors should be normally distributed:* If this is not true, statistical tests and confidence intervals are incorrect for small samples though the OLS coefficient estimates are unbiased. Figure 5.3 which illustrates this behavior has already been discussed. This problem can be avoided by having large samples, and verifying that the model is properly specified.

(c) *Errors should have constant variance:* This violation of the basic OLS assumption results in increasing the standard errors of the estimates and widening the model prediction confidence intervals (though the OLS estimates themselves are unbiased). In this sense, there is a loss in statistical power. This condition is expressed mathematically as, $\text{var}\,(y_i)=\sigma^2$. This issue is discussed further in Sect. 5.6.3.

(d) *Errors should not be serially correlated:* This violation is equivalent to have less *independent* data, and also results in a loss in statistical power with the same consequences as (c) above. Serial correlations may occur due to the manner in which the experiment is carried out. Extraneous factors, i.e., factors beyond our control (such as the weather, for example) may leave little or no choice as to how the experiments are executed. An example of a reversible experiment is the classic pipe-friction experiment where the flow through a pipe is varied so as to cover both laminar and turbulent flows, and the associated friction drops are observed. Gradually increasing the flow one way (or decreasing it the other way) may introduce biases in the data which will subsequently also bias the model parameter estimates. In other circumstances, certain experiments are irreversible. For example, the loading on a steel sample to produce a stress-strain plot has to be performed by gradually increasing the loading till the sample breaks, one cannot proceed in the other direction. Usually the biases brought about by the test sequence are small, and this may not be crucial. In mathematical terms, this condition, for a first order case, can be written as $E(\varepsilon_i.\varepsilon_{i+1})=0$. This assumption, which is said to be hardest to verify, is further discussed in Sect. 5.6.4.

(e) *Errors should be uncorrelated with the regressors:* The consequences of this violation result in OLS coefficient estimates being biased and the predicted OLS confidence intervals understated, i.e., narrower. This violation is a very important one, and is often due to "mis-specification error" or underfitting. Omission of influential regressor variables and improper model formulation (assuming a linear relationship when it is not) are likely causes. This issue is discussed at more length in Sect. 10.4.1.

(f) *Regressors should not have any measurement error:* Violation of this assumption in some (or all) regressors will result in biased OLS coefficient estimates for those (or all) regressors. The model can be used for prediction but the confidence limits will be understated. Strictly speaking, this assumption is hardly ever satisfied since there is always some measurement error. However, in most engineering studies, measurement errors in the regressors are not large compared to the random errors in the response, and so this violation may not have important consequences. As a rough rule of thumb, this violation becomes important when the errors in x reach about a fifth of the random errors in y, and when multi-collinearity is present. If the errors in x are known, there are procedures which allow unbiased coefficient estimates to be determined (see Sect. 10.4.2). Mathematically, this condition is expressed as $\text{var}\,(x_i)=0$.

(g) *Regressor variables should be independent of each other:* This violation applies to models identified by multiple regression when the regressor variables are correlated among each other (called multicollinearity). This is true even if the model provides an excellent fit to the data. Estimated regression coefficients, though unbiased, will tend to be unstable (their values tend to change greatly when a data point is dropped or added), and the OLS standard errors and the prediction intervals will be understated. Multicollinearity is likely to be problem only when one (or more) of the correlation coefficients among the regressors exceeds 0.85 or so. Sect. 10.3 deals with this issue at more length.

5.5.2 Sources of Errors During Regression

Perhaps the most crucial issue during parameter identification is the type of measurement inaccuracy present. This has a direct influence on the estimation method to be used. Though statistical theory has more or less neatly classified this behavior into a finite number of groups, the data analyst is often stymied by data which does not fit into any one category. Remedial action advocated does not seem to entirely remove the adverse data conditioning. A certain amount of experience is required to surmount this type of adversity, which, further, is circumstance-specific. As discussed earlier, there can be two types of errors:

(a) **measurement error.** The following sub-cases can be identified depending on whether the error occurs:

(i) in the dependent variable, in which case the model form is:

$$y_i + \delta_i = \beta_0 + \beta_1 x_i \tag{5.41a}$$

(ii) in the regressor variable, in which case the model form is:

$$y_i = \beta_0 + \beta_1 (x_i + \gamma_i) \tag{5.41b}$$

(iii) in both dependent and regressor variables:

$$y_i + \delta_i = \beta_0 + \beta_1 (x_i + \gamma_i) \tag{5.41c}$$

Further, the errors δ and γ (which will be jointly represented by ε) can have an additive error, in which case, $\varepsilon_i \neq f(y_i, x_i)$, or a multiplicative error: $\varepsilon_i = f(y_i, x_i)$, or worst still, a combination of both. Section 10.4.1 discusses this issue further.

(b) **model misspecification error.** How this would affect the model residuals ε_i is difficult to predict, and is extremely circumstance-specific. Misspecification could be due to several factors, for example, one or more important variables have been left out of the model, or the functional form of the model is incorrect. Even if the physics of the phenomenon or of the system is well understood and can be cast in mathematical terms, identifiability constraints may require that a simplified or macroscopic model be used for parameter identification rather than the detailed model (see Sect. 10.2). This is likely to introduce both bias and random noise in the parameter estimation process except when model R^2 is very high ($R^2 > 0.9$). This issue is further discussed in Sect. 5.6. Formal statistical procedures do not explicitly treat this case but limit themselves to type (a) errors and more specifically to case (i) assuming purely additive or multiplicative errors. The implicit assumptions in OLS and their implications, if violated, are described below.

5.6 Model Residual Analysis[3]

5.6.1 Detection of Ill-Conditioned Model Residual Behavior

The availability of statistical software has resulted in routine and easy application of OLS to multiple linear models. However, there are several underlying assumptions that affect the individual parameter estimates of the model as well as the overall model itself. Once a model has been identified, the general tendency of the analyst is to hasten and use the model for whatever purpose intended. However, it is extremely important (and this phase in often overlooked) that an assessment of the model be done to determine whether the OLS assumptions are met, otherwise the model is likely to be deficient or misspecified, and yield misleading results. In the last few decades, there has been much progress made on how to screen model residual behavior so as to provide diagnostics insight into model deficiency or misspecification.

A few idealized plots illustrate some basic patterns of improper residual behavior which are addressed in more detail in the later sections of this chapter. Figure 5.12 illustrates the effect of omitting an important dependence which suggests that an additional variable is to be introduced in the model

Fig. 5.12 The residuals can be separated into two distinct groups (shown as *crosses* and *dots*) which suggest that the response variable is related to another regressor not considered in the regression model. This residual pattern can be overcome by reformulating the model by including this additional variable. One example of such a time-based event system change is shown in Fig. 9.15 of Chap. 9.

[3] Herschel: "… almost all of the greatest discoveries in astronomy have resulted from the consideration of what … (was) termed residual phenomena".

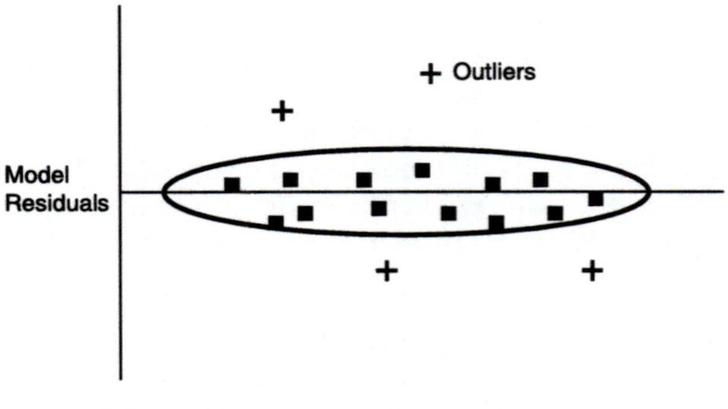

Fig. 5.13 Outliers indicated by crosses suggest that data should be checked and/or robust regression used instead of OLS

which distinguishes between the two groups. The presence of outliers and the need for more robust regression schemes which are immune to such outliers is illustrated in Fig. 5.13. The presence of non-constant variance (or heteroscedasticity) in the residuals is a very common violation and one of several possible manifestations is shown in Fig. 5.14. This particular residual behavior is likely to be remedied by using a log transform of the response variable instead of the variable itself. Another approach is to use weighted least squares estimation procedures described later in this chapter. Though non-constant variance is easy to detect visually, its cause is difficult to identify. Figure 5.15 illustrates a typical behavior which arises when a linear function is used to model a quadratic variation. The proper corrective action will increase the predictive accuracy of the model (RMSE will be lower), result in the estimated parameters being more efficient (i.e., lower standard errors), and most importantly, allow more sound and realistic interpretation of the model prediction uncertainty bounds.

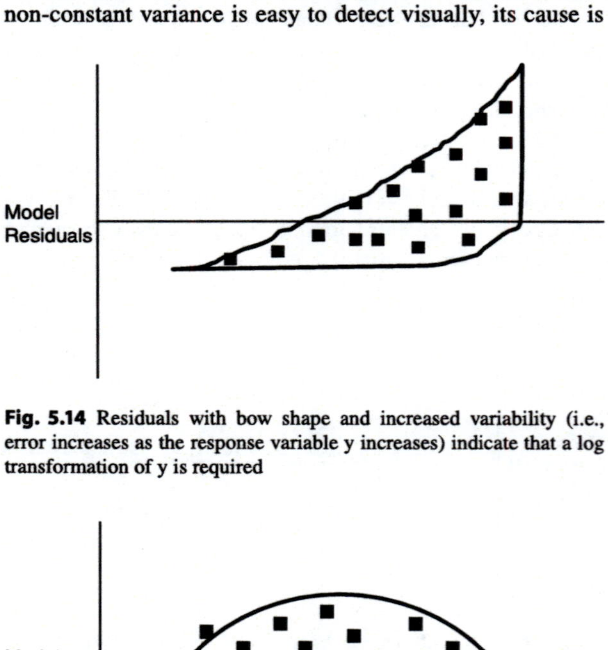

Fig. 5.14 Residuals with bow shape and increased variability (i.e., error increases as the response variable y increases) indicate that a log transformation of y is required

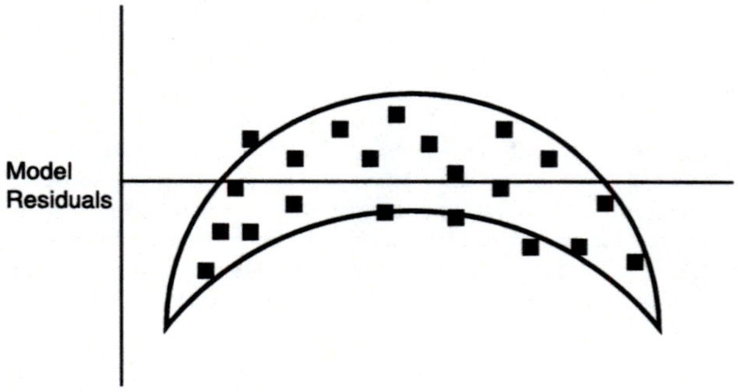

Fig. 5.15 Bow-shaped residuals suggest that a non-linear model, i.e. a model with a square term in the regressor variable to be evaluated

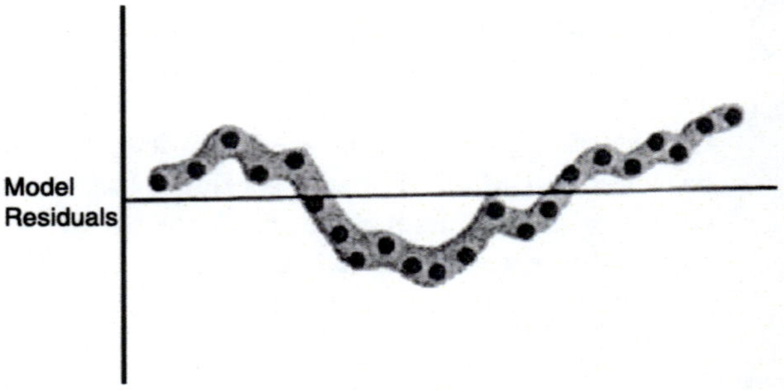

Fig. 5.16 Serial correlation is indicated by a pattern in the residuals when plotted in the sequence the data was collected, i.e., when plotted against time even though time may not be a regressor in the model

Figure 5.16 illustrates the occurrence of serial correlations in time series data which arises when the error terms are not independent. Such patterned residuals occur commonly during model development and provide useful insights into model deficiency. Serial correlation (or autocorrelation) has special pertinence to time series data (or data ordered in time) collected from in-situ performance of mechanical and thermal systems and equipment. Autocorrelation is present if adjacent model residuals, i.e., residuals show a trend or a pattern of clusters above or below the zero value that can be discerned visually. Such correlations can either suggest that additional variables have been left out of the model (model-misspecification error), or could be due to the nature of the process itself (called pure or "pseudo" autocorrelation). The latter is due to the fact that equipment loading over a day would follow an overall cyclic curve (as against random jumps from say full load to half load) consistent with the diurnal cycle and the way the system is operated. In such cases, positive residuals would tend to be followed by positive residuals, and vice versa. Time series data and models are treated further in Chap. 9.

Problems associated with *model underfitting and overfitting* are usually the result of a failure to identify the non-random pattern in time series data. Underfitting does not capture enough of the variation in the response variable which the corresponding set of regressor variables can possibly explain. For example, all four models fit to their respective sets of data as shown in Fig. 5.17, have identical R^2 values and t-statistics but are distinctly different in how they capture the data variation. Only plot (a) can be described by a linear model. The data in (b) needs to be fitted by a higher order

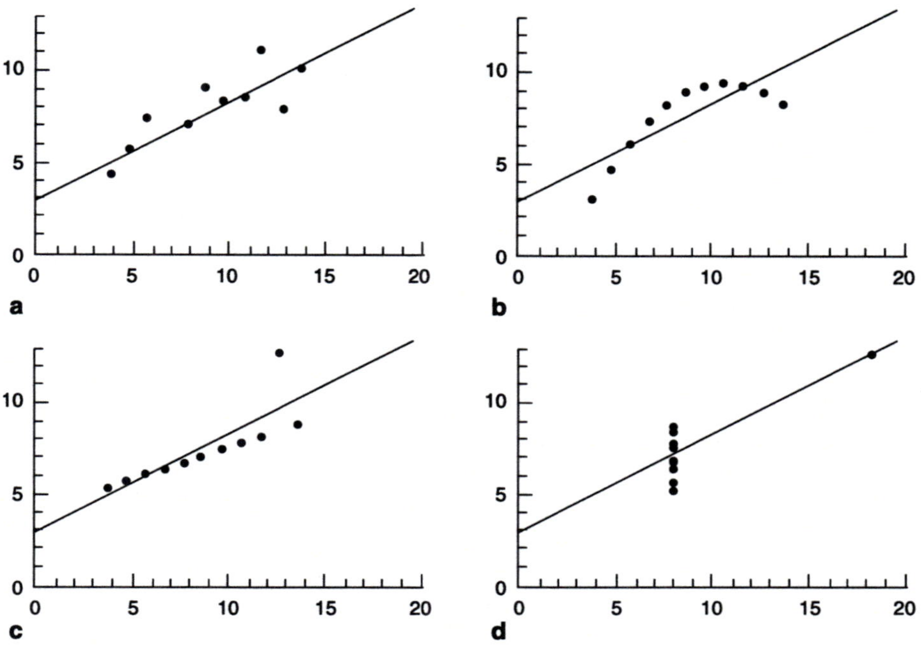

Fig. 5.17 Plot of the data (x, y) with the fitted lines for four data sets. The models have identical R^2 and t-statistics but only the first model is a realistic model. (From Chatterjee and Price 1991 by permission of John Wiley and Sons)

model, while one data point in (c) and (d) distorts the entire model. Blind model fitting (i.e., relying only on model statistics) is, thus, inadvisable.

Overfitting implies capturing randomness in the model, i.e., attempting to fit the noise in the data. A rather extreme example is when one attempts to fit a model with six parameters to six data points which have some inherent experimental error. The model has zero degrees of freedom and the set of six equations can be solved without error (i.e., RMSE=0). This is clearly unphysical because the model parameters have also "explained" the random noise in the observations in a deterministic manner.

Both underfitting and overfitting can be detected by performing certain statistical tests on the residuals. The most commonly used test for white noise (i.e., uncorrelated residuals) involving model residuals is the **Durbin-Watson (DW)** statistic defined by:

$$DW = \sum_{i=2}^{n} \left(\varepsilon_i - \varepsilon_{i-1}\right)^2 \Big/ \sum_{i=1}^{n} \varepsilon_i^2 \qquad (5.42)$$

where ε_i is the residual at time interval i, defined as $\varepsilon_i = y_i - \hat{y}_i$.

If there is no serial or autocorrelation present, the expected value of DW is 2. If the model underfits, DW would be less than 2 while it would be greater than 2 for an overfitted model, the limiting range being 0–4. Tables are available for approximate significance tests with different numbers of regressor variables and number of data points. Table A.13 assembles lower and upper critical values of DW statistics to test autocorrelation. For example, if n=20, and the model has three variables (p=3), the null hypothesis that the correlation coefficient is equal to zero can be rejected at the 0.05 significance level if its value is either below 1.00 or above 1.68. Note that the critical values in the table are one-sided, i.e., apply to one tailed distributions.

It is important to note that the DW statistic is only sensitive to correlated errors in adjacent observations, i.e., when only first-order autocorrelation is present. For example, if the time series has seasonal patterns, then higher autocorrelations may be present which the DW statistic will be unable to detect. More advanced concepts and modeling are discussed in Sect. 9.5 while treating stochastic time series data.

5.6.2 Leverage and Influence Data Points

Most of the aspects discussed above relate to identifying general patterns in the residuals of the entire data set. Another issue is the ability to identify subsets of data that have an unusual or disproportionate influence on the estimated model in terms of parameter estimation. Being able to flag such influential subsets of individual points allows one to investigate their validity, or to glean insights for better experimental design since they may contain the most interesting system behavioral information. Note that such points are not necessarily "bad" data points which should be omitted, but should be viewed as being "distinctive" observations in the overall data set. Scatter plots reveal such outliers easily for single regressor situations, but are inappropriate for multivariate cases. Hence, several statistical measures have been proposed to deal with multivariate situations, the influence and leverage indices being widely used (Belsley et al. 1980; Cook and Weisberg 1982).

The *leverage* of a point quantifies the extent to which that point is "isolated" in the x-space, i.e., its distinctiveness in terms of the regressor variables. It has a large impact on the numerical values of the model parameters being estimated. Consider the following matrix (called the hat matrix):

$$H = X(X'X)^{-1}X' \tag{5.43}$$

If one has a data set with two regressors, the order of the H matrix would be (3×3), i.e, equal to the number of parameters in the model (constant plus the two regressor coefficients). The diagonal element p_{ii} can be related to the distance between x_i and $\bar{x}$, and is defined as the *leverage* of the ith data point. Since the diagonal elements have values between 0 and 1, their average value is equal to (p/n) where n is the number of observation sets. Points with $p_{ii} > 3$ (p/n) are regarded as points with high leverage (sometimes the threshold is taken as 2 (p/n).

Large residuals are traditionally used to highlight suspect data points or data points unduly affecting the regression model. Instead of looking at residuals ε_i, it is more meaningful to study a normalized or scaled value, namely the standardized residuals or R-student residuals, where

$$\text{R-student} = \frac{\varepsilon_i}{\text{RMSE} \cdot [1-p_{ii}]^{1/2}} \tag{5.44}$$

Points with |R-student|>3 can be said to be influence points which corresponds to a significance level of 0.01. Sometimes a less conservative value of 2 is used corresponding to the 0.05 significance level, with the underlying assumption that residuals or errors are Gaussian.

A data point is said to be *influential* if its deletion, singly or in combination with a relatively few others, cause statistically significant changes in the fitted model coefficients. See Sect. 3.5.3 for a discussion based on graphical considerations of this concept. There are several measures used to describe influence, a common one is DFITS:

$$\text{DFITS}_i = \frac{\varepsilon_i (p_{ii})^{1/2}}{s_i (1 - p_{ii})^{1/2}} \tag{5.45}$$

where ε_i is the residual error of observation i, and s_i is the standard deviation of the residuals without considering the ith residual. Points with $DFITS \geq 2[p/(n-p)]^{1/2}$ are flagged as influential points.

Both the R-student statistic and the DFITS indices are often used to detect influence points. In summary, just because a point has high leverage does not make it influential. It is advisable to identify points with high leverage, and, then, examine them to determine whether they are influential as well.

Influential observations can impact the final regression model in different ways (Hair et al. 1998). For example, in Fig. 5.18a, the model residuals are not significant and the two influential observations shown as filled dots reinforce the general pattern in the model and lower the standard error of the parameters and of the model prediction. Thus, the two points would be considered to be leverage points which are beneficial to our model building. Influential points which adversely impact model building are illustrated in Fig. 5.18b and c. In the former, the two influential points almost totally account for the observed relationship but would not have been identified as outlier points. In Fig. 5.18c, the two influential points have totally altered the model identified, and the actual data points would have shown up as points with large residuals which the analyst would probably have identified as spurious.

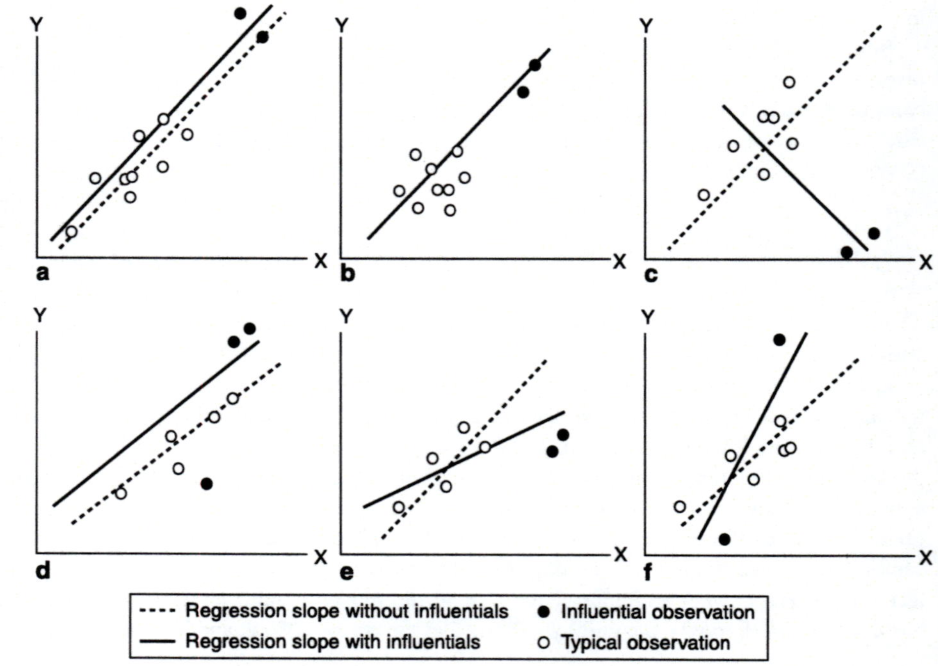

Fig. 5.18a–f Common patterns of influential observations. (From Hair et al. 1998 by © permission of Pearson Education)

The next frame (d) illustrates the instance when an influential point changes the intercept of the model but leaves the slope unaltered. The two final frames, Fig. 5.18e and f, illustrate two, hard to identify and rectify, cases when two influential points reinforce each other in altering both the slope and the intercept of the model though their relative positions are very much different. Note that data points that satisfy both these statistical criteria, i.e., are both influential and have high leverage, are the ones worthy of closer scrutiny. Most statistical programs have the ability to flag such points, and hence performing this analysis is fairly straightforward.

Thus, in conclusion, individual data points can be outliers, leverage or influential points. Outliers are relatively simple to detect and to interpret using the R-student statistic. Leverage of a point is a measure of how unusual the point lies in the x-space. An influence point is one which has an important affect on the regression model when that particular point were to be removed from the data set. Influential points are the ones which need particular attention since they provide insights about the robustness of the fit. In any case, all three measures (leverage p_{ii}, DFITS and R-student) provide indications as to the role played by different observations towards the overall model fit. Ultimately, the decision of deciding whether to retain or reject such points is somewhat based on judgment.

Example 5.6.1: *Example highlighting different characteristic of outliers or residuals versus influence points.*
Consider the following made-up data (Table 5.5) where x ranges from 1 to 10, and the model is $y=10+1.5*x$ to which random normal noise $\varepsilon=[0, \sigma=1]$ has been added to give y (second column). The last observation has been intentionally corrupted to a value of 50 as shown.

How well a linear model fits the data is depicted in Fig. 5.19. The table of unusual residuals shown below lists all observations which have Studentized residuals greater than 2.0 in absolute value. Note that observation 10 is flagged as an unusual residual (not surprising since this was intentionally corrupted) and no observation has been identified as influential despite it being very much of an outlier (the studentized value is very large—recall that a value of 3.0 would indicate a 99% CL). Thus, the error in one point seems to be overwhelmed by the well-behaved nature of the other nine points. This example serves to highlight the different characteristic of outliers versus influence points. ■

Row	x	y	Predicted y	Residual	Studentized residual
10	10.0	50.0	*37.2572*	12.743	*11.43*

Table 5.5 Data table for Example 5.6.1

x	y[0,1]	y1
1	11.69977	11.69977
2	12.72232	12.72232
3	16.24426	16.24426
4	19.27647	19.27647
5	21.19835	21.19835
6	23.73313	23.73313
7	21.81641	21.81641
8	25.76582	25.76582
9	29.09502	29.09502
10	28.9133	*50*

5.6.3 Remedies for Non-uniform Model Residuals

Non-uniform model residuals or heteroscedasticity can be due to: (i) the nature of the process investigated, (ii) noise in the data, or (iii) the method of data collection from samples which are known to have different variances. Three possible generic remedies for non-constant variance are to (Chatterjee and Price 1991):

(a) **introduce additional variables into the model and collect new data:** The physics of the problem along with model residual behavior can shed light into whether certain key variables, left out in the original fit, need to be introduced or not. This aspect is further discussed in Sect. 5.6.5;
(b) **transform the dependent variable:** This is appropriate when the errors in measuring the dependent variable may follow a probability distribution whose variance is a function of the mean of the distribution. In such cases, the model residuals are likely to exhibit heterosce-

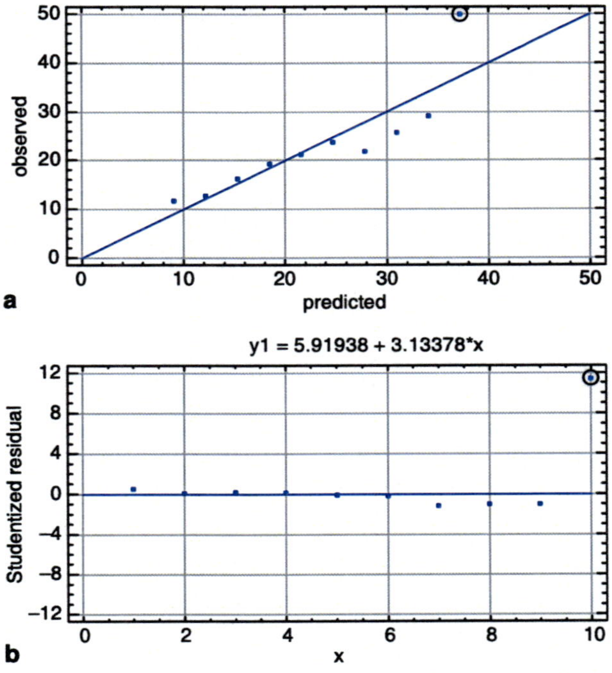

Fig. 5.19 **a** Observed vs predicted plot. **b** Residual plot versus regressor

Table 5.6. Transformations in dependent variable y likely to stabilize non-uniform model variance

	Variance of y in terms of its mean μ	Transformation
Poisson	μ	$y^{1/2}$
Binomial	$\mu(1-\mu)/n$	$\sin^{-1}(y)^{1/2}$

dasticity which can be removed by using exponential, Poisson or Binomial transformations. For example, a variable which is distributed Binomially with parameters "n and p" has mean (n.p.) and variance [n.p.(1−p)] (Sect. 2.4.2). For a Poisson variable, the mean and variance are equal. The transformations shown in Table 5.6 will stabilize variance, and the distribution of the transformed variable will be closer to the normal distribution.

The logarithmic transformation is also widely used in certain cases to transform a non-linear model into a linear one (see Sect. 9.5.1). When the variables have a large standard deviation compared to the mean, working with the data on a log scale often has the effect of dampening variability and reducing asymmetry. This is often an effective means of removing heteroscedascity as well. However, this approach is valid only when the magnitude of the residuals increase (or decrease) with that of one of the variables.

Example 5.6.2: *Example of variable transformation to remedy improper residual behavior*

The following example serves to illustrate the use of variable transformation. Table 5.7 shows data from 27 departments in a university with y as the number of faculty and staff and x the number of students.

A simple linear regression yields a model with R-squared=77.6% and a RMSE=21.7293. However, the residuals reveal an unacceptable behavior with a strong funnel behavior (see Fig. 5.20a).

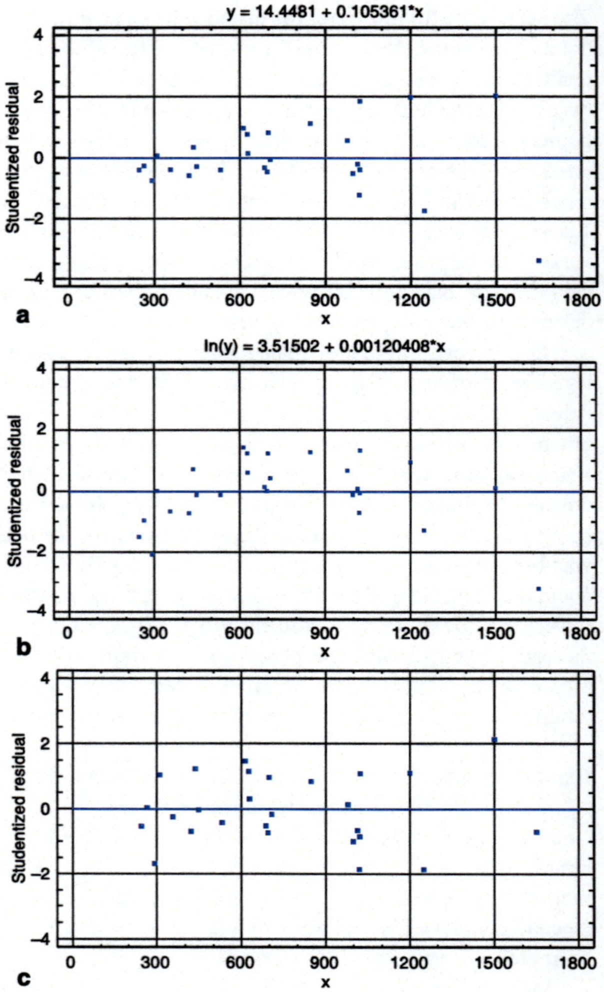

Fig. 5.20 **a** Residual plot of linear model. **b** Residual plot of log transformed linear model. **c** Residual plot of log transformed linear model

Table 5.7 Data table for Example 5.6.2

	x	y		x	y
1	294	30	15	615	100
2	247	32	16	999	109
3	267	37	17	1,022	114
4	358	44	18	1,015	117
5	423	47	19	700	106
6	311	49	20	850	128
7	450	56	21	980	130
8	534	62	22	1,025	160
9	438	68	23	1,021	97
10	697	78	24	1,200	180
11	688	80	25	1,250	112
12	630	84	26	1,500	210
13	709	88	27	1,650	135
14	627	97			

Instead of a linear model in y, a linear model in ln(y) is investigated. In this case, the model R-squared=76.1% and RMSE=0.252396. However, these statistics should NOT be compared directly since the y variable is no longer the same (in one case, it is "y"; in the other "ln y").

Let us not look into this aspect, but rather study the residual behavior. Notice that a linear model does reduce some of the improper residual variance but the inverted u shape behavior is indicative of model mis-specification (see Fig. 5.20b).

Finally, using a quadratic model along with the ln transformation results in a model:

$$\ln(y) = 2.8516 + 0.00311267^{*}x - 0.00000110226^{*}x^2$$

The residuals shown in Fig. 5.20c are now quite well behaved as a result of such a transformation. ■

(c) **perform weighted least squares.** This approach is more flexible and several variants exist (Chatterjee and Price 1991). As described earlier, OLS model residual behavior can exhibit non-uniform variance (called heteroscedasticity) even if the model is structurally complete, i.e., the model is not mis-specified. This violates one of the standard OLS assumptions. In a multiple regression model, detection of heteroscedasticity may not be very straight-forward since only one or two variables may be the culprits. Examination of the residuals versus each variable in turn along with intuition and understanding of the physical phenomenon being modeled can be of great help. Otherwise, the OLS estimates will lack precision, and the estimated standard errors of the model parameters will be wider. If this phenomenon occurs, the model identification should be redone with explicit recognition of this fact.

During OLS, the sum of the model residuals of all points are minimized with no regard to the values of the individual points or to points from different domains of the range of variability of the regressors. The basic concept of weighted least squares (WLS) is to simply assign different weights to different points according to a certain scheme. Thus, the general formulation of WLS is that the following function should be minimized:

$$\text{WLS function} = \sum w_i\left(y_i - \beta_0 - \beta_1 x_{1i} \cdots - \beta_p x_{pi}\right)^2 \tag{5.46}$$

where w_i are the weights of individual points. These are formulated differently depending on the weighting scheme selected which, in turn, depends on prior knowledge about the process generating the data.

(c-i) Errors Are Proportional to x Resulting in Funnel-Shaped Residuals Consider the simple model $y=\alpha+\beta x+\varepsilon$ whose residuals ε have a standard deviation which increases as the regressor variable (resulting in the funnel-like shape in Fig. 5.21). Assuming a weighting scheme such as $\text{var}(\varepsilon_i) = k^2 x_i^2$, transforms the model into:

$$\frac{y}{x} = \frac{\alpha}{x} + \beta + \frac{\varepsilon}{x} \text{ or } y' = \alpha x' + \beta + \varepsilon' \tag{5.47}$$

Note that the variance of ε' is constant and equals k^2. If the assumption about the weighting scheme is correct, the transformed model will be homoscedastic, and the model parameters α and β will be efficiently estimated by OLS (i.e., the standard errors of the estimates will be optimal).

The above transformation is only valid when the model residuals behave as shown in Fig. 5.21. If residuals behave differently, then different transformations or weighting schemes will have to be explored. Whether a particular transformation is adequate or not can only be gauged by the behavior of the variance of the residuals. Note that the analyst has to perform two separate regressions: one an OLS regression in order to

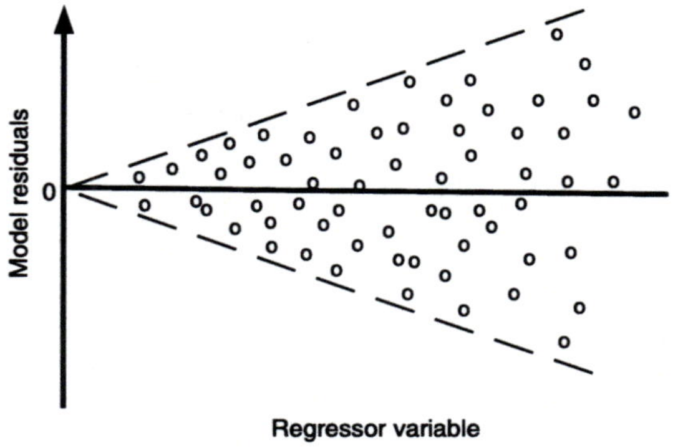

Fig. 5.21 Type of heteroscedastic model residual behavior which arises when errors are proportional to the magnitude of the x variable

determine the residual amounts of the individual data points, and then a WLS regression for final parameter identification. This is often referred to as *two-stage estimation.*

(c-ii) Replicated Measurements with Different Variance It could happen, especially with models involving one regressor variable only and when the data is obtained in the framework of a designed experimental study (as against observational or non-experimental data), that one obtains replicated measurements on the response variable corresponding to a set of fixed values of the explanatory variables. For example, consider the case when the regressor variable x takes several discrete values. If the physics of the phenomenon cannot provide any theoretical basis on how to select a particular weighty scheme, then this has to be determined experimentally from studying the data. If there is an increasing pattern in the heteroscedascity present in the data, this could be modeled either by a logarithmic transform (as illustrated in Example 5.6.2) or a suitable variable transformation. Here, another more versatile approach which can be applied to any pattern of the residuals is illustrated. Each observed residual ε_{ij} (where the index for discrete x values is i, and the number of observations at each discrete x value is $j=1, 2, \ldots n_i$) is made up of two parts, i.e., $\varepsilon_{ij} = (y_{ij} - \bar{y}_i) + (\bar{y}_i - \hat{y}_{ij})$. The first part is referred to as pure error while the second part measures lack of fit. An assessment of heteroscedasticity is based on pure error. Thus, the WLS weight may be estimated as $w_i = 1/s_i^2$ where the mean square error is:

$$s_i^2 = \frac{\sum (y_{ij} - \bar{y}_i)^2}{(n_i - 1)} \tag{5.48}$$

Alternatively a model can be fit to the mean values of x and the s_i^2 values in order to smoothen out the weighting function, and this function used instead. Thus, this approach would also qualify as a two-stage estimation process. The following example illustrates this approach.

Table 5.8 Measured data, OLS residuals deduced from Eq. 5.49a and the weights calculated from Eq. 5.49b

x	y	Residual ε_i	w_i	x	y	Residual ε_i	w_i
1.15	0.99	0.26329	0.9882	9.03	9.47	−0.20366	0.4694
1.90	0.98	−0.59826	1.7083	9.07	11.45	1.730922	0.4614
3.00	2.60	−0.2272	6.1489	9.11	12.14	2.375506	0.4535
3.00	2.67	−0.1572	6.1489	9.14	11.50	1.701444	0.4477
3.00	2.66	−0.1672	6.1489	9.16	10.65	0.828736	0.4440
3.00	2.78	−0.0472	6.1489	9.37	10.64	0.580302	0.4070
3.00	2.80	−0.0272	6.1489	10.17	9.78	−1.18802	0.3015
5.34	5.92	0.435964	15.2439	10.18	12.39	1.410628	0.3004
5.38	5.35	−0.17945	13.6185	10.22	11.03	0.005212	0.2963
5.40	4.33	−1.22216	12.9092	10.22	8.00	−3.02479	0.2963
5.40	4.89	-0.66216	12.9092	10.22	11.90	0.875212	0.2963
5.45	5.21	−0.39893	11.3767	10.18	8.68	−2.29937	0.3004
7.70	7.68	−0.48358	0.9318	10.50	7.25	−4.0927	0.2696
7.80	9.81	1.53288	0.8768	10.23	13.46	2.423858	0.2953
7.81	6.52	−1.76847	0.8716	10.03	10.19	−0.61906	0.3167
7.85	9.71	1.37611	0.8512	10.23	9.93	−1.10614	0.2953
7.87	9.82	1.463402	0.8413				
7.91	9.81	1.407986	0.8219				
7.94	8.50	0.063924	0.8078				

Example 5.6.3:[4] *Example of weighted regression for replicate measurements*

Consider the data given in Table 5.8 of replicate measurements of y taken at different values of x (which vary slightly).

A scatter plot of this data and the simple OLS linear model are shown in Fig. 5.22a. The regressed model is:

$$y = -0.578954 + 1.1354^{*}x \quad \text{with} \quad R^2 = 0.841 \quad \text{and} \quad \text{RMSE} = 1.4566 \tag{5.49a}$$

Note that the intercept term in the model is not statistically significant (p-value=0.4 for the t-statistic), while the overall model fit given by the F-ratio is significant. The model residuals of a simple OLS fit are shown in Fig. 5.22b.

Coefficients

Parameter	Least squares estimate	Standard error	t-statistic	P-value
Intercept	−0.578954	0.679186	−0.852423	0.4001
Slope	1.1354	0.086218	13.169	0.0000

Analysis of Variance

Source	Sum of squares	Df	Mean square	F-ratio	P-value
Model	367.948	1	367.948	173.42	0.0000
Residual	70.0157	33	2.12169		
Total (Corr.)	437.964	34			

[4] From Draper and Smith (1981) by permission of John Wiley and Sons.

The residuals of a simple linear OLS model shown in Fig. 5.22b reveal, as expected, marked heteroscadascity. Hence, the OLS model is bound to lead to misleading uncertainty bands even if the model predictions themselves are not biased. The model residuals from the above model are also shown in the table. Subsequently, the mean and the mean square error s_i^2 are calculated following Eq. 5.48 to yield the following table:

$\bar{x}$	s_i^2
3	0.0072
5.39	0.373
7.84	1.6482
9.15	0.8802
10.22	4.1152

Then, because of the pattern exhibited, a second order polynomial OLS model is regressed to this data (see Fig. 5.22c):

$$s_i^2 = 1.887 - 0.8727.\bar{x} + 0.9967.\bar{x}^2 \text{ with } R^2 = 0.743 \tag{5.49b}$$

The regression weights w_i can thus be deduced by using individual values of x_i instead of $\bar{x}$ in the above equation. The values of the weights are also shown in the data table. Finally, a weighted regression is performed following Eq. 5.46 (most statistical packages have this capability) resulting in:

$$y = -0.942228 + 1.16252^{*}x \quad \text{with} \quad R^2 = 0.896 \quad \text{and} \quad \text{RMSE} = 1.2725. \tag{5.49c}$$

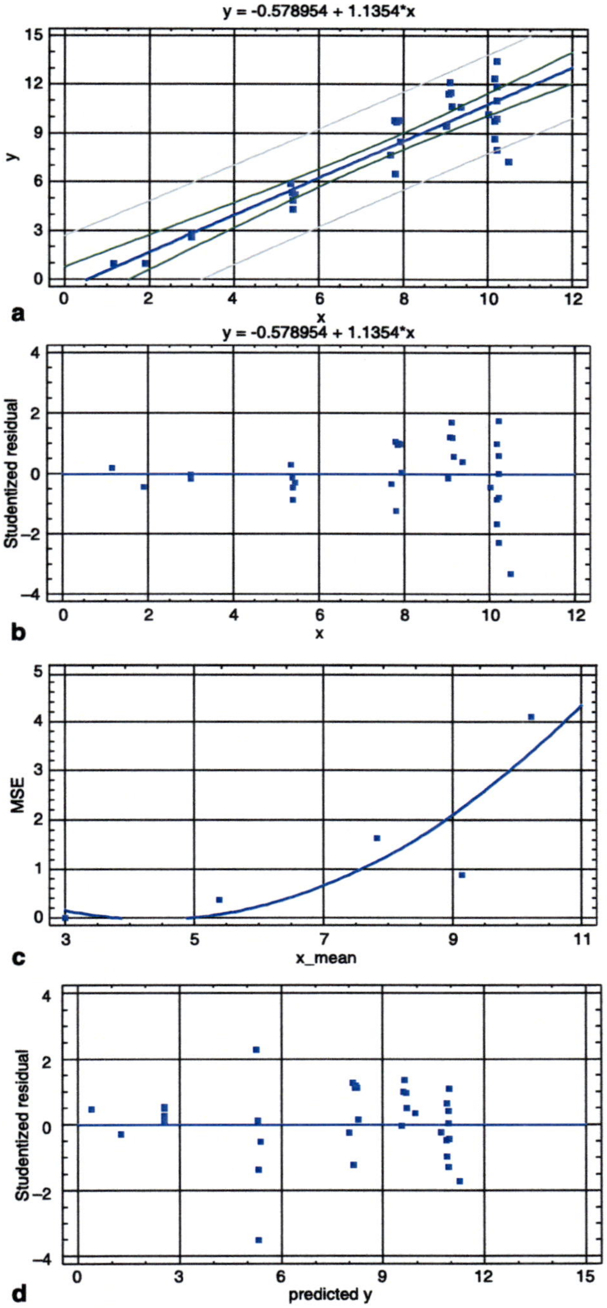

Fig. 5.22 **a** Data set and OLS regression line of observations with non-constant variance and replicated observations in x. **b** Residuals of a simple linear OLS model fit (Eq. 5.49a). **c** Residuals of a second order polynomial OLS fit to the mean x and mean square error (MSE) of the replicate values (Eq. 5.49b). **d** Residuals of the weighted regression model (Eq. 5.49c)

The residual plots are shown as Fig. 5.22d. Though the goodness of fit is only slightly better than the OLS model, the real advantage is that this model will have better prediction accuracy and realistic prediction errors. ■

(c-iii) Non-patterned Variance in the Residuals A third type of non-constant residual variance is one when no pattern is discerned with respect to the regressors which can be discrete or vary continuously. In this case, a practical approach is to look at a plot of the model residuals against the response variable, divide the range in the response variable into as many regions as seem to have different variances, and calculate the standard deviation of the residuals for each of these regions. In that sense, the general approach parallels the one adopted in case (c-ii) when dealing with replicated values with non-constant variance; however, now, no model such as 5.49b is needed. The general approach would involve the following steps:

- First, fit an OLS model to the data;
- Next, discretize the domain of the regressor variables into a finite number of groups and determine ε_i^2 from which the weights w_i for each of these groups can be deduced;
- Finally, perform a WLS regression in order to estimate the efficient model parameters.

Though this two-stage estimation approach is conceptually easy and appealing for simple models, it may become rather complex for multivariate models, and moreover, there is no guarantee that heteroscedasticity will be removed entirely.

5.6.4 Serially Correlated Residuals

Another manifestation of improper residual behavior is serial correlation (discussed in Sect. 5.6.1). As stated earlier, one should distinguish between the two different types of autocorrelation, namely pure autocorrelation and model-misspecification, though often it is difficult to discern between them. The latter is usually addressed using the weight matrix approach (Pindyck and Rubinfeld 1981) which is fairly formal and general, but somewhat demanding. Pure autocorrelation relates to the case of "pseudo" patterned residual behavior which arises because the regressor variables have strong serial correlation. This serial correlation behavior is subsequently transferred over to the model, and thence to its residuals, even when the regression model functional form is close to "perfect". The remedial approach to be adopted is to transform the original data set prior to regression itself. There are several techniques of doing so, and the widely-used Cochrane-Orcutt (CO) procedure is described. It involves the use of generalized differencing to alter the linear model into one in which the errors are independent. The *two stage first-order CO procedure* involves:

(i) fitting an OLS model to the original variables;
(ii) computing the first-order serial correlation coefficient ρ of the model residuals;
(iii) transforming the original variables y and x into a new set of pseudo-variables:

$$y_t^* = y_t - \rho \cdot y_{t-1} \quad \text{and} \quad x_t^* = x_t - \rho \cdot x_{t-1} \qquad (5.50)$$

(iv) OLS regressing of the pseudo variables y* and x* to re-estimate the parameters of the model;
(v) Finally, obtaining the fitted regression model in the original variables by a back transformation of the pseudo regression coefficients:

$$b_0 = b_0{}^*/(1-\rho) \quad \text{and} \quad b_1 = b_1{}^* \tag{5.51}$$

Though two estimation steps are involved, the entire process is simple to implement. This approach, when originally proposed, advocated that this process be continued till the residuals become random (say, based on the Durbin-Watson test). However, the current recommendation is that alternative estimation methods should be attempted if one iteration proves inadequate. This approach can be used during parameter estimation of MLR models provided only one of the regressor variables is the cause of the pseudo-correlation. Also, a more sophisticated version of the CO method has been suggested by Hildreth and Lu (Chatterjee and Price 1991) involving only one estimation process where the optimal value of ρ is determined along with the parameters. This, however, requires non-linear estimation methods.

Example 5.6.4: *Using the Cochrane-Orcutt procedure to remove first-order autocorrelation*

Consider the case when observed pre-retrofit data of energy consumption in a commercial building support a linear regression model as follows:

$$E_i = a_o + a_1 T_i \tag{5.52}$$

where
T = daily average outdoor dry-bulb temperature,
E_i = daily total energy use predicted by the model,
i = subscript representing a particular day, and,
a_o and a_1 are the least-square regression coefficients

How the above transformation yields a regression model different from OLS estimation is illustrated in Fig. 5.23 with year-long daily cooling energy use from a large institutional building in central Texas. The first-order autocorrelation coefficients of cooling energy and average daily temperature were both equal to 0.92, while that of the OLS residuals was 0.60. The Durbin-Watson statistic for the OLS residuals (i.e. untransformed data) was DW=3 indicating strong residual autocorrelation, while that of the CO transform was 1.89 indicating little or no autocorrelation. Note that the CO transform is inadequate in cases of model mis-specification and/or seasonal operational changes. ■

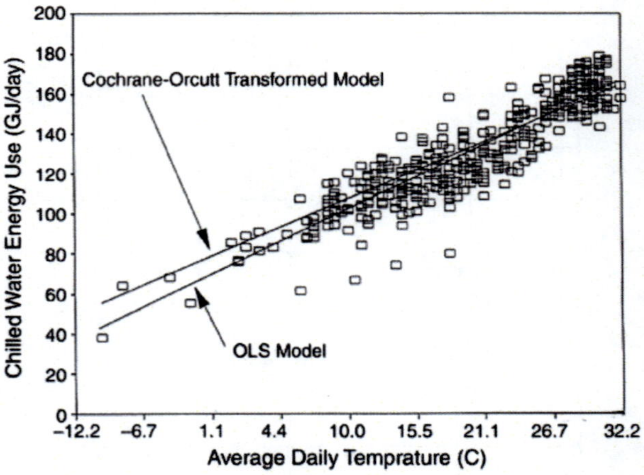

Fig. 5.23 How serial correlation in the residuals affects model identification (Example 5.6.4)

5.6.5 Dealing with Misspecified Models

An important source of error during model identification is *model misspecification* error. This is unrelated to measurement error, and arises when the functional form of the model is not appropriate. This can occur due to:

(i) *inclusion of irrelevant variables:* This does not bias the estimation of the intercept and slope parameters, but generally reduces the efficiency of the slope parameters, i.e., their standard errors will be larger. This source of error can be eliminated by, say, step-wise regression or simple tests such as t-tests;
(ii) *exclusion of an important variable:* This case will result in the slope parameters being both biased and inconsistent.
(iii) *assumption of a linear model:* This arises when a linear model is erroneously assumed, and
(iv) *incorrect model order:* This corresponds to the case when one assumes a lower or higher model than what the data warrants.

The latter three sources of errors are very likely to manifest themselves in improper residual behavior (the residuals will show sequential or non-constant variance behavior). The residual analysis may not identify the exact cause, and several attempts at model reformulations may be required to overcome this problem. Even if the physics of the phenomenon or of the system is well understood and can be cast in mathematical terms, experimental or identifiability constraints may require that a simplified or macroscopic model be used for parameter identification rather than the detailed model. This could cause model misspecification, especially so if the model is poor.

Example 5.6.5: *Example to illustrate how inclusion of additional regressors can remedy improper model residual behavior*

Energy use in commercial buildings accounts for about 18% of the total energy use in the United States and consequently, it is a prime area of energy conservation efforts. For this purpose, the development of baseline models, i.e., models of energy use for a specific end-use before energy conservation measures are implemented, is an important modeling activity for monitoring and verification studies.

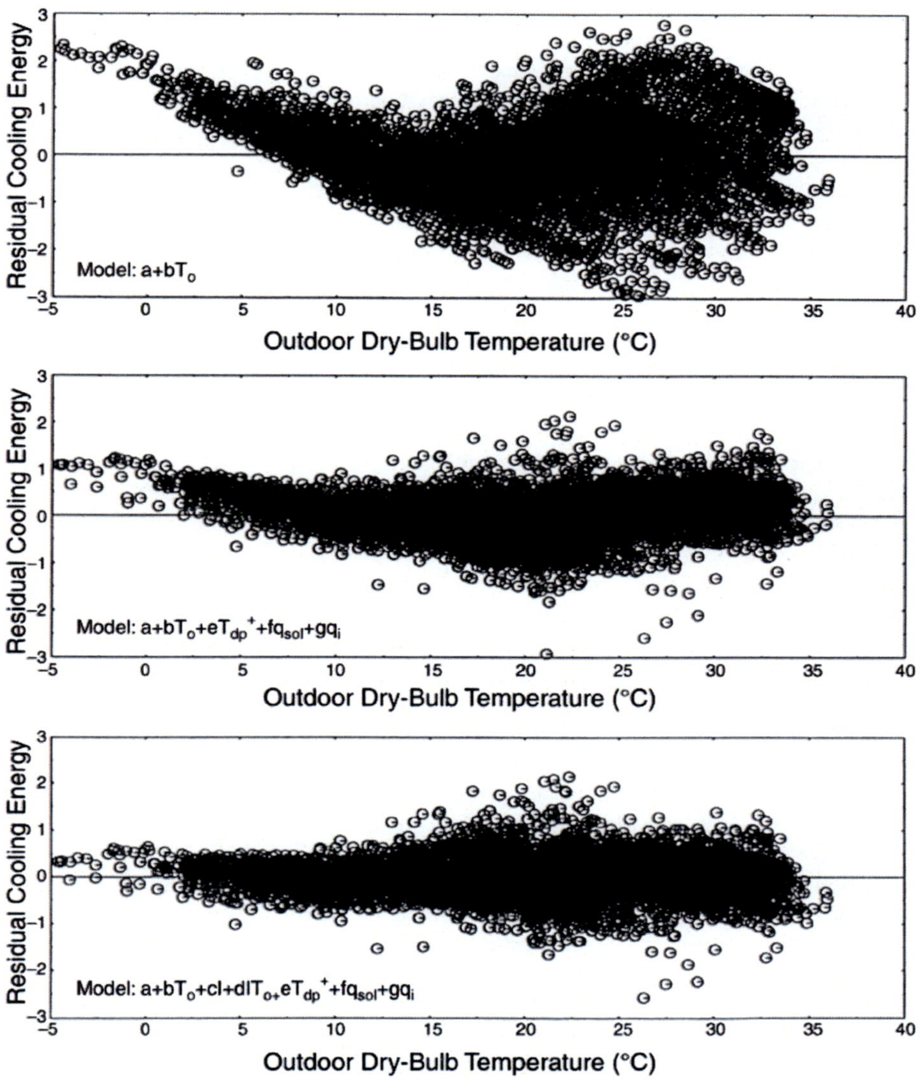

Fig. 5.24 Improvement in residual behavior for a model of hourly energy use of a variable air volume HVAC system in a commercial building as influential regressors are incrementally added to the model. (From Katipamula et al. 1998)

Let us illustrate the effect of improper selection of regressor variables or model misspecification for modeling measured thermal cooling energy use of a large commercial building operating 24 hours a day under a variable air volume HVAC system (Katipamula et al. 1998). Figure 5.24 illustrates the residual pattern when hourly energy use is modeled with only the outdoor dry-bulb temperature (T_o). The residual pattern is blatantly poor exhibiting both non-constant variance as well as systematic bias in the low range of the x-variable. Once the outdoor dew point temperature (T_{dp}^+)[5], the global horizontal solar radiation (q_{sol}) and the internal building heat loads q_i (such as lights and equipment) are introduced in the model, the residual behavior improves significantly but the lower tail is still present. Finally, when additional terms involving indicator variables I to both intercept (T_o) are introduced, (described in Sect. 5.7.2), an acceptable residual behavior is achieved. ■

[5] Actually the outdoor humidity impacts energy use only when the dew point temperature exceeds a certain threshold which many studies have identified to be about 55°F (this is related to how the HVAC is controlled in response to human comfort). This type of conditional variable is indicated as a + superscript.

5.7 Other OLS Parameter Estimation Methods

5.7.1 Zero-Intercept Models

Sometimes the physics of the system dictates that the regression line pass through the origin. For the linear case, the model assumes the form:

$$y = \beta_1 x + \varepsilon \tag{5.53}$$

The interpretation of R^2 under such a case is not the same as for the model with an intercept, and this statistic cannot be used to compare the two types of models directly. Recall that the R^2 value designated the percentage variation of the response variable *about its mean* explained by that of the regressor variable. For the no-intercept case, the R^2 value explains the percentage variation of the response variable *about the origin* explained by that of the regressor variable. Thus, when comparing both models, one should decide on which is the better model based on their RMSE values.

5.7.2 Indicator Variables for Local Piecewise Models—Spline Fits

Spline functions are an important class of functions, described in numerical analysis textbooks in the framework of interpolation, which allow distinct functions to be used over different ranges while maintaining continuity in the function. They are extremely flexible functions in that they allow a wide range of locally different behavior to be captured within one elegant functional framework. Thus, a globally nonlinear function can be decomposed into simpler local patterns. Two cases arise.

(a) The simpler case is one where it is known which points lie on which trend, i.e., when the physics of the system is such that the location of the structural break or "hinge point" x_c of the regressor is known. The simplest type is the piece-wise *linear* spline (as shown in Fig. 5.25), with higher order polynomial splines up to the third degree being also used often to capture non-linear trends. The objective here is to formulate a linear model and identify its parameters which best describe data points in Fig. 5.25. One cannot simply divide the data into two, and fit each region with a separate linear model since the constraint that the model be continuous at the hinge point would be violated. A model of the following form would be acceptable:

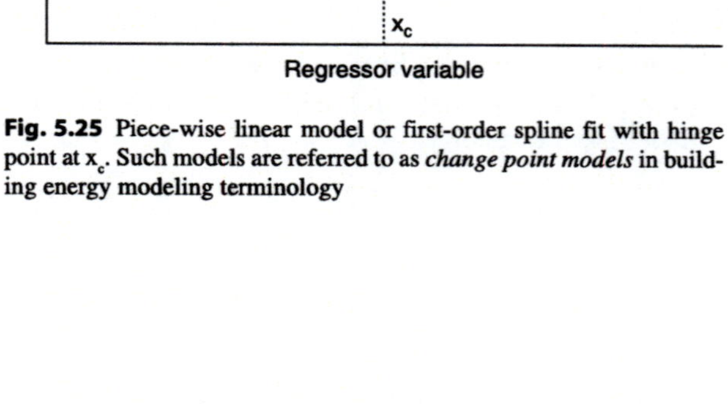

Fig. 5.25 Piece-wise linear model or first-order spline fit with hinge point at x_c. Such models are referred to as *change point models* in building energy modeling terminology

$$y = \beta_0 + \beta_1 x + \beta_2 (x - x_c) I \tag{5.54a}$$

where the indicator variable

$$I = \begin{cases} 1 & \text{if } x > x_c \\ 0 & \text{otherwise} \end{cases} \tag{5.54b}$$

Hence, for the region $x \leq x_c$, , the model is:

$$y = \beta_0 + \beta_1 x \tag{5.55}$$

and for the region $x > x_c$ $y = (\beta_0 - \beta_2 x_c) + (\beta_1 + \beta_2)x$. Thus, the slope of the model is β_1 before the break and $(\beta_1 + \beta_2)$ afterwards. The intercept term changes as well from β_0 before the break to $(\beta_0 - \beta_2 x_c)$ after the break. The logical extensions to linear spline models with two structural breaks or to higher order splines involving quadratic and cubic terms are fairly straightforward.

(b) The second case arises when the change point is not known. A simple approach is to look at the data, identify a "ball-park" range for the change point, perform numerous regression fits with the data set divided according to each possible value of the change point in this ball-park range, and pick that value which yields the best overall R-square or RMSE. Alternatively, the more accurate but more complex approach is to cast the problem as a nonlinear estimation method with the change point variable as one of the parameters.

Example 5.7.1: *Change point models for building utility bill analysis*

The theoretical basis of modeling monthly energy use in buildings is discussed in several papers (for example, Reddy et al. 1997). The interest in this particular time scale is obvious—such information is easily obtained from utility bills which are usually on a monthly time scale. The models suitable for this application are similar to linear spline models, and are referred to as **change point models** by building energy analysts. A simple example is shown below to illustrate the above equations. Electricity utility bills of a residence in Houston, TX have been normalized by the number of days in the month and assembled in Table 5.9 along with the corresponding month and monthly mean outdoor temperature values for Houston (the first three columns of the table). The intent is to use Eq. 5.54 to model this behavior.

The scatter plot and the trend lines drawn in Fig. 5.26 suggest that the change point is in the range 17–19°C. Let us perform the calculation assuming a value of 17°C. Defining an indicator variable:

$$I = \begin{cases} 1 & \text{if } x > 17^\circ\text{C} \\ 0 & \text{otherwise} \end{cases}$$

Table 5.9 Measured monthly energy use data and calculation step for deducing the change point independent variable assuming a base value of 17°C

Month	Mean outdoor temperature (°C)	Monthly mean daily electric use (kWh/m²/day)	x (°C)	$(x-17°C)I$ (°C)
Jan	11	0.1669	11	0
Feb	13	0.1866	13	0
Mar	16	0.1988	16	0
Apr	21	0.2575	21	4
May	24	0.3152	24	7
Jun	27	0.3518	27	10
Jul	29	0.3898	29	12
Aug	29	0.3872	29	12
Sept	26	0.3315	26	9
Oct	22	0.2789	22	5
Nov	16	0.2051	16	0
Dec	13	0.1790	13	0

Based on this assumption, the last two columns of the table have been generated to correspond to the two regressor variables in Eq. 5.54. A linear multiple regression yields:

$$y = 0.1046 + 0.005904x + 0.00905(x - 17)I$$
$$\text{with} \quad R^2 = 0.996 \quad \text{and} \quad \text{RMSE} = 0.0055$$

with all three parameters being significant. The reader can repeat this analysis assuming a different value for the change point (say $x_c=18°C$) in order to study the sensitivity of the model to the choice of the change point value. Though only three parameters are determined by regression, this is an example of a four parameter (or 4-P) model in building science terminology. The fourth parameter is the change point x_c which also needs to be determined. Software programs have been developed to determine the optimal value of x_c (i.e., that which results in minimum RMSE of different possible choices of x_c) following a numerical search process akin to the one described in this example. ■

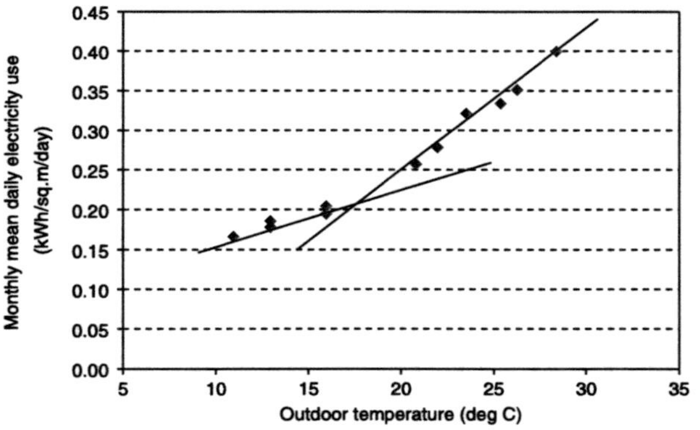

Fig. 5.26 Piece-wise linear regression lines for building electric use with outdoor temperature. The change point is the point of intersection of the two lines. The combined model is called a change point model, which, in this case, is a four parameter model given by Eq. 5.54

5.7.3 Indicator Variables for Categorical Regressor Models

The use of indicator (also called dummy) variables has been illustrated in the previous section when dealing with spline models. They are also used in cases when shifts in either the intercept or the slope are to be modeled with the *condition of continuity now being relaxed.* The majority of variables encountered in mechanistic models are quantitative, i.e., the variables are measured on a numerical scale. Some examples are temperature, pressure, distance, energy use and age. Occasionally, the analyst comes across models involving qualitative variables, i.e., regressor data that belong in one of two (or more) possible categories. One would like to evaluate whether differences in intercept and slope between categories are significant enough to warrant two separate models or not. This concept is illustrated by the following example.

Whether the annual energy use of a regular commercial buildings is markedly higher than that of another certified as being energy efficient is to be determined. Data from several buildings which fall in each group is gathered to ascertain whether the presumption is supported by the actual data. Factors which affect the *normalized energy use* (variable y) of both experimental groups are conditioned floor area (variable x_1) and outdoor temperature (variable x_2). Suppose that a linear relationship can be assumed with the same intercept for both groups. One approach would be to separate the data into two groups: one for regular buildings and one for efficient buildings, and develop regression models for each group separately. Subsequently, one could perform a t-test to determine whether the slope terms of the two models are significantly different or not. However, the assumption of constant intercept term for both models may be erroneous, and this may confound the analysis. A better approach is to use the entire data and adopt a modeling approach involving indicator variables.

Let model 1 be for regular buildings:

$$y = a + b_1 x_1 + c_1 x_2$$

and, model 2 be for energy efficient buildings:

$$y = a + b_2 x_1 + c_2 x_2 \tag{5.56}$$

The complete model (or model 3) would be formulated as:

$$y = a + b_1 x_1 + c_1 x_2 + b_2 (I \cdot x_1) + c_2 (I \cdot x_2) \tag{5.57}$$

where I is an indicator variable such that

$$I = \begin{cases} 1 & \text{for energy efficient buildings} \\ 0 & \text{for regular buildings} \end{cases}$$

Note that a basic assumption in formulating this model is that the intercept is unaffected by the building group. Formally, one would like to test the null hypothesis H: $b_2=c_2=0$. The hypothesis is tested by constructing an F statistic for the comparison of the two models. Note that model 3 is referred to as the *full model* (FM) or as the pooled model. Model 1, when the null hypothesis holds, is the *reduced model* (RM). The idea is to compare the goodness-of-fit of the FM and that of the RM. If the RM provides as good a fit as the FM, then the null hypothesis is valid. Let SSE(FM) and SSE(RM) be the corresponding model sum of square error or squared model residuals. Then, the following F-test statistic is defined:

$$F = \frac{[SSE(RM) - SSE(FM)]/(k\text{-}m)}{SSE(FM)/(n\text{-}k)} \tag{5.58}$$

where n is the number of data sets, k is the number of parameters of the FM, and m the number of parameters of the RM. If the observed F value is larger than the tabulated value of F with (n-k) and (k-m) degrees of freedom at the pre-specified significance level (provided by Table A.6), the RM is unsatisfactory and the full model has to be retained. As a cautionary note, this test is strictly valid only if the OLS assumptions for the model residuals hold.

Example 5.7.2: *Combined modeling of energy use in regular and energy efficient buildings*

Consider the data assembled in Table 5.10. Let us designate the regular buildings by group (A) and the energy efficient buildings by group (B), with the problem simplified by assuming both types of buildings to be located in the same geographic location. Hence, the model has *only one regressor variable* involving floor area. The complete model with the indicator variable term given by Eq. 5.57 is used to verify whether group B buildings consume less energy than group A buildings.

The full model (FM) given by Eq. 5.57 reduces to the following form since only one regressor is involved:

Table 5.10 Data table for Example 5.7.2

Energy use (y)	Floor area (x_1)	Bldg type	Energy use (y)	Floor area (x_1)	Bldg type
45.44	225	A	32.13	224	B
42.03	200	A	35.47	251	B
50.1	250	A	33.49	232	B
48.75	245	A	32.29	216	B
47.92	235	A	33.5	224	B
47.79	237	A	31.23	212	B
52.26	265	A	37.52	248	B
50.52	259	A	37.13	260	B
45.58	221	A	34.7	243	B
44.78	218	A	33.92	238	B

$y=a+b_1x_1+b_2I\cdot x_2$ where the variable I is an indicator variable such that it is 0 for group A and 1 for group B. The null hypothesis is that H_0: $b_2=0$. The reduced model (RM) is $y=a+b\,x_1$.

The estimated model is $y=14.2762+0.14115\ x_1-13.2802\ (I\cdot x_2)$. The analysis of variance shows that the SSR(FM)=7.7943 and SSR(RM)=889.245. The F statistic in this case is:

$$F = \frac{(889.245 - 7.7943)/1}{7.7943/(20-3)} = 1922.5$$

One can thus safely reject the null hypothesis, and state with confidence that buildings built as energy-efficient ones consume energy which is statistically lower than those which are not.

It is also possible to extend the analysis and test whether both slope and intercept are affected by the type of building. The FM in this case is $y=a_1+b_1\ x_1+c(I)+d(I\cdot x_1)$ where I is an indicator variable which is, say 0 for Building A and 1 for Building B. The null hypothesis in this case is that $c=d=0$. This is left for the interested reader to solve. ■

5.7.4 Assuring Model Parsimony—Stepwise Regression

Perhaps the major problem with multivariate regression is that the "independent" variables are not really independent but collinear to some extent (how to deal with collinear regressors by transformation is discussed in Sect. 9.3). In multivariate regression, a thumb rule is that the number of variables should be less than four times the number of observations (Chatfield 1995). Hence, with n=12, the number of variables should be at most 3 or less. Moreover, some authors go so far as stating that multivariate regression models with more than 4–5 variables are suspect. There is, thus, a big benefit in identifying models that are parsimonious. The more straightforward approach is to use the simpler (but formal) methods to identify/construct the "best" model linear in the parameters if the comprehensive set of all feasible/possible regressors of the model is known (Draper and Smith 1981; Chatterjee and Price 1991):

(a) *All possible regression models*: This method involves: (i) constructing models of different basic forms (single variate with various degrees of polynomials and multi-variate), (ii) estimating parameters that correspond to all possible predictor variable combinations, and (iii) then selecting one considered most desirable based on some criterion. While this approach is thorough, the computational effort involved may be significant. For example, with p possible parameters, the number of model combinations would be p^2. However, this may be moot if the statistical analysis program being used contains such a capability. The only real drawback is that blind curve fitting may suggest a model with no physical justification which in certain applications may have undesirable consequences. Further, it is advised that the cross-validation scheme should be used to avoid overfitting (see Sect. 5.3.2-d).

In any case, one needs a statistical criterion to determine, if not the "best[6]" model, then, at least a subset of desirable models from which one can be chosen based on the physics of the problem. One could use the adjusted R-square given by Eq. 5.7b which includes the number of model parameters. Another criterion for model selection is the *Mallows* C_p *statistic* which gives a normalized estimate of the total expected estimation error for all observations in the data set and takes account of both bias and variance:

$$C_p = \frac{SSE}{\sigma^2} + (2p - n) \quad (5.59)$$

where SSE is the sum of square errors (see Eq. 5.2), σ^2 is the variance of the residuals with the full set of variables, n is the number of data points, and p is the number of parameters in the specific model. It can be shown that the expected value of C_p is p when there is no bias in the fitted equation containing p terms. Thus "good" or desirable model possibilities are those whose C_p values are close to the corresponding number of parameters of the model.

Another automatic selection approach to handling models with large number of possible parameters is the *iterative approach* which comes in three variants.

(b-1) Backward Elimination Method: One begins with selecting an initial model that includes the full set of possible predictor variables from the candidate pool, and then successively dropping one variable at a time on the basis of their contribution to the reduction of SSE. The OLS method is used to estimate all model parameters along with t-values for each model parameter. If all model parameters are statistically significant, the model building process stops. If some model parameters are not significant, the model parameter of least significance (lowest t-value) is omitted from the regression equation, and the reduced model is refit. This process continues until all parameters that remain in the model are statistically significant.

(b-2) Forward Selection Method: One begins with an equation containing no regressors (i.e., a constant model). The model is then augmented by including the regressor variable with the highest simple correlation with the response variable. If this regression coefficient is significantly different from zero, it is retained, and the search for a second variable is made. This process of adding regressors one-by-one is terminated when the last variable entering the equation has an insignificant regression coefficient or when all the variables are included in the model. Clearly, this approach involves fitting many more models than in the backward elimination method.

(b-3) Stepwise Regression Method: This is one of the more powerful model building approaches and combines both the above procedures. Stepwise regression begins by computing correlation coefficients between the response and each predictor variable. The variable most highly correlated with the response is then allowed to "enter the regression equation". The parameter for the single-variable regression equation is then estimated along with a measure of the goodness of fit. The next most highly correlated predictor variable is identified, given the current variable already in the regression equation. This variable is then allowed to enter the equation and the parameters re-estimated along with the goodness of fit. Following each parameter estimation, t-values for each parameter are calculated and compared to t-critical to determine whether all parameters are still statistically significant. Any parameter that is not statistically significant is removed from the regression equation. This process continues until no more variables "enter" or "leave" the regression equation. In general, it is best to select the model that yields a reasonably high "goodness of fit" for the fewest parameters in the model (referred to as *model parsimony*). The final decision on model selection requires the judgment of the model builder, and on mechanistic insights into the problem. Again, one has to guard against the danger of overfitting by performing a cross-validation check.

When a black-box model is used containing several regressors, step-wise regression would improve the robustness of the model by reducing the number of regressors in the model, and thus hopefully reduce the adverse effects of multicollinearity between the remaining regressors. Many packages use the F-test indicative of the overall model instead of the t-test on individual parameters to perform the step-wise regression. A value of F=4 is often chosen. It is suggested that step-wise regression not be used in case the regressors are highly correlated since it may result in non-robust models. However, the backward procedure is said to better handle such situations than the forward selection procedure.

A note of caution is warranted in using stepwise regression for engineering models based on mechanistic considerations. In certain cases, stepwise regression may omit a regressor which ought to be influential when using a particular data set, while the regressor is picked up when another data set is used. This may be a dilemma when the model is to be used for subsequent predictions. In such cases, discretion based on physical considerations should trump purely statistical model building.

[6] Actually, there is no "best" model since random variables are involved. A better term would be "most plausible" and should include mechanistic considerations, if appropriate.

Example 5.7.3:[7] *Proper model identification with multivariate regression models*

An example of multivariate regression is the development of model equations to characterize the performance of refrigeration compressors. It is possible to regress compressor manufacturer's tabular data of compressor performance using the following simple bi-quadratic formulation (see Fig. 5.11 for nomenclature).

$$y = C_0 + C_1 \cdot T_{cho} + C_2 \cdot T_{cdi} + C_3 \cdot T_{cho}^2 + C_4 \cdot T_{cdi}^2 + C_5 \cdot T_{cho} \cdot T_{cdi} \quad (5.60)$$

where y represents either the compressor power (P_{comp}) or the cooling capacity (Q_{ch}).

OLS is then used to develop estimates of the six model parameters, C_0–C_5, based on the compressor manufacturer's data. The biquadratic model was used to estimate the parameters for compressor cooling capacity (in Tons) for a screw compressor. The model and its corresponding parameter estimates are given below. Although the overall curve fit for the data was excellent (R^2=99.96%), the t-values of two parameter estimates (C_2 and C_4) are clearly insignificant.

A second stage regression is done omitting these regressors resulting in the following model and coefficient t-values shown in Table 5.11.

$$y = C_0 + C_1 \cdot T_{cho} + C_3 \cdot T_{cho}^2 + C_5 \cdot T_{cho} \cdot T_{cdi}$$

All of the parameters in the simplified model are significant and the overall model fit remains excellent: R^2=99.5%. ■

5.8 Case Study Example: Effect of Refrigerant Additive on Chiller Performance[8]

The objective of this analysis was to verify the claim of a company which had developed a refrigerant additive to improve chiller COP. The performance of a chiller before (called pre-retrofit period) and after (called post-retrofit period) addition of this additive was monitored for several months to determine whether the additive results in an improvement in chiller performance, and if so, by how much. The same four variables described in Example 5.4.3, namely two temperatures (T_{cho} and T_{cdi}), the chiller thermal cooling load (Q_{ch}) and the electrical power consumed (P_{comp}) were measured in intervals of 15 min. Note that the chiller COP can be deduced from the last two variables. Altogether, there were 4,607 and 5,078 data points for the pre-and post periods respectively.

Step 1: Perform Exploratory Data Analysis At the onset, an exploratory data analysis should be performed to determine the spread of the variables, and their occurrence frequencies during the pre- and post-periods, i.e., before and after addition of the refrigerant additive. Further, it is important to ascertain whether the operating conditions during both periods are similar or not. The eight frames in Fig. 5.27 summarize the spread and frequency of the important variables. It is noted that though the spreads in the operating conditions are similar, the frequencies are different during both periods especially in the condenser temperatures and the chiller load variables. Figure 5.28 suggests that $COP_{post} > COP_{pre}$. An ANOVA test with results shown in Table 5.12 and Fig. 5.29 also indicates that the mean of post-retrofit power use is statistically different at 95% confidence level as compared to the pre-retrofit power.

t Test to Compare Means

Null hypothesis: mean (COP_{post}) = mean (COP_{pre})

Alternative hypothesis: mean (COP_{post}) ≠ mean (COP_{pre}) assuming equal variances:

$$t = 38.8828, \text{ p-value} = 0.0$$

The null hypothesis is rejected at α=0.05.

Of particular interest is the confidence interval for the difference between the means, which extends from 0.678 to 0.750. Since the interval does not contain the value 0.0, there is a statistically significant difference between the means of the two samples at the 95.0% confidence level. However, it would be incorrect to infer that $COP_{post} > COP_{pre}$ *since the operating conditions are different, and thus one should not use this approach to draw any conclusions.* Hence, a regression model based approach is warranted.

Step 2: Use the Entire Pre-retrofit Data to Identify a Model The GN chiller models (Gordon and Ng 2000) are described in Pr. 5.13. The monitored data is first used to compute the variables of the model given by the regression model Eqs. 5.70 and 5.71. Then, a linear regression is performed which is given below along with standard errors of the coefficients shown within parenthesis:

Table 5.11 Results of the first and second stage model building

With all parameters			With significant parameters only	
Coefficient	Value	t-value	Value	t-value
C_0	152.50	6.27	114.80	73.91
C_1	3.71	36.14	3.91	11.17
C_2	−0.335	−0.62	–	–
C_3	0.0279	52.35	0.027	14.82
C_4	−0.000940	−0.32	–	–
C_5	−0.00683	−6.13	−0.00892	−2.34

[7] From ASHRAE (2005) © American Society of Heating, Refrigerating and Air-conditioning Engineers, Inc., www.ashrae.org.

[8] The monitored data was provided by Ken Gillespie for which we are grateful.

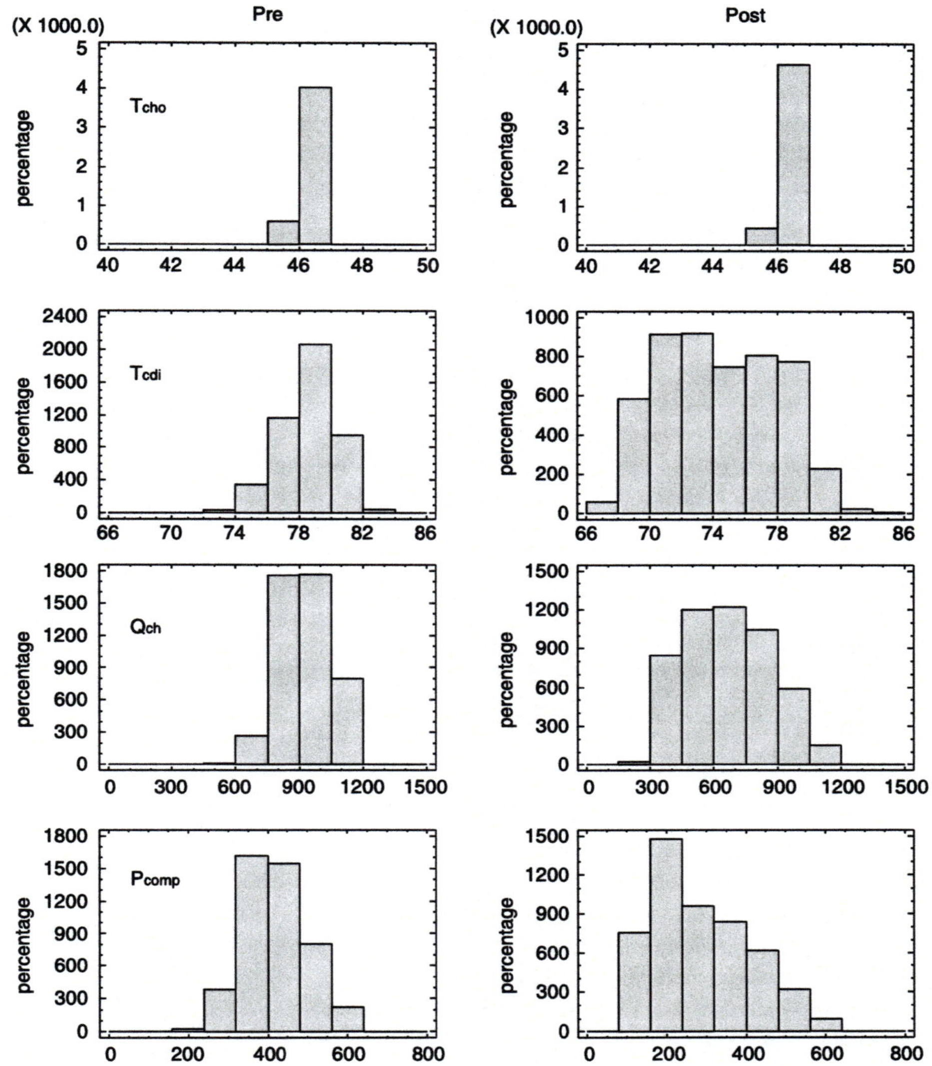

Fig. 5.27 Histograms depicting the range of variation and frequency of the four important variables before and after the retrofit (pre=4,607 data points, post=5,078 data points). The condenser water temperature and the chiller load show much larger variability during the post period

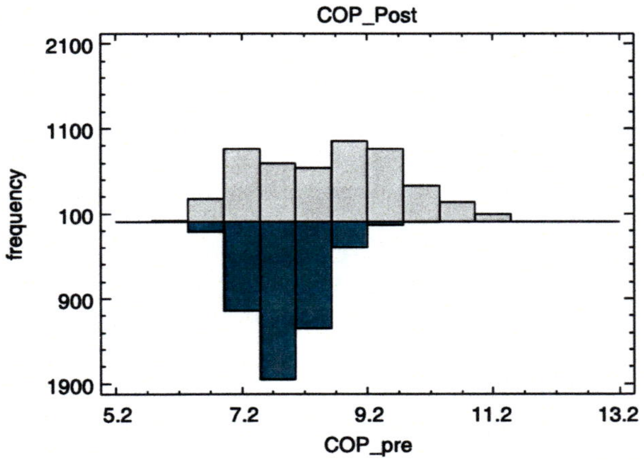

Fig. 5.28 Histogram plots of Coefficient of Performance (COP) of chiller before and after retrofit. Clearly, there are several instances when $COP_{post} > COP_{pre}$ but that could be due to operating conditions. Hence, a regression modeling approach is clearly warranted

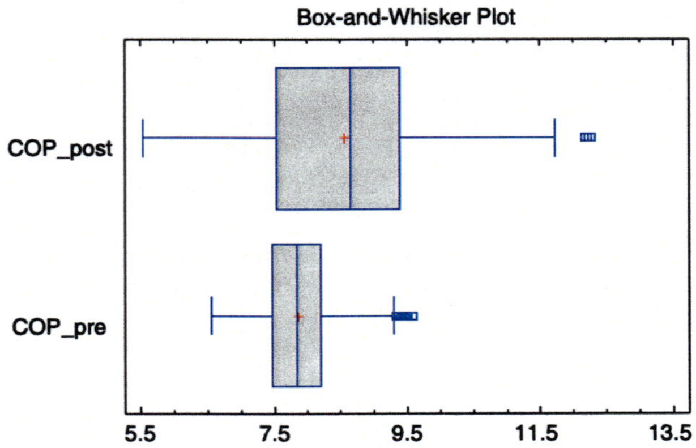

Fig. 5.29 ANOVA test results in the form of box-and-whisker plots for chiller COP before and after addition of refrigerant additive

Table 5.12 Results of the ANOVA Test of comparison of means at significance level of 0.05

95.0% confidence interval for mean of COP_{post}:	8.573 ±0.03142 = [8.542, 8.605]
95.0% confidence interval for mean of COP_{pre}:	7.859 ±0.01512 = [7.844, 7.874]
95.0% confidence interval for the difference between the means assuming equal variances:	0.714 ±0.03599 = [0.678, 0.750]

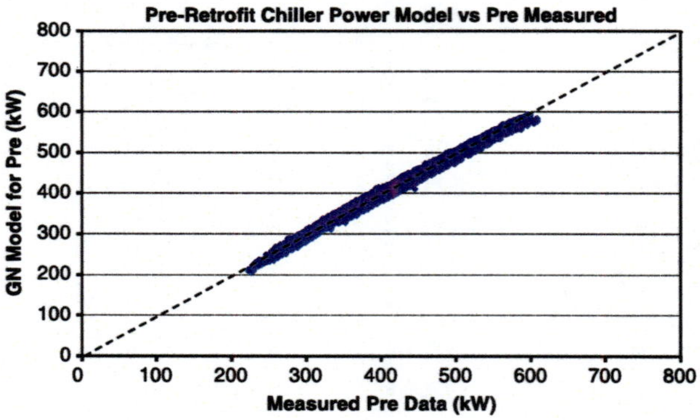

Fig. 5.30 X–Y plot of chiller power during pre-retrofit period. The overall fit is excellent (RMSE=9.36 kW and CV=2.24%), and except for a few data points, the data seems well behaved. Total number of data points=4,607

$$y = \underset{(0.00163)}{-0.00187} \cdot x_1 + \underset{(15.925)}{261.2885} \cdot x_2 + \underset{(0.000111)}{0.022461} \cdot x_3$$
$$\text{with adjusted } R^2 = 0.998 \qquad (5.61)$$

This model is then re-transformed into a model for power using Eq. 5.76, and the error statistics using the pre-retrofit data are found to be: RMSE=9.36 kW and CV=2.24%. Figure 5.30 shows the x–y plot from which one can visually evaluate the goodness of fit of the model. Note that the mean power use=418.7 kW while the mean model residuals=0.017 kW (negligibly close to zero, as it should be. This step validates the fact that the spreadsheet cells have been coded correctly with the right formulas).

Step 3: Calculate Savings in Electrical Power The above chiller model representative of thermal performance of the chiller without refrigerant additive is used to estimate savings by first predicting power use for each 15 min interval using the two operating temperatures and the load corresponding to the 5,078 post-retrofit data points. Subsequently, savings in chiller power are deduced for each of the 5,078 data points:

$$\begin{aligned} \textit{Power savings} &= \textit{Model-predicted pre-retrofit use} \\ &\quad - \textit{measured post-retrofit use} \end{aligned} \qquad (5.62)$$

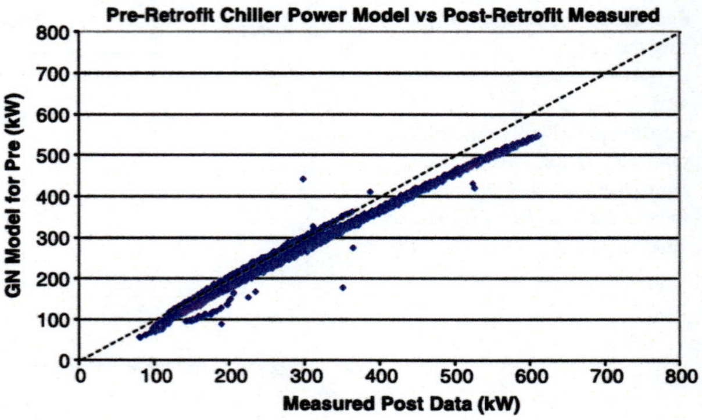

Fig. 5.31 Difference in X–Y plots of chiller power indicating that post-retrofit values are higher than those during pre-retrofit period (mean increase=21 kW or 7.88%). One can clearly distinguish two operating patterns in the data suggesting some intrinsic behavioral change in chiller operation. Entire data set for the post-period consisting of 5,078 observations has been used in this analysis

It is found that mean power savings=−21.0 kW (i.e., **an increase in power use**) or a decrease of 7.88% in the measured mean power use of 287.5 kW. Figure 5.31 visually illustrates the extent to which power use during the post-retrofit period has increased as compared to the pre-retrofit model. Overlooking the few outliers, one notes that there are two patterns: a larger number of data points indicating that post-retrofit electricity power use was much higher and a smaller set when the difference is little to nil. The reason for the onset of two distinct patterns in operation is worthy of a subsequent investigation.

Step 4: Calculate Uncertainty in Savings and Draw Conclusions The uncertainty arises from two sources: prediction model and power measurement errors. The latter are usually small, about 0.1% of the reading, which in this particular case is less than 1 kW. Hence, this contribution can be neglected during an initial investigation such as this one. The model uncertainty is given by:

$$\begin{aligned} &\text{absolute uncertainty in power use savings or reduction} \\ &\quad = (\text{t\_value x RMSE}) \end{aligned} \qquad (5.63)$$

The t-value at 90% confidence level=1.65 and RMSE of model (for pre-retrofit period)=9.36 kW.

Hence the calculated increase in power due to refrigerant additive = −21.0 kW ± 15.44 kW at 90% CL. Thus, one would conclude that the refrigerant additive is actually penalizing chiller performance by 7.88% since electric power use is increased.

Note: The entire analysis was redone by cleaning the post-retrofit data so as to remove the dual sets of data (see Fig. 5.31). Even then, the same conclusion was reached.

Table 5.13 Data table for Problem 5.1

Temperature t (°C)	0	10	20	30	40	50	60	70	80	90	100
Specific volume v (m^3/kg)	206.3	106.4	57.84	32.93	19.55	12.05	7.679	5.046	3.409	2.361	1.673
Sat. vapor enthalpy kJ/kg	2501.6	2519.9	2538.2	2556.4	2574.4	2592.2	2609.7	2626.9	2643.8	2660.1	2676

Problems

Pr. 5.1 Table 5.13 lists various properties of saturated water in the temperature range 0–100°C.

(a) Investigate first order and second-order polynomials that fit saturated vapor enthalpy to temperature in °C. Identify the better model by looking at R^2, RMSE and CV values for both models. Predict the value of saturated vapor enthalpy at 30°C along with 95% confidence intervals and 95% prediction intervals.

(b) Repeat the above analysis for specific volume but investigate third-order polynomial fits as well. Predict the value of specific volume at 30°C along with 95% confidence intervals and 95% prediction intervals.

Pr. 5.2 Tensile tests on a steel specimen yielded the results shown in Table 5.14.

(a) Assuming the regression of y on x to be linear, estimate the parameters of the regression line and determine the 95% confidence limits for x=4.5

(b) Now regress x on y, and estimate the parameters of the regression line. For the same value of y predicted in (a) above, determine the value of x. Compare this value with the value of 4.5 assumed in (a). If different, discuss why.

(c) Compare the R^2 and CV values of both models.

(d) Plot the residuals of both models

(e) Of the two models, which is preferable for OLS estimation.

Pr. 5.3 The yield of a chemical process was measured at three temperatures (in °C), each with two concentrations of a particular reactant, as recorded in Table 5.15.

(a) Use OLS to find the best values of the coefficients a, b, and c in the equation: $y=a+bt+cx$.

(b) Calculate the R^2, RMSE, and CV of the overall model as well as the SE of the parameters

(c) Using the β coefficient concept described in Sect. 5.4.5, determine the relative importance of the two independent variables on the yield.

Table 5.14 Data table for Problem 5.2

Tensile force x	1	2	3	4	5	6
Elongation y	15	35	41	63	77	84

Table 5.15 Data table for Problem 5.3

Temperature, t	40	40	50	50	60	60
Concentration, x	0.2	0.4	0.2	0.4	0.2	0.4
Yield y	38	42	41	46	46	49

Pr. 5.4 *Cost of electric power generation versus load factor and cost of coal*

The cost to an electric utility of producing power (C_{Ele}) in mills per kilowatt-hr ($\$10^{-3}$/kWh) is a function of the load factor (LF) in % and the cost of coal (C_{coal}) in cents per million Btu. Relevant data is assembled in Table 5.16.

(a) Investigate different models (first order and second order with and without interaction terms) and identify the best model for predicting C_{Ele} vs LF and C_{Coal}. Use stepwise regression if appropriate. (Hint: plot the data and look for trends first).

(b) Perform residual analysis

(c) Calculate the R^2, RMSE, and CV of the overall model as well as the SE of the parameters

Pr. 5.5 *Modeling of cooling tower performance*

Manufacturers of cooling towers often present catalog data showing outlet-water temperature T_{co} as a function of ambient air wet-bulb temperature (T_{wb}) and range (which is the difference between inlet and outlet water temperatures). Table 5.17 assembles data for a specific cooling tower. Identify an appropriate model (investigate first order and second order polynomial models for T_{co}) by looking at R^2, RMSE and CV values, the individual t-values of the parameters as well as the behavior of the overall model residuals.

Pr. 5.6 *Steady-state performance testing of solar thermal flat plate collector*

Solar thermal collectors are devices which convert the radiant energy from the sun into useful thermal energy that goes to heating, say, water for domestic or for industrial applications. Because of low collector time constants, heat capacity effects are usually small compared to the hourly time step

Table 5.16 Data table for Problem 5.4

LF	85	80	70	74	67	87	78	73	72	69	82	89
C_{Coal}	15	17	27	23	20	29	25	14	26	29	24	23
C_{Ele}	4.1	4.5	5.6	5.1	5.0	5.2	5.3	4.3	5.8	5.7	4.9	4.8

Table 5.17 Data table for Problem 5.5

	T_{wb} (°C)				
Range (°C)	20	21.5	23	23.5	26
10	25.89	26.65	27.49	27.78	29.38
13	26.40	27.11	27.90	28.18	29.75
16	26.99	27.64	28.38	28.66	30.18
19	27.65	28.24	28.94	29.20	30.69
22	28.38	28.92	29.58	29.83	31.28

used to drive the model. The steady-state useful energy q_C delivered by a solar flat-plate collector of surface area A_C is given by the Hottel-Whillier-Bliss equation (Reddy 1987):

$$q_c = A_c F_R \left[I_T \eta_n - U_L (T_{Ci} - T_a) \right]^+ \tag{5.64}$$

where F_R is called the heat removal factor and is a measure of the solar collector performance as a heat exchanger (since it can be interpreted as the ratio of actual heat transfer to the maximum possible heat transfer); η_n is the optical efficiency or the product of the transmittance and absorptance of the cover and absorber of the collector at normal solar incidence; U_L is the overall heat loss coefficient of the collector which is dependent on collector design only, I_T is the radiation intensity on the plane of the collector, T_{ci} is the temperature of the fluid entering the collector, and T_a is the ambient temperature. The $^+$ sign denotes that only positive values are to be used, which physically implies that the collector should not be operated if q_C is negative i.e., when the collector loses more heat than it can collect (which can happen under low radiation and high T_{ci} conditions).

Steady-state collector testing is the best manner for a manufacturer to rate his product. From an overall heat balance on the collector fluid and from Eq. 5.64, the expressions for the instantaneous collector efficiency η_c under normal solar incidence are:

$$\eta_C \equiv \frac{q_C}{A_C I_T} = \frac{(mc_p)_C (T_{Co} - T_{Ci})}{A_C I_T} = \left[F_R \eta_n - F_R U_L \left(\frac{T_{Ci} - T_a}{I_T} \right) \right] \tag{5.65}$$

where m_c is the total fluid flow rate through the collectors, c_{pc} is the specific heat of the fluid flowing through the collector, and T_{ci} and T_{co} are the inlet and exit temperatures of the fluid to the collector. Thus, measurements (of course done as per the standard protocol, ASHRAE 1978) of I_T, T_{ci} and T_{co} are done under a pre-specified and controlled value of fluid flow rate. The test data are plotted as η_c against reduced temperature $[(T_{Ci}-T_a)/I_T]$ as shown in Fig. 5.32. A linear fit is made to these data points by regression, from which the values of $F_R\ \eta_n$ and $F_R\ U_L$ are easily deduced.

If the same collector is testing during different days, slightly different numerical values are obtained for the two

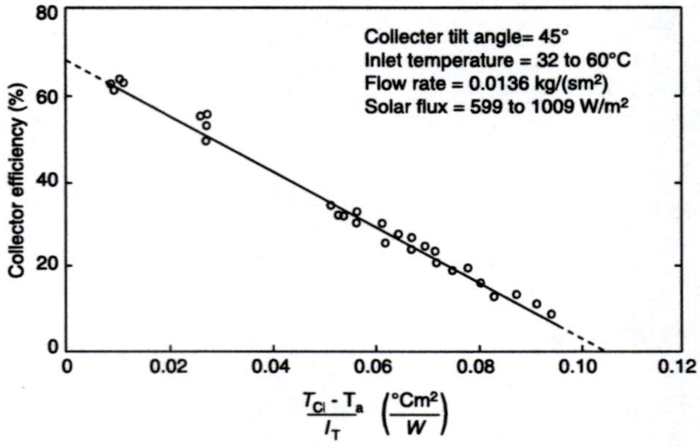

Fig. 5.32 Test data points of thermal efficiency of a double glazed flat-plate liquid collector with reduced temperature. The regression line of the model given by Eq. 5.65 is also shown. (From ASHRAE (1978) © American Society of Heating, Refrigerating and Air-conditioning Engineers, Inc., www.ashrae.org)

parameters $F_R\eta_n$ and F_RU_L which are often, but not always, within the uncertainty bands of the estimates. Model misspecification (i.e., the model is not perfect, for example, it is known that the collector heat losses are not strictly linear) is partly the cause of such variability. This is somewhat disconcerting to a manufacturer since this introduces ambiguity as to which values of the parameters to present in his product specification sheet.

The data points of Fig. 5.32 are assembled in Table 5.18. Assume that water is the working fluid.

(a) Perform OLS regression using Eq. 5.65 and identify the two parameters $F_R\eta_n$ and F_RU_L along with their standard errors. Plot the model residuals, and study their behavior.
(b) Draw a straight line visually through the data points and determine the x-axis and y-axis intercepts. Estimate the $F_{R\,n}$ and F_RU_L parameters and compare them with those determined from (a).
(c) Calculate the R^2, RMSE and CV values of the model
(d) Calculate the F-statistic to test for overall model significance of the model
(e) Perform t-tests on the individual model parameters
(f) Use the model to predict collector efficiency when I_T=800 W/m², T_{ci}=35°C and T_a=10°C

Table 5.18 Data table for Problem 5.6

x	y (%)	x	y (%)	x	y (%)	x	y (%)
0.009	64	0.051	30	0.064	27	0.077	20
0.011	65	0.052	30	0.065	26	0.080	16
0.025	56	0.053	31	0.065	24	0.083	14
0.025	56	0.056	29	0.069	24	0.086	14
0.025	52.5	0.056	29	0.071	23	0.091	12
0.025	49	0.061	29	0.071	21	0.094	10
0.050	35	0.062	25	0.075	20		

(g) Determine the 95% CL intervals for the mean and individual responses for (f) above.
(h) The steady-state model of the solar thermal collector assumes the heat loss term given by $[UA(T_{ci}-T_a]$ to be linear with the temperature difference between collector inlet temperature and the ambient temperature. One wishes to investigate whether the model improves if the loss term is to include an additional second order term:
  (i) Derive the resulting expression for collector efficiency analogous to Eq. 5.65?
  (Hint: start with the fundamental heat balance equation—Eq. 5.64)
  (ii) Does the data justify the use of such a model?

Pr. 5.7[9] *Dimensionless model for fans or pumps*
The performance of a fan or pump is characterized in terms of the head or the pressure rise across the device and the flow rate for a given shaft power. The use of dimensionless variables simplifies and generalizes the model. Dimensional analysis (consistent with fan affinity laws for changes in speed, diameter and air density) suggests that the performance of a centrifugal fan can be expressed as a function of two dimensionless groups representing flow coefficient and pressure head respectively:

$$\Psi = \frac{SP}{D^2\omega^2\rho} \quad \text{and} \quad \Phi = \frac{Q}{D^3\omega} \tag{5.66}$$

where SP is the static pressure, Pa; D the diameter of wheel, m; ω the rotative speed, rad/s; ρ the density, kg/m^3 and Q the volume flow rate of air, m^3/s.

For a fan operating at constant density, it should be possible to plot one curve Ψ vs Φ that represents the performance at all speeds. The performance of a certain 0.3 m diameter fan is shown in Table 5.19.

Table 5.19 Data table for Problem 5.7

Rotation ω (Rad/s)	Flow rate Q (m^3/s)	Static pressure SP (Pa)	Rotation ω (Rad/s)	Flow rate Q (m^3/s)	Static pressure SP (Pa)
157	1.42	861	94	0.94	304
157	1.89	861	94	1.27	299
157	2.36	796	94	1.89	219
157	2.83	694	94	2.22	134
157	3.02	635	94	2.36	100
157	3.30	525	63	0.80	134
126	1.42	548	63	1.04	122
126	1.79	530	63	1.42	70
126	2.17	473	63	1.51	55
126	2.36	428			
126	2.60	351			
126	3.30	114			

[9] From Stoecker (1989) by permission of McGraw-Hill.

(a) First, plot the data and formulate two or three promising functions.
(b) Identify the best function by looking at the R^2, RMSE and CV values and also at the residuals.

Assume density of air at STP conditions to be 1.204 kg/m^3

Pr. 5.8 Consider the data used in Example 5.6.3 meant to illustrate the use of weighted regression for replicate measurements with non-constant variance. For the same data set, identify a model using the logarithmic transform approach similar to that shown in Example 5.6.2

Pr. 5.9 *Spline models for solar radiation*
This problem involves using splines for functions with abrupt hinge points. Several studies have proposed correlations to predict different components of solar radiation from more routinely measured components. One such correlation relates the fraction of hourly diffuse solar radiation on a horizontal radiation (I_d) and the global radiation on a horizontal surface (I) to a quantity known as the hourly atmospheric clearness index ($k_T=I/I_0$) where I_0 is the extraterrestrial hourly radiation on a horizontal surface at the same latitude and time and day of the year (Reddy 1987). The latter is an astronomical quantity and can be predicted almost exactly. Data has been gathered (Table 5.20) from which a correlation between $(I_d/I)=f(k_T)$ needs to be identified.

(a) Plot the data and visually determine likely locations of hinge points. (Hint: there should be two points, one at either extreme).
(b) Previous studies have suggested the following three functional forms: a constant model for the lower range, a second order for the middle range, and a constant model for the higher range. Evaluate with the data provided whether this functional form still holds, and report pertinent models and relevant goodness-of-fit indices.

Table 5.20 Data table for Problem 5.9

k_T	(I_d/I)	k_T	(I_d/I)
0.1	0.991	0.5	0.658
0.15	0.987	0.55	0.55
0.2	0.982	0.6	0.439
0.25	0.978	0.65	0.333
0.3	0.947	0.7	0.244
0.35	0.903	0.75	0.183
0.4	0.839	0.8	0.164
0.45	0.756	0.85	0.166
		0.9	0.165

Table 5.21 Data table for Problem 5.10

Balance point temp. (°C)	25	20	15	10	5	0	–5
VBDD (°C-Days)	4,750	3,900	2,000	1,100	500	100	0

Pr. 5.10 *Modeling variable base degree-days with balance point temperature at a specific location*

Degree-day methods provide a simple means of determining annual energy use in envelope-dominated buildings operated constantly and with simple HVAC systems which can be characterized by a constant efficiency. Such simple single-measure methods capture the severity of the climate in a particular location. The variable base degree day (VBDD) is conceptually similar to the simple degree-day method but is an improvement since it is based on the actual balance point of the house instead of the outdated default value of 65°F or 18.3°C (ASHRAE 2009). Table 5.21 assembles the VBDD values for New York City, NY from actual climatic data over several years at this location.

Identify a suitable regression curve for VBDD versus balance point temperature for this location and report all pertinent statistics.

Pr. 5.11 *Change point models of utility bills in variable occupancy buildings*

Example 5.7.1 illustrated the use of linear spline models to model monthly energy use in a commercial building versus outdoor dry-bulb temperature. Such models are useful for several purposes, one of which is for energy conservation. For example, the energy manager may wish to track the extent to which energy use has been increasing over the years, or the effect of a recently implemented energy conservation measure (such as a new chiller). For such purposes, one would like to correct, or normalize, for any changes in weather since an abnormally hot summer could obscure the beneficial effects of a more efficient chiller. Hence, factors which change over the months or the years need to be considered explicitly in the model. Two common normalization factors include changes to the conditioned floor area (for example, an extension to an existing wing), or changes in the number of students in a school. A model regressing monthly utility energy use against outdoor temperature is appropriate for buildings with constant occupancy (such as residences) or even offices. However, buildings such as schools are practically closed during summer, and hence, the occupancy rate needs to be included as the second regressor. The functional form of the model, in such cases, is a multi-variate change point model given by:

$$y = \beta_{0,un} + \beta_0 f_{oc} + \beta_{1,un} x + \beta_1 f_{oc} x + \beta_{2,un}(x - x_c)I + \beta_2 f_{oc}(x - x_c)I \quad (5.67)$$

where x and y are the monthly mean outdoor temperature (T_o) and the electricity use per square foot of the school (E) respectively, and $f_{oc} = N_{oc}/N_{total}$ represents the fraction of days in the month when the school is in session (N_{oc}) to the total number of days in that particular month (N_{total}). The factor f_{oc} can be determined from the school calendar. Clearly, the unoccupied fraction $f_{un} = 1 - f_{oc}$.

The term I represents an indicator variable whose numerical value is given by Eq. 5.54b. Note that the change point temperatures for occupied and unoccupied periods are assumed to be identical since the monthly data does not allow this separation to be identified.

Consider the monthly data assembled (shown in Table 5.22).

(a) Plot the data and look for change points in the data. Note that the model given by Eq. 5.67 has 7 parameters of which x_c (the change point temperature) is the one which makes the estimation non-linear. By inspection of the scatter plot, you will assume a reasonable value for this variable, and proceed to perform a linear regression as illustrated in Example 5.7.1. The search for the best value of x_c (one with minimum RMSE) would require several OLS regressions assuming different values of the change point temperature.

Table 5.22 Data table for Example 5.11

Year	Month	E (W/ft²)	T_o (°F)	f_{oc}	Year	Month	E (W/ft²)	T_o (°F)	f_{oc}
94	Aug	1.006	78.233	0.41	95	Aug	1.351	81.766	0.39
94	Sep	1.123	73.686	0.68	95	Sep	1.337	76.341	0.71
94	Oct	0.987	66.784	0.67	95	Oct	0.987	65.805	0.68
94	Nov	0.962	61.037	0.65	95	Nov	0.938	56.714	0.66
94	Dec	0.751	52.475	0.42	95	Dec	0.751	52.839	0.41
95	Jan	0.921	49.373	0.65	96	Jan	0.921	49.270	0.65
95	Feb	0.947	53.764	0.68	96	Feb	0.947	55.873	0.66
95	Mar	0.876	59.197	0.58	96	Mar	0.873	55.200	0.57
95	Apr	0.918	65.711	0.66	96	Apr	0.993	66.221	0.65
95	May	1.123	73.891	0.65	96	May	1.427	78.719	0.64
95	Jun	0.539	77.840	0	96	Jun	0.567	78.382	0.1
95	Jul	0.869	81.742	0	96	Jul	1.005	82.992	0.2

(b) Identify the parsimonious model, and estimate the appropriate parameters of the model. Note that of the six parameters appearing in Eq. 5.67, some of the parameters may be statistically insignificant, and appropriate care should be exercised in this regard. Report appropriate model and parameter statistics.
(c) Perform a residual analysis and discuss results.

Pr. 5.12 *Determining energy savings from monitoring and verification (M&V) projects*
A crucial element in any energy conservation program is the ability to verify savings from measured energy use data—this is referred to as monitoring and verification (M&V). Energy service companies (ESCOs) are required, in most cases, to perform this as part of their services. Figure 5.33 depicts how energy savings are estimated. A common M&V protocol involves measuring the monthly total energy use at the facility for whole year before the retrofit (this is the baseline period or the pre-retrofit period) and a whole year after the retrofit (called the post-retrofit period). The time taken for implementing the energy saving measures (called the "construction period") is neglected in this simple example. One first identifies a baseline regression model of energy use against ambient dry-bulb temperature T_o during the pre-retrofit period $E_{pre} = f(T_o)$. This model is then used to predict energy use during each month of the post-retrofit period by using the corresponding ambient temperature values. The difference between model predicted and measured monthly energy use is the energy savings during that month.

$$\textit{Energy savings} = \textit{Model-predicted pre-retrofit use} - \textit{measured post-retrofit use} \quad (5.68)$$

The determination of the annual savings resulting from the energy retrofit and its uncertainty are finally determined. It is very important that the uncertainty associated with the savings estimates be determined as well for meaningful conclusions to be reached regarding the impact of the retrofit on energy use.

You are given monthly data of outdoor dry bulb temperature (T_o) and area-normalized whole building electricity use WB_e) for two years (Table 5.23). The first year is the pre-retrofit period before a new energy management and control system (EMCS) for the building is installed, and the second is the post-retrofit period. Construction period, i.e., the period it takes to implement the conservation measures is taken to be negligible.
(a) Plot time series and x–y plots and see whether you can visually distinguish the change in energy use as a result of installing the EMCS (similar to Fig. 5.33);
(b) Evaluate at least two different models (with one of them being a model with indicator variables) for the pre-retrofit period, and select the better model;

Table 5.23 Data table for Problem 5.12

Pre-retrofit period			Post-retrofit period		
Month	T_o (°F)	WB_e (W/ft²)	Month	T_o (°F)	WB_e (W/ft²)
1994-Jul	84.04	3.289	1995-Jul	83.63	2.362
Aug	81.26	2.827	Aug	83.69	2.732
Sep	77.98	2.675	Sep	80.99	2.695
Oct	71.94	1.908	Oct	72.04	1.524
Nov	66.80	1.514	Nov	62.75	1.109
Dec	58.68	1.073	Dec	57.81	0.937
1995-Jan	56.57	1.237	1996-Jan	54.32	1.015
Feb	60.35	1.253	Feb	59.53	1.119
Mar	62.70	1.318	Mar	58.70	1.016
Apr	69.29	1.584	Apr	68.28	1.364
May	77.14	2.474	May	78.12	2.208
Jun	80.54	2.356	Jun	80.91	2.070

(c) Use this baseline model to determine month-by-month energy use during the post-retrofit period representative of energy use had not the conservation measure been implemented;
(d) Determine the month-by-month as well as the annual energy savings (this is the "model-predicted pre-retrofit energy use" of Eq. 5.68);
(e) The ESCO which suggested and implemented the ECM claims a savings of 15%. You have been retained by the building owner as an independent M&V consultant to verify this claim. Prepare a short report describing your analysis methodology, results and conclusions. (Note: you should also calculate the 90% uncertainty in the savings estimated assuming zero measurement uncertainty. Only the cumulative annual savings and their uncertainty are required, not month-by-month values).

Pr. 5.13[10] *Grey-box and black-box models of centrifugal chiller using field data*
You are asked to evaluate two types of models: physical or gray-box models versus polynomial or black-box models. A brief overview of these is provided below.

(a) Gray-Box Models The *Universal Thermodynamic Model* proposed by Gordon and Ng (2000) is to be used. The GN model is a simple, analytical, universal model for chiller performance based on first principles of thermodynamics and linearized heat losses. The model predicts the dependent chiller COP (defined as the ratio of chiller (or evaporator) thermal cooling capacity Q_{ch} by the electrical power P_{comp} consumed by the chiller (or compressor) with specially chosen independent (and easily measurable) parameters such as the fluid (water or air) inlet temperature from the condenser

[10] Data for this problem is given in Appendix B.

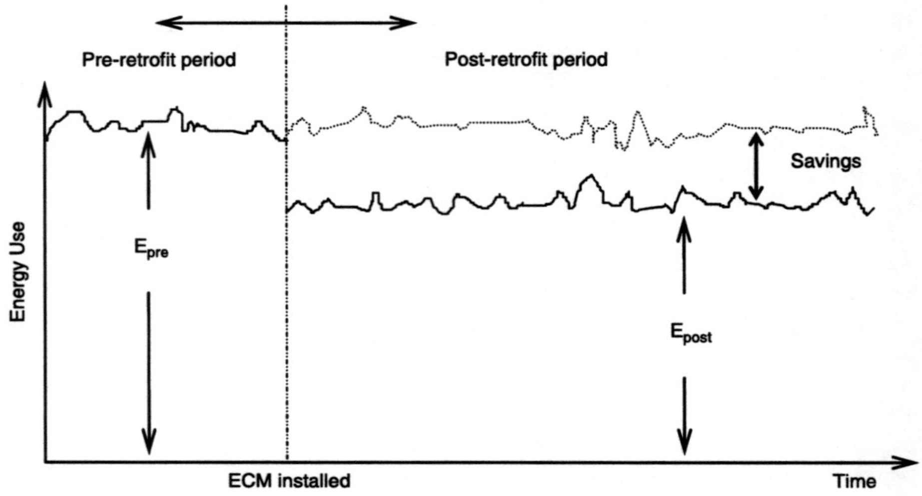

Fig. 5.33 Schematic representation of energy use prior to and after installing energy conservation measures (ECM) and of the resulting energy savings

T_{cdi}, fluid temperature leaving the evaporator (or the chilled water return temperature from the building) T_{cho}, and the thermal cooling capacity of the evaporator (similar to the figure for Example 5.4.3). The GN model is a three-parameter model which, for parameter identification, takes the following form:

$$\left(\frac{1}{COP}+1\right)\frac{T_{cho}}{T_{cdi}}-1 = a_1\frac{T_{cho}}{Q_{ch}}+a_2\frac{(T_{cdi}-T_{cho})}{T_{cdi}Q_{ch}}+a_3\frac{(1/COP+1)Q_{ch}}{T_{cdi}} \tag{5.69}$$

where the **temperatures are in absolute units,** and the parameters of the model have physical meaning in terms of irreversibilities:

$a_1=\Delta s$, the total internal entropy production rate in the chiller due to internal irreversibilities,

$a_2=Q_{leak}$, the rate of heat losses (or gains) from (or in to) the chiller,

$a_3 = R = \dfrac{1}{(mCE)_{cond}} + \dfrac{1-E_{evap}}{(mCE)_{evap}}$ i.e., the total heat exchanger thermal resistance which represents the irreversibility due to finite-rate heat exchanger, and m is the mass flow rate, C the specific heat of water, and E is the heat exchanger effectiveness.

The model applies both to unitary and large chillers operating under steady state conditions. Evaluations by several researchers have shown this model to be very accurate for a large number of chiller types and sizes. If one introduces:

$$x_1=\frac{T_{cho}}{Q_{ch}}, x_2=\frac{(T_{cdi}-T_{cho})}{T_{cdi}Q_{ch}}, x_3=\frac{(1/COP+1)Q_{ch}}{T_{cdi}}$$
$$\text{and} \quad y=\left(\frac{1}{COP}+1\right)\frac{T_{cho}}{T_{cdi}}-1 \tag{5.70}$$

Eq. 5.69 assumes the following linear form:

$$y=a_1x_1+a_2x_2+a_3x_3 \tag{5.71}$$

Although most commercial chillers are designed and installed to operate at constant coolant flow rates, *variable condenser water flow operation* (as well as evaporator flow rate) is being increasingly used to improve overall cooling plant efficiency especially at low loads. In order to accurately correlate chiller model performance under variable condenser flow, an analytic model as follows was developed:

$$\frac{T_{cho}(1+1/COP)}{T_{cdi}}-1-\frac{1}{(V\rho C)_{cond}}\frac{(1/COP+1)Q_{ch}}{T_{cdi}} = c_1\frac{T_{cho}}{Q_{ch}}+c_2\left(\frac{T_{cdi}-T_{cho}}{Q_{ch}T_{cdi}}\right)+c_3\frac{Q_{ch}(1+1/COP)}{T_{cdi}} \tag{5.72}$$

If one introduces

$$x_1=\frac{T_{cho}}{Q_{ch}},\quad x_2=\frac{T_{cdi}-T_{cho}}{Q_{ch}T_{cdi}},\quad x_3=\frac{(1/COP+1)Q_{ch}}{T_{cdi}}$$

and

$$y=\frac{T_{cho}(1/COP+1)}{T_{cdi}}-1-\frac{1}{(V\rho C)_{cond}}\frac{(1/COP+1)Q_{ch}}{T_{cdi}} \tag{5.73}$$

where V, ρ and c are the volumetric flow rate, the density and specific heat of the condenser water.

For the variable condenser flow rate, Eq. 5.72 becomes

$$y=c_1x_1+c_2x_2+c_3x_3 \tag{5.74}$$

(b) Black-Box Models Whereas the structure of a gray box model, like the GN model, is determined from the underlying physics, the black box model is characterized as having no (or sparse) information about the physical problem incorporated in the model structure. The model is regarded as a black box and describes an empirical relationship between input and output variables. The commercially available DOE-2 building energy simulation model (DOE-2 1993) relies on the same parameters as those for the physical model, but uses a second order linear polynomial model instead. This "standard" empirical model (also called a *multivariate polynomial linear model or MLR*) has 10 coefficients which need to be identified from monitored data:

$$\begin{aligned} COP = b_0 &+ b_1 T_{cdi} + b_2 T_{cho} \\ &+ b_3 Q_{ch} + b_4 T_{cdi}{}^2 + b_5 T_{cho}^2 + b_6 Q_{ch}^2 \\ &+ b_7 T_{cdi} T_{cho} + b_8 T_{cdi} Q_{ch} + b_9 T_{cho} Q_{ch} \end{aligned} \quad (5.75)$$

These coefficients, unlike the three coefficients appearing in the GN model, have no physical meaning and their magnitude cannot be interpreted in physical terms. Collinearity in regressors and ill-behaved residual behavior are also problematic issues. Usually one needs to retain in the model only those parameters which are statistically significant, and this is best done by step-wise regression.

Table B.3 in Appendix B assembles data consisting of 52 sets of observations from a 387 ton centrifugal chiller with variable condenser flow data. A sample hold-out cross-validation scheme will be used to guard against over-fitting. Though this is a severe type of split, **use the first 36 data points as training data and the rest (shown in italics) as testing data.**

(a) You will use the three models described above (Eqs. 5.71, 5.74 and 5.75) to identify suitable regression models. Study residual behavior as well as collinearity issues between regressors. Identify the best forms of the GN and the MLR model formulations.

(b) Evaluate which of these models is superior in terms of their external prediction accuracy The GN and MLR models have different y-values and so you cannot use the statistics provided by the regression package directly. You need to perform subsequent calculations in a spreadsheet using the power as the basis of comparing model accuracy and reporting internal and external prediction accuracies. For the MLR model, this is easily deduced from the model predicted COP values. For the GN model with constant flow, rearranging terms of Eq. 5.71 yields the following expression for the chiller electric power P_{ch}:

$$P_{comp} = \frac{Q_{ch}(T_{cdi} - T_{cho}) + a_1 T_{cdi} T_{cho} + a_2 (T_{cdi} - T_{cho}) + a_3 Q^2{}_{ch}}{T_{cho} - a_3 Q_{ch}} \quad (5.76)$$

(c) Report all pertinent steps performed in your analysis and present your results succinctly.

Helpful tips:

(i) Convert temperatures into degrees Celsius, Q_{ch} into kW and volumetric flow rate V into L/s for unit consistency (work in SI units)

(ii) For the GN model, all temperatures should be in absolute units

(iii) Degrees of freedom (d.f.) have to be estimated correctly in order to compute RMSE and CV. For internal prediction, d.f.=n−p where n is the number of data points and p the number of model parameters. For external prediction accuracy, d.f.=m where m is the number of data points.

Pr. 5.14[11] Effect of tube cleaning in reducing chiller fouling

A widespread problem with liquid-cooled chillers is condenser fouling which increases heat transfer resistance in the condenser and results in reduced chiller COP. A common remedy is to periodically (every year or so) brush-clean the insides of the condenser tubes. Some practitioners question the efficacy of this process though this is widely adopted in the chiller service industry. In an effort to clarify this ambiguity, an actual large chiller (with refrigerant R11) was monitored during normal operation for 3 days before (9/11-9/13-2000) and 3 days after (1/17-1/19-2001) tube cleaning was done. Table B.4 (in Appendix B) assembles the entire data set of 72 observations for each period. This chiller is similar to the figure for Example 5.4.3.

Analyze, using the GN model described in Pr. 5.13, the two data sets, and determine the extent to which the COP of the chiller has improved as a result of this action. Prepare a report describing your analysis methodology, your analysis results, the uncertainty in your results, your conclusions, and any suggestions for future analysis work.

References

ASHRAE, 1978, Standard 93-77: Methods of Testing to Determine the Thermal Performance of Solar Collectors, American Society of Heating, Refrigerating and Air-Conditioning Engineers, Atlanta, GA.

ASHRAE, 2005. *Guideline 2-2005: Engineering Analysis of Experimental Data*, American Society of Heating, Refrigerating and Air-Conditioning Engineers, Atlanta, GA.

ASHRAE, 2009. *Fundamentals Handbook*, American Society of Heating, Refrigerating and Air-Conditioning Engineers, Atlanta, GA.

Belsley, D.A., E. Kuh, and R.E Welsch, 1980, *Regression Diagnostics*, John Wiley & Sons, New York.

Chatfield, C., 1995. *Problem Solving: A Statistician's Guide*, 2nd Ed., Chapman and Hall, London, U.K.

Chatterjee, S. and B. Price, 1991. *Regression Analysis by Example*, 2nd Edition, John Wiley & Sons, New York.

[11] Data for this problem is given in Appendix B.

Cook, R.D. and S. Weisberg, 1982. *Residuals and Influence in Regression*, Chapman and Hall, New York.
DOE-2, 1993. Building Energy simulation software developed by Lawrence Berkeley National Laboratory with funding from U.S. Department of Energy, http://simulationresearch.lbl.gov/
Draper, N.R. and H. Smith, 1981. *Applied Regression Analysis*, 2nd Ed., John Wiley and Sons, New York.
Gordon, J.M. and K.C. Ng, 2000. *Cool Thermodynamics*, Cambridge International Science Publishing, Cambridge, UK
Hair, J.F., R.E. Anderson, R.L. Tatham, and W.C. Black, 1998. *Multivariate Data Analysis*, 5th Ed., Prentice Hall, Upper Saddle River, NJ,
Katipamula, S., T. A. Reddy and D. E. Claridge, 1998. "Multivariate regression modeling", *ASME Journal of Solar Energy Engineering,* vol. 120, p.177, August.
Pindyck, R.S. and D.L. Rubinfeld, 1981. *Econometric Models and Economic Forecasts*, 2nd Edition, McGraw-Hill, New York, NY.
Reddy, T.A., 1987. *The Design and Sizing of Active Solar Thermal Systems*, Oxford University Press, Clarendon Press, U.K., September.
Reddy, T.A., N.F. Saman, D.E. Claridge, J.S. Haberl, W.D. Turner W.D., and A.T. Chalifoux, 1997. Baselining methodology for facility-level monthly energy use- part 1: Theoretical aspects, *ASHRAE Transactions*, v.103 (2), American Society of Heating, Refrigerating and Air-Conditioning Engineers, Atlanta, GA.
Schenck, H., 1969. *Theories of Engineering Experimentation*, Second Edition, McGraw-Hill, New York.
Shannon, R.E., 1975. *System Simulation: The Art and Science*, Prentice-Hall, Englewood Cliffs, NJ.
Stoecker, W.F., 1989. *Design of Thermal Systems*, 3rd Edition, McGraw-Hill, New York.
Walpole, R.E., R.H. Myers, and S.L. Myers, 1998. *Probability and Statistics for Engineers and Scientists*, 6th Ed., Prentice Hall, Upper Saddle River, NJ

Design of Experiments

6

The data from which performance models are identified may originate either from planned experiments or from non-intrusive (or observational) data gathered while the system is in normal operation. A large body of well accepted practices is available for the former which falls under the general terminology of "Design of Experiments" (DOE). This is the process of defining the structural framework, i.e., prescribing the exact manner in which samples for testing need to be selected, and the conditions and sequence under which the testing needs to be performed. This would provide the "richness" in the data set necessary for statistically sound performance models to be identified between the response variable and the several categorical factors. Experimental design methods, which allow extending hypothesis testing to multiple variables as well as identifying sound performance models, are presented. Selected experimental design methods are discussed such as randomized block, Latin Squares and 2^k factorial designs. The parallel between model building in a DOE framework and linear multiple regression is illustrated. Finally, this chapter addresses response surface methods (RSM) which allow accelerating the search towards optimizing a process or towards finding the conditions under which a desirable behavior of a product is optimized. RSM is a sequential approach where one starts with test conditions in a plausible area of the search space, analyzes test results to determine the optimal direction to move, performs a second set of test conditions, and so on till the required optimum is reached.

6.1 Background

The two previous chapters dealt with statistical techniques for analyzing data which was already gathered. However, no amount of "creative" statistical data analysis can reveal information not available in the data itself. Thus, the process by which this data is gathered is itself an equally important field of study. The process of proper planning and execution of experiments, intentionally designed to provide data rich in information especially suited for the intended objective, is referred to as *experimental design*. Optimal experimental design is one which stipulates the conditions under which each observation should be taken so as to minimize/maximize certain optimal constraints (say, the bias and variance of the parameter estimators). Practical considerations and constraints often complicate the design of optimal experiments and these factors should also be explicitly factored in.

One needs to differentiate between two conditions under which data can be collected. On one hand, one can have a controlled setting where the various variables of interest can be altered by the experimenter. In such a case, referred to as *intrusive testing*, one can plan an "optimal" experiment where one can adjust the inputs and boundary or initial conditions as well as choose the number and location of the sensors so as to minimize the effect of errors on estimated values of the parameters. On the other hand, one may be in a situation where one is a mere "spectator", i.e., the system or phenomenon cannot be controlled, and the data is collected under non-experimental conditions (as is the case of astronomical observations). Such an experimental protocol, known as *non-intrusive* identification, is usually not the best approach. In certain cases, the driving forces may be so weak or repetitive that even when a "long" data set is used for identification, a strong enough or varied output signal cannot be elicited for proper statistical treatment (see Chap. 10). An *intrusive* or controlled experimental protocol, wherein the system is artificially stressed to elicit a strong response, is more likely to yield robust and accurate models and their parameter estimates. However, in some cases, the type and operation of the system may not allow such intrusive experiments to be performed.

One should appreciate differences between measurements made in a laboratory setting and in the field. The potential for errors, both bias and random, is usually much greater in the latter. Not only can measurements made on a piece of laboratory equipment be better designed and closely controlled, but they will be more accurate as well because more expensive sensors can be selected and placed correctly in the system. For example, proper flow measurement requires that

T. Agami Reddy, *Applied Data Analysis and Modeling for Energy Engineers and Scientists*,
DOI 10.1007/978-1-4419-9613-8_6, © Springer Science+Business Media, LLC 2011

the flowmeter be placed 30 pipe diameters after a bend. A laboratory set-up can be designed accordingly, while field conditions may not allow such conditions to be met satisfactorily. Further, systems being operated in the field may not allow controlled tests to be performed, and one has to develop a model or make decisions based on what one can observe.

Experimental design techniques were developed about 100 years back primarily in the context of agricultural research, subsequently migrating to industrial engineering, and then on to other fields. The historic reason for their development was to allow ascertaining, by hypothesis testing, whether a certain treatment, which could be an additive nutrient such as a fertilizer or a design change such as an alloy modification for a machine part, increased the yield or improved the product. The statistical techniques which stipulate how each of the independent variables or *factors* have to be varied so as to obtain the most information about system behavior quantified by a response variable, and do so with a minimum of tests (and hence, least effort and expense), are called experimental designs or *design of experiments* (*DOE*). To rephrase, DOE involves the complete reasoning process of defining the structural framework, i.e., prescribing the exact manner in which samples for testing need to be selected, and the conditions and sequence under which the testing needs to be performed under specific restrictions imposed by space, time and nature of the process (Mandel 1964).

The applications of DOE have expanded to the area of *model building* as well. It is now used to identify which subsets among several variables influence the response variable, and to determine a quantitative relationship between them. Generally, DOE and model building involve three issues:

(a) to "screen" a large number of possible candidates or likely variables, and select the dominant variables, which are referred to as *factors* in experimental design terminology. These possible candidate factors are then subject to more extensive investigation;
(b) to formulate how the tests need to be carried out so that sources of unsuspecting and uncontrollable/extraneous errors can be minimized, while eliciting the necessary "richness" in system behavior. The richness of the data set cannot be ascertained by the *amount of data* but by the extent to which all possible states of the system are represented in the data set. This is especially true for observational data sets where data is collected while the system is under routine day-to-day operation without any external intervention by the observer. However, under controlled test operation, such as in a laboratory, DOE allows optimal model identification to be achieved with the least effort and expense;
(c) to build a suitable model between the factors and the response variable using the data set acquired previously. This involves both hypothesis testing so as to identify the significant factors as well as the model building and residual diagnostic checking phases.

The relative importance of the three issues depends on the specific circumstance. For example, often, and especially so in engineering model building, the dominant regressor set is know beforehand, acquired either from mechanistic insights or prior experimentation, and so the screening phase may be redundant. In the context of calibrating a detailed simulation program with monitored data, the problem involves dozens, if not hundreds, of input parameters. Which parameters are best tuned and which left alone can be determined from a sensitivity analysis which is directly based on the principles of screening tests in DOE.

6.2 Complete and Incomplete Block Designs

6.2.1 Randomized Complete Block Designs

Recall that the various hypothesis tests presented in Chap. 4 dealt mainly with problems involving one variable or factor only. The design of experiments can be extended to include:

(i) *several factors*, but only one, two and three factors will be discussed for the sake of simplicity. The factors could be either controllable by the experimenter or cannot be controlled (also called, extraneous or nuisance factors). The source of variation of controllable variables can be reduced by fixing them at preselected levels, while that of the uncontrollable variables requires suitable experimental procedures-this process is called *blocking*, and is described below.
(ii) *several levels* (or *treatments*, a term widely used in DOE terminology), but only two-four levels will be considered for conceptual simplicity. The levels of a factor are the different values or categories which the factor can assume. This can be dictated by the type of variable (which can be continuous or discrete or categorical) or selected by the experimenter. Different combinations of the levels of the factors are called *experimental units*.

Note that though the factors can be either continuous or categorical, the initial design of experiments treats them in the same fashion. In case of continuous variables, their range of variation is discretized into a relatively small set of numerical values; as a result, the levels have a magnitude associated with them. This is not the case with categorical variables, such as say, male or female where there is no magnitude involved; the grouping is done based on some classification criterion such as a level or treatment.

It is important to keep in mind that the intent of the experimental design process, as stated earlier, is to provide the necessary "richness" in the data set in order to: (i) perform hypothesis testing of several factors at different levels, and (ii) to identify statistically sound performance models between the response variable and the several factors. This is done using ANOVA techniques. If a statistical model describing the impact of a single variable or factor on the dependent or response variable y is to be identified, the *one-way ANOVA* (described in Sect. 4.3.1) allows one to ascertain whether there are statistically significant differences between the mean at different levels of x if x is a continuous variable, or at different categories of x if x is a categorical variable. If two or more variables are involved, the procedure is called *multifactor ANOVA*.

The simplest type of design is the *randomized unrestricted block design* which involves selecting at random the combinations of the factors or levels under which to perform the experiments. This type of design, if done naively, is not very efficient and may require an unnecessarily large number of experiments to be performed. The concept is illustrated with a simple example, from the agricultural area, from which DOE emerged. Say, one wishes to evaluate the yield of four newly developed varieties of wheat (labeled x_1, x_2, x_3, x_4). Four plots of land or field stations (or experimental units) are available, which unfortunately, are located in 4 different geographic regions which differ in climate. The geographic location is likely to affect yield. The simplest way of assigning which station will be planted by which variety of wheat is to do so randomly—one such result (among many possible ones) is shown in Table 6.1. Note that the intent of this investigation is to determine the effect of wheat variety on yield. The location of the field or station is a "nuisance" variable, but in this case can be controlled; this is achieved by suitable blocking. However, there may be other variables which are uncontrollable (say, excessive rainfall in one of regions during the test) and this can be partially compensated for by assuming them to be random and replicating or repeating the tests more than once for each combination. The above example serves to illustrate the principle of *restricted randomization* by blocking the variability in an uncontrollable effect.

Such an unrestricted randomization leads to needless replication (for example, wheat type x_1 is tested twice in Region 1 and not at all in Region 4) and may not be very efficient. Since the intention is to reduce variability in the uncontrolled variable, in this case, the "region" variable, (note that the station to station difference in a region also exists but it will be less important and can be viewed as the error or random effect), one can insist that each variety of wheat be tested in each region. There are again several possibilities, with one being shown in Table 6.2.

Table 6.1 Example of *unrestricted* randomized block design (one of several possibilities) for one factor of interest at levels x_1, x_2, x_3, x_4 and one nuisance variable (Regions 1, 2, 3, 4). This is not an efficient design since x_1 appears twice under Region 1 and not at all in Region 4

Region 1	Region 2	Region 3	Region 4
x_1	x_1	x_2	x_3
x_3	x_2	x_1	x_4
x_1	x_3	x_3	x_2
x_2	x_1	x_3	x_4

Table 6.2 Example of *restricted* randomized block design (one of several possibilities) for the same example of Table 6.1

Region 1	Region 2	Region 3	Region 4
x_1	x_2	x_3	x_4
x_2	x_3	x_4	x_1
x_3	x_4	x_1	x_2
x_4	x_1	x_2	x_3

The approach can be extended to the treatment of two or multiple factors, and are called *factorial designs* (Devore and Farnum 2005). Consider two factors (labeled A and B) which are to be studied at "a and b" levels respectively. This is often referred to as *$(a \times b)$ design*, and the standard manner of representing the results of the tests is by assembling them as shown in Table 6.3. Each combination of factor-level can be tested more than once in order to minimize the effect of random errors, and this is called *replication*. Though more tests are done, replication reduces experimental errors introduced by extraneous factors not explicitly controlled during the experiments that can bias the results. Often for mathematical convenience, each combination is tested at the same replication level, and this is called a *balanced design*. Thus, Table 6.3 is an example of a (3×2) balanced design with replication $r=2$.

The above terms are perhaps better understood in the context of regression analysis (treated in Chap. 5). Let Z be the response variable which is linear in regressor variable X, and a model needs to be identified. Further, say, another variable Y is known to influence Z which may smear the sought after relation (such as the field station variable in the above example). Selecting three specific values is akin to selecting three levels for the factor X (say, x_1, x_2, x_3). The nuisance effect of variable Y can be "blocked" by performing the tests at preselected *fixed* levels or values of Y (say, y_1 and y_2). The corresponding scatter plot is shown in Fig. 6.1. Repeat testing at each of the six combinations in order to reduce experimental errors is akin to replication; in this example, replication $r=3$. Finally, if the 18 tests are performed in random sequence, the experimental design would qualify as a full factorial random design.

Table 6.3 Standard method of assembling test results for a balanced (3×2) design with two replication levels

		Factor B		Average
		Level 1	Level 2	
Factor A	Level 1	10,14	18,14	14
	Level 2	23,21	16,20	20
	Level 3	31,27	21,25	26
Average		21	19	20

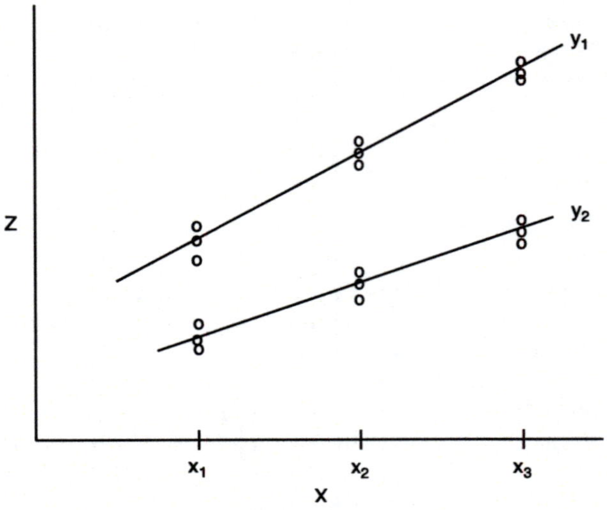

Fig. 6.1 Correspondence between block design approach and multiple regression analysis

The averages shown in Table 6.3 correspond to those of either the associated row or the associated column. Thus, the average of the first row, i.e. {10, 14, 18, 14}, is shown as 14, and so on. Plots of the average response versus the levels of a factor yield a graph which depicts the trend, called *main effect* of the factor. Thus, Fig. 6.2a suggests that the average response tends to increase linearly as factor A changes from A_1 to A_3, while that of factor B decreases a little as factor B changes from B_1 to B_2. The effect of the factors on the response may not be purely additive, a multiplicative component may be included as well. In such cases, the two factors are said to *interact* with each other. Whether this interaction effect is statistically significant or not can be determined by performing the calculations shown in Table 6.4.

Thus, the effect of going from A_1 to A_3 is 17 under B_1 and only 7 under B_2. This suggests interaction effects. A simpler and more direct approach is to graph the two factor interaction plot as shown in Fig. 6.2b. Since the lines are not parallel (in this case they cross each other), one would infer strong interaction between the two factors. However, in many instances, such plots are not conclusive enough, and one needs to perform statistical tests to determine whether the main or the interaction effects are significant or not. Figure 6.3 shows the type of interaction plots one would obtain for the case when the interaction effects are not significant.

ANOVA decompositions allow breaking up the observed total sum of square variation (SST) into its various contributing causes (the one factor ANOVA was described in Sect. 4.3.1). For a two-factor ANOVA decomposition (Devore and Farnum 2005):

$$SST = SSA + SSB + SS\,(AB) + SSE \tag{6.1}$$

where observed sum of squares:

$$\begin{aligned} SST &= \sum_{i=1}^{a}\sum_{j=1}^{b}\sum_{m=1}^{r}(y_{ijm} - \langle y \rangle)^2 \\ &= (stdev)^2(abr-1) \end{aligned}$$

sum of squares associated with factor A:

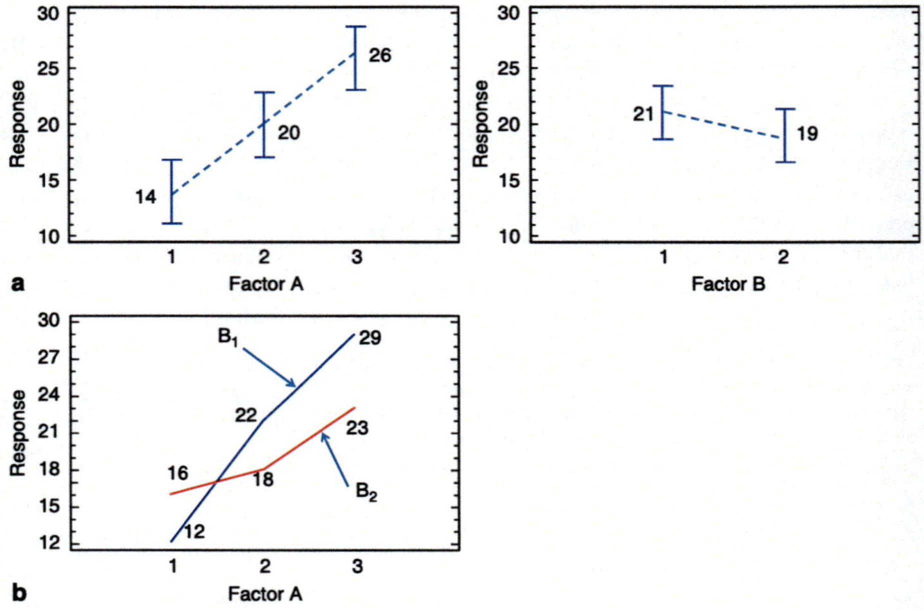

Fig. 6.2 Plots for the (3×2) balanced factorial design. **a** Main effects of factors A and B with mean and 95% intervals (data from Table 6.3). **b** Two-factor interaction plot (data from Table 6.4)

Table 6.4 Interaction effect calculations

Effect of changing A (B fixed at B_1)			Effect of changing A (B fixed at B_2)		
A_1 and B_1	(10+14)/2=12		A_1 and B_2	(18+14)/2=16	
A_2 and B_1	(23+21)/2=22	29−12=17	A_2 and B_2	(16+20)/2=18	23−16=7
A_3 and B_1	(31+27)/2=29		A_3 and B_2	(21+25)/2=23	

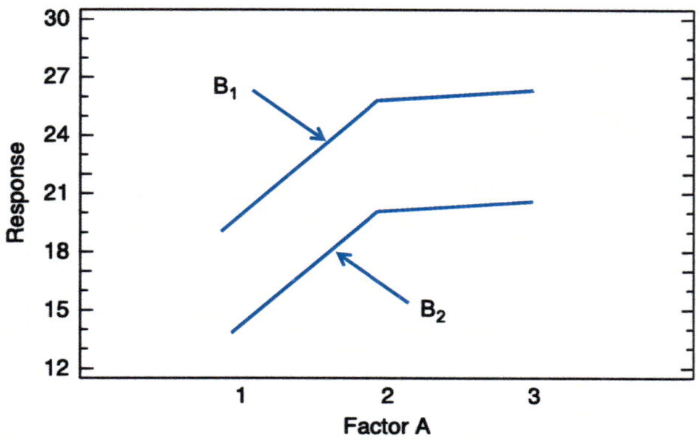

Fig. 6.3 An example of a two-factor interaction plot when the factors have no interaction

$$SSA = b.r.\sum_{i=1}^{a}(\bar{A}_i - <y>)^2$$

sum of squares associated with factor B:

$$SSB = a.r.\sum_{j=1}^{b}(\bar{B}_j - <y>)^2$$

error or residual sum of squares:

$$SSE = \sum_{i=1}^{a}\sum_{j=1}^{b}\sum_{m=1}^{r}(y_{ijm} - \bar{y}_{ij})^2$$

sum of square associated with the AB interaction is SST(AB)

with y_{ijm}=observation under m[th] replication when A is at level i and B is at level j
a=number of levels of factor A
b=number of levels of factor B
r=number of replications per cell
$\bar{A}_i$ =average of all response values at i[th] level of factor A
$\bar{B}_j$=average of all response values at j[th] level of factor B
$\bar{y}_{ij}$ =average for each cell (i.e., across replications)
<y>=grand average
i=1, … a is the index for levels of factor A
j=1, … b is the index for levels of factor B
m=1, … r is the index for replicate.

Note that SS(AB) is deduced from Eq. 6.1 since all other quantities can be calculated. A linear statistical model, referred to as a *random effects model*, between the response and the two factors which includes the interaction term between factors A and B can be deduced. More specifically, this is called a *non-additive two-factor model* (it is non-additive because of the interaction term present) which assumes the following form since one starts with the grand average and adds individual effects of the factors, the interaction terms and the noise or error term:

$$y_{ij} = <y> + \alpha_i + \beta_j + (\alpha\beta)_{ij} + \varepsilon_{ij} \qquad (6.2)$$

where
α_i represents the main effect of factor A at the ith level $= \bar{A}_i - <y>$ and $= \sum_{i=1}^{a}\alpha_i = 0$
β_j the main effect of factor B at the jth level $= \bar{B}_j - <y>$ and $= \sum_{j=1}^{b}\beta_j = 0$
$(\alpha\beta)_{ij}$ the interaction between factors A and B $= \bar{y}_{ij} - (<y> + \bar{A}_i + \bar{B}_j)$ and $= \sum_{j=1}^{b}\sum_{i=1}^{a}(\alpha\beta)_{ij} = 0$
and ε_{ij} is the error (or residuals) assumed uncorrelated with mean zero and variance $\sigma^2 = MSE$.

The analysis of variance is done as described earlier, but care must be taken to use the correct degrees of freedom to calculate the mean squares (refer to Table 6.5). The analysis of variance model (Eq. 6.2) can be viewed as a special case of multiple linear regression (or more specifically to one with indicator variables—see Sect. 5.7.3). This is illustrated in the example below.

Example 6.2.1: *Two-factor ANOVA analysis and random effect model fitting*
Using the data from Table 6.3, determine whether the main effect of factor A, main effect of factor B, and the interaction effect of AB are statistically significant at $\alpha=0.05$. Subsequently, identify the random effects model.

First, using all 12 observations, one computes the grand average <y>=20 and the standard deviation stdev=6.015. Then, following Eq. 6.1:

Table 6.5 Computational procedure for a two-factor ANOVA design

Source of Variation	Sum of Squares	Degrees of Freedom	Mean square	Computed F statistic	Degrees of Freedom for p-value
Factor A	SSA	a−1	MSA=SSA/(a−1)	MSA/MSE	a−1, ab(r−1)
Factor B	SSB	b−1	MSB=SSB/(b−1)	MSB/MSE	b−1, ab(r−1)
AB interaction	SS(AB)	(a−1)(b−1)	MS(AB)=SS(AB)/(a−1)(b−1)	MS(AB)/MSE	(a−1)(b−1), ab(r−1)
Error	SSE	ab(r−1)	MSE=SSE/[ab(r−1)]	–	–
Total variation	SST	abr−1	–	–	–

$$\begin{aligned} SST &= stdev^2.(abr-1) = 6.015^2[(3)(2)(2)-1] \\ &= 398 \\ SSA &= (2).(2)[(14-20)^2+(20-20)^2+(26-20)^2] \\ &= 288 \\ SSB &= (3).(2)[(21-20)^2+(19-20)^2] = 12 \\ SSE &= [(10-12)^2+(14-12)^2+(18-16)^2 \\ &\quad +(14-16)^2+(23-22)^2+(21-22)^2 \\ &\quad +(16-18)^2+(20-18)^2+(31-29)^2 \\ &\quad +(27-29)^2+(21-23)^2+(23-25)^2 \\ &= 42 \end{aligned}$$

Then, from

$$\begin{aligned} SST &= SSA+SSB+SS(AB)+SSE \\ SS(AB) &= SST-SSA-SSB-SSE \\ &= 398-288-12-42 = 56 \end{aligned}$$

Next, the expressions shown in Table 6.5 result in:

$$\begin{aligned} MSA &= \frac{SSA}{a-1} = \frac{288}{3-1} = 144 \\ MSB &= \frac{SSB}{b-1} = \frac{12}{2-1} = 12 \\ MS(AB) &= \frac{SS(AB)}{(a-1)(b-1)} = \frac{56}{(2)(1)} = 28 \\ MSE &= \frac{SSE}{ab(r-1)} = \frac{42}{(3)(2)(1)} = 7 \end{aligned}$$

The statistical significance of the factors can now be evaluated.

- Factor A: $F-value = \frac{MSA}{MSE} = \frac{144}{7} = 20.57$. Since critical F value for degrees of freedom (2, 6)=$F_c(2,6)$ @ 0.05 significance level=5.14, and because calculated F>F_c, one concludes at the 95% confidence level that this factor is significant.
- Factor B: $F-value = \frac{MSB}{MSE} = \frac{12}{7} = 1.71$. Since $F_c(1,6)$ @ 0.05=5.99; this factor is **not** significant.
- Factor AB: $F-value = \frac{MS(AB)}{MSE} = \frac{28}{7} = 4$. Since $F_c(2,6)$ @ 0.05=5.14; this factor is not significant.

The use of Eq. 6.2 can also be illustrated in terms of this example. The main effect of A and B are given by the differences between the cell averages and the grand average $\langle y \rangle=20$ (see Table 6.3):

$$\begin{aligned} &\alpha_1 = (14-20) = -6; \alpha_2 = (20-20) = 0; \\ &\alpha_3 = (26-20) = 6; \\ &\beta_1 = (21-20) = 1; \beta_2 = (19-20) = -1; \end{aligned}$$

and, the interaction terms by (refer to Table 6.4):

$$\begin{aligned} (\alpha\beta)_{11} &= 12-(20-6+1) = -3; \\ (\alpha\beta)_{21} &= 22-(20+0+1) = 1; \\ (\alpha\beta)_{31} &= 29-(20+6+1) = 2; \\ (\alpha\beta)_{12} &= 16-(20-6-1) = 3; \\ (\alpha\beta)_{22} &= 18-(20+0-1) = -1; \\ (\alpha\beta)_{32} &= 23-(20+6-1) = -2; \end{aligned}$$

Following Eq. 6.2, the random effects model is:

$$\begin{aligned} \hat{y}_{ij} &= 20+\{-6,0,6\}_i+\{1,-1\}_j \\ &\quad +\{-3,1,2,3,-1,-2\}_{ij} \\ &\text{with} \quad i=1,2,3 \quad \text{and} \quad j=1,2 \end{aligned} \tag{6.3}$$

For example, the cell corresponding to (A1, B1) has a mean value of 12 which is predicted by the above model as: $\hat{y}_{ij} = 20-6+1-3 = 12$, and so on. Finally, the prediction error of the model has a variance $\sigma^2 = MSE = 7$. Recasting the above model as a regression model with indicator variables may be insightful (though cumbersome) to those more familiar with regression analysis methods:

$$\begin{aligned} \hat{y}_{ij} &= 20+(-6)I_1+(0)I_2+(6)I_3+(1)J_1+(-1)J_2 \\ &\quad +(-3)I_1J_1+(1)I_1J_2+(2)I_2J_1+(3)I_2J_2 \\ &\quad +(-1)I_3J_1+(-2)I_3J_2 \end{aligned}$$

where I_i and J_i are indicator variables. ■

Table 6.6 Machining time (in minutes) for Example 6.2.2.

Machine	Operator 1	2	3	4	5	6	Average
1	42.5	39.3	39.6	39.9	42.9	43.6	41.300
2	39.8	40.1	40.5	42.3	42.5	43.1	41.383
3	40.2	40.5	41.3	43.4	44.9	45.1	42.567
4	41.3	42.2	43.5	44.2	45.9	42.3	43.233
Average	40.950	40.525	41.225	42.450	44.050	43.525	42.121

Table 6.7 ANOVA table for Example 6.2.2.

Source of variation	Sum of Squares	Degrees of Freedom	Mean square	Computed F statistic	Probability
Machines	15.92	3	5.31	3.34	0.048
Operators	42.09	5	8.42		
Error	23.84	15	1.59		
Total	81.86	23	–		

Example 6.2.2:[1] *Evaluating performance of four machines while blocking effect of operator dexterity*

This example will illustrate the concept of randomized complete block design with one factor. The performance of four different machines M_1, M_2, M_3 and M_4 are to be evaluated in terms of speed in making a widget. It is decided that the same widget will be manufactured on these machines by six different machinists or operators in a randomized block experiment. The machines are assigned in a random order to each operator. Since dexterity is involved, there will be a difference among the operators in the time needed to machine the widget. Table 6.6 assembles the time in minutes to manufacture a widget.

Here, machine type is the treatment, while the uncontrollable factor is the operator. The effect of this factor is blocked or taken into consideration (or its effect minimized) by the randomized complete block design where all operators use all 4 machines. The analysis calls for testing the hypothesis at the 0.05 level of significance that the performance of the machines is identical.

Let Factor A correspond to the machine type and B to the operator. Thus a=4 and b=6, with replication r=1. Then, Eq. 6.1 reduces to:

$$\text{SST} = \text{SSA} + \text{SSB} + \text{SSE}$$

$$\begin{aligned} SSA &= (6)[(41.3 - 42.121)^2 \\ &\quad + (41.383 - 42.121)^2 + \cdots] \\ &= 15.92 \end{aligned}$$

$$\begin{aligned} SSB &= (4)[(40.95 - 42.121)^2 \\ &\quad + (40.525 - 42.121)^2 + \cdots] \\ &= 42.09 \end{aligned}$$

Total variation=SST

$$= (abr - 1).stdev^2 = (23)(1.8865^2) = 81.86$$

Subsequently, SSE=81.86−15.92−42.09=23.84

The ANOVA table can then be generated as given by Table 6.7. The value of F=3.34 is significant at p=0.048. One would conclude that the performance of the machines cannot be taken to be similar at the 0.05 significance level (this is a close call though!).

[1] From Walpole et al. (2007) by © permission of Pearson Education.

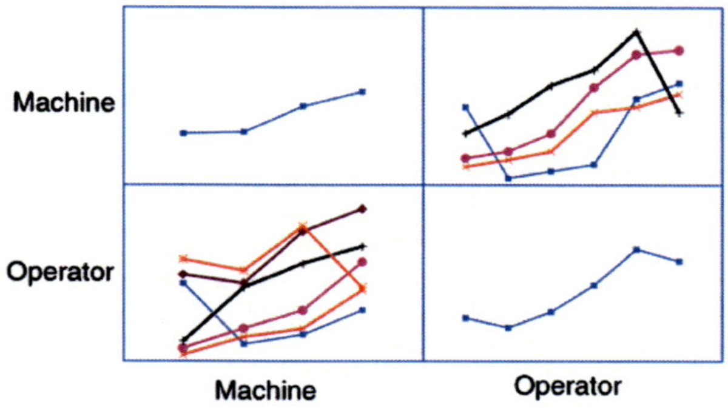

Fig. 6.4 Factor mean plots of the two factors with six levels for Operator variable and four for Machine variable

As illustrated earlier, graphical display of data can provide useful diagnostic insights in ANOVA type of problems as well. For example, a simple plotting of the raw observations around each treatment mean can provide a feel for variability between sample means and within samples. Figure 6.4 depicts all the data as well as the mean variation. One notices that there are two unusually different values which stand out, and it may be wise to go back and study the experimental conditions which produced these results. Without these, the interaction effects seem small.

A random effects model can also be identified. In this case, an additive linear model is appropriate such as:

$$y_{ij} = <y> + \alpha_i + \beta_j + \varepsilon_{ij} \tag{6.4}$$

Inspection of the residuals can provide diagnostic insights with regard to violation of normality and non-uniform variance akin to regression analysis. Since model predictions are given by:

$$\begin{aligned} \hat{y}_{ij} &= <y> + \left(\bar{A}_i - <y>\right) + \left(\bar{B}_j - <y>\right) \\ &= \bar{A}_i + \bar{B}_j - <y> \end{aligned} \tag{6.5a}$$

the residuals of the (i,j) observation are:

$$\begin{aligned} &\varepsilon_{ij} \equiv y_{ij} - \hat{y}_{ij} = y_{ij} - (\bar{A}_i + \bar{B}_j - <y>) \\ &i = 1, \ldots, 4 \quad \text{and} \quad j = 1, \ldots, 6 \end{aligned} \tag{6.5b}$$

Two different residual plots have been generated. Figure 6.5 and 6.6 reveal that the variance of the errors versus operators

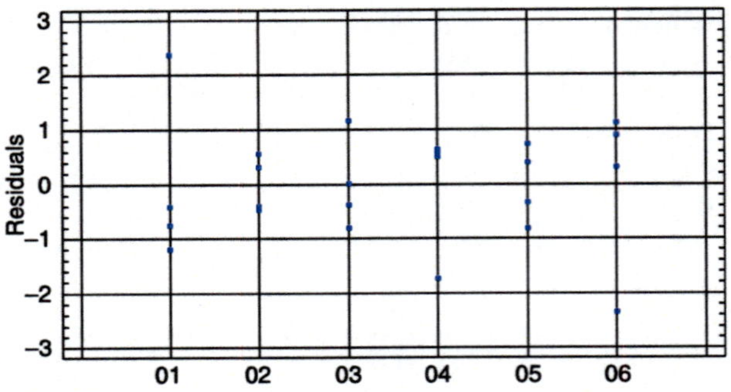

Fig. 6.5 Scatter plot of the residuals versus the six operators

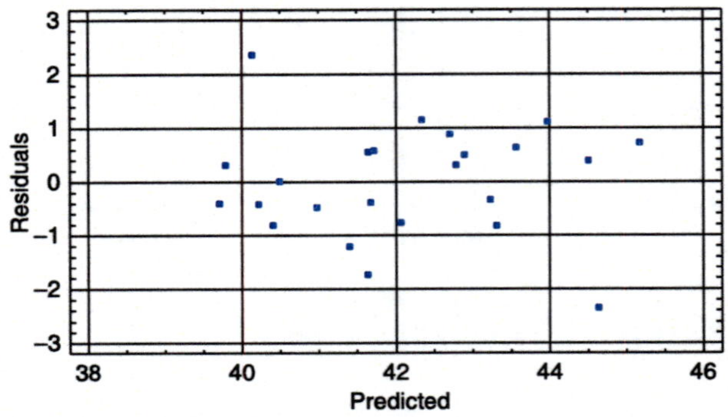

Fig. 6.6 Scatter plot of residuals versus predicted values

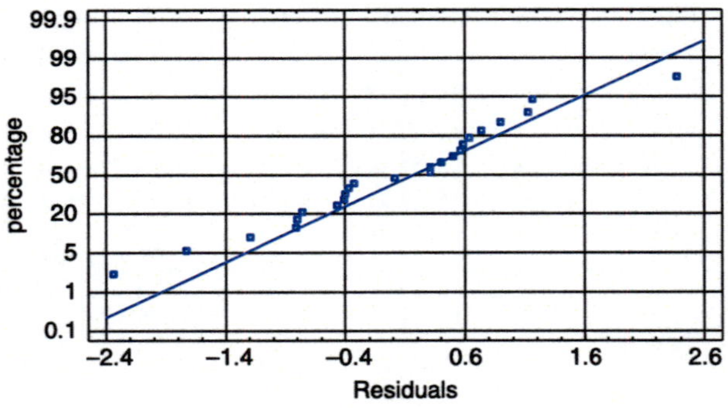

Fig. 6.7 Normal probability plot of the residuals

and versus model predicted values are fairly random except for two large residuals (as noted earlier). Further, a normal probability plot of the model residuals seems to be normally distributed except for the two outliers (Fig. 6.7).

An implicit and important assumption in the above model design is that the treatment and block effects are additive, i.e. negligible interaction effects. In the context of Example 6.2.2, it means that if, say, Operator 3 is 0.5 min faster on the average than Operator 2 on machine 1, the same difference also holds for machines 2, 3, and 4. This pattern would be akin to that depicted in Fig. 6.3 where the mean responses of different blocks differ by the same amount from one treatment to the next. In many experiments, this assumption of additivity does not hold, and the treatment and block effects interact (as illustrated in Fig. 6.2b). For example, Operator 1 may be faster by 0.5 min on the average than Operator 2 when machine 1 is used, but he may be slower by, say, 0.3 min on the average than Operator 2 when machine 2 is used. In such a case, the operators and the machines are said to be interacting. ■

The above treatment of full factorial designs was limited to two factors. The treatment can be extended to more number of factors, but the analysis gets messier though the extension is quite straightforward. The interested reader can refer to the Box et al. (1978) or Montgomery (2009) for such an analysis.

6.2.2 Incomplete Factorial Designs—Latin Squares

The previous section covered two important concepts (randomization and blocking) which, performed together, allow sounder conclusions to be drawn from fewer tests. After the experimental design is formulated, the sequence of the tests, i.e., the selection of the combinations of different levels and different factors should be done in a random manner. This *randomization* would reduce (maybe, even eliminate) unforeseen biases in experimental data which could be due to the effect of subtle factors not considered in the experiment. The other concept, namely *blocking,* is a form of stratified sampling whereby subjects or items in the sample of data are grouped into blocks according to some "matching" criterion so that the similarity of subjects within each block or group is maximized while those from block to block are minimized. Pharmaceutical companies wishing to test the effectiveness of a new drug adopt the above concepts extensively. Since different people react differently, grouping of subjects is done according to some criteria (such as age, gender, body fat percentage,…). Such blocking would result in more uniformity among groups. Subsequently, a random administration of the drugs to half of the people within each block with a placebo to the other half would constitute randomization. Thus, any differences between each block taken separately would be more pronounced than if randomization was done without blocking.

When multiple factors are studied at multiple levels, the number of experiments required for full-factorial design can increase dramatically:

$$\text{Number of Experiments for Full Factorial} = \prod_{i=1}^{k} Levels_i \tag{6.6}$$

where i is the index for the factors which total k. For the special case when all factors have the same number of levels, the number of experiments necessary for a complete factorial design which includes all main effects and interactions is n^k where k is the number of factors and n the number of levels. If certain assumptions are made, this number can be reduced considerably. Such methods are referred to as *incomplete or fractional factorial designs*. The *Latin squares* is one such special design meant for problems: (i) involving *three factors or more,* (ii) that allows *blocking in two directions,* i.e., eliminating two sources of nuisance variability, (iii) where *the number of levels for each factor is the same,* and (iv) *where interaction terms among factors are negligible* (i.e., the interaction terms $(\alpha\beta)_{ij}$ in the statistical effects model given by Eq. 6.2 are dropped). This allows a large reduction in the number of experimental runs especially when several levels are involved.

A Latin Square for n levels denoted by (n×n) is a square of n rows and n columns with each of the n^2 cells containing *one specific treatment that appears once, and only once, in each row and column.* Consider a three factor experiment at three different levels each. The number of experiments required for full factorial, i.e., to map out the entire experimental space would be $3^3=27$. For incomplete factorials, the number of experiments reduces to 3^2 or 9 experiments. The (3×3) Latin square design with three factors (A, B, C) is shown in Table 6.8. While levels A and B are laid down as rows and columns, the level of the third factor is displayed in each cell. Thus, the first cell requires a test with all three factors set at level 1, while the last cell requires A and B to be set at level 3 and C at level 2. A simple manner of generating Latin Square designs for higher values of n is to simply write them in order of level in the first row with the subsequent rows generated by simply shifting the sequence of levels one space to the left.

Note that the Latin Square design shown in Table 6.8 is not unique. There are 12 different combinations of (3×3) Latin squares for the three level case but each design only needs 9 as against 27 experiments required for the full factorial design. Thus, Latin square designs reduce the required number of experiments from n^3 to n^2 (where n is the number of levels), thereby saving cost and time. In general, the fractional factorial design requires n^{k-1} experiments, while the full factorial requires n^k.

Table 6.9 assembles the analysis of variance equations for a Latin Square design, which will be illustrated in Example 6.2.3. It is said (Montgomery 2009) that Latin Square designs usually have a small number of error degrees of freedom (for example, 2 for a 3×3 and 6 for a 4×4 design), which allows a measure of model variance to be deduced.

Table 6.8 A (3×3) Latin Square Design with three factors (A, B, C) with 3 levels each denoted by (1, 2, 3)

C ↘		B		
		1	2	3
A	1	1	2	3
	2	2	3	1
	3	3	1	2

Table 6.9 The analysis of variance equations for (n×n) Latin square design

Source of variation	Sum of squares	Degrees of Freedom	Mean square	Computed F statistic
Row	SSR	$n-1$	$SSR/(n-1)$	$F_R=(MSR/MSE)$
Column	SSC	$n-1$	$SSC/(n-1)$	$F_C=(MSC/MSE)$
Treatment	SSTr	$n-1$	$SSTr/(n-1)$	$F_{Tr}=(MSTr/MSE)$
Error	SSE	$(n-1)(n-2)$	$SSE/(n-1)(n-2)$	–
Total	SST	n^2-1	–	–

In conclusion, while the randomized block design allows blocking of one source of variation, the Latin square design allows systematic blocking of two sources of variability for problems involving three or more factors or variables. The concept, under the same assumptions as those for Latin Square design, can be extended to problems with four factors or more where three sources of variability need to be blocked; this is done using Graeco-Latin square designs (see, Box et al. 1978; Montgomery 2009).

Example 6.2.3: *Evaluating impact of three factors (school, air filter type and season) on breathing complaints*
In an effort to reduce breathing related complaints from students, four different types of air cleaning filters (labeled A, B, C and D, which are viewed as treatments in DOE terminology) are being considered for all schools in a district. Since seasonal effects are important, tests are to be performed under each of the four seasons (and correct for the days when the school is in session for each of these seasons). Further, it is decided that tests should be conducted in four schools (labeled 1 through 4). Because of the potential for differences between schools, it is logical to insist that each filter type be tested at each school during each season of the year.

(a) Develop a DOE This is a three factor problem with four levels in each, i.e., (4×4). The total number of treatment combinations for a completely randomized design would be $4^3=64$. The selection of the same number of categories for all three criteria of classification could be done following a Latin square design, and the analysis of variance performed using the results of only 16 treatment combinations. A typical Latin square, selected at random from all possible (4×4) squares, is given in Table 6.10.

Table 6.10 Experimental design

School	Season			
	Fall	Winter	Spring	Summer
1	A	B	C	D
2	D	A	B	C
3	C	D	A	B
4	B	C	D	A

Table 6.11 Data table showing number of breathing complaints. A, B, C and D are four different types of air filters being evaluated.

School	Fall	Winter	Spring	Summer	Average
1	A 70	B 75	C 68	D 81	73.5
2	D 66	A 59	B 55	C 63	60.75
3	C 59	D 66	A 39	B 42	51.50
4	B 41	C 57	D 39	A 55	48.00
Average	59.00	64.25	50.25	60.25	58.4375

The rows and columns represent the two sources of variation one wishes to control. One notes that in this design, each treatment occurs exactly once in each row and in each column. Such a *balanced arrangement* allows the effect of the air cleaning filter to be separated from that of the season variable. Note that if interaction between the sources of variation is present, the Latin square model cannot be used; this assessment ought to be made based on previous studies or expert opinion.

(b) Perform an ANOVA analysis Table 6.11 summarizes the data collected under such an experimental protocol, where the numerical values shown are the number of breathing-related complaints per season corrected for the number of days when the school is in session and for changes in number of student population.

Assuming that the various sources of variation do not interact, the objective is to statistically determine whether any (and, if so, which) of the three factors (school, season and filter type) affect the number of breathing complaints.

Generating scatter plots such as that shown in Fig. 6.8 for filter type is a good start. In this case, one would make a fair guess based on the intra and within variation that filter type is probably not an influential factor on the number of complaints. The averages of the four treatments or filter types are:

$$\bar{A} = 55.75, \bar{B} = 53.25, \bar{C} = 61.75, \bar{D} = 63.00$$

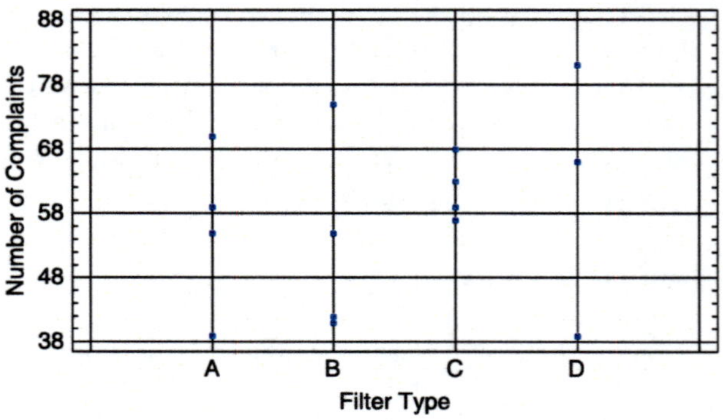

Fig. 6.8 Scatter plot of filter type on number of complaints suggests a lack of correlation. This is supported by the ANOVA analysis.

Table 6.12 ANOVA results following equations shown in Table 6.9

Source of variation	Sum of squares	Degrees of freedom	Mean square	Computed F statistic	Probability
School	1557.2	3	519.06	11.92	0.006
Season	417.69	3	139.23	3.20	0.105
Filter type	263.69	3	87.90	2.02	0.213
Error	261.37	6	43.56	–	–
Total	2499.94	15	–	–	–

The standard deviation is also determined as stdev=12.91. The analysis of variance approach is likely to be more convincing because of its statistical rigor.

From the probability values in the last column of Table 6.12, it can be concluded that the number of complaints is strongly dependent on the school variable, statistically significant at the 0.10 level on the season, and not statistically significant on filter type. ■

6.3 Factorial Designs

6.3.1 2^k Factorial Designs

The above treatment of full and incomplete factorial designs can lead to a prohibitive number of runs when numerous levels need to be considered. As pointed out by Box et al. (1978), it is wise to design a DOE investigation in stages, with each successive iteration providing incremental insight into important issues and suggesting subsequent investigations. Factorial designs, primarily 2^k and 3^k, are of great value at the early stages of an investigation, where a large number of possible factors are investigated with the intention of either narrowing down the number (as in screening design), or to get a preliminary understanding of the mathematical relationship between factors and the response variable. These are, thus, viewed as logical lead-in to the response surface method discussed in Sect. 6.4. The associated mathematics and interpretation of 2^k designs are simple, and can provide insights into the framing of more sophisticated and complete experimental designs. They are popular in R&D of products and processes, and are used extensively.

The 2^k factorial design derives its terminology from the fact that only two levels for each factor k are presumed, one indicative of the lower level or range of variation (coded as –) and the other representing the higher level (coded as +). The factors can be categorical or continuous; if the latter, they are discretized into categories or levels. Figure 6.9a illustrates

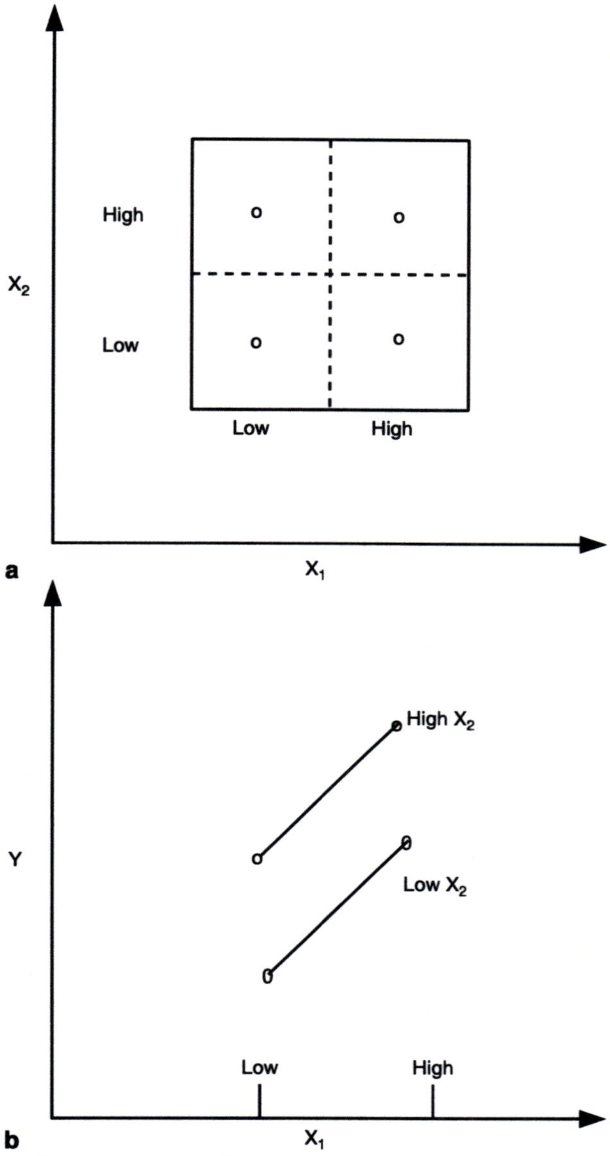

Fig. 6.9 Illustration of how models are built from factorial design data. A 2^2 factorial design is assumed. **a** Discretize the range of variation of the regressors x_1 and x_2 into "low" and "high" ranges, and **b** regress the system performance data as they appear on a scatter plot (no factor interaction is assumed since the two lines are shown parallel)

how the continuous regressors x_1 and x_2 are discretized depending on their range of variation into four system states, while Fig. 6.9b depicts how these four observations would appear in a scatter plot should they exhibit no factor interaction (that is why the lines are parallel). For two factors, the number of trials (without any replication) would be $2^2=4$; for three factors, this would be $2^3=8$, and so on. The formalism of coding the low and high levels of the factors −1 and +1 respectively is most widespread though other ways of coding variables have been proposed.

Table 6.13 The standard form (suggested by Yates) for setting up the two-level three-factor (or 2^3) design

	Level of Factors			
Trial	A	B	C	Response
1	−	−	−	y_1
2	+	−	−	y_2
3	−	+	−	y_3
4	+	+	−	y_4
5	−	−	+	y_5
6	+	−	+	y_6
7	−	+	+	y_7
8	+	+	+	y_8

Table 6.13 depicts a quick and easy way of setting up a two-level three-factor design (following the standard form suggested by Yates). Notice that the last but one column has four (−) followed by four (+), the last but two column by successive pairs of (−) and (+), and the second has alternating (−) and (+). The Yates algorithm is easily extended to higher number of factors. However, the sequence in which the runs are to be performed should be randomized; a good way is to simply sample the set of trials {1, …, 8} in random fashion without replacement.

The approach can be modified to treat the case of parameter interaction, i.e., when the factors interact. Table 6.13 is simply modified by including separate columns for the three interaction terms, as shown in Table 6.14. The sign coding for the interactions is determined by multiplying the signs of each of the two corresponding terms. For example, interaction AB for trial 1, would be (−)(−)=(+); and so on.

The main effect of, say, factor C can be determined simply as:

$$\text{Main effect of } C = \bar{C}_+ - \bar{C}_- = \frac{(y_5 + y_6 + y_7 + y_8)}{4} - \frac{(y_1 + y_2 + y_3 + y_4)}{4} \quad (6.7)$$

Statistical text books on DOE provide elaborate details of how to obtain estimates of all main and interaction effects

Table 6.14 The standard form of the two-level three-factor (or 2^3) design with interactions

	Level of Factors			Interactions				
Trial	A	B	C	AB	AC	BC	ABC	Response
1	−	−	−	+	+	+	−	y_1
2	+	−	−	−	−	+	+	y_2
3	−	+	−	−	+	−	+	y_3
4	+	+	−	+	−	−	−	y_4
5	−	−	+	+	−	−	+	y_5
6	+	−	+	−	+	−	−	y_6
7	−	+	+	−	−	+	−	y_7
8	+	+	+	+	+	+	+	y_8

Table 6.15 Response table for the 2^k design with interactions (omitting the ABC term)

Trial	Resp.	A_+	A_-	B_+	B_-	C_+	C_-	AB_+	AB_-	AC_+	AC_-	BC_+	BC_-
1	y_1		y_1		y_1		y_1	y_1		y_1		y_1	
2	y_2	y_2			y_2		y_2		y_2		y_2	y_2	
3	y_3		y_3	y_3			y_3		y_3	y_3			y_3
4	y_4	y_4		y_4			y_4	y_4			y_4		y_4
5	y_5		y_5		y_5	y_5		y_5			y_5		y_5
6	y_6	y_6			y_6	y_6			y_6	y_6			y_6
7	y_7		y_7	y_7		y_7			y_7		y_7	y_7	
8	y_8	y_8		y_8		y_8		y_8		y_8		y_8	
Sum													
No.	8	4	4	4	4	4	4	4	4	4	4	4	4
Avg		$\bar{A}_+$	$\bar{A}_-$	$\bar{B}_+$	$\bar{B}_-$	$\bar{C}_+$	$\bar{C}_-$	$\bar{AB}_+$	$\bar{AB}_-$	$\bar{AC}_+$	$\bar{AC}_-$	$\bar{BC}_+$	$\bar{BC}_-$
Effect		$\bar{A}_+ - \bar{A}_-$		$\bar{B}_+ - \bar{B}_-$		$\bar{C}_+ - \bar{C}_-$		$\bar{AB}_+ - \bar{AB}_-$		$\bar{AC}_+ - \bar{AC}_-$		$\bar{BC}_+ - \bar{BC}_-$	

when more factors are to be considered, and then how to use statistical procedures such as ANOVA to identify the significant ones. The standard form shown in Table 6.14 can be rewritten as shown in Table 6.15 for the 2^3 design with interactions. This is referred to as the *response table* form, and is advantageous in that it allows the analysis to be done in a clear and modular manner. The interpretation of what is implied by the interaction terms appearing in the table needs to be clarified. For example, AB_+ denotes the effect of A when B is held fixed at the B_+ or higher level. On the other hand, AB_- denotes the effect of A when B is held fixed at the B_- or lower level.

For example, the main effect of A is conveniently determined as:

$$= (\bar{A}_+ - \bar{A}_-) = \frac{1}{4}\left[(y_2 + y_4 + y_6 + y_8) - (y_1 + y_3 + y_5 + y_7)\right] \tag{6.8a}$$

while the interaction effect of, say, BC can be determined by the average of the B effect when C is held constant at +1 minus the B effect when C is held constant at –1. Thus:

$$\text{Interaction effect of BC} = \left(\bar{BC}_+ - \bar{BC}_-\right) = \frac{1}{4}\left[(y_1 + y_2 + y_7 + y_8) - (y_3 + y_4 + y_5 + y_6)\right] \tag{6.8b}$$

These individual and interaction effects directly provide a prediction model of the form:

$$\hat{y} = \underbrace{b_0 + b_1A + b_2B + b_3C}_{\text{Main effects}} + \underbrace{b_{12}AB + b_{13}AC + b_{23}BC + b_{123}ABC}_{\text{Interaction terms}} \tag{6.9a}$$

The intercept term is given by the grand average of all the response values y. This model is analogous to Eq. 5.18c which is one form of the additive multiple linear models discussed in Chap. 5. Note that Eq. 6.9 has eight parameters and with eight experimental runs, the model fit will be perfect with no variance. A measure of the random error can only be deduced if the degrees of freedom (d.f.)>0, and so replication (i.e., repeats of runs) is necessary. Another option, relevant when interaction effects are known to be negligible, is to adopt a model with only main effects such as:

$$\hat{y} = b_0 + b_1A + b_2B + b_3C \tag{6.9b}$$

In this case, d.f.=4, and so a measure of random error of the model can be determined.

Example 6.3.1: ***Deducing a prediction model for a 2^3 factorial design***

Consider a problem where three factors {A, B, C} are presumed to influence a response variable y. The problem is to perform a DOE, collect data, ascertain the statistical importance of the factors and then identify a prediction model. The numerical values of the factors or regressors corresponding to the high and low levels are assembled in Table 6.16 (while the numerical value of A is a fraction about its mean value, those of B and C are not).

It was decided to use two replicate tests for each of the 8 combinations to enhance accuracy. Thus 16 runs were per-

Table 6.16 Assumed low and high levels for the three factors (Example 6.3.1)

Factor	Low level	High level
A	0.9	1.1
B	1.20	1.30
C	20	30

Table 6.17 Standard table (Example 6.3.1)

	Level of Factors			
Trial	A	B	C	Responses
1	0.9	1.2	20	34,40
2	1.1	1.2	20	26,29
3	0.9	1.3	20	33,35
4	1.1	1.3	20	21,22
5	0.9	1.2	30	24,23
6	1.1	1.2	30	23,22
7	0.9	1.3	30	19,18
8	1.1	1.3	30	18,18

formed and the results are tabulated in the standard form as suggested by Yates (Table 6.13) and shown in Table 6.17.

(a) Identify statistically significant terms

This tabular data can be used to create a table similar to Table 6.15, which is left to the reader. Then, the main effects and interaction terms can be calculated following Eq. 6.8. Thus:

Main effect of factor A:

$$\begin{aligned}&\frac{1}{(2)(4)}[(26+29)+(21+22)+(23+22)\\&\quad+(18+18)-(34+40)-(33+35)\\&\quad-(24+23)-(19+18)]\\&=-\frac{47}{8}=-5.875\end{aligned}$$

while the effect sum of squares SSA $=(-47.0)^2/16=$ 138.063

Similarly, B: −4.625; C: −9.375; AB: −0.625; AC: 5.125; BC: −0.125; ABC: 0.875. The results of the ANOVA analysis are assembled in Table 6.18. One concludes that the main effects A, B and C and the interaction effect AC are significant at the 0.01 level. The main effect and interaction effect plots are shown in Figs. 6.10 and 6.11. These plots do confirm that interaction effects are present only for factors A and C since the lines are clearly not parallel.

Table 6.18 Results of the ANOVA analysis. Interaction effects AB, BC and ABC are not significant (Example 6.3.1)

Source	Sum of Squares	D.f.	Mean Square	F-Ratio	p-Value
Main effects					
Factor A	138.063	1	138.063	41.68	0.0002
Factor B	85.5625	1	85.5625	25.83	0.0010
Factor C	351.563	1	351.563	106.13	0.0000
Interactions					
AB	1.5625	1	1.5625	0.47	0.5116
AC	105.063	1	105.063	31.72	0.0005
BC	0.0625	1	0.0625	0.02	0.8941
ABC	3.063	1	3.063	0.92	0.3640
Residual or error	26.5	8	3.3125		
Total (corrected)	711.438	15			

(b) Identify prediction model

Only four terms, namely A, B, C and AC interaction were found to be statistically significant at the 0.05 level. (See Table 6.18) In such a case, the functional form of the prediction model given by Eq. 6.9 reduces to:

$$\hat{y}=b_0+b_1x_A+b_2x_B+b_3x_c+b_4x_Ax_C$$

Substituting the values of the effect estimates determined earlier results in

$$\begin{aligned}\hat{y}=25.313-2.938x_A-2.313x_B\\-4.688x_c+2.563x_Ax_C\end{aligned}\qquad(6.10)$$

where coefficient b_0 is the mean of all observations. Also, note that the values of the model coefficients are half the values of the main and interaction effects determined in part (a). For example, the main effect of factor A was calculated to be (−5.875) which is twice the (−2.938) coefficient for the x_A factor shown in the equation above. The division by 2 is needed because of the manner in which the factors were coded, i.e. the high and low levels, coded as +1 and −1, are separated by 2 units.

The performance equation thus determined can be used for predictions. For example, when $x_A=+1$, $x_B=-1$, $x_C=-1$, one gets $\hat{y}=26.813$ which agrees reasonably well with the average of the two replicates performed (26 and 29).

(c) Comparison with linear multiple regression approach

The parallel between this approach and regression modeling involving indicator variables (described in Sect. 5.7.3) is obvious but note worthy. For example, if one were to perform a multiple regression to the above data with the three regressors coded as −1 and +1 for low and high values respectively, one obtains the following results (Table 6.19).

Note that the same four variables (A, B, C and AC interaction) are statistically significant while the model coefficients are identical to the ones determined by the ANOVA analysis. If the regression were to be redone with only these four variables present, the *model coefficients would be identical.* This is a great advantage with factorial designs in that one could include additional variables incrementally in the model without impacting the model coefficients of variables already identified. Why this is so is explained in the next section. Table 6.20 assembles pertinent goodness-of-fit indices for the complete model and the one with only the four significant regressors. Note that while the R^2 value of the former is higher (a misleading statistic to consider when dealing with multivariate model building), the adjusted R^2 and the RMSE of the reduced model are superior. Finally, Figs. 6.12 and 6.13 are model predicted versus observed plots which allow one to ascertain how well the model has fared; in this case, there seems to be larger scatter at higher values indicative of non-additive errors. This suggests that a linear additive model may not be the best. ■

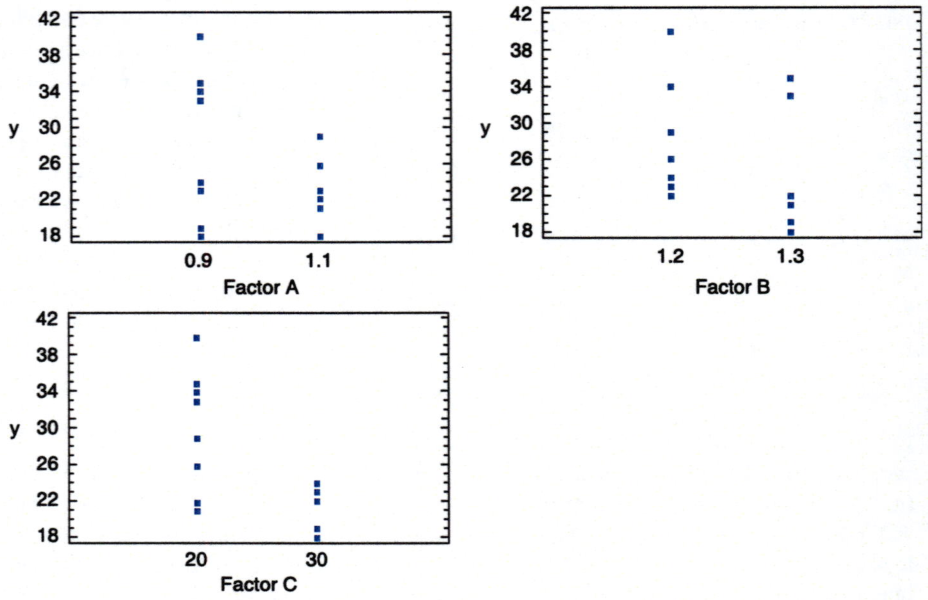

Fig. 6.10 Main effect scatter plots for the three factors

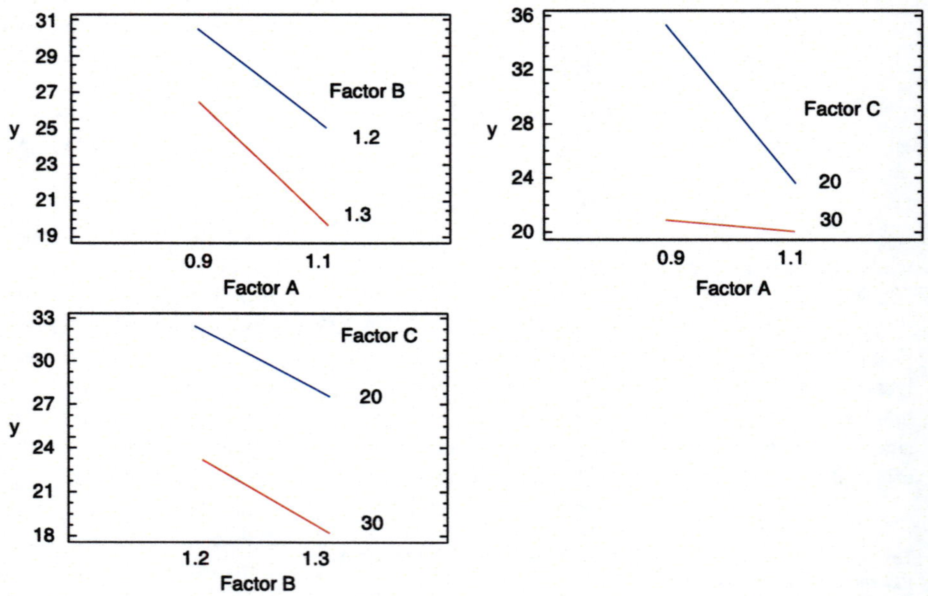

Fig. 6.11 Interaction plots

6.3.2 Concept of Orthogonality

An important concept in DOE is *orthogonality* by which is implied that trials should be framed such that the data matrix X[2] results in $(X'X) = -1$.[3] In such a case, the off-diagonal terms of the matrix $(X'X)$ will be zero, i.e., the regressors are uncorrelated. This would lead to the best designs since it would minimize the variance of the regression coefficients. For example, consider Table 6.13 where the standard form for the two-level three-factor design is shown. Replacing low and high values (i.e., – and +) by -1 and $+1$, and noting that an extra column of 1 needs to be introduced to take care of the constant term in the model (see Eq. 5.26b) results in the regressor matrix being defined by:

[2] Refer to Sect. 5.4.3 for refresher.

[3] Recall from basic geometry that two straight lines are perpendicular when the product of their slopes is equal to -1. Orthogonality is an extension of this concept to multi-dimensions.

Table 6.19 Results of performing a multiple linear regression to the same data with coded regressors (Example 6.3.1)

Parameter	Parameter estimate	Standard error	t-statistic	p-value
Constant	25.3125	0.455007	55.631	0.0000
Factor A	−2.9375	0.455007	−6.45595	0.0002
Factor B	−2.3125	0.455007	−5.08234	0.0010
Factor C	−4.6875	0.455007	−10.302	0.0000
Factor A*Factor B	−0.3125	0.455007	−0.686803	0.5116
Factor A*Factor C	2.5625	0.455007	5.63178	0.0005
Factor B*Factor C	−0.0625	0.455007	−0.137361	0.8941
Factor A*Factor B*Factor C	0.4375	0.455007	0.961524	0.3644

Table 6.20 Goodness-of-fit statistics of multiple linear regression models (Example 6.3.1)

Regression model	Model R^2	Adjusted R^2	RMSE
With all terms	0.963	0.930	1.820
With only four significant terms	0.956	0.940	1.684

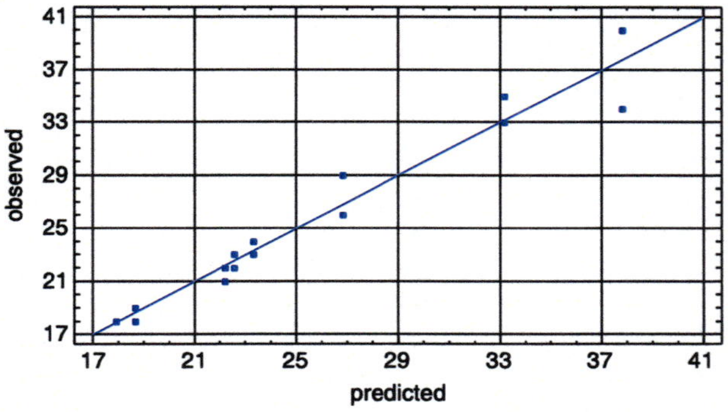

Fig. 6.12 Observed versus predicted values for the regression model indicate larger scatter at high values (Example 6.3.1)

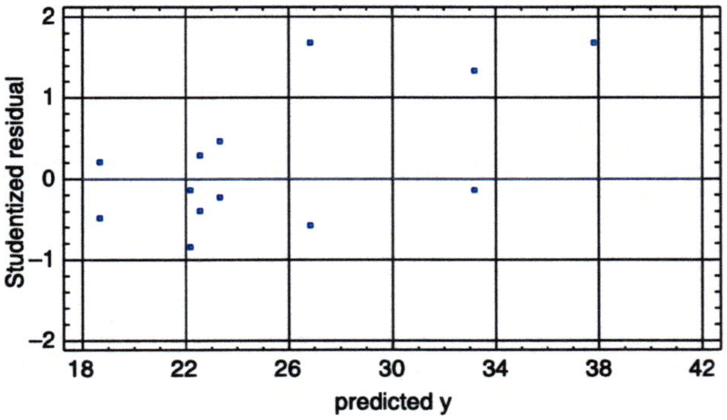

Fig. 6.13 Model residuals versus model predicted values highlight the larger scatter present at higher values indicative of non-additive errors (Example 6.3.1)

$$X = \begin{bmatrix} 1 & -1 & -1 & -1 \\ 1 & +1 & -1 & -1 \\ 1 & -1 & +1 & -1 \\ 1 & +1 & +1 & -1 \\ 1 & -1 & -1 & +1 \\ 1 & +1 & -1 & +1 \\ 1 & -1 & +1 & +1 \\ 1 & +1 & +1 & +1 \end{bmatrix} \qquad (6.11)$$

The reader can verify that the off-diagonal terms of the matrix $(X'X)$ are indeed zero. All n^k factorial designs are thus orthogonal, i.e., $(X'X)^{-1}$ is a diagonal matrix with nonzero diagonal components. This leads to the *most sound parameter estimation* (as discussed in Sect. 11.2). Another benefit of orthogonal designs is that parameters of regressors already identified remain unchanged as additional regressors are added to the model; thereby allowing the model to be developed incrementally. Thus, the effect of each term of the model can be examined independently. These are two great benefits when factorial designs are adopted for model identification.

Example 6.3.2:[4] *Matrix approach to a inferring prediction model for a 2^3 design*
This example will illustrate the analysis procedure for a complete 2^k factorial design with three factors similar to Example 6.3.1 but following the matrix formulation. The model, assuming a linear form, is given by Eq. 6.9a and includes individual or main and interaction effects. Denoting the three factors by x_1, x_2 and x_3, the regressor matrix X will have four parameters (the intercept term is the first additional term) as well as the four interaction terms as shown below (refer to Table 6.14 for ease in understanding the matrix):

$$X = \begin{bmatrix} 1 & -1 & -1 & -1 & 1 & 1 & 1 & -1 \\ 1 & 1 & -1 & -1 & -1 & -1 & 1 & 1 \\ 1 & -1 & 1 & -1 & -1 & 1 & -1 & 1 \\ 1 & 1 & 1 & -1 & 1 & -1 & -1 & -1 \\ 1 & -1 & -1 & 1 & 1 & -1 & -1 & 1 \\ 1 & 1 & -1 & 1 & -1 & 1 & -1 & -1 \\ 1 & -1 & 1 & 1 & -1 & -1 & 1 & -1 \\ 1 & 1 & 1 & 1 & 1 & 1 & 1 & 1 \end{bmatrix}$$

Main effects Interaction effects

Let us assume that a DOE has yielded the following values for the response variable:

$$Y^T = [49 \quad 62 \quad 44 \quad 58 \quad 42 \quad 73 \quad 35 \quad 69]$$

The intention is to identify a parsimonious model, i.e., one in which only the statistically significant terms following Eq. 6.9a are retained in the model.

[4] From Beck and Arnold (1977) by permission of Beck.

The inverse of $\mathbf{X^TX} = \left(\frac{1}{8}\right)\mathbf{I}$ and the $\mathbf{X^TX}$ terms can be deduced by taking the sums of the y_i terms multiplied by either + or – as indicated in $\mathbf{X^T}$. The coefficient $b_0=54$ (average of all eight values of y), b_1 following Eq. 6.8a is: $b_1=[(62+58+73+69)-(49+44+42+35)]/(4\times 2)=11.5$, and so on. The resulting model is:

$$\hat{y}_i = 54 + 11.5x_{1i} - 2.5x_{2i} + 0.75x_{3i} + 0.5x_{1i}x_{2i} + 4.75x_{1i}x_{3i} - 0.25x_{2i}x_{3i} + 0.25x_{1i}x_{2i}x_{3i} \quad (6.12a)$$

With eight parameters and eight observations (and no replication), the model will be perfect with zero degrees of freedom; this is referred to as a *saturated model*. This is not a prudent situation since a model variance cannot be computed nor can the p-values of the various terms inferred. Had replication been adopted, an estimate of the variance in the model could have been conveniently estimated and some measure of the goodness-of-fit of the model deduced (as in Example 6.3.1). In this case, the simplest recourse is to drop one of the terms from the model (say the $(x_1x_2x_3)$ interaction term) and then perform the ANOVA analysis. Because of the orthogonal behavior, the significance of the dropped term can be evaluated at a later stage without affecting the model terms already identified.

The effect of individual terms is now investigated in a manner similar to the previous example. The ANOVA analysis shown in Table 6.21 suggests that only the terms x_1 and (x_1x_3) are statistically significant at the 0.05 level. However, the p value for x_2 is close, and so it would be advisable to keep this term. Thus, the parsimonious model assumes the form:

$$\hat{y}_i = 54 + 11.5x_{1i} - 2.5x_{2i} + 4.75x_{1i}x_{3i} \quad (6.12b)$$ ■

The above example illustrates how data gathered within a DOE framework and analyzed following the ANOVA method can yield an efficient functional predictive model of the data. It is left to the reader to repeat the analysis illustrated in Example 6.3.1 where an identical model was obtained

Table 6.21 Results of the ANOVA analysis (Example 6.3.2)

Source	Sum of squares	D.f.	Mean square	F-ratio	p-value
Main effects					
Factor x_1	1058	1	1058	2116	0.0138
Factor x_2	50.0	1	50.0	100	0.0635
Factor x_3	4.50	1	4.50	9.00	0.2050
Interactions					
x_1x_2	2.00	1	2.00	4.00	0.2950
x_1x_3	180.5	1	180.5	361	0.0335
x_2x_3	0.50	1	0.50	1.00	0.5000
Residual or error	0.50	1	0.50		
Total (Corrected)	1296	7			

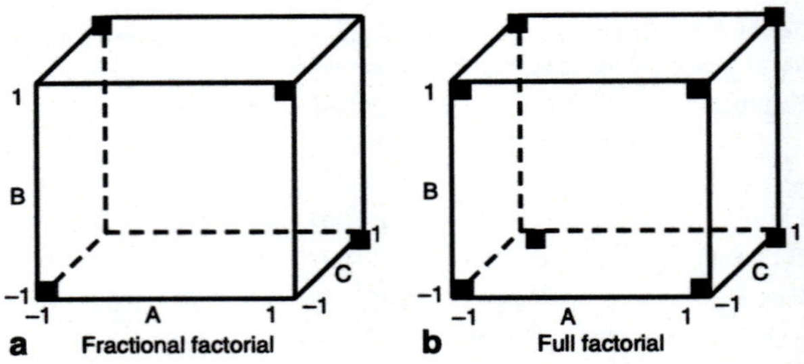

Fig. 6.14 Illustration of the differences between fractional and full factorials runs for a 2^3 DOE experiment. See Table 6.14 for a specification of the full factorial design. Several different combinations of fractional factorials designs are possible; only one such combination is shown

by straightforward use of multiple linear regression. Note that orthogonality is maintained only if the analysis is done with coded variables (–1 and +1), and not with the original ones.

Recall that a 2^2 factorial design implies two regressors or factors, each at two levels; say "low" and "high". Since there are only two states, one can only frame a first order functional model to the data such as Eq. 6.9. Thus, a 2^2 factorial design is inherently constrained to identifying a first order linear model between the regressors and the response variable. If the mathematical relationship requires higher order terms, multi-level factorial designs are more appropriate. Such designs would allow a model of the form given by Eq. 5.23 (which is a full second-order polynomial model). For example, the 3^k design will require the range of variation of the factors to be aggregated into three levels, such as "low", "medium" and "high". If the situation is one with three factors (i.e., k=3), one needs to perform 27 experiments even if no replication tests are considered. This is more than three times the number of tests needed for the 2^3 design. Thus, the added higher order insight can only be gained at the expense of a larger number of runs which, for higher number of factors, may become prohibitive.

One way of greatly reducing the number of runs, provided interaction effects are known to be negligible, is to adopt incomplete or fractional factorial designs (described in Sect. 6.2). The 27 tests needed for a full 3^3 factorial design can be reduced to 9 tests only. Thus, instead of 3^k tests, an incomplete block design would only require (3^{k-1}) tests. A graphical illustration of how a fractional factorial design differs from a full factorial one for a 2^3 instance is illustrated in Fig. 6.14. Three factors are involved (A, B and C) at two levels each (–1, 1). While 8 test runs are performed corresponding to each of the 8 corners of the cube for the full factorial, only 4 runs are required for the fractional factorial as shown. The interested reader can refer to the Box et al. (1978) or Montgomery (2009) for detailed treatment of higher level factorial design methods both complete and incomplete.

6.4 Response Surface Designs

6.4.1 Applications

Recall that the factorial methods described in the previous section can be applied to either continuous or discrete categorical variables. The 2^k factorial methods allow both for screening to identify dominant factors, and for identifying a robust linear predictive model. In a historic timeline, these techniques were then extended to optimizing a process or product by Box and Wilson in the early 1950s. A special class of mathematical and statistical techniques were developed meant to identify models and analyze data between a response and a set of continuous variables with the intent of determining the conditions under which a maximum (or a minimum) of the response variable is obtained. For example, the optimal mix of two alloys which would result in the product having maximum strength can be deduced by fitting the data from factorial experiments with a model from which the optimum is determined either by calculus or search methods (described in Chap. 7 under optimization methods). These models, called *response surface models* (RSM), can be framed as either first order or second order models, linear in the parameters, depending on whether one is far or close to the desired optimum. Response surface designs involve not just the modeling aspect, but also recommendations on how to perform the sequential search involving several DOE steps.

The reader may wonder why most of the DOE models treated in this chapter assume empirical polynomial models. This was because of historic reasons where the types of applications which triggered the development of DOE were not understood well enough to adopt mechanistic functional forms. Empirical polynomial models are linear in the parameters but can be non-linear in their functional form (such as Eq. 6.9a). Recall that a function is strictly linear only when it contains first-order regressors (i.e., main effects of the factors) with no interacting terms (such as Eq. 6.9b). Non-linear functional models can arise by introducing interaction terms in the first order linear model, as well as using higher order terms, such as quadratic terms for the regressors (see Eq. 5.21 and 5.23).

A second class of problems to which RSM can be used involves *simplifying the search for an optimum* when detailed computer simulation programs of physical systems requiring long-run times are to be used. The similarity of such problems to DOE experiments on physical processes is easily made, since the former requires: (i) performing a sensitivity analysis to determine a subset of dominant model input parameter combinations (akin to screening), (ii) defining a suitable approximation for the true functional relationship between response and the set of independent variables (akin to determining the number of necessary levels), (iii) making multiple runs of the computer model using specific values and pairings of these input parameters (akin to performing factorial experiments), and (iv) fitting an appropriate mathematical model to the data. This fitted response-surface is then used as a replacement or proxy for the computer model, and all inferences related to optimization/uncertainty analysis requiring several thousands of simulations for the original model are derived from this fitted model. The validity of this approach is of course contingent on the fact that the computer simulation is an accurate representation of the physical system. Thus, this application is very similar to the intent behind process optimization except that model simulations are done to predict system response instead of actual experiments.

6.4.2 Phases Involved

A typical RS experimental design involves three general phases performed with the specific intention of limiting the number of experiments required to achieve a rich data set. This will be illustrated using the following example. The R&D staff of a steel company wants to improve the strength of the metal sheets sold. They have identified a preliminary list of factors that might impact the strength of their metal sheets including the concentrations of chemical A and chemical B, the annealing temperature, the time to anneal and the thickness of the sheet casting. The **first phase** is to run a screening design to identify the main factors influencing the metal sheet strength. Thus, those factors that are not important contributors to the metal sheet strength are eliminated from further study. How to perform such screening tests have been discussed in Sect. 6.3.1.

As an illustration, it is concluded that the chemical concentrations A and B are the main factors that survive the screening design. To *optimize* the mechanical strength of the metal sheets, one needs to know the relationship between the strength of the metal sheet and the concentration of chemicals A and B in the formula; this is done in the **second phase** *which requires a sequential search process.* The following steps are undertaken:

(i) Identify the levels of the amount of chemicals A and B to study. Use 2 levels for linear relationships and 3 or more levels for non-linear relationships.
(ii) Generate the experimental design using one of several factorial methods.
(iii) Run the experiments.
(iv) Analyze the data using ANOVA.
(v) Draw conclusions and develop a model for the response variable. Unfortunately, this is likely to be an approximate model representative of the behavior of the metal in the *local search space* only; and usually, one cannot simply use this model to identify the global maximum.
(vi) Using optimization methods (such as calculus based methods or search methods such as steepest descent), move in the direction of the search space where the overall optimum is likely to lie (refer to Example 7.4.2).

(vii) Repeat steps (i) through (vi) till the global minimum is reached.

Once the optimum has been identified, the R&D staff would want to *confirm* that the new, improved metal sheets have higher strength; this is the **third phase**. They would resort to hypothesis tests involving running experiments to support the alternate hypothesis that the strength of the new, improved metal sheet is greater than the strength of the existing metal sheet. In summary, the goals of the second and third phases of the RS design are to determine and then confirm, with the needed statistical confidence, the optimum levels of Chemicals A and B that maximize the metal sheet strength.

6.4.3 First and Second Order Models

In most RS problems, the form of the relationship between the response and the regressors is unknown. Consider the case where the yield (Y) of a chemical process is to be maximized with temperature (T) and pressure (P) being the two independent variables (from Montgomery 2009). The 3-D plot (called the *response surface* in DOE terminology) is shown in Fig. 6.15, while its projection of a 2-D plane, known as a *contour plot*, is also shown. The maximum yield is achieved under T=138 and P=18, at which the maximum yield Y=70. If one did not know the shape of this curve, one simple approach would be to assume a starting point (say, T=117 and P=20, as shown) and repeatedly perform experiments in an effort to reach the maximum point. This is akin to a univariate optimization search (see Sect. 7.4) which is not very efficient. In this example involving a chemical process, varying one variable at a time may work because of the symmetry of the RS. However, in case (and this is often so) when the RS is asymmetrical or when the search location is far away from the optimum, such a univariate search may erroneously indicate a non-optimal maximum. A superior manner, and the one adopted in most numerical methods is the steepest gradient method which involves adjusting all the variables together (see Sect. 7.4). As shown in Fig. 6.16, if the responses Y at each of the four corners of the square are known by experimentation, a suitable model is identified (in the figure, a linear model is assumed and so the set of lines for different values of Y are parallel). The steepest gradient method involves moving along a direction perpendicular to the sets of lines (indicated by the "steepest descent" direction in the figure) to another point where the next set of experiments ought to be performed. Repeated use of this testing, modeling and stepping is likely to lead one close to the sought-after maximum or minimum (provided one is not caught in a local peak or valley or a saddle point).

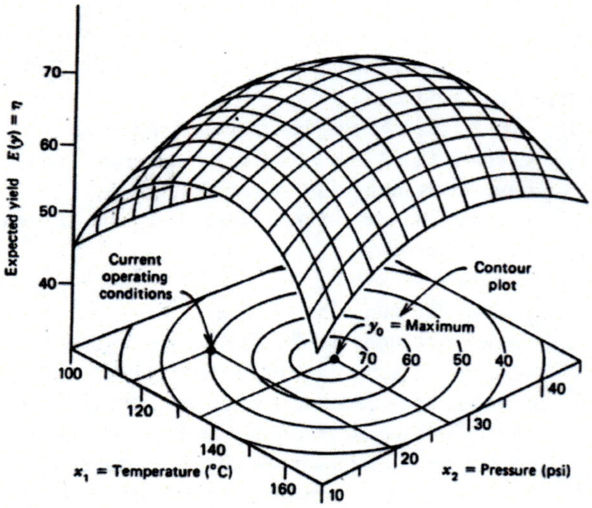

Fig. 6.15 A three-dimensional response surface between the response variable (the expected yield) and two regressors (temperature and pressure) with the associate contour plots indicating the optimal value. (From Montgomery 2009 by permission of John Wiley and Sons)

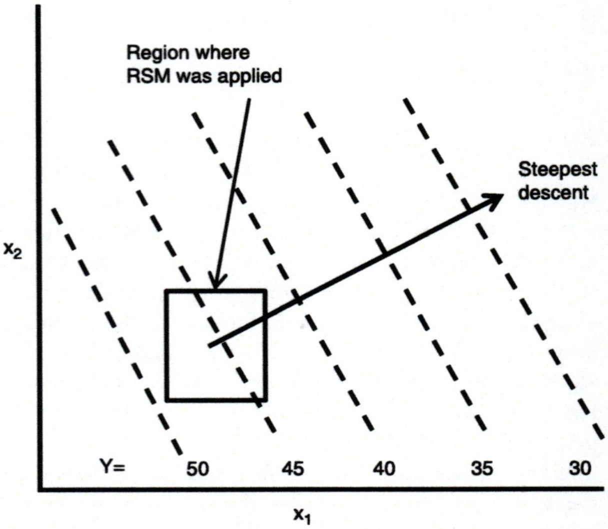

Fig. 6.16 Figure illustrating how the first-order response surface model (RSM) fit to a local region can progressively lead to the global optimum using the steepest descent search method

The following recommendations are noteworthy so as to minimize the number of experiments to be performed:

(a) *During the initial stages*, of the investigation, a *first-order polynomial model* in some region of the range of variation of the regressors is usually adequate. Such models have been extensively covered in Chap. 5 with Eq. 5.25 being the linear first order model form in vector notation. 2^k factorial designs are good choices at the preliminary stage of the RS investigation. As stated earlier, due to the benefit of *orthogonality*, these designs are best since they would minimize the variance of the regression coefficients.

(b) Once *close to the optimal region*, polynomial models higher than first order are advised. This could be a first-order polynomial (involving just main effects) or a higher-order polynomial which also includes quadratic effects and interactions between pairs of factors (two-factor interactions) to account for curvature. *Quadratic models are usually sufficient for most engineering applications*, though increasing the order of approximation to higher orders could, sometimes, further reduce model errors. Of course, it is unlikely that a polynomial model will be a reasonable approximation of the true functional relationship over the entire space of the independent variables, but for a relatively small region they usually work quite well (Montgomery 2009). Note that rarely would all of the terms of the quadratic model be needed; and how to identify a parsimonious model has been illustrated in Examples 6.3.1 and 6.3.2.

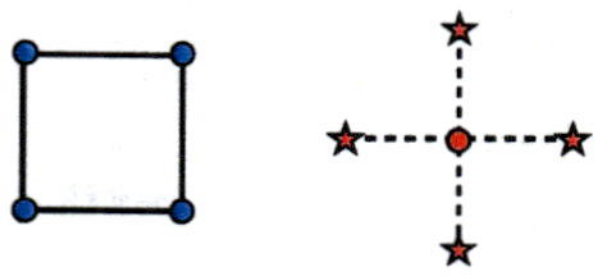

Fig. 6.17 A Central Composite Design (CCD) for two factors contains two sets of trials: a fractional factorial or "cube" portion which serve as a preliminary stage where one can fit a first-order (linear) model, and a group of axial or "star" points that allow estimation of curvature. A CCD always contains twice as many axial (or star) points as there are factors in the design. In addition, a certain number of center points are also used so as to capture inherent random variability in the process or system behavior

6.4.4 Central Composite Design and the Concept of Rotation

For a 3^k factorial design with the number of factors k=3, one needs 27 experiments with no replication, which, for k=4 grows to 81 experiments. Thus, number of trials at each iteration point increase geometrically. Hence, 3^k designs become impractical as k gets much above 3. A more efficient manner of designing experiments is to use the concept of *rotation,* also referred to as axi-symmetric. An experimental design is said to be rotatable if the trials are selected such that they are equi-distant from the center. Since the location of the optimum point is unknown, such a design would result in equal precision of estimation in all directions. In other words, the variance of the response variable at any point in the regressor space is function of only the distance of the point from the design center.

Several rotatable as well as non-rotatable designs can be found in the published literature. Of these, the *Central composite design* (CCD) is probably the most widely used for fitting a second order response surface. A CCD contains a fractional factorial design that is augmented with a group of *axial points* that allow estimation of curvature. Center point runs (which are essentially random repeats of the center point) are included to provide a measure of process stability (i.e., reduces model prediction errors) and capture any inherent variability. They, also, provide a check for curvature, i.e., if the response surface is curved, the center points will be lower or higher than predicted by the design points. The factorial or "cube" portion and center points (shown as circles in Fig. 6.17) may aid in fitting a first-order (linear) model during the preliminary stage while still providing evidence regarding the importance of a second-order contribution or curvature. A CCD always contains twice as many axial (or star) points as there are factors in the design. The star points represent new extreme values (low and high) for each factor in the design. Thus, the total number of experimental runs for a CCD with k factors$=2^k+2k+c$ where c is the number of center points.

CCDs allow for efficient estimation of the quadratic terms in the second-order model since they inherently satisfy the desirable design properties of orthogonal blocking and rotatability. A central composite design with two and three factors is shown in Fig. 6.18. For a two-factor experiment design, the CCD generates 4 factorial points and 4 axial points; for a three-factor experiment design, the CCD generates 8 factorial points and 6 axial points. The number of center points for some useful CCDs have also been suggested. Sometimes, more center points than the numbers suggested are introduced; nothing will be lost by this except the cost of performing the additional runs. For a two-factor CCD, at least two center points should be used, while many researchers routinely use as many as 6–8 points.

If the distance from the center of the design space to a factorial point is ± 1 unit for each factor, the distance from the center of the design space to an axial point is $\pm \alpha$ with $|\alpha|>1$. The precise value of α depends on certain properties desired for the design (for example, whether or not the design is orthogonally blocked) and on the number of factors involved. Similarly, the number of center point runs needed for the design also depends on certain properties required for the design. To maintain rotatability, the value of a for CCD is chosen such that (Montgomery 2009):

$$\alpha = (n_f)^{1/4} \tag{6.13}$$

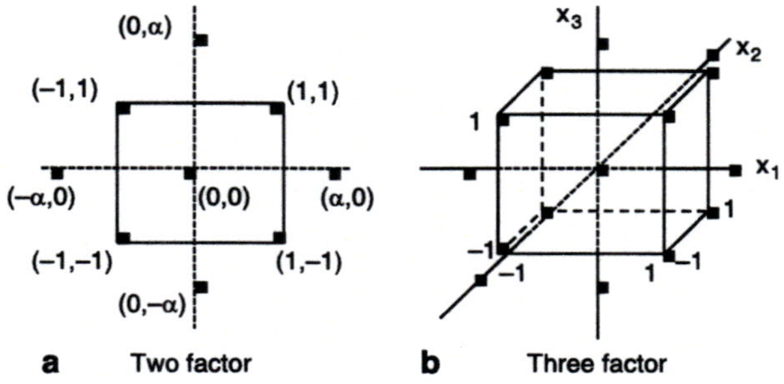

Fig. 6.18 Rotatable central composite designs for two factors and three factors during RSM. The *black dots* indicate locations of experimental runs

where n_f is the number of experimental runs in factorial portion. For example: if the experiment has 2 factors, the full factorial portion would contain $2^2=4$ points; the value of α for rotatability would be $\alpha=(2^2)^{1/4}=1.414$; if the experiment has 3 factors, $\alpha=(2^3)^{1/4}=1.682$; if the experiment has 4 factors, $\alpha=(2^4)^{1/4}=2$; and so on. As shown in Fig. 6.18, CCDs usually have axial points outside the "cube", (unless one intentionally specifies $\alpha \leq 1$ due to, say, safety concerns in performing the experiments). Finally, since the design points describe a circle *circumscribed* about the factorial square, the optimum values must fall within this experimental region. If not, suitable constraints must be imposed on the function to be optimized. This is illustrated in the example below. For further reading on RSD, the texts by Box et al. (1978) and Montgomery (2009) are recommended.

Example 6.4.1:[5] *Optimizing the deposition rate for a tungsten film on silicon wafer.*
A two-factor rotatable central composite design (CCD) was run so as to optimize the deposition rate for a tungsten film on silicon wafer. The two factors are the process pressure (in Torr) and the ratio of H_2 to WF_6 in the reaction atmosphere. The ranges for these factors are given in Table 6.22.

Let x_1 be the pressure factor and x_2 the ratio factor. The rotatable CCD design with three center points was performed, with the experimental results assembled in Table 6.23.

A second order linear regression with all 11 data points results in a model with Adj-R^2=0.969 and RMSE=608.9. The model coefficients assembled in Table 6.24 indicate that coefficients (x_1*x_2), and (x_1^2*x_2^2) are not statistically significant. Dropping these terms results in a better model with Adj-R^2=98.3, and RMSE=578.8. The corresponding values of the model coefficients are shown in Table 6.25. In determining whether the model can be further simplified, one notes that the highest p-value on the independent variables is 0.0549, belonging to (x_2^2). Since the p-value is greater or equal to 0.05, that term is not statistically significant at the 95.0% or higher confidence level. Consequently, one could consider removing this term from the model; this, however, was not done here since the value is close to 0.05.

Thus, the final model is:

$$y = 8972.6 + 3454.4x_1 + 1566.8x_2 - 762x_1^2 - 579.5x_2^2 \tag{6.14}$$

Table 6.22 Assumed low and high levels for the two factors (Example 6.4.1)

Factor	Low level	High level
Pressure	4	80
Ratio H_2/WF_6	2	10

[5] From Buckner et al. (1993).

Table 6.23 Results of the CCD rotatable design for two factors with 3 center points (Example 6.4.1)

x_1	x_2	y
−1	−1	3663
1	−1	9393
−1	1	5602
1	1	12488
−1.414	0	1984
1.414	0	12603
0	−1.414	5007
0	1.414	10310
0	0	8979
0	0	8960
0	0	8979

Table 6.24 Model coefficients for the second order complete model (Example 6.4.1)

Parameter	Estimate	Standard error	t-statistic	p-value
Constant	8972.6	351.53	25.5246	0.0000
x_1	3454.43	215.284	16.046	0.0001
x_2	1566.79	215.284	7.27781	0.0019
x_1*x_2	289.0	304.434	0.949303	0.3962
x_1^2	−839.837	277.993	−3.02107	0.0391
x_2^2	−657.282	277.993	−2.36438	0.0773

Table 6.25 Model coefficients for the reduced model (Example 6.4.1)

Parameter	Estimate	Standard error	t-statistic	p-value
Constant	8972.6	334.19	26.8488	0.0000
x_1	3454.43	204.664	16.8785	0.0000
x_2	1566.79	204.664	7.65544	0.0003
x_1^2	−762.044	243.63	−3.12787	0.0204
x_2^2	−579.489	243.63	−2.37856	0.0549

Table 6.24 Model coefficients for the second order complete model (Example 6.4.1)

$x_1^{2*}x_2$^2	310.952	430.6	0.722137	0.5102

It would be wise to look at the residuals and if there are any unusual ones. Figures 6.19 and 6.20 clearly indicate that one of the points (the first row, i.e., y=3663) is unusual with very high studentized residuals (recall that studentized residuals measure how many standard deviations each observed value of y deviates from a model fitted using all of the data except that observation—Sect. 5.6.2). Those greater than 3 in absolute value warrant a close look and if necessary removed prior to model fitting.

The optimal values of the two regressors associated with the maximum response are determined by taking partial derivatives and setting them to zero. This yields: x_1=2.267 and x_2=1.353. However, this optimum lies outside the experimental region and does not satisfy the sphe-

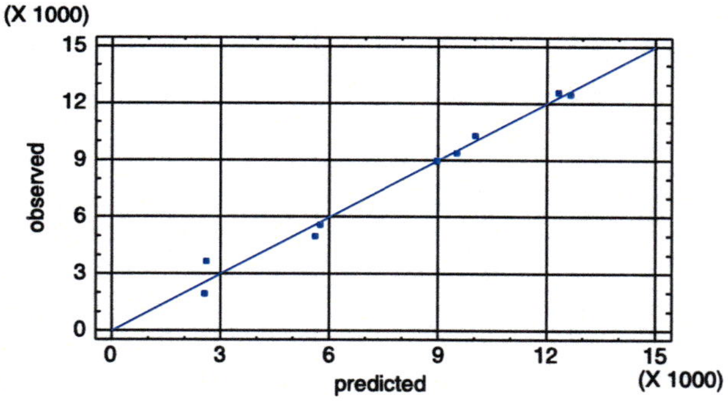

Fig. 6.19 Observed versus model predicted values

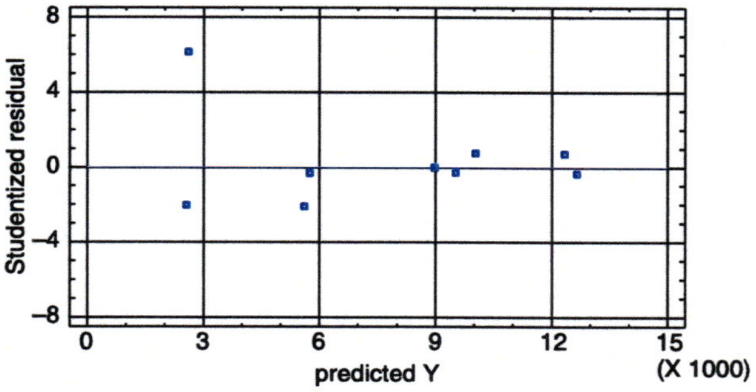

Fig. 6.20 Studentized residuals versus model predicted values (Example 6.4.1)

rical constraint, which in this case is $x_1^2 + x_2^2 \leq 2$. This is illustrated in the contour plot of Fig. 6.21. Resorting to a constrained optimization (see Sect. 7.3) results in the optimal values of the regressors: $x_1^* = 1.253$ and $x_2^* = 0.656$

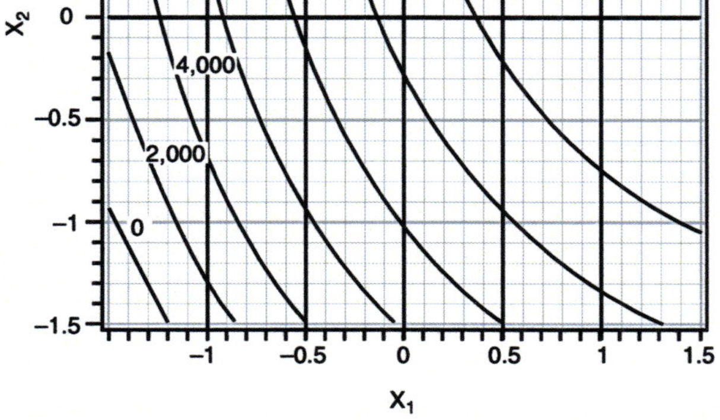

Fig. 6.21 Contour plot of Eq. 6.14

representing a maximum deposition rate $y^* = 12{,}883$. In terms of the original variables, these correspond to a pressure=[(80−4)/2+1.253*(80−4)/2)=85.6 torr and a Ratio $H_2/WF_6 = 8.0$. Finally, a confirmatory experiment would have to be conducted in the neighborhood of this optimum. ■

Problems

Pr. 6.1 Consider Example 6.2.2 where the performance of four machines was analyzed in terms of machining time with operator dexterity being a factor to be blocked. How to identify an additive linear model was also illustrated. Figure 6.2a suggests that interaction effects may be important. You will re-analyze the data to determine whether interaction terms are statistically significant or not.

Pr. 6.2[6] *Full-factorial design for evaluating three different missile systems*

A full-factorial experiment is conducted to determine which of 3 different missile systems is preferable. The propellant burning rate for 24 static firings was measured using four different propellant types. The experiment performed duplicate observations (replicate=2) of burning rates (in minutes) at each combination of the treatments. The data, after coding, is given in Table 6.26.

The following hypotheses tests are to be studied:

(a) There is no difference in the mean propellant burning rates when different missile systems are used,
(b) there is no difference in the mean propellant burning rates of the 4 propellant types,
(c) there is no interaction between the different missile systems and the different propellant types,

Pr. 6.3 *Random effects model for worker productivity*

A full-factorial experiment was conducted to study the effect of indoor environment condition (depending on such factors as dry bulb temperature, relative humidity...) on the productivity of workers manufacturing widgets. Four groups of workers were selected distinguished by such traits as age, gender,... called G1, G2, G3 and G4. The number of widgets produced over a day by two members of each group

Table 6.26 Burning rates in minutes for the (3×4) case with two replicates (Problem 6.2)

Missile System	Propellant type			
	b_1	b_2	b_3	b_4
A_1	34.0, 32.7	30.1, 32.8	29.8, 26.7	29.0, 28.9
A_2	32.0, 33.2	30.2, 29.8	28.7, 28.1	27.6, 27.8
A_3	28.4, 29.3	27.3, 28.9	29.7, 27.3	28.8, 29.1

[6] From Walpole et al. (2007) by © permission of Pearson Education.

Table 6.27 Showing the number of widgets produced by day using a replicate r=2 (Problem 6.3)

Environmental Conditions	Group number G 1	G 2	G 3	G 4
E1	227, 221	214, 259	225, 236	260, 229
E2	187, 208	181, 179	232, 198	246, 273
E3	174, 202	198, 194	178, 213	206, 219

under three different environmental conditions (E1, E2 and E3) was recorded. These results are assembled in Table 6.27.

Using 0.05 significance level, test the hypothesis that:

(a) different environmental conditions have no effect on number of widgets produced,
(b) different worker groups have no effect on number of widgets produced,
(c) there is no interaction effects between both factors,

Subsequently, identify a suitable random effects model, study model residual behavior and draw relevant conclusions.

Pr. 6.4 The thermal efficiency of solar thermal collectors decreases as their average operating temperatures increase. One of the means of improving the thermal performance is to use selective surfaces for the absorber plates which have the special property that the absorption coefficient is high for the solar radiation and low for the infrared radiative heat losses. Two collectors, one without a selective surface and another with, were tested at four different operating temperatures under replication r=4. The experimental results of thermal efficiency in % are tabulated In Table 6.28.

(a) Perform an analysis of variance to test for significant main and interaction effects,
(b) Identify a suitable random effects model,
(c) Identify a linear regression model and compare your results with those from part (b),
(d) Study model residual behavior and draw relevant conclusions.

Pr. 6.5 The close similarity between a factorial design model and a multiple linear regression model was illustrated in Example 6.3.1. You will repeat this exercise with data from Example 6.3.2.

Table 6.28 Thermal efficiencies (%) of the two solar thermal collectors (Problem 6.4)

	Mean operating temperature (°C) 80	70	60	50
Without selective surface	28, 29, 31, 32	34, 33, 35, 34	38, 39, 41, 38	40, 42, 41, 41
With selective surface	33, 36, 33, 34	38, 38, 36, 35	41, 40, 43, 42	43, 45, 44, 45

(a) Identify a multiple linear regression model and verify that the parameters of all regressors are identical to the factorial design model,
(b) Verify that model coefficients do not change when multiple linear regression is redone with the reduced model using variables coded as −1 and +1,
(c) Perform a forward step-wise linear regression and verify that you get back the same reduced model with the same coefficients.

Pr. 6.6 2^3 *factorial analysis for strength of concrete mix*
A civil construction company wishes to maximize the strength of its concrete mix with three factors or variables: A—water content, B—coarse aggregate, and C—silica. A 2^3 full factorial set of experimental runs, consistent with the nomenclature of Table 6.13, was performed. These results are assembled below:

[58.27, 55.06, 58.73, 52.55, 54.88, 58.07, 56.60, 59.57]

(a) You are asked to analyze this data so as to identify statistically meaningful terms,
(b) If the minimum and maximum range of the three factors are: A(0.3576, 0.4392), B(0.4071, 0.4353) and C(0.0153, 0.0247), develop a prediction model for this problem,
(c) Identify a multiple linear regression model and verify that the parameters of all regressors are identical to the factorial design model,
(d) Verify that model coefficients do not change when multiple linear regression is redone with the reduced model.

Pr. 6.7 As part of the first step of a response surface (RS) approach, the following linear model was identified from preliminary experimentation using two coded variables

$$y = 55 - 2.5x_1 + 1.2x_2 \quad \text{with} \quad -1 \le x_i \le +1$$

Determine the path of steepest ascent, and draw this path on a contour plot.

Pr. 6.8[7] *Predictive model inferred from* 2^3 *factorial design on a large laboratory chiller*
Table 6.29 assembles steady state data of a 2^3 factorial series of laboratory tests conducted on a 90 Ton centrifugal chiller. There are three response variables (T_{cho}—chilled water leaving the evaporator, T_{cdi}—cooling water entering the condenser, and Q_{ch}—chiller cooling load) with two levels each, thereby resulting in 8 data points without any replication. Note that there are small differences in the high and low levels of each of the factors because of operational control variability

[7] Adapted from a more extensive table from data collected by Comstock and Braun (1999). We are thankful to James Braun for providing this data.

Table 6.29 Laboratory tests from a centrifugal chiller (Problem 6.8)

Test #	Data for model development T_{cho} (°C)	T_{cdi} (°C)	Q_{ch} (kW)	COP	Data for cross-validation T_{cho} (°C)	T_{cdi} (°C)	Q_{ch} (kW)	COP
1	10.940	29.816	315.011	3.765	7.940	29.628	286.284	3.593
2	10.403	29.559	103.140	2.425	7.528	24.403	348.387	4.274
3	10.038	21.537	289.625	4.748	6.699	24.288	188.940	3.678
4	9.967	18.086	122.884	3.503	7.306	24.202	93.798	2.517
5	4.930	27.056	292.052	3.763				
6	4.541	26.783	109.822	2.526				
7	4.793	21.523	354.936	4.411				
8	4.426	18.666	114.394	3.151				

during testing. The chiller Coefficient of Performance (COP) is the response variable.

(a) Perform an ANOVA analysis, and check the importance of the main and interaction terms using the 8 data points indicated in the table,
(b) Identify the parsimonious predictive model from the above ANOVA analysis,
(c) Identify a least square regression model with coded variables and compare the model coefficients with those from the model identified in part (b),
(d) Generate model residuals and study their behavior (influential outliers, constant variance and near-normal distribution),
(e) Reframe both models in terms of the original variables and compare the internal prediction errors,
(f) Using the four data sets indicated in the table as holdout points meant for cross-validation, compute the NMSE, RMSE and CV values of both models. Draw relevant conclusions.

References

Beck, J.V. and K.J., Arnold, 1977. *Parameter Estimation in Engineering and Science*, John Wiley and Sons, New York

Box, G.E.P., W.G. Hunter, and J.S. Hunter, 1978. *Statistics for Experimenters*, John Wiley & Sons, New York.

Buckner, J., D.J. Cammenga and A. Weber, 1993. Elimination of TiN peeling during exposure to CVD tungsten deposition process using designed experiments, *Statistics in the Semiconductor Industry*, Austin, Texas: SEMATECH, Technology Transfer No. 92051125A-GEN, vol.I, 4-445-3-71.

Comstock, M.C. and J.E. Braun, 1999. Development of Analysis Tools for the Evaluation of Fault Detection and Diagnostics in Chillers, ASHRAE Research Project 1043-RP; also, Ray W. Herrick Laboratories. Purdue University. HL 99-20: Report #4036-3, December.

Devore J., and N. Farnum, 2005. *Applied Statistics for Engineers and Scientists*, 2nd Ed., Thomson Brooks/Cole, Australia.

Mandel, J., 1964. *The Statistical Analysisof Experimental Data*, Dover Publications, New York.

Montgomery, D.C., 2009, *Design and Analysis of Experiments*, 7th Edition, John Wiley & Sons, New York.

Walpole, R.E., R.H. Myers, S.L. Myers, and K. Ye, 2007, *Probability and Statistics for Engineers and Scientists*, 8th Ed., Prentice-Hall, Upper Saddle River, NJ.

Optimization Methods

7

This chapter provides an introductory overview of traditional optimization techniques as applied to engineering applications. These apply to situations where the impact of uncertainties is relatively minor, and can be viewed as a subset of decision-making problems which are treated in Chap. 12. After defining the various terms used in the optimization literature, calculus based methods covering both analytical as well as numerical techniques are reviewed. Subsequently, different solutions to problems which can be grouped as linear, quadratic or non-linear programming are described, while highlighting the differences between them and the methods used to solve such problems. A complete illustrative example of how to set up an optimization problem for a combined heat and power system is presented. Methods that allow global solutions as against local ones are described. Finally, the important topic of dynamic optimization is covered which applies to optimizing a trajectory, i.e., to discrete situations when a series of decisions have to be made to define or operate a system composed of distinct stages, such that a decision is made at each stage with the decisions at later stages not affecting the performance of earlier ones. There is a vast amount of published material on the subject of optimization, and this chapter is simply meant as a brief overview.

7.1 Background

One of the most important tools for both design and operation of engineering systems is *optimization* which corresponds to the case of decision-making under low uncertainty. This branch of applied mathematics, also studied under "operations research" (OR), is the use of specific methods where one tries to minimize or maximize a global characteristic (say, the cost or the benefit) whose variation is modeled by an objective function. The set-up of the optimization problem involves both the formulation of the objective function but as importantly, the explicit and complete consideration of a set of constraints. Optimization problems arise in almost all branches of industry or society, *e.g.*, in product and engineering process design, production scheduling, logistics, traffic control and even strategic planning.

Optimization in an engineering context involves certain basic elements consisting of some or all of the following: (i) the framing of a situation or problem (for which a solution or a course of action is sought) in terms of a mathematical model often called the objective function; this could be a simple expression, or framed as a decision tree model in case of multiple outcomes (deterministic or probabilistic) or sequential decision making stages, (ii) defining the range constraints to the problem in terms of input parameters which may be dictated by physical considerations, (iii) placing bounds on the solution space of the output variables in terms of some practical or physical constraints, (iv) defining or introducing uncertainties in the input parameters and in the types of parameters appearing in the model, (v) mathematical techniques which can solve such models efficiently (short execution times) and accurately (unbiased solutions); and (vi) sensitivity analysis to gauge the robustness of the optimal solution to various uncertainties.

Framing of the mathematical model involves two types of uncertainties: *epistemic* or lack of complete knowledge of the process or system which can be reduced as more data is acquired, and *aleotory uncertainty* which has to do with the stochasticity of the process, and cannot be reduced by collecting more data. These notions are discussed at some length in Chap. 12 while dealing with decision analysis. This chapter deals with traditional optimization techniques as applied to engineering applications which are characterized by low aleatory and epistemic uncertainty. Further, recall the concept of *abstraction* presented in Sect. 1.2.3 in the context of formulating models. It pertains to the process of deciding on the level of detail appropriate for the problem at hand without, on one hand, over-simplification which may result in loss of important system behavior predictability, while on the other hand, avoiding the formulation of an overly-detailed model which may result in undue data and computational resources as well as time spent in understanding the model assumptions and results generated. The same concept of abstraction

T. Agami Reddy, *Applied Data Analysis and Modeling for Energy Engineers and Scientists*,
DOI 10.1007/978-1-4419-9613-8_7,

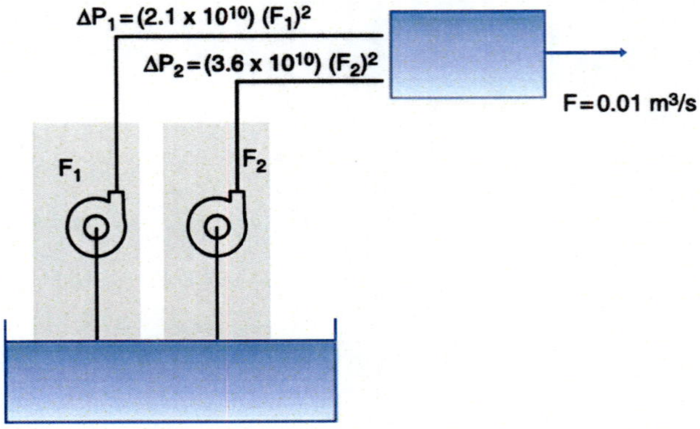

Fig. 7.1 Pumping system whose operational cost is to be optimized (Example 7.1.1)

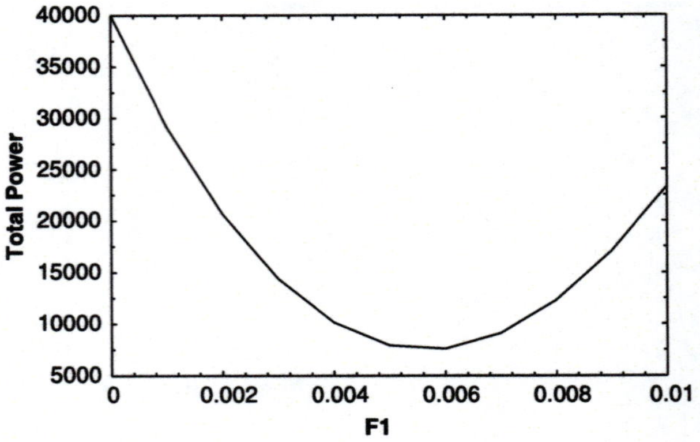

Fig. 7.2 One type of post-optimality analysis involves plotting the objective function for Total Power to evaluate the shape of the curve near the optimum. In this case, there is a broad optimum indicating that the system can be operated near-optimally over this range without much corresponding power penalty

also applies to the science of optimization. One has to set a level of abstraction commensurate with the complexity of the problem at hand and the accuracy of the solution sought.

Let us consider a problem framed as finding the optimum of a continuous function. There could, of course, be the added complexity of considering several discrete options; but each option has one or more continuous variables requiring proper control so as to achieve a global optimum. A simple example (Pr. 1.7 from Chap. 1) will be used to illustrate this case.

Example 7.1.1: *Simple example involving function minimization*

Two pumps with parallel networks (Fig. 7.1) deliver a volumetric flow rate F=0.01 m³/s of water from a reservoir to the destination. The pressure drops in Pascals (Pa) of each network are given by: $\Delta p_1 = (2.1).10^{10}.F_1^2$ and $\Delta p_2 = (3.6).10^{10}.F_2^2$ where F_1 and F_2 are the flow rates through each branch in m³/s. Assume that both the pumps and their motor assemblies have equal efficiencies $\eta_1 = \eta_2 = 0.9$. Let P_1 and P_2 be the electric power in Watts (W) consumed by the two pump-motor assemblies.

Since, power consumed is equal to volume flow rate times the pressure drop, the objective function to be minimized is the sum of the power consumed by both pumps:

$$J = \left[\frac{\Delta p_1.F_1}{\eta_1} + \frac{\Delta p_2 F_2}{\eta_2}\right]$$

$$\text{or } J = \left[\frac{(2.1).10^{10}.F_1^3}{0.9} + \frac{(3.6).10^{10}.F_2^3}{0.9}\right] \tag{7.1}$$

The sum of both flows is equal to 0.01 m³/s, and so F_2 can be eliminated in Eq. 7.1. Thus, the sought-after solution is the value of F_1 which minimizes the objective function J:

$$J^* = \text{Min }\{J\} = \text{Min}\left\{\frac{(2.1).10^{10}.F_1^3}{0.9} + \frac{(3.6).10^{10}.(0.01 - F_1^3)}{0.9}\right\} \tag{7.2}$$

subject to the constraint that $F_1 > 0$.

From basic calculus, $dJ/dF_1 = 0$ would provide the optimum solution from where F_1=0.00567 m³/s and F_2=0.00433 m³/s, and the total power of both pumps=7501 W. The extent to which non-optimal performance is likely to lead to excess power can be gauged (referred to as *post-optimality analysis*) by simply plotting the function J vs F_1 (Fig. 7.2) In this case, the optima is rather broad; the system can be operated such that F_1 is in the range of 0.005–0.006 m³/s without much power penalty. On the other hand, sensitivity analysis would involve a study of how the optimum value is affected by certain parameters. For example, Fig. 7.3 shows that varying the efficiency of pump 1 in the range of 0.85–0.95 has negligible impact on the optimal result. However, this may not be the case for some other variable. A systematic study of how various parameters impact the optimal value falls under sensitivity analysis, and there exist formal methods of inves-

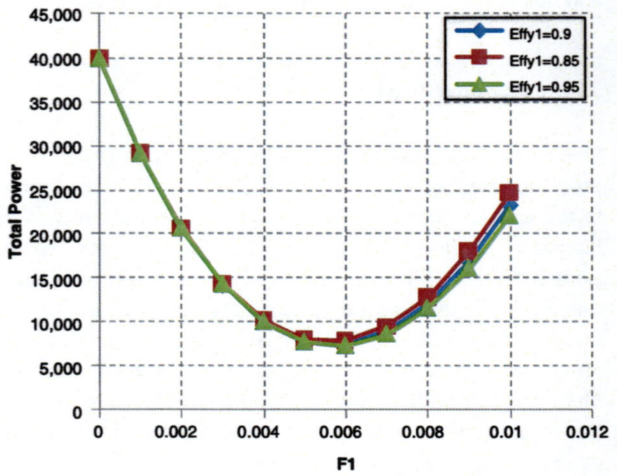

Fig. 7.3 Sensitivity analysis with respect to efficiency of pump 1 on the overall optimum

tigating this aspect of the problem. Finally, note that this is a very simple optimization problem with a simple imposed constraint which was not even considered during the optimization. ■

7.2 Terminology and Classification

7.2.1 Basic Terminology and Notation

A mathematical formulation of an optimization problem involving control of an engineering system consists of the following terms or categories:

(i) *Decision variables* or process variables, say (x_1, x_2 … x_n) whose respective values are to be determined. These can be either discrete or continuous variables;
(ii) *Control variables*, which are the physical quantities which can be varied by hardware according to the numerical values of the decision variables sought. Determining the "best" numerical values of these variables is the basic intent of optimization;
(iii) *Objective function*, which is an analytical formulation of an appropriate measure of performance of the system (or characteristic of the design problem) in terms of decision variables;
(iv) *Constraints* or restrictions on the values of the decision variables. These can be of two types: *non-negative constraints*, for example, flow rates cannot be negative; and *functional constraints* (also called structural constraints), which can be equality, non-equality or range constraints that specify a range of variation over which the decision variables can be varied. These can be based on direct considerations (such as not exceeding capacity of energy equipment, limitations of temperature & pressure control values,…) or on indirect ones (when mass and energy balances have to be satisfied).
(v) *Model parameters* are constants appearing in constraints and objective equations.

Establishing the objective function is often simple. The real challenge is usually in specifying the complete set of constraints. A *feasible solution* is one which satisfies all the stated constraints, while an infeasible solution is one where at least one constraint is violated. The *optimal solution* is a feasible solution that has the most favorable value (either maximum or minimum) of the objective function, and it is this solution which is being sought after. The optimal solutions can be a single point or even several points. Also, some problems may have no optimal solutions at all. Figure 7.4 shows a function to be maximized subject to several constraints (six in this case). Note that there is no feasible solution and one of the constraints has to be relaxed or the problem reframed. In some optimization problems, one can obtain several feasible solutions. This is illustrated in Fig. 7.5 where several combinations of the two variables, which define the line segment shown, are possible optima.

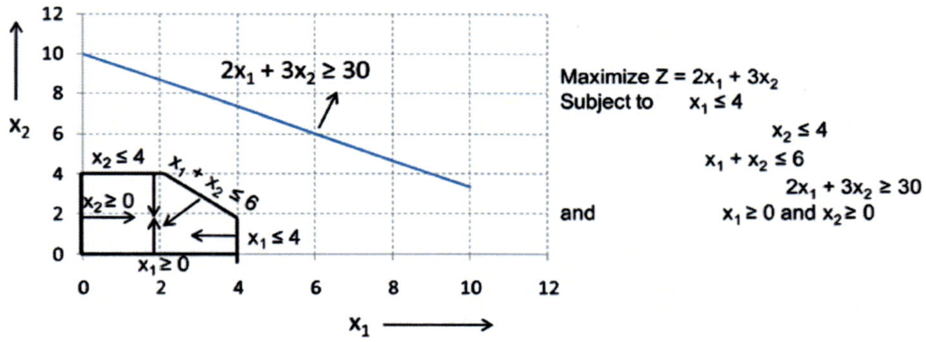

Fig. 7.4 An example of a constrained optimization problem with no feasible solution

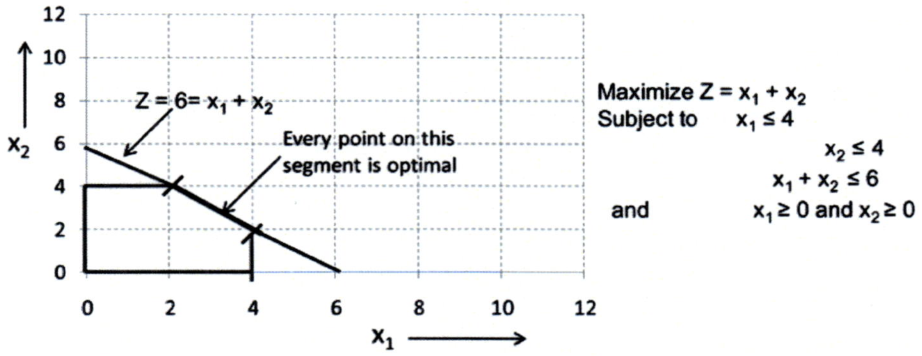

Fig. 7.5 An example of a constrained optimization problem with more than one feasible solution

Sometimes, an optimal solution may not necessarily be the one selected for implementation. A "*satisficing*" solution (combination of words "satisfactory" and "optimizing") may be the solution which is selected for actual implementation and reflects the difference between theory (which yields an optimal solution) and reality faced (due to actual implementation issues, heuristic constraints which cannot be expressed mathematically, the need to treat unpredictable occurrences, risk attitudes of the owner/operator,....). Some practitioners also refer to such solutions as "near-optimal" though this has a sort of negative connotation.

7.2.2 Traditional Optimization Methods

Optimization methods can be categorized in a number of ways:

(i) *Linear vs. non-linear*. Linear optimization problems involve linear models and a linear objective function and linear constraints. The theory is well developed, and solutions can be found relatively quickly and robustly. There is an enormous amount of published literature on this type of problem, and it has found numerous practical applications involving upto several thousands of independent variables. There are several well-know techniques to solve them (the Simplex method in Operations Research used to solve a large set of linear equations being the best known). However, many problems from engineering to economics require the use of non-linear models or constraints, in which case, non-linear programming techniques have to be used. In some cases, non-linear models (for example, equipment models such as chillers, fans, pumps and cooling towers) can be expressed as quadratic models, and algorithms more efficient than non-linear programming ones have been developed; this falls under quadratic programming methods. Calculus-based methods are often used for finding the solution of the model equations.

(ii) *Continuous vs. discontinuous*. When the objective functions are discontinuous, calculus based methods can break down. In such cases, one could use non-gradient based methods or even heuristic based computational methods such as simulated annealing or genetic algorithms. The latter are very powerful in that they can overcome problems associated with local minima and discontinuous functions, but they need long computing times and a certain amount of knowledge of such techniques on the part of the analyst. Another form of discontinuity arises when one or more of the variables are discrete as against continuous. Such cases fall under the classification known as *integer or discrete programming*.

(iii) *Static vs. dynamic*. If the optimization is done with time not being a factor, then the procedure is called static. However, if optimization has to be done over a time period where decisions can be made at several sub-intervals of that period, then a dynamic optimization method is warranted. Two such examples are when one needs to optimize the route taken by a salesman visiting different cities as part of his road trip, or when the operation of a thermal ice storage supplying cooling to a building has to be optimized during several hours of the day during which high electric demand charges prevail. Whenever possible, analysts make simplifying assumptions to make the optimization problem static.

(iv) *Deterministic vs stochastic*. This depends on whether one neglects or considers the uncertainty associated with various parameters of the objective function and the constraints. The need to treat these uncertainties together, and in a probabilistic manner, rather than one at a time (as is done in a sensitivity analysis) has led to the development of several numerical techniques, the Monte Carlo technique being the most widely used (Sect. 12.2.7).

7.2.3 Types of Objective Functions

Single criterion optimization is one where a single over-riding objective function can be formulated. It is used in the majority of optimization problems. For example, an industrialist is considering starting a factory to assemble photovoltaic (PV) cells into PV modules. Whether to invest or not, and if yes, at what capacity level are issues which can both be framed as a single criterion optimization problem. However, if maximizing the number of jobs created is another (altruistic) objective, then the problem must be treated as a multi-criteria decision problem. Such cases are discussed in Sect. 12.2.6.

7.2.4 Sensitivity Analysis or Post Optimality Analysis

Model parameters are often not known with certainty, and could be based on models identified from partial or incomplete observations, or they could even be guess-estimates. The optimum is only correct insofar as the model is accurate, and the model parameters and constraints reflective of the actual situation. Hence, the optimal solution determined needs to be reevaluated in terms of how the various types of uncertainties affect it. This is done by sensitivity analysis which determines range of values:

(i) of the parameters over which the optimal solutions will remain unchanged (allowable range to stay near- optimal). This would flag critical parameters which may require closer investigation, refinement and monitoring;

(ii) over which the optimal solution will remain feasible with adjusted values for the basic variables (allowable range to stay feasible, i.e. the constraints are satisfied). This would help identify influential constraints.

Further, the above evaluations can be performed by adopting:

(i) individual parameter sensitivity, where one parameter at a time in the original model is varied (or *perturbed*) to check its effect on the optimal solution;
(ii) total sensitivity (also called parametric programming) involves the study of how the optimal solution changes as many parameters change simultaneously over some range. Thus, it provides insight into "correlated" parameters and trade off in parameter values. Such evaluations are conveniently done using Monte Carlo methods (Sect. 12.2.7).

7.3 Calculus-Based Analytical and Search Solutions

Calculus-based solution methods can be applied to both linear and non-linear problems, and are the ones to which undergraduate students are most likely to be exposed to. They can be used for problems where the objective function and the constraints are differentiable. These methods are also referred to as *classical or traditional optimization* methods as distinct from machine-learning methods. A brief review of calculus-based analytical and search methods is presented below.

7.3.1 Simple Unconstrained Problems

The basic calculus of the univariate unconstrained optimization problem can be extended to the multivariate case of dimension n by introducing the gradient vector ∇ and by recalling that the gradient of a scalar y is defined as:

$$\nabla y = \frac{\partial y}{\partial x_1}\mathbf{i}_1 + \frac{\partial y}{\partial x_2}\mathbf{i}_2 + \ldots + \frac{\partial y}{\partial x_n}\mathbf{i}_n \quad (7.3)$$

where $\mathbf{i}_1, \mathbf{i}_2 \ldots \mathbf{i}_n$ are unit vectors, and y is the objective function and is a function of n variables: $y = y(x_1, x_2, \ldots, x_n)$

With this terminology, the condition for optimality of a continuous function y is simply:

$$\nabla y = 0 \quad (7.4)$$

However, the optimality may be associated with stationary points which could be minimum, maximum, saddle or ridge points. Since objective functions are conventionally expressed as a minimization problem, one seeks the minimum of the objective function. Recall that for the univariate case, assuring that the optimal value found is a minimum (and not a maximum or a saddle point) involves computing the numerical value of the second derivative at this optimal point, and checking that its value is positive. Graphically, a minimum for a continuous function is found (or exists) when the function is convex, while a saddle point is found for a combination function (see Fig. 7.6). In the multivariate optimization case, one checks whether the Hessian matrix (i.e., the second derivative matrix which is symmetrical) is positive definite or not. It is tedious to check this condition by hand for any matrix whose dimensionality is greater 2, and so computer programs are invariably used for such problems. A simple hand calculation method (which works well for low dimension problems) for ascertaining whether the optimal point is a minimum (or maximum) is to simply perturb the optimal solution vector obtained, compute the objective function and determine whether this value is higher (or lower) than the optimal value found.

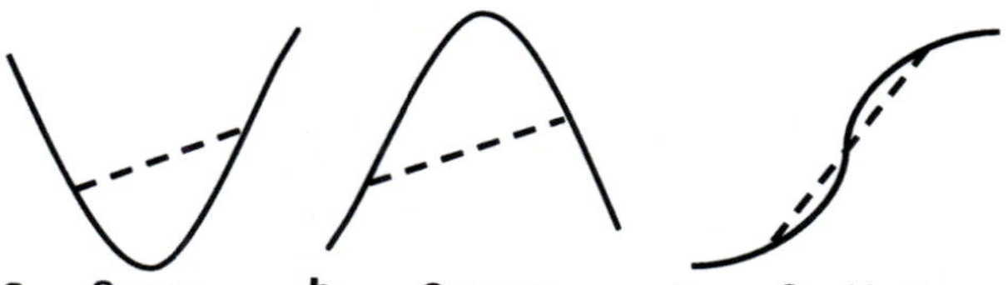

Fig. 7.6 Illustrations of convex, concave and combination functions. A convex function is one where every point on the line joining any two points on the graph does not lie below the graph at any point. A combination function is one which exhibits both convex and concave behavior during different portions with the switch-over being the saddle point

Example 7.3.1: Determine the minimum value of the following function: $y = \frac{1}{4x_1} + 8x_1^2 x_2 + \frac{1}{x_2^2}$.

First, the two first order derivatives are found:

$$\frac{\partial y}{\partial x_1} = -\frac{1}{4x_1^2} + 16x_1x_2 \quad \text{and} \quad \frac{\partial y}{\partial x_2} = 8x_1^2 - \frac{2}{x_2^3}$$

Setting the above two expressions to zero and solving results in: $x_1^* = 0.2051$ and $x_2^* = 1.8114$ at which condition $y^* = 2.133$. It is left to the reader to verify these results, and check whether this is indeed the minimum. ■

7.3.2 Problems with Equality Constraints

Most practical problems have constraints in terms of the independent variables, and often these assume the form of *equality constraints* only. There are several semi-analytical techniques which allow the constrained optimization problem to be reformulated into an unconstrained one, and the manner in which this is done is what differentiates these methods. In such a case, one does not need generalized optimization solver approaches requiring software programs which need greater skill to use properly and may take longer to solve.

The simplest approach is the *direct substitution method* where for a problem involving "n" variables and "m" equality constraints, one tries to eliminate the m constraints by direct substitution, and solve the objective function using the unconstrained solution method described above. This approach was used in Example 7.1.1.

Example 7.3.2:[1] *Direct substitution method*
Consider the simple optimization problem stated as:

$$\text{Minimize} \quad f(\mathbf{x}) = 4x_1^2 + 5x_2^2 \tag{7.5a}$$

$$\text{subject to:} \quad 2x_1 + 3x_2 = 6 \tag{7.5b}$$

Either x_1 or x_2 can be eliminated without difficulty. Say, the constraint equation is used to solve for x_1, and then substituted into the objective function. This yields the unconstrained objective function:

$$f(x_2) = 14x_2^2 - 36x_2 + 36.$$

The optimal value of $x_2^* = 1.286$ from which, by substitution, $x_1^* = 1.071$. The resulting value of the objective function is: $f(\mathbf{x})^* = 12.857$.

This simple problem allows a geometric visualization to better illustrate the approach. As shown in Fig. 7.7, the objective function is a paraboloid shown on the z axis with x_1 and x_2 being the other two axes. The constraint is represented by a plane surface which intersects the paraboloid as shown.

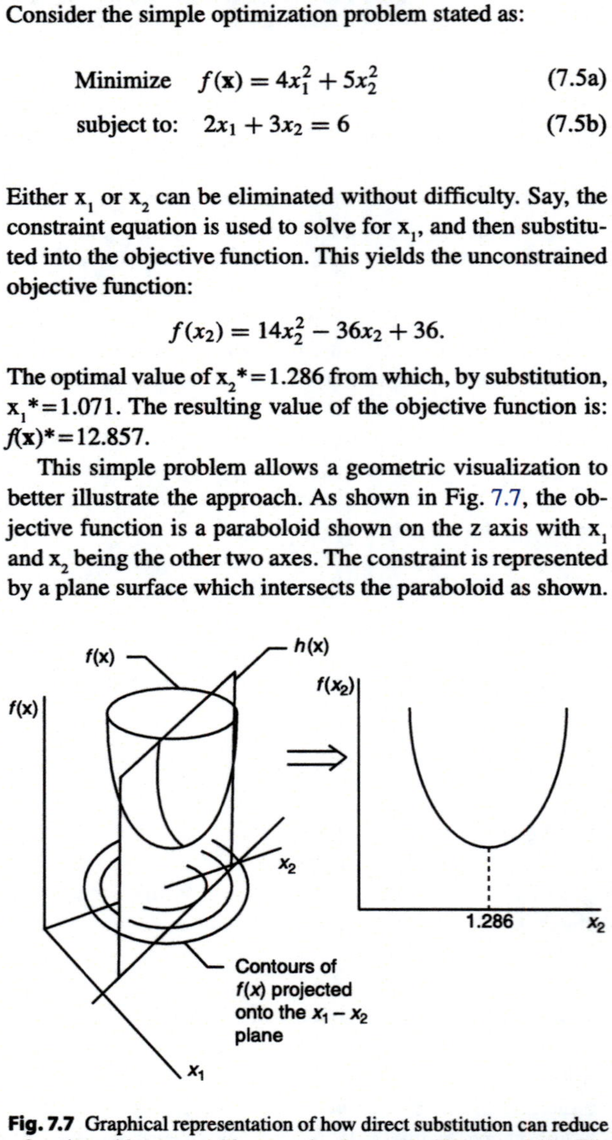

Fig. 7.7 Graphical representation of how direct substitution can reduce a function with two variables x_1 and x_2 into one with one variable. The unconstrained optimum is at (0, 0) at the center of the contours. (From Edgar et al. 2001 by permission of McGraw-Hill)

[1] From Edgar et al. (2001) by permission of McGraw-Hill.

The resulting intersection is a parabola whose optimum is the solution of the objective function being sought after. Notice how this constrained optimum is different from the unconstrained optimum which occurs at (0, 0) (Fig. 7.7). ■

The above approach requires that one variable be first explicitly expressed as a function of the remaining variables, and then eliminated from all equations; this procedure is continued till there are no more constraints. Unfortunately, this is not an approach which is likely to be of general applicability in most problems.

7.3.3 Lagrange Multiplier Method

A more versatile and widely used approach which allows the constrained problem to be reformulated into an unconstraint one is the *Lagrange multiplier approach*. Consider an optimization problem involving an objective function y, a set of n decision variable **x** and a set of m equality constraints h(**x**):

$$\text{Minimize } y = y(\mathbf{x}) \quad \text{objective function} \tag{7.6a}$$

$$\text{subject to } h(\mathbf{x}) = 0 \quad \text{equality constraints} \tag{7.6b}$$

The Lagrange multiplier method simply absorbs the equality constraints into the objective function, and states that the optimum occurs when the following modified objective function is minimized:

$$\begin{aligned} J^* &\equiv \min\{y(\mathbf{x})\} \\ &= y(\mathbf{x}) - \lambda_1.h_1(\mathbf{x}) - \lambda_2.h_2(\mathbf{x}) - \ldots = 0 \end{aligned} \tag{7.7}$$

where the quantities $\lambda_1, \lambda_2, \ldots$ are called the Lagrange multipliers. The optimization problem, thus, involves minimizing y with respect to both **x** and the Lagrange multipliers.

The cost of eliminating the constraints comes at the price of increasing the dimensionality of the problem from n to (n+m), or stated differently, one is now seeking the values of (n+m) variables as against n only which will optimize the function y.

A simple example with one equality constraint serves to illustrate this approach. The objective function $y = 2x_1 + 3x_2$ is to be optimized subject to the constraint $x_1 x_2^2 = 48$. Figure 7.8 depicts this problem visually with the two variables being the two axes and the objective function being represented by a series of parallel lines for different assumed values of y. Since the constraint is a curved line, the optimal solution is obviously the point where the tangent vector of the curve (shown as a dotted line) is parallel to these lines (shown as point A).

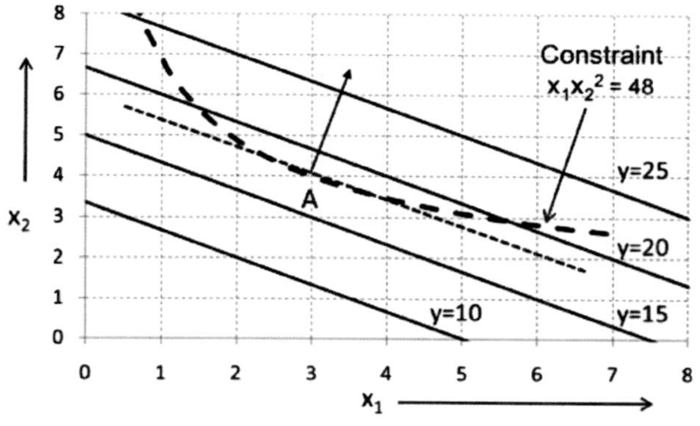

Fig. 7.8 Optimization of the linear function $y=2x_1+3x_2$ subject to the constraint shown. The problem is easily solved using the Lagrange multiplier method to yield optimal values of ($x_1^*=3$, $x_2^*=4$). Graphically, the optimum point A occurs where the constraint function and the lines of constant y (which, in this case, are linear) have a common normal indicated by the arrow at A

Example 7.3.3: *Optimizing a solar water heater system using the Lagrange method*
A solar water heater consisting of a solar collector and storage tank is to be optimized for lowest first cost consistent with the following specified system performance. During the day the storage temperature is to be raised from 30°C (equal to the ambient temperature T_a) to a temperature T_{max}, while during the night heat is to be withdrawn from storage such that the storage temperature drops back to 30°C for the next day's operation. The system should be able to store 20 MJ of thermal heat over a typical day of the year during which H_T, the incident radiation over the collector operating time (assumed to be 10 h) is 12 MJ/m². The collector performance characteristics[2] are $F_R\eta_0=0.8$ and $F_RU_L=4.0$ W/m².°C. The costs of the solar subsystem components are fixed cost C_b=$ 600, collector area proportional cost C_a=$ 200/m² of collector area, and storage volume proportional cost C_s=$ 200/m³ of storage volume.

Assume that the average inlet temperature to the collector is equal to the arithmetic mean of T_{max} and T_a.

Let A_C (m²) and V_S (m³) be the collector area and storage volume respectively. The objective function is:

$$J = 600 + 200.A_C + 200.V_S \tag{7.8}$$

The constraint of the daily amount of solar energy collected is expressed as:

$$Q_C = A_C[H_T F_R \eta_0 - U_L(T_{Ci} - T_a).\Delta t] \tag{7.9a}$$

where Δt is the number of seconds during which the collector operates.

[2] See Pr. 5.6 for a description of the solar collector model.

or

$$(20)(10^6) = A_C\left\{(12)(10^6)(0.8) - (4)(3600)(10).\left(\frac{T_{max}+30}{2} - 30\right)\right\}$$

or

$$20.10^6 = (11.76 - 0.072.T_{max})A_C \tag{7.9b}$$

A heat balance on the storage over the day yields

$$Q_C = Mc_p(T_{max} - T_{initial})$$

or

$$20.10^6 = V_S(1000).(4190)(T_{max} - 30) \tag{7.10}$$

$$\text{from which } T_{max} = \frac{20}{(4.19)V_S} + 30$$

Substituting this back into the constraint Eq. 7.9b results in

$$A_C\left(9.6 - \frac{0.344}{V_S}\right) = 20$$

This allows the combined Lagrangian objective function to be deduced as:

$$J = 600 + 200.A_C + 200.V_S - \lambda\left\{A_C\left(9.6 - \frac{0.344}{V_S}\right) - 20\right\} \tag{7.11}$$

The resulting set of Lagrangian equations are:

$$\frac{\delta J}{\delta A_C} = 0 = 200 - \left(9.6 - \frac{0.344}{V_S}\right)\lambda$$

$$\frac{\delta J}{\delta V_S} = 0 = 200 - \lambda A_C \frac{0.344}{V_2^2}$$

$$\frac{\delta J}{\delta \lambda} = 0 = A_C\left(9.6 - \frac{0.344}{V_S}\right) - 20$$

Solving this set yields the sought-after optimal values: $A_C^*=2.36$ m² and $V_S^*=0.308$ m³. The value of the Lagrangian multiplier is $\lambda=23.58$, and the corresponding initial cost J*=$ 1134. The Lagrangian multiplier can be interpreted as the sensitivity coefficient, which in this example corresponds to the marginal cost of solar thermal energy. In other words, increasing the thermal requirements by 1 MJ would lead to an increase of λ=$23.58 in the initial cost of the optimal solar system. ■

7.3.4 Penalty Function Method

Another widely used method for constrained optimizing is the use of *penalty factors* where the problem is converted to an unconstrained one. Consider the problem defined by

Eq. 7.6 with the possibility that the constraints can be inequality constraints as well. Then, a new unconstrained function is framed as:

$$J^* \equiv \min\{y(\mathbf{x})\} = \min\left\{y(\mathbf{x}) + \sum_{i=1}^{k} P_i(h_1)^2\right\} \quad (7.12)$$

where P_i is called the penalty factor for condition i with k being the number of constraints. The choice of this penalty factor provides the relative weighting of the constraint compared to the function. For high P_i values, the search will satisfy the constraints but move more slowly in optimizing the function. If P_i is too small, the search may terminate without satisfying the constraints adequately. The penalty factor can assume any function, but the nature of the problem may often influence the selection. For example, when a forward model is being calibrated with experimental data, one has some prior knowledge of the numerical values of the model parameters. Instead of simply performing a calibration based on minimizing the least square errors, one could frame the problem as an unconstrained penalty factor problem where the function to be minimized consists of a term representing the root sum of square errors, and of the penalty factor term which may be the square deviations of the model parameters from their respective estimates. The following example illustrates this approach while it is further described in Sect. 10.5.2 when dealing with non-linear parameter estimation.

Example 7.3.4: Minimize the following problem using the penalty function approach:

$$y = 5x_1^2 + 4x_2^2 \quad \text{s.t.} \quad 3x_1 + 2x_2 = 6 \quad (7.13)$$

Let us assume a simple form of the penalty factor and frame the problem as:

$$\begin{aligned} J^* &= \min(J) = \min[y + P(h)^2] \\ &= \min[5x_1^2 + 4x_2^2 + P(3x_1 + 2x_2 - 6)^2] \end{aligned} \quad (7.14)$$

$$\text{Then:} \quad \frac{\partial J}{\partial x_1} = 10x_1 + 6P(3x_1 + 2x_2 - 6) = 0 \quad (7.15a)$$

$$\text{and} \quad \frac{\partial J}{\partial x_2} = 8x_2 + 4P(3x_1 + 2x_2 - 6) = 0 \quad (7.15b)$$

Solving these results in $x_1 = \dfrac{6x_2}{5}$

which, when substituted back into the constraint of Eq. 7.13, yields:

$$x_2 = \frac{36}{\dfrac{12}{P} + \dfrac{108}{5} + 12}$$

The optimal values of the variables are found as the limiting values when P becomes very large. In this case, $x_2^* = 1.071$ and, subsequently, from Eq. 7.15b $x_1^* = 1.286$; these are the optimal solutions sought. ■

7.4 Numerical Search Methods

Most practical optimization problems will have to be solved using numerical search methods. This implies that the search towards an optimum is done systematically and progressively using an iterative approach to gradually zoom onto the optimum. Because the search is performed at discrete points, the precise optimum will not be known. The best that can be achieved is to specify an *interval of uncertainty* which is the range of x values in which the optimum is known to exist. The search methods differ depending on whether the problem is univariate/multivariate or unconstrained/constrained or have continuous/discontinuous functions. Numerous search methods have been proposed ranging from the exhaustive search method to genetic algorithms.

A general method is the *lattice search* (Fig. 7.9) where one starts at one point in the search space (shown as point 1), calculates values of the function in a number of points around the initial point (points 2–9), and moves to the point which has the lowest value (shown as point 5). This process is repeated till the overall minimum is found. Sometimes, one may use a coarse grid search initially; find the optimum within an interval of uncertainty, then repeat the search using a finer

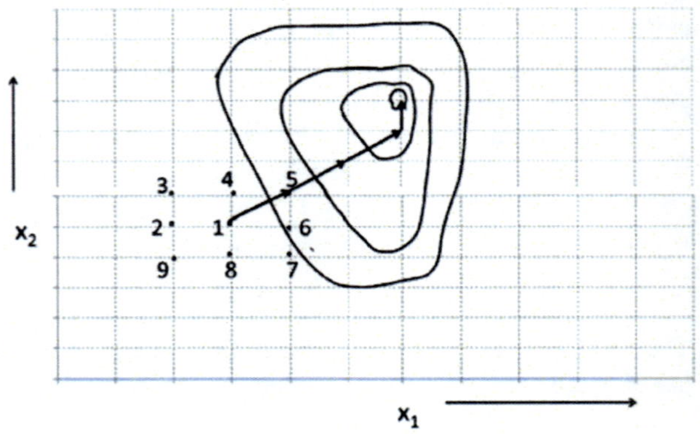

Fig. 7.9 Conceptual illustration of finding a minimum point of a bi-variate function using a lattice search method. From an initial point 1, the best subsequent move involves determining the function values around that point at discrete grid points (points 2 through 9) and moving to the point with the lowest function value

grid. Note that this is not a calculus-based method, and is not very efficient computationally, especially for higher dimension problems. However, it is robust and simple to implement.

Calculus-based methods are generally efficient for problems with continuous functions; these are referred to as *hill-climbing* methods. Strictly speaking, hill-climbing leads one to a maximum; the algorithm is easily modified for valley-descending as needed for minimization. Two general approaches are described below while other methods can be found in Stoecker (1989) or Beveridge and Schechter (1970).

(a) *Univariate search* (Fig. 7.10): this is a calculus based method where the function is optimized with respect to one variable at a time. One starts by using some preliminary values for all variables other than the one being optimized, and finds the optimum value for the selected variable (shown as x_1). One then selects a second variable to optimize while retaining this optimal value of the first variable, and finds the optimal value of the second variable, and so on for all remaining variables. Though the problem reduces to a simple univariate search, it may be computationally inefficient for higher dimension problems, and worse, may not converge to the global optimum if the search space is not symmetrical. The entire process often requires more than one iteration, as shown in Fig. 7.10.

(b) *Steepest-descent search* (Fig. 7.11): this is a widely used approach because of its efficiency. The computational algorithm involves three steps: one starts with a guess value (represented by point 1) which is selected somewhat arbitrarily but, if possible, close to the optimal value; one then evaluates the gradient of the function at the current point by computing the partial derivatives either analytically or numerically; and, finally, one moves along this gradient (hence, the terminology "steepest") by deciding, somewhat arbitrarily, on the step size. The relationship between the step sizes Δx_i and the partial derivatives $(\partial y/\partial x_i)$ is:

$$\frac{\Delta x_1}{\partial y/\partial x_1} = \frac{\Delta x_2}{\partial y/\partial x_2} = \ldots = \frac{\Delta x_i}{\partial y/\partial x_i} \quad (7.16)$$

Steps 2–3 are performed iteratively till the minimum (or maximum) point is reached. A note of caution is that too large a step size can result in numerical instability; while too small a step size increases computation time.

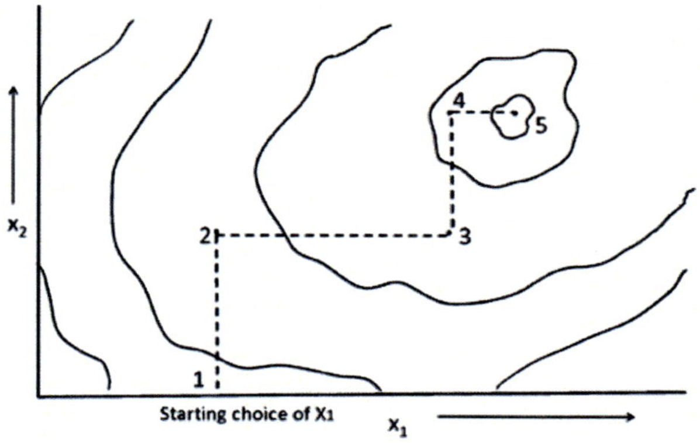

Fig. 7.10 Conceptual illustration of finding a minimum point of a bivariate function using the univariate search method. From an initial point 1, the gradient of the function is used to find the optimal point value of x_2 keeping x_1 fixed, and so on till the optimal point 5 is found

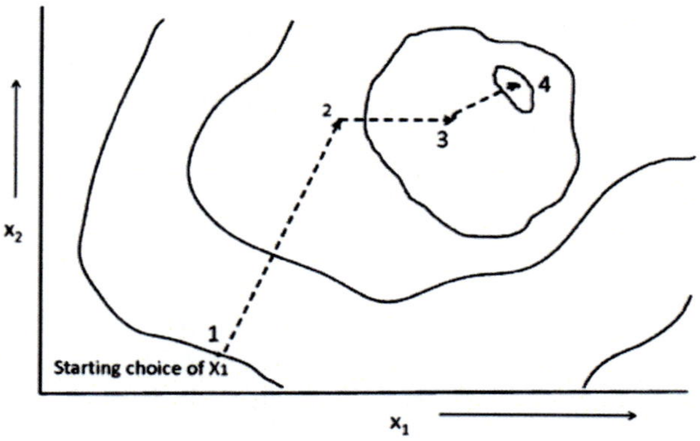

Fig. 7.11 Conceptual illustration of finding a minimum point of a bivariate function using the steepest descent search method. From an initial point 1, the gradient of the function is determined and the next search point determined by moving in that direction, and so on till the optimal point 4 is found

Extensions of this general approach to optimization problems with non-linear constraints and with inequality constraints are also well developed, and will not be discussed here. The reader can refer, for example, to Stoecker (1989) or to Beveridge and Schechter (1970) for the widely used non-linear hemstitching method which have its roots in the Lagrangian approach.

Example 7.4.1: *Illustration of the univariate search method*
Consider the following function with two variables which is to be minimized using the univariate search process starting with an initial value of x_2=3.

$$y = x_1 + \frac{16}{x_1.x_2} + \frac{x_2}{2}$$

First, the partial derivatives are derived:

$$\frac{\partial y}{\partial x_1} = 1 - \frac{16}{x_2.x_1^2} \quad \text{and} \quad \frac{\partial y}{\partial x_2} = -\frac{16}{x_1.x_2^2} + \frac{1}{2}$$

Next, the initial value of x_2=3 is used to find the next iterative value of x_1 from the $(\partial y/\partial x_1)$ function as follows:

$$\frac{\partial y}{\partial x_1} = 1 - \frac{16}{(3).x_1^2} = 0 \text{ from where } x_1 = \frac{4\sqrt{3}}{3} = 2.309$$

The other partial derivative is finally used with this value of x_1 to yield:

$$-\frac{16}{\frac{4\sqrt{3}}{3}.x_2^2}. + \frac{1}{2} = 0 \text{ from where } x_2 = 3.722$$

The new value of x_2 is now used for the next cycle, and the iterative process repeated until consecutive improvements turn out to be sufficiently small to suggest convergence. It is left to the reader to verify that the optimal values are $(x_1^* = 2, x_2^* = 4)$. ■

Example 7.4.2: *Illustration of the steepest descent method* Consider the following function with three variables to be minimized:

$$y = \frac{72x_1}{x_2} + \frac{360}{x_1x_3} + x_1x_2 + 2x_3$$

Assume a starting point of (x_1=5, x_2=6, x_3=8). At this point, the value of the function is y=115.

These numerical values are inserted in the expressions for the partial derivatives:

$$\frac{\partial y}{\partial x_1} = \frac{72}{x_2} - \frac{360}{x_3.x_1^2} + x_2 = \frac{72}{6} - \frac{360}{(8).(5)^2} + 6 = 16.2$$
$$\frac{\partial y}{\partial x_2} = -\frac{72x_1}{x_2^2} + x_1 = -\frac{72(5)}{(6)^2} + 5 = -5$$
$$\frac{\partial y}{\partial x_3} = -\frac{360}{x_1x_3^2} + 2 = -\frac{360}{(5)(8)^2} + 2 = 0.875$$

In order to compute the next point, a step size has to be assumed. Let us arbitrarily assume Δx_1=−1 (whether to take a positive or a negative value for the spatial step is not obvious—in this case a minimum is sought, and the reader can verify that taking a negative value results in a decrease in the function value y).

Applying Eq. 7.16 results in $\frac{-1}{16.2} = \frac{\Delta x_2}{-5} = \frac{\Delta x_3}{0.875}$ from where Δx_2=0.309, Δx_3=−0.054. Thus, the new point is (x_1=4, x_2=6.309, x_3=7.946). The reader can verify that the new point has resulted in a decrease in the functional value from 115 to 98.1. Repeated use of the search method will gradually result in the optimal value being found. ■

7.5 Linear Programming

The above sections were meant to provide a basic background to optimization problems using rather simplistic examples. Most practical problems would be much more complex than these, and would require a formal mathematically-based numerical approach to find the optimal solution. *Numerical efficiency* (or power) of a method of solution involves both robustness of the solution and fast execution times. Optimization problems which can be framed into a linear problem (even at the expense of a little loss in accuracy) have great numerical efficiency. Only if the objective function and the constraints are *both* linear functions is the problem designated as a linear optimization problem; otherwise it is deemed a non-linear function. The objective function can involve one or more functions to be either minimized or maximized (either objective can be treated identically since it is easy to convert one into the other).

There is a great deal of literature on methods to solve linear programs, which are referred to as linear programming methods. The *Simplex algorithm* is the most popular technique for solving linear problems and involves matrix inversion along with directional iterations; it also provides the necessary information for performing a sensitivity analysis at the same time. Hence, formulating problems as linear problems (even when they are not strictly so) has a great advantage in the solving phase.

The standard form of linear programming problems is:

$$\text{minimize } f(\mathbf{x}) = \mathbf{c}^T\mathbf{x} \quad (7.17a)$$

$$\text{subject to : } g(\mathbf{x}) : \mathbf{A}\mathbf{x} = \mathbf{b} \quad (7.17b)$$

where **x** is the column vector of variables of dimension n, **b** that of the constraint limits of dimension m, **c** that of the cost coefficients of dimension n, and **A** is the (m x n) matrix of constraint coefficients. Notice that no inequality constraints appear in the above formulation. This is because inequality constraints can be re-expressed as equality constraints by introducing additional variables, called ***slack variables***. The order of the optimization problem will increase, but the efficiency in the subsequent numerical solution approach outweighs this drawback. The following simple example serves to illustrate this approach.

Example 7.5.1: Express the following linear two-dimensional problem into standard matrix notation:

$$\text{Maximize } f(\mathbf{x}) : 3186 + 620x_1 + 420x_2 \quad (7.18a)$$

$$\text{subject to} \quad \begin{aligned} &g_1(\mathbf{x}) : 0.5x_1 + 0.7x_2 \le 6.5 \\ &g_2(\mathbf{x}) : 4.5x_1 - x_2 \le 35 \\ &g_3(\mathbf{x}) : 2.1x_1 + 5.2x_2 \le 60 \end{aligned} \quad (7.18b)$$

with range constraints on the variables x_1 and x_2 being that these should not be negative.

This is a problem with two variables (x_1 and x_2). However, three slack variables need to be introduced in order to reframe the three inequality constraints as equality constraints.

This makes the problem into one with five unknown variables. The three inequality constraints are rewritten as:

$$\begin{aligned} g_1(\mathbf{x}) &: 0.5x_1 + 0.7x_2 + x_3 = 6.5 \\ g_2(\mathbf{x}) &: 4.5x_1 - x_2 + x_4 = 35 \\ g_3(\mathbf{x}) &: 2.1x_1 + 5.2x_2 + x_5 = 60 \end{aligned} \quad (7.19)$$

Hence, the terms appearing in the standard form (eq. 7.17) are:

$$\mathbf{c} = [\ -620 \quad -420 \quad 0 \quad 0 \quad 0]^T,$$

$$\mathbf{x} = [\ x_1 \quad x_2 \quad x_3 \quad x_4 \quad x_5\ ]^T,$$

$$A = \begin{bmatrix} 0.5 & 0.7 & 1 & 0 & 0 \\ 4.5 & -1 & 0 & 1 & 0 \\ 2.1 & 5.2 & 0 & 0 & 1 \end{bmatrix},$$

$$\mathbf{b} = [\ 6.5 \quad 35 \quad 60\ ]^T$$

Note that the objective function is recast as a minimization problem simply by reversing the signs of the coefficients. Also, the constant does not appear in the optimization since it can be simply added to the optimal value of the function at the end. Step-by-step solutions of such optimization problems are given in several textbooks such as Edgar et al. (2001), Hillier and Liberman (2001) and Stoecker (1989). A commercial optimization software program was used to determine the optimal value of the above objective function:

$$f^*(x) = 9803.8$$

Note that in this case, since the inequalities are "less than or equal to zero", the numerical values of the slack variables (x_3, x_4, x_5) will be positive. The optimal values for the primary variables are: $x_1^* = 8.493, x_2^* = 3.219$, while those for the slack variables are $x_3^* = 0, x_4^* = 0, x_5^* = 25.424$ (implying that constraints 1 and 2 in Eq. 7.18b have turned out to be equality constraints). ■

7.6 Quadratic Programming

A function of dimension n (i.e., there are n variables) is said to be quadratic when:

$$f(\mathbf{x}) = a_{11}x_1^2 + a_{12}x_1x_2 + \ldots + a_{ij}x_ix_j + \ldots + a_{nn}x_n^2 = \sum_{i=1}^{n}\sum_{j=1}^{n} a_{ij}x_ix_j \quad (7.20)$$

where the coefficients are constants. Consider the function which is quadratic in two variables:

$$f(x_1, x_2) = 4x_1^2 + 12x_1x_2 - 6x_2x_1 - 8x_2^2 \quad (7.21)$$

It can be written in matrix form as:

$$f(x_1, x_2) = [x_1 \quad x_2]\begin{bmatrix} 4 & -6 \\ 12 & -8 \end{bmatrix}\begin{bmatrix} x_1 \\ x_2 \end{bmatrix}$$

Because $12x_1x_2 - 6x_2x_1 = 6x_1x_2$, the function can also be written as:

$$f(x_1, x_2) = [x_1 \quad x_2\ ]\begin{bmatrix} 4 & 6 \\ 6 & -8 \end{bmatrix}\begin{bmatrix} x_1 \\ x_2 \end{bmatrix}$$

Thus, the coefficient matrix of any quadratic function can be written in symmetric form.

Quadratic programming problems differ from the linear ones in only one aspect: the objective function is quadratic in its terms (while constraints and equalities must be linear). Even though such problems can be treated as non-linear problems, formulating the problem as a quadratic one allows for greater numerical efficiency in finding the solutions. Numerical algorithms to solve such problems are similar to the linear programming ones; a modified Simplex method has been developed which is quite popular.

The standard notation is:

$$\text{Minimize } f(\mathbf{x}) = \mathbf{cx} + \frac{1}{2}\mathbf{x}^T\mathbf{Qx} \quad (7.22)$$

$$\text{Subject to : } g(\mathbf{x}) : \mathbf{Ax} = \mathbf{b} \quad (7.23)$$

Note that the coefficient matrix Q is symmetric, as explained above.

Example 7.6.1: Express the following problem in standard quadratic programming formulation:

$$\text{Min } J = 4x_1^2 + 4x_2^2 + 8x_1x_2 - 60x_1 - 45x_2 \quad (7.24)$$

subject to $2x_1 + 3x_2 = 30$

In this case:

$$\mathbf{c} = [-60 \quad -45],\ \mathbf{x} = [x_1 \quad x_2]^T,$$

$$\mathbf{Q} = \begin{bmatrix} 8 & -8 \\ -8 & 8 \end{bmatrix}$$

$$\mathbf{A} = [2 \quad 3],\ \mathbf{b} = [30]$$

It is easy to verify that:

$$\begin{aligned} \mathbf{x}^T\mathbf{Qx} &= [x_1 \quad x_2]\begin{bmatrix} 8 & -8 \\ -8 & 8 \end{bmatrix}\begin{bmatrix} x_1 \\ x_2 \end{bmatrix} \\ &= [(8x_1 - 8x_2)(-8x_1 + 8x_2)]\begin{bmatrix} x_1 \\ x_2 \end{bmatrix} \\ &= 8x_1^2 - 16x_1x_2 + 8x_2^2 \end{aligned}$$

The reader can verify that the optimal solution corresponds to: $x_1^* = 3.75, x_2^* = 7.5$ which results in a value of J*=−56.25 for the objective function. ■

7.7 Non-linear Programming

Non-linear problems are those where either the objective function or any of the constraints are non-linear. Such problems represent the general case and are of great interest. A widely used notation to describe the *complete nonlinear optimization problem* is to frame the problem as:

$$\text{Minimize} \quad y = y(\mathbf{x}) \quad \text{objective function} \tag{7.25}$$

$$\text{subject to} \quad h(\mathbf{x}) = 0 \quad \text{equality constraints} \tag{7.26}$$

$$g(\mathbf{x}) \leq 0 \quad \text{inequality constraints} \tag{7.27}$$

$$l_j \leq x_j \leq u_j \quad \text{range or boundary constraints} \tag{7.28}$$

where **x** is a vector of p variables.

The constraints $h(\mathbf{x})$ and $g(\mathbf{x})$ are vectors of independent equations of dimension m_1 and m_2 respectively. If these constraints are linear, then the problem is said to have linear constraints; otherwise it is said to have non-linear constraints. The constraints l_j & u_j are lower and upper bounds of the decision variables of dimension m_3. Thus, the *total number of constraints* is $m=m_1+m_2+m_3$.

Several software codes have been developed to solve constrained non-linear problems which involve rather sophisticated numerical methods. The two most widely used techniques are: (i) the sequential quadratic programming (SQP) method which uses a Taylor-series quadratic expansion of the functions around the current search point, and the solution is then found successively; and (ii) the generalized reduced gradient (GRG) method which is a sophisticated search method whose basic approach was previously described in Sect. 7.4. The interested reader can refer to Venkataraman (2002) for a detailed description of these as well as other numerical optimization methods.

A note of caution is needed at this stage. Quite often, the optimal point is such that some of the constraints turn out to be redundant (but one has no way of knowing that from before), and even worse that the problem is found to have either no solution or an infinite number of solutions. In such cases, for a unique solution to be found, the optimization problem may have to be reformulated in such a manner that, while being faithful to the physical problem being solved, some of the constraints are relaxed or reframed. This is easier said than done, and even the experienced analyst may have to evaluate alternative formulations before deciding on the most appropriate one.

7.8 Illustrative Example: Combined Heat and Power System

Combined Heat and Power (CHP) components and systems are described in several books and technical papers (for example, see Petchers 2003). Such systems meant for commercial/institutional buildings (BCHP) involve multiple prime movers, chillers and boilers and require more careful and sophisticated equipment scheduling and control methods as compared to those in industrial CHP. This is due to the large variability in building thermal and electric loads as well as the equipment scheduling issue. *Equipment scheduling* involves determining which of the numerous equipment combinations to operate, i.e., is concerned with starting or stopping prime movers, boilers and chillers. The second and lower level type of control is called *supervisory control* which involves determining the optimal values of the control parameters (such as loading of primemovers, boilers and chillers) under a specific combination of equipment schedule. The complete optimization problem, for a given hour, would qualify as a *mixed-integer programming problem* because of the fact that different discrete pieces of equipment may be on or off. The problem can be tackled by using algorithms appropriate for *mixed integer programming* where certain variables can only assume integer values (for example, for the combinatorial problem, a certain piece of equipment can be on or off—which can be designated as 0 or 1 respectively). Usually, such algorithms are not too efficient, and, a typical approach in engineering problems when faced with mixed integer problems is to treat integer variables as continuous, and solve the continuous problem. The optimal values of these variables are then simply found by rounding to the nearest integer. In the particular case of the BCHP optimization problem, another approach which works well for medium sized situations (involving, say, up to about 100 combinations) is to proceed as follows. For a given hour specified by the climatic variables and the building loads, all the feasible combinations of equipment are first generated. Subsequently, a lower level optimization is done for each of these feasible equipment combinations, from which the best combination of equipment to meet the current load can be selected.

Currently, little optimization of the interactions among systems is done in buildings. Heuristic control normally used by plant operators often results in off-optimal operation due to the numerous control options available to them as well as due to dynamic, time-varying rate structures and relative changes in gas and electricity prices. Though reliable estimates are lacking in the technical literature, the consensus is that 5–15% of cost savings can be realized if these multiple-equipment BCHP plants were operated more rationally and optimally.

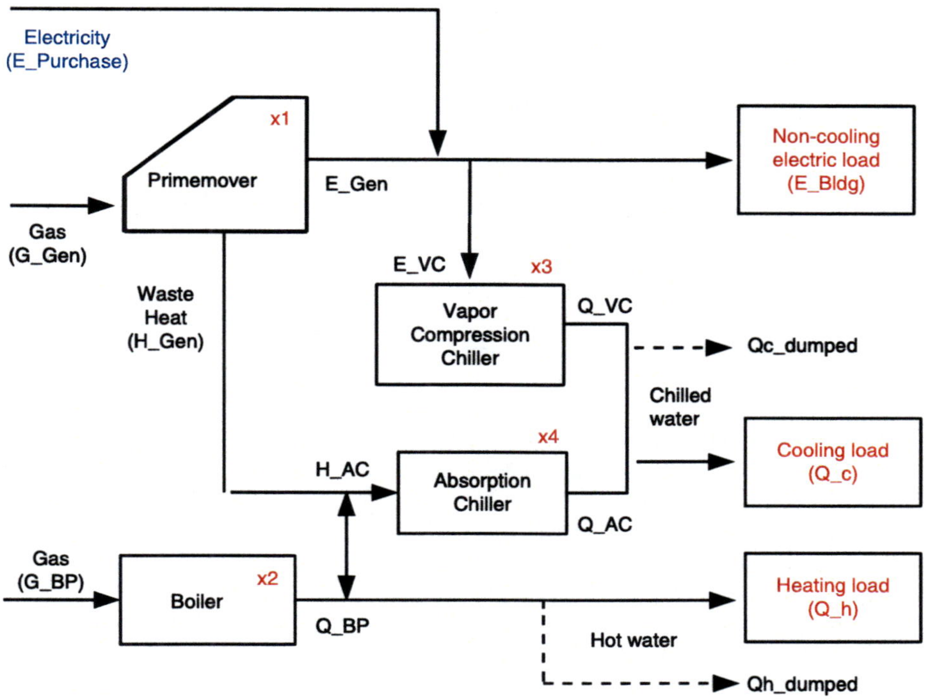

Fig. 7.12 Generic schematic of a combined heat and power (CHP) system meant to supply cooling, heating and part of the electricity needs of a building. Sub-system interactions and nomenclature used are also shown. The terms x1, x2, x3 and x4 are control variables which represent the loading fractions of the prime-mover, boiler, vapor compression chiller and the absorption chiller respectively. (From Reddy and Maor 2009)

Figure 7.12 is a generic schematic of how the important subsystems of a BCHP system (namely, primemovers, vapor compression chillers, absorption chillers and boilers) are often coupled to serve the building loads, (Reddy and Maor 2009). The static optimization case, *without utility sell-back,* involves optimizing the operating cost of the BCHP system for each time step, i.e. each hour, while it meets the building loads: the non-cooling electric load (E_{Bldg}),the thermal cooling load (Q_c) and the building thermal heating load (Q_h). Assume that the cost components only include steady state hourly energy costs for electricity and gas. So, the quantity to be minimized, J, is the total cost of energy consumption, summed over all components that are operating plus the equipment operation and maintenance (O&M) costs.

The **objective function** to be optimized for a particular time step (or hour) and for a specific BCHP system combination:

$$J^* = \min\{J\} = \min\{J_1 + J_2 + J_3\} \tag{7.29}$$

where

- the cost associated with gas use is

$$J_1 = (G_{Gen} + G_{BP}).C_g$$

- the cost associated with electric use is

$$J_2 = E_{Purchase}.C_e \tag{7.30}$$

- the O&M cost is

$$J_3 = M_{OM}$$

subject to the **inequality constraints** that the building loads must be met (called functional constraints[3]):

- building thermal cooling load

$$Q_{AC} + Q_{VC} \geq Q_c \tag{7.31}$$

- building thermal heating load

$$Q_{BP} + H_{Gen} - H_{AC} \geq Q_h$$

- building non-cooling electric load

$$(E_{Purchase} + E_{Gen} - E_{VC} - E_p) \geq E_{Bldg}$$

and subject to boundary or **range constraints** that

- the primemover part load ratio

$$x_{Gen,\min} \leq x_{Gen} \leq 1.0$$

- the vapor compression chiller part load ratio

$$x_{VC,\min} \leq x_{VC} \leq 1.0 \tag{7.32}$$

- the absorption chiller part load ratio

$$x_{AC.,\min} \leq x_{AC} \leq 1.0$$

- the boiler plant part load ratio

$$x_{BP,\min} \leq x_{BP} \leq 1.0$$

[3] The inequality constraints allow for energy dumping. In practice, this is almost never done, in which case, one could recast the three constraints of Eq. 7.31 as equality constraints.

where:

C_e unit energy cost of electricity use
C_g unit energy cost of natural gas
E_{Gen} actual electric power output of primemover
$E_{Purchase}$ amount of purchased electricity
E_p parasitic electric use of the BCHP plant (pumps, fans, etc.)
E_{VC} electricity consumed by the vapor compression chiller
G_{BP} amount of natural gas heat consumed by the boiler plant
G_{Gen} amount of natural gas heat consumed by the primemover
H_{AC} heat supplied to the absorption chiller
H_{Gen} total recovered waste heat from the primemover
M_{OM} operation and maintenance costs of the BCHP equipment which are operated
Q_{AC} amount of cooling supplied by the absorption chiller
Q_{BP} amount of heating supplied by the boiler plant
Q_{VC} amount of cooling supplied by the vapor compression chiller
x part load ratios of the four major pieces of equipment, i.e., actual load divided by their rated capacity. The subscripts: AC—absorption chiller, BP—boiler plant, Gen—generator, VC—vapor compression chiller. Note that the lower bound values of the range constraints in Eq. 7.32 are specific to the type of equipment and are limits below which such equipment should not be operated.

One also needs to model the energy consumption for each of the components as a function of the component's characteristics and of the controlled variables. The models to be used for optimization can be of three types:

(a) *detailed simulation models* originally developed for providing insights into design issues and which are most appropriate for research purposes;
(b) *semi-empirical component models* that combine deterministic modeling involving thermodynamic and heat transfer considerations with some empirical curve-fit models so as to provide some degree of modeling detail of sub-components of the major equipment, such as the effect of back-pressure on turbine performance, individual heat exchanger performance, power for gas compression,…;
(c) *semi-empirical inverse models*, which can be either grey-box or black-box depending on whether the underlying physics is used during model development. The traditional black-box approach using rated equipment performance along with polynomial models to capture part load performance is illustrated below in view of its conceptual simplicity.

A simple manner of modeling part-load performance of various equipment is given below. Part-load *electrical efficiency of reciprocating engines and microturbines* can be modeled as:

$$y_{Gen} = a_0 + a_1.x_{Gen} + a_2.x_{Gen}{}^2 \tag{7.33}$$

where y_{Gen} is the relative electrical efficiency=(actual efficiency/rated efficiency),

x_{Gen} is the relative power output = (actual power/rated power)

$$= (E_{Gen}/E''_{Gen}) \tag{7.34}$$

with the supbscript (″) denoting rated conditions.

The numerical values of the part-load model coefficients are known from manufacturer data. Since electrical efficiency of a prime mover is taken to be the electrical power output divided by the gas heat input, the expression for the *natural gas heat input* is:

$$G_{Gen} = E_{Gen}.\frac{G''_{Gen}}{E''_{Gen}}.\frac{1}{y_{Gen}} \tag{7.35a}$$

or

$$G_{Gen} = G''_{Gen}.x_{Gen}.\frac{1}{(a_0 + a_1.x_{Gen} + a_2.x_{Gen}{}^2)} \tag{7.35b}$$

The amount of waste heat which can be recovered from the primemover under part-load conditions is also needed during the simulation. Under part-load, primemover electrical efficiency degrades, and consequently a larger fraction of the supplied gas energy will appear as waste thermal heat. If one assumes that the primemover is designed such that the ratio of recovered waste heat to total waste heat is constant during its entire operation, then to a good approximation, one can model the *recovered thermal energy* under part-load operation H_{Gen} as:

$$H_{Gen} = \frac{H''_{Gen}}{G''_{Gen}}.\frac{G_{Gen}}{y_{Gen}} \tag{7.36}$$

where y_{Gen} is the relative efficiency defined earlier by Eq. 7.33.

Chiller part-load performance factor (PLF) can be modeled as (Braun 2006):

$$\begin{aligned} PLF &= b_0 + b_1.PTR + b_2.PTR^2 \\ &\quad + b_3PLR + b_4PLR^2 + b_5.PLR.PTR \end{aligned} \tag{7.37}$$

where the numerical values of the model coefficients b_i can be found from manufacturer data. Since the type of power input to the vapor compression and absorption chillers are different, the PLF for vapor compression and absorption chillers are defined as:

$$PLF_{VC} \equiv y_{VC} = \frac{E_{VC}}{E^*_{VC}} \text{ and } PLF_{AC} \equiv y_{AC} = \frac{H_{AC}}{H^*_{AC}} \tag{7.38}$$

where E_{VC} is the electric power consumed by the vapor compression chiller, and H_{AC} is the thermal heat input to the absorption chiller.

Instead of using the symbol PLR which is the part-load ratio=(actual thermal cooling load / rated thermal cooling load), the symbols x_{VC} and x_{AC} can be used to denote the part load ratios of the vapor compression and absorption chillers respectively.

$$x_{VC}(\text{or } x_{AC}) = \frac{Q_{VC}}{Q^*_{VC}}\left(\text{or } \frac{Q_{AC}}{Q^*_{AC}}\right) \tag{7.39}$$

Finally, PTR is the part-load temperature ratio of the entering condenser water temperature= $\frac{T_{cdi}}{T^*_{cdi}}$

The *boiler efficiency* is also conveniently modeled following polynomial relations:

$$y_{BP} = c_0 + c_1 x_{BP} + c_2 x_{BP}{}^2 \tag{7.40a}$$

where

x_{BP} is the part load ratio=boiler heat output by its rated value

$$\equiv \frac{Q_{BP}}{Q^*_{BP}} \tag{7.40b}$$

y_{BP} is the heat input ratio or the ratio of fuel energy input to heat output under operating condition to that under rated condition

$$= \left(\frac{G_{BP}}{Q_{BP}}\right) / \left(\frac{G^*_{BP}}{Q^*_{BP}}\right) \tag{7.40c}$$

Rearranging, one gets

$$G_{BP} = G^*_{BP} \cdot \left(\frac{Q_{BP}}{Q^*_{BP}}\right)(d_1 + d_2 x_{BP} + d_3 x_{BP}{}^2) \tag{7.41}$$

The optimization for a given period (say, a hour) would involve determining the numerical values of the four part load ratios (Eq. 7.32), which minimize the cost of operation while meeting the stated constraints. The four optimal part load ratios would allow $E^*_{Gen}, Q^*_{VC}, Q^*_{AC}, Q^*_{BP}$ to be determined from where $E^*_{Purchase}, G^*_{Gen}, G^*_{BP}$ can be deduced to finally yield J*. From a practical viewpoint, the BCHP plant has to be optimally controlled over a time horizon (which could be a day, or several hours during a day) and not just over a given period. This requires that the optimization be redone for each time period in order to achieve optimal control over the entire time horizon. In practice, there are a number of operational constraints in equipment scheduling (start-up losses, standby operation,...) which make the problem more complex than the simple static optimization approach described above (Reddy and Maor 2009).

7.9 Global Optimization

Certain types of optimization problems can have local (or sub-optimal) minima, and the optimization methods described earlier can converge to such local minima closest to the starting point, and never find the global solution. Further, certain optimization problems can have non-continuous first-order derivatives in certain search regions, and calculus based methods break down. Global optimization methods are those which can circumvent such limitations, but can only guarantee a close approximation to the global optimum (often this is not a major issue). Unfortunately, they generally require large computation times. These methods fall under two general categories (Edgar et al. 2001):

(a) *Exact methods* which include such methods as the branch-and-bound-methods and multistart methods. Most commercial non-linear optimization software programs have the multistart capability built-in whereby the search for the optimum solution is done automatically from many starting points. This is a conceptually simple approach though its efficient implementation requires robust methods of sampling the search space for starting points that do not converge to the same local optima, and also to implement rules for stopping the search process;

(b) *Heuristic search* methods are those which rely on some rules of thumb or "heuristics" to gradually reach an optimum, i.e. an iterative or adaptive improvement algorithm is central. They incorporate algorithms which circumvent the situation of non-improving moves and disallow previously visited states from being revisited. Again, there is no guarantee that a global optimum will be reached, and so often the computation stops after a certain number of computations have been completed. There are three well-known methods which fall in this category: Tabu search, simulated annealing (SA), and genetic algorithms (GA). Because of its increasing use in recent years, the last method is briefly described below.

While Tabu search and simulated annealing operate by transforming a single solution at a given step, GA works with a set of solutions $P(\mathbf{x}_i)$ called a population consisting of an array of individual members x_i, also called chromosomes, which are defined by a certain number of parameter values "p" (Burmeister 1998). This p-dimension problem is to be minimized with variables which could be binary or continuous. The GA algorithm is meant to work with: (i) unconstrained problems (constrained problems need to be reframed, for example, by adopting a penalty factor approach), and (ii) with binary variables (a continuous variable can be converted to higher order binary variable by discretizing its range of variability into m ranges and defining m new binary variables to replace the one continuous one). An initial population of size 2n to 4n starting vectors (or chromosomes or strings) is selected (heuristically) as starting values. An objective function or fitness function to be minimized is computed for each initial or parent string, and a subset of the strings which are "fitter" i.e., which yield a lower numerical value of the objective function to be minimized is retained. Successive iterations (or generations) are performed by either combining two or more fit individuals (called crossover) or by changing an individual (called mutation) in an effort to gradually minimize the function. This procedure is repeated several thousands of time until the solution converges. Because the random search was inspired by the process of natural selection underlying the evolution of natural organisms, this optimization method is called genetic algorithm. Clearly, the process is extremely computer intensive, especially when continuous variables are involved. Sophisticated commercial software is available, but the proper use of this method requires some understanding of the mathematical basis, and the tradeoffs available to speed convergence.

7.10 Dynamic Programming

Dynamic programming is a recursive technique developed to handle a type of problem where one is optimizing a trajectory rather than finding an optimum point. The term "dynamic" is used to reflect the fact that subsequent choices are affected by earlier ones. Thus, it involves multi-stage decision making of discrete processes or continuous functions which can be approximated or decomposed into stages. It is not a simple extension of the "static" or single-stage optimization methods discussed earlier, but one which, as shown below, involves solution methods that are much more computationally efficient. Instead of solving the entire problem at once, the sub-problems associated with individual stages are solved one after the other. The stages could be time intervals or spatial intervals. For example, determining the optimal flight path of a commercial airliner travelling from city A to city B which minimizes fuel consumption while taking into consideration vertical air density gradients (and hence, drag effects), atmospheric disturbances and other effects is a problem in dynamic programming.

Consider the following classic example of a travelling salesman, which has been simplified for easier conceptual understanding (see Fig. 7.13). A salesman starts from city A and needs to end his journey at City D but he is also required to visit two intermediate cities B and C of his choosing among several possibilities (in this problem, three possibilities: B1, B2, B3 at stage B; and C1, C2 and C3 at stage C). The travel costs to each of the cities at a given stage, from each of the cities from the previous stage, are specified. Thus, this problem consists of four stages (A,B,C and D) and three states (three different possible cities). The computational algorithm involves starting from the destination (city D) and working backwards to starting city A (see Table 7.1). The first calculation step involves adding the costs involved to travel from city D to cities C1, C2 and C3. The second calculation step involves determining costs from D through each of the cities C1, C2 and C3 and on to the three possibilities at stage B. One then identifies the paths through each of the cities C1, C2 and C3 which are the minimum (shown with an asterix). Thus, path D-C1-B2 is cheaper than paths D-C1-B1 and D-C1-B3. For the third and final step, one limits the calculation to these intermediate sub-optimal paths and performs three calculations only. The least cost path among these three is the optimal path sought (shown as path D-C3-B1-A in Table 7.1). Note that one does not need to compute the other six possible paths at the third stage, which is where the computational savings arise.

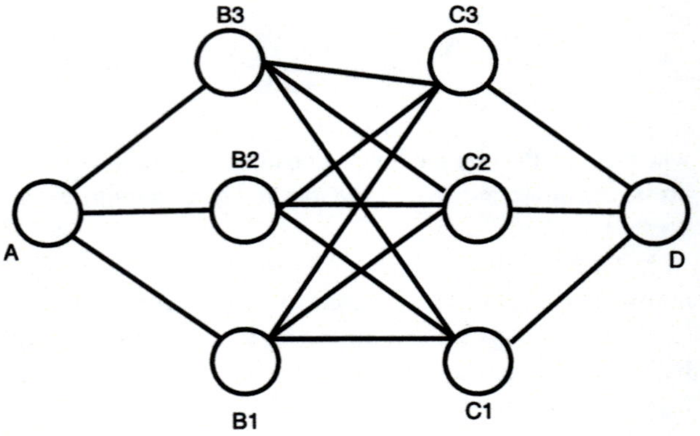

Fig. 7.13 Flow paths for the traveling salesman problem who starts from city A and needs to reach city D with the requirement that he visit one city among the three options under groups B and C

Table 7.1 Solution approach to the travelling salesman problem. Each calculation step is associated with a stage and involves determining the cumulative cost till that stage is reached

Start	First calculation step	Second calculation step	Third calculation step	Optimal path
D	D-C1	D-C1-B1		
		D-C1-B2*	D-C1-B2-A	
		D-C1-B3		
	D-C2	D-C2-B1		
		D-C2-B2*	D-C2-B2-A	
		D-C2-B3		
	D-C3	D-C3-B1*	D-C3-B1-A*	←
		D-C3-B2		
		D-C3-B3		

Note: The cells with a * are the optimal paths at each step

It is obvious that the computational savings increase for problems with increasing number of stages and states. If all possible combinations were considered for a problem involving n *intermediate* stages (excluding the start and end stages) with m states each, the total number of enumerations or possibilities would be about (m^n). On the other hand, for the dynamic programming algorithm described above, the total number would be approximately n(m x n). Thus, for n=m=4, all possible routes would require 256 calculations as against about 64 for the dynamic programming algorithm.

Basic features which characterize the dynamic programming problem are (Hillier and Lieberman 2001):

(a) the problem can be divided into stages with a policy decision made at each stage,
(b) each stage has a number of states associated with the beginning of that stage,
(c) the effect of the policy decision at each stage is to transform the current state to a state associated with the beginning of the next stage,
(d) the solution procedure is to divide the problem into stages, and given the current state at a certain stage, to find the optimal policy *for the next stage only* among all future states,
(e) the optimal policy of the remaining stages is independent of the optimal policies selected in the previous stages.

Dynamic programming has been applied to a large number of problems; to name a few, control of system operation over time, design of equipment involving multistage equipment (such as heat exchangers, reactors, distillation columns,...), equipment maintenance and replacement policy, production control, economic planning, investment. One can distinguish between deterministic and probabilistic dynamic programming methods, where the distinction arises when the next stage is not completely determined by the state and policy decisions of the current stage. *Only deterministic problems are considered in this section.* Examples of stochastic factors which may arise in probabilistic problems could be uncertainty in future demand, random equipment failures, supply of raw material,...

Example 7.10.1:[4] *Example of determining optimal equipment maintenance policy.*
Consider a facility which generates electricity from a prime-mover such as a reciprocating engine or a gas-turbine. The electricity is sold to the electric grid which results in an income to the owner. However, the efficiency of the prime-mover gradually degrades over time so that the net income generated from electricity sales reduces over time. A major service costing $ 21 (thousand) is required to bring the primemover up to its original performance, and this is needed at least once every 5 years. The net sales income (in thousands of dollars/year) is given by the following equation:

$$S = 25 - n^2 \qquad \text{for} \quad 0 \le n \le 4 \tag{7.42}$$

where n is the number of years from last major service. Hence, service is mandatory at the end of 5 years, but service may be required more frequently to maximize profits. The time value of money is not considered in this simple example.

The intent of the problem is to determine the maintenance schedule of this equipment, at a time when it is **two years old,** which maximizes *cumulative* net profit over a 4 year period. The annual profit P(n) is given by:

– at the end of the year when no servicing is done:

$$P(n) = 25 - n^2 \qquad \text{for} \quad 0 \le n \le 4 \tag{7.43}$$

– at the end of the year when servicing is done:

$$P(n=0) = 25 - 0 - 21 = 4$$

It is clear that one is seeking to maximize a sum, namely $J_N^* = \sum_{i=1}^{N} P_i(n)$ where N=4 (the time horizon for this problem) and i is the index representing the time period into this time horizon. The index i should not be confused with the index n. This is clearly a multi-stage problem with numerous combinations possible, and well suited to the dynamic programming approach. At the end of each year, one is required to make a decision whether to continue as is or whether to have service performed. Recalling one of the basic features of the dynamic programming approach, namely that the optimal policy for a certain stage is independent of the optimal policies selected in the previous stages, one can frame the problem as a recursive equation for the optimal cumulative sum J*:

$$J_i^*(n_{i+1}) = \max\{P_i(n_i) + J_{i-1}^*(n_i)\} \tag{7.44}$$

where n_i denotes the number of years after the last service corresponding to the current stage (or year) in the time horizon.

The procedure and the results of the calculation are shown in Table 7.2. Note how the calculation starts from the second column and recursively fills the subsequent cells. For example, consider the case when one is 3 years after the last service was done and 2 years of operation left. Then:

$$J_2^*(3) = \max\,[\underbrace{\{P(3) + J_1^*(4)\}}_{\text{if no service}}, \underbrace{\{P(0) + J_1^*(1)\}}_{\text{if service done}}]$$

$$\text{or} \quad J_2^*(3) = \max\,[\{16+9\}, \{4+24\}] = 28$$

This value is the one shown in the table with the indication that performing service is the preferred option.

[4] From Beveridge and Schechter (1970) by permission of McGraw-Hill.

Table 7.2 Table showing net cumulative profit ($—thousands). The bolded numbers and the arrows indicate the optimal annual policy decisions over a time horizon of 4 years of operation at a time 2 years after the last service to the equipment was performed

	Years of operation left			
Age of unit from time of service- n (yr)	1	2	3	4
1	Max(25-1^2,4) = 24 (N)	Max(24+21,4+24) =45(N)	**Max(45+16,4+24+21) =61 (N)**	Max(61+9,4+24+21+16) =70 (N)
2	Max (25-2^2, 4) =21 (N)	**Max(21+16,4+24) =37(N)**	Max(37+9,4+24+21) =49 (N or S) N selected	**Max(49+4,4+24+21+16) =65 (S)**
3	**Max(25-3^2,4) =16 (N)**	Max(16+9,4+24) =28(S)	Max(28+4,4+24+21) =49 (S)	Max(49+4,4+24+21+16) =65(S)
4	Max(25-4^2,4) =9 (N)	Max(9+4,4+24) =28(S)	Max(28+4,4+24+21) =49 (S)	Max(49+4,4+24+21+16) =65(S)
5	25-0-21 = 4 (S)	Max(4+21,4+24) =28(S)	Max(28+4,4+24+21) =49 (S)	Max(49+4,4+24+21+16) =65(S)

Note: N- no service, S- service done

The optimal policy decisions (whether to perform service or not) at each of the 4 years of the time horizon are shown bolded in Table 7.2. The path indicated by arrows is the optimal one. As the equipment is already 2 years old, one starts from the extreme right column at the second row. Since service has been done, one moves up to the third column since the age from time of service has now reduced to one. The next step should end in the cell corresponding to age of unit=2 since no service is done at this stage (year 3). Similarly, the last step takes one down to the cell corresponding to age of unit=3 since no service is again done at year 2.

This example is rather simplified and was meant to illustrate how the dynamic programming algorithm could be used to address equipment maintenance and scheduling problems. Actual cases could involve multiple equipment systems, and the inclusion of more complex engineering, financial and operational characteristics (such as varying cost of electricity generation by season and demand) as well as longer time horizons. It is in such more elaborate situations that that the greater computational efficiency of dynamic programming assumes a major consideration. ■

Example 7.10.2: *Strategy of operating a building to minimize cooling costs*
This simple example illustrates the use of dynamic programming for problems involving differential equations. The concept of lumped models was introduced in Sect. 1.2.3 and a thermal network model of heat flow through a wall was discussed in Fig. 1.5. The same concept of a thermal network can be extended to predict the dynamic thermal response of an entire building.

Many electric utilities in the U.S. have summer-peaking problems, meaning that they are hard pressed to meet the demands of their service customers during certain hot afternoon periods, referred to as *peak periods*. The air-conditioning (AC) use in residences and commercial buildings has been shown to be largely responsible for this situation. Remedial solutions undertaken by utilities involve voluntary curtailment by customers via incentives or penalties through electric rates which vary over time of day and by season (called time of day seasonal rates). Engineering solutions, also encouraged by utilities, to alleviate this situation include installing cool ice storage systems, as well as *soft* options involving dynamic control of the indoor temperature via the thermostat. This is achieved by sub-cooling the building during the night and early morning, and controlling the thermostat in a controlled manner during the peak period hours such that the "coolth" in the thermal mass of the building structure and its furnishings can partially offset the heat loads of the building, and hence, reduce the electricity demands of the AC.

Figure 7.14 illustrates a common situation where the building is occupied from 6:00 am till 7:00 pm with the peak period being from noon till 6:00 pm. The normal operation of the building is to set the thermostat to 72°F during the occupied period and at 85°F during unoccupied period (such a thermostat set-up scheme is a common energy conservation measure). Three different pre-cooling options are shown, all three involving the building to be cooled down to 70°F, representative of the lower occupant comfort level, starting from 6:00 am. *The difference in the three options lies in how the thermostat is controlled during the peak period.* The first scheme is to simply set up the thermostat to a value of 78°F representative of the high-end occupant discomfort value with the anticipation that the internal temperature T_i will not reach this value during the end of the peak period. If it does, the AC would come on and partially negate the electricity demand benefits which such a control scheme would provide. Often, the thermal mass in the structure will not be high enough for this control scheme to work satisfactorily. Another simple control scheme is to let the indoor temperature ramp us linearly which is also not optimal. The third, and optimal, scheme which would minimize the following cost function over the entire day, can be determined by solving the dynamic programming problem of this situation over time t:[5]

[5] This is a continuous path optimization problem which can be discretized to a finite sum, say a convenient time period of 1 h (the approximation improves as the number of terms in the sum increases).

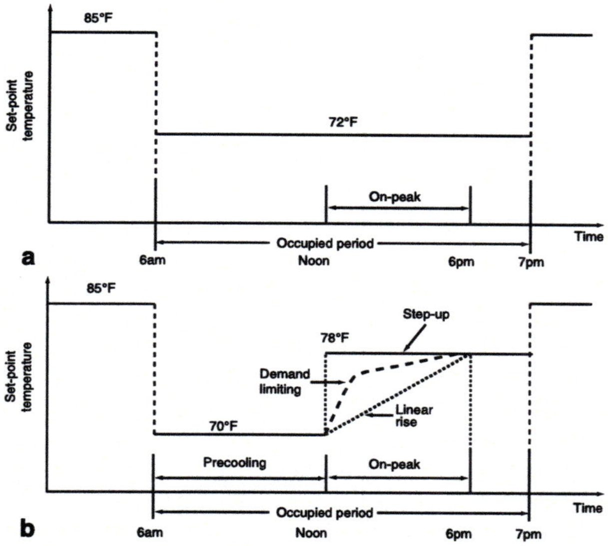

Fig. 7.14 Sketch showing the various operating periods of the building discussed in Example 7.10.2, and the three different thermostat set-point control strategies for minimizing total electric cost. **a** Normal operation. **b** Pre-cooling strategies (From Lee and Braun 2008 © American Society of Heating, Refrigerating and Air-conditioning Engineers, Inc., www.ashrae.org)

$$J^* = \min\{J\} = \min\left\{\sum_{t=1}^{24} \mathbf{c}_{e,t}.\mathbf{P}_t(T_{i,t}) + c_d \max\left[\mathbf{P}_{t,t_1-t_2}(T_{i,t})\right]\right\} \quad (7.45a)$$

$$\text{subject to:} \quad T_{i,\min} \le T_{i,t} \le T_{i,\max} \text{ and } 0 \le P_t \le P_{Rated} \quad (7.45b)$$

where

$c_{e,t}$ is the unit cost vector of electricity in \$/kWh (which can assume different values at different times of the day) as set by the electric utility,

P_t is the electric energy use during hour t, and is function of T_i which changes with time t,

c_d is the demand cost in \$/kW, also set by the electric utility, which is usually imposed on the peak hourly use during the peak hours (or there could be two demand costs, one for off-peak and one for on-peak during a given day), and

$\max(P_t)_{t1-t2}$ is the demand or maximum hourly use during the peak period t_1 to t_2.

The AC power consumed each hour represented by $\mathbf{P}_t$ is affected by $T_{i,t}$. It cannot be negative and should be less than the capacity of the AC denoted by P_{Rated}. The solution to this dynamic programming problem requires two thermal models: one to represent the thermal response of the building, and another for the performance (or efficiency) of the AC.

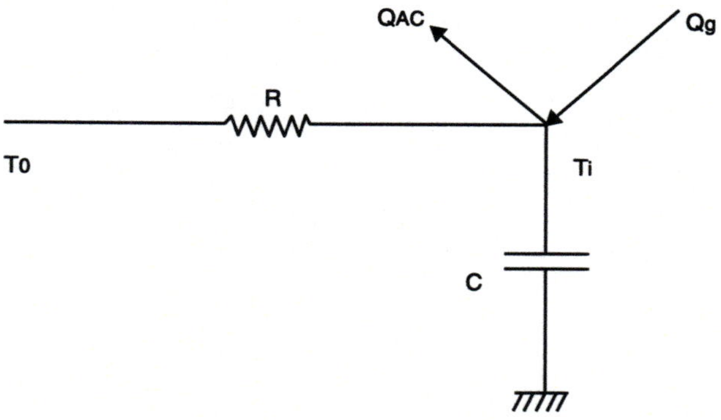

Fig. 7.15 A simplified 1R1C thermal network to model the thermal response of a building (i.e., variation of the indoor temperature T_i) subject to heat gains from the outdoor temperature T_o and from internal heat gains Q_g. The overall resistance and the capacitance of the building are R and C respectively and Q_{AC} is the thermal heat load to be removed by the air-conditioner

Models of varying complexity have been proposed in the literature. A simple model following Reddy et al. (1991) is adequate to illustrate the approach. Consider the 1R1C thermal network shown in Fig. 7.15 where the thermal mass of the building is simplistically represented by one capacitor C and an overall resistance R. The internal heat gains Q_g could include both solar heat gains coming through windows as well as thermal loads from lights, occupants and equipment generated from within the building. The simple one node lumped model for this case is:

$$C\frac{dT_i}{dt} = \frac{T_o(t) - T_i}{R} + Q_g(t) - Q_{AC}(t) \quad (7.46)$$

where T_o and T_i are the outdoor and indoor dry-bulb temperatures, and Q_{AC} is the thermal cooling provided by the AC.

For the simplified case when Q_g and T_o can be assumed constant and the AC is switched off, the transient response of this dynamic system is given by:

$$\frac{T_i - T_{i,\min}}{T_o - T_{i,\min} + R.Q_g} = 1 - e^{-\frac{\Delta t}{\tau}} \quad (7.47a)$$

where Δt is the time from when the AC is switched off. The time required for T_i to increase from $T_{i,\min}$ to $T_{i,\max}$ is then:

$$\Delta t = -\tau \ln\left[1 - \frac{T_{i,\max} - T_{i,\min}}{T_o - T_{i,\min} + R.Q_g}\right] \quad (7.47b)$$

where τ is the time constant given by (C.R).

The savings in thermal cooling energy ΔQ_{AC} which can be avoided by the linear ramp-up strategy can also be determined in a straightforward manner by performing hourly calculations over the peak period using Eq. 7.46 since the T_i values

Table 7.3 Peak AC power reduction and daily energy savings compared to base operation under different control strategies for similar days in October. (From Lee and Braun 2008 © American Society of Heating, Refrigerating and Air-Conditioning Engineers, Inc., www.ashrae.org)

Test #	$T_{out,max}$ °C	Control Strategy*	Peak power kW	Peak savings kW	Energy use kWh	Energy savings kWh
1	32.2	NS (baseline)	26.10	–	243.1	–
2	31.7	LR	23.53	2.57	226.5	16.6
3	32.8	SU	20.52	5.58	194.2	20.1
4	29.4	NS (baseline)	29.70	–	224.3	–
5	30.6	DL	20.03	9.67	219.2	5.2
6	30.6	DL	22.34	7.36	196.6	27.8
7	26.7	NS (baseline)	27.04	–	233.4	–
8	26.7	DL	16.94	10.10	190.4	43.0

**NS* baseline operation, *LR* linear ramp-up, *SU* setup, *DL* demand limiting

can be determined in advance. The total thermal cooling energy saved during the peak period is easily deduced as:

$$\Delta Q_{AC} = C.\frac{T_{i,\max} - T_{i,\min}}{\Delta t_{peak}} \quad (7.48)$$

where Δt_{peak} is the duration of the peak period.

For the dynamic optimal control strategy, the rise in $T_i(t)$ over the peak period has to be determined. For the sake of simplification, let us discretize the continuous path into say, hourly increments. Then, Eq. 7.46 with some re-arrangement and minor notational change, can be expressed as:

$$Q_{AC,t} = -C.T_{i,t+1} + T_{i,t}\left(C - \frac{1}{R}\right) + \frac{T_{o,t}}{R} + Q_{g,t} \quad (7.49a)$$

while the electric power drawn by the AC can be modeled as :

$$P_t = f(Q_{AC}, T_i, T_o) \quad (7.49b)$$

subject to conditions Eq. 7.45b.

The above problem can be solved by framing it as one with several stages (each stage corresponding to an hour into the peak period) and states representing the discretized set of possible values of T_i (say in steps of 0.5°F). One would get a set of simultaneous equations (the order being equal to the number of stages) which could be solved together to yield the optimal trajectory. Though this is conceptually appealing, it would be simpler to perform the computation using a software package given the non-linear function of P_t (Eq. 7.49b) and the need to introduce constraints on P and T_i.

The example in Sect. 7.8 assumed polynomial models though physical models using grey box models (see Pr. 5.13) are equally appropriate. Table 7.3 assembles peak AC power savings and daily energy savings for a small test building located in Palm Desert, CA which was modeled using higher order differential equations by Lee and Braun (2008). The model parameters have been deduced from experimental testing of the building and the AC equipment which were then used to evaluate different thermostat control options. The table assembles daily energy use and peak AC data for baseline operation (NS) against which other schemes can be compared. Since the energy and demand reductions would depend on the outdoor temperature, the tests have been assembled for three different conditions (tests 1–3 for very hot days, tests 4–6 for milder days, and tests 7–8 for even milder days. The optimal strategy found by dynamic programming is clearly advantageous both in demand reduction and in diurnal energy savings although the benefits show a certain amount of variability from day-to-day. This could be because of diurnal differences in the driving functions and also because of uncertainty in the determination of the model parameters. ■

Problems

Pr. 7.1[6] The collector area A_C of a solar thermal system is to be determined which minimizes the total discounted savings C_s' over n years. The solar system delivers thermal energy to an industrial process with the deficit being met by a conventional boiler system (this configuration is referred to as a solar-supplemented thermal system). Given the following expression for discounted savings:

$$C_s' = 87{,}875.66 - \frac{17{,}738.08}{1 - \exp\left(-\frac{A_C}{190.45}\right)} - 2000 - 300A_C$$

determine the optimal value of collector area A_C. Verify your solution graphically, and estimate a satisficing range of collector area values which are within 5% of the optimal value of C_s'.

[6] From Reddy (1987).

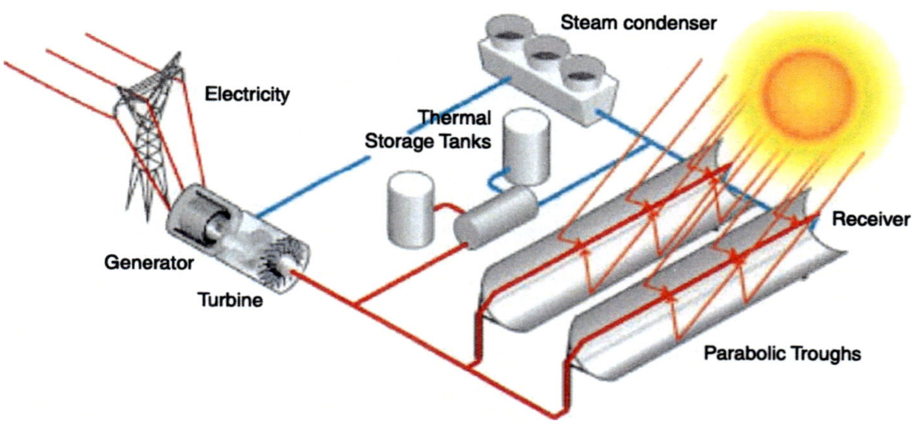

Fig. 7.16 Sketch of a parabolic trough solar power system. The energy collected from the collectors can be stored in the thermal storage tanks which is used to produce steam to operate a Rankine power engine. (Downloaded from http://www1.eere.energy.gov/solar/)

Pr. 7.2
(a) Use a graphical approach to determine the values of x_1 and x_2 which maximize the following function:

$$\begin{aligned} J &= 0.4x_1 + 0.5x_2 \\ \text{s.t.} \quad & 0.3x_1 + 0.1x_2 \leq 2.7 \\ & 0.5x_1 + 0.5x_2 = 6 \\ & 0.6x_1 + 0.4x_2 \geq 6 \\ \text{and} \quad & x_1 \geq 0 \text{ and } x_2 \geq 0 \end{aligned}$$

(b) Solve the problem analytically using the slack variable approach, and verify your results.
(c) Perform a sensitivity analysis of the optimum

Pr. 7.3 Use the Lagrange multiplier approach to minimize the following constrained optimization problem:

$$J = (x_1^2 + x_2^2 + x_3^2)/2$$
$$\text{s.t.} \quad x_1 - x_2 = 0, \quad x_1 + x_2 + x_3 = 0$$

Pr. 7.4 *Solar thermal power system optimization*
Generating electricity via thermal energy collected from solar collectors is a mature technology which is much cheaper than that from photovoltaic systems (if no rebates and financial incentives are considered). Such solar thermal power systems in essence comprise of a solar collector field, a thermal storage, a heat exchanger to transfer the heat collected from the solar collectors to a steam boiler and the conventional Rankine steam power plant (Fig. 7.16). A simper system without storage tank will be analyzed here such that the fluid temperature leaving the solar collector array (T_{co}) will directly enter the Rankine engine to produce electricity.

Problem 5.6 described the performance model of a flat-plate solar collector while usually concentrating collectors are required for power generation. A simplified expression analogous to Eq. 5.65 for concentrating collectors with concentration ratio C is:

$$\eta_C = \left[F_R \eta_{opt} - \frac{F_R U_L}{C} \left(\frac{T_{Ci} - T_a}{I_T} \right) \right]$$

The efficiency of the solar system decreases with higher values of T_{ci} while that of the Rankine power cycle increases with higher value of T_{co} such that the optimal operating point is the product of both efficiency curves (Fig. 7.17). The problem is to determine the optimal value of T_{ci} which maximizes the overall system efficiency given the following information:
(i) concentration ratio C=30
(ii) beam solar irradiation I_T=800 W/m²
(iii) trough collectors with: $F_R\eta_{opt}$ =0.75 and F_RU_L= 10.0 W/m²-°C
(iv) ambient temperature T_a=20°C
(v) Assume that the Rankine efficiency of the steam power cycle is half of that of the Carnot cycle operating bet-

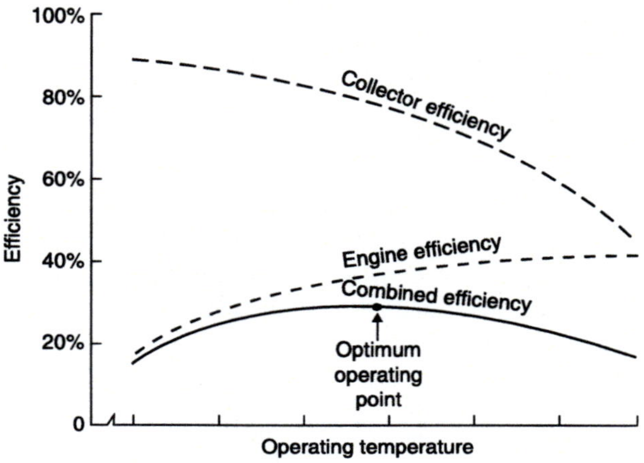

Fig. 7.17 Combined solar collector and Rankine engine efficiencies dictate the optimum operating temperature of the system

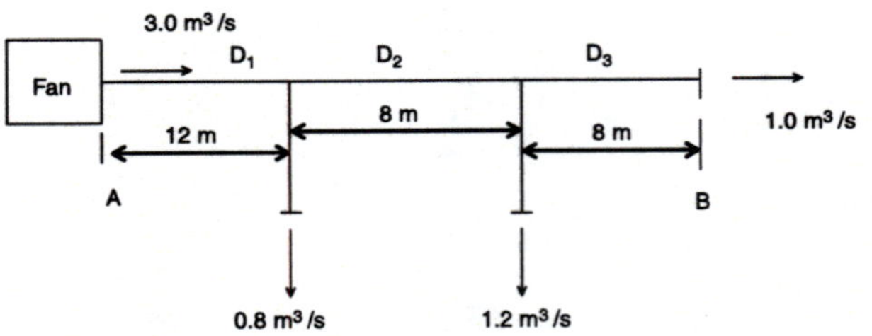

Fig. 7.18 Ducting layout with pertinent information for Problem Pr. 7.5

ween the same high and low temperatures (take the low-end temperature to be 10°C above T_a and the high temperature is to be determined from an energy balance on the solar collector: $T_{co} = T_{ci} + \frac{Q_c}{m\ c_p}$ where Q_c is the useful energy collected per unit collector area (W/m²), the mass flow rate per unit collector area m=8 kg/m² h, and the specific heat of the heat transfer fluid c_p=3.26 kJ/kg°C).

Solve this unconstrained problem analytically and verify your result graphically. Perform a post optimality analysis and identify influential variables and parameters in this problem.

Pr. 7.5[7] *Minimizing pressure drop in ducts*
Using the method of Lagrange multipliers, determine the diameters of the circular duct in the system shown in Fig. 7.18 so that the drop in the static pressure between points A and B will be a minimum. Use the following additional information:

- Quantity of sheet metal available=60 m²
- Pressure drop in a section of straight duct of diameter D (m) and length L (m) with fluid flowing at velocity v (m/s), $\Delta p = f.\left(\frac{L}{D}\right).\left(\frac{V^2}{2}\right)\rho$

where f is the friction factor=0.02 and air density ρ= 1.2 kg/m³

Neglect pressure drop in the straight-through section past the outlets and the influence of changes in velocity pressure. Use pertinent information from Fig. 7.18.

Pr. 7.6 *Replacement of filters in HVAC ducts*
The HVAC air supply in hospitals has to meet high standards in terms of biological and dust-free cleanliness for which purpose high quality filters are used in the air ducts. Fouling by way of dust build-up on these filters causes additional pressure drop which translates into an increased electricity consumption of the fan-motor circulating the air. Hence, the maintenance staff is supposed to replace these filters on a regular basis. Changing them too frequently results in undue expense due to the high cost of these filters, while not replacing them in a timely manner also increases the expense due to that associated with the pumping power. Determine the optimal filter replacement schedule under the following operating conditions (neglecting time value of money):

- The HVAC system operates 24 h/day and 7 days/week and circulates Q=100 m³/s of air
- The pressure drop in the HVAC duct when the filters are new is 5 cm of water or H=0.05 m
- The pressure drop across the filters increase in a linear fashion by 0.1 m of water gauge for every 1000 h of operation (this is a simplification—actual increase is likely to be exponential)
- The total cost of replacing all the filters is $ 800
- The efficiency of the pump is 65% and that of the motor is 90%
- The levelized cost of electricity is $ 0.10 per kW h.

The electric power consumed in kW by the pump-motor is given by: $E(kW) = \frac{Q(L/s)H(m)}{(102)\eta_{pump}\eta_{motor}}$

(Hint: The problem is better visualized by plotting the energy cost function versus hours of operation)

Pr. 7.7 *Relative loading of two chillers*
Thermal performance models for chillers have been described in Pr. 5.13 of Chap. 5. The black box model given by Eq. 5.75 for the COP often appears in a modified form with the chiller electric power consumption P being expressed as:

$$P = a_0 + a_1(T_{cdo} - T_{chi}) + a_2(T_{cdo} - T_{chi})^2 + a_3 Q_{ch} + a_4 Q_{ch}^2 + a_5(T_{cdo} - T_{chi})Q_{ch} \quad (7.50)$$

where T_{cdo} and T_{chi} are the leaving condenser water and supply chilled water temperatures respectively, Q_{ch} is the chiller thermal load, and a_i are the model parameters. Consider a situation where two chillers, denoted by Chiller A and Chiller B, are available to meet a thermal cooling load. The chillers are to be operated such that T_{cdo}=85°F and T_{chi}=45°F. Chiller B is more efficient than Chiller A at low relative load fractions and vice versa. Their model coefficients from performance data supplied by the chiller manufacturer are given in Table 7.4.

(a) The loading fraction of a chiller is the ratio of the actual thermal load supplied by the chiller to its rated value $Q_{ch\text{-}rated}$. Use the method of Lagrange multipliers to prove that the optimum loading fractions y_1* and y_2* occur when the slopes of the curves are equal, i.e., when $\frac{\partial P_A}{\partial Q_{ch,A}} = \frac{\partial P_B}{\partial Q_{ch,B}}$

(b) Determine the optimal loading (which minimizes the total power draw) of both chillers at three different values of Q_{ch}: 800, 1200, 1600 tons, and calculate the corre-

[7] From Stoecker (1989) by permission of McGraw-Hill.

Table 7.4 Values of coefficients in Eq. 7.50. (from ASHRAE 1999 © American Society of Heating, Refrigerating and Air-Conditioning Engineers, Inc., www.ashrae.org)

	Units	Chiller A	Chiller B
$Q_{ch\text{-}rated}$	Tons (cooling)	1250	550
a_0	kW	106.4	119.7
a_1	kW/°F	6.147	0.1875
a_2	kW/°F²	0.1792	0.04789
a_3	kW/ton	–0.0735	–0.3673
a_4	kW/ton²	0.0001324	0.0005324
a_5	kW/ton.°F	–0.001009	0.008526

sponding power draw. Investigate the effect of near-optimal operation, and plot your results in a fashion useful for the operator of this cooling plant.

Pr. 7.8[8] *Maintenance scheduling for plant*
The maintenance schedule for a plant is to be planned to maximize its 4-year profit. The income level of the plant at any given year is a function of the condition of the plant carried over from the previous year and the maintenance expenditure at the beginning of the year. Table 7.5 shows the necessary maintenance expenditures that result in a certain income level during the year for various income levels carried over from the previous year. The income level at the beginning of year 1 before the maintenance expenses are made is $ 36,000 and the income level specified during, and at the end of year 4, is to be $ 34,000. The profit for any one year will be the income during the year minus the expenditure made for maintenance at the beginning of the year. Use dynamic programming to determine the plan for maintenance expenditures that result in maximum profit for the 4 years.

Pr. 7.9 *Comparing different thermostat control strategies*
The three thermostat pre-cooling strategies whereby air-conditioner (AC) electrical energy use in commercial buildings can be reduced have been discussed in Example 7.10.2. You will compare these three strategies for the following small commercial building and specified control limits:

Assume that the RC network shown in Fig. 7.15 is a good representation of the actual building. The building time constant is 6 h and its overall heat transfer resistance R=2.5°C/kW. The internal loads of space can be assumed constant at Q_g=1.5 kW. The peak period lasts for 8 h and the ambient temperature can be assume constant at T_0=32°C. The minimum and maximum thermostat control set points are $T_{i,min}$ =22°C and $T_{i,max}$=28°C.

A very simple model for the AC is to be used (Kreider et al. 2009):

$$P_{AC} = \frac{Q_{AC,Rated}}{COP_{Rated}}(0.023 + 1.429{*}PLR - 0.471{*}PLR^2) \qquad (7.51)$$

where PLR=part load ration=($Q_{AC}/Q_{AC,Rated}$) and COP_{Rated} is the Coefficient of Performance of the reciprocating chiller. Assume COP_{Rated}=4.0 and $Q_{AC,Rated}$=4.0 kW.

(a) Usually, $T_{i,max}$ is not the set point temperature of the space. For pre-cooling strategies to work better, a higher temperature is often selected since occupants can tolerate this increased temperature for a couple of hours without adverse effects. Calculate the AC electricity consumed, as well as the demand, during the peak period if T_r=26°C; this will serve as the baseline electricity consumption scenario.
(b) For the simple-minded setup strategy, first compute the number of hours in which the selected $T_{i,max}$ is reached starting from $T_{i,min}$. Then, calculate the electricity consumed and the demand by the AC during the remaining number of hours left in the peak period if the space is kept at $T_{i,max}$.
(c) For the ramp-up strategy, calculate electricity consumed and the demand by the AC during the peak period (by summing those at hourly time intervals).
(d) Assuming that the thermostat is being controlled at hourly intervals, determine the optimal trajectory of the indoor temperature? What is the corresponding AC electricity use and demand?
(e) Summarize your results in a table similar to Table 7.3 and discuss your findings.

Table 7.5 Maintenance expenditures made at the beginning of year, thousands of dollars

Income level carried over from previous years	Income level during year $ 30	$ 32	$ 34	$ 36	$ 38	$ 40
$ 30	$ 2	$ 4	$ 7	$ 11	$ 16	$ 23
32	2	3	5	9	13	18
34	1	2	4	7	10	14
36	0	1	2	5	8	10
38	x	0	1	2	6	9
40	x	x	0	1	4	8

[8] From Stoecker (1989) by permission of McGraw-Hill.

References

ASHRAE,1999. ASHRAE Applications Handbook, Chap. 40, *Supervisory Control Strategies and Applications,* American Society of Heating, Refrigerating and Air-Conditioning Engineers, Atlanta.
Beveridge, G.S.G and S. Schechter,1970. *Optimization: Theory and Practice,* McGraw-Hill Kogakusha, Tokyo.
Braun, J.E., 2006. Optimized Operation of Chiller Equipment in Hybrid Machinery Rooms and Associated Operating and Control Strategies- 1200 RP, Final report of ASHRAE Research Project RP-1200

RP, Braun Consulting, 87 pages, American Society of Heating, Refrigerating and Air-Conditioning Engineers, Atlanta, August.

Burmeister, L.C.,1998. *Elements of Thermal-Fluid System Design,* Prentice-Hall, Upper Saddle River, NJ.

Edgar, T.F., D.M. Himmelblau and L.S. Lasdon, 2001. *Optimization of Chemical Processes,* 2nd Ed., McGraw Hill, Boston.

Hillier, F.S. and G.J. Lieberman, 2001, *Introduction to Operations Research,* 7th Edition, McGraw-Hill, New York,

Kreider, J.K., P.S. Curtiss and A. Rabl, 2009. *Heating and Cooling of Buildings,* 2nd Ed., CRC Press, Boca Raton, FL.

Lee, K.H. and Braun, J.E. 2008. A data-driven method for determining zone temperature trajectories that minimize peak electrical demand, *ASHRAE Trans.* 114(2), American Society of Heating, Refrigerating and Air-Conditioning Engineers, Atlanta, June.

Petchers, N., 2003. *Combined Heating, Cooling and Power Handbook: Technologies and Applications,* Fairmont Press Inc., Marcel Dekker Inc., Lilburn, GA

Reddy, T.A., 1987. *The Design and Sizing of Active Solar Thermal Systems,* Oxford University Press, Oxford, U.K.

Reddy, T.A., L.K. Norford and W.Kempton, 1991. Shaving residential air-conditioner electricity peaks by intelligent use of the building thermal mass, *Energy*, vol.16(7), pp.1010-1010.

Reddy, T.A. and I. Maor, 2009. Cost penalties of near-optimal scheduling control of BCHP systems: Part II: Modeling, optimization and analysis results, *ASHRAE Trans,* vol.116 (1), American Society of Heating, Refrigerating and Air-Conditioning Engineers, Atlanta, January.

Stoecker, W.F., 1989. *Design of Thermal Systems,* 3rd Edition, McGraw-Hill, New York.

Venkataraman, P. 2002. *Applied Optimization with Matlab Programming,* John Wiley and Sons, New York.

Classification and Clustering Methods

8

This chapter covers two widely used classes of multivariate data analysis methods, classification and clustering methods. Classification methods are meant: (i) to statistically distinguish or "discriminate" between differences in two or more groups when one knows beforehand that such groupings exist in the data set of measurements provided, and (ii) subsequently assign or allocate a future unclassified observation into a specific group with the smallest misclassification error. Numerous classification techniques, divided into three groups: parametric, heuristic and regression trees, are described and illustrated by way of examples. Clustering involves situations when the number of clusters or groups is not known beforehand, and the intent is to allocate a set of observation sets into groups which are similar or "close" to one another with respect to certain attribute(s) or characteristic(s). In general, the number of clusters is not predefined and has to be gleaned from the data set. This and the fact that one does not have a training data set to build a model make clustering a much more difficult problem than classification. Two types of clustering techniques, namely partitional and hierarchical, are described. This chapter provides a non-mathematical overview of these numerous techniques with conceptual understanding enhanced by way of simple examples as well as actual case study examples.

8.1 Introduction

Certain topics relevant to multivariate data analysis have been previously treated: ANOVA analysis (Sect. 4.3), tests of significance (Sect. 4.4), multiple OLS regression without (Sect. 5.4) and with collinear regressors (Sect. 10.3). This chapter complements these by covering two important statistical classes of problems dealing with multivariate data analysis, namely classification and clustering methods.

Clustering analysis involves several procedures by which a group of samples (or multivariate observations) can be clustered or partitioned or separated into sub-sets of greater homogeneity, i.e., those based on some pre-determined similarity criteria. Examples include clustering individuals based on their similarities with respect to physical attributes or mental attitudes or medical problems; or multivariate performance data of a mechanical piece of equipment can be separated into those which represent normal operation as against faulty operation. Thus, clustering analysis reveals inter-relationships between samples which can serve to group them under situations *where one does not know the number of sub-groups beforehand.* There are a large number of methods which have been developed, and some of the classical ones are described in Sect. 8.5. As a note of caution, certain authors (for example, Chatfield 1995) opine that the clustering results depend to a large extent on the clustering method used, and that inexperienced users are very likely to end up with misleading results. This lack of clear-cut results has somewhat dampened the use of sophisticated clustering methods, with analysts tending to rely rather on simpler and more intuitive methods.

Classification analysis, on the other hand, applies to situations when the groups are known beforehand. The purview here is to identify models which best characterize the boundaries between groups, so that future objects can be allocated into the appropriate group. Since the groups are predetermined, classification problems are somewhat simpler to analyze than clustering problems. The challenge in classification modeling is dealing with the *misclassification rate* of objects, a dilemma faced in most practical situations. Three types of classification methods are briefly treated: parametric methods (involving statistical, ordinary least squares, discriminant analysis and Bayesian techniques), heuristic classification methods (rule-based, descision-tree and k nearest neighbors), and parametric models involving classification and regression trees.

8.2 Parametric Classification Approaches

8.2.1 Distance Between Measurements

Similarity between objects or samples can be geometrically likened to a distance or a trend. The correlation coefficient

T. Agami Reddy, *Applied Data Analysis and Modeling for Energy Engineers and Scientists*,
DOI 10.1007/978-1-4419-9613-8_8,

between two variables can be used as a measure of trend and similarity. The distinctiveness of two objects can be visually ascertained by plotting them. For example, the Euclidian distance between two objects (x_1, y_1) and (x_2, y_2) plotted on Cartesian coordinates is characterized in two-dimensions by:

$$d = [(x_2 - x_1)^2 + (y_2 - y_1)^2]^{1/2} \tag{8.1}$$

In general, for p variables, the generalized Euclidean distance from object i to object j is (Manly 2005):

$$d_{ij} = \left[\sum_{k=1}^{p} (x_{ik} - x_{jk})^2 \right]^{1/2} \tag{8.2}$$

where x_{ik} is the value of the variable X_k for object i and x_{jk} is the value of the same variable for object j.

The distance term will be affected by the magnitude of the variables, i.e., physical quantities such as temperature, air flow rates, efficiency, have different scales and range of variation, and the one with largest numerical values will overwhelm the variation in the others. Thus, some sort of normalization is warranted. One common approach is to equalize the variances by defining a new variable (x_1/s_1) where s_1^2 is an estimate of the variance of variable x_1. Other ways are by min-max scaling or by standard deviation scaling as given by Eqs. 3.12 and 3.13.

Example 8.2.1: *Using distance measures for evaluating canine samples*

Consider a problem where a biologist wishes to evaluate whether the modern dog in Thailand descended from prehistoric ones from the same region or were inbred with similar dogs which migrated from nearby China or India. The basis of this evaluation will be six measurements all related to the mandible or lower jaw of the dog: x_1—breadth of mandible, x_2—height of mandible below the first molar, x_3—length of first molar, x_4—breadth of first molar, x_5—length from first to third molar, x_6—length from first to fourth premolar. Relevant data are assembled in Table 8.1.

The measurements have to be standardized, and so, the mean and standard deviations for each variable across groups is determined (shown in Table 8.1). This allows the standardized values to be computed following Eq. 3.13 as shown in Table 8.2. For example, the standardized value for the modern dog: z_1=(9.7−11.25)/1.68=−0.922, and so on.

Table 8.1 Mean mandible (or lower jaw) measurements of six variables for four canine groups. (Modified example from Higham et al. 1980)

	x_1	x_2	x_3	x_4	x_5	x_6
Modern dog	9.7	21	19.4	7.7	32	36.5
Chinese wolf	13.5	27.3	26.8	10.6	41.9	48.1
Indian wolf	11.5	24.3	24.5	9.3	40	44.6
Prehistoric dog	10.3	22.1	19.1	8.1	32.2	35
Mean	*11.25*	*23.675*	*22.45*	*8.925*	*36.525*	*41.05*
Standard deviation	*1.68*	*2.78*	*3.81*	*1.31*	*5.17*	*6.31*

Table 8.2 Standardized measurements

	z_1	z_2	z_3	z_4	z_5	z_6
Modern dog	−0.922	−0.963	−0.800	−0.937	−0.875	−0.721
Chinese wolf	1.342	1.304	1.140	1.281	1.040	1.117
Indian wolf	0.149	0.225	0.537	0.287	0.672	0.562
Prehistoric dog	−0.567	−0.567	−0.878	−0.631	−0.837	−0.958

Table 8.3 Euclidean distances between canine groups

	Modern dog	Chinese wolf	Indian wolf	Prehistoric dog
Modern dog	–			
Chinese wolf	5.100	–		
Indian wolf	3.145	2.094	–	
Prehistoric dog	0.665	4.765	2.928	–

Finally, using Eq. 8.2, the Euclidean distances among all groups are computed as shown in Table 8.3. It is clear that prehistoric dogs are similar to modern ones because their distances are much smaller than those of others. Next in terms of similarity are the Chinese and Indian wolfs, and so on. ■

Other measures of distance have been proposed for discrimination; one such measure is the *Manhattan measure* which uses absolute differences rather than squared distances. However, this is used only under special circumstances. A widely used distance measure is the *Mahanabolis measure* which is superior to the Euclidean when the variables are correlated. If $\bar{\mathbf{x}}_i' = (\bar{x}_{1i}, \bar{x}_{2i}, \ldots, \bar{x}_{pi,})'$ denotes the vector of mean values for the ith group, and **C** the variance-covariance matrix given by Eq. 5.31 in Sect. 5.4.2, then, the Mahanabolis distance from an observation **x**′ to the center of group i is computed as:

$$D_i^2 = (\mathbf{x} - \bar{\mathbf{x}}_i)' \mathbf{C}^{-1} (\mathbf{x} - \bar{\mathbf{x}}_i) \tag{8.3}$$

The observation **x** is then assigned to the group for which the value of D_i is smallest.

8.2.2 Statistical Classification

As stated earlier, classification methods provide the necessary methodology to: (i) statistically distinguish or "discriminate" between differences in two or more groups when one knows beforehand that such groupings exist in the data set of measurements provided, and (ii) subsequently assign or allocate a future unclassified observation into a specific group with the smallest probability of error.

At the onset, a conceptual understanding of the general classification problem is useful. Consider the simplest clas-

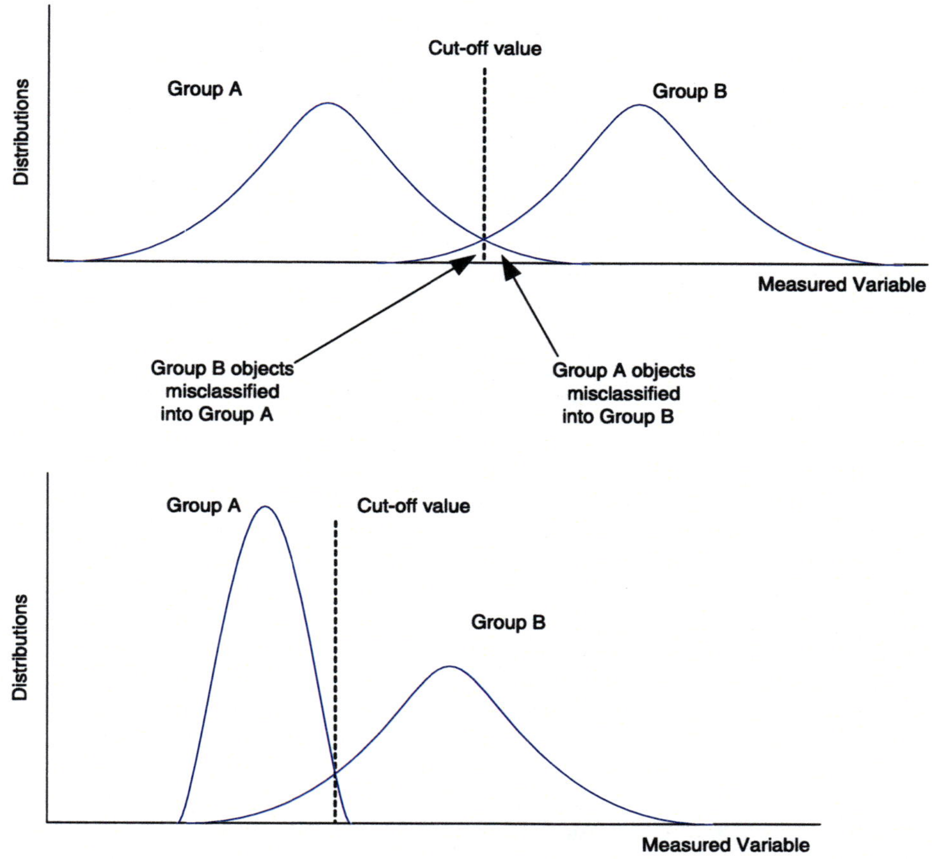

Fig. 8.1 Errors of misclassification for the univariate case. **a** When the two distributions are similar and if equal misclassification rates are sought, then the cutoff value or score is selected at the intersection point of the two distributions. **b** When the two distributions are not similar, a similar cut-off value will result in different misclassification errors

sification problem where two groups (Group A and Group B) are to be distinguished based on a single variable x. Both the groups have the same variance but the means are different. If the variable x is plotted on a single axis for both groups, the trivial problem is when there is no overlap. In this situation, determining the threshold or boundary or cut-off score is obvious. On the other hand, one could obtain a certain amount of overlap as shown in Fig. 8.1a. The objective of classification in this univariate instance is to determine the cut-off value of x which would yield fewest errors of false classification. If the two groups have equal variance in the variable x, then the best cut-off value is the point of intersection of the two distributions (as shown in Fig. 8.1a). Note that the areas represented by the tails of the distributions on either side of the cut-off value represent the misclassification probabilities or rates. An obvious extension to this simple problem is when the two distributions are different. There is no obvious best cut-off value since any choice of cut-off value would result in different misclassification error rates for both groups (see Fig. 8.1b). In such a case, the choice of the cut-off value is dictated by other issues such as the extent to which misclassification of one group is more critical (i.e., errors in one group have more severe adverse implications in cost, for example) than that of the other. The following example will illustrate the general approach for the univariate case.

Example 8.2.2: *Statistical classification of ordinary and energy efficient office buildings*

The objective is to distinguish between medium-sized office buildings which are ordinary (type O) or energy efficient (type E) judged by their "energy use index" (or EUI) or the energy used per square foot per year. Table 8.4 lists the EUIs for 14 buildings in the Phoenix, AZ area, half of which are type O and the other half type E. The first ten values (C1–C10) will be used to train or develop the cut-off score, while the last four will be used for testing, i.e., in order to determine the misclassification rate more representative of future misclassification rates than the one found during training.

This is a simple example meant for conceptual understanding. If the cut-off value is to be determined with no special weighting on misclassification rates, a first attempt at determining this value is to take it as being the mid-point of both the means. The training data set consists of five values in each category. The average value for the first five build-

Table 8.4 Data table specifying type and associated EUI and the results of the classification analysis (only misclassified ones are indicated)

Building #	Type	EUI (kBtu/ft²/yr)		Assuming cut-off score of 38.2
C1	O	40.1	Training	
C2	O	41.4	Training	
C3	O	38.7	Training	
C4	O	37.5	Training	Misclassified
C5	O	43.0	Training	
C6	E	37.4	Training	
C7	E	38.3	Training	Misclassified
C8	E	36.9	Training	
C9	E	35.3	Training	
C10	E	36.1	Training	
C11	O	37.2	Testing	Misclassified
C12	O	39.2	Testing	
C13	E	37.2	Testing	
C14	E	36.3	Testing	

ings (C1–C5 are type O) is 39.7 while that of the second five (C6–C10 are type E) is 36.8. If a cut-off value of 38.2 which is the mid-point or mean of the two averages is selected, one should expect the EUI for ordinary buildings to be higher than this value and that for energy efficient ones to be lower. The results listed in the last column of Table 8.4 indicate one misclassification in each category during the training phase. Thus, this simple-minded cutoff value is acceptable since it leads to equal misclassification rates among both categories. The table also indicates that among the last four buildings (C11–C14) used for testing the selected cut-off score, one of the ordinary buildings is improperly classified.

It is left to the reader to extend this type of analysis to the case when the misclassification rates are stipulated as not being equal. A case in point would be to deduce the cut-off value where no misclassification for category O is allowed. A cut-off value of 37.2 would fit this criterion. ■

How classification is done using the simple distance measure is conceptually illustrated in Fig. 8.1. The extension of the distance measure to bivariate (and multivariate) cases is intuitively straightforward. Two groups are shown in Fig. 8.2 with normalized variables. During training, the centers of each class can be determined as well as the separating boundaries around them. The two circles (shown continuous) encircle all the points. A future observation will be classified in that group whose class center is closest to that group. However, a deficiency is that some of the points are misclassified. One solution to decreasing, but not eliminating, the misclassification rates is to reduce the boundaries (shown by dotted circles). However, some of the observations now fall outside the dotted circles, and these points cannot be classified into either Class A or Class B. Hence, the option of reducing the boundary diameters is acceptable only if one is willing to group certain points into a third class called "unable to classify".

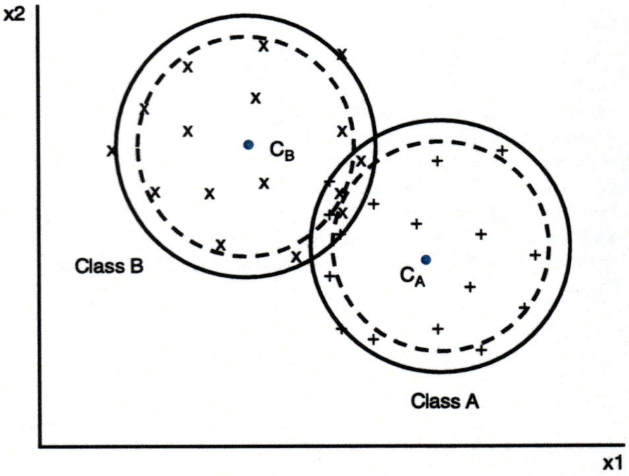

Fig. 8.2 Bivariate classification using the distance approach showing the centers and boundaries of the two groups. Data points are assigned to the class whose center (C_A or C_B) is closest to the point of interest. Note that reducing the boundaries (from *continuous* to *dotted circles*) reduces the misclassification rates but then some of the points fall outside the boundaries, and hence the need to be classified into additional groups

8.2.3 Ordinary Least Squares Regression Method

The logical ultimate extension of the univariate two-group situation is the multivariate multi-group case where the classification is based on a number of measurements or variables characterizing different attributes of each group. Instead of a single cut-off value, the multivariate case would require several functions to "separate" the groups.

Regression methods (discussed in Chap. 5) were traditionally developed to deal with model building problems and prediction. They can also be used to model differences between groups and thereby assign a future observation to a particular group. While the response variable during regression is a continuous one, that relevant to classification is a categorical variable representing the class from which the regressor set was gathered. The type of variables best suited for model building is continuous. In the case of classification models, the attributes or regressor variables can be categorical or continuous, but the *response variable has to be categorical.* Several regression methods have been described in previous chapters such as least square linear multivariate models, least square nonlinear models, maximum likelihood estimation (MLE) and neural network based multi-layer perceptron (MLP). All these methods can also be used in the classification context by simply assigning arbitrary numerical values to the different classes (called *coding*) which will serve as the dependent variable while training the classification model.

Let us consider the case of distinguishing between two groups A and B based on measurements of two predictor va-

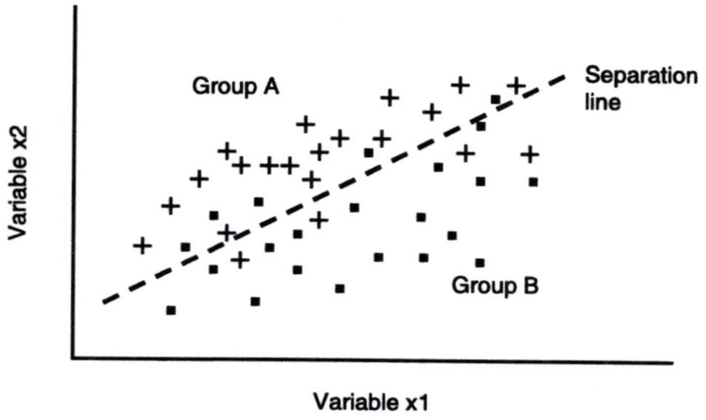

Fig. 8.3 Classification involves identifying a separating boundary between two known groups (shown as *dots* and *crosses*), based on two attributes (or variables or regressors) x_1 and x_2, which will minimize the misclassification of the points

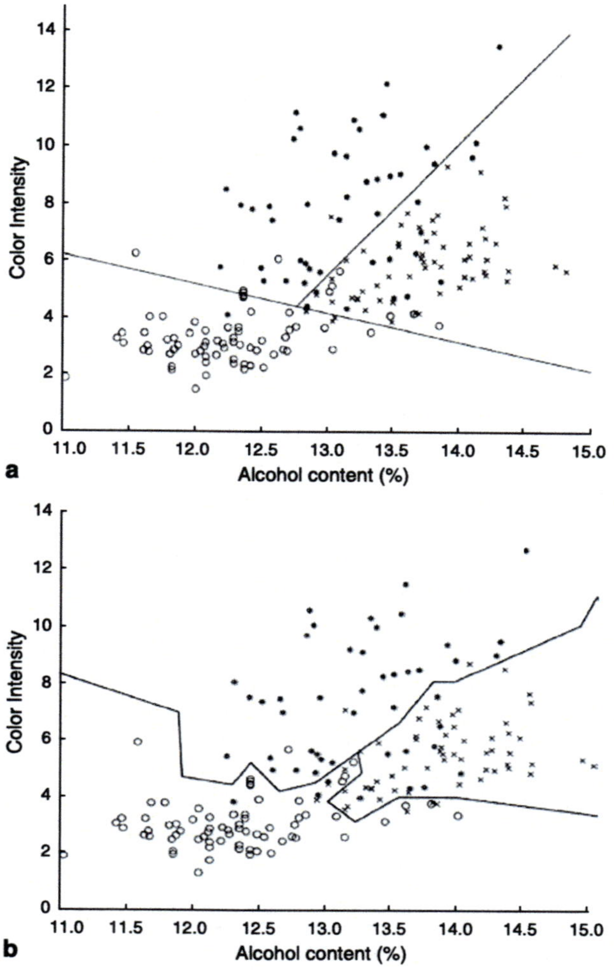

Fig. 8.4 Linear and piecewise linear decision boundaries for two-dimensional data (color intensity and alcohol content) used to classify the type of wine into one of three classes. **a** Linear decision boundaries. **b** Piecewise linear decision boundaries. (From Hair et al. 1998 by © permission of Pearson Education)

riables x_1 and x_2 (Fig. 8.3). The groups overlap and the two variables are moderately correlated; somewhat complicating the separation. One can simply use these two measurements as the regressor variables, and arbitrarily assign, say 0 and 1 to Group A and Group B respectively. Ordinary least-square (OLS) regression will directly yield the necessary model for the boundary between the two sets of data points. This model can be used to predict (or assign or classify) a future set of measurements into either Group A or Group B depending on whether the data point falls above or below the model line respectively. Just like the indices R^2 or RMSE are used to evaluate the goodness of fit of a regression model, the accuracy of classification (or the misclassification rate) of all the data points used to identify the regression model can serve as an indicator of the performance of the classification model. A better approach is to adopt the sample hold-out cross-validation approach (described in Sect. 5.3.2) where a portion of the data is used for training, and the rest for testing or evaluating the identified model. The corresponding indices are likely to be more representative of the actual model performance.

Figure 8.4a illustrates an example of a linear decision boundary for a set of data points from three known groups or classes. While separation of two classes needed a single linear model, two linear models are needed to separate three classes, and so on. The boundaries need not be linear, piecewise linear models could also be identified (Fig. 8.4b). An approach similar to using indicator variables in OLS regression can also be adopted which allows the flexibility of capturing local piece-wise behavior. An extension of such models to non-linear boundaries in more complex situations leads to multi-layer perceptron artificial neural network (ANN) models described in Sect. 11.3.3. However, it will be shown in the next section that OLS models are not really meant to be used for classification problems though they can provide good results in some cases. A better modeling approach is described in the next section.

8.2.4 Discriminant Function Analysis

Linear discriminant function analysis (LDA), originally proposed by Fisher in 1936, is similar to multiple linear regression analysis but approaches the problem differently. The similarity lies in that both approaches are based on identifying a linear model from a set of p observed quantitative variables x_i such as $z = w_0 + w_1x_1 + w_2x_2 + \cdots w_px_p$ where z is called the discriminant score and w_i are the model weights. A model such as this allows multivariate observations $\mathbf{X}_i$ to be converted into univariate observations z_i. However, the determination of the weights, which are similar to the model parameters during regression, is done differently. LDA seeks to determine weights that maximize the ratio of between-class scatter to the within-class scatter, or

$$\text{Max}\left\{\frac{\text{squared distance between means of z}}{\text{Variance of z}}\right\} = \frac{(\mu_1 - \mu_2)^2}{\sigma^2{}_z} \quad (8.4)$$

where μ_1 and μ_2 are the average values of z_i for Group A and Group B respectively, and the variance of z is that of any one group, with the assumption that the variances of both groups are equal.

Just like in OLS regression, the loss function need not necessarily be one that penalizes squared differences, but this is a form which is widely adopted. This approach allows two or more groups of data to be distinguished as best as possible. More weight is given to those regressors that discriminate well and vice versa. The method is optimal when the two classes are normally distributed with equal covariance matrices; even when they are not, the method is said to give satisfactory results.

The model, once identified, can be used for discrimination, i.e., to classify new observations as belonging to one or another group (in case of two groups only). This is done by determining the threshold or the *separating score,* with new objects having scores larger than this score being assigned to one class and those with lower scores assigned to the other class. If z_A and z_B are the mean discriminant scores of preclassified samples from groups A and B, the optimal choice for the threshold score z_{thres} when the two classes are of equal size, are distributed with similar variance and for equal misclassification rates $z_{thres} = (z_A + z_B)/2$. A new sample will be classified to one group or another depending on whether z is larger than or less than z_{thres}. Misclassification errors pertaining to one class can be reduced by appropriate weighting if the resulting consequences are more severe in one group as compared to the other.

Several studies use standardized regressors (with zero mean and unit variance, as usually done in principle component analysis or PCA) to identify the discriminant function. Others argue that an advantage of the LDA is that data need not be standardized since results are not affected by scaling of the individual variables. One difference is that, when using standardized variables, the discriminant function would not need an intercept term, while this is needed when using untransformed variables.

In constructing the discriminant functions under a multivariate case, one can include all of the regressor variables or adopt a stepwise selection procedure that includes only those variables that are statistically significant discriminators amongst the groups. A number of statistical software are available which perform such stepwise procedures and provide useful summaries and tests of significance for the number of discriminant functions. If the dependent variable is dichotomous (i.e., can only assume two values), there is only one discriminant function. If there are k levels or categories, upto (k-1) functions can be extracted. Just like in PCA (Sect. 10.3.2), successive discriminant functions are orthogonal to one another and one can test or determine how many are worth extracting. The interested reader can refer to pertinent texts such as Manly (2005), Hand (1981) or Duda et al. (2001) for a mathematical treatment of LDA, for establishing statistical significance of group differences, and for more robust methods such as quadratic discriminant analysis.

Though LDA is widely used for classification problems, it is increasingly being replaced by logistic regression (see Sect. 10.4.4) since the latter makes fewer assumptions, and hence, is more flexible (for example, discriminant analysis is based on normally distributed variables), and more robust statistically when dealing with actual data. Logistic regression is also said to be more parsimonious and the value of the weights easier to interpret. A drawback (if it is one!) is that logistic regression requires model weights to be estimated by maximum likelihood method (Sect. 10.4.3), which some analysts are not as familiar with as OLS regression.

Example 8.2.3:[1] *Using discriminant analysis to model fault-free and faulty behavior of chillers*

This example illustrates the use of multiple linear regression and linear discrimination analysis to classify two data sets representative of normal and faulty operation of a large centrifugal chiller. The faulty operation corresponds to chiller performance in which non-condensable (namely nitrogen gas) was intentionally introduced in the refrigerant loop. The three discerning variables are $T_{cd\text{-}sub}$ is the amount of refrigerant subcooling in the condenser (°C), $T_{cd\text{-}app}$ is the condenser approach temperature (°C) i.e., difference in condenser refrigerant temperature and that of the cooling water leaving the condenser, and COP is the coefficient of performance of the chiller. The data is assembled in Table 8.5 and shows the assigned grouping code (0 for fault-free and 1 for faulty), and the numerical values for the three regressors. The two data sets consist of 27 operating points each; however, 21 points are used for training the model while the remaining points are used for evaluating the classification model.

The 3-D scatter plot of the three variables used for model training is shown in Fig. 8.5 with the two groups distinguished by different symbols. One notes that the two groups are fairly distinct and that no misclassification should occur (in short, not a very challenging problem).

The following models were identified (with all coefficients being statistically significant):

- OLS regression model $z = 1.03467 - 0.33752 * \text{COP} - 0.3446 * T_{cd\text{-}sub} + 0.74493 * T_{cd\text{-}app}$
- LDA model $z = 2.1803 - 2.51329 * \text{COP} - 1.63738 * T_{cd\text{-}sub} + 4.19767 * T_{cd\text{-}app}$

[1] From Reddy (2007) with data provided by James Braun for which we are grateful.

Table 8.5 Analysis results for chiller fault detection example using two different methods, namely the OLS regression method and the linear discriminant analysis (LDA) method. Fault-free data is assigned a class value of 0 while faulty data a value of 1. The models are identified from training data and then used to predict class membership for testing data. The cut-off score is 0.5 for both approaches, i.e., if calculated score is less than 0.5, then the observation is deemed to belong to the fault-free behavior, and vice versa. LDA model scores also fall on either side of 0.5 but are magnified as compared to the OLS scores leading to more robust classification. In this example, there are no misclassification data points for either model during both training and testing periods

	Assigned grouping	Variables			OLS model score	LDA model score
	Class	COP	$T_{cd\text{-}sub}$ °C	$T_{cd\text{-}app}$ °C		
Training	0	3.765	4.911	2.319	−0.08	−5.59
	0	3.405	3.778	1.822	−0.01	−4.91
	0	2.425	2.611	1.009	0.09	−3.95
	0	4.512	5.800	3.376	0.04	−4.49
	0	4.748	4.589	2.752	−0.09	−5.71
	0	4.513	3.356	1.892	−0.19	−6.71
	0	3.503	2.244	1.272	−0.01	−4.96
	0	3.593	4.878	2.706	0.14	−3.48
	0	3.252	3.700	1.720	0.00	−4.83
	0	2.463	2.578	1.102	0.13	−3.60
	0	4.274	5.422	3.323	0.14	−3.49
	0	4.684	4.989	3.140	0.03	−4.58
	0	4.641	3.589	2.188	−0.14	−6.18
	0	3.038	1.989	1.061	0.06	−4.26
	0	3.763	4.656	2.687	0.13	−3.62
	0	3.342	3.456	1.926	0.11	−3.79
	0	2.526	2.600	1.108	0.11	−3.77
	0	4.411	5.411	3.383	0.13	−3.56
	0	4.029	3.844	2.128	−0.05	−5.31
	0	4.443	3.556	2.121	−0.11	−5.91
	0	3.151	2.333	1.224	0.04	−4.42
	1	3.587	6.656	4.497	0.62	1.14
	1	3.198	5.767	3.881	0.60	0.99
	1	2.416	4.333	2.793	0.58	0.74
	1	2.414	3.811	2.722	0.63	1.30
	1	4.525	7.256	5.359	0.65	1.42
	1	4.232	6.022	4.557	0.58	0.81
	1	3.424	4.544	3.538	0.60	0.99
	1	3.382	6.533	4.602	0.74	2.30
	1	3.017	5.667	3.907	0.68	1.72
	1	3.730	5.933	4.372	0.65	1.44
	1	2.395	3.989	2.884	0.68	1.74
	1	4.460	7.356	5.567	0.74	2.29
	1	4.166	6.044	4.697	0.66	1.53
	1	2.974	4.456	3.339	0.65	1.43
	1	3.568	7.033	4.710	0.65	1.47
	1	3.162	6.044	4.070	0.65	1.42
	1	2.382	4.544	3.116	0.69	1.83
	1	4.263	7.567	5.672	0.80	2.89
	1	3.757	5.967	4.368	0.63	1.30
	1	4.132	6.589	4.793	0.62	1.13
	1	2.944	4.811	3.470	0.65	1.47

Table 8.5 (continued)

	Assigned grouping	Variables			OLS model score	LDA model score
	Class	COP	$T_{cd\text{-}sub}$ °C	$T_{cd\text{-}app}$ °C		
Testing	0	3.947	3.567	1.914	−0.07	−5.54
	0	2.434	1.967	0.873	0.14	−3.49
	0	3.678	3.389	1.907	0.02	−4.61
	0	2.517	2.133	1.039	0.16	−3.28
	0	2.815	2.122	0.946	0.04	−4.40
	0	4.785	5.100	3.052	−0.06	−5.38
	1	4.330	7.656	5.513	0.70	1.91
	1	3.716	5.633	4.082	0.58	0.75
	1	2.309	4.489	2.954	0.65	1.43
	1	4.059	7.467	5.501	0.79	2.84
	1	2.615	4.511	3.239	0.69	1.82
	1	4.539	7.667	5.550	0.66	1.52

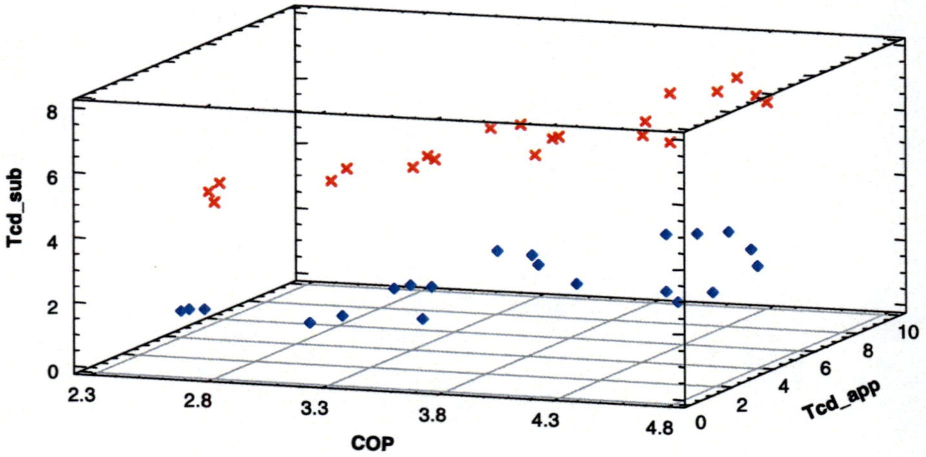

Fig. 8.5 Scatter plot of the three variables. Fault-free data (coded 0) is shown as diamonds while faulty data (coded 1) is shown as crosses. Clearly there is no overlap between the data sets and this is supported by the analysis which indicates no misclassified data points

The corresponding scores for both OLS and LDA are shown in the last two columns of the table. The threshold score z_{thres} is simply 0.5 which is the average of the two groups (coded as 0 and 1). A calculated z score less than 0.5 would suggest that the data point came from fault-free operation, while a score greater than 0.5 would suggest faulty operation. Note that there are no misclassification data points during either training or testing periods for either approach. However, note that for OLS there are several instances when the score is very close to the threshold value of 0.5. Figure 8.6 shows the predicted (using the OLS model) versus the "measured" values of the two groups which can only assume values of either 0 or 1 only. This figure clearly indicates the poor modeling capability of the OLS suggestive of the fact that OLS, though used by some for classification problems, is not really meant for this purpose. ■

8.2.5 Bayesian Classification

The Bayesian approach was addressed in Sect. 2.5 and also in Sect. 4.6. Bayesian statistics provide the formal manner by which prior opinion expressed as probabilities can be revised in the light of new information (from additional data collected) to yield posterior probabilities. The general approach can be recast into a framework which also allows classification tasks to be performed. The simplified or *Naïve Bayes* method assumes that the predictors are statistically independent and uses prior probabilities for training the model, which subsequently can be used along with the likelihood function of a new sample to classify the sample into the most likely group. The training and classification are easily interpreted. It is said to be most appropriate when the number of predictors is very high. Further, it is easy to use, handles missing data well, and

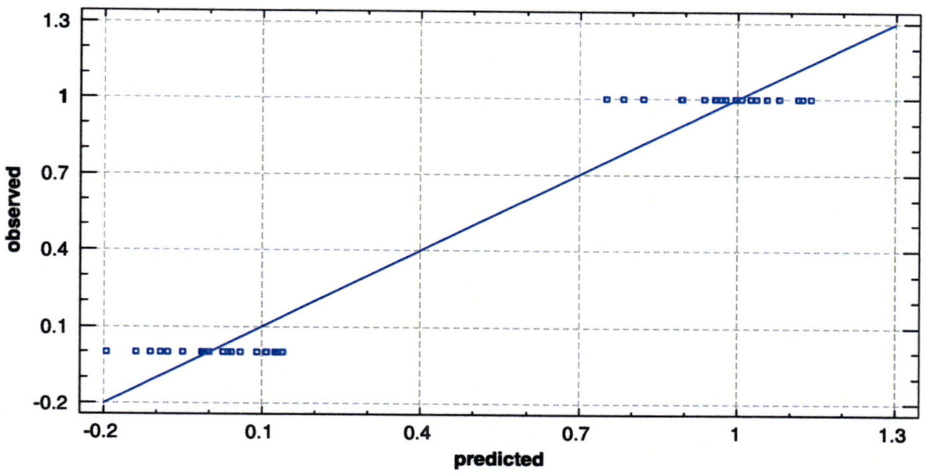

Fig. 8.6 Though an OLS model can be used for classification, it is not really meant for this purpose. This is illustrated by the poor correspondence between observed vs predicted values of the coded "class" variables. The observed values can assume numerical values of either 0 or 1 only, while the values predicted by the model range from −0.2 to about 1.2. See Example 8.2.3

Table 8.6 Count data of the 200 samples collected and calculated probabilities of attributes (Example 8.2.4)

Attribute		Number of samples			Calculated probabilities of attributes		
		Poor	Average	Good	Poor	Average	Good
Type	Bituminous	72	20	8	0.72	0.20	0.08
	Anthracite	0	44	56	0.0	0.44	0.56
Carbon %	50–60%	41	0	0	41/72	0	0
	60–70%	31	42	0	31/72	42/64	0
	70–80%	0	22	28	0	22/64	28/64
	80–90%	0	0	36	0	0	36/64
	Total	72	64	64	–	–	–

requires little computational effort. However, it does not handle continuous data well, and does not always yield satisfactory results because of its inherent assumption of predictor independence which is very often not the case. Despite these limitations, it is a useful analysis approach to have in one's toolkit.

Example 8.2.4: *Bayesian classification of coal sample based on carbon content*

There are several types of coal used in thermal power plants to produce electricity. A power plant gets two types of coal: bituminous and anthracite. Each of these two types can contain different fixed carbon content depending on the time and the location from where the sample was mined. Further, each of these types of coal can be assigned into one of three categories: Poor, Average and Good depending on the carbon content whose thresholds are different for the two types of coal. For example, a bituminous sample can be graded as "good" while an anthracite sample can be graded as "average" even though both samples may have the same carbon content. Table 8.6 shows the prior data (of 200 samples) and the associated probabilities of the attributes. This corresponds to the training data set.

These values are used to determine the prior probabilities as shown in the second column of Table 8.7. The power plant operator wants to classify a new sample of bituminous carbon which is found to contain 70–80% carbon content. The sample probabilities are shown in the third column of Table 8.7. The values in the likelihood column add up to 0.0332 which is used to determine the actual posterior probabilities shown in the last column. The category which has the highest posterior probability can then be identified. Thus, the new sample will be classified as "average". This is a simple contrived example meant to illustrate the concept and to show the vari-

Table 8.7 Calculation of the prior, sample and posterior probabilities

	Prior probabilities	Sample probabilities	Likelihood	Posterior probabilities
Poor (p)	p(p)=(41+31)/200=0.36	p(s/p)=0.72×0=0	0.36×0=0	0/0.091=0
Average (a)	p(a)=(42+22)/200=0.32	p(s/a)=0.20×(22/64)=0.0687	0.32×0.0687=0.022	0.022/0.0332=0.663
Good (g)	p(g)=(28+36)/200=0.32	p(s/g)=0.08×(28/64)=0.035	0.32×0.035=0.0112	0.0112/0.0332=0.337
			Sum=0.0332	

ous calculation steps which are straightforward to interpret by those with a basic understanding of Bayesian statistics. ■

8.3 Heuristic Classification Methods

8.3.1 Rule-Based Methods

The simplest type of rule-based method is the one involving "if-then" rules. Such classification rules consist of the "if" or antecedent part, and the "then" or consequent part of the rule (Dunham 2006). These rules must cover all the possibilities, and every instance must be uniquely assigned to a particular group. Such a heuristic approach is widely used in several fields because of the ease of interpretation and implementation of the algorithm. The following example illustrates this approach.

Example 8.3.1:[2] *Rule-based admission policy into the Yale medical school*

The selection committee framed the following set of rules for interviewing applicants into the school based on undergraduate (UG) GPA and MCAT verbal (V) and MCAT quantitative (Q) scores

- If UA GPA <3.47 and MCAT-V <555, then Class A- reject
- If UA GPA <3.47 and MCAT-V ≥ 555 and MCAT-Q <655, then Group B, reject
- If UA GPA <3.47 and MCAT-V ≥ 555 and MCAT-Q ≥ 655, then Group C, interview
- If UA GPA ≥ 3.47 and MCAT-V <535, then Group D, reject
- If UA GPA ≥ 3.47 and MCAT-V ≥ 535, then Group E, interview.

It is clear that the set of rules is comprehensive and would cover every eventuality. For example, an applicant with UA GPA $=3.6$ and MCAT-V $=525$ would fall under group D and be rejected without an interview. Thus, the pre-determined threshold or selection criteria of GPA, MCAT-V and MCAT-Q are in essence the classification model, while classification of a future applicant is straightforward. ■

8.3.2 Decision Trees

Probability trees were introduced in Sect. 2.2.4 as a means of dividing a decision problem into a hierarchical structure for easier understanding and analysis. Very similar in concept are directed graphs or decision trees which are predictive modeling approaches that can be used for classification, clustering as well as for regression model building. As stated earlier, classification problems differ from regression problems in that the response variable is categorical in the former, and continuous in the latter. Treed regression is addressed in Sect. 8.4, while this section limits itself to classification problems. Decision trees essentially divide the spatial space such that each branch can be associated with a different sub-region. A rule is associated with each node of the tree, and observations which satisfy the rule are assigned to the corresponding branch of the tree. Terminal nodes are the end-nodes of the tree. Though similar to if-then rules in their structure, decision trees are easier to comprehend in more complex situations, and are more efficient computationally.

Example 8.3.2: Consider the same problem as that of Example 8.3.1. These if-then rules can be represented by a tree diagram as shown in Fig. 8.7. The top node contains the entire data set, while each node further down contains a subset of data till the branches end in a unique branch representing one of the five possible groups (Groups A-E). In many ways, this diagram is easier to comprehend than the if-then rules, and further allows intuitive tweaking of the rules as necessary. The numbers in parenthesis shown in Fig. 8.7 represent the numbers of applicants at different stages of the tree (out of a total of 727). Note that 481 applicants made the cut (sum of those who ended up in Group C or Group E) and would be called for an interview. If the school decides to cut this number down in future years, it can use the set of 727 applicants, and iteratively evaluate the effect of modifying the rules at different levels till the intended target reduction is achieved. This can be programmed into a computer to facilitate the search for an optimum set of rules. A simpler and more intuitive manner is to study Fig. 8.7, and evaluate certain heuristic modifications. For example, more successful applicants fall in Group E than Group C, and so increasing the MCAT-V score threshold from 535 to something a little higher may be an option worth evaluating.

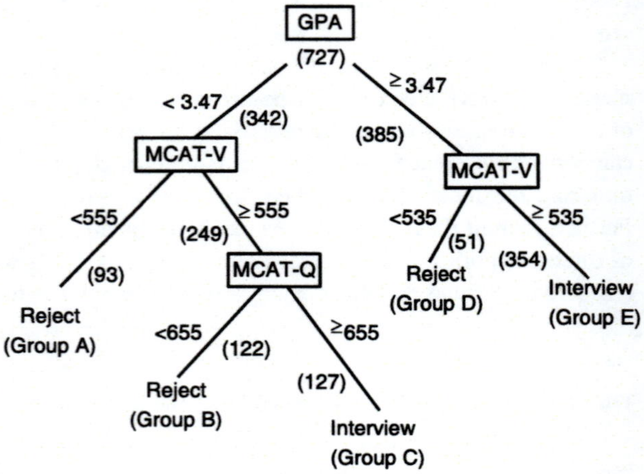

Fig. 8.7 Tree diagram for the medical school admission process with five terminal nodes, each representing a different group (Example 8.3.2). This is a binary tree with three levels

[2] From Milstein et al. (1975)

8.3.3 k Nearest Neighbors

The k nearest neighbor (kNN) method is a conceptually simple pattern recognition approach that is widely used for classification. It is based on the distance measure, and requires a training data set of observations from different groups identified as such. If a future object needs to be classified, one determines the point closest to this new object, and simply assigns the new object to the group to which the closest point belongs. Thus, no training as such is needed. The classification is more robust if a few points are used rather than a single closest neighbor. This, however, leads to the following issues which complicate the classification:

(i) how many closest points "k" should be used for the classification, and
(ii) how to reconcile differences when the nearest neighbors come from different groups.

Because of different ways by which the above issues can be addressed, kNN is more of an algorithm than a clear-cut analytical procedure. An allied classification method is the *closest neighborhood* scheme, where an object is classified in that group for which its distance from the *center* of that group happens to be the smallest as compared to its distances from the centers of other possible groups. Training would involve computing the centers of each group and distances of individual objects from this center.

A redeeming feature of kNN is that it does not impose a priori any assumptions about the distribution from which the modeling sample is drawn. Stated differently, kNN has the great advantage that it is asymptotically convergent, i.e., as the size of the training set increases, misclassification errors will be minimized if the observations are independent regardless of the distribution from which the sample is drawn. kNN can be adapted to a wide range of applications, with the distance measure modified to suit the particular application. The following example illustrates one such application.

Example 8.3.3: *Using k nearest neighborhood to calculate uncertainty in building energy savings*

This example illustrates how the nearest neighbor approach can be used to estimate the uncertainty in building energy savings after energy conservation measures (ECM) have been installed (adapted from Subbarao et al. 2011). Example 5.7.1 and Problem Pr. 5.12 describe the analysis methodology which consists of four steps:

(i) identify a baseline multivariate regression model for energy use against climatic and operating variables before the retrofits were implemented,
(ii) use this baseline model along with post-retrofit climatic and operating variables data to predict energy use $E_{\text{pre, model}}$ reflective of consumption during the pre-retrofit stage,
(iii) compute energy savings as the difference between the model predicted baseline energy use and the actual measured energy use during the post-retrofit period, $E_{\text{savings}} = (E_{\text{pre, model}} - E_{\text{post, meas}})$ and,
(iv) determine the uncertainty in the energy savings based on the multivariate baseline model goodness-of-fit (such as the RMSE) and the uncertainty in the post-retrofit measured energy use.

Unfortunately, step (iv) is not straightforward. Energy use models identified by global or year-long data do not adequately capture seasonal changes in energy use due to control and operation changes done to the various building systems since these variables do not appear explicitly as regressor variables. Hence, classical models identified from whole-year data are handicapped in this respect, and this often leads to model residuals that have different patterns during different times of the year. Such improper residual behavior results in improper estimates of the uncertainty surrounding the measured savings. An alternative is to use the nearest neighbors approach which relies on "local" model behavior as against global estimates such as the overall RMSE. However, the k-nearest neighbor approach requires two aspects to be defined specific to the problem at hand: (i) definition of the distance measure, (ii) and deciding on the number of neighbor points to select.

Let us assume that a statistical model with p regressor parameters has been identified from the pre-retrofit period based on daily variables. Any day, for specificity, a pre-retrofit day j, can be represented as a point in this p-dimensional space. If data for a whole year is available, the days are represented by 365 points in this p-dimensional space. The uncertainty in this estimate is better characterized by identifying a certain number of days in the pre-retrofit period which closely match the specific values of the regressor set for the post-retrofit day j, and then determining the error distribution from this set of days. Thus, the method is applicable regardless of the type of model residual behavior encountered.

All regressors do not have the same effect on the response variable; hence, those that are more influential need to be weighted more, and vice versa. The definition of the *distance* d_{ij} between two given days i and j specified by the set of regressor variables $x_{k,i}$ and $x_{k,j}$ is defined as:

$$d_{ij} = \sqrt{\sum_{k=1}^{p} w_k^2 (x_{k,i} - x_{k,j})^2 / p} \tag{8.5}$$

where the weights w_k are given in terms of the derivative of energy use with respect to the regressors:

$$w_k = \left(\frac{\partial E_{pre,\,model}}{\partial x_k} \right) \tag{8.6}$$

The partial derivatives can be determined numerically by perturbation, as discussed in Sect. 3.7.1. Days that are at a

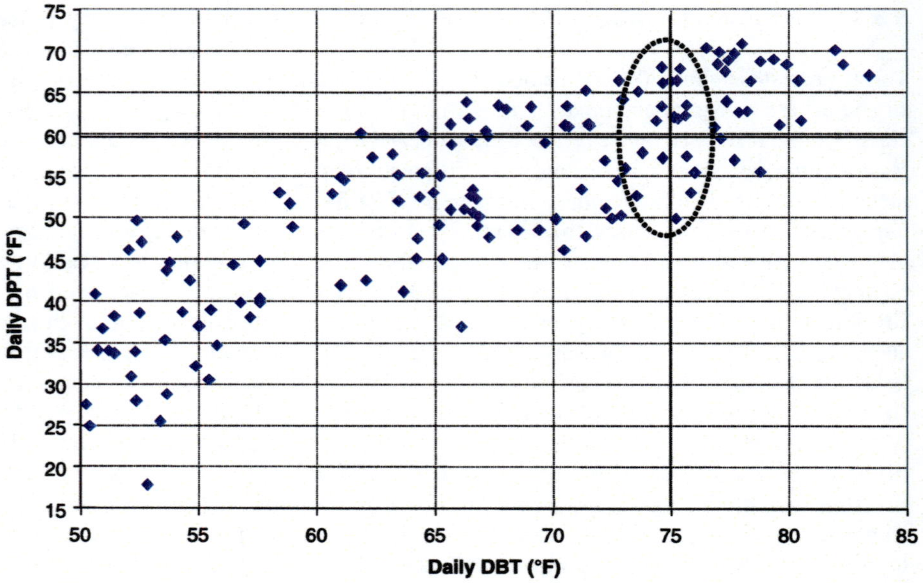

Fig. 8.8 Illustration of the neighborhood concept for a baseline with two regressors (dry-bulb temperature DBT and dew point temperature DPT) for Example 8.3.3. If the DBT variable has more "weight" than DPT on the variation of the response variable, this would translate geometrically into an elliptic domain as shown. The data set of "neighbor points" to the post datum point (75, 60) would consist of all points contained within the ellipse. Further, a given point within this ellipse may be assigned more "influence" the closer it is to the center of the ellipse. (From Subbarao et al. 2011)

given "energy distance" from a given day lie on an ellipsoid whose axis in the k-direction is proportional to $(1/\sqrt{w_k})$. This concept is illustrated in Fig. 8.8.

The selection of the number of neighbor points is somewhat arbitrary and can be done either by deciding on a maximum distance, or selecting the number of points based on the confidence level sought; the latter approach has been adopted below. One can associate an ellipsoid with each post-retrofit day in the parameter space; pre-retrofit days that lie inside this ellipsoid contribute to the determination of uncertainty in the estimation of the savings for this particular post-retrofit day. The distribution of uncertainties in the estimate $E_{\text{post, meas, j}}$ is given by the distribution of the estimates of $(E_{\text{pre, model, j}} - E_{\text{post, meas, j}})$ inside the ellipsoid. The overall size of the ellipsoid is determined by the requirements of making it as small as possible (so that variations in the daily energy use are small) while having a sufficient number of pre-retrofit days within the ellipsoid.

The proposed approach is illustrated with a simple example involving synthetic daily data of building energy use. Though the simulation involves numerous variables, only the following variables are considered as they relate to a subsequent statistical model: (i) regressors: ambient air dry-bulb temperature (DBT) and dew point temperature (DPT), and (ii) response: cooling coil thermal load (Q_c). The hourly data has first been separated in weekdays and weekends and then averaged/summed to represent daily values. Only the weekday data set consisting of 249 values have been used to illustrate the concept of the proposed methodology. Figure 8.9 is a scatter plot of cooling load Q_c versus DBT. This is a typical

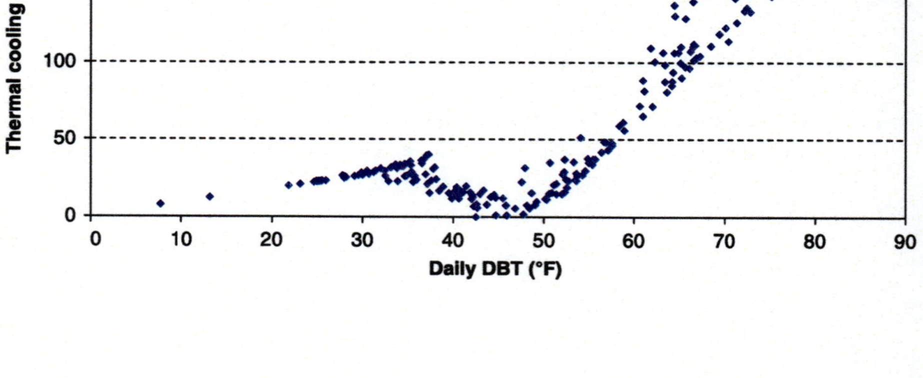

Fig. 8.9 Scatter plot of thermal cooling load Q_c versus DBT for Example 8.3.3

plot showing strong change point non-linear behavior and a large cloud at the high DBT range due to humidity loads. It would very difficult to model this behavior using traditional regression methods that would yield realistic uncertainty estimates at specific local ranges.

Previous studies have demonstrated that only when the outdoor air dew point temperature is higher than about 55 °F (which is usually close to the cold air deck temperature of the air handler unit) do humidity loads appear on the cooling coil due to the ventilation air brought into the building. Hence, the variable $(DPT\text{-}55)^+$ rather than DPT is used as the regressor variable in the model which is such that: $(DPT\text{-}55)^+ = 0$ when DPT < 55 °F, and $(DPT\text{-}55)^+ = (DPT\text{-}55)$ when DPT ≥ 55 °F.

A more critical issue with multivariate models in general is the collinear behavior between regressors; in the case of building energy use models, the most critical is the significant correlation between DBT and DPT. If one ignores this, then the energy use model may have physically unreasonable internal parameter values, but continue to give reasonable predictions. However, the derivatives of the response variable with respect to the regressor variables can be very misleading. One variable may "steal" the dependence from another variable, which will affect the weights assigned to the different regressors. To mitigate this problem, a model of $(DPT\text{-}55°F)^+$ vs DBT is identified, and the residuals of this model (ResDPT) are used instead of DPT in the regressor set for energy use modeling. Though this procedure assigns more influence to DBT, the collinearity effect is reduced. The model could be a simple linear model or could be an artificial neural network (ANN) multi-layer perceptron model, depending on the preference of the analyst. Consider the case when one wishes to determine the uncertainty in the response variable corresponding to a set of operating conditions specified by DBT=75°F and ResDPT=5°F which results in Q_c=233.88 MBtu/day. The ANN 3-10-1 model was used to numerically determine the gradients of these two regressors:

$$\frac{\partial Q_c}{\partial (DBT)} = 5.0685 \quad \text{and} \quad \frac{\partial Q_c}{\partial (\mathrm{Res}DPT)} = 7.606$$

The "distance" statistic for each of the 249 days in our synthetic data set has been computed following Eqs. 8.5 and 8.6, and the data sorted by this statistic. The top 20 data points (with smallest distance) are shown in Table 8.8, as are the regressor values, the measured and predicted values, and their residuals. The last column assembles the "distance" variable. Note that this statistic varies from 1.78 to 23.06. In case the 90% confidence intervals are to be determined, a distribution-free approach is to use the corresponding values of the 5th and the 95th percentiles of the residuals. Since there are 20 points, the two extreme values of the residuals shown in Table 8.8, which yields the 90% limits (−8.43 and 8.29 which are bolded) around the model predicted value of 233.88 MBtu/day for the cooling energy use. In this case, the distribution is fairly symmetric, and one could report a local prediction value of (233.88 ± 8.3 MBtu/day) at the 90% confidence level. If the traditional method of reporting uncertainty were to be adopted, the RMSE for the 2-10-1 ANN model, found to be 5.7414 (or a CV=6.9%), would result in (±9.44 MBtu/day) at the 90% confidence level. Thus, using the k-nearest neighbors approach has led to some reduction in the uncertainty interval around the local prediction value; but more importantly, this estimate of uncertainty is more realistic and robust since it better represents the local behavior of the relationship between energy use and the regressor variables. Needless, to say, the advantage of this entire method is that even when the residuals are not normally distributed, the data itself can be used to ascertain statistical limits.

8.4 Classification and Regression Trees (CART) and Treed Regression

Classification and regression trees (CART) approach is a non-parametric decision tree technique that can be applied either to classification or regression problems, depending on whether the dependent variable is categorical or numeric respectively. Recall that nonparametric methods are those which do not rely on assumptions about the data distribution. In Sect. 8.3.2 dealing with decision trees, the model-building step was not needed because the tree structure, the attributes and their decision rules were specified explicitly. This will not be the case in most classification problems. Constructing a tree is analogous to training in a model-building context, but here, it involves deciding on the following choices or collection of rules (Dunham 2006):

(i) choosing the splitting attributes, i.e., the set of important variables to perform the splitting; in many engineering problems, this is a moot step,
(ii) ordering the splitting attributes, i.e., ranking them by order of importance in terms of being able to explain the variation in the dependent variable,
(iii) deciding on the number of splits of the splitting attributes which is dictated by the domain or range of variation of that particular attribute,
(iv) defining the tree structure, i.e., number of nodes and branches,
(v) selecting stopping criteria which are a set of pre-defined rules meant to reveal that no further gain is being made in the model; this involves a trade-off between accuracy of classification and performance, and
(vi) pruning a tree which involves making modifications to the tree constructed using the training data so that it applies well to the testing data.

Table 8.8 Table showing how the model residual values can be used to ascertain pre-specified confidence levels of the response adopting a non-parametric approach (Example 8.3.3). Values shown of the regressor and response variables are for the 20 closest neighborhood points from a data set of 249 points. Reference point of DBT=75°F and DPT-55°=5°F determined using ANN model 2-10-1 are shown, as are the "distance" and the model residual values. The residual values shown bolded are the 5 and 95% values

	DBT (°F)	ResDPT (°F)	Q_c_Meas (10^6 Btu/day)	Q_c_Model (10^6 Btu/day)	Residuals (10^6 Btu/day)	Distance
1	74.67	4.76	225.59	233.88	**8.29**	1.78
2	75.42	4.22	236.36	231.39	−4.97	4.47
3	77.08	5.45	240.21	248.19	7.98	7.85
4	76.54	6.23	251.99	252.94	0.95	8.62
5	77.00	4.08	239.01	238.55	−0.46	8.71
6	77.46	4.32	241.97	242.60	0.64	9.54
7	72.83	3.88	224.54	217.61	−6.93	9.82
8	77.75	5.00	240.63	247.13	6.50	9.86
9	75.04	2.88	221.23	223.84	2.61	11.40
10	75.29	2.85	224.36	222.07	−2.29	11.61
11	74.71	2.80	214.56	220.89	6.33	11.89
12	78.08	6.08	252.04	256.14	4.10	12.49
13	77.33	3.07	231.57	234.58	3.00	13.32
14	71.42	3.37	210.83	204.44	−6.39	15.53
15	78.83	3.61	238.85	242.55	3.70	15.63
16	73.67	2.19	200.35	209.02	8.66	15.87
17	79.42	3.60	250.10	241.68	**−8.43**	17.54
18	73.00	1.62	213.35	204.23	−9.12	19.55
19	79.96	2.74	240.66	239.73	−0.92	21.53
20	78.42	1.37	224.75	230.06	5.32	23.06

Classification and regression trees (CART) is one of an increasing number of computer intensive methods which perform an exhaustive search to determine best tree size and configuration in multivariate data. While being a fully automatic method, it is flexible, powerful and *parsimonious, i.e., it identifies a tree with the fewest number of branches.* Another appeal of CART is that it chooses the splitting variables and splitting points that best discriminate between the outcome classes. The algorithm, however, suffers from the danger of over-fitting, and hence, a cross-validation data set is essential. This would assure that the best tree configuration is selected which minimizes misclassification rate, and also give realistic estimates of the misclassification rate of the final tree. Most trees, including CART are binary decision trees (i.e., the tree splits into two branches at each node), though they do not necessarily have to be so. Also, each branch of the tree ends in a terminal node while each observation falls into one and exactly one terminal node. The tree is created by an exhaustive search performed at each node to determine the best split. The computation stops when any further split does not improve the classification. *Treed regression* is very similar to CART except that the latter fits the mean of the dependent variable in each terminal node, while treed regression can assume any functional form.

CART and treed regression are robust methods which are ideally suited for the analysis of complex data which can be numeric or categorical, involving nonlinear relationships, high-order interactions, and missing values in either response or regressor variables. Despite such difficulties, the methods are simple to understand and give easily interpretable results. Trees explain variation of a single response variable by repeatedly splitting the data into more homogeneous groups or spatial ranges, using combinations of explanatory variables that may be categorical and/or numeric. Each group is characterized by a typical value of the response variable, the number of observations in the group, and the values of the explanatory variables that define it. The tree is represented graphically, and this aids exploration and understanding. Classification and regression have a wide range of applications, including scientific experiments, medical diagnosis, fraud detection, credit approval, and target marketing (Hand 1981). The book by Breiman et al. (1984) is recommended for those interested in a more detailed understanding of CART and its computational algorithms. Even though the following example illustrates the use of treed regression in a regression context with continuous variables, an identical approach would be used with categorical data.

Example 8.4.1: *Using treed regression to model atmospheric ozone variation with climatic variables.*

Cleveland (1994) presents data from 111 days in the New York City metropolitan region in the early 1970s consisting

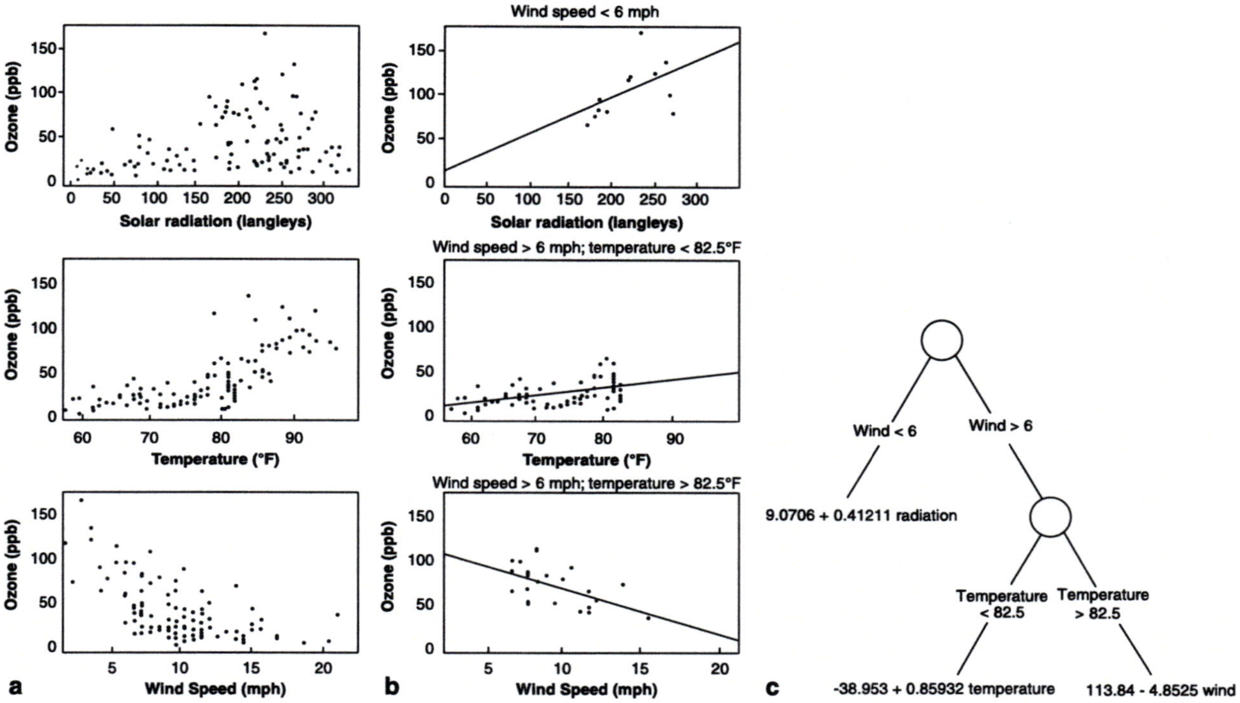

Fig. 8.10 **a** Scatter plots of ozone versus climatic data. **b** Scatter plots and linear regression models for the three terminal nodes of the treed regression model. (From Kotz 1997 by permission of John Wiley and Sons). **c** Treed regression model for predicting ozone level against climatic variables. (From Kotz 1997 by permission of John Wiley and Sons)

of the ozone concentration (an index for air pollutant) in parts per billion (ppb) and three climatic variables: ambient temperature (in°F), wind speed (in mph) and solar radiation (in langleys). It is the intent to develop a regression model for predicting ozone levels against the three variables. The pairwise scatter plots of ozone (the dependent variable) and the other three variables are shown in Fig. 8.10a. One notes that though some sort of correlation exits, the scatter is fairly important. An obvious way is to use multiple regression with inclusion of higher order terms as necessary.

An alternative, and in many cases superior, approach is to use treed regression. This involves, partitioning the spatial region into sub-regions, and identifying models for each region or terminal node separately. Kotz (1997) used a treed regression approach to identify three terminal nodes as shown in Fig. 8.10c: (i) wind speed<6 mph (representative of stagnant air conditions), (ii) wind speed>6 mph and ambient temperature <82.5°F, and (iii) wind speed>6 mph and ambient temperature >82.5°F. The corresponding pairwise scatter plots are shown in Fig. 8.10b while the individual models are shown in Fig. 8.10c. One notes that though there is some scatter around a straight line for the three terminal nodes, it is much less than adopting a straightforward multiple regression approach. ■

8.5 Clustering Methods

8.5.1 Types of Clustering Methods

The aim of cluster analysis is to allocate a set of observation sets into groups which are similar or "close" to one another with respect to certain attribute(s) or characteristic(s). Thus, an observation can be placed in one and only one cluster. For example, performance data collected from mechanical equipment could be classified as representing good, faulty or uncertain operation.

In general, the number of clusters is not predefined and has to be gleaned from the data set. This and the fact that one does not have a training data set to build a model make clustering a much more difficult problem than classification. A wide variety of clustering techniques and algorithms has been proposed, and there is no generally accepted best method. Some authors (for example, Chatfield 1995) point out that, except when the clusters are clear-cut, the resulting clusters often depend on the analysis approach used and somewhat subjective. Thus, there is often no one single best result, and there exists the distinct possibility that different analysts will arrive at different results.

Broadly speaking, there are two types of clustering methods both of which are based on distance-algorithms where objects are clustered into groups depending on their relative closeness to each other. Such distance measures have been described earlier: the Euclidian distance given by Eq. 8.2 or the Mahanabolis distance given by Eq. 8.3. One clustering approach involves partitional clustering where non-overlapping clusters are identified. The second involves hierarchic clustering which allows one to identify closeness of different objects at different levels of aggregation. Thus, one starts by identifying several lower-level clusters or groups, and then gradually merging these in a sequential manner depending on their relative closeness, so that finally only one group results. Both approaches rely, in essence, in identifying those which exhibit *small within-cluster variation* as against *large between-cluster variation.* Several algorithms are available for cluster analysis, and the intent here is to provide a conceptual understanding.

8.5.2 Partitional Clustering Methods

Partitional clustering (or disjoint clusters) determines the optimal number of clusters by performing the analysis with different pre-selected number of clusters. For example, if a visual inspection of the data (which is impossible in more than three dimensions) suggests, say, 2, 3, or 4 clusters, the analysis is performed separately for all three cases. The analysis would require specifying a criterion function used to assess the goodness-of-fit. A widely used criterion is the *within-cluster variation,* i.e., squared error metric which measures the square distance from each point within the cluster to the centroid of the cluster (see Fig. 8.11). Similarly, a between-cluster variation can be computed representative of the distance from one cluster center to another. The ratio of the between-cluster variation to the average within clusters is analogous to the F-ratio used in ANOVA tests. Thus, one starts with an arbitrary number of cluster centers, assigns objects to what is deemed to be the nearest cluster center, computes the F-ratio of the resulting cluster, and then jiggles the objects back and forth between the clusters each time re-calculating the mean so that the F ratio is maximized or is sufficiently large. It is recommended that this process be repeated with different seeds or initial centers since their initial selection may result in cluster formations which are localized. This tedious process can only be done by computers for most practical problem. A slight deviant of the above algorithm is the widely used *k-means algorithm* where instead of a F-test, the sum of the squared errors is directly used for clustering. This is best illustrated with a simple two-dimension sample.

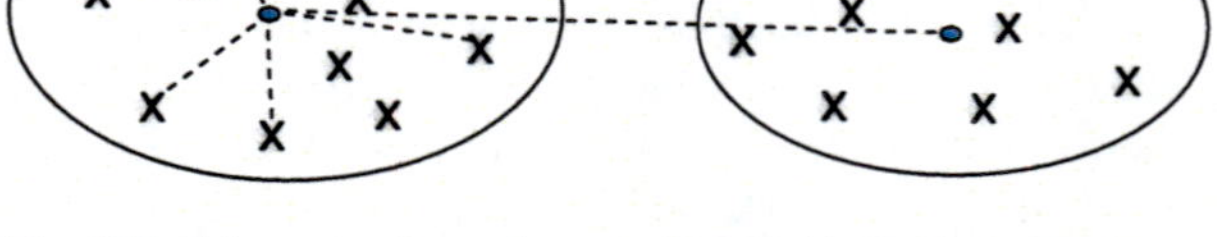

Fig. 8.11 Schematic of two clusters with individual points shown as x. The within-cluster variation is the sum of the individual distances from the centroid to the points within the cluster, while the between-cluster variation is the distance between the two centroids

Example 8.5.1:[3] *Simple example of the k-means clustering algorithm.*

Consider five objects or points characterized by two Cartesian coordinates: $x_1=(0,2)$; $x_2=(0,0)$, $x_3=(1.5,0)$, $x_4=(5,0)$, and $x_5=(5,2)$. The process of clustering these five objects is described below.

(a) *Select an initial partition of k clusters containing randomly chosen samples and compute their centroids*
Say, one selects two clusters and assigns to cluster $C_1 = (x_1, x_2, x_4)$ and $C_2 = (x_3, x_5)$. Next, the centroids of the two clusters are determined:

$$\begin{aligned} M_1 &= \{(0+0+5)/3, (2+0+0)/3\} \\ &= \{1.66,\ 0.66\} \\ M_2 &= \{(1.5+5)/2, (0+2)/2\} \\ &= \{3.25, 1.0\} \end{aligned}$$

(b) *Compute the within-cluster variations:*

$$\begin{aligned} e_1^2 &= [(0-1.66)^2 + (2-0.66)^2] + [(0-1.66)^2 \\ &\quad + (0-0.66)^2] + [(5-1.66)^2 \\ &\quad + (0-0.66)^2] = 19.36 \\ e_2^2 &= [(1.5-3.25)^2 + (0-1)^2] \\ &\quad + [(5-3.25)^2 + (2-1)^2] \\ &= 8.12 \end{aligned}$$

and the total error $E^2 = e_1^2 + e_2^2 = 19.36 + 8.12 = 27.48$

(c) *Generate a new partition by assigning each sample to the closest cluster center*
For example, the distance of x_1 from the centroid M_1 is $d(M_1, x_1) = (1.66^2 + 1.34^2)^{1/2} = 2.14$, while that for $d(M_2, x_1) = 3.40$. Thus, object x_1 will be assigned to the group which has the smaller distance, namely C_1. Similarly, one can compute distance measures of all other objects, and assign each object as shown in Table 8.9.

(d) *Compute new cluster centers as centroids of the clusters*
The new cluster centers are $M_1=\{0.5,\ 0.67\}$ and $M_2=\{5.0,\ 1.0\}$

(e) *Repeat steps (b) and (c) until an optimum value is found or until the cluster membership stabilizes*

[3] From Kantardzic (2003) by permission of John Wiley and Sons.

Table 8.9 Distance measures of the five objects with respect to the two groups

$d(M_1, x_1) = 2.14$	$d(M_2, x_1) = 3.40$	So assign $\Rightarrow x_1 \in C_1$
$d(M_1, x_2) = 1.79$	$d(M_2, x_2) = 3.40$	So assign $\Rightarrow x_2 \in C_1$
$d(M_1, x_3) = 0.83$	$d(M_2, x_3) = 2.01$	So assign $\Rightarrow x_3 \in C_1$
$d(M_1, x_4) = 3.41$	$d(M_2, x_4) = 2.01$	So assign $\Rightarrow x_4 \in C_2$
$d(M_1, x_5) = 3.60$	$d(M_2, x_5) = 2.01$	So assign $\Rightarrow x_5 \in C_2$

For the new clusters $C_1=(x_1, x_2, x_3)$ and $C_2=(x_4, x_5)$, the within-cluster variation and the total square errors are: $e_1^2 = 4.17$, $e_2^2 = 2.00$, $E^2 = 6.17$. Thus, the total error has decreased significantly just after one iteration. ■

It is recommended that the data be plotted so that starting values of the cluster centers could be visually determined. Though this is a good strategy in general, there are instances when this is not optimal. Consider the data set in Fig. 8.12a, where one would intuitively draw the two clusters (dotted circles) as shown. However, it turns out that the split depicted in Fig. 8.12b results in lower sum of squared error which is the better manner of performing the clustering. Thus, initial definition of cluster centers done visually have to be verified by analytical measures. Though the k-means clustering method is very popular, it is said to be sensitive to noise and outlier points.

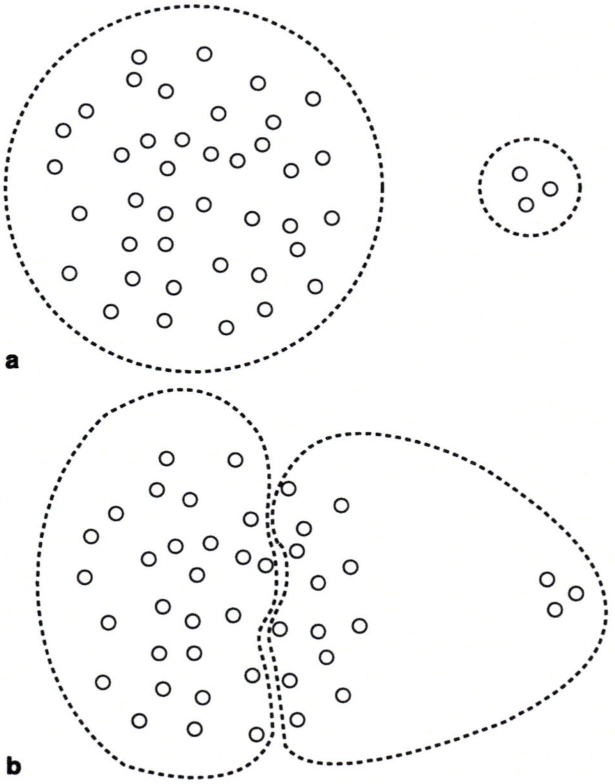

Fig. 8.12 Visual clustering may not always lead to an optimal splitting. **a** Obvious manner of clustering. **b** The better way of clustering which results in lower sum of squared error. (From Duda et al. 2001 by permission of John Wiley and Sons)

Example 8.5.2:[4] *Clustering residences based on their air-conditioner electricity use during peak summer days*

Electricity used by residential air conditioners in the U.S. has been identified as being largely responsible for the high electric demand faced by electric utilities during hot summer afternoons. Being able to classify residences based on their diurnal profiles during such critical days would be advantageous to electric utilities. For example, they would be able to better design and implement cost-effective peak shaving strategies (such as direct load control, cool storage, offering financial incentives, …). Hourly data for 73 residential homes was collected during an entire summer. Data corresponding to six of the peak days were extracted with the intent to classify residences based on their similarity in diurnal profiles during these days. Clustering would require two distinct phases: first, a process whereby the variability of the patterns can be quantified in terms of relatively few statistical parameters, and second, a process whereby objects are assigned to specific groups for which both the nuclei and the boundaries need to be determined.

First, the six diurnal profiles for each of the customers were averaged so as to obtain a single diurnal profile. The peak period occurs only for a certain portion of the day, in this case, from 2:00–8:00 pm, and hence the diurnal profile during this period is of greater importance than that outside this period. The most logical manner of quantifying the hourly variation during this period is to compute a measure representative of the mean and one of the standard deviation. Hence, a "peak size" was defined as the fraction of the air-conditioner use during the peak period divided by the total daily usage, and a "peak variation" as the standard deviation of the hourly air-conditioning values during the same peak period.

The two-dimensional data is plotted in Fig. 8.13a, while a partitional disjoint clustering approach was used to identify the nine clusters (Fig. 8.13b). While points shown as 0 are outliers, one does detect a logical and well-behaved clustering of the rest of the objects. Each of the individual clusters can be interpreted based on its peak pattern; only three of which are shown in Fig. 8.13. It can be noted that Cluster 1 includes individuals with a flat usage pattern, while Cluster 4 comprises of customers with properly sized and well-operated air-conditioners, and Cluster 7 those with a much higher peak load probably because of varying their thermostat setting during the day. ■

[4] From Hull and Reddy (1990).

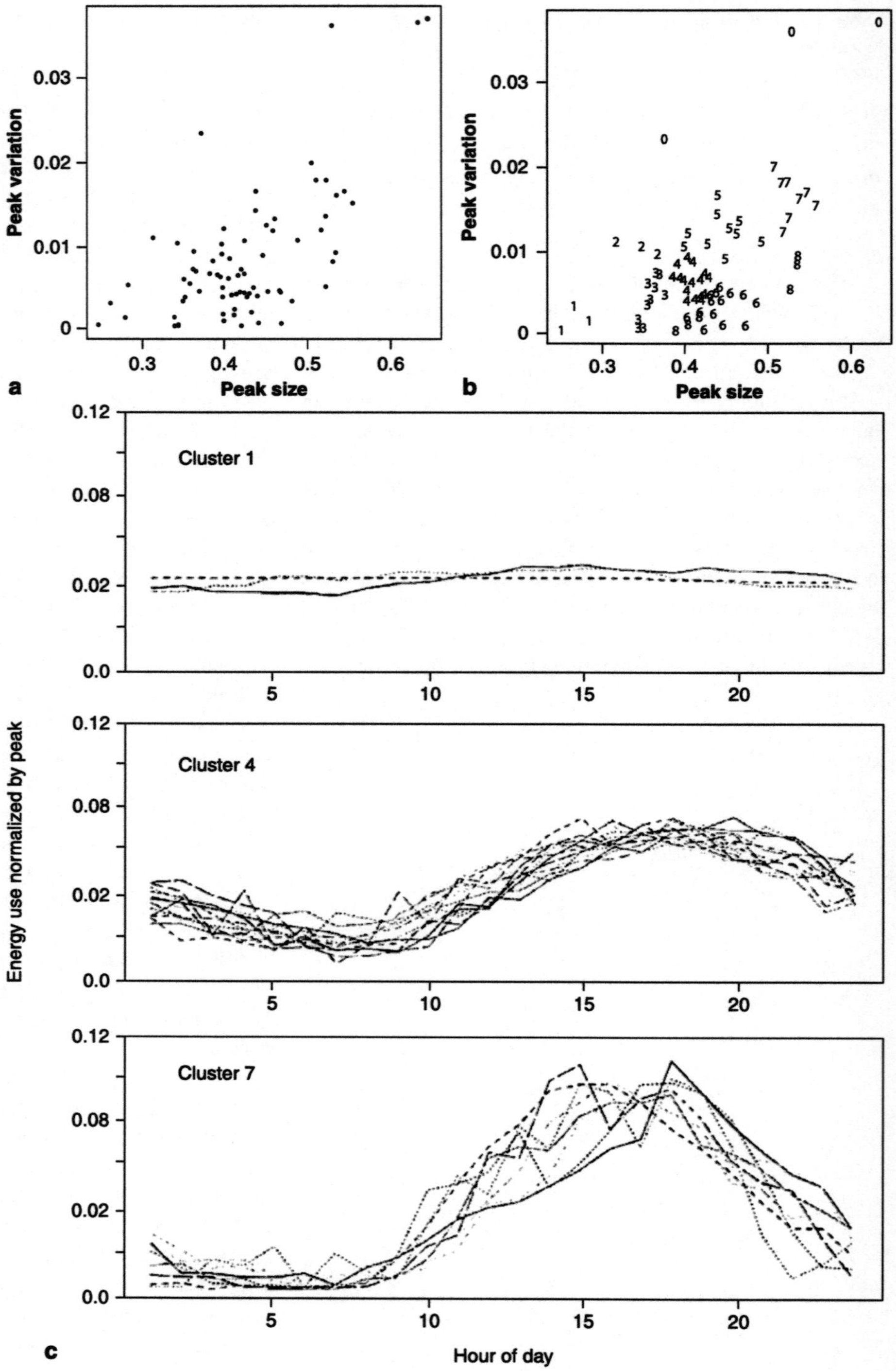

Fig. 8.13 Clustering of the 73 residences based on their air-conditioner use during peak hours. **a** Normalized two dimensional data. **b** Clustered data. **c** Normalized profiles of three of the eight clusters of homeowners identified. (From Hull and Reddy 1990)

8.5.3 Hierarchical Clustering Methods

Another cluster identification algorithm, called hierarchical clustering, does not start by partitioning a set of objects into mutually exclusive clusters, but forms them sequentially in a nested fashion. For example, the eight objects shown at the left of the tree diagram (also called dendrogram) in Fig. 8.14a are merged into clusters at different stages depending on their relative similarity. This allows one to identify objects which are close to each other at different levels. The sets of objects

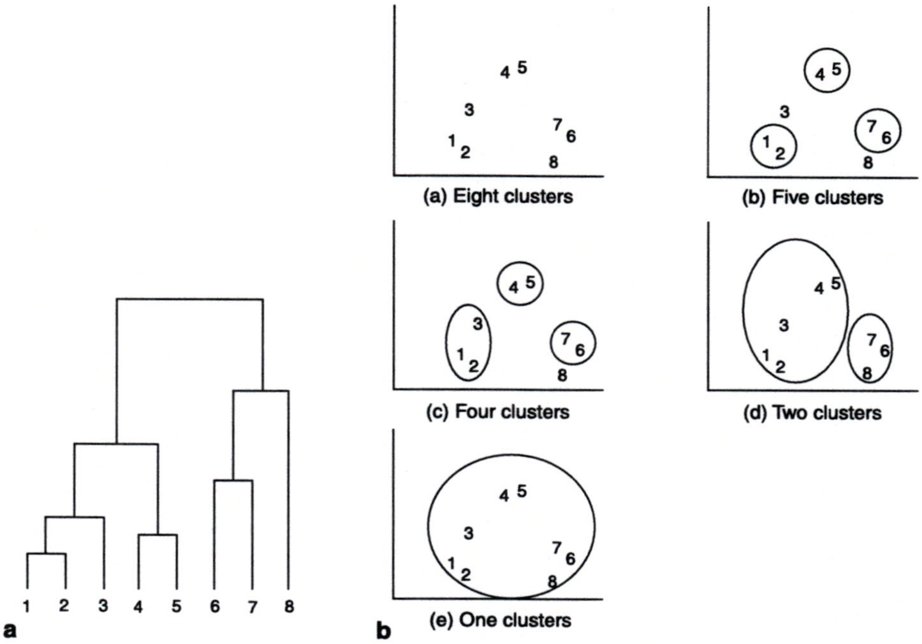

Fig. 8.14 Example of hierarchical agglomerative clustering. **a** Tree diagram or dendrogram. **b** Different levels of clustering the tree diagram

(O_1, O_2), (O_4, O_5), and (O_6, O_7), are the most similar to each other, and are merged together resulting in the five cluster diagram in Fig. 8.14b. If one wishes to form four clusters, it is best to merge object O_3 with the first set (Fig. 8.14b). This merging is continued till all objects have been combined into a single undifferentiated group. Though the last step has little value, it is the sub-levels which provide insights into the extent that different objects are close to one another. This process of starting with individual objects and repeatedly merging nearest objects into clusters till one is left with a single cluster is referred to as *agglomerative clustering*.

Another approach called *divisive clustering* tackles the problem in the other direction, namely starts by placing all objects in a single cluster and repeatedly splitting the clusters in two until all objects are placed in their own cluster. These two somewhat complementary approaches are akin to the forward and backward stepwise regression approaches. Note that both approaches are not always consistent in the way they cluster a set of data.

Hierarchical techniques are appropriate for instances when the data set has naturally-occurring or physically-based nested relationships, such as plant or animal taxonomies (Dunham 2006). The partitional clustering algorithm is advantageous in applications involving large data sets for which hierarchical clustering is computationally complex. A very basic overview of clustering methods has been given in this chapter; the interested reader can refer to pertinent texts such as Dunham (2006), Manly (2005), Hand (1981) or Duda et al. (2001) for in-depth mathematical treatment, description of the clustering algorithms and various applications.

Problems

Pr. 8.1 Rework Example 8.2.4 for a new sample of anthracite coal sample found to contain 60–70% carbon content.

Pr. 8.2 Consider the two dimensional data of three groups shown in Table 8.10. Using the standardized Euclidian distance:

(i) Identify boundaries for the three groups so as to make the misclassification rates more or less equal among all three groups. State the misclassification rates.
(ii) Classify the following four points into one of the three groups: (35,12), (18,20), (28,16) and (12,35)

Pr. 8.3 The intent is to cluster the ten cities shown in Table 8.11 into two or three groups.

(a) Perform hierarchical clustering and generate the dendrogram. Identify the different levels.

Table 8.10 Data table for Problem 8.2

Group A		Group B		Group C	
x_1	x_2	x_1	x_2	x_1	x_2
39	14	25	14	13	17
47	8	22	16	22	26
42	10	23	17	19	23
32	12	22	16	11	15
43	13	31	15	12	19
35	12	27	14	20	15
41	12	24	19	16	24
44	8	31	12	18	23

Table 8.11 Data table for Problem 8.3

City	Average horizontal annual radiation (MJ/m²-day)	Average annual ambient temperature (°C)
Miami, USA	16.764	23.7
New York, USA	12.515	12.6
Phoenix, USA	21.281	20.0
Kabul, Afghanistan	17.439	12.0
Melbourne, Australia	15.203	14.8
Beijing, China	14.598	12.0
Cairo, Egypt	19.979	22.0
New Delhi, India	19.698	25.3
Sede Boqer, Israel	19.880	18.3
Bangkok, Thailand	17.053	31.8
London, UK	9.127	10.5

(b) Perform a partitional clustering along the lines shown in Example 8.5.1.
(c) Compare both the approaches in terms of ease of use, interpretation and simplicity.

Pr. 8.4 The Human Development Index (HDI) is a composite statistic used to rank countries by level of "human development" which includes life expectance, education level and per-capita gross national product which is an indicator of standard of living. Table 8.12 assembles values of HDI and energy use per capita for 30 countries classified as Group A, B and C.

(a) Develop classification models for the three groups (plot the data to detect any trends). You can evaluate the statistical classification method or the discriminant analysis as appropriate. Report misclassification rates if any.
(b) Classify the three countries shown at the end (France, Israel and Greece).

References

Breiman, L., J.H. Freidman, R.A. Olshen and C.I. Stone, 1984. *Classification and Regression Trees*, Belmont, California, Wadsworth.

Chatfield, C., 1995. *Problem Solving: A Statistician's Guide*, 2nd Ed., Chapman and Gall, London, U.K.

Cleveland, W.S., 1994. *The Elements of Graphing Data*, revised edition, Hobart Press, Summit, NJ.

Duda R.O., P.E. Hart, and D.G. Stork, 2001. *Pattern Classification*, 2nd Ed., John Wiley & Sons, New York, NY.

Dunham, M., 2006. *Data Mining: Introductory and Advanced Topics*, Pearson Education Inc.

Hair, J.F., R.E. Anderson, R.L. Tatham, and W.C. Black, 1998. *Multivariate Data Analysis*, 5th Ed., Prentice Hall, Upper Saddle River, NJ.

Hand, D.J., 1981. *Discrimination and Classification*, John Wiley and Sons, Chichester.

Table 8.12 Data table for Problem 8.4

Group	Country	HDI (2010)	Energy (W)/capita (2003)	Group	Country	HDI (2010)	Energy (W)/capita (2003)
A	Norway	0.938	7,902	B	Chile	0.783	2,200
	Australia	0.937	7,622		Argentina	0.775	2,097
	New Zealand	0.907	5,831		Libya	0.755	4,266
	United States	0.902	10,381		Saudi Arabia	0.752	7,434
	Ireland	0.895	5,009		Mexico	0.75	2,041
	Netherlands	0.89	6,675		Russia	0.719	5,890
	Canada	0.888	11,055		Iran	0.702	2,709
	Sweden	0.885	7,677		Brazil	0.699	1,422
	Germany	0.885	5,598		Venezuela	0.696	2,739
	Japan	0.884	5,381		Algeria	0.677	1,382
C	Indonesia	0.6	1,009				
	South Africa	0.597	3,459				
	Syria	0.589	1,307				
	Vietnam	0.572	718.2				
	Morocco	0.567	476				
	India	0.519	682.4				
	Pakistan	0.49	608.2				
	Congo	0.489	363.1				
	Kenya	0.469	640.9				
	Bangladesh	0.469	214.4				
To be Classified	France	0.872	6,018				
	Israel	0.795	3,156				
	Greece	0.855	3,594				

Higham, C.F.W., A. Kijngam and B.F.J. Manly, 1980. An analysis of prehistoric canid remains from Thailand, *J. Archeological Science*, vol.7, pp.149–165.

Hull, D.A. and T.A. Reddy, 1990. A procedure to group residential air-conditioning load profiles during the hottest days in summer, *Energy*, vol. 12, pp.1085–1097.

Kantardzic, M., 2003. *Data Mining: Concepts, Models, Methods and Algorithms*, John Wiley and Sons, NY.

Kotz, S., 1997. *Encyclopedia of Statistical Sciences: Treed Regression*, vol.2, pp.677–679., John Wiley and Sons, New York, NY.

Manly, B.J.F., 2005. *Multivariate Statistical Methods: A Primer*, 3rd Ed., Chapman & Hall/CRC, Boca Raton, FL

Milstein, R.M., G.N. Burrow, L. Wilkinson and W. Kessen, 1975. Prediction of screening decisions in a medical school admission process, *J. Medical Education,* vol., 51, pp. 626–633.

Reddy, T.A., 2007. Application of a generic evaluation methodology to assess four different chiller FDD methods, *HVAC&R Research Journal*, vol.13, no.5, pp 711–729, American Society of Heating, Refrigerating and Air-Conditioning Engineers, Atlanta September.

Subbarao, K., Y. Lei and T.A. Reddy, 2011. The nearest neighborhood method to improve uncertainty estimates in statistical building energy models, *ASHRAE Trans*, vol. 117, Part 2, American Society of Heating, Refrigerating and Air-Conditioning Engineers, Atlanta.

Analysis of Time Series Data

9

This chapter introduces several methods to analyze time series data in the time domain; an area rich in theoretical development and in practical applications. What constitutes time series data and some of the common trends encountered are first presented. This is followed by a description of three types of time-domain modeling and forecasting models. The first general class of models involves moving average smoothing techniques which are methods for removing rapid fluctuations in time series so that the general secular trend can be seen. The second class of models is similar to classical OLS regression models, but here the time variable appears as a regressor, thereby allowing the trend and the seasonal behavior in the data series to be captured by the model. Its strength lies in its ability to model the deterministic or structural trend of the data in a relatively simple manner. The third class of models called ARIMA models allows separating and modeling the systematic component of the model residuals from the purely random white noise element, thereby enhancing the prediction accuracy of the overall model. ARMAX models, which are extensions of the univariate ARIMA models to multivariate problems, and their ability to model dynamic systems are also discussed with illustrative examples. Finally, an overview is provided of a practical application involving control chart techniques which are extensively used for process and condition monitoring of engineered systems.

9.1 Basic Concepts

9.1.1 Introduction

Time series data is not merely data collected over time. If this definition were true, then almost any data set would qualify as time series data. There must be some sort of ordering, i.e. a relation between successive data observations. In other words, successive observations in time-series data are usually not independent and their order or sequence needs to be maintained during the analysis. *A collection of numerical observations arranged in a natural order with each observation associated with a particular instant of time or interval of time which provides the ordering would qualify as time series data* (Bloomfield 1976). One example is temperature measurements of an iron casting as it cools over time. The hourly variation of electricity use in a commercial building during the day and over a year would also qualify as time series data. Thus, the inherent behavior or the response of the system is affected by the time variable (either directly such as the cooling of a billet, or indirectly such as the electricity use in a building). Note that "time" need not necessarily mean time in the physical sense, but any variable to which an ordering can be associated. Another more practical way of ascertaining whether the data is to be treated as time series data or not, is to determine if the analysis results would change if the sequence of the data observations were to be scrambled. The importance of time series analysis is that it provides insights and more accurate modeling and prediction to time series data than do classical statistical analysis because of the explicit manner in which the systematic residual behavior of the data is accounted for in the model.

Consider Fig. 9.1 where the hourly loads during a week of an electric utility are shown. The loads are highly influenced by those of residential and commercial buildings and industrial facilities within the utility's service territory. Because of distinct occupied and unoccupied schedules, there are both strong diurnal and weekly variations in the load. These loads are also affected by such variables as outdoor temperature and humidity. Hence, developing models which can predict or forecast, say a day ahead will allow electric utilities to better plan their operations. Figure 9.2 is another representation where the cyclic pattern (shown as dots) indicates the electric demand of an electric utility during each quarter of a 3-year period. Using traditional ordinary least squares (OLS) covered in Chap. 5, one would obtain the mean of future predictions (or forecasts) and the upper and lower confidence limits (CL) as shown. Analyzing data points in a time series framework would improve both the forecasts as well as reduce the confidence interval. Time series methods have been applied

T. Agami Reddy, *Applied Data Analysis and Modeling for Energy Engineers and Scientists*,
DOI 10.1007/978-1-4419-9613-8_9, © Springer Science+Business Media, LLC 2011

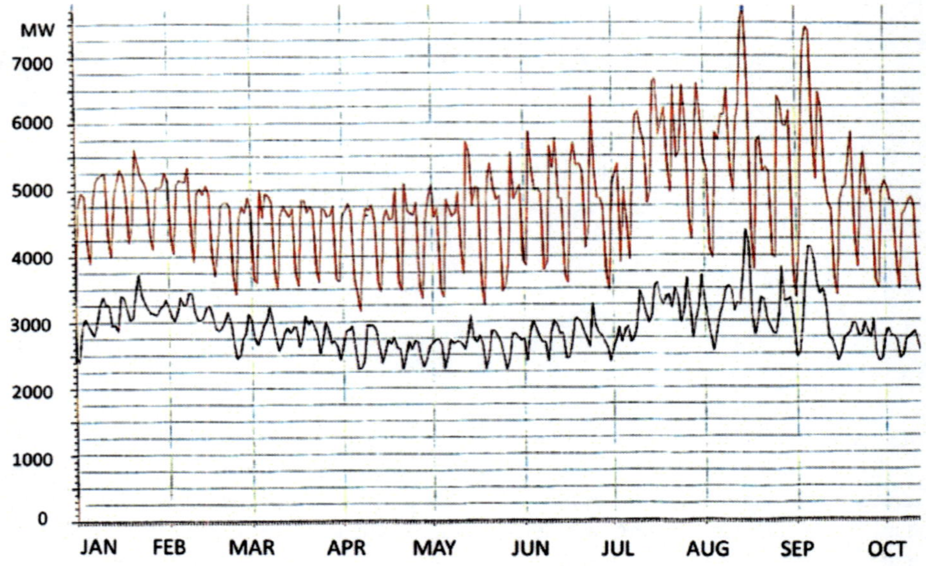

Fig. 9.1 Daily peak and minimum hourly loads over several months for a large electric utility to illustrate the diurnal, the weekday/weekend and the seasonal fluctuations and trends

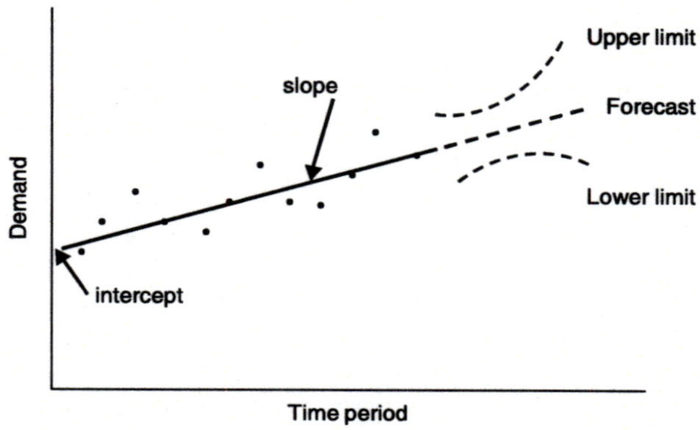

Fig. 9.2 Forecasts and prediction intervals of the quarterly electric demand of an electric utility using ordinary least square fit to historic data

to numerous applications; to name a few, to data which exhibit periodicities, for condition monitoring of industrial processes (using control chart techniques), as well as allowing a systematic way of modeling dynamic systems.

Similar to traditional data analysis, there are two kinds of time series analysis: (i) *descriptive* which uses graphical and numerical techniques to provide the necessary understanding, and (ii) *inferential* which allows future values to be forecast (or predicted) along with a measure of their confidence intervals. Both these aspects are complementary. Usually time series data (either from natural phenomenon such as, say, occurrence of sun spots, or for industrial process monitoring) need to be understood and modeled first prior to forecasting and, if possible, control. The forecast is not an end in itself, it is part of a larger issue such as taking corrective action.

Time series analysis has several features in common with classical statistical analysis. If the physics of the system is not well understood and curve fitting is resorted to, the subjective element involved in trying to select an appropriate time series model for a given set of data is a major issue. Other practical problems include missing observations, outliers or interruptions in the series due to a moment impulse acting on the system. The analysis of time series data is further complicated by the possible presence of trend and seasonal variation which can be hard to estimate and/or remove. Finally, inferences involving the non-stochastic[1] or deterministic trend are based on OLS where errors are assumed to be independent and uncorrelated (white noise), while in time series data analysis the errors are treated as being correlated. It is important to note that the OLS regression parameters identified from the data series data are unbiased per se, i.e., in the long run, an OLS regression model will yield the right average values of the parameters. However, the statistical significance of these parameters, i.e., the standard errors of the parameters will be improper, often resulting in an underestimation of the confidence intervals.

Time series can be analyzed in one of two ways:

(a) *Time domain* analysis in which the behavior of a series is described in terms of the manner in which observations at different times are related statistically. This approach is usually more intuitive to beginners and is used almost exclusively in disciplines such as econometrics and social science whose models are less deterministic; and
(b) *Frequency domain* analysis which seeks to describe the fluctuations in one or more series in terms of sinusoidal behavior at various frequencies. This approach has been

[1] A stochastic process is one which is described by a set of time indexed observations subject to probabilistic laws.

used extensively in the physical sciences especially engineering, physics, and astronomy. One can distinguish four sub-categories:

(b1) *Fourier series* analysis, in its narrow sense, is the decomposition or approximation of a function into a sum of sinusoidal components (Bloomfield 1976). In its wider sense, Fourier series analysis is a procedure that describes or measures the fluctuations in time series by comparing them with sinusoids when data exhibits *clear periodic components.*

(b2) *Harmonic analysis* extends the capability of Fourier series analysis by allowing detection of periodic components or hidden periodicities in cases when the data do not appear periodic.

(b3) *Complex demodulation* is a more flexible approach than harmonic analysis and is used to describe features in the data that would be missed by harmonic analysis, and also to verify, in some cases, that no such features exist. The price of this flexibility is a loss of precision in describing pure frequencies for which harmonic analysis is more exact.

(b4) *Spectral analysis* describes the *tendency* for oscillations of a given frequency to appear in the data, rather than the oscillations themselves. It is a modification of Fourier analysis so as to make it suitable for stochastic rather than deterministic functions of time.

Frequency domain methods will not be treated in this book, and the interested reader can refer to several good texts such as Bloomfield (1976) and Chatfield (1989).

9.1.2 Terminology

Terminology and notations used in time series analysis differ somewhat from classical statistical analysis (Montgomery and Johnson 1976).

(a) *Types of data:* A time series is *continuous* when observations are made continuously in time, even if the measured variable take on only discrete set of values. A time series is said to be *discrete* when observations are taken only at specific times, usually equally spaced, even if the variable is continuous in nature (such as say, outdoor temperature). Other types of data (such as dividend paid by a company to the shareholders) are inherently discrete. Further, one distinguishes between two types of discrete data. *Period or sampled* data represent aggregate values of a parameter *over* a period of time, such as average temperature during a day. *Point or instantaneous* data represent the value of a variable *at* specific time points, such as the temperature at noon. The difference between these two types of data has implications primarily for the type of data collection system to be used, and on the effect of measurement and data processing errors on the results.

(b) *Types of forecast:* Forecasting (or prediction) is the technique used to predict future values based upon past and present values (i.e. the estimation or model identification period) of the parameter in question. It is useful to distinguish between two types of forecast: *point forecasts* where a single number is predicted in each forecast period, and *interval forecasts* where an interval or range is deduced over which the realized value is expected to lie. The latter provides a means of ascertaining prediction intervals (see Fig. 9.2).

(c) Another type of distinction is *expost* and *exante.* In the expost forecast, the forecast period is such that observations of both the driving variables and the response variable are known with certainty. Thus, expost forecasts can be checked with existing data and provide a means of evaluating the model. An exante forecast predicts values of the response variable when those of the driving variables are: (i) known with certainty, referred to as *conditional exante forecast,* or (ii) not known with certainty, denoted as *unconditional exante forecast.* Thus, the unconditional forecast is more demanding than conditional forecasting since the driving variables need also to be predicted into the future (along with the associated uncertainty which it entails).

(d) *Types of forecast time elements.* One needs also to distinguish between the following three time periods. The *forecasting period* is the basic unit of time for which forecasts are made. For example, one may wish to forecast electricity use of a building on an hourly basis, i.e., the period is an hour. The forecasting *horizon or lead time* is the number of periods into the future covered by the forecast. If electricity use in a building over the next day is to be forecast, then the horizon is 24 h, broken down by hour. Finally, the forecasting *interval* is the frequency with which new forecasts are prepared. Often this is the same as the forecasting period, so that forecasts are revised each period using the most recent period's value and other current information as the basis for revision. If the horizon is always the same length and the forecast is revised each period, one is then operating on a *moving horizon* basis.

9.1.3 Basic Behavior Patterns

Time series data can exhibit different types of behavior patterns. One can envision various patterns, four of which are:

(i) variations around a constant mean value, referred to as a constant process such as Fig. 9.3a,
(ii) trend or secular behavior i.e. a long-term change in the mean level which may be linear, (such as Fig. 9.3b) or non-linear,
(iii) cyclic or periodic (or seasonal) behavior such as Fig. 9.3c, and

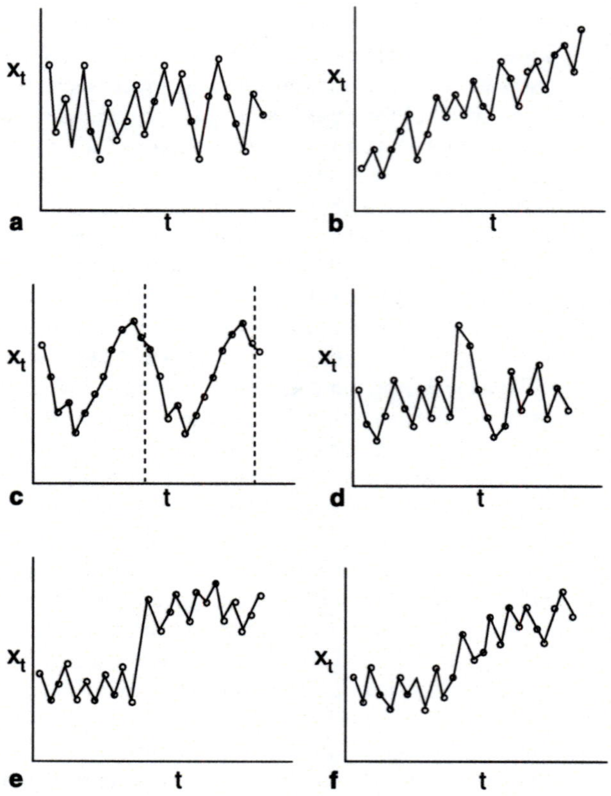

Fig. 9.3 Different characteristics of time series. **a** Constant process, **b** linear trend, **c** cyclic variation, **d** impulse, **e** step function, **f** ramp. (From Montgomery and Johnson 1976 by permission of McGraw-Hill)

(iv) transient behavior where one can have momentary impulse, such as Fig. 9.3d, or a step change (as in Fig. 9.3e) or a ramp up, i.e., where the increase is more gradual (as in Fig. 9.3f).

Much of the challenge in time series analysis is distinguishing these basic behavior patterns when they occur in conjunction. Untangling the data into these patterns requires a certain amount of experience and skill (specially because the decomposition is often not unique), only after which can model identification and forecasting be done with confidence. The problem is compounded by the fact that processes may exhibit these patterns at different times. For example, the growth of bacteria in a pond may experience an exponential growth followed by a stable constant regime, and finally, a declining trend phase.

9.1.4 Illustrative Data Set

Transient behavior in time series is indicative of a change in the basic dynamics of the phenomenon or process and has to be dealt with in a separate fashion. Removing the secular trend and the cyclic variation are the first steps in rendering the data stationary[2], only after which can a time-domain model be developed. In the frequency-domain approach, only the transient and trend patterns have to be removed from the basic time series data.

The following is a simple example of time series data exhibiting both a linear trend and seasonal periodicities. This data set will be used to illustrate many of the modeling approaches covered later in this chapter.

Table 9.1 Demand data for an electric utility. (From McClave and Benson 1988 by © permission of Pearson Education)

Year	Quarter	MW	Year	Quarter	MW
1974	1	68.8	1980	1	130.6
	2	65		2	116.8
	3	88.4		3	144.2
	4	69		4	123.3
1975	1	83.6	1981	1	142.3
	2	69.7		2	124
	3	90.2		3	146.1
	4	72.5		4	135.5
1976	1	106.8	1982	1	147.1
	2	89.2		2	119.3
	3	110.7		3	138.2
	4	91.7		4	127.6
1977	1	108.6	1983	1	143.4
	2	98.9		2	134
	3	120.1		3	159.6
	4	102.1		4	135.1
1978	1	113.1	1984	1	149.5
	2	94.2		2	123.3
	3	120.5		3	154.4
	4	107.4		4	139.4
1979	1	116.2	1985	1	151.6
	2	104.4		2	133.7
	3	131.7		3	154.5
	4	117.9		4	135.1

Example 9.1.1: *Modeling peak demand of an electric utility* Table 9.1 assembles time series data of the peak demand or load for an electric utility for 12 years at quarterly levels (48 data points in total). The data is shown graphically in Fig. 9.4 revealing distinct long-term and seasonal trends. Most time series data sets may not exhibit such well-behaved patterns (and that is why this area of time series modeling is so rich and extensive); but this example is meant as a simple illustration. ■

[2] Stationarity in a time series strictly requires that all statistical descriptors, such as the mean, variance, correlation coefficients of the data series be invariant in time. Due to the simplified treatment in this text, the discussion is geared primarily towards stabilizing the mean, i.e., removing the long-term and seasonal trends.

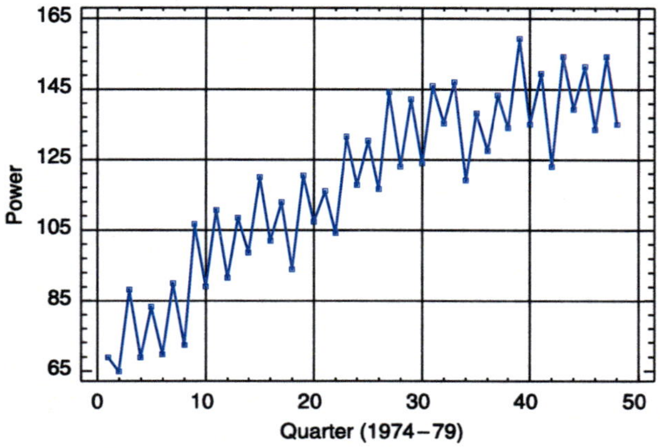

Fig. 9.4 Time series data of electric power demand by quarter (data from Table 9.1)

9.2 General Model Formulations

How does one model the behavior of the data shown in Example 9.1.1 and use it for extrapolation purposes? There are three general time domain approaches:

(a) *Smoothing methods,* which are really meant to filter the data in a computationally simple manner. However, they can also be used for extrapolation purposes (covered in Sect. 9.3);
(b) *OLS models,* which treat time series data as sectional data but with the time variable accounted for in an explicit manner as an independent variable (this is addressed in Sect. 9.4), and
(c) The *stochastic time series modeling,* approach which explicitly treats the model residual errors of (b) by adding a layer of sophistication; this is described briefly below, and at more length in Sect. 9.6).

A basic distinguishing trait is that while approach (b) uses the observations directly, stochastic time series modeling deals with stationary data series which have been made so either by removing the trend by OLS modeling or by temporal differencing, or by normalizing and stabilizing the variance by suitable transformations if necessary. Thus, the first step is to remove the deterministic trend and periodic components of the time series; this is referred to as making the series *stationary.* Let us illustrate the differences in approaches (b) and (c) in terms of an *additive model.* Heuristically:

- For approach (b): Current value at time t = [deterministic component] + [residual random error] = [constant + long-term trend + cyclic (or seasonal) trend] + [residual random error]
- For approach (c): Current value at time t = [deterministic component] + [stochastic component] = [constant + long-term trend + cyclic (or seasonal) trend] + [systematic stochastic component + white noise]

where

(a) the *deterministic component* includes the long-term (or secular) and seasonal trends taken to be independent of the error structure in the observations, and are usually identified by say, standard OLS regression. The parameters of the model which explain the deterministic behavior of the time series yield the "long run" or average or expected values of the underlying process along with the cyclic variations, though one should expect a certain deviation of any given observation from the expected value. It is the residual error (quantified, say as the standard error or the RMSE) which determines the uncertainty of the predictions;
(b) the *stochastic component* treats the residual errors in a more refined sense by separating: (i) the *systematic part* in the errors responsible for the autocorrelation in time series data (if they exist). It is the determination of the structure of this systematic part which is the novelty in time series analysis and adds to the more accurate determination of the uncertainty bands of prediction; and (ii) the *white noise* or purely random part of the stochastic component that cannot be captured in a model, and implicitly appears in the determination of prediction of the uncertainty bands. Thus, the stochastic methods exploit the dependency in successive observations to produce superior results in the prediction uncertainty determination. Note the *distinction made between random errors and white noise* in the residuals.

The deterministic models, called, trend and seasonal models are simple, easy to interpret, fairly robust and especially suitable for data with pronounced trends and/or large seasonal effect. Moreover, these models are usually simpler to use and can be applied to relatively short data series (less than, say, 50 observations). Section 9.3 describes two widely used smoothing approaches: moving average and exponential smoothing, both of which rely on data smoothening methods, i.e. methods for removing rapid fluctuations in time series so that the general trend can be seen. Subsequently, how to adapt the OLS regression approach to model trend behavior as well as seasonal behavior is discussed in Sect. 9.4. These two general classes of models involve the time series observations themselves. The third, namely the stochastic approach addressed in Sect. 9.5, builds on the latter. In other words, it is brought to bear only after the trend and seasonal behavior is removed from the data.

9.3 Smoothing Methods

Time series data often exhibits local irregularities or rapid fluctuations resulting in trends that are hard to describe. Moving average modeling is a way to smoothen out these fluctuations, thus making it easier to discern longer time trends and, thereby, allowing future or trend predictions to be made,

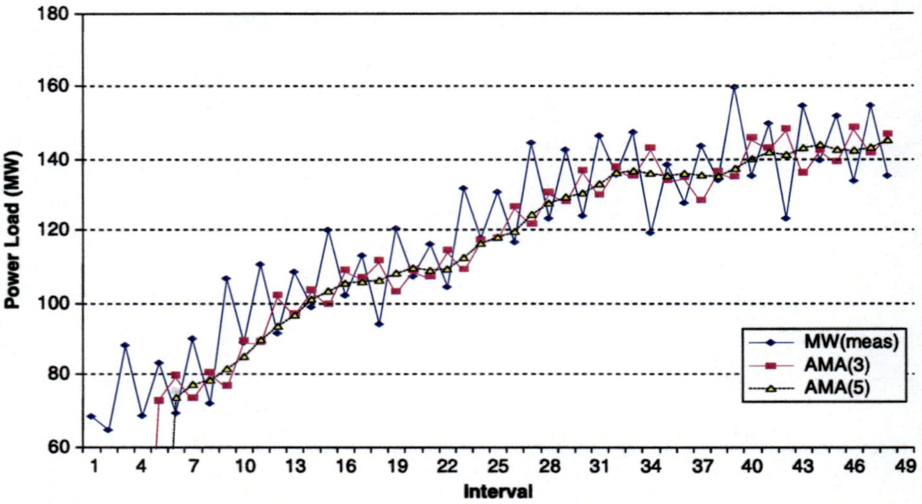

Fig. 9.5 Plots illustrating how two different AMA smoothing methods capture the electric utility load data denoted by MW(meas) (data from Table 9.1)

albeit in a simple manner. However, though they are useful in predicting mean future values, *they do not provide any information about the uncertainty of these predictions* since no modeling per se is involved, and so standard errors (which are the cause for forecast errors) cannot be estimated. The inability to quantify forecast errors is a serious deficiency.

Two types of models are often used: arithmetic moving averages and exponential moving averages, which are described below. Both of these models are recursive in that previous observations are used to predict future values. Further, they are linear in their parameters.

9.3.1 Arithmetic Moving Average (AMA)

Let Y(t), t={1,..., N...n} be time series observations at discrete intervals of time at n time intervals. Instead of the functional notation, let us adhere to subscripts so that $Y(t) \equiv Y_t$. AMA models of order N (where N<n) denoted by AMA(N) combine N number of past, current and future values of the time series in order to perform a simple arithmetic average which slides on a moving horizon basis. A time series which is a constant process, i.e., without trend or cyclic behavior, is the simplest case one can consider. In this case, one can assume a model such as:

$$Y_t = b_0 + \varepsilon_t \tag{9.1}$$

where $\varepsilon_t(0, \sigma_\varepsilon^2)$ is a random variable with mean 0 and variance σ_ε^2, and b_0 is an unknown parameter. To forecast future values of the time series, the unknown parameter b_0 is to be estimated. If all observations are equally important in estimating b_0, then the least-square criterion involves determining the value which minimizes the sum of squares:

$$SS = \sum_{t=1}^{N} (Y_t - b_0)^2, \quad \text{i.e.,} \quad b_0 = \frac{1}{N}\sum_{t=1}^{N} Y_t, \tag{9.2}$$

i.e. the arithmetic mean.

Since the model is used for forecasting purposes with AMA (N), one cannot obviously use future values. Let Y_t and $\hat{Y}_t$ denote observed and model-predicted values of Y at time interval t. Then, the following recursive equation for the average M_t of the N most recent observations is used:

$$\hat{Y}_{t+1} \equiv M_t = (Y_t + Y_{t-1} + \cdots + Y_{t-N+1})/N \tag{9.3a}$$

or

$$M_t = M_{t-1} + (Y_t - Y_{t-N})/N \tag{9.3b}$$

where for better clarity the notation M_t is used instead of $\hat{Y}_{t+1}$ to denote model predicted forecasts one time interval ahead from time t. At each period, the oldest observation is discarded and the newest one added to the set; hence its name "*N-period simple moving average*". The choice of N, though important, is of course circumstance specific, but more importantly, largely subjective or adhoc. Large values of N produce a smoother moving average but more points at the beginning of the series are lost and the data series may wind up so smooth that incremental changes are lost. How simple 3-point and 5-point AMA schemes capture the overall trend in the electric utility data is shown in Fig. 9.5, while Fig. 9.6 shows the residuals pattern with time. As expected, one notes that AMA(3) smoothening has larger residuals spikes, i.e., is not as smooth as the AMA(5) but it is quicker to reflect changes in the series. Which model is "better" cannot be ascertained since it depends on the intent of the analysis: whether to

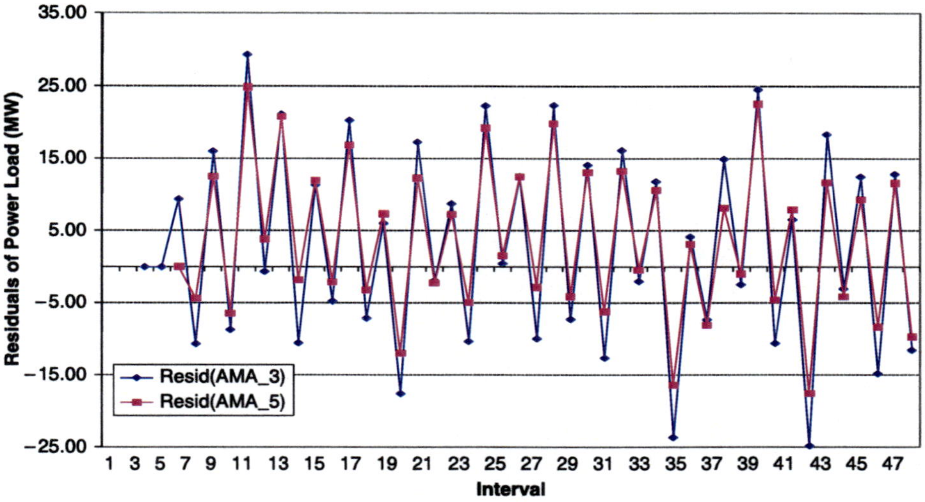

Fig. 9.6 Residual plots illustrating how two different AMA smoothing methods capture the electric utility load data (data from Table 9.1)

capture longer-term trends or shorter ones. A cross-validation evaluation can be done as illustrated later by Example 9.5.3.

It is clear that by taking simple moving averages, one overlooks the stochastic element in the data series. Its use results in a lag in $\hat{Y}_{t+1}$ or M_t in the predicted values. Further, cyclic or seasonal behavior is not well treated unless the sliding window length is selected to be a multiple of the basic frequency. For the Example 9.1.4 data, a 4-point or a 8-point sliding window would give proper weight to the forecasts. This lag, as well as higher order trends in the data, can be corrected by taking higher order MA methods, such as the moving average of moving averages, called a *double moving average*[3]. Such techniques as well as more sophisticated variants of AMA are available, but the same degree of forecast accuracy can be obtained by using exponentially weighted moving average models or the trend and seasonal models described below.

9.3.2 Exponentially Weighted Moving Average (EWA)

AMA is useful if the data trend indicates that future values are simply averages of the past values. The AMA model is appropriate if one has reason to believe that the one-step forecast is likely to be *equally* influenced by the N previous observations. However, in case recent values influence future values more strongly than do past values, a weighting scheme needs to be adopted, and this is the basis of EWA. Such smoothing models are widely used, their popularity stemming not only from their simplicity and computational efficiency but also from their ease of self-adjustment to changes in the process being forecast. Thus, they provide a means to adapt to changes in data trends which is superior to AMA. Modifying, Eq. 9.3a results in:

$$\begin{aligned}\hat{Y}_{t+1} &\equiv M_t = \alpha Y_t + (1-\alpha)M_{t-1} \\ &= \alpha Y_t + \alpha(1-\alpha)Y_{t-1} + \alpha(1-\alpha)^2 Y_{t-2}\ldots\end{aligned} \tag{9.4}$$

where $0<\alpha<1$ is the exponential smoothing fraction. If $\alpha=0.2$, then the weights of the previous observations are 0.16, 0.128, 0.1024 and so on. Note that normalization requires that the weights sum to unity, which holds true since:

$$\begin{aligned}\alpha\sum_{t=0}^{\infty}(1-\alpha)^t &= (1-\alpha)\left[1+\sum_{t=0}^{\infty}(1-\alpha)^b\right] \\ &= \frac{\alpha}{1-(1-\alpha)} = 1\end{aligned} \tag{9.5}$$

A major drawback in exponential smoothing is that it is difficult to select an "optimum" value of α without making some restrictive assumptions about the behavior of the time series data. Like many averages, the EWA series changes less rapidly than the time series itself. Note that the choice of α is critical, with smaller values giving more weight to the past values of the time series resulting in a smooth, slow changing series of forecasts. Conversely choosing α closer to 1 yields an EMA series closer to the original. Several values of α should be tried in order to determine how sensitive the forecast series is to choice of α. In practice, α is chosen such that $0.01<\alpha<0.30$. Figure 9.7 illustrates how the two EWA schemes capture the overall trend, and how the choice of α affects the smoothing of the electric load data shown in Table 9.1. The residual plots are shown in Fig. 9.8 and, as expected, residuals for EWA(0.2) are lower in magnitude than those of EWA(0.5).

[3] See for example McClave and Benson (1988) or Montgomery and Johnson (1976).

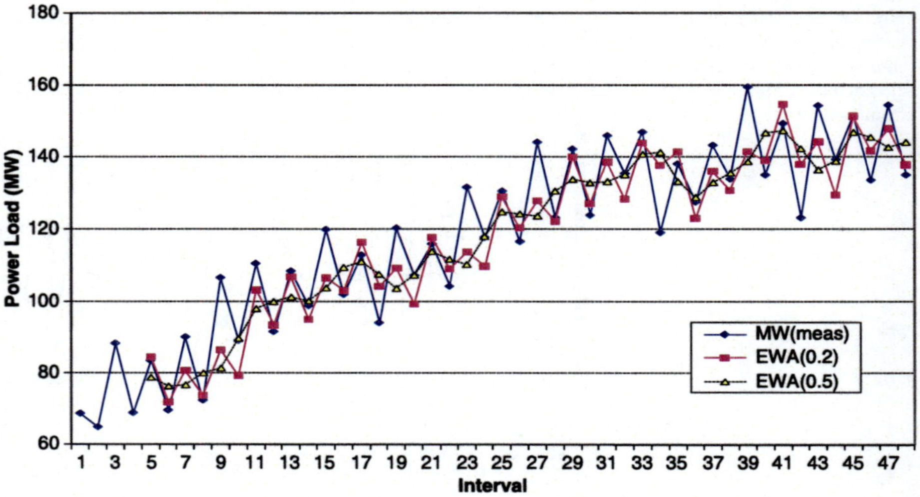

Fig. 9.7 Plots illustrating how two different EWA smoothing methods capture the electric utility load data denoted by MW(meas) (data from Table 9.1)

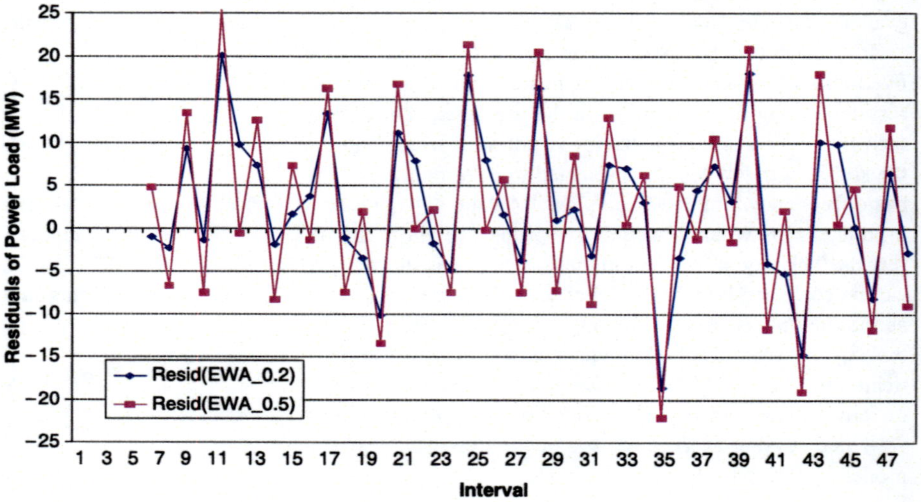

Fig. 9.8 Residual plots illustrating how two different EWA smoothing methods capture the electric utility load data (data from Table 9.1)

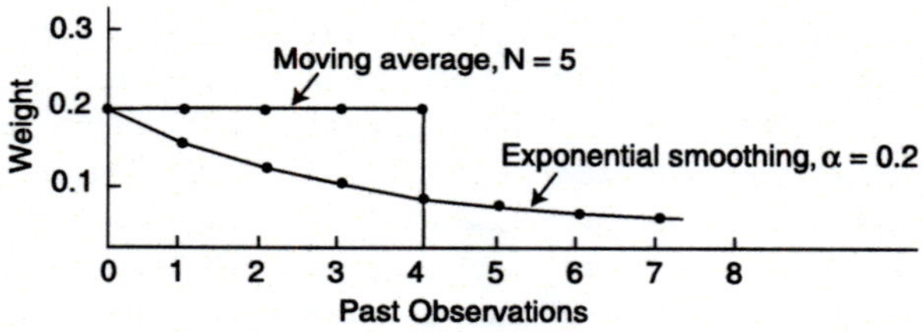

Fig. 9.9 Assumed weights for AMA5 and EWA (0.2) used to generate Figs. 9.5 and 9.7

Exponential smoothing can be used to estimate the coefficients in polynomial models of any degree. Though this approach could be used for higher order models, they get increasingly complex, and it is simpler and more convenient to use trend and seasonal models as well as stochastic models. Note that both the AMA and EWA models are procedures that adjust the future predicted value by an amount that is proportional to the most recent forecast error (see Fig. 9.9). This is the reason why forecasts based on these smoothing models fall under the class sometimes referred to as *adaptive* forecasting methods.

Filtering is a generic term used to denote an operation where time series data is modified in a pre-determined manner so as to produce an output with emphasis on variation at particular frequencies. For example, a low pass filter is used to remove local fluctuations made up of high frequency variations. Thus, AMA and EWA can be viewed as different types of low-pass filters.

9.4 OLS Regression Models

9.4.1 Trend Modeling

Pure regression models can often be used for forecasting. OLS assumes that the model residuals or errors are independent random variables, implying that successive observations are also considered to be independent. This is not strictly true in time series data, but assuming it to be so leads to a class of models which are often referred to as *trend and seasonal* time series models. In such a case, the time series is just a sophisticated method of extrapolation. For example, the simplest extrapolation model is given by the linear trend:

$$Y_t = b_0 + b_1 t \tag{9.6}$$

where t is time, and Y_t is the value of Y at time t. This is akin to a simple linear regression model with time as a regressor. The interpretation of the coefficients b_0 and b_1 for both classical regression and trend and seasonal time series analysis are identical. What differentiates both approaches is that, while OLS models are used to predict future movements in a variable by relating it to a set of other variables in a causal framework, the "pure" time series models are used to predict future movements using "time" as a surrogate variable. Model parameters are sensitive to outliers, and to the first and last observations of the time series.

To model a series whose rate of growth is proportional to its current value, the exponential growth model is more appropriate:

$$Y_t = b_0 \exp(b_1 t) \tag{9.7a}$$

which can be linearized by taking logs:

$$\ln Y_t = \ln b_0 + b_1 \ln(t) \tag{9.7b}$$

The model coefficients b_0 and b_1 can then be identified by least squares regression.

Example 9.4.1: *Modeling peak electric demand by a linear trend model*
The electric peak load data consisting of 48 data points given in Table 9.1 can be regressed following a linear trend model (given by Eq. 9.6). The corresponding OLS model is:

$$Y_t = 77.906 + 1.624\, t$$

with $R^2 = 0.783$, RMSE = 12.10 and CV = 10.3%.

Figure 9.10 depicts the model residuals, from which one notes that the residuals are patterned, but more importantly, that there is a clear quadratic trend in the residuals as indicated by the trend line drawn. This suggests that the trend is not linear and that alternative functional forms should be investigated. ■

9.4.2 Trend and Seasonal Models

In order to capture the deterministic seasonal trends in the data, a general *additive* time series model formulation analogous to a classical OLS regression model can be heuristically expressed as:

$$Y_t = b_0 + b_1 f_1 + b_2 f_2 + b_3 f_3 + \cdots + b_n t + \varepsilon \tag{9.8}$$

where t is the time index, f is the cyclic or seasonal frequency and "b"s are the model coefficients. Note that the residual effect should ideally consist of a white noise element and a structured residual pattern (which is extremely difficult to remove). Modeling the structured portion of this residual pattern is the objective of stochastic modeling of time series data and is considered in the next section.

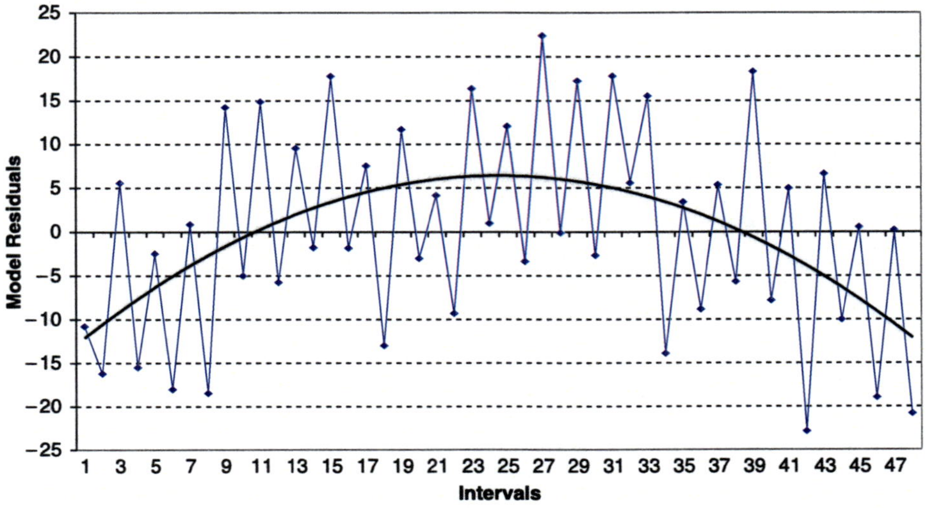

Fig. 9.10 Figure illustrating that residuals for the linear trend model (Eq. 9.6) are not random (see Example 9.4.1). They exhibit both local systematic scatter as well as an overall pattern as shown by the quadratic trend line. They seem to exhibit larger scatter than the AMA residuals shown in Fig. 9.6

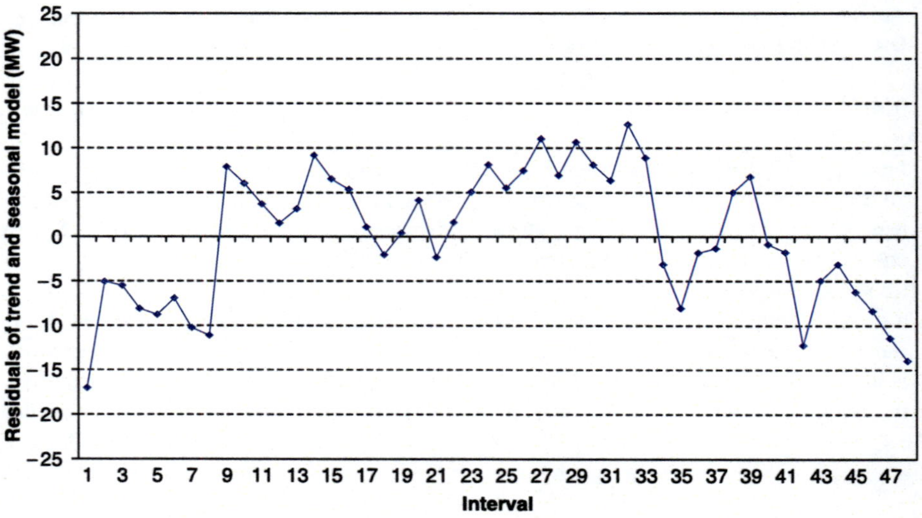

Fig. 9.11 Residuals for the linear and seasonal model (see Example 9.4.2). Note that the residuals still exhibit a pattern

Many time series data exhibit cyclic behavior, and one can use the indicator variable modeling approach (described in Sects. 5.7.2 and 5.7.3) in such cases. Consider the demand data for the electric utility shown in Table 9.1 and Fig. 9.4 which has seasonal differences. Since quarterly data are available, the following model can be assumed to capture such seasonal behavior:

$$Y_t = b_0 + b_1 I_1 + b_2 I_2 + b_3 I_3 + b_4 t \tag{9.9}$$

where t=time ranging from t=1 (first quarter of 1974) to t=48 (last quarter of 1985),

$$I_1 = 1 \text{ for quarter} = 1$$
$$= 0 \text{ for quarter} = 2,3,4$$

$$I_2 = 1 \text{ for quarter} = 2$$
$$= 0 \text{ for quarter} = 1,3,4$$

$$I_3 = 1 \text{ for quarter} = 3$$
$$= 0 \text{ for quarter} = 1,2,4.$$

When the time series is changing at an increasing rate over time, the multiplicative model is more appropriate:

$$Y_t = \exp\,(b_0 + b_1 I_1 + b_2 I_2 + b_3 I_3 + b_4 t + \varepsilon) \tag{9.10a}$$

or

$$\ln Y_t = b_0 + b_1 I_1 + b_2 I_2 + b_3 I_3 + b_4 t + \varepsilon \tag{9.10b}$$

Example 9.4.2: *Modeling peak electric demand by a linear trend plus seasonal model*

The linear trend plus seasonal model given by Eq. 9.9 has been fit to the electric utility load data yielding the following model:

$$Y_t = 70.5085 + 13.6586 I_1 - 3.7359 I_2 + 18.4695 I_3 + 1.6362\,t$$

with the following statistics: R^2=0.914, RMSE=7.86 and CV=6.68%.

Thus, the model R^2 has clearly improved from 0.783 for the linear trend model to 0.914 for the linear-seasonal model, while the RMSE has decreased from 12.10 to 7.86. However, as shown in Fig. 9.11, the residuals are still not entirely random (the residuals of the data series at either end are lower), and one would investigate other models. ■

Unlike the case of smoothing methods (such as the AMA and EWA), a model is now being used, and this allows standard model errors to be computed. Thus, one is able to estimate the errors associated with predicting the *individual as well as the mean value* of future values at a forecasting period of one time step under a moving horizon basis. Since this approach does not attempt to deal with the systematic stochastic component of the residuals, the prediction uncertainty bands are similar to those of the regression models. However, instead of using the complete expression for individual prediction bands given in Sect. 5.3.4 (specifically, Eqs. 5.15 and 5.16), the term associated with the slope parameter uncertainty is usually dropped (note that one will be underpredicting the uncertainty a little). This would result in the following simplified expressions which can be assumed constant irrespective of the number of m future time steps of predictions:

(a) ***individual predictions:***

$$\text{Uncorrelated residuals:}\ \ \text{var}(\hat{Y}_{t+m}) = \sigma_\varepsilon^2\left(1 + \frac{1}{N}\right) \tag{9.11}$$

Table 9.2 Accuracy of the various modeling approaches when applied to the electric utility load data given in Table 9.1 (48 data points covering 1974–1985). The RMSE correspond to internal prediction accuracy

	AMA(3)	AMA(5)	EWA(0.2)	EWA(0.5)	Linear	Linear+ Seasonal
RMSE	7.68	9.02	8.59	11.53	12.10	7.86

(b) ***mean values of say m future values:***

$$\text{Uncorrelated residuals: var}(<\hat{Y}_{t+m}>) = \frac{\sigma_\varepsilon^2}{m^{1/2}}\left(1+\frac{1}{N}\right) \tag{9.12}$$

Example 9.4.3: *Comparison of different models for peak electric demand*

Let us use the time series data of the electric utility demand to compare the internal predictive accuracy of the various models. Since the model identification was done using OLS, there are no bias errors to within rounding errors of the computer program used. Hence, it is logical to base this evaluation on the RMSE statistics, which are given in Table 9.2. AMA(3) is surprisingly the best in that it has the lowest RMSE of all models followed very closely by the (linear+seasonal) model. The simple linear model has the highest residuals error. This illustrates that a blind OLS fit to the data is not recommended. However, the *internal prediction accuracy* is of limited use per se, our intention being to use the model to forecast future values (to be illustrated in Example 9.5.2 below). ■

9.4.3 Fourier Series Models for Periodic Behavior

It is clear that both the smoothing models (AMA and EWA) operate directly on the observation values $\{Y_i\}$ unlike the stochastic time series models discussed in the next section. The trend and seasonal models using the OLS approach are analogous in that they model the structural component of the data. Another OLS modeling approach which achieves the same objective is to use basic Fourier series models.

Note that the word "seasonal" used to describe time series data really implies "periodic". Thus, the Fourier series modeling approach applies to data which exhibit distinct periodic behavior which are known beforehand from the nature of the system, or which can be gleaned from plotting the data. For example, the data in Fig. 9.1 exhibits strong weekly cycles while Fig. 9.12 has strong diurnal cycles. Recall that a periodic function is one which can be expressed as:

$$f(t) = f(t+T) \tag{9.13}$$

where T is a constant called the period and is related to the frequency (or cycles/time) $f = 1/T$, and to angular frequency $\omega = 2\pi/T$. This applies to any waveform such as sinusoidal, square, saw-tooth,.... For example, in Fig. 9.1, the period is

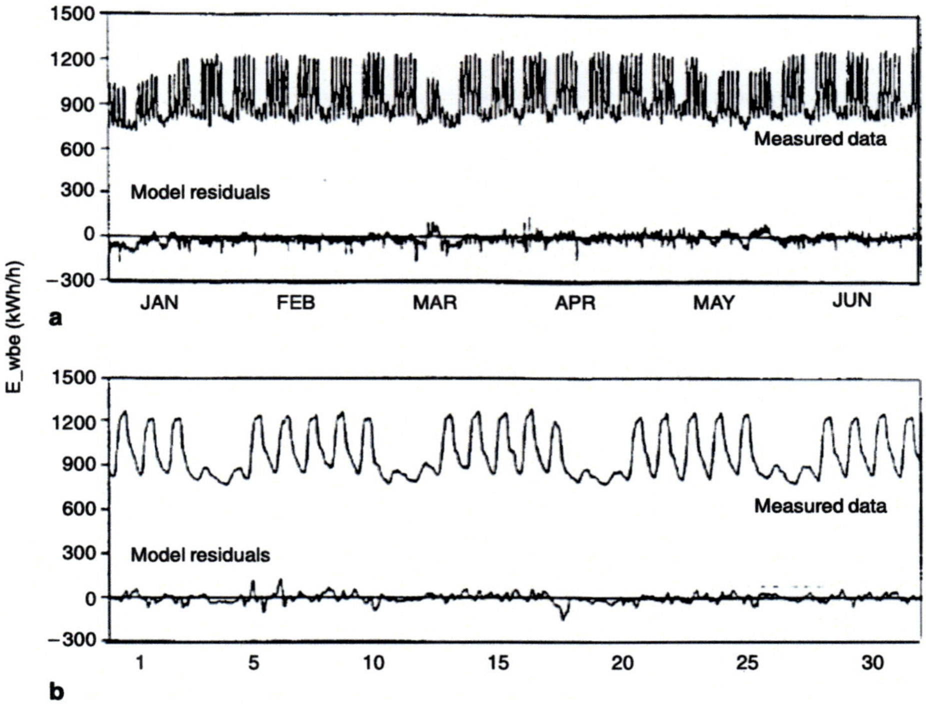

Fig. 9.12 Measured hourly whole building electric use (excluding cooling and heating related energy) for a large university building in central Texas (from Dhar et al. 1999) from January to June. The data shows distinct diurnal and weekly periodicities but no seasonal trend. Such behavior is referred to as weather-independent data. The residual data series using a pure sinusoidal model (Eq. 9.16) are also shown. **a** January–June. **b** April

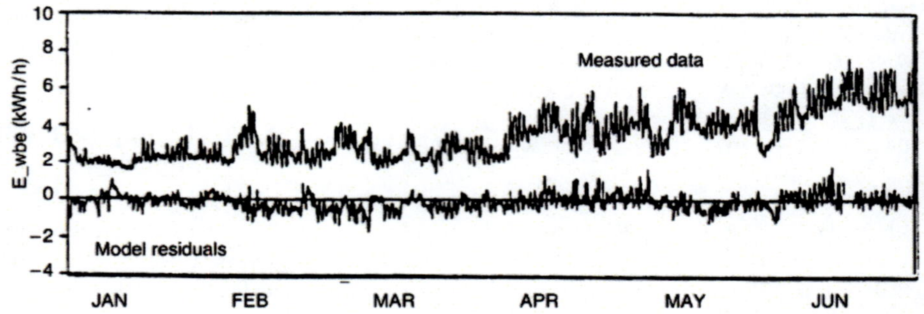

Fig. 9.13 Measured hourly whole building cooling thermal energy use for the same building as in Fig. 9.12 (from Dhar et al. 1999) from January to June. The data shows distinct diurnal and weekly periodicities as well as weather-dependency. The residual data series using a sinusoidal model with weather variables (Eq. 9.18) are also shown

1 week while the frequency would be 52 year^{-1}. A special case is the simple sinusoid function expressed as:

$$y(t) = a_0 + a_1 \cdot \cos(\omega t) + a_2 \cdot \sin(\omega t) \quad (9.14)$$

where a_0, a_1 and a_2 are model parameters to be determined by OLS regression to data. Note that if the frequency ω is known in advance, then, one sets $x_1=\cos(\omega t)$ and $x_2=\sin(\omega t)$ which allows reducing Eq. 9.14 to a linear model:

$$y(t) = a_0 + a_1 \cdot x_1 + a_2 \cdot x_2 \quad (9.15)$$

The extension of this model to more than one frequency (denoted by subscript j) is:

$$y(t) = b_0 + \sum_{j=1}^{n} [a_j \cdot \cos(j\omega t) + b_j \cdot \sin(j\omega t)] \quad (9.16)$$

The above formulation is general, and can be modified as dictated by the specific situation. One such instance is the Fourier series model meant to describe hourly energy use in commercial buildings, as illustrated by the following example taken from Dhar et al. (1999). Whole building hourly energy use E_{wbe} for a large university building in central Texas is shown in Fig. 9.12a for six months (from January to June) and in Fig. 9.12b for the month of April only in order to better illustrate the periodicities in the data. The data channel includes building internal electric loads (lights and equipment) and electricity to operate the air-handlers but does not include any cooling or heating energy use. Hence, the data shows no long-term trend but distinct diurnal periodicities (occupied vs unoccupied hours) as well as weekly periodicities (weekday vs weekends). There are also abrupt drops in usage during certain times of year reflective of events such as semester breaks. On the other hand, Fig. 9.13 is a time series plot of cooling thermal energy use for the same building. This energy use exhibits clear seasonal trend since more cooling is required as the weather warms up from January till June. Such loads are referred to as *weather dependent loads* by building energy professionals. Residual effects, i.e., differences between measured data and predictions by a Fourier series model identified by OLS are also shown in both figures to show the accuracy with which such models capture actual measured behavior. It must be noted that the time series data have been separated into three day-types: weekdays, weekends and holidays/semester break periods, and separate models have been fit to each of these periods. How to identify such periods statistically falls under the purview of classification, an issue which was addressed in Chap. 8.

A general model formulation which can capture the diurnal and seasonal periodicities as well as their interaction such as varying amplitude (see Fig. 9.14) is as follows:

$$E_{d,h} = X(d) + Y(h) + Z(d,h) + \varepsilon_{d,h} \quad (9.17)$$

where

$$X(d) = \sum_{i=0}^{i_{max}} \gamma_i . \sin\frac{2\pi}{P_i}d + \delta_i . \cos\frac{2\pi}{P_i}d$$

$$Y(h) = \sum_{j=0}^{j_{max}} \alpha_j . \sin\frac{2\pi}{P_j}h + \beta_j . \cos\frac{2\pi}{P_j}h$$

$$Z(d,h) = \sum_{i=0}^{i_{max}} \sum_{j=0}^{j_{max}} \left(\phi_i . \sin\frac{2\pi}{P_i}d + \psi_i . \cos\frac{2\pi}{P_i}d\right) \cdot \left(\eta_j . \sin\frac{2\pi}{P_j}h + \zeta_j . \cos\frac{2\pi}{P_j}h\right)$$

where h and d denote hourly and daily respectively, and $P_i=(365/i)$ and $P_j=(24/j)$ for the annual and daily cycles respectively. Also, there are 24 hourly observations in a daily cycle and 365 days in a year which means that $i_{max}=(365/2-1)=181$ and $j_{max}=(24/2-1)=11$. These restrictions would avoid the number of parameters in the model exceeding the number of observations. Obviously, one would not use so many frequencies since much fewer frequencies should provide adequate fits to the data.

Note that Y and X represent diurnal and seasonal periodicities respectively, while Z accounts for their interaction. In other words, Y alone will represent a load shape of constant mean and amplitude (shown in Fig. 9.14a). When term X is

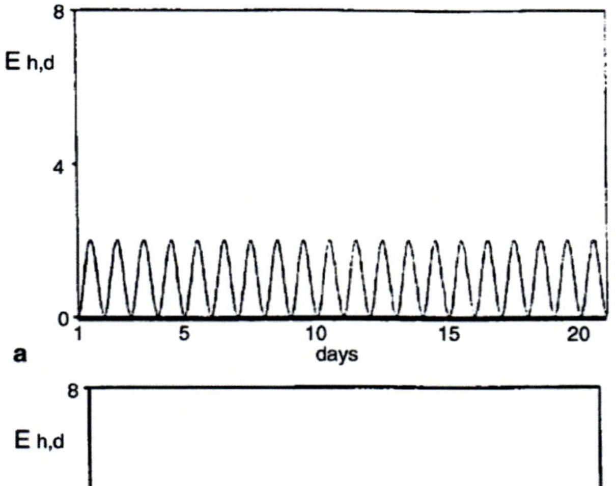

Fig. 9.14 Three general types of periodic profiles applicable to model the pronounced diurnal and annual behavior of energy use in buildings (from Dhar et al. 1999). The energy use of a given hour h and day d has been modeled following functional forms given by Eq. 9.17. **a** Constant mean and amplitude, **b** varying mean but constant amplitude, **c** varying mean and amplitude. Specifically:
(*a*) $E_{h,d} = Y(h) = 1 - \cos((2\pi/24)h)$
(*b*) $E_{h,d} = Y(h) + X(d) = [1 - \cos((2\pi/24)h)] + [1 - \cos((2\pi/24)d)]$
(*c*) $E_{h,d} = Y(h) + X(d) + Y(h).X(d)$

added to the model, variation in the mean energy use can be modeled (see Fig. 9.14b). Addition of the term Z enables the model to represent shapes with varying mean and amplitude (Fig. 9.14c). Equation 9.17 is the general "seasonal" model for capturing effects which are related to time (for example, time of day or time of year). In the case of building energy use, these reflect the manner in which the building is scheduled for operation. Energy control systems automatically switch on and off certain equipment consistent with how the building is occupied. However, there are other building loads (such as cooling energy use shown in Fig. 9.13) which are affected by other variables. Typical variables are outdoor dry-bulb temperature (T_{db}), absolute humidity of outdoor air (W) and solar radiation (S). Humidity affects the latent cooling loads in a building only at higher humidity levels. Previous studies (supported by theoretical considerations) have shown that this effect is linear, not with W but with humidity potential $W^+ = (W - 0.0092)^+$ indicating that the absolute humidity effect is zero when W<0.0092 and linear above that threshold value. In such a case, the most general formulation of the equivalent "linear and seasonal trend" model is:

$$E_{d,h} = \sum_k k \cdot [X_k + Y_k + Z_k] \tag{9.18}$$

where k = 1, T_{db}, W^+ and S.

The choice of which terms to retain in the model is done based on statistical tests or by stepwise regression (discussed in Sect. 5.7.4). Usually only a few terms are adequate. Table 9.3 assembles the results of fitting a Fourier series model to the whole building electric hourly data for a whole year shown in Fig. 9.12 and to the cooling thermal energy use for the first six months of the year (shown in Fig. 9.13). The progressive improvement of the model R^2 as stepwise regression is performed is indicative of the added contribution of the associated variable in describing the variation in the dependent variable. The weather independent model is very good with high R^2 and low CV. In the case of temperature-dependent loads, the R^2 is very good but the CV is rather poor (12%) with the residual values being about ± 1 GJ/h (from Fig. 9.13). The ambient temperature variable is by far the most influential variable.

The above case study is intended to demonstrate how the general Fourier series modeling framework can be modified or tailored to capture structural trend in time series data with distinct and clear periodicities. For other types of applications, one would expect different variables as well as different periodicities to be influential, and the model structure would have to be suitably altered. Spectral analysis is an extension of Fourier analysis as applied to stochastic processes with no obvious trend or seasonality; if there are obvious trends, these should be removed prior to spectral analysis. The spectrum of a data series is a plot of the amplitude versus the angular frequency. Inspection of the spectrum allows one to detect hidden periodicities or features that show statistical regularity (see for example, Bloomfield 1976).

Table 9.3 Fourier series model results for building loads during the school-in-session weekdays (from Dhar et al. 1998). CHi (and SHi) and CDi (and SDi) represent the ith frequency of the cosine (and sine) terms of the diurnal and seasonal cycles respectively. The climatic variables are ambient dry-bulb temperature (T_{db}), absolute humidity potential (W^+) and horizontal solar radiation (S)

No. of parameters	Weather independent (one year)—Fig. 9.12				Weather dependent (6 months)—Fig. 9.13			
	Variable	Partial R^2	Cumulative R^2	CV (%)	Variable	Partial R^2	Cumulative R^2	CV (%)
2	CH1	0.609	0.609	–	T_{db}	0.804	0.804	–
3	SH1	0.267	0.876	–	W^+	0.062	0.866	–
4	CH2	0.041	0.918	–	T_{db} * CH1	0.179	0.888	–
5	SH4	0.012	0.927	–	S * CH1	0.008	0.892	12.08
6	SH3	0.007	0.937	–				
7	SD1	0.006	0.943	3.8				

9.4.4 Interrupted Time Series

The time series data considered above was representative of a system which behaved predictably, albeit with some noise or unexplained variation, but without any abrupt changes in the system dynamics. Since the systems under interest are often dynamic entities, their behavior changes in time due to some specific cause (or *intervention*), and this gives rise to *interrupted time series*. Often, such changes can be detected fairly easily, and should be removed as part of the trend and seasonal modeling phase. In many cases, this can be done via OLS modeling or some simple transformation; two simple cases will be briefly discussed below. More complex interventions cannot be removed by such simple measures and should be treated in the framework of transfer functions (Sect. 9.6).

9.4.4.1 Abrupt One-Time Constant Change in Time

This is the simplest type of intervention arising when the time series data abruptly changes its mean value and assumes a new mean level. If the time at which the intervention took place is known, one can simply recast the trend and seasonal model to include an indicator variable (see Sect. 5.7.3) such that I=0 before the intervention and I=1 afterwards (or vice versa).

The case of an abrupt operational change in the heating and ventilation air-conditioning (HVAC) system of a commercial building is illustrated by the following example (Ruch et al. 1999). Synthetic energy data (E_t) for day t was generated using 91 days of real daily outdoor dry-bulb temperature data T_t from central Texas according to the following model:

$$E_t = 2{,}000 + 100T_t + 1{,}500I + \varepsilon_t \tag{9.19}$$

where the indicator variable
I=1 for days t=1, 2,..., 61, and
I=0 for days t=62, 63,..., 91
and ε_t is the error term assumed normally distributed.

This hypothetical building has been assumed to undergo an energy saving operational change on the 62nd day, the result being a 1,500 unit shift (or decrease) in energy use at the change point. In this case, since the shift is in the mean value only, the slope of the model is constant over the entire period. A superficial glance at a scatter plot of the data (Fig. 9.15a) suggests a single linear model, though a closer look at the residuals against time would have revealed a shift. A simple OLS fit gave a reasonable fit (R^2=0.84) but the residual autocorrelation was significant (ρ=0.80). The model with indicator variables to account for the change-point in time behavior, resulted in a much better fit of R^2=0.98 and negligible autocorrelation (see Fig. 9.15b).

9.4.4.2 Gradual and Constant Change over Time

Another type of common intervention is when the change is not abrupt but gradual and constant. In the framework of the example given above, the energy use in the building "creeps" up in time due to such causes as increase of equipment (more computers, printers,...) or gradual degradation in performance of the HVAC system. Such *energy creep* is widely observed in buildings and has been well-documented in several publications. Again, one has to remove the structural portion of the time series so as to de-trend it. A simple approach is to simply modify the model given by Eq. 9.19 when the degradation is related to temperature as follows:

$$E_t = a + bT_t + cIT_t + \varepsilon_t \tag{9.20}$$

where the indicator variable is now applied to the slope of the model, and not to the intercept, such that:
I = 0 before the onset of energy creep, and
I = 1 after the onset.

Interrupted time series can arise due to various types of interventions. This section presented some simple cases when the interventions were one-time occurrences and the time of their onset was known. There are more complex types of interventions which are due to known forcing functions which change in time, and such cases are treated using

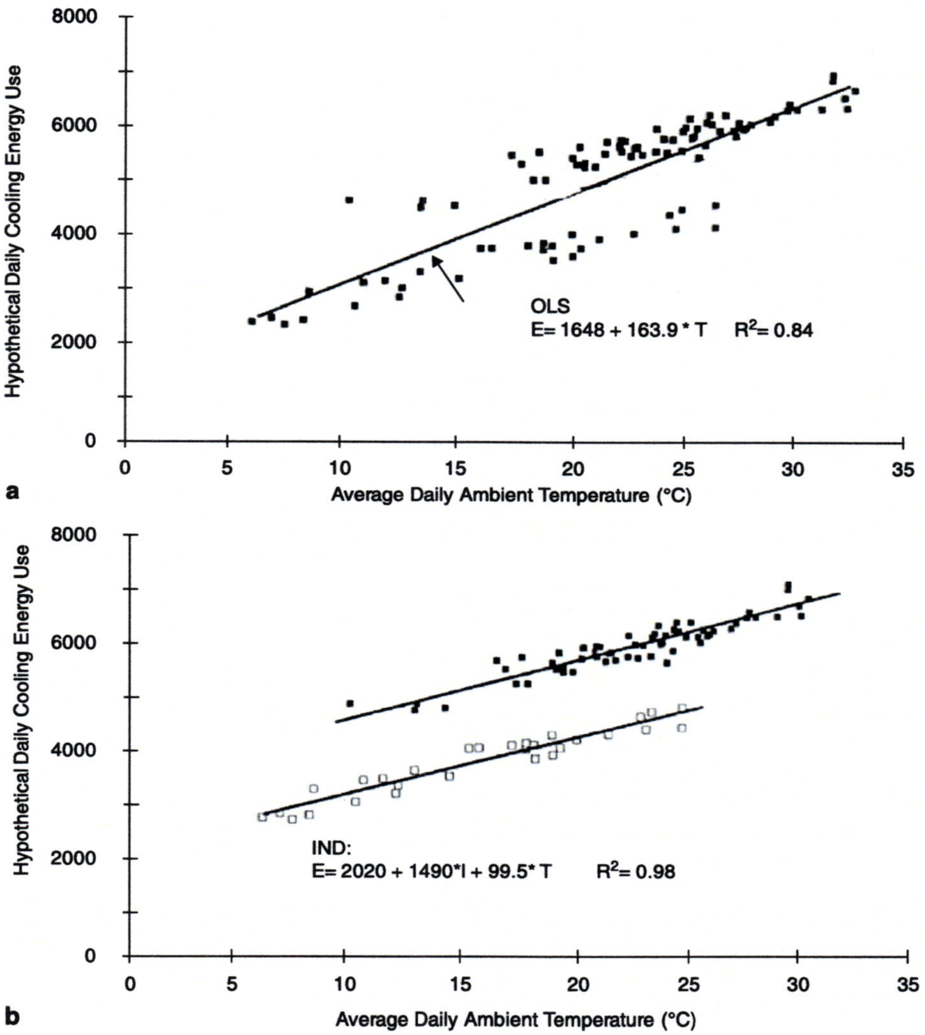

Fig. 9.15 Improvement in OLS model fit when an indicator variable is introduced to capture abrupt one-time change in energy use in a building. **a** Ordinary least square (OLS) model. **b** Indicator variable model (IND). (From Ruch et al. 1999)

transfer function models (Sect. 9.6). The reader can also refer to McCain and McCleary (1979) for more discussion on interrupted time series analysis.

9.5 Stochastic Time Series Models

9.5.1 Introduction

The strength of the classical regression techniques (such as OLS discussed in Chap. 5) is the ability to capture the deterministic or structural trend of the data, while the model residuals are treated as random errors which impact the uncertainty limits of future predictions. In time series data, the residuals are often patterned, i.e., have a serial correlation which is difficult to remove using classical regression methods. The strength of the time series techniques is their ability to first detect whether the model residuals are purely random or not, i.e., whether white noise. If they are, classical regression methods are adequate. If not, the residual errors are separated into a systematic stochastic component and white noise. The former is treated by stochastic time series models such as AR, MA, ARMA, ARIMA and ARMAX (these terms will be explained below) which are *linear* in both model and parameters, and hence, simplify the parameter estimation process. Such an approach usually allows more accurate predictions than classical regression (i.e., narrower uncertainty bands around the model predictions), and therein lies their appeal. Once it is deemed that a time series modeling approach is appropriate for the situation at hand, three separate issues are involved similar to OLS modeling: (i) identification of the order of the model (i.e., model structure), (ii) estimation of the model parameters (parameter estimation), and (iii) ascertaining uncertainty in the forecasts.

It must be noted that time series models may not always be superior to the standard OLS methods even when dealing with time series data; hence, the analyst should evaluate various models in terms of their predictive ability by performing a sample holdout cross-validation evaluation (described in Sect. 5.3.2d).

There is a rich literature on stochastic time series models as pertinent to various disciplines, and this section should be regarded as an introduction to this area (see texts such as Box and Jenkins 1976 dealing with engineering applications and Pindyck and Rubinfeld 1981 dealing with econometric applications). Some professionals view stochastic time series analysis as being too mathematically involved to be of much use for practical applications; this is no more true than any statistical method. An understanding of the basic principles and of the functionality of the different forms of ARMAX models, the willingness to learn by way of practice along with familiarity in using an appropriate statistical software are all that is needed to be able to add time series analysis to one's toolbox.

9.5.2 ACF, PACF and Data Detrending

9.5.2.1 Autocorrelation Function (ACF)

The concept of the ordinary correlation coefficient between two variables has already been introduced in Sect. 3.4.2. This concept can be extended to time series data to ascertain if successive observations are correlated. Let $\{Y_1, \ldots Y_n\}$ be a *discrete de-trended time series data* where the long-term trend and seasonal variation has been removed. Then $(n-1)$ pairs of observations can be formed, namely (Y_1, Y_2), (Y_2, Y_3),…(Y_{n-1}, Y_n). An *autocorrelation* (or serial correlation) coefficient r_1 measures the extent to which successive observations are interdependent and is given by:

$$r_1 = \frac{\sum_{t=1}^{n-1} (Y_t - \bar{Y})(Y_{t+1} - \bar{Y})}{\sum_{t=1}^{n} (Y_t - \bar{Y})^2} \tag{9.21}$$

where $\bar{Y}$ is the overall mean.

If the data is cyclic, this behavior will not be captured by the first order autocorrelation coefficient. One needs to introduce, by extension, the serial correlation at lag k, i.e., between observations k apart:

$$r_k = \frac{\left(\frac{1}{n-k}\right)\sum_{t=1}^{n-k} (Y_t - \bar{Y})(Y_{t+k} - \bar{Y})}{\left(\frac{1}{n-1}\right)\sum_{t=1}^{n} (Y_t - \bar{Y})^2} = \frac{c_k}{c_0} \tag{9.22}$$

where n is the number of data points and c_k and c_o are the autocovariance coefficients at lag k and lag zero respectively. Though it can be calculated for lags of any size, usually it is inadvisable to calculate r_k for values of k greater than about (n/4) (Chatfield 1989). A value of r_k close to zero would imply little or no relationship between observations k lags apart. The *autocorrelation function* (ACF) is a function which represents the variation of r_k with lag k. Usually, there is no need to fit a functional equation, but a graphical representation called the *correlogram* is a useful means to provide insights both into model development and to evaluate whether *stationarity* (i.e., detrending by removal of long-term trend and periodic/cyclic variation) has been achieved or not. It is clear that the ACF is an extension of the Durbin-Watson (DB) statistic presented in Sect. 5.6.1 which relates to one lag only, i.e., k=1.

Figure 9.16 illustrates how the ACF varies for four different numerical values of r_k (for both positive and negative values) assuming only first-order autocorrelation to be present. One notes that all curves are asymptotic, dropping to zero faster for the weaker correlations, while plots with negative correlation fluctuate on either side of zero as they drop towards zero. The close similarity between these plots and those of the Pearson correlation coefficient (Sect. 3.4.2 and Fig. 3.13) should be noted. Because r_k is normalized with the value at lag k=0, ACF at lag 0 is unity, i.e., $r_0=1$. If the data series were non-stationary, these plots would not asymptote towards zero (as illustrated in Fig. 9.17). Thus, stationarity is easily verified via the ACF.

The standard error for r_k is calculated under the assumption that the autocorrelations have died out till lag k using:

$$\sigma(r_k) = \left[\frac{1}{n}\left(1 + 2\sum_{i=1}^{k-1} r_i^2\right)\right]^{1/2} \tag{9.23}$$

where n is the number of observations. The corresponding confidence intervals at the selected significance level α are given by: $0 \pm z_{\alpha/2} \cdot \sigma(r_k)$. Any sample correlation which falls outside these limits is deemed to be statistically insignificant at the selected significant level. The primary value of the correlogram is that it is a convenient graphical way of determining the number of lags after which correlation coefficients are insignificant. This is useful in identifying the appropriate MA model (discussed in Sect. 9.5.3).

9.5.2.2 Partial Autocorrelation Function (PACF)

The limitation of the ACF as a statistical measure suggestive of the order of the model meant to fit the detrended data is that the magnitude of r_1 carries over to subsequent r_i values. In other words, the autocorrelation at small lags carries over to larger lags. Consider a first-order model with $r_1=0.8$ (see Fig. 9.16c). One would expect the ACF to decay exponentially. Thus, $r_2=0.8^2=0.64$ and $r_3=0.8^3=0.512$ and so on, even

Fig. 9.16 Correlograms of the first-order ACF for different magnitudes of the correlation coefficient. **a** Weak positive. **b** Moderate positive. **c** Strong positive. **d** Strong negative

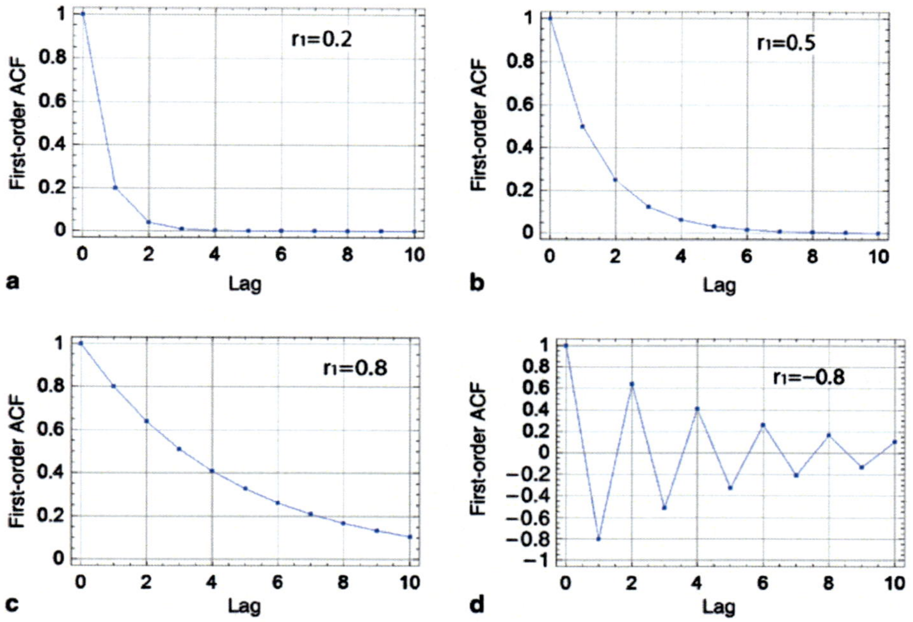

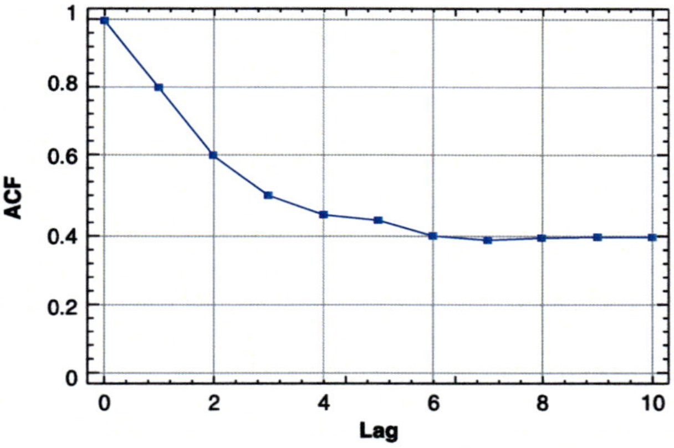

Fig. 9.17 Sample correlogram for a time series which is non-stationary since the ACF does not seem to asymptote to zero

if no second order lags are present. When second order effects are present, the decay is also close to asymptotic, which clouds the identification of the order of the model describing the time series process. A statistical quantity, by which one can measure the *excess correlation* that remains at lag k after a model of order (k-1) is fit to data, is called the *partial autocorrelation function* (PACF) ϕ_{kk}. Thus, the ACF of a time series tends to taper off as lag k increases, while the PACF cuts off after a certain lag is reached.

The standard error for ϕ_{kk} is given by:

$$\sigma(\phi_{kk}) = \left(\frac{1}{n}\right)^{1/2} \tag{9.24}$$

where n is the number of observations, while the corresponding confidence intervals at significance level α are given by $0 \pm z_{\alpha/2} \cdot \sigma(\phi_{kk})$.

The PACF is particularly useful as an aid in determining the order of the AR model as discussed in Sect. 9.5.3. It finds application in problems where it is necessary to determine the order of the ODE needed to model dynamic system behavior.

Example 9.5.1: *The ACF function applied to peak electric demand*

Consider the data assumed earlier in Example 9.1.1 and shown in Table 9.1. One would speculate that four lags are important because the data is taken quarterly (four times a year). How the ACF function is able to detect this effect will be illustrated below.

The data is plotted in Fig. 9.1. A commercial software package has been used to generate the ACF shown in Fig. 9.18 while Table 9.4 shows the estimated autocorrelations between values of electric power at various lags (only till 8 lags are shown) along with their standard errors and the 95% probability limits around 0.0. The lag k autocorrelation coefficient measures the correlation between values of electric power at time t and time (t−k). If the probability limits at a particular lag do not contain the estimated coefficient, there is a statistically significant correlation at that lag at the 95% confidence level. In this case, 4 of the 24 autocorrelation coefficients are statistically significant at the 95% confidence level (shown in italics in the table), implying that the time series is not completely random (white noise)—this is consistent with the result expected. ■

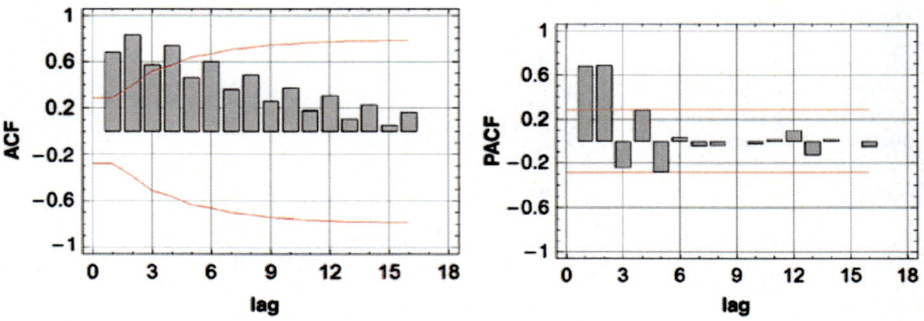

Fig. 9.18 ACF and PACF plots for the time series data given in Table 9.1 along with their 95% uncertainty bands. Note the asymptotic behavior of the ACF and the abrupt cutoff of the PACF after a finite number of lags

Table 9.4 Estimated autocorrelation coefficients till lag 8 and associated uncertainty

Lag	Autocorrelation	Stnd. error	Lower 95.0% prob. limit	Upper 95.0% prob. limit
1	*0.679357*	*0.144338*	*−0.282897*	*0.282897*
2	*0.829781*	*0.200159*	*−0.392305*	*0.392305*
3	*0.571588*	*0.262207*	*−0.513918*	*0.513918*
4	*0.737873*	*0.286994*	*−0.562499*	*0.562499*
5	0.462624	0.324116	−0.635257	0.635257
6	0.600009	0.337593	−0.661671	0.661671
7	0.358554	0.359123	−0.70387	0.70387
8	0.48272	0.366505	−0.718338	0.718338

9.5.2.3 Detrending Data by Differencing

Detrending is the process by which the deterministic trend in the data can be removed or filtered out. One could use a regression model to achieve this, but a simpler method, and one which yields insight into the order of the time series model, is *differencing*. For data series that do not have cyclic variation (i.e. non-seasonal data), differencing can make a non-stationary time series stationary. A backward first-order difference that can remove a linear trend is:

$$\nabla Y_{t+1} = Y_{t+1} - Y_t \tag{9.25a}$$

where ∇ is the backward difference operator.

Similarly, the second order differencing to remove a quadratic trend is:

$$\begin{aligned} \nabla Y_{t+2} &= (Y_{t+1} - Y_t) - (Y_t - Y_{t-1}) \\ &= Y_{t+1} - 2Y_t + Y_{t-1}) \end{aligned} \tag{9.25b}$$

Thus, differencing a time series is akin to finite differencing a derivative. The time series data in Fig. 9.19a is quadratic, with the first differencing making it linear (see Fig. 9.19b) and the second differencing (Fig. 9.19c) making it constant, i.e. totally without any trend. This is, of course, a simple example, and actual data will not detrend so cleanly.

Usually, not more than a second order sequencing is needed to make a time series stationary provided no seasonal trend is present. In case this is not so, a log transform should be investigated. Let us illustrate the above concepts using very simple examples (McCain and McCleary 1979). Consider the series $\{1, 2, 3, 4, 5\ldots, N\}$. Differencing this sequence results in $\{1, 1, 1, 1, 1,\ldots 1\}$ which is stationary. Hence, a first order sequencing is adequate. Now consider the sequence $\{2, 4, 8, 16, 32,\ldots 2^N\}$. No matter how many times this sequence is differenced, it will remain nonstationary. Let us log-transform the sequence as follows $\{1(\ln 2), 2(\ln 2), 3(\ln 2), 4(\ln 2),\ldots N(\ln 2)\}$. If this sequence is differenced just once, one gets a stationary sequence. Series which require a log-transform to make them stationary are called "explosively" nonstationary. They are not as common in practice as linear time series data, but when they do appear, they are easy to detect.

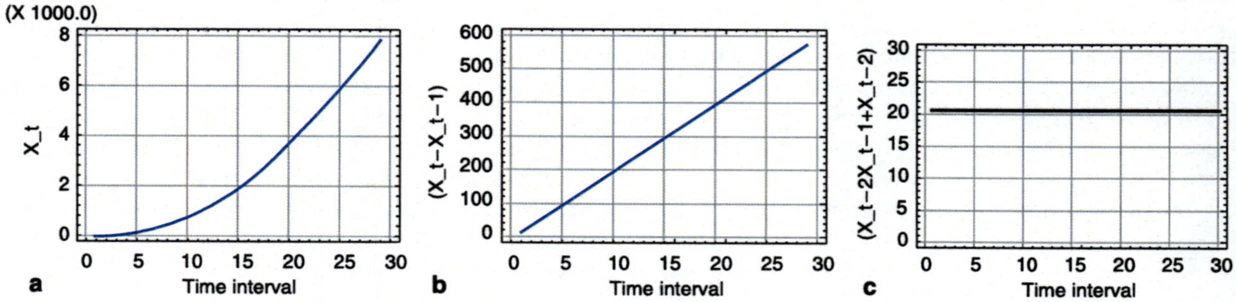

Fig. 9.19 **a–c** Simple illustration of how successive differencing can reduce a non-stationary time series to a stationary one (function assumed: $x_t = 10t^2$)

For time series data that have periodic (or seasonal) variability, *seasonal differencing* has to be done, albeit appropriately. For example, time series data of hourly electricity use in a building exhibits strong diurnal variability which is, however, fairly repetitive from one day to the next. Differencing hourly data 24 h apart would be an obvious way of making the series close to stationary. Thus, one employs the operator ∇_{24} and obtains $\nabla_{24}Y_t=Y_t-Y_{t-24}$ which is likely to detrend the data series. Though visual inspection is one way of determining whether a data has become stationary or not, a better way is to use statistical tools such as the correlogram.

Another question is whether detrending a time series by differencing it changes any of the deterministic parameters that describe the processes. The answer is that differencing does not affect the parameters but only affects the manner in which they are represented in the model. Consider the sequence {2, 4, 6, 8…, 2N}. Clearly this has a linear secular trend which can be modeled by the equation: $Y_t=2.t$, i.e., a linear model with slope 2. However, one can also represent the same series by the equation:

$$Y_t = Y_{t-1} + 2$$

Thus, the explicit trend model and the first-order linear difference equations both retain the parameter as is, but the parameter appears in different forms in both equations.

9.5.3 ARIMA Models

The ARIMA (Auto Regressive Integrated Moving Average) model formulation is a general *linear* framework which actually consists of three sub-models: the autoregressive (AR), the integrated (I), and the moving average (MA). It is a model linear in its parameters which simplifies the estimation. The integrated component is meant to render the time series data stationary, while the MA and AR components are meant to address the stochastic element. Often, time series data have mean values that are dependent upon time (drift upwards or downwards), have non-constant variance in the random shocks that drive the process, or they possess a seasonal trend. A seasonal trend is reflective of autocorrelation at a large lag, and if the cyclic patterns are known, they can be detrended as described earlier. The non-constant variance violation is often handled by transforming the data in some fashion (refer to Chatfield 1989). ARIMA models are expressed as ARIMA (p, d, q) where p is the order of AR component, q is the order of the MA component and d is the number of differencing taken to make the time series data stationary[4]. Thus, ARIMA (p, 0, q) implies that the time series is already stationary and that no differencing need be done. Similarly, ARIMA(p, 2, q) denotes that differencing needed to be done twice to make the time series data stationary.

Consider a system which could be an engineering system with well-defined behavior, or a complex system (such as the stock market). The current response of a dynamic system depends on its past value. Think of the cooling of a hot water tank where the temperature at time t is necessarily a function of the temperature at time (t−1) as well as on the "shocks" or random perturbations to which it is subjected to. The AR model captures the former element, i.e., the "memory" or past behavior of the system by expressing the series residuals at the current time as a linear function of p past residuals. The MA models capture the random shocks or perturbation on the system (which do not persist) via a linear function of q past white noise errors. The AR component is dominant in systems which are fairly deterministic and with direct relationship between adjacent observations (such as the tank cool-down example). The order of p is directly related to the order of the differential equation of the white-box model which will adequately model the system behavior (see Sect. 9.6.3). The MA component is a special type of low-pass filter which is a generalization of the EWA smoothing approach described in Sect. 9.3.2.

The MA and AR models are obviously special cases of the ARMA formulation. Unlike OLS type models, ARMA models require relatively long data series for parameter estimation (about a minimum of 50 data points and preferably 100 data points or more) and are based on *linear* model formulation. However, they provide very accurate forecasts and offer both a formal and a structured approach to model building and analysis. Several texts suggest that, in most cases, it is not necessary to include both the AR and the MA elements; one of these two, depending on system behavior, should suffice. Further, it is recommended that analysts new to this field limit themselves to low order models of 3 or less provided the seasonality has been properly filtered out.

9.5.3.1 ARMA Models

Let us consider a discrete random time series data which is stationary but serially dependent, and is represented by the series $\{Z_t\}$. Let $\{a_t\}$ be a white noise or "purely" random series, also referred to as random shocks or innovations. The ARIMA (p, 0, q) model is then written as:

$$\begin{aligned}\phi'_0 Z_t &= \phi'_1 Z_{t-1} + \phi'_2 Z_{t-2} + \cdots + \phi'_p Z_{t-p} \\ &\quad + \omega'_0 a_t + \omega'_0 a_{t-1} + \omega'_2 a_{t-2} + \cdots + \omega'_q a_{t-q}\end{aligned} \tag{9.26a}$$

where $\{\phi'_i\}$ and $\{\omega'_i\}$ are the weights on the $\{Z_t\}$ and $\{a_t\}$ series respectively. The weights are usually scaled by setting $\phi'_0 = 1$ and $\omega'_0 = 1$, so that the general ARIMA(p, 0, q) formulation is:

$$\begin{aligned}Z_t &= \phi_1 Z_{t-1} + \phi_2 Z_{t-2} + \cdots + \phi_p Z_{t-p} \\ &\quad + a_t + \omega_1 a_{t-1} + \omega_2 a_{t-2} + \cdots + \omega_q a_{t-q}\end{aligned} \tag{9.26b}$$

[4] Stationarity of a stochastic process can be interpreted qualitatively as a process which is in statistical equilibrium.

9.5.3.2 MA Models

The first order moving average model or MA(1) represents a linear system subjected to a shock in the first time interval only and which does not persist over subsequent time periods. Following Box and Luceno (1997), it is written as:

$$Z_t = a_t + \omega_1 a_{t-1} = a_t - \theta \cdot a_{t-1} \tag{9.27}$$

where the coefficient $\theta = -\omega_1$ is introduced by convention and represents the weighted portion of the previous shock at time (t−1).

Thus:

$$Z_{t-1} = a_{t-1} - \theta \cdot a_{t-2}, \quad Z_{t-2} = a_{t-2} - \theta \cdot a_{t-3}$$

and so on.

The general expression for a MA model of order q, i.e., MA (q) model, is expressed as:

$$Z_t = a_t - \theta_1 \cdot a_{t-1} - \theta_2 \cdot a_{t-2} - \ldots - \theta_q \cdot a_{t-q} \tag{9.28}$$

The white noise terms a_i are often modeled as a normal distribution with zero mean and standard deviation σ, or $N(0,\sigma)$ and, hence, the process given by Eq. 9.28 will fluctuate around zero. If a process with a non-zero mean μ but without any trend or seasonality is to be modeled, a constant term c is introduced in the model such that $c=\mu$. An example of a MA(1) process with mean 10 and $\theta_1=-0.9$ is depicted in Fig. 9.20a where a set of 100 data points have been generated in a spreadsheet program using the model shown with a random number generator $N(0,1)$ for the white noise term. Since this is a first order model, the ACF should have only one significant value (this is seen in Fig. 9.20b where ACF for greater lags fall inside the 95% confidence intervals). Ideally, there should only be one spike at lag k=1, but because random noise was introduced in the synthetic data, this obfuscates the estimation, and spikes at other lags appear which, however, are statistically insignificant.

For MA models, the ACF can be deduced from the model coefficients (Montgomery and Johnson 1976):

$$\text{For MA(1)}: \; r_k = \frac{-\theta_1}{1+\theta_1^2} \text{ for k} = 1 \text{ and } r_k = 0 \text{ for k} > 1 \tag{9.29}$$

and

$$\text{for MA(2)}: \; r_1 = \frac{-\theta_1(1-\theta_1)}{1+\theta_1^2+\theta_2^2} \text{ and}$$

$$r_2 = \frac{-\theta_2}{1+\theta_1^2+\theta_2^2} \text{ and } r_k = 0 \text{ for k} > 2$$

For the above MA(1) example, $r_1 = (0.9)/1+(0.9)^2 = 0.497$ which is what is indicated in Fig. 9.20b.

The PACF function alternates with lag term but damps out exponentially (Fig. 9.20c). MA processes are not very common in engineering, but they are often used in areas where the origin of the shocks is unexpected. For example, in econometrics, events such as strikes and government decisions are modeled as white noise.

9.5.3.3 AR Models

Autoregressive models of the order p or AR(p) models are often adopted in engineering applications. The first-order AR(1) model is written as (Box and Luceno 1997):

$$Z_t = \phi_1 \cdot Z_{t-1} + a_t \tag{9.30}$$

where ϕ_1 is the ACF at lag 1 such that $-1 \le \phi_1 \le 1$. The AR(1) model, also called a *Markov process* is often used to model physical processes. A special case is when $\phi_1 = 1$, which represents another well known process called the *random walk* model. If a process with a non-zero mean μ is to be modeled, a constant term c is introduced in the model such that $c = \mu(1-\phi_1)$.

At time (t−1): $Z_{t-1} = \phi_1 . Z_{t-2} + a_{t-1}$ which, when substituted back into Eq. 9.30, yields

$$Z_t = a_t + \phi_1 \cdot a_{t-1} + \phi_1{}^2 \cdot Z_{t-2} \tag{9.31}$$

Eventually, by successive substitution, Z_t can be expressed as an infinite-order MA. From the viewpoint of a compact model, an AR model approach is, thus, superior to the MA modeling approach.

An AR(2) process is written as:

$$Z_t = \phi_1 \cdot Z_{t-1} + \phi_2 \cdot Z_{t-2} + a_t \tag{9.32}$$

with the conditions that: $(\phi_1 + \phi_2) < 1, (\phi_2 - \phi_1) < 1$, $-1 < \phi_2 < 1$

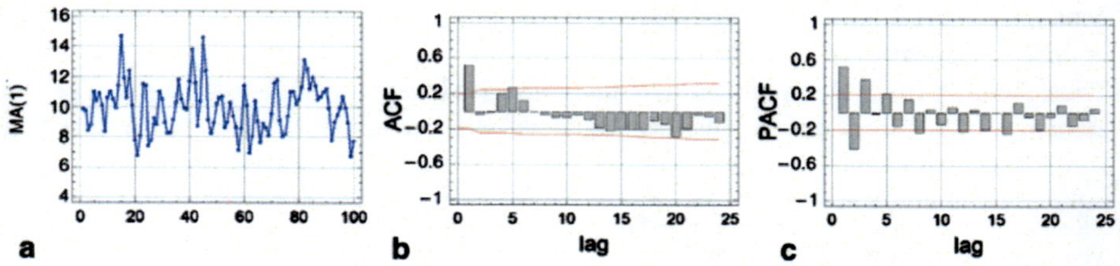

Fig. 9.20 a–c One realization of a MA(1) process for $Z_t = 10 + \varepsilon_t + 0.9\varepsilon_{t-1}$ along with corresponding ACF and PACF with error term being Normal(0,1)

Again, if a process with a non-zero mean μ is to be modeled, a constant term c is introduced in the model such that $c=\mu(1-\phi_1-\phi_2)$.

By extension, an AR(p) model will assume the form:

$$Z_t = \phi_1 \cdot Z_{t-1} + \phi_2 \cdot Z_{t-2} + \cdots + \phi_p \cdot Z_{t-p} + a_t \quad (9.33)$$

$$\text{For AR(1): ACF is } r_k = \phi_1^k \text{ (exponential decay)} \\ \text{and PACF is } \phi_{11} = r_1 \quad (9.34)$$

and

$$\text{AR(2): ACF is } r_0 = 1, \quad r_1 = \frac{\phi_1}{1-\phi_2}, \\ r_k = \phi_1 r_{k-1} + \phi_2 r_{k-2} \text{ for k} > 0$$

while PACF is $\phi_{11}=r_1$ and ϕ_{22} requires an iterative solution.

There are no simple formulae to derive the PACF for orders higher than 2, and hence software programs involving iterative equations, known as the Yule-Walker equations, are used to estimate the parameters of the AR model (see for example, Box and Luceno 1997).

Two processes, one for AR(1) with a positive coefficient and the other for AR(2) with one positive and one negative coefficients are shown in Figs. 9.21 and 9.22 along with their respective ACF and PACF plots. The ACF function, though it is a model of order one, dies down exponentially, and this is where the PACF is useful. Only one PACF term is statistically significant at the 95% significance level for AR1 in Fig. 9.21 while two terms are so in Fig. 9.22 (as it should be). The process mean line for AR(1) and the constant term appearing in the model are related: $\mu = c/(1-\phi) = 5/(1-0.8) = 25$ which is consistent with the process behavior shown in Fig. 9.21a.

For AR(2), the process mean line is $\mu = c/(1-\phi_1-\phi_2) = 25/(1-0.8+0.8)$ also consistent with Fig. 9.22a. For the ACF: $r_1 = 0.8/(1-(-0.8)) = 0.44$ and $r_2 = -0.8 = (0.8)^2/(1-(-0.8)) = -0.44$. The latter value is, slightly different from the value of about -0.5 shown in Fig. 9.22b which is due to the white noise introduced in the synthetic sequence.

Finally, Fig. 9.23 illustrates an ARMA (1,1) process where elements of both MA(1) and AR(1) processes are present: exponential damping of both the ACF and the PACF.

These four sets of figures (Figs. 9.20–9.23) partially illustrate the fact that one can model a stochastic process using different models, a dilemma which one faces even when identifying classical OLS models. Hence, evaluation of competing models using a cross-validation sample is high advisable as well as investigating whether there is any correlation structure left in the residuals of the series after the stochastic model effect has been removed. These tests closely parallel those which one would perform during OLS regression.

9.5.3.4 Identification and Forecasting

One could use the entire ARIMA model structure as described above to identify a complete model. However, with the

Fig. 9.21 a–c One realization of a AR(1) process for $Z_t=5+0.8Z_{t-1}+\varepsilon_t$ along with corresponding ACF and PACF with error term being Normal(0,1)

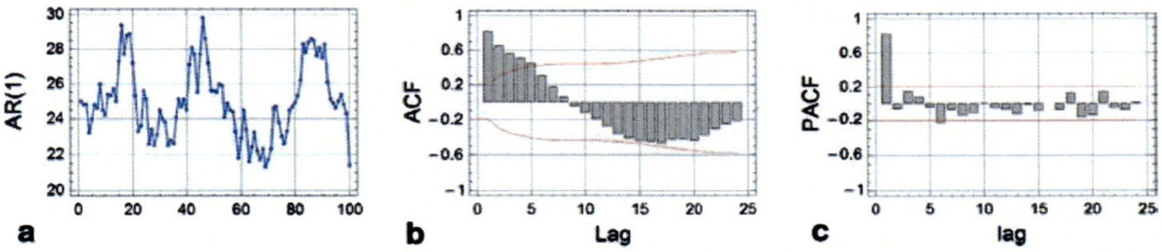

Fig. 9.22 a–c One realization of a AR(2) process for $Z_t=25+0.8Z_{t-1}-0.8Z_{t-2}+\varepsilon_t$ along with corresponding ACF and PACF with error term being Normal(0,1)

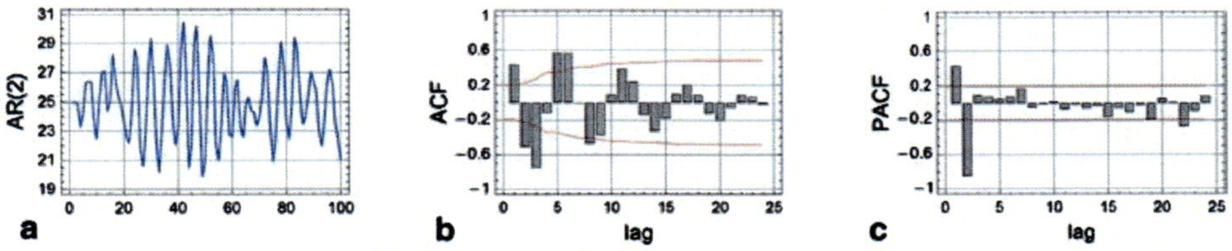

Fig. 9.23 One realization of a ARMA(1,1) process for $Z_t=15+0.8Z_{t-1}+\varepsilon_t+0.9\varepsilon_{t-1}$ along with corresponding ACF and PACF with error term being Normal(0,1)

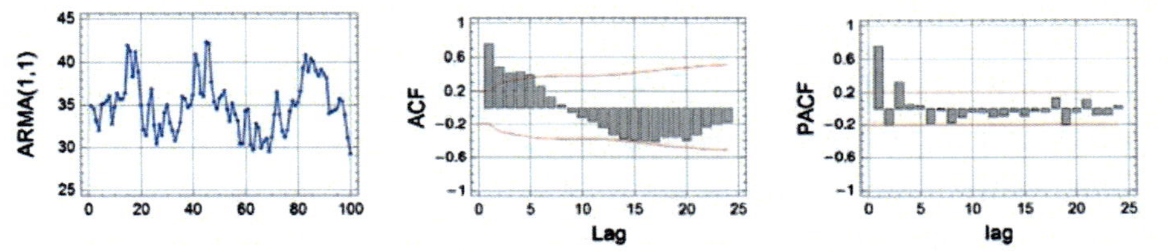

intention of simplifying the process of model identification, and in recognition of the importance of AR models, let us limit the scope to AR models only. The procedure for identifying an AR(p) model with n data points, and then using it for forecasting purposes are summarized below.

To Identify a AR(p) Model:

(i) Evaluate different trend and seasonal models using OLS, and identify the best one based on internal and external predictive accuracies:

$$Y_t = b_0 + b_1 \cdot x_{1,t} + \cdots + b_k \cdot x_{k,t} \qquad (9.35a)$$

where the x terms are regressors which account for the trend and seasonal variation.

(ii) Calculate the residual series as the difference between observed and predicted: $\{Z_t\} = \{Y_t - \hat{Y}_t\}$

(iii) Determine the ACFs of the residual series for different lags: $r_1, r_2, \ldots r_p$

(iv) Determine the PACF function of the residual series for different lags: $\phi_{11}, \phi_{2,2}, \ldots \phi_{pp}$

(v) Generate correlograms for the ACF and PACF, and make sure that series is stationary

(vi) Evaluate different AR models based on their internal and external predictive accuracies, and select the most parsimonious AR model (often, 1 or 2 terms should suffice)—see Sect. 9.5.4 for general recommendations:

$$\begin{aligned}\hat{Y}_t &= (b_0 + b_1 \cdot x_{1,t} + \cdots + b_k \cdot x_{k,t}) \\ &+ (\phi_1 Z_{t-1} + \phi_2 Z_{t-2} + \cdots + \phi_p Z_{t-p})\end{aligned} \qquad (9.35b)$$

(vii) Calculate the RMSE of the overall model (trend plus seasonal plus stochastic) thus identified.

To Forecast a Future Value Y_{t+1} When Updating Is Possible (i.e., the Value Y_{t+1} Is Known):

(i) Compute the series: $\hat{Z}_t, \hat{Z}_{t-1}, \ldots \hat{Z}_{t-p}$

(ii) Estimate $\hat{Z}_{t+1} = \phi_1 \cdot \hat{Z}_t + \phi_2 \cdot \hat{Z}_{t-1} + \cdots + \phi_p \cdot \hat{Z}_{t-p+1}$

(iii) Finally, use the overall model (Eq. 9.35b) modified to time step (t+1) as follows:

$$\hat{Y}_{t+1} = (b_0 + b_1 \cdot x_{1,t+1} + \cdots + b_k \cdot x_{k,t+1}) + \hat{Z}_{t+1} \qquad (9.35c)$$

(iv) Calculate approximate 95% prediction limits for the forecast given by:

$$\hat{Y}_{t+1} \pm 2 \cdot RMSE \qquad (9.36)$$

(v) Re-initialize the series by setting Z_t as the residual of the most recent period, and repeat steps (i) to (iv).

To Forecast a Future Value When Updating Is Not Possible: In case, one lacks observed values for future forecasts (such as having to make forecasts over a horizon involving several time steps ahead), these are to be estimated in a recursive manner as follows. The first forecast is made as before, but now, one is unable to compute the model error which is to be used for predicting the second forecast, and the subsequent accumulation of errors widens the confidence intervals as one predicts further into the future. Then:

(i) Future forecast Y_{t+2} for the case when no updating is possible (i.e., Y_{t-1} is not known and so one cannot determine Z_{t+1}):

$$\begin{aligned}\hat{Z}_{t+2} &= \phi^2{}_1 \cdot \hat{Z}_t + \phi^2{}_2 \cdot \hat{Z}_{t-1} + \cdots + \phi_p \cdot \hat{Z}_{t-p+1} \\ \hat{Y}_{t+2} &= (b_0 + b_1 \cdot x_{1,t+2} + \cdots + b_k \cdot x_{k,t+2}) + \hat{Z}_{t+2}\end{aligned} \qquad (9.37)$$

and so forth…

(ii) An approximate 95% prediction interval for m time steps should be determined, and this is provided by the software program used. For the simple case of AR(1), for forecasts m time-steps ahead :

$$\hat{Y}_{t+m} \pm 2 \cdot RMSE\,[1 + {\phi_1}^2 + \cdots + {\phi_1}^{2(m-1)}]^{1/2} \qquad (9.38)$$

Note that the ARMA models are usually written as equations with fixed estimated parameters representing a stochastic structure that does not change with time. Hence, such models are not adaptive. This is the reason why some researchers caution the use of these models for forecasting several time steps ahead when updating is not possible.

Example 9.5.2: *AR model for peak electric demand*
Consider the same data shown in Table 9.1 for the electric utility which consists of four quarterly observations per year for 12 years (from 1974–1986). Let us illustrate the use of the AR(1) model with this data set and highlight its importance as compared to the various models described earlier.

The trend and seasonal model is given in Example 9.4.2. This model is used to calculate the residuals $\{Z_t\}$ for each of the 44 data points. The ACF and PACF functions for $\{Z_t\}$ are shown in Fig. 9.24. Since the PACF cuts off abruptly after lag 1, it is concluded that an AR(1) model is adequate to model the stochastic residual series. The corresponding model was identified by OLS:

$$Z_t = 0.657 \cdot Z_{t-1} + a_t$$

Note that the value ϕ_1=0.657 is consistent with the value shown in the PACF plot of Fig. 9.24. Table 9.5 assembles the RMSE values of various models used in previous examples as well as for the AR(1) model. The internal RMSE prediction error for the AR(1) model has decreased to 4.98 which is a great improvement compared to 7.86 for the (linear+seasonal) regression model assumed earlier. ■

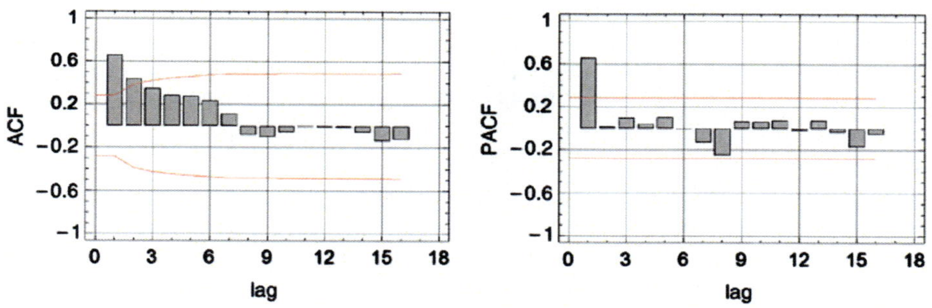

Fig. 9.24 The ACF and PACF functions for the residuals in the time series data after removing the linear trend and seasonal behavior

Table 9.5 Accuracy of the various modeling approaches when applied to the electric utility load data given in Table 9.1 (48 data points covering 1974–1985). The RMSE correspond to internal prediction accuracy of the various models evaluated

	AMA(3)	AMA(5)	EWA(0.2)	EWA(0.5)	Linear	Linear+Seasonal	Linear+Seasonal+AR(1)
RMSE	7.68	9.02	8.59	11.53	12.10	7.86	4.98

Table 9.6 Forecast accuracy of the various models applied to the electric load data. The actual values correspond to the recorded values for the four quarters for 1986

	Actual Load (1986)	AMA(3)	AMA(5)	EWA(0.2)	EWA(0.5)	Linear	Linear+Seasonal	Linear+Seasonal+AR(1)
Quarter 1	151.3	141.1	142.9	150.6	144.8	157.5	164.3	155.2
Quarter 2	132.9	143.6	143.6	138.2	142.9	159.1	148.6	140.0
Quarter 3	160.5	140.0	141.9	148.1	144.4	160.7	172.4	162.1
Quarter 4	161.0	141.5	143.6	140.2	143.2	162.4	155.6	147.8
Average	151.4	141.5	143.0	144.3	143.8	159.9	160.2	151.3
%Diff. in average	–	*−6.6*	*−5.6*	*−4.7*	*−5.0*	*5.6*	*5.8*	*0.1*
RMSE	–	*15.96*	*14.44*	*12.40*	*13.40*	*13.48*	*12.11*	*7.78*

Example 9.5.3: *Comparison of the external prediction error of different models for peak electric demand*
The various models illustrated in previous sections have been compared in terms of their internal prediction errors. The more appropriate manner of comparing them is in terms of their external prediction errors such as bias and RMSE. The peak loads for the next four quarters for 1986 will be used as the basis of comparison.

The AR model can also be used to forecast the future values for the four quarters of 1986 (Y_{49} to Y_{52}). First, one determines the residual for the last quarter of 1985 (Z_{48}) using the trend and seasonal model to forecast $\hat{Y}_{LS,48}$. The AR(1) correction is subsequently determined, and finally, the forecast for the first quarter of 1986 or Y_{49} is computed from:

$$\hat{Z}_{48} = \hat{Y}_{48} - (\hat{Y}_{LS,48}) = 135.1 - (149.05) = -139.5$$
$$\hat{Z}_{49} = r_1.\hat{Z}_{48} = (0.657)(-13.95) = -9.16$$
$$\hat{Y}_{49} = (\hat{Y}_{LS,49}) + \hat{Z}_{49} = (164.34) - 9.16 = 155.2$$

Finally, the 95% confidence limits are: $\hat{Y}_{t+1} \pm 2.RMSE = 155.2 \pm 2.(4.98) = (145.6, 164.8)$

The individual and mean forecasts for all methods are shown in Table 9.6. One can compare the various models in how accurately they are able to predict these four individual values. They indicate that the mean differences in forecasts are consistent across models, about 5–6% except for AMA(3) which is closer to 7%. Note that the forecast errors for AMA and EWA are negative; this is because the inherent lags in these smoothing techniques result in forecasts being lower than actual. EWA(0.2) is the most accurate among all models. The (linear + seasonal) model turns out to be quite poor with an average bias of 5.8%.On the other hand, the predictions are almost perfect for the AR model (the bias is only 0.1%) while the external prediction RMSE is also the lowest at 7.78. This is closely followed by the (linear+seasonal) model with RMSE=12.11. The others models have higher RMSE values. This example clearly illustrates the added benefit brought in by the AR(1) term. ■

9.5.4 Recommendations on Model Identification

As stated earlier, several researchers suggest that, in most cases, it is not necessary to include both the AR and the MA elements; one of these two, depending on system behavior, should suffice. Further, it is recommended that low order models of 3 or less should be adequate in most instances provided the seasonality has been properly filtered out. Other texts state that adopting an ARMA model is likely to result in a model with fewer terms than those of a pure MA or AR process by itself (Chatfield 1989). Some recommendations on how to identify and evaluate a stochastic model are summarized below.

(a) *Stationarity check:* Whether the series is stationary or not in its mean trend is easily identified (one has also got to verify that the variance is stable, for which transformations such as taking logarithms may be necessary). A non-constant trend in the underlying mean value of a process will result in the ACF not dying out rapidly. If seasonal behavior is present, the ACF will be able to detect it as well since it will exhibit a decay with cyclic behavior. The seasonality effect needs to be removed by using an appropriate regression model (for example, the traditional OLS or even a Fourier Series model) or by differencing. Figure 9.18 illustrates how a seasonal trend shows up in the ACF of the time series data of Table 9.1. The seasonal nature of the time series is reflected in the clamped sinusoidal behavior of the ACF. Differencing is another way of detrending the series which is especially useful when the cyclic behavior is known (such as 24 h lag differencing for electricity use in buildings). If more than twice differencing does not remove seasonality, consider a transformation of the time series data using natural logarithms.

(b) *Model selection:* The correlograms of both the ACF and the PACF are the appropriate means for identifying the model type (whether ARIMA, ARMA, AR or MA) and the model order. The identification procedure can be summarized as follows (McCain and McCleary 1979):
   (i) For AR(1): ACF decays exponentially, PACF has a spike a lag 1, and other spikes are not statistically significant, i.e., are contained within the 95% confidence intervals
   (ii) For AR(2): ACF decays exponentially (indicative of positive model coefficients) or with sinusoidal-exponential decay (indicative of a positive and a negative coefficient), and PACF has two statistically significant spikes
   (iii) For MA(1): ACF has one statistically significant spike at lag 1 and PACF damps down exponentially
   (iv) For MA(2): ACF has two statistically significant spikes (one at lag 1 and one at lag 2), and PACF has an exponential decay or a sinusoidal-exponential decay
   (v) For ARMA (1,1): ACF and PACF have spikes at lag 1 with exponential decay.

   Usually, it is better to start with the lowest values of p and q for an ARMA(p, q) process. Subsequently, the model order is increased until no systematic patterns are evident in the residuals of the model. Most time series data from engineering experiments or from physical systems or processes should be adequately modeled by low orders, i.e., about 1–3 terms. If higher orders are required, the analyst should check his data for bias or unduly large noise effects. Cross-validation using the sample handout approach is strongly recommended for model selection since this avoids over-fitting, and would better reflect the predictive capability of the model. The model selection is somewhat subjective as described above. In an effort to circumvent this arbitrariness, objective criteria have been proposed for model selection. Wei (1990) describes several such criteria; the Akaike Information Criteria (AIC), the Bayesian Information Criteria (BIC) and the Criterion for Autoregressive Transfer function (CAT) to name three of several indices.

(c) *Model evaluation:* After a tentative time series model has been identified and its parameters estimated, a diagnostic check must be made to evaluate its adequacy. This check could consist of two steps as described below:
   (i) the autocorrelated function of the simulated series (i.e., the time series generated by the model) and that of the original series must be close;
   (ii) the residuals from a satisfactory model should be white noise. This would be reflected by the sample autocorrelation function of the residuals being close or equal to zero. Since it is assumed that the random error terms in the actual process are normally distributed and independent of each other (i.e., white noise), the model residuals should also behave similarly. This is tested by computing the sample autocorrelation function for lag k of the residuals. If the model is correctly specified, the residual autocorrelations r_k (upto about K=15 or so) are themselves uncorrelated, normally distributed random variables with mean 0 and variance (1/n), where n is the number of observations in the time series. Finally, the sum of the squared independent normal random variables denoted by the Q statistic is computed as:

$$Q = n\sum_{k=1}^{K} r_k^2 \tag{9.39}$$

   Q must be approximately distributed as chi-square χ^2 with (K-p-q) degrees of freedom. Lookup Table A.5 provides the critical value to determine whether or not to accept the hypothesis that the model is acceptable.

Despite their obvious appeal, a note of caution on ARMA models is warranted. Fitting reliable multi-variate time series models is difficult. For example, in case of non-experimental data which is not controlled, there may be high correlation between and within series which may or may not be real (there may be mutual correlation with time). An apparent good fit to the data may not necessarily result in better forecasting accuracy than using a simpler univariate model.

Though ARMA models usually provide very good fits to the data series, often, a much simpler method may give results just as good.

An implied assumption of ARMA models is that the data series is stationary and normally distributed. If the data series is not, it is important to find a suitable transformation to make it normally distributed prior to OLS model fitting. Further, if the data series has non-random disturbance terms, the maximum likelihood estimation (MLE) method described in Sect. 10.4.3 is said to be statistically more efficient than OLS estimation. The reader can refer to Wei (1990), Box and Jenkins (1976), or Montgomery and Johnson (1976) for more details. The autocorrelation function and the spectral method are closely related; the latter can provide insight into the appropriate order of the ARMA model (Chatfield 1989).

9.6 ARMAX or Transfer Function Models

9.6.1 Conceptual Approach and Benefit

The ARIMA models presented above involve detrending the data (via the "Integrated" component) prior to modeling. The systematic stochastic component is modeled by ARMA models which are univariate by definition since they only consist of lagged variables of the detrended series $\{Z_t\}$. An alternate form of detrending is to use OLS models such as described in Sect. 9.4 which can involve indicator variables for seasonality as well as the time variable explicitly. One can even have other "independent" variables **X** appear in the OLS model if necessary; for example: $Y_t = f(\mathbf{X}_t)$. However, such models do not contain lagged variables in **X,** as shown in Fig. 9.25a. That is why the ARMA models are said to be basically univariate since they relate to the detrended series $\{Z_t\}$. This series is taken to be the detrended response of white noise plus a feedback loop whose effect is taken into consideration via the variation of the lagged variables. Thus, the stochastic time series data points are in equilibrium over time and fluctuate about a mean value.

However, there are systems whose response cannot be satisfactorily modeled using ARMA models alone since their mean values vary greatly over time. The obvious case is of dynamic systems which have some sort of feedback in the independent or regressor variables, and explicit recognition of such effects need to be considered. Traditional ARMA models would then be of limited use since the error term would include some of the structural variation which one could directly attribute to the variation of the regressor variables. Thus, the model predictions could be biased with uncertainty bands so large as to make predictions very poor, and often useless. In such cases, a model relating the dependent variable with lagged values of itself *plus current and lagged values of the independent variables*, plus the error term captured by the time-series model, is likely to be superior to the ARMA models alone (see Fig. 9.25b). Such a model formulation is called "Multivariate ARMA" (or MARMA) or ARMAX models or *transfer function models*. Such models have found extensive applications is engineering, econometrics and other disciplines as well, and are briefly described below.

9.6.2 Transfer Function Modeling of Linear Dynamic Systems

Dynamic systems are modeled by differential equations. A simple example taken from Kreider et al. (2009) will illustrate how linear differential equations can be recast as ARMAX models such that the order of the differential equation is equal to the number of lag terms in a time series. Consider a plane wall represented by a 1C1R thermal network as shown in Fig. 1.5 (see Sect. 1.2.4 for an introductory discussion about representing the transient heat conduction through a plane wall by electrical network analogues). The internal node is the indoor air temperature T_i which is assumed to be closely coupled to the thermal mass of the building or room. This node is impacted by internal heat loads generated from people and various equipment (Q) and also by heat conduc-

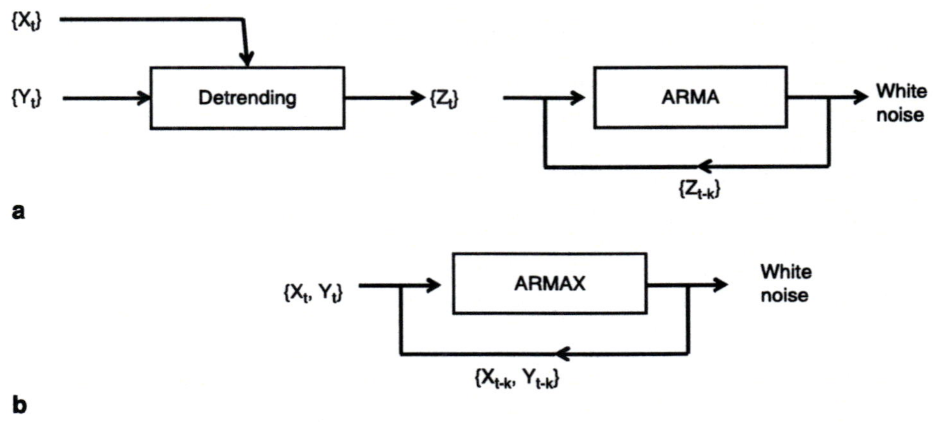

Fig. 9.25 Conceptual difference between the single-variate ARMA approach and the multivariate ARMAX approach applied to dynamic systems. **a** Traditional ARMA approach. **b** ARMAX approach

tion from the outdoors at temperature T_o through the outdoor wall with an effective resistance R.

The thermal performance of such a system is modeled by:

$$C\dot{T}_i = \frac{T_o - T_i}{R} + Q \tag{9.40a}$$

where $\dot{T}_i$ is the time derivative of T_i.

Introducing the time constant $\tau = RC$, the above equation can be re-written as:

$$\tau\dot{T}_i + T_i = T_o + RQ \tag{9.40b}$$

For the simplifying case when both the driving terms T_o and Q are constant, one gets

$$\tau\,\dot{T}(t) + T(t) = 0 \quad \text{where} \quad T(t) = T_i(t) - T_o - RQ \tag{9.41}$$

The solution is

$$\begin{aligned} T(t+1) &= T(0)\cdot\exp\left(-\frac{t+1}{\tau}\right) \\ &= T(0)\cdot\exp\left(-\frac{t}{\tau}\right)\cdot\exp\left(-\frac{1}{\tau}\right) \\ &= T(t)\cdot\exp\left(-\frac{1}{\tau}\right) \end{aligned}$$

which can be expressed as:

$$T(t) + a_1 T(t-1) = 0 \quad \text{where} \quad a_1 = -\exp\left(-\frac{1}{\tau}\right) \tag{9.42}$$

The first order ODE is, thus, recast as the traditional single-variate AR(1) model with one-lag term. In this case, there is a clear interpretation of the coefficient a_1 in terms of the time constant of the system. Example 7.10.2 illustrates the use of such models in the context of operating a building so as to minimize the cooling energy use and demand during the peak period of the day.

Kreider et al. (2009) also give another example of a network with two nodes (i.e., 2R2C network) where, for a similar assumption of constant driving terms T_o and Q, one obtains a second order ODE which can be cast as a time series model with two lag terms, with the time series coefficients (or transfer function coefficients) still retaining a clear relation with the resistances and the two time constants of the system. For more complex models and for cases when the driving terms are not constant, such clear interpretation of the time series model coefficients in terms of resistances and capacitances would not exist since the same time series model can apply to different RC networks, and so uniqueness is lost.

For the general case of the free response of a non-airconditioned room or building represented by indoor air temperature T_i and which is acted upon by two driving terms T_o and Q which are time variant, the general form of the transfer function or ARMAX model of order n is:

$$\begin{aligned} T_{i,t} &+ a_1 T_{i,t-1} + a_2 T_{i,t-2} + \cdots + a_n T_{i,t-n} \\ &= b_0 T_{o,t} + b_1 T_{o,t-1} + b_2 T_{o,t-2} + \cdots + b_n T_{o,t-n} \\ &\quad + c_0 Q_t + c_1 Q_{t-1} + c_2 Q_{t-2} + \cdots + c_n Q_{t-n} \end{aligned} \tag{9.43}$$

The model identification process, when applied to the observed time series of these three variables, would determine how many coefficients or weighting factors to retain in the final model for each of the variables. Once such a model has been identified, it can be used for accurate forecasting purposes. In some cases, physical considerations can impose certain restrictions on the transfer function coefficients, and it is urged that these be considered since it would result in sounder models. For the above example, it can be shown that at the limit of steady state operation when present and lagged values of each the three variables are constant, heat loss would be expressed in terms of the overall heat loss coefficient U times the cross-sectional area A perpendicular to heat flow: $Q_{loss} = UA(T_o - T_i)$. This would require that the following condition be met:

$$(1 + a_1 + a_2 + \cdots + a_n) = (b_0 + b_1 + b_2 + \cdots + b_n) \tag{9.44}$$

The transfer function approach[5] has been widely used in several detailed building energy simulation software programs developed during the last 30 years to model unsteady state heat transfer and thermal mass storage effects such as wall and roof conduction, solar heat gains and internal heat gains. It is a widely understood and accepted modeling approach among building energy professionals (see for example, Kreider et al. 2009). The following example illustrates the approach.

Example 9.6.1: *Transfer function model to represent unsteady state heat transfer through a wall*

For the case when indoor air temperature T_i is kept constant by air-conditioning, Eq. 9.43 can be re-expressed as follows consistent with industry practice:

$$\begin{aligned} Q_{cond,t} = &- d_1 Q_{cond,t-1\cdot\Delta t} - d_2 Q_{cond,t-2\cdot\Delta t} - \cdots \\ &+ b_0 T_{solair,t} + b_1 T_{solair,t-1\cdot\Delta t} + b_2 T_{solair,t-2\cdot\Delta t} + \cdots \\ &- T_i \sum_{n\geq 0} c_n \end{aligned} \tag{9.45a}$$

[5] Strictly, this formulation should be called discrete transfer function or z-transform since it uses discrete time intervals (of one hour).

Table 9.7 Conduction transfer function coefficients for a 4″ concrete wall with 2″ insulation

	n=0	n=1	n=2	n=3	n=4
b_n	0.00099	0.00836	0.00361	0.00007	0.00
d_n	1.00	−0.93970	0.04664	0.00	0.00
$\sum c_n$	0.01303				

or

$$Q_{cond,t} = -\sum_{n>1} d_n Q_{cond,t-n\cdot\Delta t} + \sum_{n\geq 0} b_0 T_{solair,t-n\cdot\Delta t} - T_i \sum_{n\geq 0} c_n \quad (9.45b)$$

where

Q_{cond} is the conduction heat gain through the wall

T_{solair} is the sol-air temperature (a variable which includes the combined effect of outdoor dry-bulb air temperature and the solar radiation incident on the wall)

T_i is the indoor air temperature

Δt is the time step (usually 1 h)

and b_n, c_n and d_n are the transfer function coefficients

Table 9.7 assembles values of the transfer function coefficients for a 4″ concrete wall with 2″ insulation. For a given hour, say 10:00 am, Eq. 9.45 can be expressed as:

$$\begin{aligned} Q_{cond,10} &= (0.93970)Q_{cond,9} - (0.04664)Q_{cond,8} \\ &\quad + (0.00099)T_{solair,10} + (0.00836)T_{solair,9} \\ &\quad + (0.00361)T_{solair,8} + (0.00007)T_{solair,7} - (0.01303)T_i \end{aligned} \quad (9.46)$$

First, values of the driving terms T_{solair} have to be computed for all the hours over which the computation is to be performed. To start the calculation, initial guess values are assumed for $Q_{cond,9}$ and $Q_{cond,8}$. For a specified T_i, one can then calculate $Q_{cond,10}$ and repeat the recursive calculation for each subsequent hour. The heat gains are periodic because of the diurnal periodicity of T_{sol}. The effect of the initial guess values soon dies out and the calculations attain the desired accuracy after a few iterations. ■

9.7 Quality Control and Process Monitoring Using Control Chart Methods

9.7.1 Background and Approach

The concept of statistical quality control and quality assurance was proposed by Shewart in the 1920s (and, hence, many of these techniques bear his name) with the intent of using sampling and statistical analysis techniques to improve and maintain quality during industrial production. Process monitoring, using control chart techniques provides an ongoing check on the stability of the process and points to problems whose elimination can reduce variation and permanently improve the system (Box and Luceno 1997). It has been extended to include condition monitoring and performance degradation of various equipment and systems, and for control of industrial processes involving process adjustments using feedback control to compensate for sources of drift variation. The basic concept is that variation in any production process is unavoidable the causes of which can be categorized into:

(i) *Common causes* or random fluctuations due to the overall process itself, such as variation in quality of raw materials and in consistency of equipment performance—these lead to random variability and statistical concepts apply;

(ii) *Special or assignable causes* (or non-random or sporadic changes) due to specific deterministic circumstances, such as operator error, machine fault, faulty sensors, or performance degradation of the measurement and control equipment.

When the variability is due to random or common causes, the process is said to be in statistical control. The normal curve is assumed to describe the process measurements with the confidence limits indicated as the *upper and lower control limits (UCL and LCL)* as shown in Fig. 9.26. The practice of plotting the attributes or characteristics of the process over time on a plot is called monitoring via *control charts*. It consists of a horizontal plot which locates the process mean (called "*centerline*") and two lines (the UCL and LCL limits), as shown in Fig. 9.27. The intent of statistical process monitoring using control charts is to detect the occurrence of non-random events which impact the central tendency and

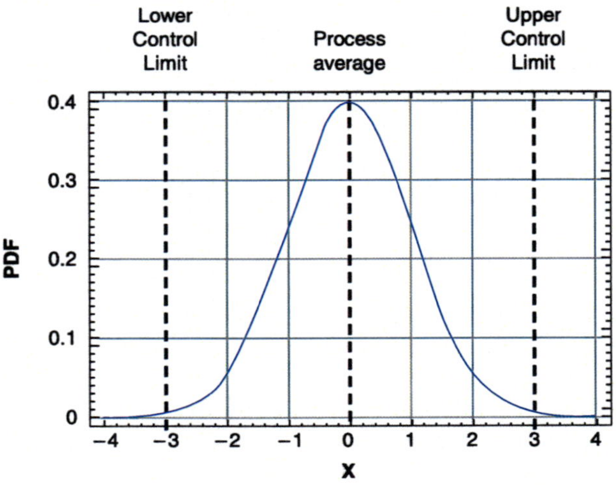

Fig. 9.26 The upper and lower three-sigma limits indicative of the UCL and LCL limits shown on a normal distribution

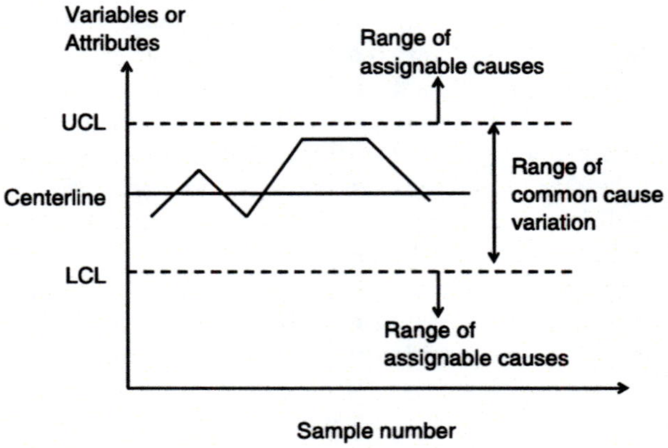

Fig. 9.27 The Shewhart control chart with primary limits

the variability of the process, and thereby take corrective action in order to eliminate them. Thus, the process can again be brought back to stable and statistical control as quickly as possible. These limits are often drawn to correspond to 3 times the standard deviation so that one can infer that there is strong evidence that points outside the limits are faulty or indicate an unstable process. The decision as to whether to deem an incoming sample of size n at a particular point in time to be in or out of control is, thus, akin to a two-tailed hypothesis test where:

$$\begin{array}{ll}\text{Null hypothesis} & H_0: \text{Process is in control, } \bar{X} = \mu_0 \\ \text{Alternative hypothesis} & H_a: \text{process out of control } \bar{X} \neq \mu_0\end{array} \tag{9.47}$$

where $\bar{X}$ denotes the sample mean of n observations and μ_0 the expected mean deduced from an in-control process. Note that the convention in statistical quality control literature is to use upper case letters for the mean value. Just as in hypothesis testing (Sect. 4.2), type I and type II errors can result. For example, type I error (or false positive error) arises when a sample (or point on the chart) of an in-line control process falls outside the control bands. As before, the probability of occurrence is reduced by making appropriate choices of sample size and control limits.

9.7.2 Shewart Control Charts for Variables and Attributes

The Shewart control chart method is a generic name which includes a number of different charts. It is primarily meant to monitor a process, while it is said to be of limited usefulness for adjusting the process.

(a) ***Shewart chart for variables:*** for continuous measurements such as diameter, temperature, flow, as well as derived parameters or quantities such as overall heat loss coefficient, efficiency,…

(i) **mean or $\bar{X}$ chart** is used to detect the onset of bias in measured quantities or estimated parameters. This detection is based on a two-tailed hypothesis test assuming a normal error distribution. The control chart plots are deduced as upper and lower control limits about the centerline where the norm is to use the 3-sigma confidence limits for the z-value:

$$\textit{when s is known}: \quad \{UCL, LCL\}_{\bar{x}} = \bar{X} \pm 3 \cdot \frac{s}{n^{1/2}} \tag{9.48}$$

where s is the process standard deviation and $(s/n^{1/2})$ is the standard error of the sample means. Recall that the 3-sigma limits include 99.74% of the area under the normal curve, and hence, that the probability of a type I error (or false positive) is only 0.26%. When the standard deviation of the process is not known, it is suggested that the average range $\bar{R}$ of the numerous samples be used for 3-sigma limits as follows:

$$\textit{when s is not known}: \quad \{UCL, LCL\}_{\bar{x}} = \bar{X} \pm A_2 \cdot \bar{R} \tag{9.49}$$

where the factor A_2 is given in Table 9.8. Note that this factor decreases as the number of samples increases. Recall that the *range* for a given sample is simply the difference between the highest and lowest values of the given sample.

Devore and Farnum (2005) cite a study which demonstrated that the use of medians and the interquartile range (IQR) was superior to the traditional means and range control charts. The former were found to be more robust, i.e., less influenced by spurious outliers. The suggested control limits were:

$$\{UCL, LCL\}_{\tilde{x}} = \tilde{X} \pm 3 \cdot \frac{IQR}{k_n(n)^{1/2}} \tag{9.50}$$

where $\tilde{X}$ is the median and the values of k_n are selected based on the sample size n given by the Table 9.9.

(ii) **range or R charts** to control variation to detect uniformity or consistency of a process. The range is a rough measure of the "rate of change" of the observed variable which is a more sensitive measure than the mean. Hence, a point which is out of control on the range chart may be flagged as an abnormality before the mean chart does. Consider the case of drawing k samples each with sample size n (i.e., each sample consists of drawing n

Table 9.8 Numerical values of the three coefficients to be used in Eqs. 9.49 and 9.51 for constructing the three-sigma limits for the mean and range charts. (Source: Adapted from "1950 ASTM Manual on Quality Control of Materials," *American Society for Testing and Materials*, in J. M. Juran, ed., *Quality Control Handbook* (New York: McGraw-Hill Book Company, 1974), Appendix II, p. 39.)

Number of observations in each sample, n	Factor for determining control limits, control chart for the mean, A_2	Factors for determining control limits, control chart for the range	
		D_3	D_4
2	1.880	0	3.268
3	1.023	0	2.574
4	0.729	0	2.282
5	0.577	0	2.114
6	0.483	0	2.004
7	0.419	0.076	1.924
8	0.373	0.136	1.864
9	0.337	0.184	1.816
10	0.308	0.223	1.777
11	0.285	0.256	1.744
12	0.266	0.284	1.717
13	0.249	0.308	1.692
14	0.235	0.329	1.671
15	0.223	0.348	1.652

Table 9.9 Values of the factor k_n to be used in Eq. 9.50

N	4	5	6	7	8
k_n	0.596	0.990	1.282	1.512	0.942

items or taking n individual measurements)[6]. The 3-sigma limits for the range chart are given by:

$$\begin{aligned} &\text{mean line:} && \bar{R} \\ &\text{lower control limit:} && LCL_{\bar{R}} = D_3\bar{R} \\ &\text{upper control limit:} && UCL_{\bar{R}} = D_4\bar{R} \end{aligned} \tag{9.51}$$

where $\bar{R}$ is the mean of the ranges of the k samples, and the numerical values of the coefficients D_3 and D_4 are given in Table 9.8 for different number of sample sizes.

It is suggested that the mean and range chart be used together since their complementary properties allow better monitoring of a process. Figure 9.28 illustrates two instances where the benefit of using both charts reveal behavior which one chart alone would have missed.

There are several variants of the above mean and range charts since several statistical indices are available to measure the central tendency and the variability. One common chart is the *standard deviation or s charts*, while other types of charts involve s^2 charts. Which chart to use depends to some extent on personal preference. Often, the sample size for control chart monitoring is low (around 5 according to Himmelblau 1978) which results is standard deviation being not a very robust statistic. One suggestion is to use the s charts when the sample size is around 8–10 or larger, and to use range charts for smaller sample sizes. Further, the range is easier to visualize and interpret, and is more easily determined than the standard deviation.

[6] Note the distinction between the number of samples (k) and the sample size (n).

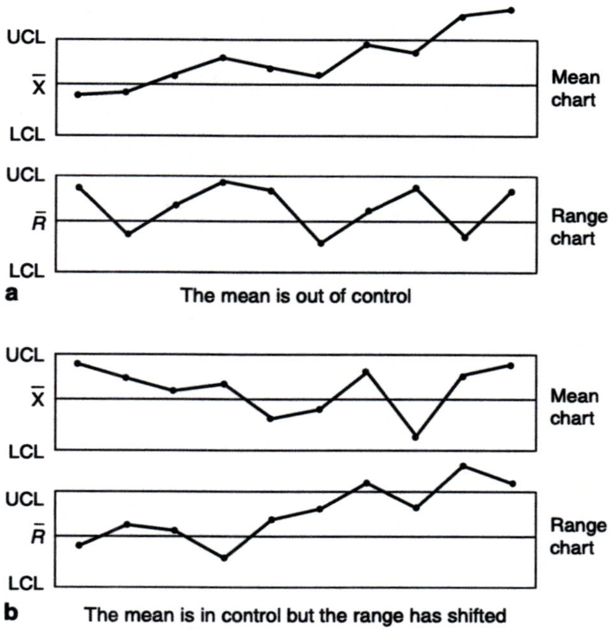

Fig. 9.28 The combined advantage provided by the mean and range charts in detecting out-of-control processes. Two instances are shown: **a** where the variability is within limits but the mean is out of control which is detected by the mean chart, and **b** where the mean is in control but not the variability which is detected by the range chart

Example 9.7.1: *Illustration of the mean and range charts*
Consider a process where 20 samples, each consisting of 4 items, are gathered as shown in Table 9.10. The mean and range charts will be used to illustrate how to assess whether the process is in control or not.

The X-bar and R charts are shown in Fig. 9.29. Note that no point is beyond the control limits in either plot indicating that the process is in statistical control.

The data in Table 9.10 has been intentionally corrupted such that the four items of one sample only have a higher mean value (about 1.6). How the mean and range plots flag this occurrence is shown in Fig. 9.30 illustrates a case where the reverse holds. The process fault is detected by the X-bar but not by R chart. Therefore, it is recommended that the process be monitored using both the mean and range charts which can provide additional insights not provided by each control chart technique alone. ■

(b) ***Shewart control charts for attributes*** In complex assembly operations (or during condition monitoring involving several sensors as in many thermal systems), numerous qua-

Table 9.10 Data table for the 20 samples consisting of four items and associated mean and range statistics (Example 9.7.1)

Sample#	Item 1	Item 2	Item 3	Item 4	X-bar ($\bar{X}$)	Range (R)
1	1.405	1.419	1.377	1.400	1.400	0.042
2	1.407	1.397	1.377	1.393	1.394	0.030
3	1.385	1.392	1.399	1.392	1.392	0.014
4	1.386	1.419	1.387	1.417	1.402	0.033
5	1.382	1.391	1.390	1.397	1.390	0.015
6	1.404	1.406	1.404	1.402	1.404	0.004
7	1.409	1.386	1.399	1.403	1.399	0.023
8	1.399	1.382	1.389	1.410	1.395	0.028
9	1.408	1.411	1.394	1.388	1.400	0.023
10	1.399	1.421	1.400	1.407	1.407	0.022
11	1.394	1.397	1.396	1.409	1.399	0.015
12	1.409	1.389	1.398	1.399	1.399	0.020
13	1.405	1.387	1.399	1.393	1.396	0.018
14	1.390	1.410	1.388	1.384	1.393	0.026
15	1.393	1.403	1.387	1.415	1.400	0.028
16	1.413	1.390	1.395	1.411	1.402	0.023
17	1.410	1.415	1.392	1.397	1.404	0.023
18	1.407	1.386	1.396	1.393	1.396	0.021
19	1.411	1.406	1.392	1.387	1.399	0.024
20	1.404	1.396	1.391	1.390	1.395	0.014
Grand Mean					1.398	0.022

lity variables would need to be monitored, and in principle, each one could be monitored separately. A simpler procedure is to inspect n finished products, denoting a sample, at regular intervals and to simply flag the proportion of products in the sample found to be defective or non-defective. Thus, analysis using attributes would only differentiate between two possibilities: acceptable or not acceptable. Different types of charts have been proposed, two important ones are listed below:

(a) *p-chart for fraction or proportion* of defective items in a sample (it is recommended that typically n=100 or so). An analogous chart for tracking the *number of defectives* in a sample, i.e., the variable (n.p.) is also widely used;
(b) *c-chart for rate of defects* or minor flaws or number of nonconformities per unit time. This is a more sophisticated type of chart where an item may not be defective so as to render it useless, but would nevertheless compromise the quality of the product. An item can have non-conformities but still be able to function as intended. It is based on the Poisson distribution rather than the Binomial distribution which is the basis for method (a) above. The reader can refer to texts such as Devore and Farnum (2005) or Walpole et al. (2007) for more detailed description of this approach.

Method (a) is briefly described below. Let p be the probability that any particular item is defective. One manner of determining probability p is to infer it as the long run proportion of defective items taken from a previous in-control period. If the process is assumed to be independent between samples, then the expected value and the variance of a binomial random variable X in a random sample n with p being the fraction of defectives is given by (see Sect. 2.4.2):

$$E(\hat{p}) = \bar{p} \qquad \operatorname{var}(\hat{p}) = \frac{\bar{p}(1-\bar{p})}{n} \tag{9.52}$$

Thus, the 3 sigma upper and lower limits are given by:

$$\{UCL, LCL\}_{\bar{p}} = \bar{p} \pm 3\left[\frac{\bar{p}(1-\bar{p})}{n}\right]^{1/2} \tag{9.53}$$

In case the LCL is negative, it has to be set to zero, since negative values are physically impossible.

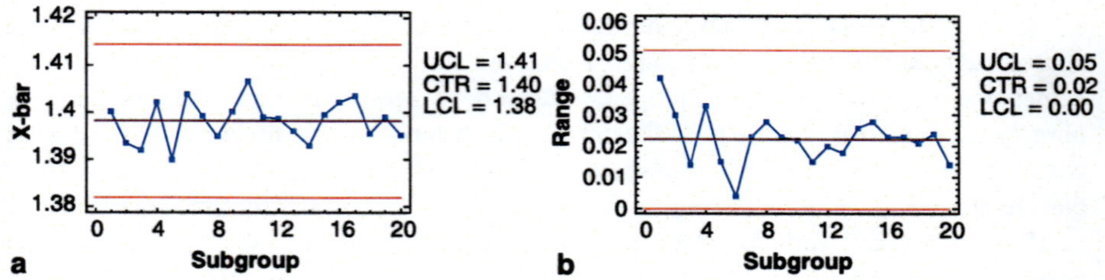

Fig. 9.29 Shewart charts for mean and range using data from Table 9.10. **a** X-bar chart. **b** Range chart

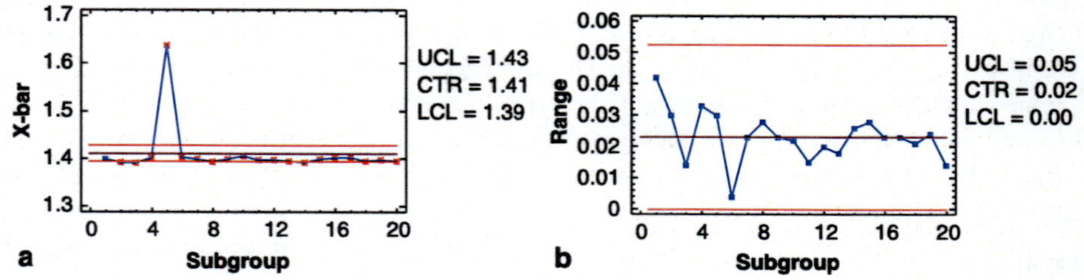

Fig. 9.30 Shewart charts when one of the samples in Table 9.10 has been intentionally corrupted. **a** X-bar chart. **b** Range chart

Table 9.11 Data table for Example 9.7.2

Sample	Number of defective components	Fraction defective, p
1	8	0.16
2	6	0.12
3	5	0.10
4	7	0.14
5	2	0.04
6	5	0.10
7	3	0.06
8	8	0.16
9	4	0.08
10	4	0.08
11	3	0.06
12	1	0.02
13	5	0.10
14	4	0.08
15	4	0.08
16	2	0.04
17	3	0.06
18	5	0.10
19	6	0.12
20	3	0.06
Mean		0.088

Example 9.7.2: *Illustration of the p-chart method*
Consider the data shown in Table 9.11 collected from a process where 20 samples are gathered with each sample size being n=50. If prior knowledge is available as to the expected defective proportion p, then that value should be used. In case it is not, and provided the process is generally fault-free, it can be computed from the data itself as shown. From the table, the mean p value=0.088. This is used as the baseline for comparison in this example.

The centerline and the UCL and LCL values for the p chart following Eqs. 9.52 and 9.53 are shown in Fig. 9.31. The process can be taken to be in control since the individual points are contained within the UCL and LCL bands. Since the p value cannot be negative, the LCL is forced to zero; this is the reason for the asymmetry in the UCL and LCL bands around the CTR. The analogous chart for the number of defectives is also shown. Note that the two types of charts look very similar except for the numerical values of the UCL and LCL; this is not surprising since the number of samples n (taken as 50 in this example) is a constant multiplier. ■

(c) ***Practical implementation issues*** The basic process for constructing control charts is to first gather at least k=25–30 samples of data with a fixed number of objects or observations of size n from a production process known to be working properly, i.e., one in statistical control. A typical value of n for X-bar and R charts is n=5. As the value of n is increased, one can detect smaller changes but at the expense of more time and money.

The mean, UCL and LCL values could be preset and unchanging during the course of operation, or they could be estimated anew at each updating period. Say, the analysis is done once a day, with four observations (n=4) taken hourly, and the process operates 24 h/day. The limits for each day could be updated based on the statistics of the 24 samples taken the previous day or kept fixed at some pre-set value. Such choices are best done based on physical insights into the specific process or equipment being monitored. A practical consideration is that process operators do not like frequent adjustments made to the control limits. Not only can this lead to errors in resetting the limits, but this may led to psychological skepticism on the reliability of the entire statistical control approach.

When a process is in control, the points from each sample plotted on the control chart should fluctuate in a random manner between the UCL and the LCL with no clear pattern. Several "rules" have been proposed to increase the sensitivity of Shewhart charts. Other than "no points outside the control limits", one could check for such effects as: (i) the number of points above and below the centerline are about equal, (ii) there is no steady rise or decrease in a sequence of points, (iii) most of the points are close to the centerline rather than hugging the limits, (iv) there is a sudden shift in the process mean, (v) cyclic behavior…

Devore and Farnum (2005) present an extended list of "out-of-control" rules involving counting the number of points falling within different bounds corresponding to one, two and three sigma lines. Eight different types of out-of-control behavior patterns are shown to illustrate that several possible schemes can be devised for process monitoring. Others have developed similar types of rules in order to increase the sensitivity of the monitoring process. However, using such types of extended rules also increases the possibility of false alarms (or type I errors), and so, rather than being ad hoc, there should be some statistical basis to these rules.

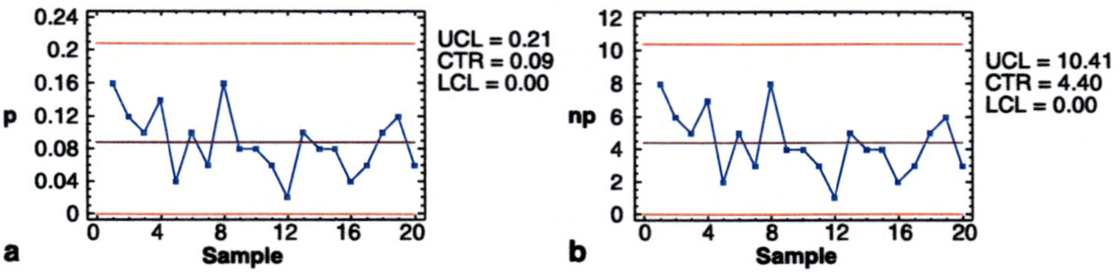

Fig. 9.31 Shewart p-charts **a** Chart for the proportion of defectives. **b** Chart for the number of defectives (n.p)

9.7.3 Statistical Process Control Using Time Weighted Charts

Traditional Shewhart chart methods are based on investigating statistics (mean or range, for example) of an individual sample data of n items. Time weighted procedures allow more sensitive detection by basing the inferences, not on one individual sample statistics, but on the cumulative sum or the moving sum of a series of successive observations. This is somewhat similar to the rather heuristic "out-of-control" rules stated by Devore and Farnum (2005), but now this is done in a more statistically sound manner. Typical time weighted approaches are those using Cusum and moving average methods since they have a shorter average run length in detecting small to moderate process shifts. In short, they incorporate past history of process and are more sensitive to small gradual changes. However, non-normality and serial correlation in the data has an important effect on conclusions drawn from Cusum plots (Himmelblau 1978).

9.7.3.1 Cusum Charts

Cusum or cumulative sum charts are similar to the Shewart charts in that they are diagnostic tools which indicate whether a process has gone out of control or not due to the onset of special non-random causes. However, the cusum approach makes the inference based on a *sum of deviations* rather than individual samples. They damp down random noise while amplifying true process changes. They can indicate *when* and *by how much* the mean of the process has shifted. Consider a control chart for the mean with a reference or target level established at μ_0. Let the sample means be given by $(\bar{X}_1, \bar{X}_2 ... \bar{X}_r)$. Then, the first r cusums are computed as:

$$\begin{aligned} S_1 &= \bar{X}_1 - \mu_0 \\ S_2 &= S_1 + (\bar{X}_2 - \mu_0) = (\bar{X}_1 - \mu_0) + (\bar{X}_2 - \mu_0) \\ &\dots \\ S_r &= S_{r-1} + (\bar{X}_r - \mu_0) = \sum_{i=1}^{r} (\bar{X}_i - \mu_0) \end{aligned} \tag{9.54}$$

The above discussion applied to the mean residuals, i.e., the difference between the measured value and its expected value. Such charts could be based on other statistics such as the range, the variable itself, absolute differences, or successive differences between observations.

The cusum chart is simply a plot of S_r over time like the Shewart charts, but they provide a different type of visual record. Since the deviations add up i.e., cumulate, an increase (or decrease) in the process mean will result in an upward (or downward) slope of the value S_r. The magnitude of the slope is indicative of the size of the change in the mean. Special templates or overlays are generated according to certain specific rules for constructing them (see for example, Walpole et al. 2007). Often, these overlays are shaped as a Vee (since the slope is indicative of a change in the mean of the process) which is placed over the most recent point. If the data points fall within the opening of the Vee, then the process is considered to be in control, otherwise it is not. If there is no shift in the mean, the cusum chart should fluctuate around the horizontal line. Even a moderate change in the mean, however, would result in the cusum chart exhibiting a slope with each new observation highlighting the slope more distinctly. Cusum charts are drawn with pre-established limits set by the user which apply to the mean:

AQL acceptable quality level i.e., when the process is in-control $S_r = \mu_0$

RQL rejectable quality level, i.e., when the process is out-of-control, $S_r \neq \mu_0$

Clearly, these limits are similar to the concepts of null and alternate mean values during hypothesis testing. The practical advantages and disadvantages of this approach are discussed by Himmelblau (1978). The following example should clarify this approach.

Example 9.7.3: *Illustration of Cusum plots*

The data shown in Table 9.10 represents a process known to be in control. Two tests will be performed:

(a) Use the Cusum approach to verify that indeed this data is in control

(b) Corrupt only one of the 20 sample data as done in Example 9.7.1 (i.e., the numerical values of four items forming the sample) and illustrate how the Cusum chart behaves under this situation.

Figure 9.32a shows the Cusum plot with the Vee mask. Since no point is outside the opening, the process is deemed to be in control. However, when the corruption of one data point (case b above) is introduced, one notes from Fig. 9.32b that the Cusum chart signals this effect quite dramatically since several points are outside the opening bounded by the Vee mask. ■

9.7.3.2 EWMA Monitoring Process

Moving average control charts can provide greatest sensitivity in process monitoring and control since information from past samples is combined with that of the current sample. There is, however, the danger that an incipient trend which gradually appears in the past observations may submerge any small shifts in the process. The exponential weighted moving average (EWMA) process (discussed in Sect. 9.3.2) which has direct links to the AR1 model (see Sect. 9.5.3) has redeeming qualities which make it attractive as a statistical quality control tool. One can apply this monitoring approach to either the sample mean of a set of observations forming a sample, or to individual observations taken from a system while in operation.

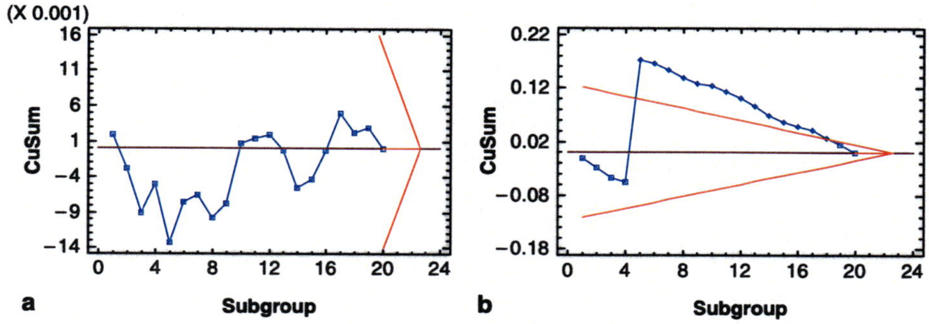

Fig. 9.32 **a** The Cusum chart with data from Table 9.10. **b** The Cusum chart with one data sample intentionally corrupted

An exponential weighted average with a discount factor θ such that $-1 \leq \theta \leq +1$ would be (Box and Luceno 1997)

$$\overset{\approx}{Y}_t = (1-\theta)(Y_t + \theta \cdot Y_{t-1} + \theta^2 \cdot Y_{t-2} + \cdots) \quad (9.55)$$

where the constant $(1-\theta)$ is introduced in order to normalize the sum of the series to unity since

$$(1 + \theta + \theta^2 + \cdots) = (1-\theta)^{-1}$$

Instead of using Eq. 9.55 to repeatedly recalculate $\overset{\approx}{Y}_t$ with each fresh observation, a convenient updating formula is:

$$\overset{\approx}{Y}_t = \lambda Y_t + \theta \cdot \overset{\approx}{Y}_{t-1} \quad (9.56)$$

where the new variable (seemingly redundant) is introduced by convention such that $\lambda = 1-\theta$.

If $\lambda = 1$, all the weight is placed in the latest observation, and one gets the Shewhart chart.

Consider the following sequence of observations from an operating system (Box and Luceno 1997):

Observation		1	2	3	4	5	6	7	8
Y	10	6	9	12	11	5	6	4	10

Note that the starting value of 10 is taken to be the target value. If $\lambda = 0.4$, then:

$$\overset{\approx}{Y}_1 = (0.4x6) + (0.6x10) = 8.4$$
$$\overset{\approx}{Y}_2 = (0.4x9) + (0.6x8.4) = 8.64$$
$$\overset{\approx}{Y}_3 = (0.4x12) + (0.6x8.64) = 9.98$$

and so on.

If the process is in perfect state of control and any deviations can be taken as a random sequence with standard deviation σ_Y, it can be shown that the associated standard deviation of the EWMA process is:

$$\sigma_{\bar{Y}} = \sigma_Y \left(\frac{\lambda}{2-\lambda}\right)^{1/2} \quad (9.57)$$

Thus, if one assumes $\lambda = 0.4$, then $(\sigma_{\overset{\approx}{Y}}/\sigma_Y) = 0.5$. The benefits of both the traditional Shewhart charts and the EWMA charts can be combined by generating a co-plot (in the above case, the three-sigma bands for EWMA will be half the width of the three-sigma Shewhart mean bands). Both sets of metrics for each observation can be plotted on such a co-plot for easier visual tracking of the process. An excellent discussion on EWMA and its advantage in terms of process adjustment using feedback control is provided by Box and Luceno (1997).

9.7.4 Concluding Remarks

There are several other related analysis methods which have been described in the literature. To name a few (Devore and Farnum 2005):

(a) *Process capability analysis:* This analysis provides a means to quantify the ability of a process to meet specifications or requirements. Just because a process is in control does not mean that specified quality characteristics are being met. Process capability analysis compares the distribution of process output to specifications when only common causes determine the variation. Should any special causes be present, this entire line of enquiry is invalid, and so one needs to carefully screen data for special effects before undertaking this analysis. Process capability is measured by the proportion of output that can be produced within design specifications. By collecting data, constructing frequency distributions and histograms, and computing basic descriptive statistics (such as mean and variance), the nature of the process can be better understood.

(b) *Pareto analysis for quality assessment:* Pareto Analysis is a statistical procedure that seeks to discover from an analysis of defect reports or customer complaints which "vital few" causes are responsible for most of the reported problems. The old adage states that 80% of reported problems can usually be traced to 20% of the various underlying causes. By concentrating one's efforts on rectifying the vital 20%, one can have the greatest immediate impact on product quality. It is used with attri-

Table 9.12 Relative effectiveness of control charts in detecting a change in a process. (From Himmelblau 1978)

Cause of change	Mean ($\bar{X}$)	Range (R)	Control chart: Standard deviation (s)	Control chart: Cumulative sum (CS)
Gross error (blunder)	1	2	–	3
Shift in mean	2	–	3	1
Shift in variability	–	1	–	–
Slow fluctuation (trend)	2	–	–	1
Rapid fluctuation (cycle)	–	1	2	–

1 = most useful, 2 = next best, 3 = least useful, and – = not appropriate

bute data based on histograms/frequency of each type of fault, and reveals the most frequent defect.

There are several instances when certain products and processes can be analyzed with more than one method, and there is no clear cut choice. $\bar{X}$ and R charts are quite robust in that they yield good results even if the data is not normally distributed, while Cusum charts are adversely affected by serial correlation in the data. Table 9.12 provides useful practical tips as to the effectiveness of different control chart techniques under different situations.

Problems

Pr. 9.1 Consider the time series data given in Table 9.1 which was used to illustrate various concepts throughout this chapter. Example 9.4.2 revealed that the model was still not satisfactory since the residuals still has a distinct trend. You will investigate alternative models (such as a second order linear or an exponential) in an effort to improve the residual behavior

Perform the same types of analyses as illustrated in the text. This involves determining whether the model is more accurate when fit to the first 48 data points, whether the residuals show less pronounced patterns, and whether the forecasts of the four quarters for 1986 have become more accurate. Document your findings in a succinct manner along with your conclusions.

Pr. 9.2 Use the following time series models for forecasting purposes:

(a) $Z_t = 20 + \varepsilon_t + 0.45\varepsilon_{t-1} - 0.35\varepsilon_{t-2}$. Given the latest four observations: $\{17.50, 21.36, 18.24, 16.91\}$, compute forecasts for the next two periods

(b) $Z_t = 15 + 0.86Z_{t-1} - 0.32Z_{t-2} + \varepsilon_t$. Given the latest two values of Z {32, 30}, determine the next four forecasts.

Pr. 9.3 Section 9.5.3 describes the manner in which various types of ARMA series can be synthetically generated as shown in Figs. 9.21, 9.22 and 9.23, and how one could verify different recommendations on model identification. These are useful aids for acquiring insights and confidence in the use of ARMA. You are asked to synthetically generate 50 data points using the following models and then use these data sequences to re-identify the models (because of the addition of random noise, there will be some differences in model parameters identified);

(a) $Z_t = 5 + \varepsilon_t + 0.7\varepsilon_{t-1}$ with $N(0, 0.5)$

(b) $Z_t = 5 + \varepsilon_t + 0.7\varepsilon_{t-1}$ with $N(0, 1)$

(c) $Z_t = 20 + 0.6Z_{t-1} + \varepsilon_t$ with $N(0, 1)$

(d) $Z_t = 20 + 0.8Z_{t-1} - 0.2Z_{t-1} + \varepsilon_t$ with $N(0, 1)$

(e) $Z_t = 20 + 0.8Z_{t-1} + \varepsilon_t + 0.7\varepsilon_{t-1}$ with $N(0, 1)$

Pr. 9.4 *Time series analysis of sun spot frequency per year from 1770–1869*

Data assembled in Table B.5 (in Appendix B) represents the so-called Wolf number of sunspots per year (n) over many years (from Montgomery and Johnson 1976 by permission of McGraw-Hill).

(a) First plot the data and visually note underlying patterns

(b) You will develop at least 2 alternative models using data from years 1770–1859. The models should include different trend and/or seasonal OLS models, as well as subclasses of the ARIMA models (where the trends have been removed by OLS models or by differencing). Note that you will have to compute the ACF and PACF for model identification purposes

(c) Evaluate these models using the expost approach where the data for years 1860–1869 are assumed to be known with certainty (as done in Example 9.6.1).

Pr. 9.5 *Time series of yearly atmospheric CO_2 concentrations from 1979–2005*

Table B.6 (refer to Appendix B) assembles data of yearly carbon-dioxide (CO_2) concentrations (in ppm) in the atmosphere and the temperature difference with respect to a base year (in °C) (from Andrews and Jelley 2007 by permission of Oxford University Press).

(a) Plot the data both as time series as well as scatter plots and look for underlying trends

(b) Using data from years 1979–1999, develop at least two models for the Temp. difference variable. These could be trend and /or seasonal or ARIMA type models

(c) Repeat step (b) but for the CO_2 concentration variable

(d) Using the same data from 1979–1999, develop a model for CO_2 where Temp.diff is one of the regressor variables

(e) Evaluate the models developed in (c) and (d) using data from 2000–2005 assumed known (this is the expost conditional case)

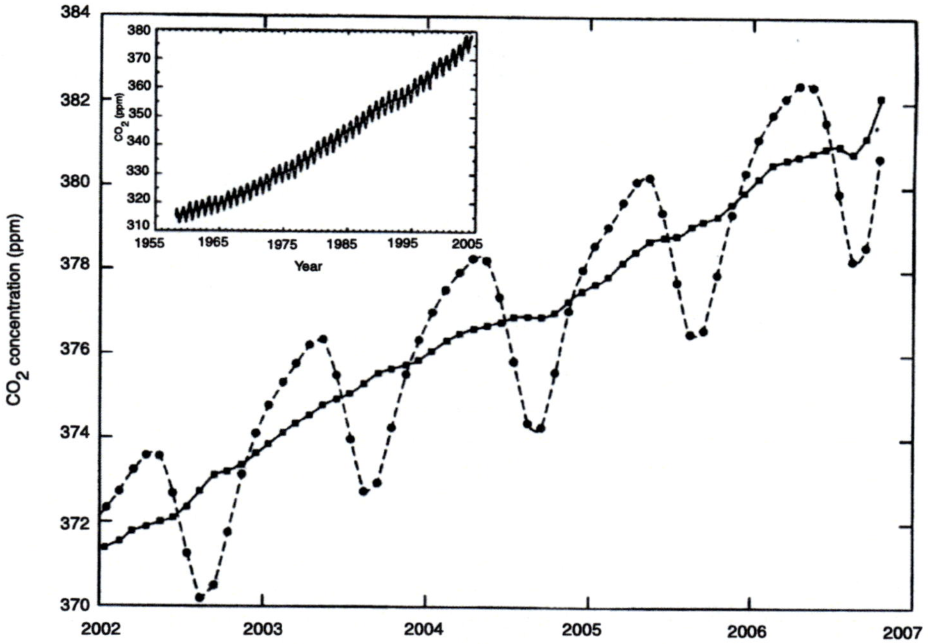

Fig. 9.33 Monthly mean global CO_2 concentration for the period 2002–2007. The smoothened line is a moving average over 10 adjacent months. (Downloaded from NOAA website http://www.cmdl.noaa.gov/ccgg/trends/index.php#mlo, 2006)

(f) Compare the above results for the exante unconditional situation. In this case, future values of temperature difference are not know, and so model developed in step (b) will be used to first predict this variable, which will then be used as an input to the model developed in step (d)
(g) Using the final model, forecast the CO_2 concentration for 2006 along with 95% CL.

Pr. 9.6 *Time series of monthly atmospheric CO_2 concentrations from 2002–2006*

Figure 9.33 represents global CO_2 levels but at monthly levels. Clearly there is both a long-term trend and a cyclic seasonal variation. The corresponding data is shown in Table B.7 (and can be found in Appendix B). You will use the first four years of data (2002–2005) to identify different moving average smoothing techniques, trend+seasonal OLS models, as well as ARIMA models as illustrated through several examples in the text. Subsequently, evaluate these models in terms of how well they predict the monthly values of the last year i.e., year 2006).

Pr. 9.7 *Transfer function analysis of unsteady state heat transfer through a wall*

You will use the conduction transfer function coefficients given in Example 9.6.1 to calculate the hourly heat gains (Q_{cond}) through the wall for a constant room temperature of 24°C and the hourly solar-air temperatures for a day given in Table 9.13 (adapted from Kreider et al. 2009). You will assume guess values to start the calculation, and repeat the diurnal calculation over as many days as needed to achieve convergence assuming the same T_{solair} values for successive days. This problem is conveniently solved on a spreadsheet.

Pr. 9.8 *Transfer function analysis using simulated hourly loads in a commercial building*

The hourly loads (total electrical, thermal cooling and thermal heating) for a large hotel in Chicago, IL have been generated for three days in August using a detailed building energy simulation program. The data shown in Table B.8 (given in Appendix B) consists of outdoor dry-bulb (T_{db}) and wet-bulb (T_{wb}) temperatures in °F as well as the internal electric loads of the building Q_{int} (these are the three regressor variables). The response variables are the total building electric power use (kWh) and the cooling and heating thermal loads (Btu/h).

(a) Plot the various variables as time series plots and note underlying patterns.
(b) Use OLS to identify a trend and seasonal model using indicator variables but with no lagged terms for Total Building electric power.
(c) For the same response variable, evaluate whether the seasonal differencing approach, i.e., $\nabla_{24}Y_t = Y_t - Y_{t-24}$ is as good as the trend and seasonal model in detrending the data series.
(d) Identify ARMAX models for all three response variables separately by using two days for model identification and the last day for model evaluation

Table 9.13 Data table for Problem 9.7

Hour ending	Solar-air temp (°C)	Hour ending	Solar-air temp (°C)	Hour ending	Solar-air temp (°C)
1	24.4	9	32.7	17	72.2
2	24.4	10	35.0	18	58.8
3	23.8	11	37.7	19	30.5
4	23.3	12	40.0	20	29.4
5	23.3	13	53.3	21	28.3
6	25.0	14	64.4	22	27.2
7	27.7	15	72.7	23	26.1
8	30.0	16	75.5	24	25.0

(e) Report all pertinent statistics and compare the results of different models. Provide reasons as to why the particular model was selected as the best one for each of the three response variables.

Pr. 9.9 Example 9.7.1 illustrated the use of Shewhart charts for variables.

(a) Reproduce the analysis results in order to gain confidence
(b) Repeat the analysis but using the Cusum and EWMA (with $\lambda=0.4$) and compare results.

Pr. 9.10 Example 9.7.2 illustrated the use of Shewhart charts for attributes variables.

(a) Reproduce the analysis results in order to gain confidence
(b) Repeat the analysis but using the Cusum and EWMA (with $\lambda=0.4$) and compare results.

References

Andrews, J. and N. Jelley, 2007. *Energy Science: Principles, Technologies and Impacts*, Oxford University Press, Oxford.

Bloomfield, P., 1976. *Fourier Analysis of Time Series: An Introduction.* John Wiley, New York.

Box, G.E.P. and G.M. Jenkins, 1976. *Time Series Analysis: Forecasting and Control,* Holden Day.

Box, G.E.P and A. Luceno, 1997. A., *Statistical Control by Monitoring and Feedback Adjustment*, John Wiley and Sons, New York.

Chatfield, C., 1989. *The Analysis of Time Series: An Introduction*, 4th Ed., Chapman and Hall, London.

Devore J., and N. Farnum, 2005. *Applied Statistics for Engineers and Scientists*, 2nd Ed., Thomson Brooks/Cole, Australia.

Dhar, A., T.A. Reddy and D.E. Claridge, 1999. Generalization of the Fourier series approach to model hourly energy use in commercial buildings, *Journal of Solar Energy Engineering*, vol. 121, p. 54, Transactions of the American Society of Mechanical Engineers, New York.

Himmelblau, D.M. 1978. *Fault Detection and Diagnosis in Chemical and Petrochemical Processes*, Chemical Engineering Monograph 8, Elsevier.

Kreider, J.K., P.S. Curtiss and A. Rabl, 2009. *Heating and Cooling of Buildings*, 2nd Ed., CRC Press, Boca Raton, Fl.

McCain L.J. and R. McCleary, 1979. The statistical analysis of the simple interrupted time series quasi-experiment, in *Quasi-experimentation: Design and analysis issues for field settings*, Houghton Mifflin Company, Boston, MA.

McClave, J.T. and P.G. Benson, 1988. *Statistics for Business and Economics*, 4th Ed., Dellen and Macmillan, London.

Montgomery, D.C. and L.A. Johnson, 1976. *Forecasting and Time Series Analysis*, McGraw-Hill, New York, NY.

Pindyck, R.S. and D.L. Rubinfeld, 1981. *Econometric Models and Economic Forecasts*, 2nd Edition, McGraw-Hill, New York, NY.

Ruch, D.K., J.K. Kissock, and T.A. Reddy, 1999. "Prediction uncertainty of linear building energy use models with autocorrelated residuals, *ASME Journal of Solar Energy Engineering*, vol.121, pp.63-68, February, Transactions of the American Society of Mechanical Engineers, New York.

Walpole, R.E., R.H. Myers, S.L. Myers, and K. Ye, 2007. *Probability and Statistics for Engineers and Scientists*, 8th Ed., Pearson Prentice Hall, Upper Saddle River, NJ.

Wei, W.W.S., 1990. *Time Series Analysis: Univariate and Multivariate Methods*, Addison-Wesley Publishing Company, Redwood City, CA.

10 Parameter Estimation Methods

This chapter covers topics related to estimation of model parameters and to model identification of univariate and multivariate problems not covered in earlier chapters. First, the fundamental notion of estimability of a model is introduced which includes both structural and numerical identifiability concepts. Recall that in Chap. 5 issues addressed were relevant to parameter estimation of linear models and to variable selection using step-wise regression in multivariate analysis. This chapter extends these basic notions by presenting the general statistical parameter estimation problem, and then presenting a few important estimation methods. Multivariate estimation methods (such as principle component analysis, ridge regression and stagewise regression) are discussed along with case study examples. Next, the error in variable (EIV) situation is treated when the errors in the regressor variables are large. Subsequently, another powerful and widely used estimation method, namely maximum likelihood estimation (MLE) is described, and its application to parameter estimation of probability functions and logistic models is presented. Also covered is parameter estimation of models non-linear in the parameters which can be separated into those that are transformable into linear ones, and those which are intrinsically non-linear. Finally, computer intensive numerical methods are discussed since such methods are being increasingly used nowadays because of the flexibility and robustness they provide. Different robust regression methods, whereby the influence of outliers on parameter estimation can be deemphasized, are discussed. This is, followed by the bootstrap resampling approach which is applicable for parameter estimation, and for ascertaining confidence limits of estimated model parameters and of model predictions.

10.1 Background

Statistical indices and residual analysis meant to evaluate the suitability of linear models, ways of identifying parsimonious models by step-wise regression, and OLS parameter estimation were addressed in Chap. 5. Basic to proper parameter estimation is the need to assess whether the data at hand is sufficiently rich and robust for the intended purpose; this aspect was addressed under design of experiments in Chap. 6. There are instances where OLS techniques, though most widely used, may not be suitable for parameter estimation. One such case occurs when regressors are correlated in a multiple linear regression (MLR) problem (recall that in Sect. 5.7.4, it was stated that step-wise regression was not recommended in such a case). Several techniques have been proposed to deal with such a situation, while allowing one to understand the underlying structure of the multivariate data and reduce the dimensionality of the problem. A basic exposure to these important statistical concepts is of some importance.

Also, parameter estimation of non-linear models relies on search methods similar to the ones discussed in Sect. 7.3 under optimization methods. In fact, the close form solutions for identifying the "best" OLS parameters are directly derived by minimizing an objective or loss function framed as the sum of square errors subject to certain inherent conditions (see Eq. 5.3). Such analytical solutions are no longer possible for non-linear models or for certain situations (such as when the measurement errors in the regressors are large). It is in such situations that estimation methods such as the maximum likelihood method or the different types of computer intensive methods discussed in this chapter (which offer great flexibility and robustness at the expense of computing resources) are appropriate, and have, thereby, gained popularity.

10.2 Concept of Estimability

The concept of estimability is an important one and relates to the ill-conditioning of the parameter matrix. It consists of two separate issues: *structural identifiability* and *numerical identifiability,* both of which are separately discussed below. Both these issues closely parallel the mathematical concepts of ill-conditioning of functions and uniqueness of

T. Agami Reddy, *Applied Data Analysis and Modeling for Energy Engineers and Scientists,*
DOI 10.1007/978-1-4419-9613-8_10, © Springer Science+Business Media, LLC 2011

equations which are covered in numerical analysis textbooks (for example, Chapra and Canale 1988), and so these are also reviewed.

10.2.1 Ill-Conditioning

The *condition* of a mathematical problem relates to its sensitivity to changes in the data. A computation is numerically unstable with respect to round-off and truncation errors if these uncertainties are grossly magnified by the numerical method. Consider the first order Taylor series:

$$f(x) = f(x_0) + f'(x_0)(x - x_0) \tag{10.1}$$

The relative error of f(x) can be defined as:

$$\varepsilon[f(x)] = \frac{f(x) - f(x_0)}{f(x_0)} \approx \frac{f'(x_0)(x - x_0)}{f(x_0)} \tag{10.2}$$

The relative error of x is given by

$$\varepsilon(x) = \frac{x - x_0}{x_0} \tag{10.3}$$

The *condition number* is the ratio of these relative errors:

$$C_d \equiv \frac{\varepsilon[f(x)]}{\varepsilon(x)} = \frac{x_0 f'(x_0)}{f(x_0)} \tag{10.4}$$

The condition number provides a measure of the extent to which an uncertainty in x is magnified by f(x). A value of 1 indicates that the function's relative error is identical to the relative error in x. Functions with very large values are said to be *ill-conditioned.*

Example 10.2.1: Calculate the condition number of the function f(x)=tan x at $x = (\pi/2)$.

The condition number $C_d = \dfrac{x_0(1/\cos^2 x_0)}{\tan x_0}$

The condition numbers at the following values of x are:

(a) at $x_0 = \pi/2 + 0.1(\pi/2)$, $C_d = \dfrac{1.7279(40.86)}{-6.314} = -11.2$

(b) at $x_0 = \pi/2 + 0.01(\pi/2)$, $C_d = \dfrac{1.5865(4053)}{-63.66} = -101$

For case (b), the major source of ill-conditioning appears in the derivative which is due to the singularity of the function close to $(\pi/2)$. ■

Let us extend this concept of ill-conditioning to sets of equations (Lipschutz 1966). The simplest case is a system of two linear equations in two unknowns in, say, x and y:

$$\begin{aligned} a_1 x + b_1 y &= c_1 \\ a_2 x + b_2 y &= c_2 \end{aligned} \tag{10.5}$$

Three cases can arise which are best described geometrically.

(a) The system has exactly one solution—where both lines intersect (see Fig. 10.1a)
(b) The system has no solution—the lines are parallel as shown in Fig. 10.1b. This will arise when the slopes of the two lines are equal but the intercept are not, i.e. when: $\dfrac{a_1}{a_2} = \dfrac{b_1}{b_2} \neq \dfrac{c_1}{c_2}$
(c) The system has an infinite number of solutions since they coincide as shown in Fig. 10.1c. This will arise

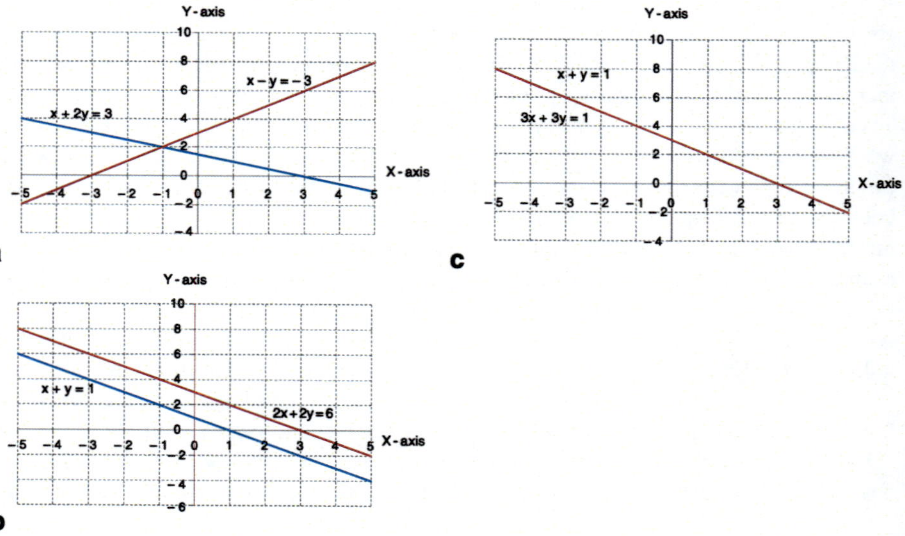

Fig. 10.1 Geometric representation of a system of two linear equations **a** the system has exactly one solution, **b** the system has no solution, **c** the system has an infinite number of equations

when both the slopes and the intercepts are equal, i.e., when $\frac{a_1}{a_2} = \frac{b_1}{b_2} = \frac{c_1}{c_2}$

Consider a set of linear equations represented by

$$Ax = b \quad \text{whose solution is} \quad x = A^{-1}b \tag{10.6}$$

Whether this set of linear equations can be solved or not is easily verified by computing the rank of the matrix A which is the integer representing the order of the highest non-vanishing integer. Consider the following matrix:

$$\begin{Bmatrix} 1 & 2 & -2 & 3 \\ 2 & -1 & 3 & -2 \\ -1 & 3 & 1 & -4 \\ 3 & 6 & -6 & 9 \end{Bmatrix}$$

Since the first and last rows are identical, or more correctly "linearly dependent", the rank is equal to 3. Computer programs would identify such deficiencies correctly and return an error message such as "matrix is not positive definite" indicating that the estimated data matrix is singular. Hence, one cannot solve for all four unknowns but only for three. However, such a test breaks down when dealing with real data which includes measurement as well as computation errors (during the computation of the inverse of the matrix). This is illustrated by using the same matrix but the last row has been corrupted by a small noise term—of the order of 5% only.

$$\begin{Bmatrix} 1 & 2 & -2 & 3 \\ 2 & -1 & 3 & -2 \\ -1 & 3 & 1 & -4 \\ \frac{63}{20} & \frac{117}{20} & \frac{-123}{20} & \frac{171}{20} \end{Bmatrix}$$

Most computer programs if faced with this problem would determine the rank of this matrix as 4! Hence, even small noise in the data can lead to misleading conclusions. The notion of condition number, introduced earlier, can be extended to include such situations. Recall that any set of linear equations of order n has n roots, either distinct or repeated, which are usually referred to as characteristic roots or *eigenvalues*. The stability or the robustness of the solution set, i.e., its closeness to singularity, can be characterized by the same concept of condition number (C_d) of the matrix A computed as:

$$C_d = \left(\frac{\text{largest eigenvalue}}{\text{smallest eigenvalue}}\right)^{1/2} \tag{10.7}$$

The value of the condition number for the above matrix is C_d=371.7. Hence, a small perturbation in b induces a relative perturbation 370 times greater in the solution of the system of equations. Thus, even a singular matrix will be signaled as an ill-conditioned (or badly conditioned) matrix due to roundoff and measurement errors, and the analyst has to select thresholds to infer whether the matrix is singular or merely ill-conditioned.

10.2.2 Structural Identifiability

Structural identifiability is defined as the problem of investigating the *conditions under* which system parameters can be uniquely estimated from experimental data, no matter how noise-free the measurements. This condition can be detected before the experiment is conducted by analyzing the basic modeling equations. Two commonly used testing techniques are the *sensitivity coefficient approach,* and one involving looking at the behavior of the poles of the *Laplace Transforms* of the basic modeling equations (Sinha and Kuszta 1983). Only the sensitivity coefficient approach to detect identifiability is introduced below.

Let us first look at some simple, almost trivial, examples of structural identifiability of models, where y is the system response and x is the input variable (or the input variable matrix).

(a) A model such as $y = (a + cb)x$ where c is a system constant will not permit unique identification of a and b when measurements of y and x are made. At best, the overall term (a+cb) can be identified.
(b) A model given by $y = (ab)x$ will not permit explicit identification of a and b, merely the product (a·b).
(c) The model $y = b_1/(b_2 + b_3 t)$ where t is the regressor variable will not allow all three parameters to be identified, only the ratios (b_2/b_1) and (b_3/b_1). This case is treated in Example 10.2.3 below.
(d) The lumped parameter differential equation model for the temperature drop with time T(t) of a cooling sphere is given by: $Mc_p\frac{dT}{dt} = hA(T - T_\infty)$ where M is the mass of the sphere, c_p its specific heat, h the heat loss coefficient from the sphere to the ambient, A the surface area of the sphere and T_∞ the ambient temperature assumed constant (see Sect. 1.2.4). If measurements over time of T are made, one can only identify the group (Mc_p/hA). Note that the reciprocal of this quantity is the time constant of the system.

The geometric interpretation of ill-conditioning (see discussion above relevant to Eq. 10.5) can be related to structural identifiability. As stated by Godfrey (1983), one can distinguish between three possible outcomes of an estimability analysis:

(i) The model parameters can be estimated uniquely, and the model is *globally identifiable,*

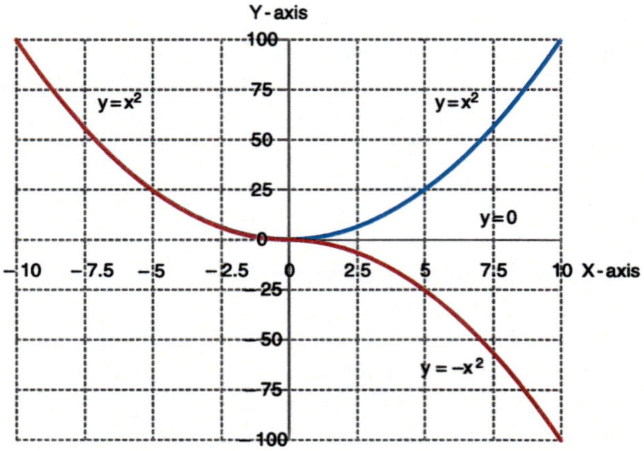

Fig. 10.2 Plot of Example 10.2.2 used to illustrate local versus global identifiability

(ii) A finite number of alternative estimates of model parameters is possible, and the model is *locally identifiable*, and
(iii) An infinite number of model parameter estimates are possible, and the model is *unidentifiable* from the data (this is the over-parameterized case).

Example 10.2.2:[1] Let us consider a more involved problem to illustrate the concept of a locally identifiable problem. Consider the first-order differential equation:

$$xy' = 2y \quad \text{whose solution is } y(x) = Cx^2 \tag{10.8}$$

where C is a constant to be determined from the initial value. If y(−1)=1, then C=1. Thus, one has a unique solution $y(x) = x^2$ on some open interval about x=−1 which passes through the origin (see Fig. 10.2). But to the right of the origin, one may choose any value for C in Eq. 10.8. Three different solutions are shown in Fig. 10.2. Hence, though one has uniqueness of the solution near some point or region, the solution may branch elsewhere, and the uniqueness may be lost. ■

The problem of identifiability is almost a non-issue for simple models; for example, models such as (a) or (b) above, and even for solved Example 10.2.2. However, more complex models demand a formal method rather than depend on adhoc manipulation and inspection. The *sensitivity coefficient* approach allows formal testing to be performed. Consider a model y(t, **b**) where t is an independent variable and **b** is the parameter vector. The first derivative of y with respect to $\mathbf{b}_j$ is the sensitivity coefficient for $\mathbf{b}_j$, and is designated by $(\partial y/\partial b_j)$. Sensitivity coefficients indicate the magnitude of change in the response y due to perturbations in the values of the parameters. Let i be the number of observation sets representing the range under which the experiment was performed. The condition for structural identifiability is that the sensitivity coefficients over the range of the observations should not be linearly dependent. Linear dependence is said to occur when, for p parameters in the model, the following relation is true for *all* i observations even if all x_j values are not zero (a formal proof is given by Beck and Arnold 1977):

$$x_1 \frac{\partial y_i}{\partial b_1} + x_2 \frac{\partial y_i}{\partial b_2} + \cdots + x_p \frac{\partial y_i}{\partial b_p} = 0 \tag{10.9}$$

Example 10.2.3: Let us apply the condition given by Eq. 10.9 to the model (c) above, namely $y = b_1/(b_2 + b_3 t)$. Though mere inspection indicates that all three parameters b_1, b_2 and b_3 cannot be individually identified, the question is "can the ratios (b_2/b_1) and (b_3/b_1) be determined under all conditions?" In this case, the sensitivity coefficients are:

$$\frac{\partial y_i}{\partial b_1} = \frac{1}{b_2 + b_3 t_i}, \quad \frac{\partial y_i}{\partial b_2} = \frac{-b_1}{(b_2 + b_3 t_i)^2} \text{ and } \quad \frac{\partial y_i}{\partial b_3} = \frac{-b_1 t_1}{(b_2 + b_3 t_i)^2} \tag{10.10}$$

It is not clear whether there is linear dependence or not. One can verify this by assuming $x_1=b_1$, $x_2=b_2$ and $x_3=b_3$. Then, the model can be expressed as

$$b_1 \frac{\partial y_i}{\partial b_1} + b_2 \frac{\partial y_i}{\partial b_2} + b_3 \frac{\partial y_i}{\partial b_3} = 0 \tag{10.11}$$

or

$$z = z_1 + z_2 + z_3 \tag{10.12}$$

where

$$z_1 = \frac{\partial y_i}{\partial b_1}, \quad z_2 = \frac{b_2}{b_1}\frac{\partial y_i}{\partial b_2}, \quad z_3 = \frac{b_3}{b_1}\frac{\partial y_i}{\partial b_3}$$

The above function can occur in various cases with linear dependence. Arbitrarily assuming $b_2=1$, the variation of the sensitivity coefficients, or more accurately those of z, z_1, z_2 and z_3, are plotted against $(b_3 \cdot t)$ in Fig. 10.3. One notes that z=0 throughout the entire range denoting therefore that identification of all three parameters is impossible. Further, inspection of Fig. 10.3 reveals that both z_2 and z_3 seem to have become constant, therefore becoming linearly dependent for $b_3 t>3$. This means that not only is it impossible to estimate all three parameters simultaneously from measurements of y and t, but that it is also impossible to estimate both b_1 and b_3 using data over the spatial range $b_3 t>3$. ■

[1] From Edwards and Penney (1996) by © permission of Pearson Education.

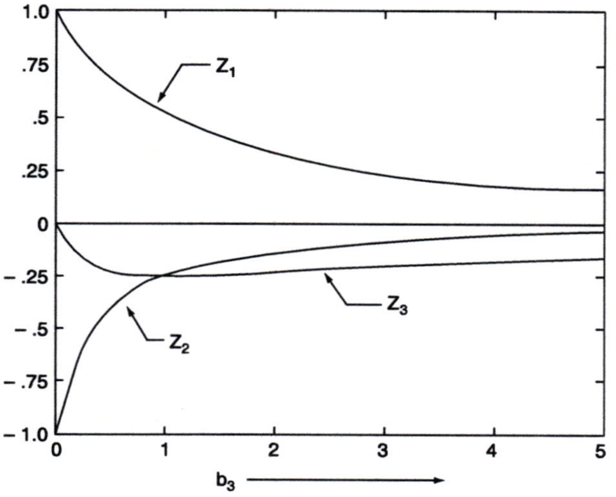

Fig. 10.3 Verifying linear dependence of model parameters of Eq. 10.12. (From Beck and Arnold 1977 by permission of Beck)

10.2.3 Numerical Identifiability

A third crucial element is the numerical scheme used for analyzing measured data. Even when the disturbing noise or experimental error in the system is low, OLS may not be adequate to identify the parameters without bias or minimum variance because of multi-collinearity effects between regressor variables. As the noise becomes more significant, more bias is introduced in the parameter estimates and so elaborate numerical schemes such as iterative methods or multi-step methods have to be used (discussed later in this chapter). Recall from Sect. 5.4.3 that the parameter estimator vector is given by $\mathbf{b} = (\mathbf{X}'\mathbf{X})^{-1}\mathbf{X}'\mathbf{Y}$ while the variance-covariance matrix by: $\mathrm{var}(\mathbf{b}) = \sigma^2(\mathbf{X}'\mathbf{X})^{-1}$ where σ^2 is the mean square error of the model error terms. *Numerical identifiability,* also called redundancy, is defined as the inability to obtain proper parameter estimates from the data even if the experiment is structurally identifiable. This can arise when the data matrix $(\mathbf{X}'\mathbf{X})$ is close to singular. Such a condition associated with inadequate richness in the data rather than with model mis-specification is referred to as *ill-conditioned* data (or, more loosely as *weak data*). If OLS estimation was used with ill-conditioned data, the parameters, though unbiased, are not efficient (unreliable with large standard errors) in the sense they no longer have minimum variance. More importantly, OLS formulae would understate both the standard errors and the models prediction uncertainty bands (even though the overall fit may be satisfactory). The only recourse is either to take additional pertinent measurements or to simplify or aggregate the model structure so as to remove some of the collinear variables from the model. How to deal with such situations is discussed in Sect. 10.3.

Data is said to be ill-conditioned when the regressor variables of a model are correlated with each other, leading to the correlation coefficient $R = \mathbf{X}'\mathbf{X}$ matrix (see Eq. 5.32) becoming close to singular, and resulting in the reciprocal becoming undetermined or very large. This may have serious effects on the estimates of the model coefficients (unstable with large variance), and on the general applicability of the estimated model. There are three commonly used diagnostic measures, all three depend on the variance-covariance matrix **C**, which can be used to evaluate the magnitude of ill-conditioning (Belsley et al. 1980).

(i) The **correlation matrix R** which is akin to matrix **C** (given by Eq. 5.31) where the diagonal elements are centered and scaled by unity (subtracting by the mean and dividing by the standard deviation). This allows one to investigate correlation between pairs of regressors in a qualitative manner, but may be of limited use in assessing the magnitude of overall multicollinearity of the regressor set.

(ii) **Variance inflation factors** which provide a better quantitative measure of the overall collinearity. The diagonal elements of the **C** matrix are:

$$C_{jj} = \frac{1}{\left(1 - R_j^2\right)} \quad j = 1, 2, \ldots k \tag{10.13}$$

where R^2_j is the coefficient of multiple determination resulting from regressing x_j on the other $(k-1)$ variables. Clearly, the stronger the linear dependency of x_j on the remaining regressors, the larger the value of R^2_j. The variance of b_j is said to be "inflated" by the quantity $(1-R^2_j)$. Thus, the variance inflation factors VIF $(b_j) = C_{jj}$. The VIF allows one to look at the joint relationship among a specified regressor and all other regressors. Its weakness, like that of the coefficient of determination R^2, is its inability to distinguish among several coexisting near-dependencies and in the inability to assign meaningful thresholds between high and low VIF values (Belsley et al. 1980). Many texts suggest rules of thumb: VIF > 10 indicate strong ill-conditioning, while 5 < VIF < 10 indicates a moderate problem.

(iii) **Condition number,** discussed above, is widely used by analysts to determine robustness of the parameter estimates, i.e., how low are their standard errors, since it provides a measure of the joint relationship among regressors. Evidence of collinearity is suggested for condition numbers > 15, and corrective action is warranted when the value exceeds 30 or so (Chatterjee and Price 1991).

To summarize, ill-conditioning of a matrix **X** is said to occur when one or more columns can be expressed as linear combinations of another column, i.e., det $(\mathbf{X}'\mathbf{X}) = 0$ or close to 0. Possible causes are either the data set is inadequate or the model is overspecified, i.e., too many parameters have been included in the model. Possible remedies one should investigate are (i) collecting more data, or (ii) dropping variables from the model based on physical insights. If these fail, one can use biased estimation methods such as ridge regression, or other transformations such as principle component analysis (PCA) which are presented below.

10.3 Dealing with Collinear Regressors During Multivariate Regression

10.3.1 Problematic Issues

Perhaps the major problem with multivariate regression is that the "independent" variables are not really independent but collinear to some extent (and hence, the suggestion that the term "regressor" be used instead). Strong collinearity has the result that the variables are "essentially" influencing or explaining the same system behavior. For linear models, the Pearson correlation coefficient (presented in Sect. 3.4.2) provides the necessary indication of the strength of this overlap. This issue of collinearity between regressors is a very common phenomenon which has important implications during model building and parameter estimation. Not only can regression coefficients be strongly biased, but they can even *have the wrong sign*. Note that this could also happen if the range of variation of the regressor variables is too small, or if some important regressor variable has been left out.

Example 10.3.1: Consider the simple example of a linear model with two regressors both of which are positively correlated with the response variable y. The data consists of six samples as shown in Table 10.1. The pairwise plots shown in Fig. 10.4 clearly depict the fairly strong relationship between the two regressors.

From the correlation matrix C for this data (Table 10.2), the correlation coefficient between the two regressors is 0.776, which can be considered to be of moderate strength. An OLS regression results in the following model:

$$y = 1.30 + 0.75x_1 - 0.05x_2 \tag{10.14}$$

The model identified suggests a negative correlation between y and x_2 which is contrary to both the correlation coefficient matrix and the graphical trend in Fig. 10.4. This irrationality is the result of the high inter-correlation between the regressor variables. What has occurred is that the inverse of the variance-covariance matrix ($\mathbf{X'X}$) of the estimated regression coefficients has become ill-conditioned and unstable. A simple layperson explanation is to say that x_1 has usurped more than its appropriate share of explicative power of y at the detriment of x_2 which, then, had to correct itself to such a degree that it ended up assuming a negative correlation. ■

Table 10.1 Data table for Example 10.3.1

y	x_1	x_2
2	1	2
2	2	3
3	2	1
3	5	5
5	4	6
6	5	4

Table 10.2 Correlation matrix for Example 10.3.1

	x_1	x_2	Y
x_1	1.000	0.776	0.742
x_2		1.000	0.553
y			1.000

Mullet (1976), discussing why regression coefficients in the physical sciences often have wrong signs, quotes: (i) Marquardt who postulated that multicollinearity is likely to be a problem only when correlation coefficients among regressor variables is higher than 0.95, and (ii) Snee who used 0.9 as the cut-off point. On the other hand, Draper and Smith (1981) state that multicollinearity is likely to be a problem if the simple correlation between two variables is larger than the correlation of one or either variable with the dependent variable.

Significant collinearity between regressor variables is likely to lead to two different problems:

(i) though the model may provide a good fit to the current data, its usefulness as a reliable predictive model is suspect. The regression coefficients and the model predictions tend to have large standard errors and uncertainty bands which makes the model unstable. It is imperative

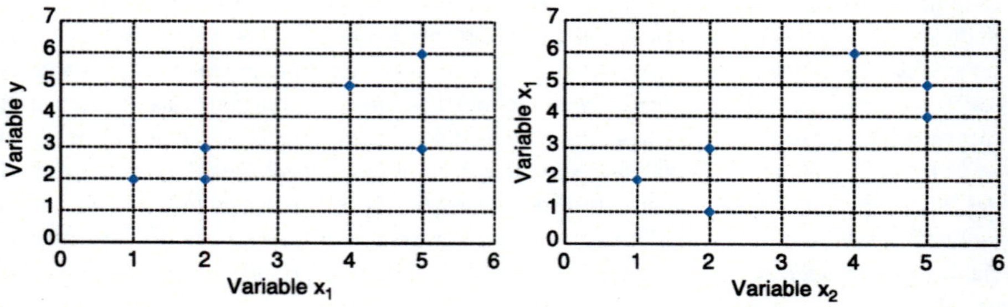

Fig. 10.4 Data for Example 10.3.1 to illustrate how multicollinearity in the regressors could result in model coefficients with wrong signs

that a sample cross-validation evaluation be performed to identify a suitable model (a case study is presented in Sect. 10.3.4);

(ii) the regression coefficients in the model are no longer proper indicators of the relative physical importance of the regressor parameters.

10.3.2 Principle Component Analysis and Regression

Principle Component Analysis (PCA) is one of the best known multivariate methods for removing the adverse effects of collinearity, while summarizing the main aspects of the variation in the regressor set (see for example, Draper and Smith 1981 or Chatterjee and Price 1991). It has a simple intuitive appeal, and though very useful in certain disciplines (such as the social sciences), its use has been rather limited in engineering applications. It is not a statistical method leading to a decision on a hypothesis, but a general method of identifying which parameters are collinear and reducing the dimension of multivariate data. This reduction in dimensionality is sometimes useful for gaining insights into the behavior of the data set. It also allows for more robust model building, an aspect which is discussed below.

The premise in PCA is that the variance in the collinear multi-dimension data comprising of the regressor variable vector **X** can be reframed in terms of a set of orthogonal (or uncorrelated) transformed variable vector **U**. This vector will then provide a means of retaining only a subset of variables which explain most of the variability in the data. Thus, the dimension of the data will be reduced without losing much of the information (reflected by the variability in the data) contained in the original data set, thereby allowing a more robust model to be subsequently identified. A simple geometric explanation of the procedure allows better conceptual understanding of the method. Consider the two-dimension data shown in Fig. 10.5. One notices that much of the variability in the data occurs along one dimension or direction. If one were to rotate the orthogonal axis such that the major u_1 axis were to lie in the direction of greatest data variability (see Fig. 10.5b), most of this variability will become uni-directional with little variability being left for the orthogonal u_2 axis to account for. The variability in the two-dimensional original data set is, thus, largely accounted for by only one variable, i.e., the transformed variable u_1.

The real power of this method is when one has a large number of dimensions; in such cases one needs to have some mathematical means of ascertaining the degree of variation in the multi-variate data along different dimensions. This is achieved by looking at the eigenvalues. The eigenvalue can be viewed as one which is indicative of the length of the axis while the eigenvector specifies the direction of rotation.

Usually PCA analysis is done with standardized variables **Z** instead of the original variables **X** such that variables **Z** have zero mean and unit variance. Recall that the eigenvalues λ (also called characteristic roots or latent roots) and the eigenvector A of a matrix **Z** are defined by:

$$\mathbf{AZ} = \lambda \mathbf{Z} \tag{10.15}$$

The eigenvalues are the solutions of the determinant of the covariance matrix of **Z**:

$$|\mathbf{Z}'\mathbf{Z} - \lambda \mathbf{I}| = 0 \tag{10.16}$$

Because the original data or regressor set **X** is standardized, an important property of the eigenvalues is that their sum is equal to the trace of the correlation matrix **C**, i.e.,

$$\lambda_1 + \lambda_2 + \cdots + \lambda_p = p \tag{10.17}$$

where p is the dimension or number of variables. This follows from the fact that the diagonal elements for a correlation matrix should sum to unity. Usually, the eigenvalues are ranked such that the first has the largest numerical value, the second the second largest, and so on. The corresponding eigenvector represents the coefficients of the principle components (PCs). Thus, the linearized transformation for the PC from the original vector of standardized variables **Z** can be represented by:

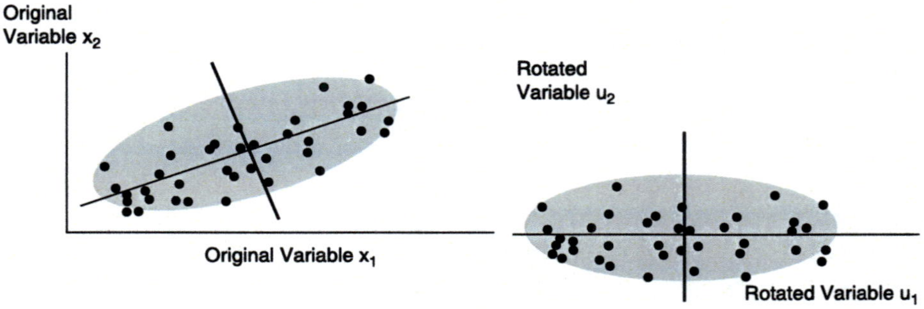

Fig. 10.5 Geometric interpretation of what a PCA analysis does in terms of variable transform for the case of two variables. The rotation has resulted in the primary axis explaining a major part of the variability in the original data with the rest explained by the second rotated axis. Reduction in dimensionality can be achieved by accepting a little loss in the information contained in the original data set and dropping the second rotated variable altogether

$$\begin{aligned} PC1: u_1 &= a_{11}z_1 + a_{12}z_2 + \cdots + a_{1p}z_p \quad \text{subject to} \\ &a_{11}^2 + a_{12}^2 + \cdots + a_{1p}^2 = 1 \\ PC2: u_2 &= a_{21}z_1 + a_{22}z_2 + \cdots + a_{2p}z_p \quad \text{subject to} \\ &a_{21}^2 + a_{22}^2 + \cdots + a_{2p}^2 = 1 \\ \ldots \text{ to } PCp \end{aligned} \tag{10.18}$$

where a_{ii} are called the component weights and are the scaled elements of the corresponding eigenvector.

Thus, the correlation matrix (given by Eq. 5.31) for the standardized and rotated variables is now transformed into:

$$\mathbf{C} = \left[\begin{pmatrix} \lambda_1 & 0 & 0 \\ 0 & \ddots & 0 \\ 0 & 0 & \lambda_p \end{pmatrix} \right] \quad \text{where} \quad \lambda_1 > \lambda_2 > \cdots > \lambda_p. \tag{10.19}$$

Note that the off-diagonal terms are zero because the variable vector **U** is orthogonal. Further, note that the eigenvalues represent the variability of the data along the principle components.

If one keeps all the PCs, nothing is really gained in terms of reduction in dimensionality, even though they are orthogonal (i.e., uncorrelated), and the model building by regression will be more robust. Model reduction is done by rejecting those transformed variables **U** which contribute little or no variance to the data. Since the eigenvalues are ranked, PC1 explains the most variability in the original data while each succeeding eigenvalue accounts for increasingly less. A typical rule of thumb to determine the cut-off is to drop any factor which explains less than (1/p) of the variability, where p is the number of parameters or the original dimension of the regressor data set.

PCA has been presented as an approach which allows the dimensionality of the multivariate data to be reduced while yielding uncorrelated regressors. Unfortunately, in most cases, the physical interpretation of the **X** variables, which often represent a physical quantity, is lost as a result of the rotation. A few textbooks (Manly 2005) provide examples where the new rotated variables retain some measure of physical interpretation, but these are the exception rather than the rule in the physical sciences. In any case, the reduced set of transformed variables can now be used to identify multivariate models that are, often but not always, more robust.

Example 10.3.2: Consider Table 10.3 where PC rotation has already been performed. The original data set contained nine variables which were first standardized, and a PCA resulted in the variance values as shown in the table. One notices that PC1 explains 41% of the variation, PC2 23%, and so on till all nine PC explain as much variation as was present in the original data. Had the nine PC been independent or orthogonal, each one would have explained on an average (1/p)=(1/9)=11% of the variance. The eigenvalues listed in the table corresponds to the number of variables which would have explained an equivalent amount of variation in the data that is attributed to the corresponding PC. For example, the first eigenvalue is determined as: 41/(100/9)=3.69 i.e., PC1 has the explicative power of 3.69 of the original variables, and so on.

Table 10.3 How to interpret PCA results and determine thresholds. (From Kachigan 1991 by permission of Kachigan)

Extracted factors	% of total variance accounted for		Eigenvalues	
	Incremental	Cumulative	Incremental	Cumulative
PC1	41%	41%	3.69	3.69
PC2	23	64	2.07	5.76
PC3	14	78	1.26	7.02
PC4	7	85	0.63	7.65
PC5	5	90	0.45	8.10
PC6	4	94	0.36	8.46
PC7	3	97	0.27	8.73
PC8	2	99	0.18	8.91
PC9	1	100	0.09	9.00

The above manner of studying the relative influence of each PC allows heuristic thresholds to be defined. The typical rule of thumb as stated above would result in all PC whose eigenvalues are less than 1.0 being omitted from the reduced multivariate data set. This choice can be defended on the grounds that an eigenvalue of 1.0 would imply that the PC explains less than would the original untransformed variable, and so retaining it would be illogical since it would be defeating the basic purpose, i.e. trying to achieve a reduction in the dimensionality of the data. However, this is to be taken as a heuristic criterion and not as a hard and fast rule. A convenient visual indication of how higher factors contribute increasingly less to the variance in the multivariate data can be obtained from a *scree plot* generated by most PCA analysis software. This is simply a plot of the eigenvalues versus the PCs (i.e., the first and fourth columns of Table 10.3), and provides a convenient visual representation as illustrated in the example below. ■

Example 10.3.3: *Reduction in dimensionality using PCA for actual chiller data*
Consider data assembled in Table 10.4 which consists of a data set of 15 possible variables or characteristic features (CFs) under 27 different operating conditions of a centrifugal chiller. With the intention of reducing the dimensionality of the data set, a PCA is performed in order to determine an optimum set of principle components. Pertinent steps will be shown and the final selection using the eigenvalues and the scree plot will be justified. Also the component weights table which assembles the PC models will be discussed.

Table 10.4 Data table with 15 regressor variables for Example 10.3.3. (From Reddy 2007 from data supplied by James Braun)

CF1	CF2	CF3	CF4	CF5	CF6	CF7	CF8	CF9	CF10	CF11	CF12	CF13	CF14	CF15
3.765	5.529	5.254	3.244	15.078	4.911	2.319	5.473	83.069	39.781	0.707	0.692	1.090	5.332	0.706
3.405	3.489	3.339	3.344	19.233	3.778	1.822	4.550	73.843	32.534	0.603	0.585	0.720	4.977	0.684
2.425	1.809	1.832	3.500	31.333	2.611	1.009	3.870	73.652	21.867	0.422	0.397	0.392	3.835	0.632
4.512	6.240	5.952	2.844	12.378	5.800	3.376	5.131	71.025	45.335	0.750	0.735	1.260	6.435	0.701
3.947	3.530	3.338	3.322	18.756	3.567	1.914	4.598	71.096	32.443	0.568	0.550	0.779	5.846	0.675
2.434	1.511	1.558	3.633	35.533	1.967	0.873	3.821	72.116	18.966	0.335	0.311	0.362	3.984	0.611
4.748	5.087	4.733	3.156	14.478	4.589	2.752	5.060	70.186	39.616	0.665	0.649	1.107	6.883	0.690
4.513	3.462	3.197	3.444	19.511	3.356	1.892	4.716	69.695	31.321	0.498	0.481	0.844	6.691	0.674
3.503	2.153	2.053	3.789	28.522	2.244	1.272	4.389	68.169	22.781	0.347	0.326	0.569	5.409	0.647
3.593	5.033	4.844	2.122	13.900	4.878	2.706	3.796	72.395	48.016	0.722	0.706	0.990	5.211	0.689
3.252	3.466	3.367	2.122	17.944	3.700	1.720	3.111	76.558	42.664	0.626	0.607	0.707	4.787	0.679
2.463	1.956	2.004	2.233	27.678	2.578	1.102	2.540	73.381	32.383	0.472	0.446	0.415	3.910	0.630
4.274	6.108	5.818	2.056	11.944	5.422	3.323	4.072	71.002	52.262	0.751	0.739	1.235	6.098	0.701
3.678	3.330	3.228	2.089	17.622	3.389	1.907	3.066	70.252	41.724	0.593	0.573	0.722	5.554	0.662
2.517	1.644	1.714	2.256	30.967	2.133	1.039	2.417	69.184	29.607	0.383	0.361	0.385	4.126	0.610
4.684	5.823	5.522	2.122	12.089	4.989	3.140	4.038	71.271	50.870	0.732	0.721	1.226	6.673	0.702
4.641	4.002	3.714	2.456	16.144	3.589	2.188	3.829	70.354	40.714	0.591	0.574	0.928	6.728	0.690
3.038	1.828	1.796	2.689	29.767	1.989	1.061	3.001	70.279	26.984	0.347	0.327	0.470	4.895	0.621
3.763	5.126	4.924	1.400	12.744	4.656	2.687	2.541	73.612	62.921	0.733	0.721	1.030	5.426	0.694
3.342	3.344	3.318	1.567	16.933	3.456	1.926	2.324	70.932	50.601	0.631	0.611	0.698	5.073	0.659
2.526	1.940	2.053	1.378	25.944	2.600	1.108	1.519	74.649	46.588	0.476	0.453	0.421	4.122	0.613
4.411	6.244	5.938	1.522	11.689	5.411	3.383	3.193	70.782	61.595	0.749	0.740	1.282	6.252	0.705
4.029	3.717	3.559	1.178	14.933	3.844	2.128	1.917	69.488	61.187	0.628	0.609	0.817	6.035	0.667
2.815	1.886	1.964	1.378	26.333	2.122	0.946	1.394	79.851	48.676	0.420	0.398	0.448	4.618	0.609
4.785	5.528	5.203	1.611	11.756	5.100	3.052	2.948	69.998	59.904	0.717	0.704	1.190	6.888	0.694
4.443	3.882	3.679	1.933	15.578	3.556	2.121	3.038	70.939	46.667	0.612	0.597	0.886	6.553	0.678
3.151	2.010	2.054	1.656	25.367	2.333	1.224	1.859	69.686	41.716	0.409	0.390	0.500	5.121	0.615

A principle component analysis is performed with the purpose of obtaining a small number of linear combinations of the 15 variables which account for most of the variability in the data. From the eigenvalue table (Table 10.5) as well as the scree plot shown (Fig. 10.6), one notes that there are three components with eigenvalues greater than or equal to 1.0, and that together they account for 95.9% of the variability in the original data. Hence, it is safe to only retain three components.

The equations of the principal components can be deduced from the table of components weights shown (Table 10.6). For example, the first principal component has the equation

$$\begin{aligned} PC1 = {} & 0.268037 * CF1 + 0.215784 * CF10 + 0.294009 \\ & * CF11 + 0.29512 * CF12 + 0.302855 * CF13 + 0.247658 \\ & * CF14 + 0.29098 * CF15 + 0.302292 * CF2 + 0.301159 \\ & * CF3 - 0.06738 * CF4 - 0.297709 * CF5 + 0.297996 \\ & * CF6 + 0.301394 * CF7 + 0.123134 * CF8 - 0.0168 * CF9 \end{aligned}$$

where the values of the variables in the equation are standardized by subtracting their means and dividing by their standard deviations. ■

Table 10.5 Eigenvalue table for Example 10.3.3

Component number	Eigenvalue	Percent of cumulative	
		Variance	Percentage
1	10.6249	70.833	70.833
2	2.41721	16.115	86.947
3	1.34933	8.996	95.943
4	0.385238	2.568	98.511
5	0.150406	1.003	99.514
6	0.0314106	0.209	99.723
7	0.0228662	0.152	99.876
8	0.00970486	0.065	99.940
9	0.00580352	0.039	99.979
10	0.00195306	0.013	99.992
11	0.000963942	0.006	99.998
12	0.000139549	0.001	99.999
13	0.0000495725	0.000	100.000
14	0.0000261991	0.000	100.000
15	0.0000201062	0.000	100.000

Thus, in summary, PCA takes a group of "n" original regressor variables and re-expresses them as another set of "n" artificial variables, each of which represents a linear combi-

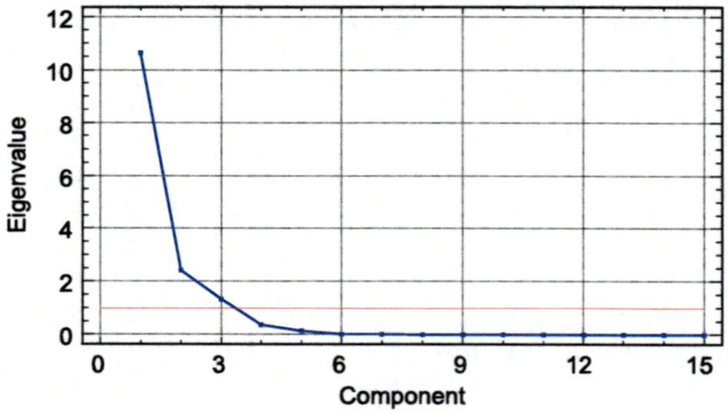

Fig. 10.6 Scree plot of Table 10.5 data

Table 10.6 Component weights for Example 10.3.3

	Component 1	Component 2	Component 3
CF1	0.268037	0.126125	−0.303013
CF10	0.215784	−0.447418	−0.0319413
CF11	0.294009	−0.0867631	0.155947
CF12	0.29512	−0.0850094	0.149828
CF13	0.302855	0.0576253	−0.0135352
CF14	0.247658	0.117187	−0.388742
CF15	0.29098	0.133689	0.08111
CF2	0.302292	0.0324667	0.0801237
CF3	0.301159	0.0140657	0.0975298
CF4	−0.06738	0.620147	0.100548
CF5	−0.297709	0.0859571	0.00911293
CF6	0.297996	0.0180763	0.130096
CF7	0.301394	0.0112407	−0.0443937
CF8	0.123134	0.576743	0.151909
CF9	−0.0168	−0.0890017	0.796502

nation of the original variables. This transformation retains all the information found in the original regressor variables. These indices, known as principal components (PCs) have several useful properties: (i) they are uncorrelated with one another, and (ii) they are ordered so that the first PC explains the largest proportion of the variation of the original data, the second PC explains the next largest proportion, and so on. When the original variables are highly correlated, the variance of many of the later PCs will be so small that they can be ignored. Consequently, the number of regressor variables in the model can be reduced with little loss in model goodness-of-fit. The same reduction in dimensionality also removes the collinearity between the regressors and would lead to more stable parameter estimation and robust model identification. The coefficients of the PCs retained in the model are said to be more stable and, when the resulting model with the PC variables is transformed back in terms of the original regressor variables, the coefficients are said to offer more realistic insight into how the individual physical variables influence the response variable.

If the principal components can be interpreted in physical terms, then it would have been an even more valuable tool. Unfortunately, this is often not the case. Though it has been shown to be useful in social sciences as a way of finding effective combinations of variables, it has had limited success in the physical and engineering sciences. Draper and Smith (1981) caution that PCA may be of limited usefulness in physical engineering sciences contrary to social sciences where models are generally weak and numerous correlated regressors tend to be included in the model. Reddy and Claridge (1994) conducted synthetic experiments in an effort to evaluate the benefits of PCA against multiple linear regression (MLR) for modeling energy use in buildings, and reached the conclusion that only when the data is poorly explained by the MLR model and when correlation strengths among regressors were high, was there a possible benefit to PCA over MLR; with, however, the caveat that injudicious use of PCA may exacerbate rather than overcome problems associated with multi-collinearity.

10.3.3 Ridge Regression

Another remedy for ill-conditioned data is to use *ridge regression* (see for example, Chatterjee and Price 1991; Draper and Smith 1981). This method results in more stable estimates than those of OLS in the sense that they are more robust, i.e., less affected by slight variations in the estimation data. There are several alternative ways of defining and computing ridge estimates; the ridge trace is perhaps the most intuitive. It is best understood in the context of a graphical representation which unifies the problems of detection and estimation. Since the determinant of (**X'X**) is close to singular, the approach involves introducing a known amount of "noise" via a variable k, *leading to the determinant becoming less sensitive to multicollinearity.* With this approach, the parameter vector for OLS is given by:

$$\mathbf{b_{Ridge}} = (\mathbf{X'X} + \mathbf{k.I})^{-1}\mathbf{X'Y} \qquad (10.20)$$

where I is the identity matrix.

Parameter variance is then given by

$$\mathrm{var}(b)_{Ridge} = \sigma^2(\mathbf{X'X} + \mathbf{k.I})^{-1}\mathbf{X'X}(\mathbf{X'X} + \mathbf{k.I})^{-1} \qquad (10.21a)$$

with prediction bands:

$$\mathrm{var}(\hat{y}_0)_{Ridge} = \sigma^2\{1 + \mathbf{X}_0'[(\mathbf{X'X} + \mathbf{k.I})^{-1}\mathbf{X'X}(\mathbf{X'X} + \mathbf{k.I})^{-1}\mathbf{X_0}] \qquad (10.21b)$$

where σ^2 is the mean square error of the residuals.

It must be pointed out that ridge regression should be performed with standardized variables (i.e., the individuals observations subtracted by the mean and divided by the standard deviation) in order to remove large differences in the numerical values of the different regressors.

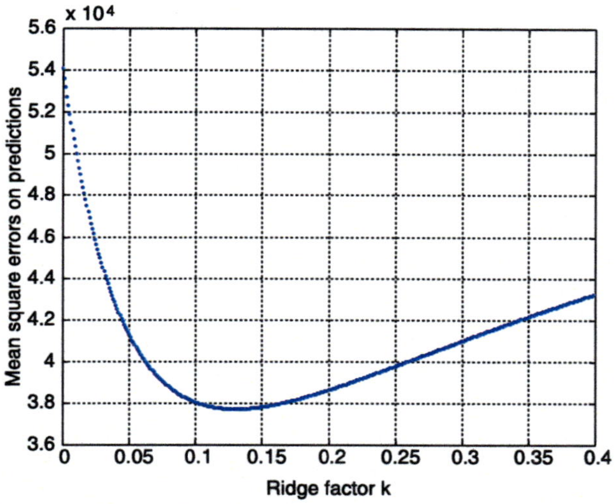

Fig 10.7 The optimal value of the ridge factor k is the value for which the mean square error (MSE) of model predictions is minimum. k=0 corresponds to OLS estimation. In this case, k=0.131 is optimal with MSE=3.78×10^4

One increases the values of the "jiggling factor" k from 0 (the OLS case) to 1.0 to determine the most optimum value of k which yields the least model mean square error (MSE). This is often based on the cross-validation or testing data set (see Sect. 5.2.3d) and not for the training data set from which the model was developed. Usually the value of k is the range 0–0.2. This is illustrated in Fig. 10.7. The ridge estimators are biased but tend to be stable and (hopefully) have smaller variance than OLS estimators despite the bias (see Fig. 10.8). Forecasts of the response variable would tend to be more accurate and the uncertainty bands more realistic. Unfortunately, many practical problems exhibit all the classical signs of multicollinear behavior but, often, applying PCA or ridge analysis does not necessarily improve the prediction accuracy of the model over the standard multi-linear OLS regression (MLR). The case study below illustrates such an instance.

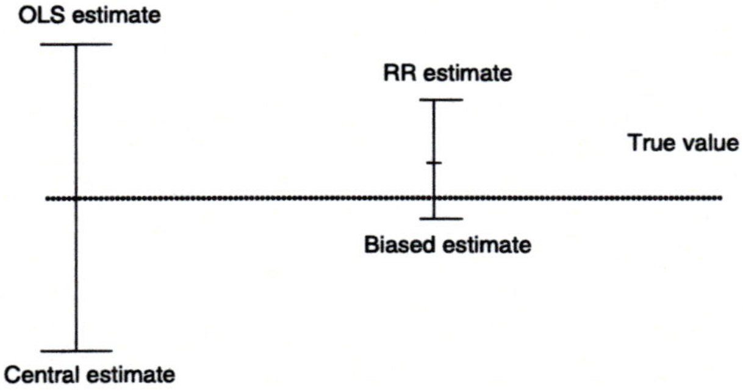

Fig. 10.8 Ridge regression (RR) estimates are biased compared to OLS estimates but the variance of the parameters will (hopefully) be smaller as shown

10.3.4 Chiller Case Study Analysis Involving Collinear Regressors

This section describes a study (Reddy and Andersen 2002) where the benefits of ridge regression compared to OLS are evaluated in terms of model prediction accuracy in the framework of steady state chiller modeling. Hourly field monitored data (consisting of 810 observations) of a centrifugal chiller included four variables: (i) Thermal cooling capacity Q_{ch} in kW; (ii) Compressor electrical power P in kW; (iii) Supply chilled water temperature T_{chi} in K, and (iv) Cooling water inlet temperature T_{cdi} in K.

The data set has to be divided into two sub-sets: (i) a training set, which is meant to compare how different model formulations fit the data, and (ii) a testing (or validating) data set, which is meant to single out the most suitable model in terms of its predictive accuracy. The intent of having training and testing data sets is to avoid over-fitting and obtaining a more accurate indication of the model prediction errors, which is why a model is being identified in the first place (see Sect. 5.3.2d).

Given a data set, there are numerous ways of selecting the training and testing data sets. The simplest is to split the data time-wise (containing the first 550 data points or about 2/3rd of the monitored data) and a testing set (containing the second 260 data points or about 1/3rd of the monitored data). It is advisable to select the training data set such that the range of variation of the individual variables is larger than that of the same variables in the testing data set, and also that the same types of cross-correlations among variables be present in both data sets. This avoids the issue of model extrapolation errors interfering with the model building process. However, if one wishes to specifically compare the various models in terms of their extrapolation accuracy, i.e., their ability to predict beyond the range of their original variation in the data used to identify the model, the training and testing data set can be selected in several ways. An extreme form of data separation is to sort the data by Q_{ch} and select the lower 2/3rd portion of the data for model development and the upper 1/3rd for model evaluation. The results of such an analysis are reported in Reddy and Andersen (2002) but omitted here since the evaluation results were similar (indicative of a robust model).

Pertinent descriptive statistics for both sets are given in Table 10.7. Note that there is relatively little variation in the two temperature variables, while the cooling load and power experience important variations. Further, as stated earlier, the range of variation of the variables in the testing data set are generally within those of the training data set. Another issue is to check the correlations and serial correlations among the variables. This is shown in Table 10.8. Note that the correlation between (P, T_{chi}) and (Q_{ch}, T_{chi}) are somewhat different during the training and testing data sets

Table 10.7 Descriptive statistics for the chiller data

	Training data set (550 data points)				Testing data set (260 data points)			
	P	Q_{ch}	T_{chi}	T_{cdi}	P	Q_{ch}	T_{chi}	T_{cdi}
Mean	222	1,108	285	302	202	1,011	288	302
Std. Dev.	37.8	282	2.43	0.73	14.7	149	2.97	0.77
Max	154	517	282	299	163	630	283	297
Min	340	1,771	292	304	230	1, 241	293	303

Table 10.8 Correlation matrices for the training and testing data sets

	Training data set (550 data points)			
	P	Q_{ch}	T_{chi}	T_{cdi}
P	–	0.98	0.52	0.57
Q_{ch}	0.99	–	0.62	0.59
T_{chi}	0.86	0.91	–	0.37
T_{cdi}	0.54	0.61	0.54	–
	Testing data set (260 data points)			

(the correlation seems to have increased from around 0.5 to about 0.9); one has to contend with this difference. In terms of collinearity, note that the correlation coefficient between (P, Q_{ch}) is very important (0.98) while the others are about 0.6 or less, which is not negligible but not very significant either. A scatter plot of the thermal load against COP is shown in Fig. 10.9.

Two different steady-state chiller thermal performance models have been evaluated. These are the black-box model (referred to as MLR) and the grey-box model referred to as GN model. These models and their functional forms prior to regression are described in Pr. 5.13 of Chap. 5. Note that

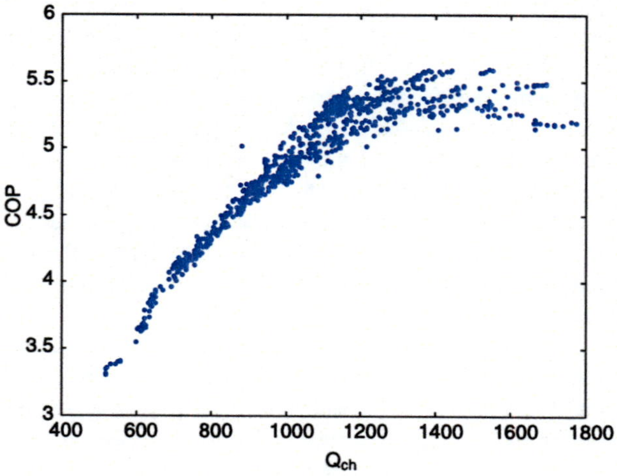

Fig. 10.9 Scatter plot of chiller COP against thermal cooling load Q_{ch} in kW. High values of Q_{ch} are often encountered during hot weather conditions at which times condenser water temperatures tend to be higher, which reduce chiller COP. This effect partly explains the leveling off and greater scatter in the COP at higher values of thermal cooling load

Table 10.9 Correlation coefficient matrix for regressors of the GN model during training

	Training data set (550 data points)			
	Y	X_1	X_2	X_3
Y	1.0	0.93	0.82	−0.79
X_a		1.0	0.96	−0.95
X_2			1.0	−0.92
X_3				1.0

while the MLR model uses the basic measurement variables, the GN model uses transformed variables (x_1, x_2, x_3). It must be pointed out that ridge regression should be performed with standardized variables in order to remove large differences in the numerical values of the different regressors.

A look at the descriptive statistics for the regressors in the physical model provides an immediate indication as to whether the data set may be ill-conditioned or not. One notes that the magnitudes of the three variables differ by several orders, but since ridge regression uses standardized variables, this is not an issue. The estimated correlation matrix for the regressors in the GN model is given in Table 10.9. It is clear that there is evidence of strong multi-collinearity between the regressors.

From the statistical analysis it is found that the GN physical model fits the field—monitored data well except perhaps at the very high end (see Fig. 10.10a). The adjusted R^2=99.1%, and the coefficient of variation (CV)=1.45%. An analysis of variance also shows that there is no statistical evidence to reduce the model order. The model residuals have constant variance, as indicated by the studentized residual plots versus time (row number of data) and by regressor variable (Fig. 10.10b, c). Further, since most of them are contained within bounds of 2.0, one need not be unduly concerned with influence points and outlier points though a few can be detected.

The regressors are strongly correlated (Table 10.9). The condition number of the matrix is close to 81 suggesting that the data is ill-conditioning, since this value is larger than the threshold value of 30 stated earlier. In an effort to overcome this adversity, ridge regression is performed with the ridge factor k varied from 0 (which is the OLS case) to k=0.2. The ridge trace for individual parameters is shown in Fig. 10.10d and the VIF factors are shown in Table 10.10.

As stated earlier, OLS estimates will be unbiased, while ridge estimation will be biased but the estimation is likely to be more efficient. The internal and external predictive accuracy of both estimation methods using the training and testing data sets respectively were evaluated. It is clear from Table 10.10 that, during model training, CV values increase as k is increased, while the VIF values of the parameters decrease. Hence, one cannot draw any inferences about whether ridge regression is better than OLS and if so, which value of ridge parameter k is optimal. Adopting the rule-of-thumb

Fig. 10.10 Analysis of chiller data using the GN model (gray-box model). **a.** x-y plot **b.** Residual plot versus time sequence in which data was collected **c.** Residual plot versus predicted response **d.** Variance inflation factors for the regressor variables

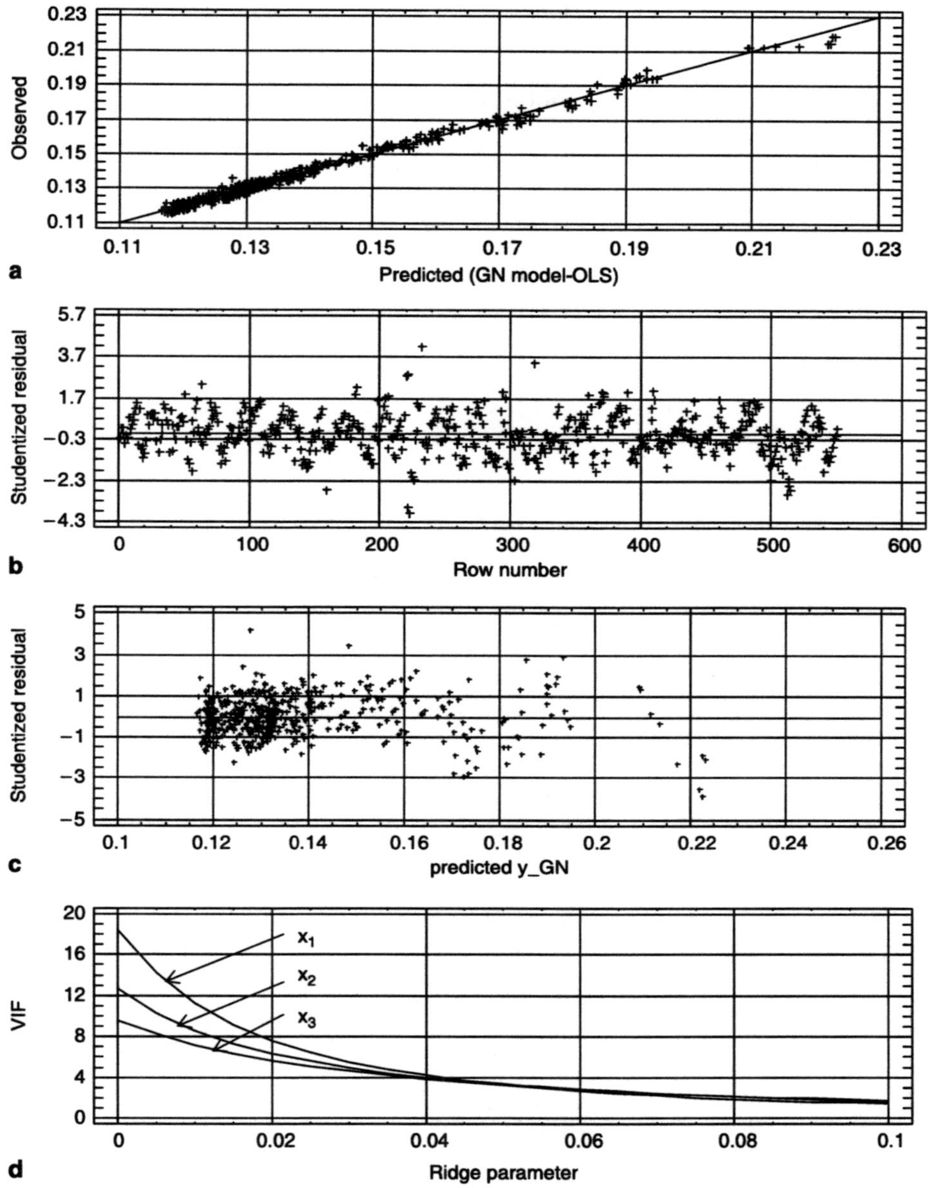

Table 10.10 Results of ridge regression applied to the GN model. Despite strong collinearity among the regressors, the OLS model has lower CV and NMBE than those from ridge regression

	Training data set					Testing data set	
	Adj-R^2	CV(%)	VIF(X_1)	VIF(X_2)	VIF(X_3)	CV(%)	NMBE(%)
k=0.0[a]	99.1	1.45	18.43	12.69	9.61	1.63	−1.01
k=0.02	89.8	3.02	7.67	6.39	5.71	2.86	1.91
k=0.04	85.1	4.16	4.22	3.99	3.87	4.35	3.18
k=0.06	82.3	4.87	2.69	2.78	2.82	5.20	3.86
k=0.08	80.2	5.32	1.88	2.07	2.17	5.73	4.27

[a] Equivalent to OLS estimation

that the lower bound for the VIF values should be 5 would suggest k=0.02 or 0.04 to be reasonable choices. The CV and normalized mean bias error (NMBE) values of the models for the testing data set are also shown in Table 10.10. For OLS (k=0), these values are 1.63 and −1.01% indicating that the identified OLS model can provide extremely good predictions. Again, both these indices increase as the value of k is increased indicating poorer predictive ability both in variability or precision and in bias. Hence, in this case, even though the data is ill-conditioned, the OLS identification is the better estimation approach if the chiller model identified is to be used for predictions only.

The MLR model is a black-box model with linear second order terms in the three regressor variables (Eq. 5.75). Some or many of the variables may be statistically insignificant, and so a step-wise OLS regression was performed. Both forward selection and backward elimination techniques were evaluated using the F-ratio of 4 as the cut-off. While the backward elimination retained seven terms (excluding the constant), forward selection only retained three. The Adjusted—R^2 and CV statistics were almost identical and so the forward selection model is retained for parsimony.

The final MLR model contains the three following variables: $[Q_{ch}{}^2, T_{cdi}{}^*Q_{chi}, T_{chi}{}^*Q_{ch}]$. The fit is again excellent with adjusted R^2=99.2% and CV=0.95% (very slightly better than those of the GN model). An analysis of variance also shows that there is no statistical evidence to reduce the model order, while the residuals are well-behaved. Unfortunately, the regressors are very strongly correlated (the correlation coefficients for all three variables are 0.99) and indicates ill-conditioned data. The condition number of the matrix is very large (C_d=76), again suggesting ill-conditioning.

The ridge regression results for the MLR model are shown in Table 10.11. How the CV values, during model training, increase as the ridge factor k is increased from 0 to 0.02 can be noted along with the VIF of the regressors. The CV and NMBE values of the models for the testing data set are also assembled. Again, despite the strong ill-conditioning of the OLS model, the OLS model (with k=0) turns out to be the best predictive model, with internal and external CV values of 1.13 and 0.62% respectively, which are excellent.

10.3.5 Stagewise Regression

Another approach that offers the promise of identifying sound parameters of a linear model whose regressors are correlated is *stagewise regression* (Draper and Smith 1981). It was extensively used prior to the advent of computers. The approach, though limited in use nowadays, is still a useful technique to know, and can provide some insights into model structure. Consider the following multivariate linear model with collinear regressors:

$$y = \beta_0 + \beta_1 x_1 \cdots + \beta_p x_p \qquad (10.22)$$

The basic idea is to perform a simple regression *with one regressor at a time* with the order in which they are selected depending on their correlation strength with the response variable. This strength is re-evaluated at each step. The algorithm describing its overall methodology consists of the following steps:

(a) Compute the correlation coefficients of the response variable y against each of the regressors, and identify the strongest one, say x_i.
(b) Perform a simple OLS regression of y vs x_i, and compute the model residuals u. This becomes the new response variable.
(c) From the remaining regressor variables, identify the one most strongly correlated with the new response variable. If this is represented by x_j, then regress u vs x_j and recompute the second stage model residuals, which become the new response variable.
(d) Repeat this process for as many remaining regressor variables as are significant.
(e) The final model is found by rearranging the terms of the final expression into the standard regression model form, i.e., with y on the left-hand side and the significant regressors on the right-hand side.

The basic difference between this method and the forward stepwise multiple regression method is that in the former method the selection of which regressor to include in the second stage depends on its correlation strength with the *residuals* of the model defined in the first stage. Stepwise regression selection is based on the strength of the regres-

Table 10.11 Results of ridge regression applied to the MLR model. Despite strong collinearity among the regressors, the OLS model has lower CV and NMBE than those from ridge regression

	Training data set					Testing data set	
	Adj-R^2	CV(%)	VIF(X_1)	VIF(X_2)	VIF(X_3)	CV(%)	NMBE(%)
k=0.0[a]	99.2	0.95	49.0	984.8	971.3	1.13	0.62
k=0.005	93.1	1.81	26.3	15.0	15.2	2.86	2.38
k=0.01	89.9	2.39	16.4	6.43	6.57	3.52	2.89
k=0.015	87.8	2.80	11.2	3.89	4.00	3.95	3.21
k=0.02	86.3	3.10	8.17	2.69	2.75	4.25	3.43

[a] Equivalent to OLS estimation

sor with the *response variable.* Stagewise regression is said to be less precise, i.e., larger RMSE; however, it minimizes, if not eliminates, the effect of correlation among variables. Simplistically interpreting the individual parameters of the final model as the relative influence which these have on the response variable is misleading since the regressors which enter the model earlier pick up more than their due share at the expense of those which enter later. Though Draper and Smith (1981) do not recommend its use for typical problems since the true OLS is said to provide better overall prediction, this approach, under certain circumstance, is said to yield realistic and physically-meaningful estimates of the individual parameters.

There are several studies in the published literature which use the basic concept behind stagewise regression without explicitly identifying their approach as such. For example, it has been used in the framework of controlled experiments where the influence of certain regressors are blocked by either conducting tests during certain times when the physical regressors have no influence (such as doing tests at night in order to eliminate solar effects). This allows a partial model to be identified which is gradually expanded to identifying parameters of the full model. For example, Saunders et al. (1994) modeled the dynamic performance of a house as a 2R1C (two resistors and one capacitor) electric network with five physical parameters. A stagewise approach was then adopted to infer these from controlled tests done during one night only. The Primary and Secondary Terms Analysis and Renormalization (PSTAR) method (Subbarao 1988) has a similar objective but is more versatile and accurate. It uses a detailed forward simulation model of the building in order to get realistic estimates of the various thermal flows, and identify the influential ones depending on the specific circumstance. In order to keep the statistical estimation robust, only the most important flows (usually three or four terms) are then corrected by introducing renormalization coefficients so that the predicted and measured thermal performance of the building are as close as possible. Again, an intrusive experimental protocol allows these renormalization parameters to be deduced in stages requiring two nights and one daytime of testing. Thus, the analyst can effectively select periods when the influence of certain variables is small-to-negligible, and identify the model parameters in stages. This allows multi-collinearity effects to be minimized, and the physical interpretation of the model parameters retained. This approach is illustrated below by way of a case study involving parameter estimation of a model for the thermal loads of large commercial buildings.

10.3.6 Case Study of Stagewise Regression Involving Building Energy Loads

The superiority of stagewise regression as compared to OLS can be illustrated by a synthetic case study example (Reddy et al. 1999). Here, the intent was to estimate building and ventilation parameters of large commercial buildings from non-intrusive monitoring of its heating and cooling energy use from which the net load can be inferred. Since the study uses synthetic data (i.e., data generated by a commercial detailed hourly building energy simulation software), one knows the "correct" values of the parameters in advance, which allows one to judge the accuracy of the estimation technique. The procedure involves first deducing a macro-model for the thermal loads of an ideal one-zone building suitable for use with monitored data, and then using a multistage linear regression approach to determine the model coefficients (along with their standard errors) which can be finally translated into estimates of the physical parameters (along with the associated errors). The evaluation was done for two different building geometries and building mass at two different climatic locations (Dallas, TX and Minneapolis, MN) using daily average and/or summed data so as to remove/minimize dynamic effects.

(a) Model Formulation First, a steady state model for the total heat gains (Q_b) was formulated in terms of variables that can be conveniently monitored. Building internal loads consist of lights and receptacle loads and occupant loads. Electricity used by lights and receptacles (q_{LR}) inside a building can be conveniently measured. Heat gains from occupants consisting of both sensible and latent portions and other types of latent loads are not amenable to direct measurement and are, thus, usually estimated. Since the schedule of lights and equipment closely follows that of building occupancy (especially at a daily time scale as presumed in this study), a convenient and logical manner to include the unmonitored sensible loads was to modify q_{LR} by a constant multiplicative correction factor k_s which accounts for the miscellaneous (i.e., unmeasurable) internal sensible loads. Also, a simple manner of treating internal latent loads was to introduce a constant multiplicative factor k_l defined as the ratio of internal latent load to the total internal sensible load ($k_s \cdot q_{LR}$) which appears *only* when outdoor specific humidity w_0 is larger than that of the conditioned space. Assuming the sign convention that energy flows are positive for heat gains and negative for heat losses, the following model was proposed:

$$Q_B = q_{LR} k_s (1 + k_l \delta) A + a'_{sol} + (b_{sol} + UA_s + m_v A c_p) \times (T_O - T_z) + m_v A h_v \delta (w_O - w_z) \quad (10.23)$$

where

- A Conditioned floor area of building
- A_S Surface area of building
- c_p Specific heat of air at constant pressure
- h_v Heat of vaporization of water
- k_l Ratio of internal latent loads to total internal sensible loads of building

k_s Multiplicative factor for converting q_{LR} to total internal sensible loads
m_v Ventilation air flow rate per unit conditioned area
Q_B Building thermal loads
q_{LR} Monitored electricity use per unit area of lights and receptacles inside the building
T_0 Outdoor air dry-bulb temperature
T_z Thermostat set point temperature
U Overall building shell heat loss coefficient
W_0 Specific humidity of outdoor air
W_z Specific humidity of air inside space
δ is an indicator variable which is 1 when $w_0 > w_z$ and 0 otherwise.

The effect of solar loads is linearized with outdoor temperature T_0 and included in the terms a'_{sol} and b_{sol}. The expression for Q_B given by Eq. 10.23 includes six physical parameters: k_s, k_l, UA_S, m_v, T_z and w_z. One could proceed to estimate these parameters in several ways.

(b) One-Step Regression Approach One way to identify these parameters is to directly resort to OLS multiple linear regression provided monitored data of q_{LR}, T_0 and w_0 is available. For such a scheme, it is more appropriate to combine solar loads into the loss coefficient U and rewrite Eq. 10.23 as:

$$Q_B/A = a + b \cdot q_{LR} + c \cdot \delta \cdot q_{LR} + d \cdot T_0 + e \cdot \delta \cdot (w_0 - w_z) \tag{10.24a}$$

where the regression coefficients are:

$$\begin{array}{lll} a = -(UA_S/A + m_v \cdot c_p) \cdot T_z & b = k_s & c = k_s \cdot k_l \\ d = (UA_S/A + m_v \cdot c_p) & e = m_v \cdot h_v & \end{array} \tag{10.24b}$$

Subsequently, the five physical parameters can be inferred from the regression coefficients as:

$$\begin{array}{lll} k_s = b & k_l = c/b & m_v = e/h_v \\ UA_S/A = d - e \cdot c_p/h_v & T_z = -a/d & \end{array} \tag{10.25}$$

The uncertainty associated with these physical parameters can be estimated from classical propagation of errors formulae discussed in Sect. 3.7, and given in Reddy et al. (1999).

The "best" value of building specific humidity w_z could be determined by a search method: select the value of w_z that yields the best goodness-of-fit to the data (i.e., highest R^2 or lowest CV). Since w_z has a more or less well known range of variation, the search is not particularly difficult. Prior studies indicated that the optimal value has a broad minimum in the range of 0.009–0.011 kg/kg. Thus, the choice of w_z was not a critical issue, and one could simply assume w_z=0.01 kg/kg without much error in subsequently estimating other parameters.

(c) Two-Step Regression Approach Earlier studies based on daily data from several buildings in central Texas indicate that for positive values of $(w_0 - w_z)$ the variables (i) q_{LR} and T_0, (ii) q_{LR} and $(w_0 - w_z)$, and (iii) T_0 and $(w_0 - w_z)$ are strongly correlated, and are likely to introduce bias in the estimation of parameters from OLS regression. It is the last set of variables which is usually the primary cause of uncertainty in the parameter estimation process. Two-step regression involves separating the data set into two groups depending on δ being 0 or 1 (with w_z assumed to be 0.01 kg/kg). During a two-month period under conditions of low outdoor humidity, $\delta = 0$, and Eq. 10.24 reduces to

$$Q_B/A = a + b \cdot q_{LR} + d \cdot T_0 \tag{10.26}$$

Since q_{LR} and T_0 are usually poorly correlated under such low outdoor humidity conditions, the coefficients b and d deduced from multiple linear regression are likely to be unbiased. For the remaining year-long data when $\delta = 1$, Eq. 10.24 can be re-written as:

$$Q_B/A = a + (b + c) \cdot q_{LR} + d \cdot T_0 + e \cdot (w_0 - w_z) \tag{10.27}$$

Now, there are two ways of proceeding. One variant is to use Eq. 10.27 as is, and determine coefficients a, (b+c), d and e from multiple regression. The previous values of a and d determined from Eq. 10.26 are rejected, and the parameter b determined from Eq. 10.26 along with a, c, d and e determined from Eq. 10.27 are retained for deducing the physical parameters. This approach, termed 2-step *variant A*, may, however, suffer from the collinearity effects between T_0 and $(w_0 - w_z)$ when Eq. 10.27 is used.

A second variant of the two-step approach, termed 2-step *variant B*, would be to retain both coefficients b and d determined from Eq. 10.26 and use the following modified equation to determine a, c and e from data when $\delta = 1$:

$$Q_B/A - d \cdot T_0 = a + (b + c) \cdot q_{LR} + e \cdot (w_0 - w_z) \tag{10.28}$$

The collinearity effects between q_{LR} and $(w_0 - w_z)$ when $\delta = 1$ are usually small, and this is likely to yield less unbiased parameter estimates than variant A.

(d) Multi-stage Regression Approach The multi-stage approach involves four steps, i.e., four regressions are performed based on Eq. 10.24 as against only two in the two-stage. The various steps involved in the multi-stage regression are shown in Table 10.12. First, (Q_B/A) is regressed against q_{LR} only, using a two-parameter (2-P) regression model to determine coefficient b. Next, the residual $Y_1 = (Q_B/A - b \cdot q_{LR})$ is regressed against T_0 using either a two parameter (2-P) model or a four parameter (4-P) change point model (see Sect. 5.7.2 for explanation of these terms) to determine co-

Table 10.12 Steps involved in the multi-stage regression approach using Eq. 10.24. Days when $\delta = 1$ correspond to days with outdoor humidity higher than that indoors

	Dependent variable	Regressor variables	Type of regression	Parameter identified from Eq. 10.24	Data set used
Step 1	Q_B/A	q_{LR}	2-P	b	Entire data
Step 2	$Y_1 = Q_B/A - b \cdot q_{LR}$	T_0	2-P or 4-P	d	Entire data
Step 3	$Y_2 = Q_B/A - b \cdot q_{LR} - d \cdot T_0$	q_{LR}	2-P	c	Data when $\delta = 1$
Step 4	$Y_2 = Q_B/A - b \cdot q_{LR} - d \cdot T_0$	$(w_0 - w_z)$	2-P or 4-P	e	Data when $\delta = 1$

efficient d. Next, for data when $\delta = 1$, the new residual is regressed against q_{LR} and $(w_0 - w_z)$ in turn to obtain coefficients c and e respectively. Note that this procedure does not allow coefficient a in Eq. 10.24 to be estimated, and so T_z cannot be identified. This, however, is not a serious limitation since the range of variation of T_z is fairly narrow for most commercial buildings. The results of evaluating whether this identification scheme is superior to the other two schemes are presented below.

(e) Evaluation A summary of how accurately the various parameter identification schemes (one-step, two-step variant A, two-step variant B, and the multistage procedures) are able to identify or recover the "true" parameters is shown in Fig. 10.11. Note that simulation runs R1–R5 contain the influence of solar radiation on building loads, while the effect of this variable has been "disabled" in the remaining four computer simulation runs. The "true" values of each of the four parameters are indicated by a solid line, while the estimated parameters along with their standard errors are shown

	R1	R2	R3	R4	R5	R6	R7	R8	R9
Identif. Scheme	1-step	2-A	2-B	Multi.	Multi.	Multi.	Multi.	Multi.	Multi.
Location	Dallas	Dallas	Dallas	Dallas	Dallas	Dallas	Minneap.	Dallas	Minneap.
Building	B1	B1	B1	B1	B1	B1	B1	B2	B2
Solar (?)	Yes	Yes	Yes	Yes	Yes	No	No	No	No

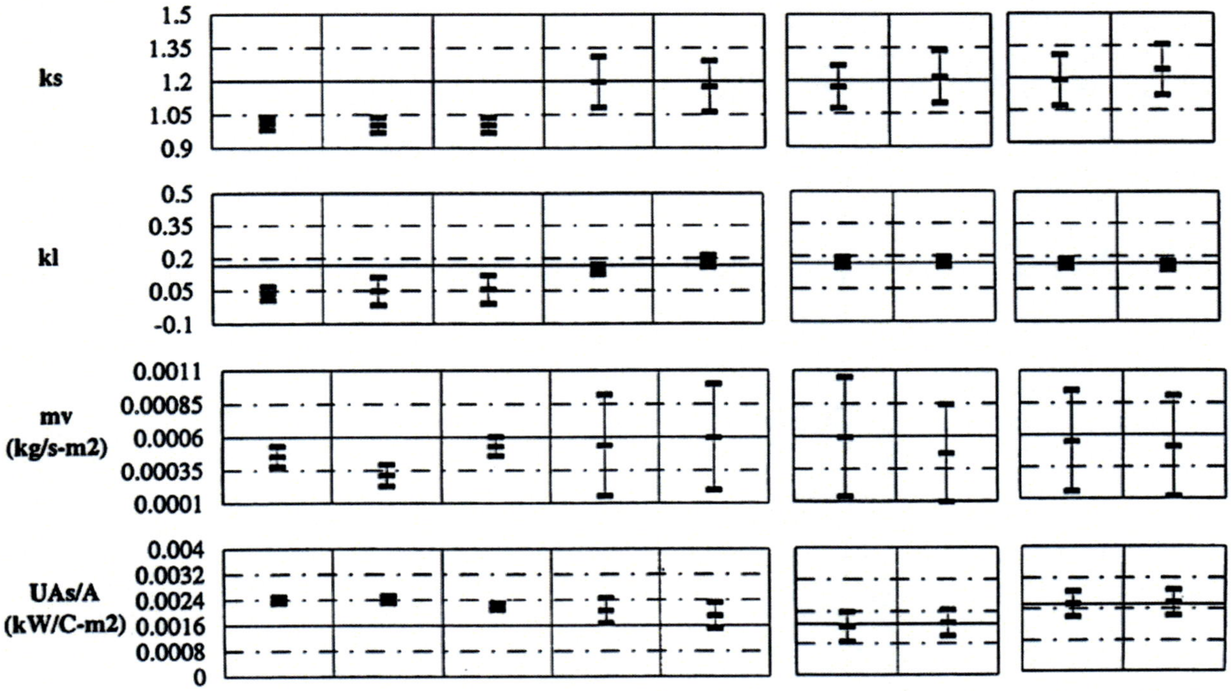

Fig. 10.11 Comparison of how the various estimation schemes (R1–R9) were able to recover the "true" values of the four physical parameters of the model given by Eq. 10.24. The solid lines depict the correct value while the mean values estimated for the various parameters and their 95% uncertainty bands are also shown

as small boxes. It is obvious that parameter identification is very poor for one-step and two-step procedures (R1, R2 and R3) while, except for (UA_S/A), the three other parameters are very accurately identified by the multistep procedure (R4). Also, noteworthy is the fact that the single step regression to daily Q_B values for buildings B1 and B2 at Dallas and Minneapolis are excellent (R^2 in the range of 0.97–0.99). *So this by itself does not assure accurate parameter identification but seems to be a necessary condition for being able to do so.*

The remaining runs (R6–R9) do not include solar effects and in such cases the multistep parameter identification scheme is accurate for both climatic types (Dallas and Minneapolis) and building geometry (B1 and B2). From Fig. 10.11, it is seen that though there is no bias in estimating the parameter m_v, there is larger uncertainty associated with this parameter than with the other four parameters. Finally, note that the bias in identifying (UA_S/A) using the multistep approach when solar is present (R4 and R5) is not really an error: simply that the steady-state overall heat loss coefficient has to be "modified" in order to implicitly account for solar interactions with the building envelope.

A physical explanation as to why the multistage identification scheme is superior to the other schemes (especially the two-step scheme) has to do with the cross-correlation of the regressor variables. Table 10.13 presents the correlation coefficients of the various variables, as well as variables Y_1 and Y_2 (see Table 10.12). Note that for both locations, q_{LR}, because of the finite number of schedules under which the building is operated (5 day-types in this case such as weekday, weekends, holidays, …) is the variable least correlated with Q_B as well as with the other regressor variables. Hence,

Table 10.13 Correlation coefficient matrix of various parameters for the two cities selected at the daily time scale for Runs #6 and #7 (R6 and R7)

	Dallas						
	$Q_{B,1\text{-}zone}$	Y_1	Y_2	q_{LR}	T_0	W_{0z}	$q_{LR}\cdot\delta$
$Q_{B,1\text{-}zone}$		0.85	0.52	0.53	0.88	0.72	0.82
Y_1	0.88		0.78	0.00	0.97	0.80	0.70
Y_2	−0.86	−0.82		−0.27	0.59	0.68	0.44
q_{LR}	0.48	0.01	−0.30		0.11	0.07	0.42
T_0	0.91	0.97	−0.93	0.13		0.75	0.72
W_{0z}	0.57	0.59	−0.40	0.10	0.54		0.66
$q_{LR}\cdot\delta$	0.66	0.60	−0.48	0.27	0.58	0.72	
	Minneapolis						

regressing Q_B with q_{LR} is least likely to result in the regression coefficient of q_{LR} (i.e., b in Eq. 10.24) picking up the influence of other regressor variables, i.e., the bias in the estimation of b is likely to be minimized. Had one adopted a scheme of regressing Q_B with T_0 first, the correlation between Q_B and T_0 as well as between T_0 and ($w_0 - w_z$) for data when $\delta = 1$ would result in coefficient d of Eq. 10.24 being assigned more than its due share of importance, thereby leading to a bias in UA_S value (see R1, R2 and R3 in Fig. 10.11), and, thus, underestimating k_S.

The regression of Q_B versus q_{LR} for R6 is shown in Fig. 10.12a. The second step involves regressing the residual Y_1 versus T_0 because of the very strong correlation between both variables (correlation coefficients of about 0.97–0.99, see Table 10.13). Equally good results were obtained by using step 3 and step 4 (see Table 10.12) in any order. Step 3 (see Fig. 10.12c) allows identi-

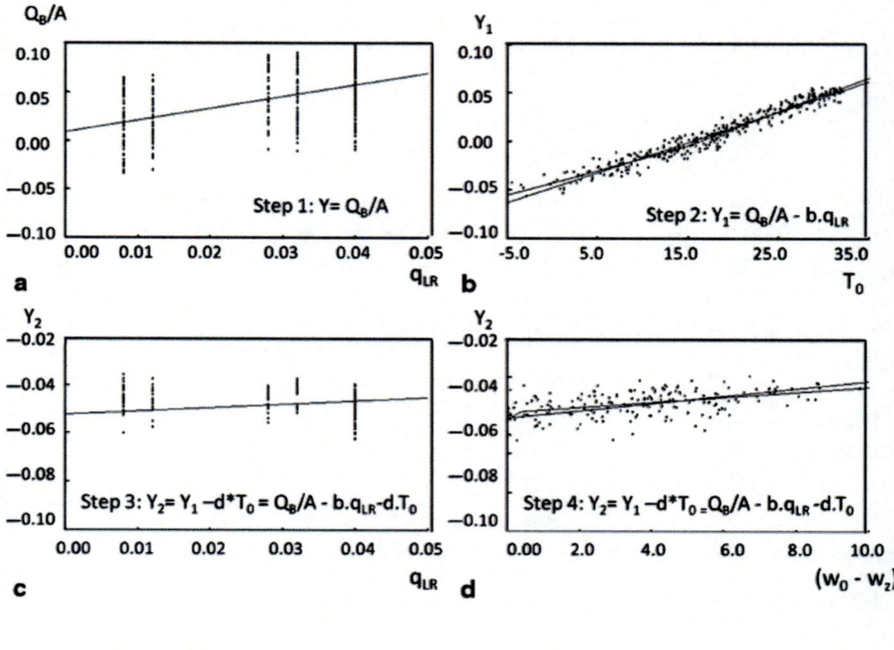

Fig. 10.12 Different steps in the stagewise regression approach to estimate the four model parameters "b, c, d, e" following Eq. 10.24 as described in Table 10.12. **a.** Estimation of parameter b **b.** Estimation of parameter d **c.** Estimation of parameter c **d.** Estimation of parameter e

fication of the regression coefficient c in Eq. 10.24 representing the building internal latent load, while step 4 (i.e., coefficient e of Eq. 10.24) identifies the corresponding regression coefficient associated with outdoor humidity (Fig. 10.12d). In conclusion, this case study illustrates how a multistage identification scheme has the potential to yield accurate parameter estimates by removing much of the bias introduced in multiple linear regression approach with correlated regressor variables.

10.3.7 Other Methods

Factor analysis (Manly 2005) is similar to PCA in that a reduction in dimensionality is sought by removing the redundancy from a set of correlated regressors. However, the difference in this approach is that each of the original regressors is now reformulated in terms of a small number of common factors which impact all of the variables, and a set of errors or specific factors which affect only a single *X* variable. Thus, the regressor is modeled as: $X_i = a_{i1}F_1 + a_{i2}F_2 + a_{i3}F_3 + \ldots + \varepsilon_i$ where a_{ii} are called the factor loadings, F_j is the value of the jth common factor, and ε_i is the part of the test result specific to the ith variable. Factor rotation can be orthogonal or oblique to yield correlated factors. One can use PCA and set an initial estimate for the number of factors (equal to the number of eigenvalues which are greater than one), and drop factors which exhibit little or no explicative power. Thus, while PCA is not based on a model, factor analysis presumes a model where the data is made up of common factors. Factor analysis is frequently used for identifying hidden or latent trends or structure in the data whose effects cannot be directly measured. Like PCA, several authors are skeptical of factor analysis since it is somewhat subjective; however, others point out its descriptive benefit as a means of understanding the causal structure of multivariate data.

Canonical correlation analysis is an extension of multiple linear regression (MLR) for systems which have several response variables (Y) and a vector of Y is to be determined. Both the regressor set (**X**) and the response set (**Y**) are first standardized, and then represented by weighted linear combinations U and V respectively, which are finally regressed against each other (akin to MLR regression) in order to yield canonical weights (or derived model parameters) which can be ranked. These canonical weights can be interpreted as beta coefficients (see Sect. 5.4.5) in that they yield insights into the relative contributions of the individual derived variables. Thus, the approach allows one to understand the relationship between and within two sets of variables X and Y. However, in many systems, the response variables Y are not all "equal" in importance, some may be deemed to be more influential than others based on physical insights of system behavior. This relative physical importance is not retained during the rotation since it is based purely on statistical criteria. Thus, this approach is said to be more relevant to the social and softer sciences than in engineering and the hard sciences (though it is widely used in hydro-climatology).

PCA and factor analyses are basically multivariate data analysis tools which involve analyzing the **X'X** matrix only; application to regression model building is secondary, if at all. Canonical regression uses both **X'X and Y'Y,** but still the regression is done only after the initial transformations are completed. A more versatile and flexible model identification approach which considers the covariance structure between predictor and response variable during model identification is called *partial least squares* (PLS) regression. It uses the cross-product matrix **Y'XX'Y** to identify the multivariate model. It has been used in instances when there are fewer observations than predictor variables, and further is useful for exploratory analysis and for outlier detection. PLS has found applications in numerous disciplines where a large number of predictors are used, such as in economics, medicine, chemistry, psychology, ... Note that PLS is not really meant to understand the underlying structure of multivariate data (as do PCA and factor analysis), but as an accurate tool for predicting system response.

10.4 Non-OLS Parameter Estimation Methods

10.4.1 General Overview

The general parameter estimation problem is discussed below, and some of the more commonly used methods are pointed out. The approach adopted in OLS was to minimize an objective function (also referred to as the *loss function*) expressed as the sum of the squared residuals (given by Eq. 5.3). One was able to derive closed form solutions for the model parameters and the uncertainties as well as the standard errors of regression (for mean response) and those for prediction (for individual response) under certain simplifying assumptions as to how noise corrupts measured system performance. Such closed form solutions cannot be obtained for many situations where the function to be minimized has to be framed differently, and these require the adoption of search methods. Thus, *parameter estimation problems are, in essence, optimization problems where the objective function is framed in accordance with what one knows about the errors in the measurements and in the model structure.*

A model can be considered to be either linear or non-linear in their parameters. The latter situation is covered in Sect. 10.5, while here the discussion is limited to linear estimation problems. Figure 10.13 is a sketch depicting how errors influence the parameter estimation process. Errors

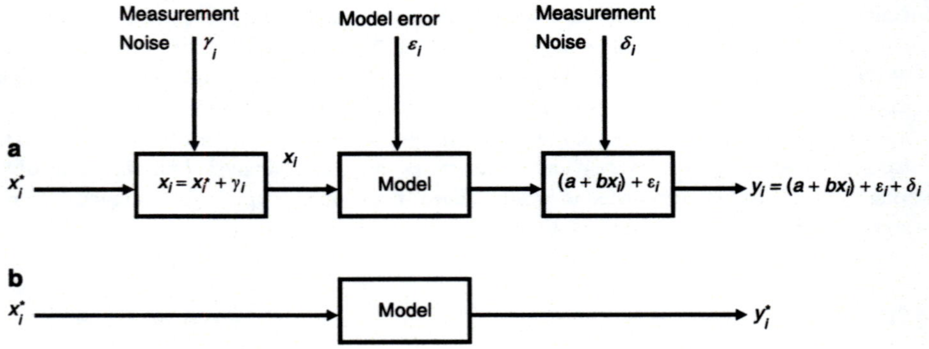

Fig. 10.13 **a.** Sketch and nomenclature to explain the different types of parameter estimation problems for the simple case of one exploratory variable (x) and one response variable (y). Noise and errors are assumed to be additive. **b.** The idealized case without any errors or noise

which can corrupt the process can be viewed as either additive, multiplicative or mixed. For the simplest case, namely the additive error situation, such errors can appear as measurement error (γ) on the regressor/exploratory variable, on the response variable (δ), and also on the postulated model (ε). Note from Fig. 10.13, that x_i^* and y_i^* are the true values of the variables while x_i and y_i are the measured values at observation i. Even when errors are assumed additive, one can distinguish between three broad types of situations:

(a) Measurement errors (γ_i, δ_i) and model error (ε_i) may be: (i) Unbiased or biased, (ii) normal or non-normally distributed along different vertical slices of x_i values (see Fig. 5.3), (iii) variance may be zero, uniform, non-uniform over the range of variation of x;
(b) Covariance effects may, or may not, exist between the errors and the regressor variable, i.e. between $(x, \delta, \gamma, \varepsilon)$;
(c) Autocorrelation or serial correlation over different temporal lags may, or may not, exist between the errors and the regressor variables, i.e., between $(x, \delta, \gamma, \varepsilon)$.

The reader is urged to refer back to Sect. 5.5.1, where these conditions as applicable to OLS were stated. The OLS situation strictly applies when there is no error in x i.e., $\gamma = 0$, when $\delta \approx N(0, \sigma_\delta^2)$ i.e., unbiased, normally distributed with constant variance, when $\text{cov}(x_i, \delta_i) = 0$ and when $\text{cov}(\delta_i, \delta_{i+1}) = 0$. Notice that it is impossible to separate the effects of δ_i from ε_i in the OLS case, and so the combined effect is to increase the error variance of the following model which will be reflected in the RMSE value:

$$y = a + bx + (\varepsilon + \delta) \tag{10.29}$$

Situations when the error in the x variable, i.e. $\gamma \approx N(0, \sigma_\gamma^2)$ is non-negligible would need the error in variable (EIV) approach presented in Sect. 10.4.2. For such cases as when δ is a known function, maximum likelihood estimation (MLE) is relevant and is discussed in Sect. 10.4.3.

10.4.2 Error in Variables (EIV) and Corrected Least Squares

Parameter estimation using OLS is based on the premise that the regressors are known without error $(\gamma = 0)$, or that their errors are very small compared to those of the response variable. There are situations when the above conditions are invalid (for example, when the variable x is a function of several basic measurements, each with their own measurement errors), and this is when the error in variable (EIV) approach is appropriate, whereas OLS results in biased estimates. If the error variances of the measurement errors are known, the bias of the OLS estimator can be removed and a consistent estimator, called *Corrected Least Squares* (CLS) can be derived. This is illustrated in Fig. 10.14, taken from Andersen and Reddy (2002), which shows how the model parameter estimation using OLS gradually increases in bias as more noise is introduced in the x variable, while no such bias is seen for CLS estimates. However, note that the 95% uncertainty bands by both methods are about the same.

CLS accounts for both the uncertainty in the regressor variables as well as that in the dependent variable by minimizing the distance given by the ratio of these two uncertainties. Let us consider the case of simple linear regression model: $y = \alpha + \beta x$, and assume that the errors in the dependent variables and the errors in the independent variables are uncorrelated. The basis of the minimization scheme is described graphically by Fig. 10.15. Let us define a parameter representing the relative weights of the variance of measurements of x and y as:

$$\lambda = \frac{\text{var}(\gamma)}{\text{var}(\delta)} \tag{10.30}$$

The loss function assumes the form (Mandel 1964):

$$S = \sum_i [(x_i - \hat{x}_i)^2 + (y_i - \hat{y}_i)^2 \lambda] \tag{10.31}$$

subject to the condition that $\hat{y}_i = a + b\hat{x}_i$ and $(\hat{y}_i, \hat{x}_i)$ are the estimates of (y_i, x_i). Omitting the derivation, the CLS estimates turn out to be:

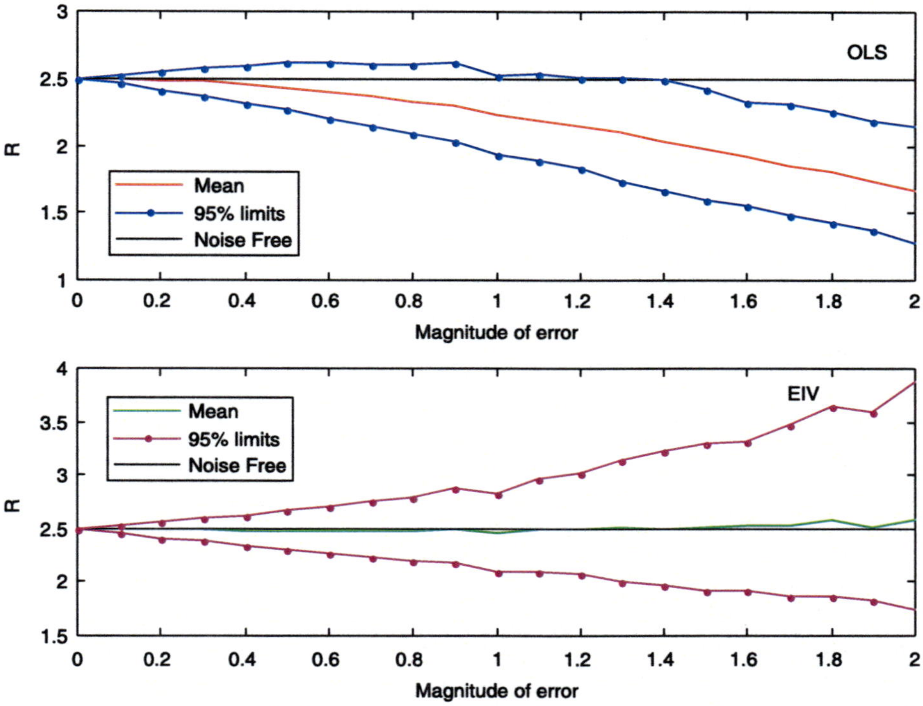

Fig. 10.14 Figure depicting how a physical parameter in a model (in this case the heat exchanger thermal resistance R of a chiller) estimated using OLS becomes gradually more biased as noise is introduced in the x variable. No such bias is seen when EIV estimation is adopted. The uncertainty bands by both methods are about the same as the magnitude of the error is increased. (From Andersen and Reddy 2002)

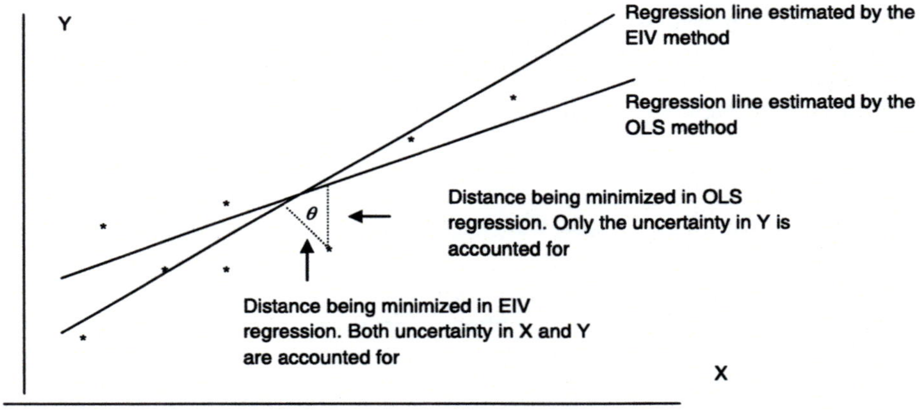

Fig. 10.15 Differences in both slope and the intercept of a linear model when the parameters are estimated under OLS or EIV. Points shown as "*" denote data points and the solid lines are the estimates of the two models. The dotted lines, which differ by angle θ indicate the distances whose squared sum is being minimized in both approaches. (From Andersen and Reddy 2002)

$$b = \frac{(\lambda S_{yy} - S_{xx}) + [(\lambda S_{yy} - S_{xx})^2 + 4\lambda S_{xy}^2]^{1/2}}{2\lambda S_{xy}} \quad (10.31a)$$

and

$$a = \bar{y} - b \cdot \bar{x} \quad (10.32b)$$

where S_{xx}, S_{yy} and S_{xy} are defined by Eq. 5.6.

The extension of the above equation to the multivariate case is given by Fuller (1987):

$$b_{CLS} = (X'X - S^2{}_{xx})^{-1}(X'Y - S^2{}_{xy}) \quad (10.32c)$$

where $S^2{}_{xx}$ is a $p \times p$ matrix with the covariance of the measurement errors and $S^2{}_{xy}$ is a $p \times 1$ vector with the covariance between the regressor variables and the dependent variable (given by Eq. 5.6).

A simple conceptual explanation is that Eq. 10.31 performs on the estimator matrix an effect essentially the opposite of what ridge regression does. While ridge regression "jiggles" or randomly enhances the dispersion in the numerical values of the X variables in order to reduce the adverse effect of multi-collinearity on the estimated parameter bias, CLS tightens the variation in an attempt to reduce the effect of random error on the x variables. For the simple regression model, the EIV approach recommends that minimization or errors be done following angle θ (see Fig. 10.15) given by:

$$\tan(\theta) = \frac{b}{(s_y/s_x)} \quad (10.33)$$

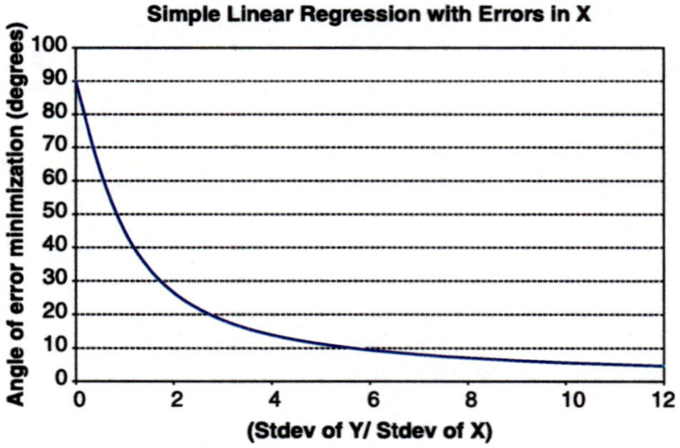

Fig. 10.16 Plot of eq. (10.33) with b = 1

where s is the standard deviation.

For the case of b = 1, one gets a curve such as Fig. 10.16. Note that when the ratio of the standard deviations is less than about 5, the angle theta varies little and is about 10°.

As a rough rule of thumb, it is advocated that if the measurement uncertainty of the x variable characterized by the standard deviation is much less, (say, less than 1/5th) than that in the response variable, then there is little benefit in applying EIV regression as compared to OLS. The 1/5th rule has been suggested for the simple regression case, and should not be used for multi-regression with correlated regressors. The interested reader can refer to Beck and Arnold (1977) and to Fuller (1987) for more in-depth treatment of the EIV approach.

10.4.3 Maximum Likelihood Estimation (MLE)

The various estimation methods discussed earlier, to deduce point estimates of a sample or model parameters from a data set of observations, are based on moments of the data (the first moment is the mean, the second is the variance, and so on); hence, this approach is referred to as the Method of Moments Estimation (MME). Maximum Likelihood Estimation (MLE) is another approach, and is generally superior to MME since it can handle any type of error distribution, while MME is limited to normally distributed errors (Pindyck and Rubinfeld 1981; Devore and Farnum 2005). MLE allows generation of estimators of unknown parameters that are generally more efficient and consistent than MME, though sometimes estimates can be biased. In many situations, the assumption of normally distributed errors is reasonable, and in such cases, MLE and MME give identical results. Thus, in that regard, MME can be viewed as a special (but important) case of MLE.

Consider the following simple example meant to illustrate the concept of MLE. Suppose that a shipment of computers is sampled for quality, and that two out of five are found defective. The common sense approach of estimating population proportion of defectives is $\pi = 2/5 = 0.40$. One could use an alternative method by considering a whole range of possible π values. For example, if $\pi = 0.1$, then the probability of s = 2 defectives out of a sample of n = 5 observations would be given by the binomial formula:

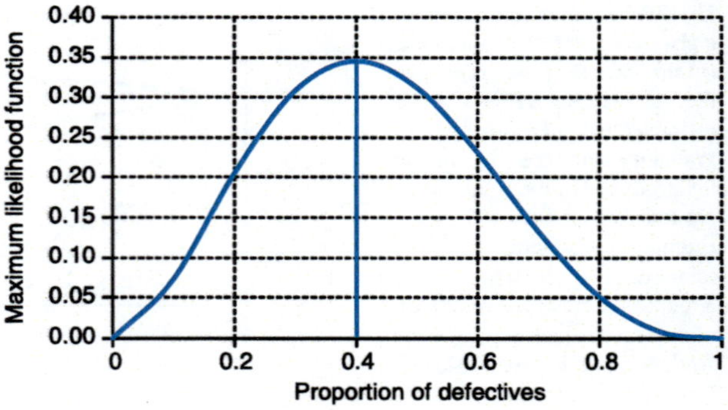

Fig. 10.17 The maximum likelihood function for the case when two computers out of a sample of five are found defective

$$\binom{n}{s}\pi^s(1-\pi)^{n-s} = \binom{5}{2}0.1^2(0.9)^3 = 0.0729$$

In other words, if $\pi = 0.1$, there is only 7.3% probability of getting the actual sample that was observed. However, if $\pi = 0.2$, the chances improve since one gets p = 20.5%. By trying out various values, one can determine the best value of π. How the maximum likelihood function varies for different values of π is shown in Fig. 10.17 from which one finds that the most likely value is 0.4, the same as the common sense approach. Thus, the MLE approach is to simply determine the value of π which maximizes the likelihood function given by the above binomial formula. In other words, the MLE is the population value that is more likely than any other value to generate the sample which was actually observed, or which maximizes the likelihood of the observed sample.

The above approach can be generalized as follows. Suppose a sample $(x_1, x_2, \ldots x_n)$ of independent observations is drawn from a population with probability function $p(x_i/\theta)$ where θ is the unknown population parameter to be estimated. If the sample is random, then the joint probability function for the whole sample is:

$$p(x_1, x_2, \ldots x_n/\theta) = p(x_1/\theta).p(x_2/\theta)\cdots p(x_n/\theta) \quad (10.34)$$

The objective is now to estimate the most likely value of θ among all its possible values which maximizes the above probability function. The likelihood function is thus:

$$L(\theta/x_1, \ldots x_n) = \prod_{i=1}^{n} p(x_i/\theta) \quad (10.35a)$$

where $\prod$ denotes the product of n factors.

The parameter θ is easily determined by taking natural logarithms, in which case,

$$\ln(L(\theta)) = \sum_{i=1}^{n} \ln p(x_i/\theta) \qquad (10.35b)$$

Thus, one could determine the MLE of θ by performing a least-squares regression in case the probability function is known or assumed. Say, one wishes to estimate the two regression parameters β_0 and β_1 of a simple linear model *assuming the error distribution to be normal* with variance σ (Sect. 2.4.3a), the probability distribution of the residuals would be given by:

$$p(y_i) = \frac{1}{(2\pi\sigma^2)^{1/2}} \exp\left[-\frac{1}{2\sigma^2}(y_i - \beta_0 - \beta_1 x_i)^2\right] \qquad (10.36)$$

where (x_i, y_i) are the individual sample observations. The maximum likelihood function is then:

$$L(y_1, y_2, \ldots, y_n, \beta_0, \beta_1, \sigma^2) = p(y_1) \cdot p(y_2) \cdots p(y_n)$$
$$= \prod_{i=1}^{n} \frac{1}{(2\pi\sigma^2)^{1/2}} \exp\left[-\frac{1}{2\sigma^2}(y_i - \beta_0 - \beta_1 x_i)^2\right] \qquad (10.37)$$

The three unknown parameters $(\beta_0, \beta_1, \sigma)$ can be determined either analytically (by setting the partial derivatives to zero) or numerically which can be done by most statistical software programs. For this case, it can be shown that MLE estimates and OLS estimates of β_0 and β_1 are identical, while those for σ^2 is biased (though consistent).

The advantages of MLE go beyond its obvious intuitive appeal:

(a) though biased for small samples, the bias reduces as the sample size increases,
(b) where MLE is not the same as MME, the former is generally superior in terms of yielding minimum variance,
(c) MLE is very straightforward and can be easily solved by computers,
(d) in addition to providing estimates, MLE is useful to show the range of plausible values for the parameters, and also for deducing confidence limits.

The main drawback is that MLE may lack robustness in dealing with a population of unknown shape, i.e., it cannot be used when one has no knowledge of the underlying error distribution. In such cases, one can evaluate the goodness of fit of different probability distributions using the Chi-square criterion (see Sect. 4.2.6), identify the best candidates and pick one based on some prior physical insights. The computation is easily done on a computer, but the final selection is to some extent at the discretion of the analyst. MLE is often used to estimate parameters appearing in probability distribution functions when sample data is available. The following examples illustrate the approach.

Table 10.14 Data table for Example 10.4.1

2,100	2,107	2,128	2,138	2,167	2,374
2,412	2,435	2,438	2,456	2,596	2,692
2,738	2,985	2,996	3,369		

Example 10.4.1: *MLE for exponential distribution*
The lifetime of several products and appliances can be described by the exponential distribution (see Sect. 2.4.3e) given by the following one parameter model:

$$E(x;\lambda) = \lambda \cdot e^{-\lambda x} \quad \text{if } x > 0$$
$$= 0 \quad \text{otherwise}$$

Sixteen appliances have been tested and operating life data in hours are assembled in Table 10.14. The parameter λ is to be estimated using MLE.

The likelihood function

$$L(\lambda/x_i) = \prod_{i=1}^{n} p(x_i/\lambda) = (\lambda e^{-\lambda x_1})(\lambda e^{-\lambda x_2})(\lambda e^{-\lambda x_3})\ldots$$
$$= \lambda^n e^{-\lambda \sum x_i}$$

Taking logs, one gets $\ln(L(\lambda)) = n\ln(\lambda) - \lambda \sum x_i$

Differentiating with respect to λ and setting it to zero yields:

$$\frac{d\ln(L(\lambda))}{d\lambda} = \frac{n}{\lambda} - \sum x_i = 0, \quad \text{from which} \quad \lambda = \frac{n}{\sum x_i} = \frac{1}{\bar{x}}$$

Thus the MLE estimate of the parameter $\lambda = (2508.188)^{-1} = 0.000399$. ■

Example 10.4.2: *MLE for Weibull distribution*
The observations assembled in Table 10.15 are values of wind speed (in m/s) at a certain location. The Weibull distribution (see Sect. 2.4.3f) with parameters (α, β) is appropriate for modeling wind distributions:

$$p(x) = \frac{\alpha}{\beta^\alpha} x^{\alpha-1} e^{-(x/\beta)^\alpha}$$

Estimate the values of the two parameters using MLE.

As previously, taking the partial derivatives of $\ln(L(\alpha, \beta))$, setting them to zero, and solving for the two equations results in (Devore and Farnum 2005):

Table 10.15 Data table for Example 10.4.2

4.7	5.8	6.5	6.9	7.2	7.4
7.7	7.9	8.0	8.1	8.2	8.4
8.6	8.9	9.1	9.5	10.1	10.4

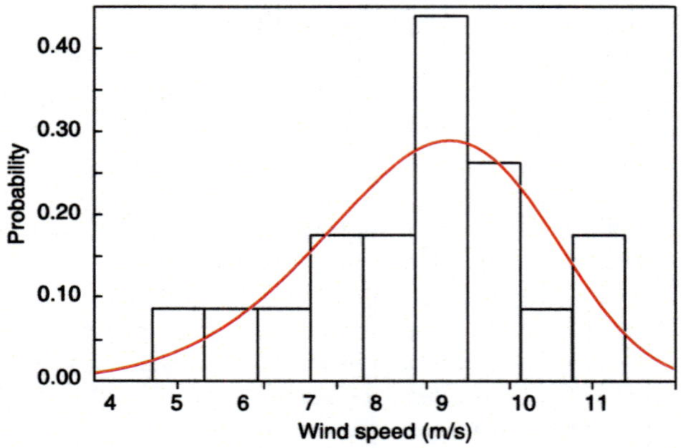

Fig. 10.18 Fit to data of the Weibull distribution with MLE parameter estimation (Example 10.4.2)

$$\alpha = \left[\frac{\sum x_i^\alpha \ln(x_i)}{\sum x_i^\alpha} - \frac{\sum \ln(x_i)}{n} \right]^{-1} \quad \text{and} \quad \beta = \left(\frac{\sum x_i^\alpha}{n} \right)^{1/\alpha}$$

This approach is tedious and error-prone, and so one would tend to use a computer program to perform MLE. Resorting to this option resulted in MLE parameter estimates of ($\alpha = 7.9686, \beta = 0.833$). The goodness-of-fit of the model can be evaluated using the Chi-square distribution (see Sect. 2.4.10) which is found to be 0.222. The resulting plot and the associated histogram of observations are jointly shown in Fig. 10.18. ■

10.4.4 Logistic Functions

The exponential model was described in Sect. 2.4.3e in terms of modeling unrestricted growth. Logistic models are extensions of the exponential model in that they apply to instances where growth is initially unrestricted, but gradually changes to restricted growth as resources get scarcer. They are an important class of equations which appear in several fields for modeling various types of growth—that of populations (humans, animal and biological) as well as energy and material use patterns (Draper and Smith 1981; Masters and Ela 2008). The non-linear shape of these models is captured by an S curve (called a sigmoid function) that reaches a steady-state value (see Fig. 10.19). One can note two phases: (i) an early phase during which the environmental conditions are optimal and allow the rate of growth to be exponential, and (ii) a second phase where the rate of growth is restricted by the amount of growth yet to be achieved, and assumed to be directly proportional to this amount. The following model captures this behavior of population N over time t:

$$\frac{dN}{dt} = rN\left(1 - \frac{N}{k}\right) \quad N(0) = N_0 \qquad (10.38)$$

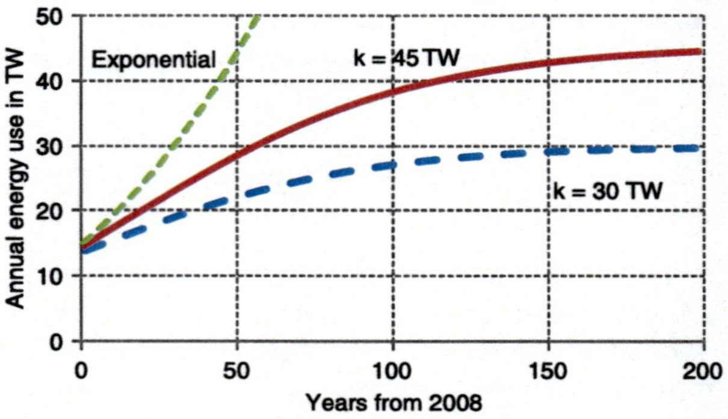

Fig. 10.19 Exponential and logistic growth curves for annual worldwide primary energy use assuming an initial annual growth rate of 2.4% and under two different carrying capacity values k (Example 10.4.3)

where N(0) is the population at time t=0, r is the growth rate constant, and k is the *carrying capacity* of the environment. The factor k can be constant or time varying; an example of the latter is the observed time variant periodic behavior of predator-prey populations in closed ecosystems. The factor [(1−N/k)] is referred to as the *environmental factor*. One way of describing this behavior in the context of biological organisms is to assume that during the early phase, food is used for both growth and sustenance, while at the saturation level it is used for sustenance only since growth has stopped. The solution to Eq. 10.38 is:

$$N(t) = \frac{k}{1 + \exp[-r(t - t^*)]} \qquad (10.39a)$$

where t^* is the time at which N=k/2, and is given by:

$$t^* = \frac{1}{r} \ln\left(\frac{k}{N_0} - 1\right) \qquad (10.40)$$

If the instantaneous growth at t=0 is represented by $R(0)=R_0$, then it can be shown that

$$r = \frac{R_0}{1 - \frac{N_0}{k}}$$

and Eq. 10.39a can be rewritten as:

$$N(t) = \frac{k}{1 + \left(\frac{k}{N_0} - 1\right) \exp(-R_0 t)} \qquad (10.39b)$$

Thus, knowing the quantities N_0 and R_0 at the start, when t=0, allows r, t^* and then N(t) to be determined.

Another useful concept in population biology is the concept of *maximum sustainable yield* of an ecosystem (Masters and Ela 2008). This corresponds to the maximum removal rate which can sustain the existing population, and would occur when $\frac{dN}{dt} = 0$, i.e. from Eq. 10.38, when N=k/2. Thus, if the fish population in a certain pond follows the logistic growth curve when there is no fish harvesting, then the maximum rate of fishing would be achieved when the actual fish population is maintained at half its carrying capacity. Many refinements to the basic logistic growth model have been proposed which allow such factors as fertility and mortality rates, population age composition, migration rates, … to be considered.

Example 10.4.3: *Use of logistic models for predicting growth of worldwide energy use*

The primary energy use in the world in 2008 was about 14 Terra Watts (TW). The annual growth rate is close to 2.4%.

(a) If the energy growth is taken to be exponential (implying unrestricted growth), in how many years would the energy use double?
The exponential model is given by: $Q(t) = Q_0 \exp(R_0 t)$ where R_0 is the growth rate (=0.024) and Q_0(=14 TW) is the annual energy use at the start, i.e. for year 2008. The doubling time would occur when $Q(t)=2Q_0$, and by simple algebra: $t_{doubling} = \frac{0.693}{R_0} = 28.9$ years, or at about year 2037.

(b) If the growth is assumed to be logistic with a carrying capacity of k=45 TW (i.e., the value of annual energy use is likely to stabilize at this value in the far future), determine the annual energy use for year 2037. With t=28.9, and R_0=0.024, Eq. 10.39b yields:

$$Q(t) = \frac{45}{1 + \left(\frac{45}{14} - 1\right) \cdot \exp[-(0.0024) \cdot (28.9)]} = 21.36 \text{ TW}$$

which is (as expected) less than that predicted by the exponential model of 28 TW. The plots in Fig. 10.19, generated using a spreadsheet, illustrate logistic growth curves for two different values of k with the exponential curve also drawn for comparison. The asymptotic behavior of the logistic curves and the fact that the curves start deviating from each other quite early on are noteworthy points. ■

Logistic models have numerous applications other than modeling restricted growth. Some of them are described below:

(a) in marketing and econometric applications for modeling the time rate at which new technologies penetrate the market place (for example, the saturation curves of new household appliances) or for modeling changes in consumer behavior (i.e., propensity to buy) when faced with certain incentives or penalties (see Pindyck and Rubinfeld 1981 for numerous examples);

(b) in neural network modeling (a form of non-linear black-box modeling discussed in Sect. 11.3.3) where logistic curves are used because of their asymptotic behavior at either end. This allow the variability in the regressors to be squashed or clamped within pre-defined limits;

(c) to model probability of occurrence of an event against one or several predictors which could be numerical or categorical. A medical application could entail a model for the probability of a heart attack for a population exposed to one or more risk factors. Another application is to model the spread of disease in epidemiological studies. A third is for dose response modeling meant to identify a non-linear relationship between the dose of a toxic agent to which a population is exposed and the response or risk of infection of the individuals stated as a probability or proportion. This is discussed further below;

(d) to model binary responses. For example, manufactured items may be defective or satisfactory; patients may respond positively or not to a new drug during clinical trials; a person exposed to a toxic agent could be infected or not. Logistic regression could be used to discriminate between the two groups or multiple groups (see Sect. 8.2.4 where classification methods are covered).

As stated in (c) above, logistic functions are widely used to model how humans are affected when exposed to different toxic load (*called dose-response*) in terms of a proportion or a probability P. For example, if a group of people is treated by a drug, not all of them are likely to be responsive. If the experiments can be performed at different dosage levels x, the percentage of responsive people is likely to change. Here the response variable is called the probability of "success" (which actually follows the binomial distribution) and can assume values between 0 and 1, while the regressor variable can assume any numerical value. Such variables can be modeled by the following two parameter model:

$$P(x) = \frac{1}{1 + \exp[-(\beta_0 + \beta_1 x)]} \tag{10.41}$$

where x is the dose to which the group is exposed and (β_0, β_1) are the two parameters to be estimated by MLE. Note that this follows from Eq. 10.39a when k=1.

One defines a new variable, called the odds ratio, as [P/(1−P)]. Then, the log of this ratio is:

$$\pi = \ln \frac{P}{1-P} \tag{10.42}$$

where π is called the *logit function*. Then, simple manipulation of Eqs. 10.39 and 10.40 leads to a linear functional form for the logit model:

$$\pi = \beta_0 + \beta_1 x \tag{10.43}$$

Thus, rather than formulating a model for P as a non-linear function of x, the approach is to model π as a linear function of x. The logit allows the number of failures and successes to be conveniently determined. For example, π=3 would imply that success P is $e^3 = 20$ times more likely than failure. Thus, unit increase in x will result in β_1 change in the logit function.

The above model can be extended to multi-regressors. The *general linear logistic* model allows the combined effect of several variables or doses to be included:

$$\pi = \beta_0 + \beta_1 x_1 + \beta_2 x_2 \cdots + \beta_k x_k \qquad (10.44)$$

Note that the regression variables can be continuous, categorical or mixed. The above models can be used to predict the dosages which induce specific levels of responses. Of particular interest is the dosage which produces a response in 50% of the population (median dose). The following example illustrates these notions.

Example 10.4.4:[2] *Fitting a logistic model to the kill rate of the fruit fly*
A toxicity experiment was conducted to model the kill rate of the common fruit fly when exposed to different levels of nicotine concentration for a pre-specified time interval (recall that concentration times duration of exposure equals the dose). Table 10.16 assembles the experimental results.

The single variate form of the logistic model (Eq. 10.41) is used to estimate the two model parameters using MLE. The regressor variable is the concentration while the response variable is the percent killed or the proportion. A MLE analysis yields the results shown in Table 10.17.

To estimate the dose or concentration which will result in p=0.5 or 50% fatality or success rate is straightforward. From Eq. 10.42, the probit value is $\pi = \ln \frac{P}{1-P} = \ln(1) = 0$. Then, using Eq. 10.43, one gets: $x_{50} = -(\beta_0/\beta_1) = 0.276$ g/100 cc. ■

Table 10.16 Data table for Example 10.4.4

Concentration (g/100 cc)	Number of insects	Number killed	Percent killed
0.10	47	8	17.0
0.15	53	14	26.4
0.20	55	24	43.6
0.30	52	32	61.5
0.50	46	38	82.6
0.70	54	50	92.6
0.95	52	50	96.2

Table 10.17 Results of maximum likelihood estimation (MLE)

Parameters	Estimate	Standard error	Chi-square	p-value
β_0	−1.7361	0.2420	51.4482	<0.0001
β_1	6.2954	0.7422	71.9399	<0.0001

[2] From Walpole et al. (2007) by © permission of Pearson Education.

Example 10.4.5: *Dose response modeling for sarin gas*
Figure 10.20a are the dose response curves for sarin gas for both casualty dose (CD), i.e. which induces an adverse reaction, and lethal dose (LD) which causes death. The CD_{50} and LD_{50} values, which will affect 50% of the population exposed, are specifically indicated because of their importance as stated earlier. Though various functions are equally plausible, the parameter estimation will be done using the logistic curve. Specifically, the LD curve is to be fitted whose data has been read off the plot and assembled in Table 10.18.

In this example, let us assume that all conditions for standard multiple regression are met (such as equal error variance across the range of the dependent variable), and so MLE and OLS will yield identical results. In case they were not, and the statistical package being used does not have the MLE capability, then the weighted least squares method could be framed to yield maximum likelihood estimates.

First, the dependent variable i.e., the fraction of people affected is transformed into its logit equivalent given by Eq. 10.43; the corresponding numerical values are given in the last row of Table 10.18 (note that the entries at either end are left out since log of 0 is undefined). A second order model in the dose level but linear in the parameters of the form: $\pi = b_0 + b_1 x + b_2 x^2$ was found to be more appropriate than the simple model given by Eq. 10.43. The model parameters were statistically significant while the model fits the observed data very well (see Fig. 10.20b) and with Adj R^2=94.3%. The numerical values of the parameters and their 95% CL listed in Table 10.19 provide a visual indication of how well the second order probit model fits the data. ■

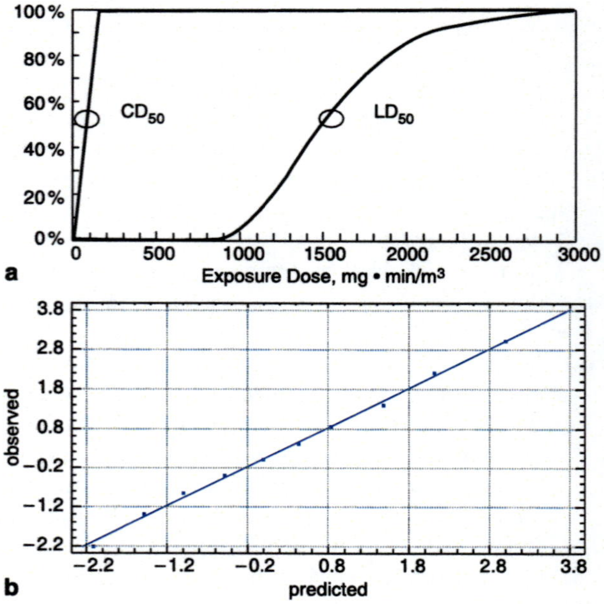

Fig. 10.20 **a** Dose-response curve for sarin gas. (From Kowalski 2002 by permission of McGraw-Hill) **b** Plot depicting the accuracy of the identified second order probit model with observed data

Table 10.18 Data used for model building (for Example 10.4.5)

LD Dose (x)	800	1,100	1, 210	1,300	1,400	1,500	1,600	1,700	1,880	2,100	3,000
Fatalities%	0	10	20	30	40	50	60	70	80	90	100
Logit values	Undefined	−2.197	−1.386	−0.847	−0.405	0	0.405	0.847	1.386	2.197	Undefined

Table 10.19 Estimated model parameters of the second order probit model

Parameter	Estimate	Standard error	95% CL limits	
			Lower limit	Upper limit
b_0	−10.6207	0.334209	−11.411	−9.83045
b_1	0.0096394	0.00035302	0.00880463	0.0104742
b_2	−0.00000169973	8.55464E-8	−0.00000190202	−0.00000149745

Generalized least squares (GLS) (not to be confused with the *general linear models* described in Sect. 5.4.1) is another widely used approach which is more flexible than OLS in the types of non-linear models and error distributions it can accommodate. It can also handle autocorrelated errors and non constant variance. The non-linearity can be overcome by transforming the response variable into another intermediate variable using a *link function*, and then identifying a linear model between the transformed variable and the response variable. Thus, GLS allows ordinary linear regression to be unified with other identification methods (such as logistic regression) while requiring MLE for parameter estimation (discussed in Sect. 10.4.3).

10.5 Non-linear Estimation

Non-linear estimation applies to instances when the model is non-linear in the parameters. This should not be confused with non-linear models wherein the function may be non-linear but the parameters may (or may not) appear in a linear fashion (an example is a polynomial model). The parameter estimation can be done by either least squares or by MLE of a suitably defined loss function. Models non-linear in their parameters are of two types: (i) those which can be made linear by a suitable variable transformation, and (ii) those which are intrinsically nonlinear. The latter is very similar to optimizing a function (discussed in Chap. 7). A short discussion of estimation under both instances is given below.

10.5.1 Models Transformable to Linear in the Parameters

Whenever appropriate it is better to convert models non-linear in the parameters to linear ones as the parameter estimation simplifies considerably. The popularity of linear estimation methods stems from the facts that the computation effort is low because of closed form solutions, the approach is intuitively appealing, and there exists a wide body of statistical knowledge supporting them. However, the transformation results in sound parameter estimation *only when certain conditions are met* regarding errors/noise, which is discussed below.

Table 10.20 gives a short list of useful transformations for nonlinear functions that result in simple linear models, while Fig. 10.21 assembles plots of such functions. For example, an exponential model is used in many fields of science, engineering, biology and numerous other fields to characterize quantities (such as population, radioactive decay, …) which increase or decrease at a rate that is directly proportional to their own magnitude. There are different forms of the exponential model; the one shown is the most general. For example, special cases of the function shown in Table 10.20 are: $y = ae^{bx}$ or $y = 1 - ae^{bx}$. How the numerical value of b affects the function is shown in Fig. 10.21a.

The power function (Fig. 10.21b) is widely used; notice the shape of the curves depending on whether the model coefficient b is positive or negative. Consider the following model:

Table 10.20 Some common transformations to make models non-linear in the parameters into linear ones

Function type	Functional model	Transformation	Transformed linear regression model
Exponential	$y = \exp(a + b_1x_1 + b_2x_2)$	$y^* = \ln y$	$y^* = a + b_1 \cdot x_1 + b_2 \cdot x_2$
Power or multiplicative	$y = a{x_1}^b{x_2}^c$	$y^* = \log y,\ x^* = \log x$	$y^* = a + b \cdot x_1^* + c \cdot x_2^*$
Logarithmic	$y = a + b \cdot \log x_1 + c.\log x_2$	$x^* = \log x$	$y = a + b \cdot x_1^* + c \cdot x_2^*$
Reciprocal	$y = (a + b_1x_1 + b_2x_2)^{-1}$	$y^* = 1/y$	$y^* = a + b_1 \cdot x_1 + b_2 \cdot x_2$
Hyperbolic	$y = x/(a + b \cdot x)$	$y^* = 1/y;\ x^* = 1/x$	$y^* = b + a \cdot x_1^*$
Saturation	$y = a \cdot x/(b + x)$	$y^* = 1/y;\ x^* = 1/x$	$y^* = 1/a + (b/a) \cdot x_1^*$
Logit	$y = \frac{\exp(a+b_1x_1+b_2x_2)}{1+\exp(a+b_1x_1+b_2x_2)}$	$y^* = \ln \frac{y}{1-y}$	$y^* = a + b_1 \cdot x_1 + b_2 \cdot x_2$

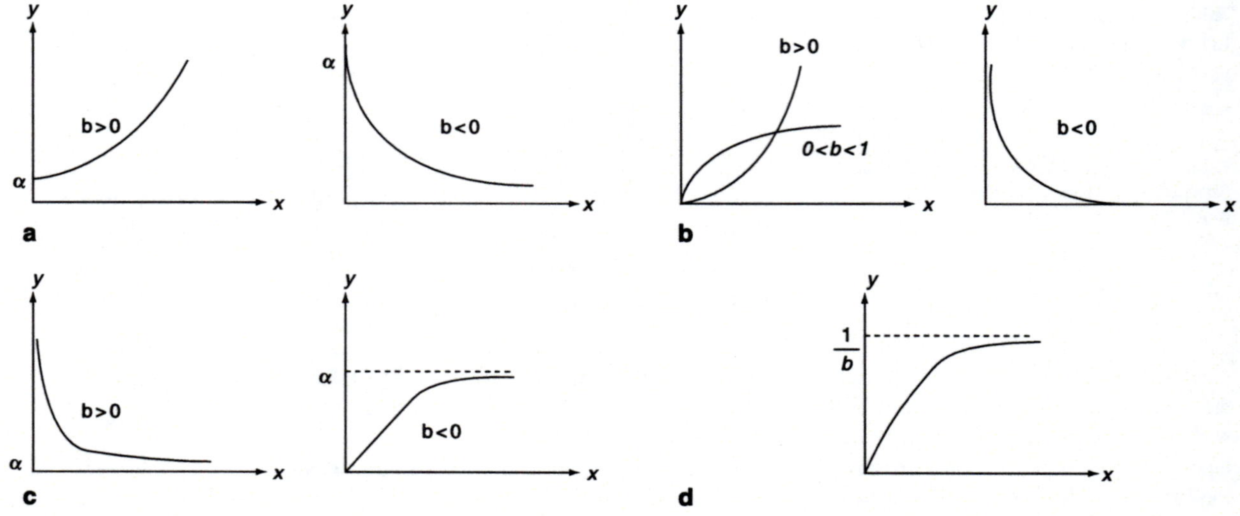

Fig. 10.21 Diagrams depicting different non-linear functions (with slope b) which can be transformed to functions linear in the parameters as shown in Table 10.20. **a** Exponential function. **b** Power function. **c** Reciprocal function. **d** Hyperbolic function (from Shannon 1975)

$$y = a\exp(bx) \tag{10.45}$$

Taking logarithms results in the simple linear model whose parameters are easily identified by OLS:

$$\ln y = \ln a + bx \tag{10.46}$$

However, two important points need to be made. The first is that it is the transformed variable which has to meet the OLS criteria (listed in Sect. 5.5.1) and not the original variable. One of the main implicit implications is that the *errors of the variable y are multiplicative.* The transformation into ln(y) has (hopefully) made the errors in the new model essentially additive, thereby allowing OLS to be performed. If the analyst believes this to be invalid, then there are two options. Adopt a non-linear estimation approach, or use weighted least squares (see Sect. 5.6.3 which presents several alternatives). The assumption of multiplicative models in certain cases, such as exponential models, is usually a good one since one would expect the magnitude of the errors to be greater as the magnitude of the variable increases. However, this is by no means obvious for other transformations. The second point is that if OLS has been adopted, the statistical goodness-of-fit indices, such as R^2 and RMSE of the regression model, as well as any residual checks apply to the transformed equation and not to the original model. Overlooking this aspect can provide misleading results in inferences being drawn about how the model explains the variation in the original response variable.

Consider another example. The solution of a first-order linear differential equation of a decay process: $T(t) = T_0 \cdot \exp(-t/\tau)$ where T_0 and τ (interpreted as the initial condition and the system time constant respectively) are the model parameters. Taking logarithms on both sides results in $\ln T(t) = \alpha + \beta t$ where $\alpha = \ln T_0$ and $\beta = 1/\tau$. The model has, thus, become linear, and OLS can be used to estimate α and β , and thence, T_0 and τ. The parameter estimates will, however, be biased; but this will not cause any prediction bias when the model is used, i.e., T(t) will be accurately predicted. However, the magnitude of the confidence and the prediction intervals provided by OLS model will be incorrect.

Natural extension of the above single variate models to multivariate ones are obvious. For example, consider the mutivariate power model:

$$y = b_0 x_1^{b_1} x_2^{b_2} \cdots x_p^{b_p} \tag{10.47}$$

If one defines:

$$z = \ln y, c = \ln b_0, w_i = \ln x_i \text{ for i} = 1, 2 \ldots \text{p} \tag{10.48}$$

Then, one gets the linear model:

$$z = c + \sum_{i=1}^{p} b_i w_i \tag{10.49}$$

Example 10.5.1: *Data shown in Table 10.21 needs to be fit using the simple power equation:* $y = ax^b$.

(a) Taking natural logarithms results in: $\ln y = \ln a + b \ln x$ which can be expressed as: $y^* = a' + bx^*$
Subsequently, a simple OLS regression yields: a'=−0.702 and b'=1.737 with R^2=0.99 (excellent) and RMSE=0.292. From here, a=0.498, and b=b'=1.737.

Table 10.21 Data table for Example 10.5.1

x	1	1.5	2	2.5	3	3.5	4	4.5	5
y	0.5	1	1.7	2.2	3.4	4.7	5.7	6.2	8.4

The goodness of fit to the original power model is illustrated in Fig. 10.22a.

(b) The scatter plot of the residuals versus x (Fig. 10.22b) reveals some amount of improper residual behavior (non-constant variance) at the high end, but a visual inspection is of limited value since it does not allow statistical inferences to be made. Nonetheless, the higher variability at the high end indicates that OLS assumptions are not fully met. Finally, the residual analysis reveals that:
   (i) the 8th observation is an unusual residual (studentized value of −4.83) as is clearly noticeable
   (ii) the 8th and 9th observations are flagged as leverage points with DFITS values of −2.99 and 1.78. This is not surprising since these are the high end, and with the lower value being close to zero, their influence on the overall fit is bound to be great.

(c) What is the predicted value for y at x=3.5. Also, determine the 95% CL for the mean prediction and that for a specific value?
The answers are summarized in the table below.

y-value from model	Standard error for forecast	Lower 95.0% CL for specific value	Upper 95.0% CL for specific value	Lower 95.0% CL for mean	Upper 95.0% CL for mean
3.856	0.31852	3.103	4.609	3.556	4.156

The prediction of y for x=3.5 is very poor. Instead of a y-value close to 4.7, the predicted value is 3.86. Even the 95% CL does not capture this value. This suggests that alternative models should be evaluated. ■

10.5.2 Intrinsically Non-linear Models

There are numerous functions which are intrinsically non-linear. Two examples of such models are:

$$y = b_0 + b_1 \exp(-b_2 x) \quad \text{and} \quad y = \exp(b_1 + b_2 x^2) \tag{10.50}$$

Nonlinear regression approach is the only recourse in such cases, as well as in cases when a transformation resulting in a model linear in the parameters still suffers from improper residual behavior. Unlike linear parameter estimation which has closed form matrix solutions, non-linear estimation requires iterative methods which closely parallel the search techniques used in optimization problems (see Sect. 7.3). Recall that the three major issues are: importance of specifying good initial or starting estimates, a robust algorithm that suggests the proper search direction and step size, and a valid stopping criterion. Hence, non-linear estimation is prone to all the pitfalls faced by optimization problems requiring search methods such as local convergence, no solution being found, and slow convergence. Either gradient-based (such as steepest-descent, Gauss-Newton's method, first or second order Taylor series,...) or gradient-free methods are used. The former are faster but sometimes may not converge to a solution, while the latter are slower but more robust. Perhaps the most widely used algorithm is the Levenburg-Markquardt algorithm which uses the desirable features of both the linearized Taylor Series and the steepest ascent methods. Its attractiveness lies in the fact that it always converges and does not slow down as do most steepest-descent methods. Most of the statistical software packages have the capability of estimating parameters of non-linear models. However, it is advisable to use them with due care; otherwise, the trial and error solution approach can lead to the program being terminated abruptly (one of the causes being ill-conditioning- see Sect. 10.2). Some authors suggest not to trust the output of a nonlinear solution until one has plotted the measured values against the predicted and looked at the residual plots The interested reader can refer to several advanced texts which deal with non-linear estimation (such as Bard 1974; Beck and Arnold 1977; and Draper and Smith 1981).

Section 7.3.4, which dealt with numerical search methods, described and illustrated the general *penalty function* approach for constrained optimization problems. Such problems also apply to non-linear parameter estimation where one has a reasonable idea beforehand of the range of variation of the individual parameters (say, based on physical considerations), and would like to constrain the search space to this range. Such problems can be converted into unconstrai-

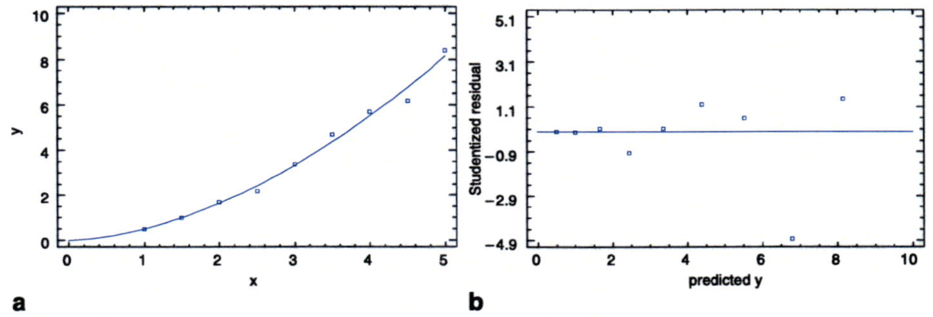

Fig. 10.22 Back transformed power model (Example 10.5.1) **a** Plot of fitted model to original data. **b** Plot of residuals

ned multi-objective problems. The objective of minimizing the squared errors δ between measured and model-predicted values is combined with another term which tries to maintain reasonable values of the parameters **p** by adversely weighting the difference between the search values of parameters and their preferred values based on prior knowledge. If the vector **p** denotes the set of parameters to be estimated, then the loss or objective function is written as the weighted square sum of the model residuals and those of the parameter deviations:

$$J(\mathbf{p}) = \sum_{j=1}^{n} w \cdot \delta_j^2 + \sum_{i=1}^{N} (1-w) \cdot (\mathbf{p}_i - \hat{\mathbf{p}}_i)^2 \quad (10.51)$$

where w is the weight (usually a fraction) associated with the model residuals, and $(1-w)$ is the weight associated with the penalty of deviating from the preferred values of the parameter set. Note that the preferred vector $\hat{\mathbf{p}}$ is not necessarily the optimal solution $\mathbf{p}^*$.

A simplistic example will make the approach clearer. One wishes to constrain the parameters "a and b" to be positive (for example, certain physical quantities cannot assume negative values). An arbitrary user-specified loss function could be of the sort:

$$J(\mathbf{p}) = \sum_{j=1}^{n} \delta_j^2 + 1{,}000^*(a<0) + 1{,}000^*(b<0) \quad (10.52)$$

where the multipliers of 1,000 are chosen simply to impose a large penalty should either a or b assume negative values. It is obvious that some care must be chosen to assign such penalties pertinent to the problem at hand, with the above example meant for conceptual purposes only.

Example 10.5.2[3]: Fit the following non-linear model to the data in Table 10.22: $y = a(1-e^{-bt}) + \varepsilon$.

First, it would be advisable to plot the data and look at the general shape. Next, statistical software is used for the non-linear estimation using least squares.

The equation of the fitted model is found to be: y=2.498* (1-exp(−0.2024*t)) with adjusted R^2=98.9%, standard error of estimate= 0.0661 and mean absolute error=0.0484. The overall model fit is thus deemed excellent. Further, the parameters have low standard errors as can be seen from the parameter estimation results shown in Table 10.23 along with the 95% intervals.

Table 10.22 Data table for Example 10.5.2

t	1	2	3	4	5	7	9	11
y	0.47	0.74	1.17	1.42	1.60	1.84	2.19	2.17

Table 10.23 Results of the non-linear parameter estimation

Parameter	Estimate	Asymptotic standard error	Asymptotic 95.0% confidence interval	
			Lower	Upper
a	2.498	0.1072	2.2357	2.7603
b	0.202	0.0180	0.1584	0.2464

How well the model predicts the observed data is illustrated in Fig. 10.23a, while Fig. 10.23b is a plot of the model residuals. Recall that studentized residuals measure how many standard deviations each observed value of y deviates from a model fitted using all of the data except that observation. In this case, there is one Studentized residual greater than 2 (point 7), but none greater than 3. DFITS is a statistic which measures how much the estimated coefficients would change if each observation was removed from the data set. Two data points (points 7 and 8) were flagged as having unusually large values of DFITS, and it would be advisable to look at these data points more carefully especially since these are the two end points. Preferably, the model function may itself have to be revised, and, if possible, collecting more data points at the high end would result in a more robust and accurate model. ■

10.6 Computer Intensive Methods

10.6.1 Robust Regression

Proper estimation of parameters of a pre-specified model depends on the assumptions one makes about the errors (see Sect. 10.4.1). Robust regression methods are parameter esti-

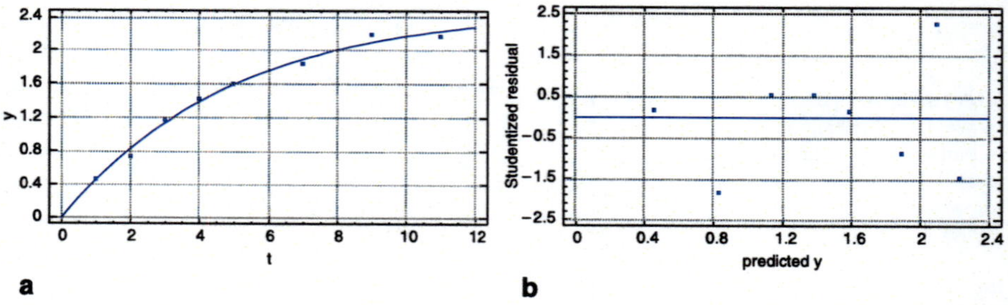

Fig. 10.23 Back transformed non-linear model (Example 10.5.2) **a** Plot of fitted model to original data. **b** Plot of residuals

[3] From Draper and Smith (1981) by permission of John Wiley and Sons.

mation methods which are not critically dependant on such assumptions (Chatfield 1995). The term is also used to describe techniques by which the influence of outlier points can be automatically down-weighted during parameter estimation. Detection of gross outlier points in the data have been addressed previously, and includes such measures as limit checks and balance checks (Sect. 3.3.2), visual means (Sect. 3.3.3), and statistical means (Sect. 3.6.6). Diagnostic methods using model residual analysis (Sect. 5.6) are a more refined means of achieving additional robustness since they allow detection of influential points. There are also automated methods that allow robust regression when faced with large data sets, and some of these are described below.

Recall that OLS assumes certain idealized conditions to hold, one of which is that the errors are normally distributed. Often such departures from normality are not serious enough to warrant any corrective action. However, under certain cases, OLS regression results are very sensitive to a few seemingly outlier data points, or when response data spans several orders of magnitude. Under such cases, the square of certain model residuals may overwhelm the regression and lead to poor fits in other regions. Common types of deficiencies include errors that may be symmetric but non-normal; they may be more peaked than the normal with lighter tails, or the converse. Even if the errors are normally distributed, certain outliers may exist. One could identify outlier points, and repeat the OLS fit by ignoring these. One could report the results of both fits to document the effect or sensitivity of the fits to the outlier points. However, the identification of outlier points is arbitrary to some extent, and rather than rejecting points, methods have been developed whereby less emphasis is placed during regression on such dubious points. Such methods are called *robust fitting methods* which assume some appropriate weighting or loss function. Figure 10.24 shows several such functions. While the two plots in the upper frame are continuous, those at the bottom are discontinuous with one function basically ignoring points which lie outside some pre-stipulated deviation value.

(a) Minimization of the least absolute deviations or MAD (Fig. 10.24a). The parameter estimation is performed with the objective function being to:

$$\text{minimize} \sum |y_i - \hat{y_i}| \qquad (10.53)$$

where y_i and $\hat{y}_i$ are the measured and modeled response variable values for observation i. This is probably the best known robust method, but is said to be generally the least powerful in terms of managing outliers.

(b) Lorentzian minimization adopts the following criterion:

$$\text{minimize} \sum \ln(1 + |y_i - \hat{y_i}|^2) \qquad (10.54)$$

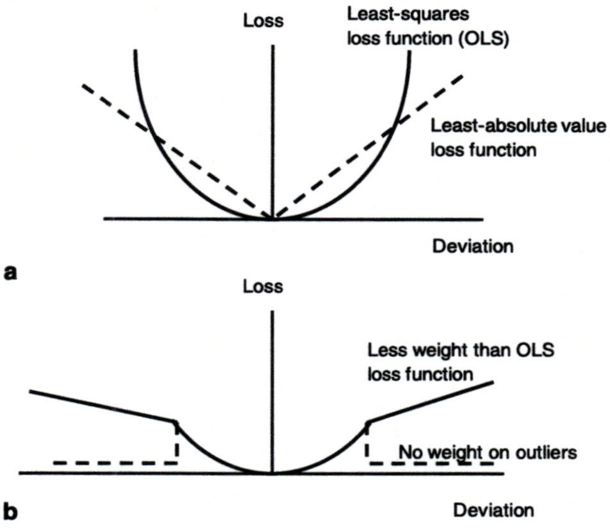

Fig. 10.24 Different weighting functions for robust regression. **a** OLS versus MAD. **b** Two different outlier weighting functions

This is said to be very effective with noisy data and data that spans several orders of magnitude. It is similar to the normal curve but with much wider tails; for example, even at 10 standard errors, the Lorenzian contains 94.9% of the points. The Gaussian, on the other hand, contains the same percentage at 2 standard errors. Thus, this function can accommodate instances of significant deviations in the data.

(c) Pearson minimization proceeds to

$$\text{minimize} \sum \ln(\sqrt{1 + |y_i - \hat{y_i}|^2} \qquad (10.55)$$

This is the most robust of the three methods with outliers having almost no impact at all on the fitted line. This minimization should be used in cases where wild and random errors are expected as a natural course.

In summary, robust regression methods are those which are less affected by outliers, and this is a seemingly advisable path to follow. However, Draper and Smith (1981) caution against the blind and indiscriminate use of robust regression since clear rules indicating the most appropriate method to use for a presumed type of error distribution do not exist. Rather, they recommend against the use of robust regression based on any one of the above functions, and suggest that maximum likelihood estimation be adopted instead. In any case, when the origin, nature, magnitude and distribution of errors are somewhat ambiguous, the cautious analyst should estimate parameters by more than one method, study the results, and then make a final decision with due diligence.

Example 10.6.1: *Consider the simple linear regression data set given in Example 5.3.1. One wishes to investigate* *the extent to which MAD estimation would differ from the standard OLS method.*

A commercial software program has been used to refit the same data using the MAD optimization criterion which is more resistant to outliers. The results are summarized below:

- OLS model identified in Example 5.3.1: y=3.8296+0.9036*x
  with the 95% CL for the intercept being {0.2131, 7.4461} and for the slope {0.8011, 1.0061}.
- Using MAD analysis: y=2.1579+0.9474*x.

One notes that the MAD parameters fall comfortably within the OLS 95% CL intervals but a closer look at Fig. 10.25 reveals that there is some deviation in the model lines especially at the low range. The difference is small, and one can conclude that the data set is such that outliers have little effect on the model estimated. Such analyses provide an additional level of confidence when estimating model parameters using the OLS approach. ■

10.6.2 Bootstrap Sampling

Recall that in Sect. 4.8.3, the use of the bootstrap method (one of the most powerful and popular methods currently in use) was illustrated for determining standard errors and confidence intervals of a parametric statistical measure in a univariate context, and also in a situation involving a nonparametric approach, where the correlation coefficient between two variables was to be deduced. Bootstrap is a statistical method where random resampling with replacement is done repeatedly from an original or initial sample, and then each bootstrapped sample is used to compute a statistic (such as the mean, median or the inter-quartile range). The resulting empirical distribution of the statistic is then examined and interpreted as an approximation to the true sampling distribution. Bootstrap is often used as a robust alternative to inference-type problems when parametric assumptions are in doubt (for example, knowledge of the probability distribution of the errors), or where parametric inference is impossible or requires very complicated formulas for the calculation of standard errors. Note, however, that the bootstrap method cannot overcome some of the limitations inherent in the original sample. The bootstrap samples are to the sample what the sample is to the population. Hence, if the sample does not adequately cover the spatial range or if the sample is not truly random, then the bootstrap results will be seriously inaccurate as well.

How the bootstrap approach can be used in a regression context is quite simple in concept. The purpose of a regression analysis can be either to develop a predictive model or to identify model parameters. In such cases, one is interested in ascertaining the uncertainty in either the model predictions or in determining the confidence intervals of the estimated parameters. Standard or classical techniques were described earlier to perform both tasks. These tasks can also be performed by numerical methods. Paraphrasing what was already stated in Sect. 4.8.1: "Efron and Tibshirami (1982) have argued that given the available power of computing, one should move away from the constraints of traditional parametric theory with its over-reliance on a small set of standard models for which theoretical solutions are available, and substitute computational power for theoretical analysis. This parallels the manner in which numerical methods have, in large part, augmented/replaced closed forms solution techniques in almost all fields of engineering and science."

Say, one has a data set of multivariate observations: $z_i = \{y_i, x_{1i}, x_{2i, \ldots}\}$ with i=1, …n (this can be viewed as a sample with n observations in the bootstrap context taken from a population of possible observations). One distinguishes between two approaches:

(a) *case resampling,* where the predictors and response observations i are random and change from sample to

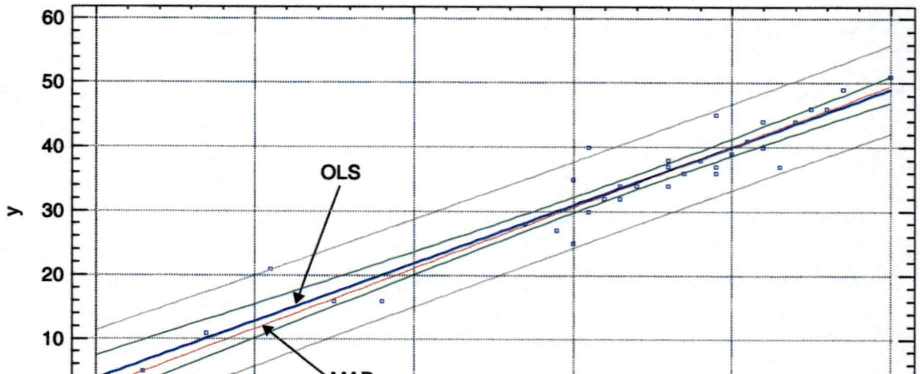

Fig. 10.25 Comparison of OLS model (shown *solid*) along with 95% confidence and prediction bands and the MAD model (Example 10.6.1)

sample. One selects a certain number of bootstrap subsamples (say 1,000) from z_i, fits the model and saves the model coefficients from each bootstrap sample. The generation of the confidence intervals for the regression coefficients is now similar to the univariate situation, and is quite straightforward. One of the benefits is that the correlation structure between the regressors is maintained;

(b) *model-based resampling* or fixed-**X** resampling, where the regressor data structure is already imposed or known with confidence. Here, the basic idea is to generate or resample the model residuals and not the observations themselves. This preserves the stochastic nature of the model structure and so the standard errors are better representative of the model's own assumption. The implementation involves attaching a random error to each y_i, and thereby producing a fixed-**X** bootstrap sample. The errors could be generated: (i) parametrically from a normal distribution with zero mean and variance equal to the estimated error variance in the regression if normal errors can be assumed (this is analogous to the concept behind the Monte Carlo approach), or (ii) non-parametrically, by resampling residuals from the original regression. One would then regress the bootstrapped values of the response variable on the fixed **X** matrix to obtain bootstrap replications of the regression coefficients. This approach is often adopted with designed experiments.

The reader can refer to Efron and Tibshirani (1982); Davison and Hinkley (1997) and other more advanced papers such as Freedman and Peters (1984) for a more complete treatment.

Problems

Pr. 10.1 Compute and interpret the condition numbers for the following:

(a) $f(x) = e^{-x}$ for $x = 10$
(b) $f(x) = [(x^2+1)^{1/2} - x]$ for $x = 1000$

Pr. 10.2 Compute the condition number of the following matrix

$$\begin{bmatrix} 21 & 7 & -1 \\ 5 & 7 & 7 \\ 4 & -4 & 20 \end{bmatrix}$$

Pr. 10.3 *Chiller data analysis using PCA*
Consider Table 10.4 of Example 10.3.3 which consists of a data set of 15 possible characteristic features (CFs) or variables under 27 different operating conditions of a centrifugal chiller (Reddy 2007). Instead of retaining all 15 variables, you will reduce the data set first by generating the correlation matrix of this data set and identifying pairs of variables which exhibit (i) the most correlation and (ii) the least correlation. It is enough if you retain only the top 5–6 variables. Subsequently repeat the PCA analysis as shown in Example 10.3.3 and compare results.

Pr. 10.4 Quality control of electronic equipment involves taking a random sample size n and determining the proportion of items which are defective. Compute and graph the likelihood function for the two following cases: (a) n=6 with 2 defectives, (b) n=8 and 3 defectives.

Pr. 10.5 Consider the data of Example 10.4.2 to which the two parameters of a Weibull distribution were estimated based on MLE. Compare the goodness of this fit (based on the Chi-square statistic) to that using a logistic model.

Pr. 10.6 Indoor air quality measurements of carbon dioxide concentration reveal how well the building is ventilated, i.e., whether adequate ventilation air is being brought in and properly distributed to meet the comfort needs of the occupants dispersed throughout the building. The following twelve measurements of CO_2 in parts per million (ppm) were taken in the twelve rooms of a building:

$$\{732, 816, 875, 932, 994, 1003, 1050, 1113, 1163, 1208, 1292, 1382\}$$

(a) Assuming normal distribution, estimate the true average concentration and the standard deviation using MLE,
(b) How are these different from classical MME values? Discuss.

Pr. 10.7 *Non-linear model fitting to thermodynamic properties of steam*
Table 10.24 lists the saturation pressure in kPascals for different values of temperature extracted from the well-known steam tables.

Two different models proposed in the literature are: $p_{v,sat} = c.\exp\left(\frac{at}{b+t}\right)$ and $\ln p_{v,sat} = a + \frac{b}{T}$ where T is the temperature in units Kelvin.

(a) You are asked to estimate the model parameters by both OLS (using variable transformation to make the esti-

Table 10.24 Data table for Problem 10.7

Temperature t (°C)	10	20	30	40	50	60	70	80	90
Pressure $p_{v,sat}$ (kPa)	1.227	2.337	4.241	7.375	12.335	19.92	31.16	47.36	70.11

Table 10.25 Data table for Problem 10.8

Type of equipment	Total number of units	Annual energy use (kWh)		Initial number of new models	Initial growth rate	Predicted saturation fraction
		Current model	New model	N_0	r %	
Central A/C	10,000	3,500	2,800	100	5	0.40
Color TV	15,000	800	600	150	8	0.60
Lights	40,000	1,000	300	500	20	0.80

mation linear) and by MLE along with standard errors of the coefficients. Comment on the differences of both methods and the implied assumption,
(b) Which of the two models is the preferred choice? Give reasons,
(c) You are asked to use the identified models to predict saturation pressure at t=75°C along with the model prediction error. Comment.

Pr. 10.8 *Logistic functions to study residential equipment penetration*
Electric utilities provide incentives to homeowners to replace appliances by high-efficiency ones—such as lights, air conditioners, dryers/washers, …. In order to plan for future load growth, the annual penetration levels of such equipment needs to be estimated with some accuracy. Logistic growth models have been found to be appropriate since market penetration rates reach a saturation level, often specified as a saturation fraction or the fractional number of the total who purchase this equipment. Table 10.25 gives a fictitious example of load estimation with three different types of residential equipment. If the start year is 2010, plot the year-to-year energy use for the next 20 years assuming a logistic growth model for each of the three pieces of equipment separately, and also for the combined effect.

Hint: The carrying capacity can be calculated from the predicted saturation fraction and the difference in annual energy use between the current model and the new one.

Pr. 10.9 *Fitting logistic models for growth of population and energy use*
Table 10.26 contains historic population data from 1970 till 2010 (current) as well as extrapolations till 2050 (from the U.S. Census Bureau's International Data Base). Primary energy use in million tons of oil equivalent (MTOE) consumed annually has also been gathered but for the same years (the last value for 2010 was partially extrapolated from 2008). You are asked to analyze this data using logistic models.
(a) Plot the population data,

Table 10.26 Data table for Problem 10.9

Year	World population (in billion)	Primary energy use in MTOE
1970	3.712	4970.2
1980	4.453	6629.7
1990	5.284	8094.7
2000	6.084	9262.6
2010	6.831	1,150[a]
2020	7.558 (projected)	–
2030	8.202	–
2040	8.748	–
2050	9.202	–

[a] Extrapolated from measured 2008 value

(b) Estimate growth rate and carrying capacity by fitting a logistic model to the data from 1970–2010. Does your analysis support the often quoted estimate that the world population would plateau at 9 billion,
(c) Predict population values for the next four decades along with 95% uncertainty estimates, and compare them with the projections by the US Census Bureau,
(d) Use a logistic model to fit the data and estimate the growth rate and the carrying capacity. Calculate the per capita annual energy use for each of the five decades from 1970 to 2010. Analyze results and draw pertinent conclusions.

Pr. 10.10 *Dose response model fitting for VX gas*
You will repeat the analysis illustrated in Example 10.4.5 using the dose response curves for VX gas which is a nerve agent. You will identify the logistic model parameters for both the causality dose (CD) and lethal dose (LD) curves and report relevant model fit and parameter statistics (Fig. 10.26).

Pr. 10.11 *Non-linear parameter estimation of a model between volume and pressure of a gas*
The pressure P of a gas corresponding to various volumes V is given in Table 10.27.

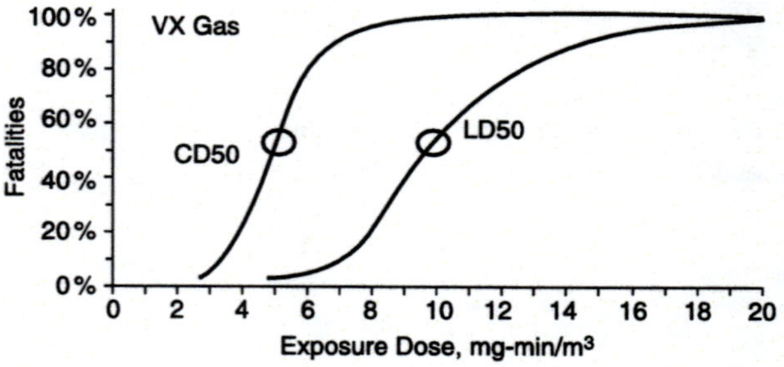

Fig. 10.26 Dose response curves for VX gas with 50% casualty dose (CD50) and 50% lethal dose (LD50) points. (From Kowalski 2002 by permission of McGraw-Hill)

Table 10.27 Data table for Problem 10.11

V (cm³)	50	60	70	90	100
P (kg/cm²)	64.7	51.3	40.5	25.9	7.8

Table 10.28 Data table for Problem 10.12

x (cm)	30	35	40	45	50	55	60	65	70	75
y	0.85	0.67	0.52	0.42	0.34	0.28	0.24	0.21	0.18	0.15

(a) Estimate the coefficients a and b assuming the ideal gas law: $PV^a = b$,

Hint: Take natural logs and re-arrange the model in a form suitable for regression,

(b) Study model residuals and draw relevant conclusions.

Pr. 10.12 *Non-linear parameter estimation of a model between light intensity and distance*

An experiment was conducted to verify the intensity of light (y) as a function of distance (x) from a light source; the results are shown in Table 10.28.

(a) Plot this data and fit a suitable polynomial model,
(b) Fit the data with an exponential model,
(c) Fit the data using a model derived from the underlying physics of the problem,
(d) Compare the results of all three models in terms of their model statistics as well as their residual behavior.

Pr. 10.13 Consider a hot water storage tank which is heated electrically. A heat balance on the storage tank yields:

$$P(t) = C\dot{T}_s(t) + L[T_s(t) - T_a(t)] \qquad (10.56)$$

where

$C = M\,c_p$ = thermal capacity of the storage tank [J/°C]
$L = U\,A$ = heat loss coefficient of the storage tank [W/°C] (A = surface area of storage)
$T_s(t)$ = temperature of storage [°C]
$T_a(t)$ = ambient temperature [°C]
$P(t)$ = heating power [W] as function of time t

Suppose one has data for $T_s(t)$ during cool-down under constant conditions: $P(t)=0$ and $T_a(t)=\text{constant}$. In that case, the energy balance can be written as (where the constant has been absorbed in T, i.e. T is now the difference between the storage temperature and the ambient temperature)

$$\tau\,\dot{T}(t) + T(t) = 0 \qquad (10.57)$$

with $\tau = \frac{C}{L}$ = time constant.

Table 10.29 assembles the test results during storage tank cool-down.

(a) First, linearize the model and estimate its parameters,
(b) Study model residuals and draw relevant conclusions,
(c) Determine the time constant of the storage tank along with standard errors

Table 10.29 Data table for Problem 10.13

t(h)	0	1	2	3	4	5	6	7	8	9	10	11
T(t)	10.1	8	6.8	5.7	4.4	3.8	3	2.4	2	1.8	1.1	1

Pr. 10.14 *Model identification for wind chill factor*

The National Weather Service generates tables of wind chill factor (WC) for different values of ambient temperature (T) in °F and wind speed (V) in mph. The WC is an equivalent temperature which has the same effect on the rate of heat loss as that of still air (an apparent wind of 4 mph). The equation used to generate the data in Table 10.30 is:

$$WC = \log V(0.26T) - 23.68(0.63T + 32.9) \qquad (10.58)$$

(a) You are given this data without knowing the model. Examine a linear model between WC=f(T,V), and point out inadequacies in the model by looking at the model fits and the residuals,
(b) Investigate, keeping T fixed, a possible relation between WC and V,
(c) Repeat with WC and T
(d) Adopt a stage-wise model building approach and evaluate suitability,
(e) Fit a model of the type: $WC = a + b\cdot T + c\cdot V + d(V)^{1/2}$ and evaluate suitability

Table 10.30 Data table for Problem 10.14. (From Chatterjee and Price 1991 by permission of John Wiley and Sons)

Wind speed	Actual air temperature (°F)											
(mph)	50	40	30	20	10	0	−10	−20	−30	−40	−50	−60
5	48	36	27	17	5	−5	−15	−25	−35	−46	−56	−66
10	40	29	18	5	−8	−20	−30	−43	−55	−68	−80	−93
15	35	23	10	−5	−18	−29	−42	−55	−70	−83	−97	−112
20	32	18	4	−10	−23	−34	−50	−64	−79	−94	−108	−121
25	30	15	−1	−15	−28	−38	−55	−72	−88	−105	−118	−130
30	28	13	−5	−18	−33	−44	−60	−76	−92	−109	−124	−134
35	27	11	−6	−20	−35	−48	−65	−80	−96	−113	−130	−137
40	26	10	−7	−21	−37	−52	−68	−83	−100	−117	−135	−140
45	25	9	−8	−22	−39	−54	−70	−86	−103	−120	−139	−143
50	25	8	−9	−23	−40	−55	−72	−88	−105	−123	−142	−145

Table 10.31 Data table for Problem 10.15

Before weather-stripping		After weather-stripping	
Δp (Pa)	Q (m^3/h)	Δp (Pa)	Q (m^3/h)
3.0	365.0	2.2	99.2
5.0	445.9	5.5	170.4
5.8	492.7	6.7	185.6
6.7	601.8	8.2	208.5
8.2	699.2	11.6	263.2
9.0	757.5	13.5	283.1
10.0	812.4	15.6	310.2
11.0	854.1	18.2	346.2

Pr. 10.15 *Parameter estimation for an air infiltration model in homes*
The most common manner of measuring exfiltration (or infiltration) rates in residences is by artificially pressurizing (or depressurizing) the home using a device called a *blower door*. This device consists of a door-insert with an rubber edge which can provide an air-tight seal against the door-lamb of one of the doors (usually the main entrance). The blower door has a variable speed fan, an air flow measuring meter and a pressure difference manometer with two plastic hoses (to allow the inside and outside pressure differential to be measured). All doors and windows are closed during the testing. The fan speed is increased incrementally and the pressure difference Δp and air flow rate Q are measured at each step. The model used to correlate air flow with pressure difference is a modified orifice flow model given by $Q = k(\Delta p)^n$ where k is called the flow coefficient (which is proportional to the effective leakage area of the building envelope) and n is the flow exponent. The latter is close to 0.5 when the flow is strictly turbulent which occurs when the flow paths through the interstices of the envelope are small and tortuous (like in a well-built "tight" house), while n is close to 1.0 when the flow is laminar (such as in a "loose" house). Values of n around 0.65 have been experimentally determined for typical residential construction in the U.S.

Table 10.31 assembles test results of an actual house where blower door tests were performed both before and after weather-stripping. Identify the two sets of coefficients k and n for tests done before and after the house tightening. Based on the uncertainty estimates of these coefficients, what can you conclude about the effect of weather-stripping the house?

References

Andersen, K.K., and Reddy, T.A., 2002. "The error in variable (EIV) regression approach as a means of identifying unbiased physical parameter estimates: Application to chiller performance data", *HVAC&R Research Journal*, vol.8, no.3, pp. 295-309, July.

Bard, Y., 1974. *Nonlinear Parameter Estimation*, Academic Press, New York.

Beck, J.V. and K.J., Arnold, 1977. *Parameter Estimation in Engineering and Science*, John Wiley and Sons, New York

Belsley, D.A., Kuh, E. and Welsch, R.E., 1980, *Regression Diagnostics*, John Wiley & Sons, New York.

Chapra, S.C. and Canale, R.P., 1988. *Numerical Methods for Engineers*, 2nd Edition, McGraw-Hill, New York.

Chatfield, C., 1995. *Problem Solving: A Statistician's Guide*, 2nd Ed., Chapman and Gall, London, U.K.

Chatterjee, S. and B. Price, 1991. *Regression Analysis by Example*, 2nd Edition, John Wiley & Sons, New York

Devore J., and N. Farnum, 2005. *Applied Statistics for Engineers and Scientists*, 2nd Ed., Thomson Brooks/Cole, Australia.

Davison, A.C. and D.V.Hinkley, 1997. *Bootstrap Methods and their Applications*, Cambridge University Press, Cambridge

Draper, N.R. and H. Smith, 1981. *Applied Regression Analysis*, 2nd Ed., John Wiley and Sons, New York.

Edwards, C.H. and D.E. Penney, 1996. *Differential Equations and Boundary Value Problems*, Prentice Hall, Englewood Cliffs, NJ

Efron, B. and J. Tibshirani, 1982. *An Introduction to the The Bootstrap*, Chapman and Hall, New York

Freedman, D.A., and S.C. Peters, 1984. Bootstrapping a regression equation: Some empirical results, *J. of the American Statistical Association,* 79(385), pp.97-106, March.

Fuller, W. A., 1987. *Measurement Error Models*, John Wiley & Sons, NY.

Godfrey, K., 1983. *Compartmental Models and Their Application*, Academic Press, New York.

Kachigan, S.K., 1991. *Multivariate Statistical Analysis*, 2nd Ed., Radius Press, New York.

Kowalski, W.J., 2002. *Immune Building Systems Technology*, McGraw-Hill, New York

Lipschutz, S., 1966. *Finite Mathematics*, Schaum's Outline Series, McGraw-Hill, New York

Mandel, J., 1964. *The Statistical Analysis of Experimental Data*, Dover Publications, New York.

Manly, B.J.F., 2005. *Multivariate Statistical Methods: A Primer*, 3rd Ed., Chapman & Hall/CRC, Boca Raton, FL

Masters, G.M. and W.P. Ela, 2008. *Introduction to Environmental Engineering and Science*,3rd Ed. Prentice Hall, Englewood Cliffs, NJ

Mullet, G.M., 1976. Why regression coefficients have the wrong sign, *J. Quality Technol.*, 8(3).

Pindyck, R.S. and D.L. Rubinfeld, 1981. *Econometric Models and Economic Forecasts*, 2nd Edition, McGraw-Hill, New York, NY.

Reddy, T.A., and D.E. Claridge, 1994 "Using synthetic data to evaluate multiple regression and principal component analysis for statistical models of daily energy consumption", *Energy and Buildings*, vol.21, pp. 35-44.

Reddy, T.A., S. Deng, and D.E. Claridge, 1999. "Development of an inverse method to estimate overall building and ventilation parameters of large commercial buildings", *ASME Journal of Solar Energy Engineering*, vol.121, pp.40-46, February.

Reddy, T.A. and K.K. Andersen, 2002. "An evaluation of classical steady-state off-line linear parameter estimation methods applied to chiller performance data", *HVAC&R Research Journal*, vol.8, no.1, pp.101-124.

Reddy, T.A., 2007. Application of a generic evaluation methodology to assess four different chiller FDD Methods (RP1275), *HVAC&R Research Journal*, vol.13, no.5, pp 711-729, September.

Saunders, D.H., J.J. Duffy, F.S. Luciani, W. Clarke, D. Sherman, L. Kinney, and N.H. Karins, 1994. "Measured performance rating method for weatherized homes", *ASHRAE Trans.*, V.100, Pt.1, paper no. 94-25-3.

Shannon, R.E., 1975. *System Simulation: The Art and Science*, Prentice-Hall, Englewood Cliffs, NJ.

Sinha, N.K. and B. Kuszta, 1983. *Modeling and Identification of Dynamic Systems*, Van Nostrand Reinhold Co., New York.

Subbarao, K., 1988. "PSTAR-Primary and Secondary Terms Analysis and Renormalization: A Unified Approach to Building Energy Simulations and Short-Term Monitoring", SERI/TR - 253- 3175, Solar Energy Research Institute, Golden CO.

Walpole, R.E., R.H. Myers, S.L. Myers, and K. Ye, 2007, *Probability and Statistics for Engineers and Scientists*, 8th Ed., Prentice-Hall, Upper Saddle River, NJ.

11 Inverse Methods

A broad description of inverse methods (introduced in Sect. 1.3.3) is that they pertain to the case when the system under study already exists, and one uses measured or observed system behavior to aid in the model building. The three types of inverse problems were classified as: (i) calibration of white-box models (which can be either relatively simple or complex-coupled simulation models) which requires that one selectively manipulate certain parameters of the model to fit observed data. Monte Carlo methods which are powerful random sampling techniques are described along with regional sensitivity analyses as a means of reducing model order of complex simulation models; (ii) model selection and parameter estimation involving positing either black-box or grey-box models, and using regression methods to identify model parameters based on some criterion of error minimization (the least squares regression method and the maximum likelihood method being the most popular); and (iii) control problems where input states and/or boundary conditions are inferred from knowledge of output states and model parameters. This chapter elaborates on these methods and illustrates their approach with case study examples. Local polynomial regression methods are also briefly described as well as the multilayer perceptron approach, which is a type of neural network modeling method that is widely used in modeling non-linear or complex phenomena. Further, the selection of a grey-box model based more on policy decisions rather than on how well a model fits the data is illustrated in the framework of dose-response models. Finally, state variable model formulation and compartmental modeling appropriate for describing dynamic behavior of linear systems are introduced along with a discussion of certain identifiability issues in practice.

11.1 Inverse Problems Revisited

Inverse problems were previously introduced and classified in Sect. 1.3.3. They are classes of problems pertaining to the case when the system under study already exists, and one uses measured or observed system behavior to aid in the model building. The three types of inverse problems were also briefly described, namely, calibration of white box models, model selection and parameter estimation (or system identification), and control problems (see Fig. 1.12). The inverse approach makes use of the additional information not available to the forward approach, viz. measured system performance data, to either tune the parameters of an existing model or identify a macroscopic model that captures the major physical interactions of the system and whose parametric values are determined by regression to the data. This "fine tuning" makes the inverse approach more suitable for diagnostics and analysis of system properties provides better conceptual insights into system behavior, and allows more accurate predictions and optimal control. The forward approach, on the other hand, is more appropriate for evaluating different alternatives during the design phase and for sensitivity studies. The inverse approach has been used in a wide variety of areas such as engineering, industrial process control, aeronautics, signal processing, biomedicine, ecology, transportation, and robotics. This chapter elaborates on these methods, and illustrates their approach with case study examples.

11.2 Calibration of White Box Models

11.2.1 Basic Notions

Calibration problems generally involve mechanistic white-box models that have a well defined model structure, i.e., the set of modeling equations can be solved uniquely. They can be of two types. One situation is when a detailed simulation program (developed for forward applications) is used to provide the model structure, with the numerous model parameters needing to be tuned so that simulated output closely matches observed system performance. Most often, such models have so many parameters that the measured data does not allow them to be identified uniquely. This is referred to as over-constrained (or over-parameterized) which has no unique solution since one is trying to identify the numerous

T. Agami Reddy, *Applied Data Analysis and Modeling for Energy Engineers and Scientists*,
DOI 10.1007/978-1-4419-9613-8_11,

model parameters with too little "information" about the behavior of the system (i.e., relatively too few data points and/or limited in "richness"). The order of the model has to be reduced mathematically by freezing certain of these parameters at their "best-guess" values. Which parameters to freeze, how many parameters can one hope to identify, what will be the uncertainty of such calibrated models are important aspects which need to be considered. The second type of situation is when the analyst only has a few data points, and adopts the approach of *developing* a simple mechanistic model which directly provides an analytical relationship between the simulation outputs and inputs consistent with the data at hand, i.e., such that the model parameters can be identified uniquely. Hence, both situations are characterized with relatively little data "richness", and parameters are identified by solving the set of equations or by search methods requiring no regression.

A trivial example of an over-constrained problem is one where the three model coefficients $\{a_1, a_2, a_3\}$ are to be determined when only two observations of $\{\Delta p, V\}$ are available:

$$\Delta p = a_1 + a_2 \cdot V + a_3 \cdot V^2$$

There are two observations and three parameters, and clearly this cannot be solved unless the model is simplified or reformulated so as to have two parameters only. Alternatively, one could collect more data and circumvent this limitation.

11.2.2 Example of Calibrated Model Development: Global Temperature Model

The atmosphere around the earth acts as a greenhouse, selectively letting in short-wave solar radiation and absorbing (and, thus trapping) the long wave infrared back-radiation from the earth's surface. This warming is beneficial to life on earth since the earth's average surface temperature has been increased from about −18°C (−0.4°F) to the current 14°C (57°F). However, the recent increase in global warming of the earth's lower atmosphere, as a result of human activity (called *anthropogenic global warming*—AGW) and associated dire consequences, has resonated with society at large, and associates spurred scientists and policymakers to find ways to mitigate against it. Very complex mechanistic circulation models have been developed, and subsequently calibrated against actual measurements. Such calibrated models are meant to make extrapolations over time so as to study the impact of different AGW mitigation strategies. However, since these mitigation measures have financial and trans-continental societal implications, coupling)them to models which capture such considerations would result in them soon becoming too complex for clear interpretation and policy-making. It is in such cases that inverse models can be useful, and have been the object of numerous studies in the published literature. A rather simplistic inverse model of the earth's radiation budget is presented below meant more to illustrate the concept of developing a calibrated white-box model than for any practical usefulness.

Consider Fig. 11.1 taken from IPCC (1996) which shows the various thermal fluxes and their numerical values (determined by direct measurement and from other complex models). Let us assume that these quantities are known to us (similar to taking measurements of an actual existing system), and can be used to develop an appropriate model. Clearly, the complexity of the model is constrained by the available measurements. The atmosphere is modeled as a single node with different upward and downward emissivities (so as to partly account for the AGW effect) interacting with the surface of

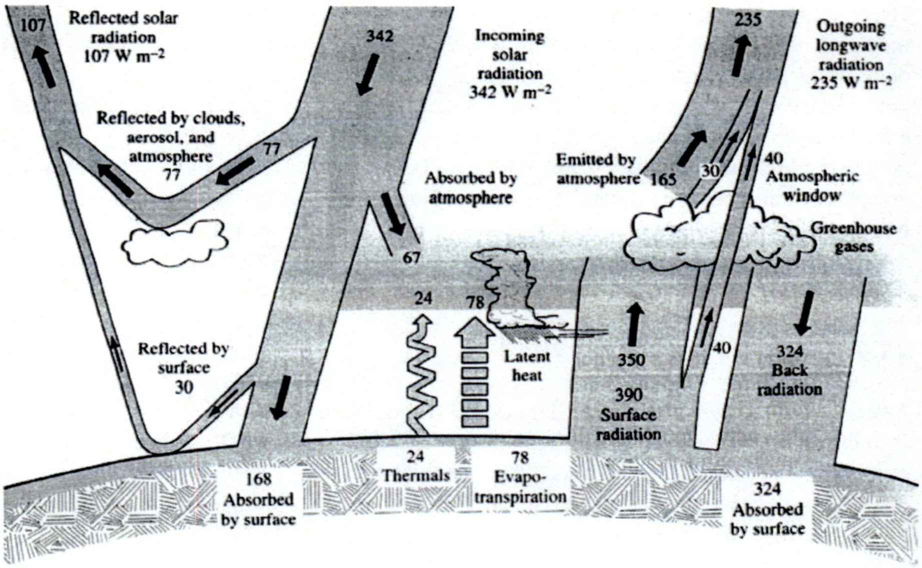

Fig. 11.1 Average energy flows between the Earth's surface, the atmosphere and space under global equilibrium. (Taken from IPCC 1996)

Fig. 11.2 Sketch of simplified global temperature model. The atmosphere and the earth's surface have been modeled as single nodes

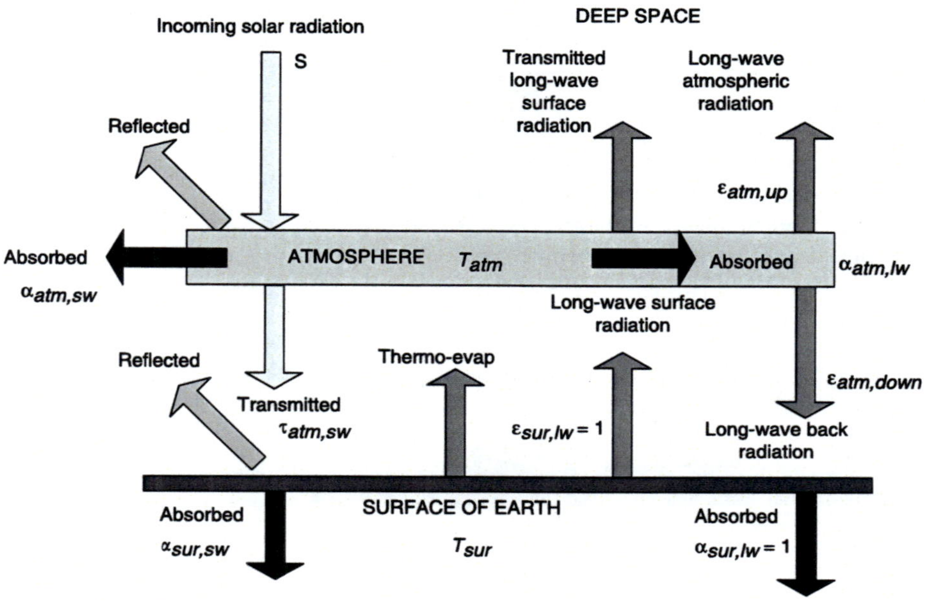

the earth, also represented by a single node. The various heat flows to be considered in the model are shown in Fig. 11.2.

Let T_{atm} be the aggregated effective atmospheric temperature (in Kelvin), T_{sur} the average effective surface temperature of the earth (in Kelvin), R the mean radius of the earth and S_{ext} the extraterrestrial incoming solar flux at the mean sun-earth distance called the solar constant (=1,367 W/m^2). The temperature of deep space is taken to be 0 K. The thickness of the atmosphere is so small compared to R that one can safely neglect it. Then, the average solar power per unit area on top of the atmospheric layer:

$$S = \frac{S_{ext}\pi R^2}{4\pi R^2} = \frac{S_{ext}}{4} = \frac{1367}{4} = 342\ \text{W/m}^2 \quad (11.1)$$

This is the average value distributed over day and night and over all latitudes (shown in Fig. 11.1 as the incoming solar radiation). The various heat flows are modeled by adopting average aggregate radiative properties such that (refer to Fig. 11.2):

(i) the incoming short wave solar radiation undergoes absorption in the atmosphere (absorptivity $\alpha_{atm,sw}$), transmission (transmittivity $\tau_{atm,sw}$) and reflection back to deep space;

(ii) most of the transmitted solar radiation is absorbed by the earth's surface (absorptivity $\alpha_{sur,sw}$) and the rest is reflected back to deep space (multiple reflection effect is overlooked);

(iii) the earth is assumed to be a black body for long wave radiation (absorptivity $\alpha_{sur,lw}$ =emissivity $\varepsilon_{sur,lw} = 1$);

(iv) thermo-evaporation accounts for a certain amount of heat transfer from the earth to the atmosphere (=24+78 W/m^2), and a simplified model is

$$q_{th-evap} = h_{c-e}(T_{sur} - T_{atm}) \quad (11.2)$$

where h_{c-e} is an effective combined coefficient;

(v) the atmosphere transmits some of the long-wave surface radiation from the earth to deep space through the atmospheric window (=40 W/m^2) and absorbs most of it (long-wave absorptivity $\alpha_{atm,lw}$) with reflectivity being negligible;

(vi) the atmosphere loses some of the heat flux (=165+30 W/m^2) by long wave radiation upwards to deep space (emissivity $\varepsilon_{atm,up}$), and the rest downwards to earth as back radiation (emissivity $\varepsilon_{atm,down}$).

Heat balance on the atmosphere:

$$\begin{aligned} &S \cdot \alpha_{atm,sw} + h_{c-e} \cdot (T_{sur} - T_{atm}) \\ &\quad + \alpha_{atm,lw} \cdot (\varepsilon_{sur,lw} \cdot \sigma \cdot T_{sur}^4) \\ &\quad = 2(\varepsilon_{atm,up} + \varepsilon_{atm,down})\sigma \cdot T_{atm}^4 \end{aligned} \quad (11.3)$$

where the multiplier of 2 is introduced since heat losses from the atmosphere occur both upwards and downwards.

Heat balance on the earth's surface:

$$\begin{aligned} &(S \cdot \tau_{atm,sw})\alpha_{sur,sw} + \alpha_{sur,lw} \cdot (\varepsilon_{atm,down} \cdot \sigma T_{atm}^4) \\ &\quad = h_{c-e} \cdot (T_{sur} - T_{atm}) + \varepsilon_{sur,lw} \cdot \sigma T_{sur}^4 \end{aligned} \quad (11.4)$$

where σ is the Stephan-Boltzmann constant $= 5.67 \times 10^{-8}$W/m^2 K^4.

A forward simulation approach would involve solving the above equations using pre-calculated or best-guess values of the numerous coefficients appearing in these equations without any consideration to actual measured fluxes and temperatures. The same model becomes an inverse calibrated model if model parameters can be estimated or tuned from actual measurements, and the value of the variables (in this case, the earth surface temperature), subsequently, predicted. Comparing the predicted and measured values provides a means of evaluating the model. In order to avoid the issue of over-parameterization, the model has been deliberately selected so as to be well specified with zero degrees of freedom.

Numerical values of most of these coefficients can be determined from the quantities shown in Fig. 11.1. Average solar flux absorbed by the atmosphere=67 $= S \cdot \alpha_{atm,sw}$ from where

$$\alpha_{atm,sw} = 67/342 = 0.196.$$

Average solar flux transmitted through the atmosphere= $(342-77-67) = S \cdot \tau_{atm,sw}$ from where

$$\tau_{atm,sw} = 198/342 = 0.579$$

Average solar flux absorbed by the earth's surface=168 $= (\tau_{atm,sw} S) \cdot \alpha_{sur,sw}$ from where

$$\alpha_{sur,sw} = 168/198 = 0.848$$

Average long wave radiation flux from earth absorbed by the atmosphere=350 $= Q_{sur,lw} \cdot \alpha_{atm,lw}$ from where

$$\alpha_{atm,lw} = 350/390 = 0.897$$

Average long wave radiation fluxes to deep space and towards the earth are:

$$\begin{aligned} 2 \cdot \varepsilon_{atm,up} \cdot \sigma T_{atm}^4 &= 165 + 30 \\ 2 \cdot \varepsilon_{atm,down} \cdot \sigma T_{atm}^4 &= 324 \end{aligned} \tag{11.5}$$

with: $\varepsilon_{atm,up} + \varepsilon_{atm,down} = 1$

Using parameter values already deduced above, Eqs. 11.3–11.5 can be solved resulting in $\varepsilon_{atm,up} = 0.376$ and $\varepsilon_{atm,down} = 0.624$ while $T_{atm} = 260.1$ K or -12.9°C and $T_{sur} = 288$ K or 15°C which is within 1°C of the measured mean surface temperature of the earth. Values of all the parameters of the model are assembled in Table. 11.1 for easy reference.

Table 11.1 Description of quantities and their numerical values appearing in Sect. 11.2.2

Quantity	Symbol	Numerical value
Average incoming solar flux	S	342 W/m²
Absorptivity of atmosphere to solar radiation	$\alpha_{atm,sw}$	0.196
Transmitivity of atmosphere to solar radiation	$\tau_{atm,sw}$	0.579
Absorptivity of earth's surface to solar radiation	$\alpha_{sur,sw}$	0.848
Absorptivity of atmosphere to long wave radiation	$\alpha_{atm,lw}$	0.897
Emissivity of earth's surface to long wave radiation	$\varepsilon_{sur,lw}$	1.0
Emissivity of atmosphere for radiation to deep space	$\varepsilon_{atm,up}$	0.376
Emissivity of atmosphere for radiation downwards to earth	$\varepsilon_{atm,down}$	0.624

The above example illustrates the use of a white box model whose parameters have been calibrated or tuned with the data (observed or estimated heat fluxes). Having such a model with parameters that can be interpreted as physical quantities allows one to evaluate the impact of different mitigation measures. One geo-engineering option is to seed the atmosphere with reflective strips so as to increase reflectivity to incoming solar radiation, and thereby reduce $\tau_{atm,sw}$. Another avenue would be to evaluate the extent to which reducing carbon dioxide levels in the atmosphere would lower surface temperatures of the Earth. This measure would impact the absorptivity, and thereby the transmittivity $\tau_{atm,lw}$ of the atmosphere to long wave re-radiation from the surface. Fig. 11.3a and b are plots indicating how the earth's surface temperature and the atmospheric mean temperature would

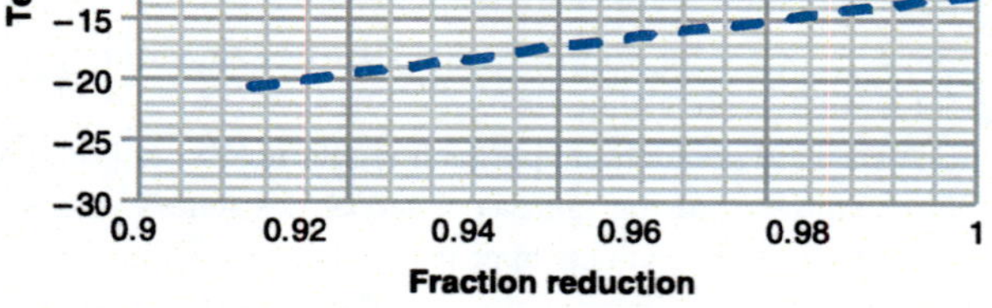

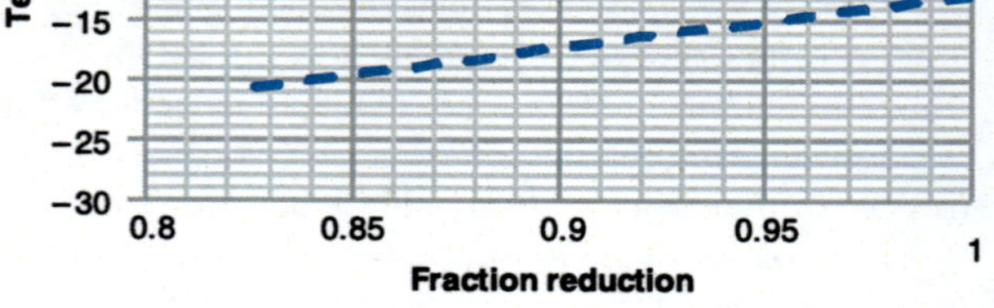

Fig. 11.3 Sensitivity results of changing key parameters of the simplistic global temperature model. **a** Effect of changing the baseline absorptivity of the atmosphere to long wave radiation from earth surface. **b** Effect of changing the baseline transmittivity of the atmosphere to solar radiation

vary under these scenarios. One notices that the trends are fairly linear indicating that a 5% reduction in $\tau_{atm,lw}$ would reduce the earth's surface temperature by 4°C while a 10% reduction would result in a decrease of 7°C.

How good or realistic are these predictions? Should one include additional terms than those shown in Fig. 11.2 which have been neglected in the above analysis? Is our approach of assuming the entire atmosphere to be one lumped node and making somewhat empirical adjustments to account for differences in long wave radiation in the upward and downward directions realistic? How valid is our approximation of assuming a mean surface temperature of the earth? The oceans and the landmass behave very differently—should these not be treated separately? Should latitude dependency and altitude effects be brought into the model; and if so, how? Detailed models have indicated that even if the man-made carbon dioxide emissions are cut to zero immediately, the effects of global warming will persist for decades to come; this suggests that dynamic effects are important and require the use of differential equations. How should feedback loops between the ocean, earth and the atmosphere be included into the model? The atmosphere has at least four distinct layers with different temperature profiles; how should these be treated? All these issues related to system identification are some of the many modeling considerations one ought to evaluate. One could also question the manner in which the various parameters were estimated (as shown in Table. 11.1). Are there other ways of estimating them which will lead to better results? Finally, consider the data itself, i.e., the fluxes shown in Fig. 11.1. How accurate are these values? How have they been determined? Can additional spatial and temporal measurements (of the ocean and the earth surface, for example) be made which can be used to improve our model? Such questions require that one try to acquire more data (either by direct observations or from detailed computer simulations) and gradually increase the complexity of the model so as to make it more realistic.

This example has been presented more to illustrate the process of framing and tuning white box models to available data. However, the overarching nagging concern in such cases is whether the model developed is realistic and complete enough that it captures the inherent complexity of the problem at hand. In such cases, a powerful argument can be made that calibrating detailed and complex models meant for forward simulation, even if the process has its own limitations, is the only rational approach for very complex problems. This approach is addressed in the next two sections.

11.2.3 Analysis Techniques Useful for Calibrating Detailed Simulation Models

The previous section dealt with developing and calibrating rather simple mechanistic models which can be algebraic or differential equations. This section pertains to coupled complex set of models with large number of parameters and where a precise relationship between outputs and inputs cannot be expressed analytically because of the complex nature of the coupling. Solving such sets of equations require computer programs with relatively sophisticated solution routines. Calibration and validation of such models has been addressed in several books and journal papers in diverse areas of engineering and science such as environmental, structural, hydrology, epidemiology and structural engineering. The crux of the problem is that the highly over-parameterized situation leads to a major difficulty aptly stated by Hornberger and Spear (1981) as: "…most simulation models will be complex, with many parameters, state-variables and non-linear relations. Under the best circumstances, such models have many degrees of freedom and, with judicious fiddling, can be made to produce virtually any desired behavior, often with both plausible structure and parameter values." This process is also referred to in the scientific community as **GIGOing** (garbage in–garbage out) where a false sense of confidence can result since precise outputs are obtained by arbitrarily restricting the input space (Saltelli 2002). Hence, given the limited monitored data available, one can at best identify only some of the numerous input parameters of the set of models. Thus, model reduction is a primary concern and several relevant analysis techniques are discussed below.

(a) Sensitivity Analysis The aim of sensitivity analysis is to determine or identify which parameters, have a significant effect on the simulated output parameters, and then, to quantify their relative importance. There are two types of sensitivities: (i) individual sensitivities which describe the influence of a single parameter on system response, and (ii) total sensitivities due to variation in all parameters together. The general approach to determining individual sensitivity coefficients is summarized below:

(i) Formulate a base case reference and its description,
(ii) Study and break down the factors into basic parameters (parameterization),
(iii) Identify parameters of interest and determine their base case values,
(iv) Determine which simulation outputs are to be investigated and their practical implications,
(v) Introduce perturbations to the selected parameters about their base case values one at a time,
(vi) Study the corresponding effects of the perturbation on the simulation outputs,
(vii) Determine the sensitivity coefficients for each selected parameter.

Sensitivity coefficients (also called, elasticity in economics, as well as influence coefficients) are defined in various ways as shown in Table. 11.2. The first group is identical to the partial derivative of the output variable OP with respect

Table 11.2 Different forms of sensitivity coefficient. (From Lam and Hui 1996)

Form	Formula[1]	Dimension	Common name(s)
1	$\frac{\Delta OP}{\Delta IP}$	With dimension	Sensitivity coefficient, influence coefficient
2a	$\frac{\Delta OP/OP_{BC}}{\Delta IP/IP_{BC}}$	$\frac{\%\ OP \text{ change}}{\%\ IP \text{ change}}$	Influence coefficient, point elasticity
2b	$\frac{\Delta OP/OP_{BC}}{\Delta IP}$	With dimension	Influence coefficient
3a	$\frac{\Delta OP/\frac{(OP_1+OP_2)}{2}}{\Delta IP/\frac{(IP_1+IP_2)}{2}}$	$\frac{\%\ OP \text{ change}}{\%\ IP \text{ change}}$	Arc mid-point elasticity
3b	$\left(\frac{\Delta OP}{\Delta IP}\right) / \left(\frac{\bar{OP}}{\bar{IP}}\right)$	$\frac{\%\ OP \text{ change}}{\%\ IP \text{ change}}$	(See note 2)

1. $\Delta OP, \Delta IP$ = changes in output and input respectively
   OP_{BC}, IP_{BC} = base case values of output and input respectively
   IP_1, IP_2 = two values of input
   OP_1, OP_2 = two values of the corresponding output
   $\bar{OP}, \bar{IP}$ = mean values of output and input respectively
2. For the form (3b), the slope of the linear regression line divided by the ratio of the mean output and mean input values is taken for determining the sensitivity coefficient

to the input parameter (IP). The second group uses the base case values to express the sensitivity in percentage change, while the third group uses the mean values to express the percentage change (this is similar to forward differencing and central differencing approaches used in numerical methods). Form (1) is used in comparative studies because the coefficient thus calculated can be used directly for error assessment. Forms (2a), (3a) and (3b) have the advantage that the sensitivity coefficients are dimensionless. However, form (3a) can only be applied to one-step change and cannot be used for multiple sets of parameters.

There are a wide range of such analysis methods as presented in a book by Saltelli et al. (2000) and in numerous technical papers. In such methods, the analyst is often faced with the difficult task of selecting the one method most appropriate for his application. A report by Iman and Helton (1985) compares different sensitivity analysis methods as applied to complex engineering systems, and summarizes current knowledge in this area. Of all the techniques, three have been found to be promising for multi-response sensitivities: (i) *Response surface*[1] replacement of the computer model where fractional factorial design is used to generate the response surface (see Sect. 6.4). This method is optimal if the models are linear; (ii) *Differential analysis* which is intended to provide information with respect to small perturbations about a point. However, this approach is not suited for complex models with large uncertainties; and (iii) *Latin hypercube Monte Carlo sampling* which offers several advantages and is described below. When the number of input parameters is large along with large uncertainty in the input parameters, and when the input parameters are interdependent and non-linear, Monte Carlo methods (though computationally more demanding) are simpler to implement and require a much lower level of mathematics while providing adequate robustness.

(b) Monte Carlo (MC) Methods The Monte Carlo (MC) approach, of which there are several types, comprise that branch of experimental mathematics which relies on experiments using random numbers to infer the response of a system (Hammersley and Handscomb 1964). MC is a general method of analysis where chance events are artificially recreated numerically (on a computer), the simulation run numerous times, and the results provide the necessary insights. MC methods provide approximate solutions to a variety of deterministic and stochastic problems, and hence their widespread appeal. The application of such methods for uncertainty propagation calculations was described in Sect. 3.7.3, and for low aleatory but high epistemic uncertainty problems in Sect. 12.2.7 in the framework of risk analysis and decision making involving stochastic system simulation. The many advantages of MC methods are: low level of mathematics, applicability to a large number of different types of problems, ability to account for correlations between inputs, and suitability to situations where model parameters have unknown distributions. They allow synthetic data to be generated from observations which have been corrupted with noise, usually additive or multiplicative of pre-selected magnitude (with or without bias) and specified probability distribution. This allows one to generate different sets of data sequences of pre-selected sample size, from which sampling distributions of the parameter estimates can be deduced and their sensitivity to the various assumptions evaluated.

MC methods are numerical methods in that all the uncertain inputs must be assigned a definite probability distribution. For each simulation, one value is selected at random for each input based on its probability of occurrence. Numerous such input sequences are generated and simulations performed[2]. Provided the number of runs is large, the simulation output values will be normally distributed irrespective of the probability distributions of the inputs. Though any non-linearity between the inputs and output are accounted for, the accuracy of the results depends on the number of runs. However, given the power of modern computers, the relatively large computational effort is not a major limitation

[1] Experimental design methods, such as 2^k factorial designs (see Sect. 6.3.1) also provide a measure of sensitivities and have been in existence for several decades. However, these methods have not been identified as promising since they only provide one-way sensitivity (i.e., the effect on the system response when only one parameter is varied at a time) rather than the multi-response sensitivity sought.

[2] There is a possibility of confusion between the bootstrap and Monte Carlo simulation approaches. The tie between them is obvious: both are based on repetitive sampling and then direct examination of the results. A key difference between the methods, however, is that bootstrapping uses the original or initial sample as the population from which to resample, whereas Monte Carlo simulation is based on setting up a sample data generation process for the inputs of the simulation or computational model.

except in very large simulation studies. The concept of "efficiency" has been used to compare different schemes of implementing MC methods. Say two methods, scheme 1 and 2, are to be compared. Method 1 calls for n_1 units of computing time (i.e., number of times that the simulation is performed), while method 2 calls for n_2 times. Also, let the resulting estimates of the response variable have variances σ_1^2 and σ_2^2. Then, the efficiency of method 2 with respect to method 1 is determined as:

$$\frac{\varepsilon_1}{\varepsilon_2} = \frac{n_1 \cdot \sigma_1^2}{n_2 \cdot \sigma_2^2} \tag{11.6}$$

where (n_1/n_2) is called the labor ratio, and (σ_1^2/σ_2^2) is called the variance ratio.

MC methods have emerged as a basic and widely used tool to quantify uncertainties associated with model predictions, and also for examining the relative importance of model parameters in affecting model performance (Spears et al. 1994). There are different types of MC methods depending on the sampling algorithm of generating the trials (Helton and Davis 2003):

(i) *Hit and miss methods* which were the historic manner of explaining MC methods. They involve using random sampling for estimating integrals (i.e., for computing areas under a curve and solving differential equations);
(ii) *Crude MC* using traditional random sampling where each sample element is generated independently following a pre-specified distribution;
(iii) *Stratified MC* (also called importance sampling) where the population is divided into groups or strata according to some pre-specified criterion, and sampling is done so that each strata is guaranteed representation (unlike the crude MC method). This method is said to be an order of magnitude more efficient than the crude MC method);
(iv) *Latin hypercube,* LHMC, uses stratified sampling without replacement, and is easiest to implement especially when the number of variables is large. It can be viewed as a compromise procedure combining many of the desirable features of random and stratified sampling. It produces more stable results than random sampling and does so more efficiently. It is easier to implement than stratified sampling for high dimension problems since it is not necessary to determine strata and strata probabilities. Because of its efficient stratification process, LHMC is primarily intended for long-running models.

LHMC is said to be one of the most promising methods for performing sensitivity studies in long-running complex models (Hofer 1999). LHMC sampling is conceptually easy to grasp. Say a sample of size n is to be generated from $\mathbf{p}=[p_1, p_2, p_3, \ldots p_n]$. The range of each parameter p_j is divided into n disjoint intervals of *equal probability* and one value is selected randomly from each interval. The n values thus obtained for p_1 are paired at random *without replacement* with similarly obtained n values for p_2. These n2-pairs are then combined in a random manner without replacement with the n values of p_3 to form n3-triples. This process is continued until a sample of np-tuples is formed. This constitutes one LHMC sample. How to modify this method to deal with correlated variables has also been proposed. The paper by Helton and Davis (2003) cites over 150 references in the area of sensitivity analysis, discusses the clear advantages of LHMC for analysis of complex systems, and enumerates the reasons for the popularity of such methods.

(c) Regional Sensitivity Analysis Once the LHMC runs have been performed, one needs to identify the strong or influential parameters and/or the weak ones appearing in the set of modeling equations; this is achieved by a process called *regional sensitivity analysis*. If the "weak" parameters can be fixed at their nominal values, and removed from further consideration in the calibration process, the parameter space would be reduced enormously, and somewhat alleviate the "curse of dimensionality". This model reduction is achieved in two steps: (i) filtering done so as to reject runs that fail to meet some prescribed criteria of model performance or goodness-of-fit of simulation outputs with measured system performance, and (ii) ascertain from the remaining runs which sets of parameters appeared more frequently; these can then be assumed to be the influential parameters.

Since running detailed simulation programs are computationally intensive and have long run-times, one cannot afford to perform separate simulation runs for identifying promising parameter vector combinations and for sensitivity analysis. Therefore, the following procedure can be adopted in order to satisfy both these objectives simultaneously. Assume that 30 "candidate" parameter vectors were identified with each parameter discretized into three states for performing the LHMC runs. One would expect that if the individual parameters were "weak" they would be randomly distributed among these 30 "candidate" vectors. Thus, the extent to which the number of occurrences of an individual parameter differs from 10 within each discrete state would indicate whether this parameter is strong or weak. This is a type of sensitivity test where the weak and strong parameters are identified using non-random pattern tests (Saltelli et al. 2000). The well-known chi-square χ^2 test for comparing distributions (see Sect. 2.4.3g) is advocated to assess statistical independence for each and every parameter. First, the χ^2 statistic is computed as:

$$\chi^2 = \sum_{s=1}^{3} \frac{(p_{obs,s} - p_{\exp})^2}{p_{\exp}} \tag{11.7}$$

where p_{obs} is the observed number of occurrences, and p_{exp} is the expected number (in this example above, this will be

Table 11.3 Critical thresholds for the Chi-square statistic with different significance levels for degrees of freedom 2

d. f	α =0.001	α =0.005	α =0.01	α =0.05	α =0.2	α =0.3	α =0.5	α =0.9
2	13.815	10.597	9.210	5.991	3.219	2.408	1.386	0.211

10), and the subscript s refers to the index of the state (in this case, there are three states). If the observed number is close to the expected number, the χ^2 value will be small indicating that the observed distribution fits the theoretical distribution closely. This would imply that the particular parameter is weak since the corresponding distribution can be viewed as being random. Note that this test requires that the degrees of freedom (d. f.) be selected as (number of states−1), i.e., in our case d. f.=2. The critical values for the χ^2 distribution for different significance levels α are given in Table. 11.3. If the χ^2 statistic for a particular parameter is greater than 9.21, one could assume it to be very strong since the associated statistical probability is greater than 99%. On the other hand, a parameter having a value of 1.386 (α=0.5) could be considered to be weak, and those in between the two values as uncertain in influence.

11.2.4 Case Study: Calibrating Detailed Building Energy Simulation Programs to Utility Bills

(a) Background The Oil shock of 1973 led to the widespread initiation of Demand Side Management (DSM) projects especially targeted at residential and small commercial building stock. Subsequently, in the 1980s, building professionals started becoming aware of the potential and magnitude of energy conservation savings in large buildings (office, commercial, hospitals, retail…). DSM measures implemented included any retrofit or operational practice, usually some sort of passive load curtailment measure during the peak hours such as installing thermal storage systems, retrofits to save energy (such as delamping, energy efficient lamping, changing constant air volume HVAC systems into variable air volume), demand meters in certain equipment (such as chillers), and energy management and control systems (EMCS) for lighting load management.

During the last decade, electric market transformation and utility de-regulation has led to a new thinking towards more pro-active load management of single and multiple buildings. The proper implementation of DSM measures involved first the identification of the appropriate energy conservation measures (ECMs), and then assessing their impact or performance once implemented. This need resulted in monitoring and verification (M&V) activities to acquire a key importance. Typically, retrofits involved rather simple energy conservation measures in numerous more or less identical buildings. The economics of these retrofits dictated that associated M&V also be low-cost. This led to utility bill analysis (involving no extra metering cost), and even analyzing only a representative sub-set of the entire number of retrofitted residences or small commercial buildings. In contrast, large commercial buildings have much higher utility costs, and the HVAC&R devices are not only more complex but more numerous as well. Hence, the retrofits were not only more extensive but the large cost associated with them justified a relatively large budget for M&V as well. The analysis tools used for DSM projects were found to be too imprecise and inadequate, which led to subsequent interest in the development of more specialized inverse modeling and analysis methods which use monitored energy data from the building along with other variables such as climatic variables and operating schedules. Most of the topics presented in this book have direct relevance to such specialized modeling and analysis methods.

A widely used technique is the calibrated simulation approach whereby the input parameters specifying the building and equipment necessary to run the detailed building energy simulation program are tuned so as to match measured data. Such a model/program would potentially allow more reliable and accurate predictions than with regression models or statistical approaches. Initial attempts, dating back to the early 1980s, involved using utility bills with which to perform the calibration. A large number of energy professionals are involved in performing calibrated simulations, and numerous more profess an active interest in this area. Further, the drastic spurt in activity by Energy Service Companies (ESCOs) led to numerous papers being published in this area (reviewed by Reddy 2006).

Calibrated simulation can be used for the following purposes:

(i) To prove/improve specific models used in a larger simulation program (Clarke 1993);
(ii) To provide insight to an owner into his building's thermal and/or electrical diurnal loadshapes using utility bill data (Sonderegger et al. 2001);
(iii) To provide an electric utility with a breakdown of baseline, cooling, and heating energy use for one or several buildings based on their utility bills in order to predict impact of different load control measures on the aggregated electrical load (Mayer et al. 2003);
(iv) To support investment-grade recommendations made by an energy auditor tasked to identify cost effective ECMs (equipment change, schedule change, control settings,..) specific to the individual building and determine their payback;
(v) For M&V under one or several of the following circumstances (ASHRAE 14 2002): (i) to identify a proper

contractual baseline energy use against which to measure energy savings due to ECM implementation; (ii) to allow making corrections to the contractual baseline under unanticipated changes (creep in plug load, changes in operating hours, changes in occupancy or conditioned area, addition of new equipment…); (iii) when the M&V requires that the effect of a end-use retrofit be verified using only whole building monitored data; (iv) when retrofits are complex and interactive (ex., lighting and chiller retrofits) and the effect of individual retrofits need to be isolated without having to monitor each sub-system individually; (v) either pre-retrofit or post-retrofit data may be inadequate or not available at all (for example, for a new building or if the monitoring equipment is installed after the ECM has been implemented), and (vi) when length of post-retrofit monitoring for verification of savings needs to be reduced;

(vi) To provide facility/building management services to owners and ESCOs the capability of implementing: (i) continuous commissioning or fault detection (FD) measures to identify equipment malfunction and take appropriate action (such as tuning/optimizing HVAC and primary equipment controls—Claridge and Liu 2001), (ii) optimal supervisory control, equipment scheduling and operation of building and its systems, either under normal operation or under active load control in response to real-time price signals.

(b) Proposed Methodology The case study described in this section is a slightly simplified version[3] of a research study fully documented in Reddy et al. (2007a, b). The calibration methodology proposed and evaluated is based on the concepts described in the previous section which can be summarized into the following phases (see Fig. 11.4):

(i) Gather relevant building, equipment and sub-system information along with performance data in the form of either utility bills and/or hourly monitored data;

(ii) Identify a building energy program which has the ability to simulate the types of building elements and systems present, and set up the simulation input file to be as realistic as possible;

(iii) Reduce the dimensionality of the parameter space by resorting to walk-thru audits and heuristics. For a given building type, identify/define a set of influential parameters and building operating schedules along with their best-guess estimates (or preferred values) and their range of variation characterized by either the minimum or the maximum range or the upper and lower 95th probability threshold values. The set of influential parameters to be selected should be such that they correspond to specific and easy-to-identify inputs to the simulation program;

(iv) Perform a "bounded" grid calibration (or unstructured or blind search) using LHMC trials or realizations with different combinations of input parameter values. Preliminary filtering or identification of a small set of the trials which meet pre-specified goodness-of-fit criteria along with regional sensitivity analysis to provide a means of identifying the weak and the strong parameters as well as determining narrower bounds of variability of these strong parameters;

(v) The conventional wisdom was that once a simulation model is calibrated with actual utility bills, the effect of different intended ECMs can be predicted with some degree of confidence by making changes to one or more of the model input parameters which characterize the ECM. Such thinking is clearly erroneous since the utility billing data is the aggregate of several end-uses within the building, each of which is affected by one or more specific and interacting parameters. While performing calibration, the many degrees of freedom may produce good calibration overall even though the individual parameters may be incorrectly identified. Subsequently, altering one or more of these incorrectly identified parameters to mimic the intended ECM is very likely to yield biased predictions. The approach proposed here involves using several *plausible predictions* on which to make inferences which partially overcome the danger of misleading predictions of "so-called" calibrated models. Thus, rather than using only the "best" calibrated solution of the input parameter set (determined on how well it fits the data) to make predictions about the effect of intended ECMs, a small number of the top *plausible* solutions are identified instead with which to make predictions. Not only is one likely to obtain a more robust prediction of the energy and demand reductions, but this would allow determining their associated prediction uncertainty as well.

(c) Description of Buildings The calibration methodology was applied to two synthetic office buildings in several locations and to one actual office building. The DOE-2 detailed public domain hourly building energy simulation program (Winkelmann et al. 1993) was used. The calibration was to be done for the prevalent case when only monthly utility billing data for a whole year were available. Moreover, it was presumed that the level and accuracy of knowledge about the building geometry, scheduling and various system equipment would be consistent with a "detailed investment grade" audit, involving equipment nameplate information as well as some limited onsite measurements (clamp-on meters,…)

[3] The original study suggested an additional phase involving refining the estimates of the strong parameters after the bounded grid search was completed. This could be done by one of several methods such as analytical optimization or genetic algorithms. This step has been intentionally left out in order not to overly burden the reader.

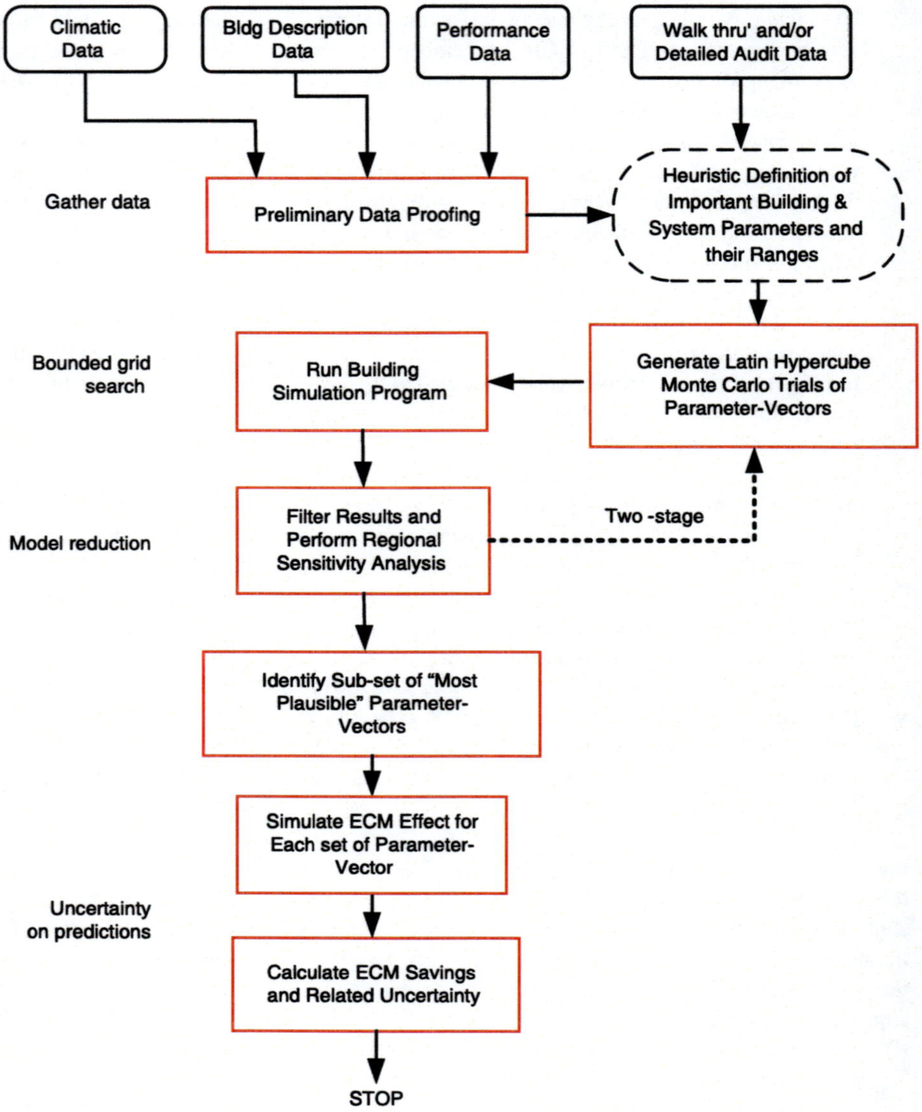

Fig. 11.4 Flowchart of the methodology for calibrating detailed building energy simulation programs. (Modified from Reddy et al. 2007a)

performed during different times of the day (morning, afternoon, night) as well as over different days of the week in order to better define operating schedules in some of the simulation inputs.

Evaluation using the synthetic buildings involved selecting a building and specifying its various construction and equipment parameters as well as its operating schedules (called reference values), and using the DOE-2 simulation program to generate "electric utility bill" data for a whole year coinciding with calendar months. The utility billing data are then assumed to be the measured data against which calibration is performed. Since the "correct" or reference parameters are known beforehand, one can evaluate the accuracy and robustness of the proposed calibration methodology by determining how correctly the calibrated models can fit the utility bill data, and also how accurately the effect of various ECMS can be predicted. The results of only one synthetic building are presented and discussed below while the interested reader can refer to the source documents for complete results.

The synthetic office building (S2) is similar to an actual building in terms of construction and mechanical equipment and is located in Atlanta, GA. It is a class A large building (20,289 m^2) with 7 floors and a penthouse with the lobby, cafeteria, service areas and mechanical/electrical rooms on the first floor, and offices in the remaining floors. Building cooling is provided by electricity while heating is met by natural gas (in units of Therms). Table. 11.4 assembles a list of heuristically-identified influential parameters which have simple and clear correspondence to specific and easy to

Table 11.4 List of influential parameters for the complex office building category S2. The minimum, base (or best guess), and maximum values characterize the assumed range while the reference values are those used to generate the synthetic utility bills used for calibration

No		Description	Variable type	Unit	Minimum value	Base value	Maximum value	Reference case value
1	Load schedules (rooms)	Lighting schedule	D	NA	OffLgt_1	OffLgt_D	OffLgt_2	OffLgt_D
2		Equipment schedule	D	NA	OffEqpt_1	OffEqpt_D	OffEqpt_2	OffEqpt_2
3		Auxiliary equipment schedule	D	NA	AuxOffEqpt_1	AuxEqpt_D	AuxOffEqpt_2	AuxEqpt_1
4	System schedules (zones)	Fans schedule	D	NA	OffFan_1	OffFan_D	OffFan_2	OffFan_2
5		Space heating temperature schedule	D	NA	OffHtT_1	OffHtT_D	OffHtT_2	OffHtT_1
6		Space cooling temperature schedule	D	NA	OffClT_1	OffClT_D	OffClT_2	OffClT_D
7		Outside air (ventilation) schedule	D	NA	OffOA_1	OffOA_D	OffOA_2	OffOA_2
8	Envelop loads	Window shading coefficient	C	Fraction	0.16	0.75	0.93	0.8
9		Window U value	C	Btu-h/°F*Sqft	0.25	0.57	1.22	0.65
10		Wall U value	C	Btu-h/°F*Sqft	0.0550	0.1000	0.5800	0.3000
11	Internal loads (rooms)	Lighting power density	C	W/Sqft	1.3	1.7	2	1.8
12		Equipment power density	C	W/Sqft	0.8	1.0	1.2	0.9
13	System variables	Supply fan power/delta_T	C	kW/CFM	0.00124	0.00145	0.00166	0.00135
14		Energy efficiency ratio (EIR)	C	Fraction	0.1564	0.1849	0.2275	0.20478
15		On hours control	D	NA	VFD	IGV	IGV	IGV
16		Off hours control	D	NA	OFF	Cycle on any Zone	Cycle on any Zone	Cycle on any Zone
17		Minimum supply air flow	C	Fraction	0.3	0.65	1.0	0.7
18		Economizer	D	NA	Yes	Yes	No	Yes
19		Minimum outside air	C	Fraction	0.1	0.3	0.5	0.4
20	Auxiliary electrical loads	Auxiliary electrical loads—non-HVAC effect	C	kW	25	50	75	65
21		Cooling tower fan power	C	BHP/GPM	0.0118	0.0184	0.0212	0.0189
22		Cooling tower fan control	D	NA	VFD	Two speed	Single speed	Two speed
23		Primary CHW & cond.pump flow	C	GPM/Ton	2.25	2.7	3.38	2.9
24		Boiler efficiency ratio	C	Fraction	1.25	1.43	1.54	1.33

C continuous, *D* discrete

The schedules (parameters 1–7) involve a vector of 24 hourly values which can be found in Reddy et al. (2007a, b)

identify inputs to the DOE-2 simulation program. The simple building category includes 24 parameters consisting of 7 discrete schedules, 13 continuous parameters, and 4 binary parameters (i.e., either on or off or one of only two possible categories). Table. 11.4 also contains information about the range of the 24 parameters (i.e., the minimum and the maximum values which the parameter can assume) as well as the base or preferred value, i.e., the values of the various parameters which the analyst deems most likely. These may be (and are) different from the reference case values which were used to generate the synthetic data. It should be pointed out that the discrete parameters P1–P7 are diurnal schedules which consist of a set of 24 hourly values (which are fully described in the source documents).

(d) Bounded Grid Search This phase, first, involves a blind LHMC search with the range of each parameter divided into three intervals of equal probability followed by a regional sensitivity analysis. Here, Monte Carlo (MC) filtering allows rejecting sets of model simulations that fail to meet some prescribed criteria of model performance which combines all three energy channels (electricity use and demand, and gas use). Such goodness of fit (GOF) criteria are based on the normalized mean bias error (NMBE) or the coefficient of variation (CV) or on a combined index which weights both of them (GOF-Total). Once a LHMC batch run of several trials is completed, the GOF_{CV} and GOF_{NMBE} indices are computed for each trial, from which those parameter vectors which result in high GOF numbers, (i.e., those whose predictions fit the utility bills poorly) can be weeded out. The information contained in these "good" or promising parameter vectors can be used to identify the weak parameters which can then be removed from further consideration in our calibration process.

For the sake of brevity only a representative selection of results are assembled in Table. 11.5 for building S2. Two

Table 11.5 Summary of various sets of runs for building S#2. The range of the 24 input parameters are given in Table. 11.4 Equal weighting factors for kWh/kW/Gas utility bills are assumed while a ratio of 3 : 1 has been selected for NMBE: CV values. Calibrated solutions are considered feasible if both GOF_NMBE and GOF_CV<15%

Run	Number of LHMC trials	Number of input parameters	Number of feasible trials	GOF_Total of top 20 solutions		
		–		Min %	Max %	Median %
1a	1,500	24	141	5.26	7.66	6.75
2a	3,000	24	326	3.75	6.57	6.00
3a	5,000	24	548	2.80	5.88	4.85
2b	3,000/3,000	24/19	391	3.34	6.45	5.94

a single-stage calibration
b two stage calibration involves freezing the weak parameters whose Chi-square value < 1.4

variants have been analyzed: one stage calibration (coded "a") and two-stage calibration (coded "b"). The two-stage calibration simply refers to a process where an additional LHMC set of runs are performed with the weak parameters identified during the first set frozen at their best guess values[4]. This variant was investigated to see whether the calibration process is improved as a result. It was found that though there was a small improvement (compare runs 2a and 2b in Table. 11.5), the improvement was not significant. The number of trials for each run, as well as the minimum, maximum and the median of the top 20 calibrated runs in terms of GOF_Total are shown in the table. Equal weighting factors for electric energy use (kWh) and demand (kW) and gas use were assumed while a ratio of 3:1 has been selected for NMBE: CV values while computing GOF-Total. Calibrated solutions are considered *feasible* if both GOF_NMBE and GOF_CV < 15% which is consistent with the recommendation of ASHRAE Guideline 14 (2002). This translated into GOF_Total of 11% when using these weights. One notes from Table. 11.5 that only about 10% of the total number of LHMC trials are deemed to contain feasible sets of input parameters.

As expected, the number of LHMC trials is a major factor; increasing this number results in better calibration. For good accuracy of fits it would seem that about 3,000–5,000 calibration trials would be needed. Figure 11.5 indicates how the GOF indices (NMBE and CV) of the three energy use quantities scatter about the origin for each of the 3,000 LHMC trials. Only the parameter vectors corresponding to those runs which cluster around the origin are selected for further analysis involving identifying strong and weak parameters. How well the best run (i.e., the one with lowest

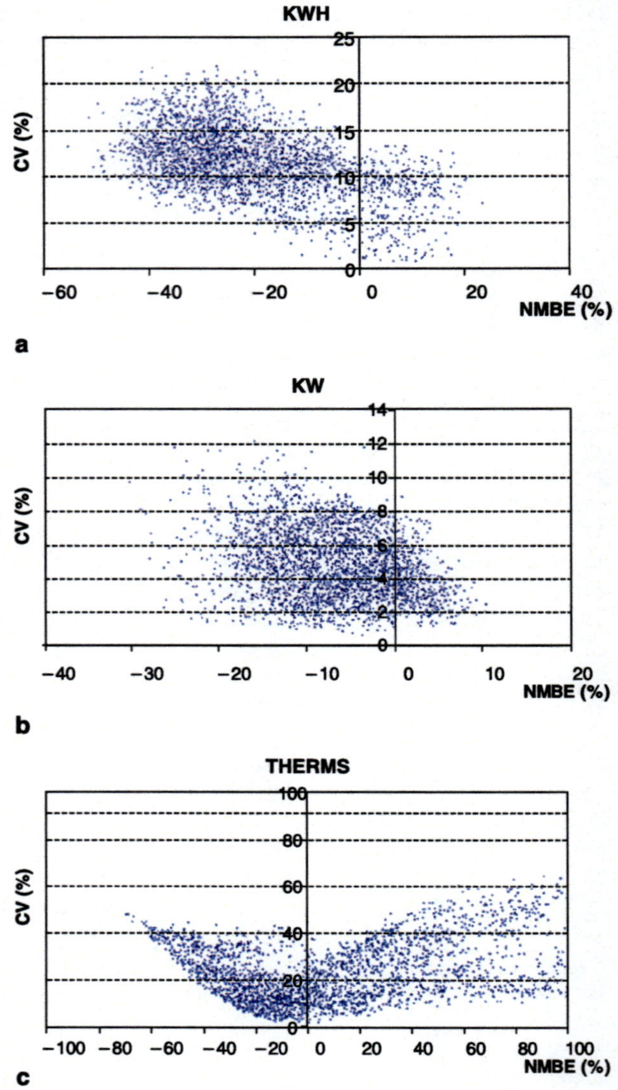

Fig. 11.5 Scatter plots depicting goodness-of-fit of the three energy use channels with 3,000 LHMC trials. Only those of the numerous trials close to the origin are then used for regional sensitivity analysis. **a** Electricity use. **b** Electricity demand. **c** Gas thermal energy

GOF_Total) for the 3,000 LHMC trials is able to track the actual utility bill data is illustrated by the three time series plots of Fig. 11.6. One notices that the fits are excellent (in fact, the top dozen or so runs are very close, and those from a LHMC run with 5,000 trials, even better). An upper limit to the calibration accuracy for the best trial seems to be about 2%, and for the median of the top 20 calibrated solutions to be around 4–7%.

(e) Uncertainty in Calibrated Model Predictions The issue of specific relevance here relates to the accuracy with which the calibrated solutions are likely to predict energy and demand

[4] There were a large number of parameters identified as strong parameters (about 8–12 out of 20 input parameters). Further, it was not always clear as to which of the three equal-probability intervals to select. Hence, for the two-stage calibration, it was more practical to freeze the weak parameters rather than the strong parameters.

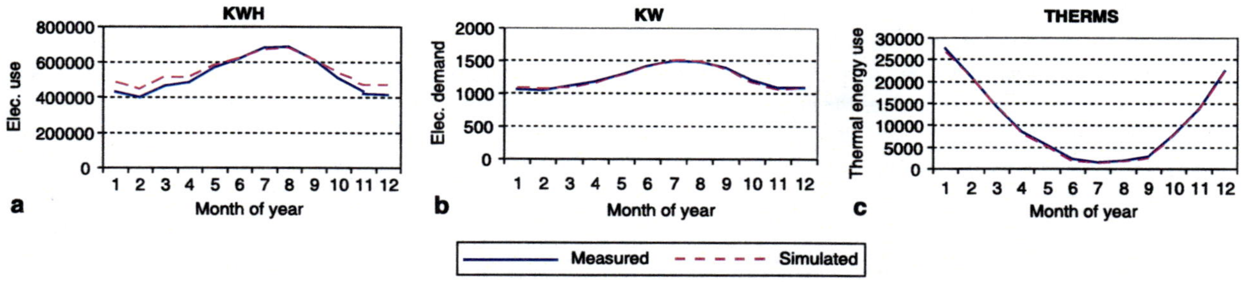

Fig. 11.6 Time series plots of the three energy use channels for the best calibrated trial corresponding to one-stage run with 3,000 LHMC trials (Run 2a of Table. 11.5). **a** Electricity use. **b** Electricity demand. **c** Gas thermal energy

savings if specific energy conservation measures (ECMs) were to be implemented. In other words, one would like to investigate the ECM savings and its associated uncertainty as predicted by the calibrated solutions. As stated earlier, relying solely on the predictions of any *one* calibration (even if it were to fit the utility bill data very accurately) is unadvised, since it provides no guarantee that individual parameters have been accurately tuned. Instead, the calibration approach is likely to be more robust if a small set of most plausible solutions are identified instead. The effect of large deviations from any individual outlier predictions can be greatly minimized (if not eliminated) by positing that the actual value is likely to be bracketed by the inter-quartile standard deviation around the median value. This would provide greater robustness by discounting the effect of outlier predictions. The top 20 calibrated solutions were selected (somewhat arbitrarily). The median and the inter-quartile standard deviation (i.e., the 10 trials whose predicted savings are between the 25 and 75% percentiles) are then calculated from these 20 predictions.

The predictive accuracy of the calibrated simulations have been investigated using four different sets of ECM measures, some of which would result in increased energy use and demand, such as ECM_C (Table. 11.6). Retrofits for ECM_A involve modifying the minimum supply air flow (parameter P17)

Table 11.6 Predictions of electricity use and monthly demand savings for synthetic building S2 by various calibration runs. The "correct" savings and the savings predicted by the inter-quartile values of the Top 20 calibrated solutions are shown for the five ECM measures simulated

ECM number	Parameters	Baseline values	New values	Run[a]	Number of LHMC trials	Median GOF_Total	Predicted savings (Top20 solutions)			
							kWh%		kW%	
							Median %	Stdev[b]	Median %	Stdev[b]
ECM_A	P17/P19	0.65/0.30	0.30/0.10	**Exact**	–	–	**18.47**	**0.00**	**8.76**	**0.00**
				2a	1,500	6.75	19.08	0.98	12.79	1.32
				4a	3,000	6.00	18.06	4.73	12.00	1.85
				4b	3,000	5.94	18.43	1.75	14.00	2.33
				6a	5,000	4.85	18.85	1.33	12.31	1.07
ECM_B	P17/P19	0.65/0.30	0.20/0.25	**Exact**	–	–	**19.69**	**0.00**	**6.06**	**0.00**
				2a	1,500	6.75	20.05	0.68	11.35	1.04
				4a	3,000	6.00	19.32	4.22	9.68	2.22
				4b	3,000	5.94	20.58	2.00	11.26	2.24
				6a	5,000	4.85	20.12	1.24	9.98	1.24
ECM_C	P11/P12	1.67/1.0	3.0/3.0	**Exact**	–	–	**–50.31**	**0.00**	**–58.63**	**0.00**
				2a	1,500	6.75	–47.79	1.26	–52.92	0.75
				4a	3,000	6.00	–50.32	2.90	–55.39	1.25
				4b	3,000	5.94	–48.56	2.57	–54.99	1.06
				6a	5,000	4.85	–45.43	1.88	–54.10	0.88
ECM_D	P8/P9/P14	0.64/0.64/0.19	0.40/0.4/0.50	**Exact**	–	–	**–30.32**	**0.00**	**–36.76**	**0.00**
				2a	1,500	6.75	–33.65	5.47	–29.74	3.28
				4a	3,000	6.00	–32.97	3.52	–35.10	2.86
				4b	3,000	5.94	–29.20	4.86	–33.82	3.64
				6a	5,000	4.85	–26.42	1.33	–29.61	1.46

[a] *a* refers to one-stage and *b* refers to two-stage calibration

[b] *Stdev* standard deviation of only those runs in the inter-quartile range (i.e., between 25 and 75%)

and the minimum outdoor air fraction (P19), both of which are strong parameters. The reference case corresponds to a value of P17=0.30 (reduced from 0.65) and a value of P19=0.1 (down from 0.3). The results are computed as the predicted % savings in energy use and demand compared to the original building. For example, the % savings in kWh have been calculated as:

$$\%\text{KWH} = 100 \times \frac{\text{Baseline annual kWh} - \text{Predicted annual kWh}}{\text{Baseline annual kWh}} \quad (11.8)$$

Table. 11.6 assembles the results of all the runs for all four ECMs considered. The base values of the parameters as well as their altered values are also indicated. For example, ECM_A entails changing the baseline values of 0.65 and 0.30 for P17 and P19 respectively to 0.3 and 0.1 respectively. The "correct" savings values of the three channels of energy data are shown bolded. For example, the above ECM_A change would lower electricity use and demand by 18.47% and 8.76% respectively. The same information is plotted in Fig. 11.7 for easier visualization where the correct values are shown as boxes without whiskers. One note that the calibration methodology seems to work satisfactorily for the most part (with the target values contained within the spread of the whiskers of the inter-quartile range of the top 20 calibrated simulation predictions) for kWh savings though there are deviations for ECM_C and ECM_D. However, there are important differences for monthly demand (kW savings), especially for ECM_A and ECM_B. This is somewhat to be expected considering that electric demand for each month is a one-time occurrence and is likely to be harder to pin down accurately. Monthly electricity use, on the other hand, is a value summed over all hours of each month and is likely to be more stable.

Generally, it would seem that LHMC trials should be of the order of 3,000 trials/run for this building. Further, it is also noted that there is little benefit in performing a two-stage calibration. Another observation from Fig. 11.7 is that, generally, the uncertainty bands for the high 5,000 trials/run cases are narrower compared to those for the 3,000 trial/run cases but the median values are not less biased (in fact they seem to be worse in a slight majority of cases). This trend suggests that the danger of biased predictions does not improve with larger number of LHMC runs, and other solutions need to be considered.

There were several ad hoc elements in the above calibration methodology. Another study by Sun and Reddy (2006) attempted to frame the problem in a general analytic framework with firmer mathematical and statistical basis. For example, the number of parameters deemed influential were selected by first normalizing them against the Euclidean norm, while an attempt was made to infer the number of parameters which one could hope to tune with the data at hand from the order of the highest order definite submatrix (or rank of the Hessian marix) whose condition number is less than a predefined threshold value (refer back to Sect. 10.2 for explanation of these terms). The overall conclusion was that trying to calibrate a detailed simulation program with only utility bill information is *never likely to be satisfactory* because of the numerical identifiability issue (described in Sect. 10.2.3). One needs to either enrich the data set by non-intrusive submonitoring, at say, hourly time scales for a few months if not the whole year, or reduce the number of strong parameters to be tuned by performing controlled experiments when the building is unoccupied, and estimate their numerical values (the LHMC process could be used to identify the strong parameters needing to be thus determined). This case study example was meant to illustrate the general approach; this whole area of calibrating detailed building energy simulation programs is still an area requiring further research before it can be used routinely by the professional community.

11.3 Model Selection and Identifiability

11.3.1 Basic Notions

Model selection is the process of identifying the functional form or model structure, while *parameter estimation* is the process of identifying the parameters in the model. These two distinct but related issues are often jointly referred to as *system identification*[5] (terms such as complete identification, structure recovery, and system characterization are also used). Note that the use of the word *estimation* has a connotation of uncertainty. This uncertainty in determining the parameters of the model which describe the physical system is unavoidable, and arises from simplification errors and noise invariably present both in the physical model and in the measurements. The challenge is to decide on how best to simplify the model and design the associated experimental protocol so as to minimize the uncertainty in our parameter estimates from data corrupted by noise. Thus, such analyses not only require knowledge of the physical behavior of the system, but also presume expertise in modeling, designing and performing experiments and in regression/search methods. Many of the chapters in this book (specifically Chaps. 1, 5, 6, 7, 9 and 10) directly pertain to such issues.

System identification is formally defined as the determination of a mathematical model of a system based on its input-output information over a relatively short period of time, usually for either understanding the system or phe-

[5] Though there is a connotational difference between the words "identification" and "estimation" in the English language, no such difference is usually made in the field of inverse modeling. Estimation is a term widely used in statistical mathematics to denote a similar effect as the term identification which appears in electrical engineering literature.

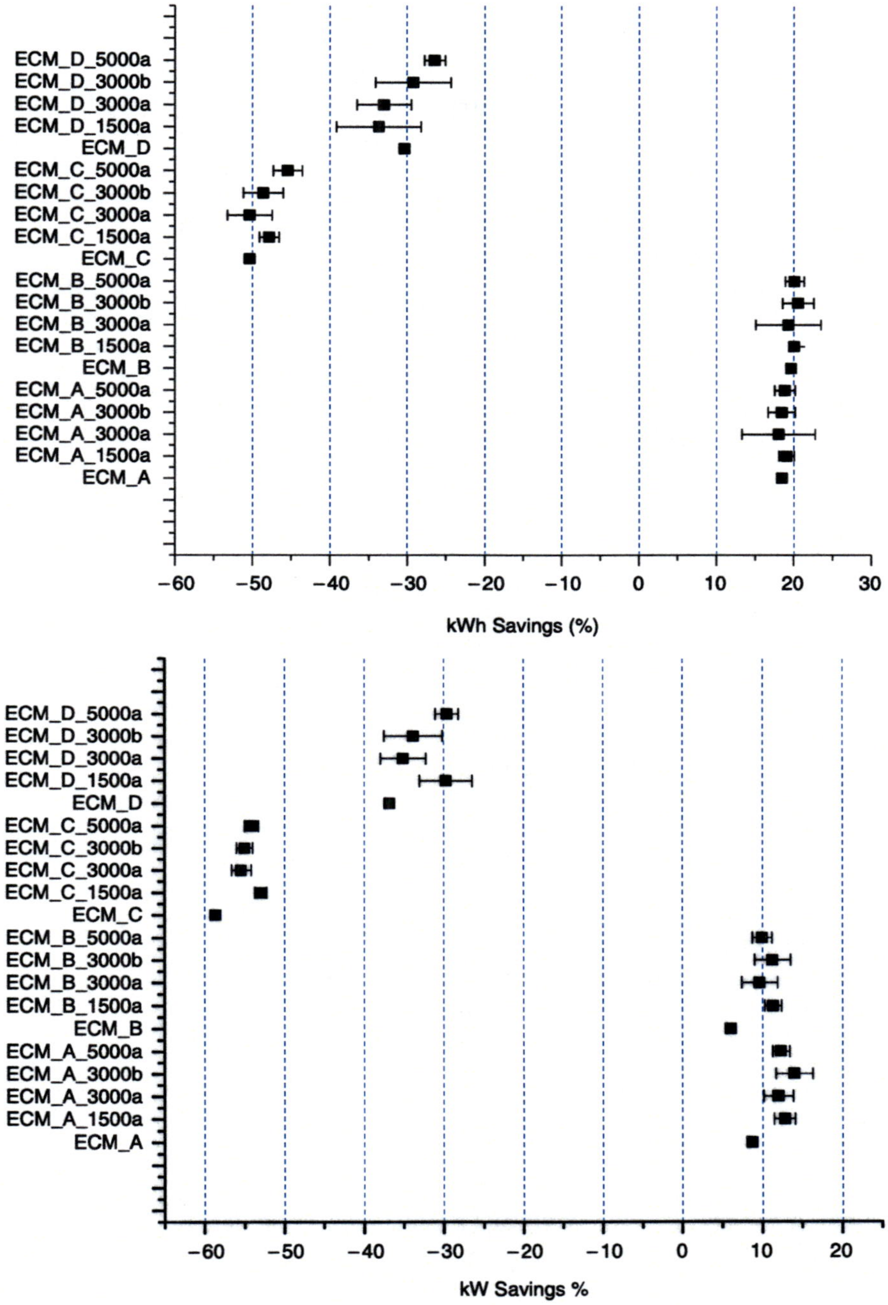

Fig. 11.7 Electricity use (kWh) and monthly demand (kW) savings (in % of the baseline building values) for the four ECM measures predicted by the top20 calibrated solutions from different number of LHMC trials (see Table. 11.6). The "correct" values are the ones without whiskers

nomenon being studied, or for making predictions of system behavior. It deals with the choice of a specific model for a class of models that are mathematically equivalent to a given physical system. Model selection problems involve under-constrained problems with degrees of freedom greater than zero where an infinite number of solutions are possible. One differentiates between (i) situations when nothing is known about the system functional behavior, sometimes referred to as *complete identification problems,* requiring the use of black-box models, and (ii) partial identification problems wherein some insights are available and allow grey-box models to be framed to analyze the data at hand. The objectives are to identify the most plausible system models and the most likely parameters/properties of the system by performing certain statistical analyses and experiments. The difficulty is that several mathematical expressions may appear to explain the input-output relationships, partially due to the presence of unavoidable errors or noise in the measurements and/or limitations in the quantity and spread of data available. Usually, the data is such that it can only support models with limited number of parameters. Hence, by necessity (as against choice) are the mathematical inverse models *macroscopic* in nature, and usually allow determination of only certain essential properties of the system. Thus, an important and key conceptual difference between forward (also called *microscopic* or microdynamic) models and system identification inverse problems (also called macrodynamic) is that one should realistically expect the latter to involve models containing only a few aggregate interactions and parameters. Note that knowledge of the physical (or structural) parameters is not absolutely necessary if internal prediction or a future (in time) forecast is the only purpose. Such forecasts can be obtained through the reduced form equations directly and exact deduction of parameters is not necessary (Pindyck and Rubinfeld 1981).

Classical inverse estimation methods can be enhanced by using the Bayesian estimation method which is more subtle in that it allows one to include prior subjective knowledge about the value of the estimate or the probability of the unknown parameters in conjunction with information provided by the sample data (see Sect. 2.5 and 4.6). The prior information can be used in the form of numerical results obtained from previous tests by other researchers or even by the same researcher on identical equipment. Including prior information allows better estimates. For larger samples, the Bayesian estimates and the classical estimates are practically the same. It is when sample are small that Bayesian estimation is particularly useful.

The following sub-sections will present the artificial neural network approach to identifying black-box models and also discuss modeling issues and techniques relevant to grey-box models.

11.3.2 Local Regression—LOWESS Smoothing Method

Sometimes the data is so noisy that the underlying trend may be obscured by the data scatter. A non-parametric black-box method called the "locally weighted scatter plot smoother" or *LOWESS* (Devore and Farnum 2005) can be used to smoothen the data scatter. Instead of using all the data (such as done in traditional ordinary least squares fitting), the intent is to fit a series of lines (usually polynomial functions) using a prespecified portion of the data. Say, one has n pairs of (x, y) observations, and one elects to use subsets of 20% of the data at a time. For each individual x_0 point, one selects 20% of the closest x-points, and fit a polynomial line with only this subset of data. One then uses this model to predict the corresponding value of the response variable $\hat{y}_0$ at the individual point x_0. This process is repeated for each of the n points so that one gets n sets of points of $(x_0, \hat{y}_0)$. The LOWESS plot is simply the plot of these n sets of data points. Figure 11.8 illustrates a case where the LOWESS plot indicated a trend which one would be hard pressed to detect in the original data. Looking at Fig. 11.8a which is a scatter plot of the characteristics of bears, with x as the chest girth of the bear, and y its weight, one can faintly detect a non-linear behavior; but it is not too clear. However, the same data when subject to the LOWESS procedure (Fig. 11.8b) assuming a pre-specified portion of 50% reveals a clear bi-linear behavior with a steeper trend when x > 38. In conclusion, LOWESS is a powerful functional estimation method which is local (as suggested by the name), non-parametric, and can potentially capture any arbitrary feature present in the data.

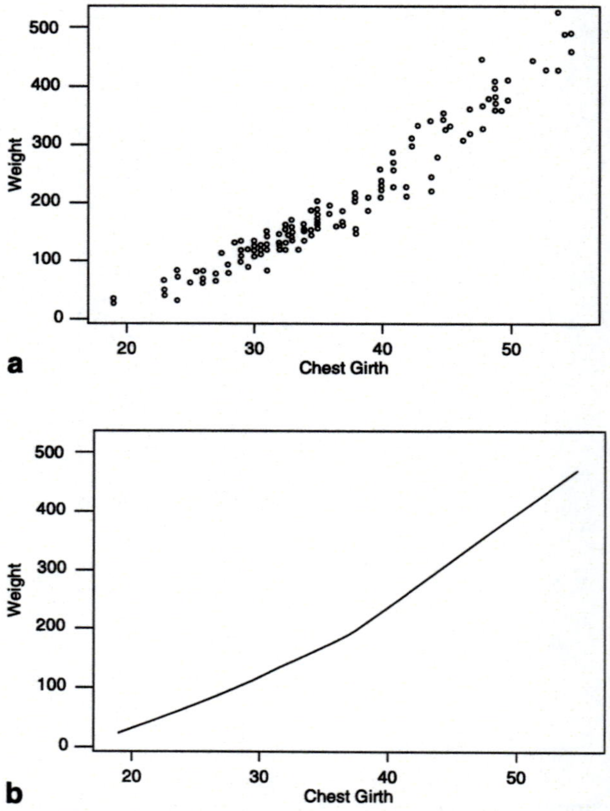

Fig. 11.8 Example to illustrate the insight which can be provided by the LOWESS smoothing procedure. The data set is meant to detect the underlying pattern between the weight of a wild bear and its chest girth. While the traditional manner of plotting the data (frame a) suggests a linear function, (frame b) assuming a 50% prespecified smoothing length, reveals a bi-linear behavior. (From Devore and Farnum 2005 by © permission of Cengage Learning)

11.3.3 Neural Networks—Multi-Layer Perceptron (MLP)

A widely used black-box approach is based on neural networks (NN) which are a class of mathematical models based on the manner in which the human brain functions while performing such activities as decision-making, pattern or speech recognition, image or signal processing, system prediction, optimization and control (Wasserman 1989). NN grew out of research in artificial intelligence, and hence, they are often referred to as artificial neural networks. NN possess several unique attributes that allow them to be superior to the traditional methods of knowledge acquisition (of which data analysis and modeling is one specific activity). They have the ability to: (i) exploit a large amount of data/information, (ii) respond quickly to varying conditions, (iii) learn from examples, and to generalize underlying rules of system behavior, (iv) map complex nonlinear behavior for which input-output variable set is known but not their structural interaction (viz, black-box models), and (v) ability to handle noisy data, i.e., have good robustness. The last 30 years have seen an explosion in the use of NN. These have been successfully applied across numerous disciplines such as engineering, physics, finance, psychology, information theory and medicine. For example in engineering, NN have been used for system modeling and control, short-term electric load forecasting, fault detection and control of complex systems. Stock market prediction and classification in terms of credit-worthiness of individuals applying for credit cards are examples of financial applications. Medical applications of NN involve predicting the probable onset of certain medical conditions based on a variety of health-related indices. In this book, let us limit ourselves to its model building capability.

There are numerous NN topologies by which the relati–onship between one or more response variable(s) and one or more regressor variable(s) can be framed (Wasserman 1989; Fausett 1993). A widely used architecture for applications involving predicting system behavior is the feed-forward multi-layer perceptron (MLP). A typical MLP consists of an input layer, one or more hidden layers and an output layer (see Fig. 11.9). The input layer is made up of discrete nodes (or neurons, or units), each of which represents a single regressor variable, while each node of the output layer represents a single response variable. Only one output node is shown but the extension to a set of response variables is obvious. Networks can consist of various topographies; for example, with no hidden layers (though this is uncommon), with one hidden layer, and with numerous hidden layers. It is the wide consensus that except in rare circumstances, a MLP architecture with only one hidden layer is usually adequate for most applications. While the number of nodes in the input and output layers are dictated by the specific problem at hand, the number of nodes in each of the hidden layers is a design choice (certain heuristics are described further below).

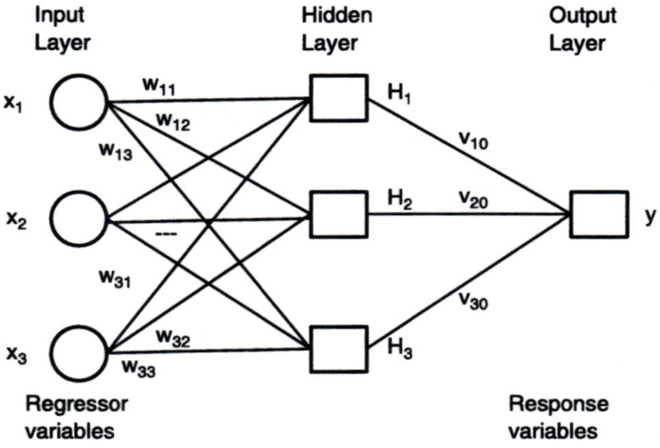

Fig. 11.9 Feed-forward multilayer perceptron (MLP) topology for a network with 3 nodes in the input layer, 3 in the hidden layer and one in the output layer denoted by MLP(3,3,1). The weights for each of the interactions between the input and hidden layer nodes are denoted by ($w_{11}, \ldots w_{33}$) while those between the hidden and the output nodes are denoted by (v_{10}, v_{20}, v_{30}). Extending the architecture to deal with more nodes in any layer and to more number of hidden layers is intuitively straightforward. The square nodes indicate those where some sort of processing is done as elaborated in Fig. 11.10

Each node of the input and first hidden layers are connected by lines to indicate information flow, and so on for each successive layer till the output layer is reached. This is why such a representation is called "feed-forward". The input nodes or the **X** vector are multiplied by an associative weight vector **W** representative of the strength of the specific connection. These are summed so that (see Figs. 11.9 and 11.10):

$$NET = (w_{11}x_1 + w_{21}x_2 + w_{31}x_3) = \sum XW \qquad (11.9)$$

For each node i, the NET_i signal is then processed further by an activation function:

$$OUT_i = f(NET_i) + b_i \qquad (11.10)$$

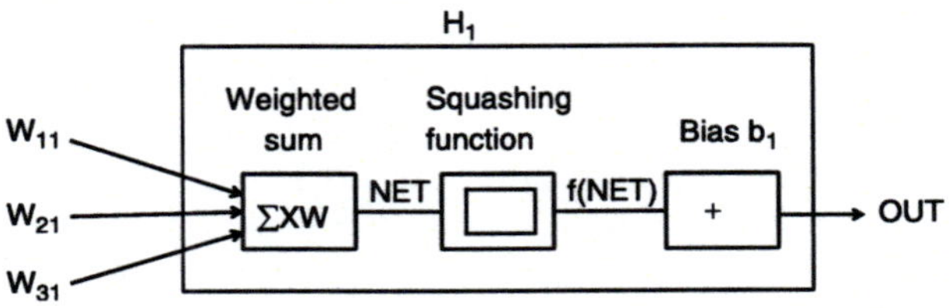

Fig. 11.10 The three computational steps done at processing node H_1 represented by square nodes in Fig. 11.9. The incoming signals are weighted and summed to yield the NET which is then transformed non-linearly by a squashing (or basis function) to which a bias term is added. The resulting OUT signal becomes the input to the next node downstream to which it is connected, and so on. These steps are done at each of the processing nodes

The activation function f(NET) (also called basis function or squashing function) can be a linear mapping function (this would lead to the simple linear regression model), but more generally, the following monotone sigmoid forms are adopted:

$$f(NET_i) \equiv [1 + \exp(-NET_i)]^{-1} \text{ or}$$
$$f(NET_i) \equiv \tanh(NET_i) \quad (11.11)$$

Logistic or hyperbolic functions are selected because of their ability to squash or limit the values of the output within certain limits such as (–1,1). It is this mapping which allows non-linearity to be introduced in the model structure. The bias term b is called the activation threshold for the corresponding node and is introduced to avoid the activation function getting stuck in the saturated or limiting tails of the function. As shown in Fig. 11.9, this process is continued till an estimate of the output variable $\hat{y}$ is found.

The weight structure determines the total network behavior. Training the MLP network is done by adjusting the network weights in an orderly manner such that each iteration (referred to as "epoch" in NN terminology) results in a step closer to the final value. Numerous epochs (of the order of 100,000 or more) are typically needed; a simple and quick task for modern personal computers. The gradual convergence is very similar to the gradient descent method used in nonlinear optimization or during estimating parameters of a non-linear model. The loss function is usually the squared error of the model residuals just as done in OLS. Adjusting the weights as to minimize this error function is called *training the network* (some use the terminology "learning by the network"). The most commonly used algorithm to perform this task is the *back-propagation training algorithm* where partial derivatives (reflective of the sensitivity coefficients) of the error surface are used to determine the new search direction. The step size is determined by a user-determined learning rate selected so as to hasten the search but not lead to over-shooting and instability (notice the similarity of the entire process with the traditional non-linear search methodology described in Sect. 7.4). As with any model identification where one has the possibility of adding a large number of model parameters, there is the distinct possibility of overfitting or over-training, i.e., fitting a structure to the noise in the data. Hence, it is essential that a sample cross-validation scheme be adopted such as that used in traditional regression model identification (see Sect. 5.3.2). In fact, during MLP modeling, the recommended approach is to sub-divide the data set used to train the MLP into three subsets:

(i) the *training data set* used to evaluate different MLP architectures and to train the weights of the nodes in the hidden layer;
(ii) *validation data set* meant to monitor the performance of the MLP during training. If the network is allowed to train too long, it tends to over-train leading to a loss in generality of the model;
(iii) *testing data set*, similar to the cross-validation data set, meant to evaluate the predictive or external accuracy of the MLP network using such indices as the CV and NMBE. These are often referred to as *generalization errors*.

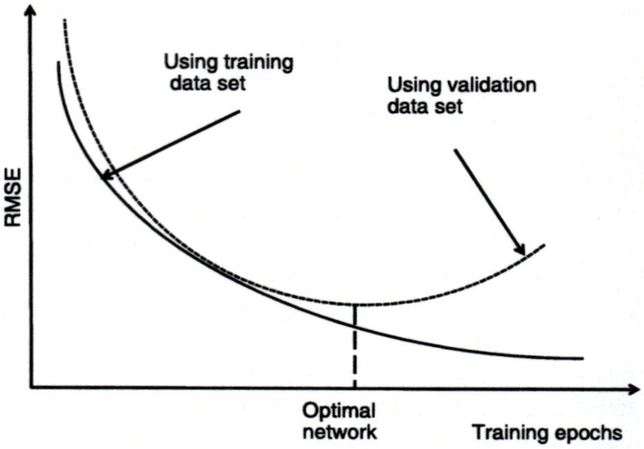

Fig. 11.11 Conceptual figure illustrating how to select the optimal MLP network weights based on the RMSE errors from the training and validation data sets

The usefulness of the validation data set during training of a specific network is illustrated in Fig. 11.11. During the early stages of training, the RMSE of both the training and validation data sets drop at the same rate. The RMSE for the training data set keeps on decreasing as more iterations (or epochs) are performed. At each epoch, the trained model is applied to the validation data set, and the corresponding RMSE error computed. The validation error eventually starts to rise; it is at this point that training ought to be stopped. The node weights at this point are the ones which correspond to the optimal model. Note, however, that this process is specific to a preselected architecture, and so different architectures need to be tried in a similar manner. Compare this process with the traditional regression model building approach wherein the optimal solution is given by closed form solutions and no search or training is needed. However, the need to discriminate between competing model functions still exists and statistical indices such as RMSE or R^2 and adjusted R^2 are used for this purpose (see Sect. 5.3.2).

MLP networks have also been used to model and forecast dynamic system behavior as well as time series data over a time horizon. There are two widely used architectures. The time series feed forward network shown in Fig. 11.12a is easier and more straightforward to understand and implement, and so is widely used. A recurrent network is one where the outputs of the nodes in the hidden layer are fed back as inputs to previous layers. This allows a higher level of non-linearity to be mapped, but they are more generally

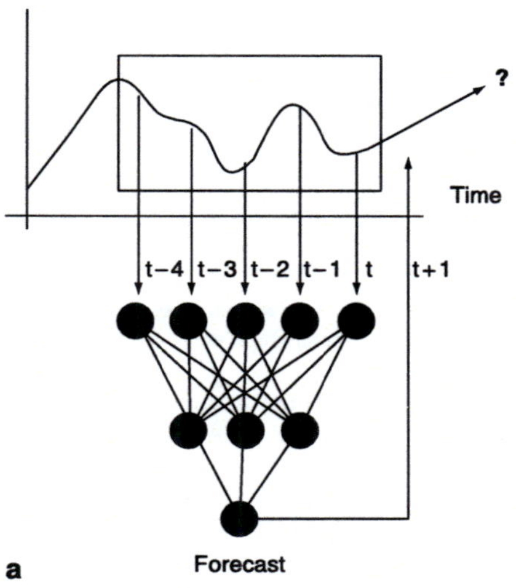

Fig. 11.12 Two different MLP model architectures appropriate for time series modeling. The simpler feed forward network uses time-lagged values till time t to predict a future value at time (t+1). Recurrent networks use internally generated past data with only the current value for prediction, and are said to be generally more powerful while, however, needing more expertise in their proper use. **a** Feedforward network. **b** Recurrent network. (From SPSS 1997)

difficult to train properly. Many types of interconnections are possible with the more straightforward network arrangement for one hidden layer shown in Fig. 11.12b. The networks assumed in these figures use the past values of the variable itself; extensions to the multivariate case would involve investigating different combinations of present and lagged values of the regressor variables as well.

Some useful heuristics with MLP training are assembled below (note that some of these closely parallel those followed by traditional model fitting):

(i) in most cases, one hidden layer should be adequate;
(ii) start with a small set of regressor variables deemed most influential, and gradually include additional variables only if the magnitude of the model residuals decrease and if this leads to better behavior of the residuals. This is illustrated in Fig. 11.13 where the use of one regressor results in very improper model residuals which is greatly reduced as two more relevant regressor variables are introduced;
(iii) the number of training data points should not be less than about 10 times the total number of weights to be tuned. Unfortunately, this criterion is not met in some published studies;
(iv) the number of nodes in the hidden layer should be directly related to the complexity/non-linearity of the problem. Often, this number should be in the range of 1–3 times the number of regressor variables;
(v) the original data set should be split such that training uses about 50–60% of the number of observations, with the validation and the testing using about 20–25% each. All three subsets should be chosen randomly;
(vi) as with non-linear parameter estimation, training should be repeated with different plausible architectures, and further, each of the architectures should be trained using different mixes of training/validation/testing subsets so as to avoid the pitfalls of local minima. Commercial software are available which automate this process, i.e., train a number of different architectures and let the analyst select the one he deems most appropriate.

Two examples of MLP modeling from the published building energy literature dealing with predicting building response over time are discussed below. Kawashima et al. (1998) evaluated several time series modeling methods for predicting the hourly thermal load of a building over a 24 h time horizon using current and lagged values of outdoor temperature and solar insolation. Comparative results in terms of the CV and NMBE statistics during several days in summer and in winter are assembled in Table. 11.7 for five methods (out of a total of seven methods in the original study), all of which used 15 regressor variables. It is quite clear that the MLP models are vastly superior in predictive accuracy to the traditional methods such as ARIMA, EWMA and Linear regression (the last used15 regressor variables). The recurrent model is slightly superior to the feed-forward MLP model. However, the MLP models, both of which have 15 nodes in the input layer and 31 nodes in the single hidden layer, imply that a rather complex model was needed to adequately capture the short-term forecasts.

Let us briefly summarize another study (Miller and Seem 1991) in order to point out that many instances only warrant simple MLP architectures, and that, even then, traditional methods may be more appropriate in practice. The study compared MLP models with traditional methods (namely, the recursive least squares) meant to predict the amount of time needed for a room to return to its desired temperature after night or weekend thermostat set-back. Only two input variables were used: the room temperature and the outdoor temperature. It was found that the best MLP was one with

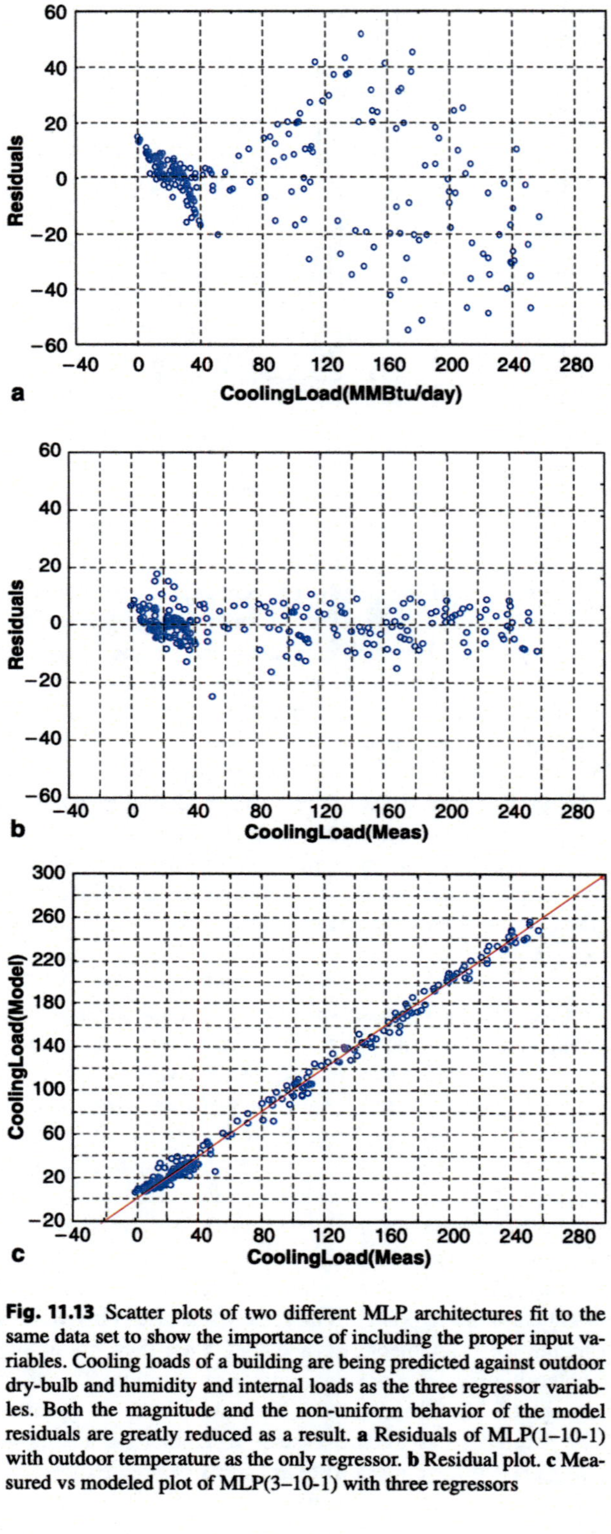

Fig. 11.13 Scatter plots of two different MLP architectures fit to the same data set to show the importance of including the proper input variables. Cooling loads of a building are being predicted against outdoor dry-bulb and humidity and internal loads as the three regressor variables. Both the magnitude and the non-uniform behavior of the model residuals are greatly reduced as a result. **a** Residuals of MLP(1–10-1) with outdoor temperature as the only regressor. **b** Residual plot. **c** Measured vs modeled plot of MLP(3–10-1) with three regressors

Table 11.7 Accuracy of different modeling approaches in predicting hourly thermal load of a building over a 24 h time horizon. (Extracted from Kawashima et al. 1998)

Model type	Coefficient of variation (%)		Normalized mean biased error (%)	
	Winter	Summer	Winter	Summer
Auto regressive integrated moving average (ARIMA)	27.7	34.4	1.6	1.0
Exponential weighted moving average (EWMA)	12.0	26.0	1.8	3.3
Linear regression with 15 regressors	27.8	21.4	−2.0	−1.3
Multi-layer Perceptron—MLP(15,31,1)	11.2	9.1	−0.5	−0.2
Recurrent Multi-layer Perceptron—MLP(15,31,1)	9.3	6.8	−0.5	−0.4

one hidden layer with two nodes even though evaluations were done with two hidden layers and up to 24 hidden nodes. The general conclusion was that even though the RMS errors and the maximum error of the MLP were slightly lower, the improvement was not significant enough to justify the more complex and expensive implementation cost of the MLP algorithm in actual HVAC controller hardware as compared to more traditional approaches.[6]

For someone with a more traditional background and outlook to model building, a sense of unease is felt when first exposed to MLP. Not only is a clear structure or functional form lacking even after the topography and weights are determined, but the "model" identification is also somewhat of an art. Further, there is the unfortunate tendency among several analysts to apply MLP to problems which can be solved more easily by traditional methods, with more transparency and allow clearer physical interpretation of the model structure and of its parameters. MLP should be viewed as another tool in the arsenal of the data analyst and, despite its power and versatility, not as the sole one. As with any new approach, repeated use and careful analysis of MLP results will gradually lead to the analyst gaining familiarity, discrimination of when to use it, and increased confidence in its proper use. The interested reader can refer to several excellent textbooks of varied level of theoretical complexity; for example, Wasserman (1989); Fausett (1993), and Haykin (1999). The MLP architecture is said to be "inspired" by how the human brain functions. This is quite a stretch, and at best a pale replication, considering that the brain typically has about 10 billion neurons with each neuron having several thousands of interconnections!

[6] This is a good example of the quote by Einstein expressing the view that ought to be followed by all good analysts: "*Everything should be as simple as possible, but not simpler*".

11.3.4 Grey-Box Models and Policy Issues Concerning Dose-Response Behavior

Example 1.3.2 in Chap. 1 discussed three methods of extrapolating dose-response curves down to low doses using observed laboratory tests performed at high doses. While the three types of models agree at high doses, they deviate substantially at low doses because the models are functionally different (Fig. 1.16). Further, such tests are done on laboratory animals, and how well they reflect actual human response is also suspect. In such cases, model selection is based more on policy decisions rather than how well a model fits the data. This aspect is illustrated below using grey-box models based on simplified but phenomenological assumptions of how biological cells become cancerous.

This section will discuss the use of inverse models to an application involving modeling risk to humans when exposed to toxins. Toxins are biological poisons usually produced by bacteria or fungi under adverse conditions such as shortage of nutrients, water or space. They are, generally, extremely deadly even in small doses. *Dose* is the variable describing the total mass of toxin which the human body ingests (either by inhalation or by food/water intake) and is a function of the toxin concentration and duration of exposure (some models are based on the rate of ingestion, not simply the dose amount). *Response* is the measurable physiological change in the body produced by the toxin which has many manifestations, but here the focus will be on human cells becoming cancerous. Since different humans (and test animals) react different to the same dose, the response is often interpreted as a probability of cancer being induced, which can be interpreted as a risk. Responses may have either no or small threshold values to the injected dose coupled with linear or non-linear behavior (see Fig. 1.16).

Dose-response curves passing through the origin are considered to apply to carcinogens, and models have been suggested to describe their behavior. The risk or probability of infection to a toxic agent with time-variant concentration C(t) from times t_1 to t_2 is provided by *Haber's law*:

$$R(C,t) = k \int_{t_1}^{t_2} C(t)dt \qquad (11.12a)$$

where k is a proportionality constant of the specific toxin and is representative of the slope of the curve.

The non-linear dose-response behavior is modeled using the *toxic load equation:*

$$R(C,t) = k \int_{t_1}^{t_2} C^n(t)dt \qquad (11.12b)$$

where n is the toxic load exponent and depends on the toxin. The value of n generally varies between 2.00 and 2.75. The implication of $n=2$ is that if a given concentration is doubled with the exposure time remaining unaltered, the response increases fourfold (and not by twice as predicted by a linear model).

The above models are somewhat empirical (or black-box) and are useful as performance models. However, they provide little understanding or insights of the basic process itself. Grey-box models based on simplified but phenomenological considerations of how biological cells become cancerous have also been proposed. Though other model types have also been suggested (such as probit and logistic models discussed in Sect. 10.4.4), the Poisson distribution (Sect. 2.4.2f) is appropriate since it describes the number of occurrences of isolated independent events when the probability of a single outcome occurring over a very short time period is proportional to the length of the time interval. The process by which a tumor spreads in a body has been modeled by multi-stage multi-hit models of which the simpler ones are shown in Fig. 11.14. The probability of getting a hit (i.e., a cancerous cell coming in contact with a normal cell) in a n-hit model is proportional to the dose level and the number n of hits necessary to cause the onset of cancer. Hence, in a one-hit model, one hit is enough to cause a toxic response in the cell with a known probability; in a two-hit model, the probabilities are treated as random independent

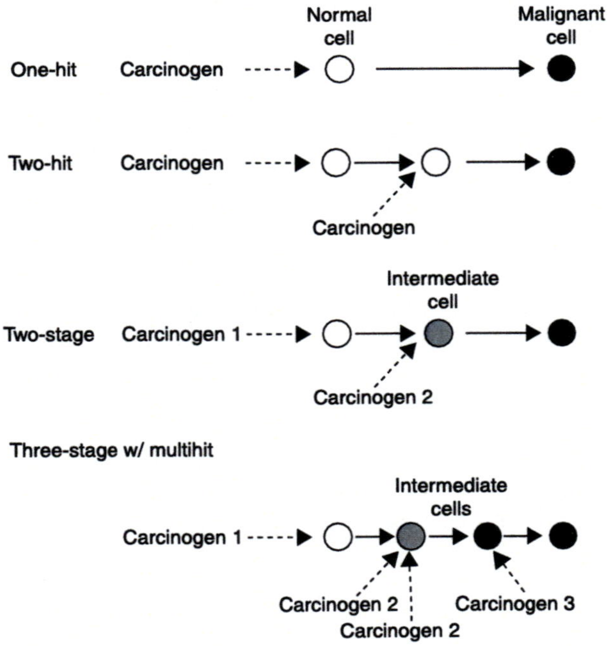

Fig. 11.14 Grey-box models for dose-response are based on different phenomenological presumptions of how human cells react to exposure. (From Kammen and Hassenzahl 1999 by permission of Princeton University Press)

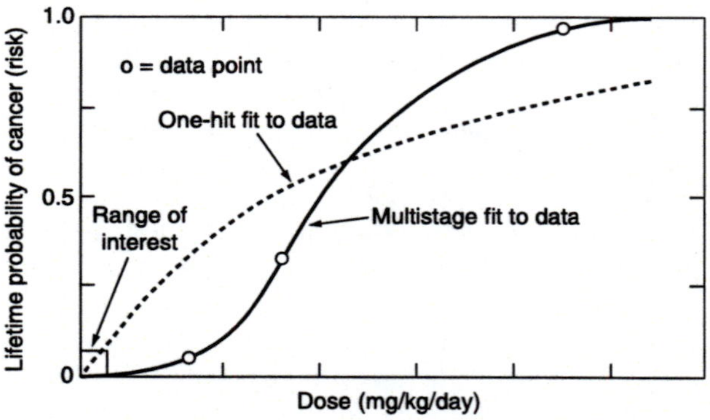

Fig. 11.15 Two very different models of dose-response behavior with different phenomenological basis. The range of interest is well below the range where data actually exist. (From Crump 1984)

events. Thus, the probabilities of the response cumulate as independent random events. The two-stage model looks superficially similar to the two-hit model; but is based on a distinctly different phenomenological premise. The process is treated as one where the cell goes through a number of distinct stages with each stage leading it gradually towards becoming carcinogenous by disabling certain specific functions of the cell (such as tumor suppression capability,…). This results in the dose effects of each successive hit cumulating non-linearly and exhibiting an upward-curving function. Thus, the multistage process is modeled as one where the accumulation of dose is not linear but includes historic information of past hits in a non-linear manner (see Fig. 11.15).

Following Masters and Ela (2008), the one-hit model expresses the probability of cancer onset P(d) as:

$$P(d) = 1 - \exp(-q_0 + q_1 d) \qquad (11.13)$$

where d is the dose, and q_0 and q_1 are empirical best fit parameters. However, cancer can be induced by other causes as well (called "background" causes). Let P(0) be the background rate of cancer incidence) corresponding to d=0. Then, since $\exp(x) \simeq 1 + x$, P(0) is found from Eq. 11.13 to be:

$$P(0) \simeq 1 - [1 - (q_0)] = q_0 \qquad (11.14)$$

Thus, model coefficient q_0 can be interpreted as the background risk. Hence, the lifetime probability of getting cancer from small doses is a linear model given by:

$$P(d) = 1 - [1 - (q_0 + q_1 d)] = q_0 + q_1 d = P(0) + q_1 d \qquad (11.15)$$

In a similar fashion, the multi-stage model of order m takes the form:

$$P(d) = 1 - \exp(-q_0 + q_1 d + q_2 d^2 + \ldots q_m d^m) \qquad (11.16)$$

Thus, though the model structure looks empirical superficially, grey box models allow some interpretation of the model coefficients. Note that the one-hit and the multi-stage models are linear only in the low dose region (which is the region to which most humans will be exposed to but where no data is available) but exponential over the entire range. Figure 11.15 illustrates how these models (Eqs. 11.13 and 11.16) capture measured data, and more importantly that there are several orders of magnitude differences in these models when applied to the low dosage region. Obviously the multi-stage model seems more accurate than the one-hit model as far as data fitting is concerned, and one's preference would be to select the former. However, there is some measure of uncertainty in these models when extrapolated downwards, and further there is no scientific evidence which indicates that one is better than the other in capturing the basic process. The U.S. Environmental Protection Agency has deliberately chosen to select the one-hit model since it is much more conservative (i.e., predicts higher risk for the same dose value) than the other at lower doses. This was a deliberate choice in view of the lack of scientific evidence in favor of one over the other model.

This example discussed one type of problem where inverse methods are used for decision making. The application of black-box and grey models to the same process was also illustrated. Instances when the scientific basis is poor and when one has to extrapolate the model well beyond the range over which it was developed qualify as one type of ill-defined problem. The mathematical form of the dose-response curve selected can provide widely different estimates of the risk at low doses. The lack of a scientific basis and the need to be conservative in drafting associated policy measures led to the selection of a model which is probably less accurate in how it fits the data collected but is, probably, preferable for its final intended purpose.

11.3.5 State Variable Representation and Compartmental Models

A classification of various dynamic modeling approaches is shown in Fig. 11.16. There is an enormous amount of knowledge in this field and on relevant inverse methods, and only a basic introduction to a narrow class of models is provided here. As described in Sect. 1.2.4, one differentiates between distributed parameter and lumped parameter system models, which can be analyzed either in time domain or frequency domain. The models, in turn, can be divided into linear and non-linear, and then into time continuous or discrete time models. This book limits itself to ARMAX or transfer function models described in Sect. 9.6.2 which are discrete time linear models with constant weight coefficients, and to the state variable formulation (described below).

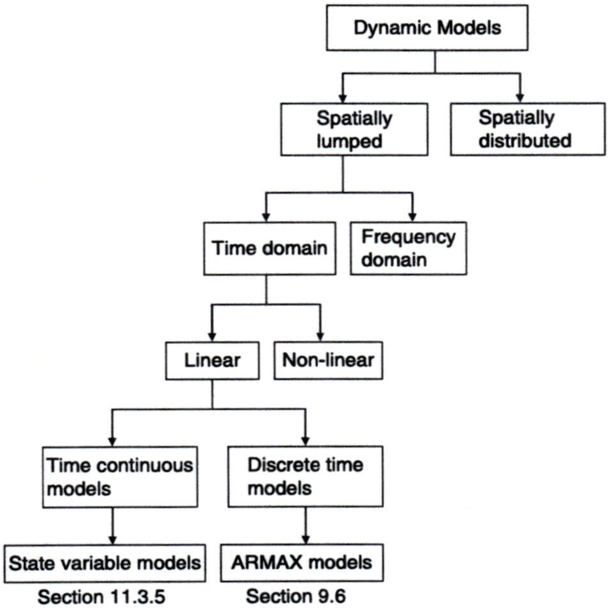

Fig. 11.16 Classification of approaches to analyze dynamic models

Dynamic models are characterized by differential equations of higher orders which are difficult to solve. Linear differential equations of order n can be represented by a set of n simultaneous first order differential equations. The standard nomenclature adopted to represent such systems is shown in Fig. 11.17. The system is acted upon by a vector of external variables or signals or influences **u** while **y** is the output or response vector (the system could have several responses). The vector **x** characterizes the state of the internal variables or elements of the system that may not be the outputs nor can they be measurable. For example, in mechanical systems, these internal elements may be positions and velocities of separate components of the system, or in thermal RC networks representing a wall, these could be the temperatures of the internal nodes within the wall. In many applications, the variables x may not have direct physical significance, nor are they necessarily unique.

The state variable representation[7] of a multi-input multi-output linear system model is framed as:

$$\dot{\mathbf{x}} = \mathbf{Ax} + \mathbf{Bu}$$
$$\mathbf{y} = \mathbf{Cx} + \mathbf{du} \tag{11.17}$$

where **A** is called the state matrix, **B** the input matrix, **C** the output matrix and **d** the direct transmission term (Franklin et al. 1994). Note that the first function relates the state vector

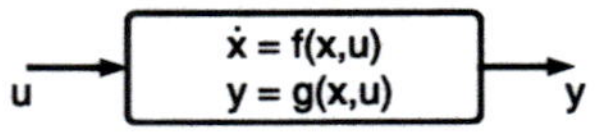

Fig. 11.17 Nomenclature adopted in modeling dynamic systems by the state variable representation

at current time with those of its previous values. This can be expanded into a linear function of p states and m inputs:

$$\dot{x}_i = a_{i1}x_1 + a_{i2}x_2 + \cdots + a_{ip}x_p + b_{i1}u_1 + b_{i2}u_2 + \cdots + b_{im}u_m \tag{11.18a}$$

Note the similarity between this formulation and that of the ARMAX model given by Eq. 9.43. Finally, the outputs themselves may or may not be the state variables. Hence, a more general representation is to express them as linear algebraic combinations:

$$y_i = c_{i1}x_1 + c_{i2}x_2 + \cdots + c_{ip}x_p + d_{i1}u_1 + d_{i2}u_2 + \cdots + d_{ip}u_p \tag{11.18b}$$

Compartmental models are a sub-category of the state variable representation appropriate when a complex process can be broken up into simpler discrete sub-systems where each can be viewed as homogeneous and well-mixed that exchange mass with each other and/or a sink/environment. This is a form of discrete linear lumped parameter modeling approach more appropriate for time invariant model parameters which is described in several textbooks (for example, Godfrey 1983). Further, several assumptions are inherent in this approach: (i) the materials or energy within a compartment get instantly fully-mixed and homogeneous, (ii) the exchange rate among compartments are related to the concentrations or densities of these compartments, (iii) the volumes of the compartments are taken to be constant over time, and (iv) usually no chemical reaction is involved as the materials flow from one cell to another. The quantity or concentration of material in each compartment can be described by first-order (linear or non-linear) constrained differential equations, the constraints being that physical quantities such as flow rates be non-negative. This type of model has been extensively used in such diverse areas as medicine (biomedicine, pharmacokinetics), science and engineering (ecology, environmental engineering, and indoor air quality), and even in social sciences. For instance, in a pharmacokinetic model, the compartments may represent different sections of a body within which the concentration of a drug is assumed to be uniform.

Though these models can be analyzed with time-variant coefficients, time invariance is usually assumed. Compartmental models are not appropriate for certain engineering applications such as closed-loop control, and even conserva-

[7] Some texts refer to this form as the "state-space" model. Control engineers retain the distinction in both terms, and refer to state space as a specific type of control design technique which is based on the state variable model formulation.

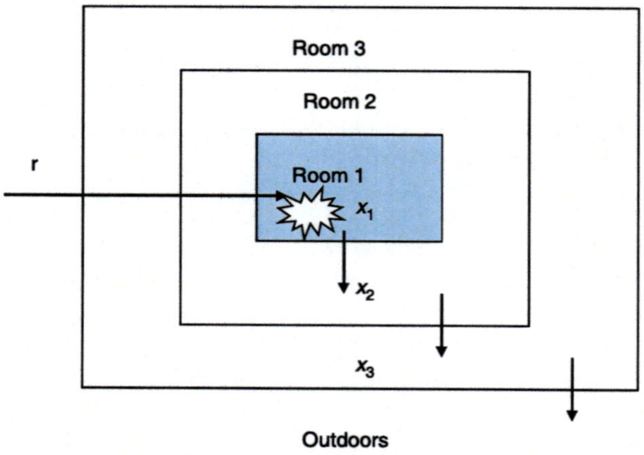

Fig. 11.18 A system of three interconnected radial rooms in which an abrupt contamination release has occurred. A quantity of outdoor air r is supplied to Room 1. The temporal variation in the concentration levels in the three rooms can be conveniently modeled following the compartmental modeling approach

tion of momentum equations that are non-compartmental in nature. Two or more dependent variables, each a function of a single independent variable (usually time for dynamic modeling of lumped physical systems) appear in such problems which lead to a system of ODEs.

Consider the three radial-room building as shown in Fig. 11.18 with volumes V_1, V_2 and V_3. A certain amount of uncontaminated outdoor air (of flow rate r) is brought into Room A from where it flows outwards to the other rooms as shown. Assume that an amount of contaminant is injected in the first room as a single burst which mixes uniformly with the air in the first room. The contaminated air then flows to the second room, mixes uniformly with the air in the second room, and on to the third room from where it escapes to the sink (or the outdoors). Let $x_1(t), x_2(t), x_3(t)$ be the volumetric concentrations of the contaminant in the three rooms, and let $k_i = r/V_i$. The entire system is modeled by a set of three ordinary differential equations (ODEs) as follows:

$$\begin{aligned}\dot{x}_1 &= -k_1 x_1 \\ \dot{x}_2 &= k_1 x_1 - k_2 x_2 \\ \dot{x}_3 &= k_2 x_2 - k_3 x_3\end{aligned} \qquad (11.19)$$

In matrix form, the above set of ODEs can be written as:

$$\begin{bmatrix}\dot{x}_1 \\ \dot{x}_2 \\ \dot{x}_3\end{bmatrix} = \begin{bmatrix}-k_1 & 0 & 0 \\ k_1 & -k_2 & 0 \\ 0 & k_2 & -k_3\end{bmatrix}\begin{bmatrix}x_1 \\ x_2 \\ x_3\end{bmatrix} \qquad (11.20a)$$

or

$$\dot{\mathbf{x}} = \mathbf{Ax} \qquad (11.20b)$$

The eigenvalue value method of solving equations of this type consist in finding values of a scalar, called the eigenvalue λ which satisfies the equation

$$(\mathbf{A} - \lambda\mathbf{I})\mathbf{x} = 0 \qquad (11.21a)$$

where $\mathbf{I}$ is the identifty matrix (Edwards and Penney 1996). For the three-room example, the expanded form of Eq. 11.21a is:

$$\begin{pmatrix}-k_1-\lambda & 0 & 0 \\ k_1 & -k_2-\lambda & 0 \\ 0 & k_2 & -k_3-\lambda\end{pmatrix}\begin{pmatrix}x_1 \\ x_2 \\ x_3\end{pmatrix} = \begin{pmatrix}0 \\ 0 \\ 0\end{pmatrix} \qquad (11.21b)$$

The eigenvalues are then determined from the following equation:

$$|\mathbf{A} - \lambda\mathbf{I}| = 0 \text{ or } (-k_1 - \lambda_1)(-k_2 - \lambda_2)(-k_3 - \lambda_3) = 0 \qquad (11.22)$$

The three distinct eigenvalues are the roots of Eq. 11.22; namely: $\lambda_1 = -k_1, \lambda_2 = -k_2, \lambda_3 = -k_3$. With each eigenvalue is associated an eigenvector $\mathbf{v}$ from where the general solution can be determined. For example, consider the case when k_1=0.5, k_2=0.25 and k_3=0.2. Then, $\lambda_1 = -0.5, \lambda_2 = -0.25, \lambda_3 = -0.2$.

The eigenvector associated with the first eigenvalue is found by substituting λ by $\lambda_1 = -0.5$ in Eq. 11.21b, to yield

$$[\mathbf{A} + (0.5)\mathbf{I}]\mathbf{v} = \begin{bmatrix}0 & 0 & 0 \\ 0.5 & 0.25 & 0 \\ 0 & 0.25 & 0.3\end{bmatrix}\begin{bmatrix}v_1 \\ v_2 \\ v_3\end{bmatrix} = \begin{bmatrix}0 \\ 0 \\ 0\end{bmatrix} \qquad (11.23)$$

Solving it results in: $\mathbf{v_1} = [3\ \text{-}6\ 5]^T$ where $[...]^T$ denotes the transpose of the vector. A similar approach is followed for the two other vectors. The general solution is finally determined as:

$$\mathbf{x}(t) = c_1\begin{bmatrix}3 \\ -6 \\ 5\end{bmatrix}e^{-0.5t} + c_2\begin{bmatrix}0 \\ 1 \\ -5\end{bmatrix}e^{-0.25t} + c_3\begin{bmatrix}0 \\ 0 \\ 1\end{bmatrix}e^{-0.2t} \qquad (11.24)$$

Finally, the three constants are determined from the initial conditions. Expanding the above equation results in:

$$\begin{aligned}x_1(t) &= 3c_1.e^{-0.5t} \\ x_2(t) &= -6c_1.e^{-0.5t} + c_2.e^{-0.25t} \\ x_3(t) &= 5c_1.e^{-0.5t} - 5c_2.e^{-0.25t} + c_3.e^{-0.2t}\end{aligned} \qquad (11.25)$$

Let the initial concentration levels in the three rooms be $x_1(0) = 900, x_2(0) = 0, x_3(0) = 0$. Inserting these in Eq. 11.25, and solving them yields $c_1 = 30$, $c_2 = 1800$, $c_3 = 7500$. Finally, the equations for the concentrations in the three rooms are given by the following tri-exponential solutions (plotted in Fig. 11.19):

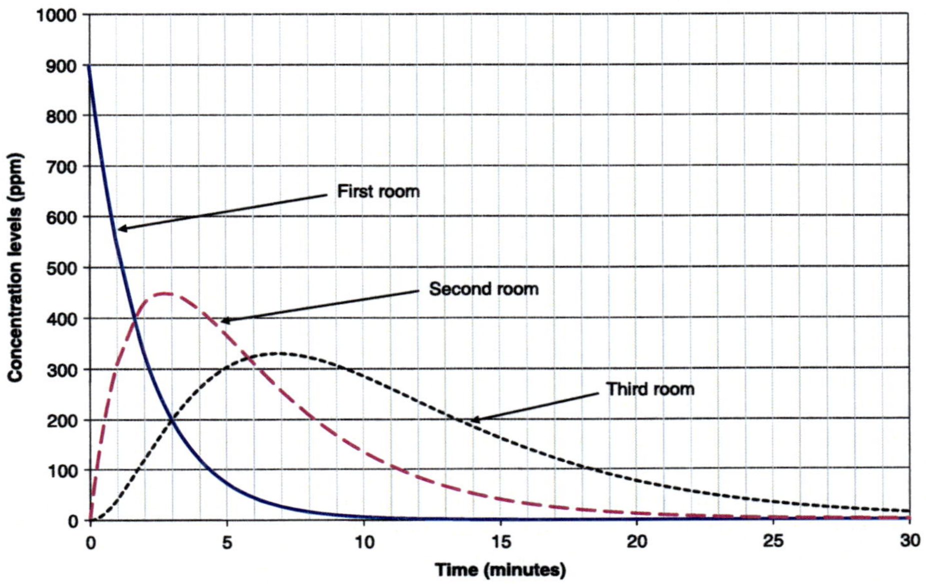

Fig. 11.19 Variation in the concentrations over time for the three interconnected rooms modeled as compartmental models. (Following Eq. 11.26)

$$\begin{aligned} x_1(t) &= 900e^{-0.5t} \\ x_2(t) &= -1800e^{-0.5t} + 1800e^{-0.25t} \\ x_3(t) &= 1500e^{-0.5t} - 9000e^{-0.25t} + 7500e^{-0.2t} \end{aligned} \tag{11.26}$$

The application of an inverse modeling approach to this problem can take several forms depending on the intent in developing the model. The basic premise is that such a building with three inter-connected rooms exists in reality from which actual concentration measurements can be taken. If an actual test similar to that assumed above were to be carried out, where should the sensors be placed (in all three rooms, or would placing sensors in Rooms 1 and 3 suffice), and what should be their sensitivities and response times? One has to account for sensor inaccuracies or even drifts, and so would some manner of fusing or combining all three data stream result in more robust model identification? Can a set of models identified under one test condition be accurate enough to predict dynamic behavior in the three rooms under other contaminant releases? What should be the sampling frequency? While longer time intervals may be adequate for routine measurements, the estimation would be better if high frequency samples were available immediately after a contaminant release was detected. Such practical issues are surveyed in the next section.

11.3.6 Practical Identifiability Issues

The complete identification problem consists of selecting an appropriate model, and then estimating the parameters of the matrices $\{\mathbf{A}, \mathbf{B}, \mathbf{C}\}$ in Eq. 11.17. The concepts of structural and numerical identifiability were also introduced in Sects. 10.2.2 and 10.2.3 respectively. *Structural identifiability*, also called "deterministic identifiability", is concerned with arriving at performance models of the system under perfect or noise free observations are available. If it is found that parameters of the assumed model structure cannot be uniquely identified, either the model has to be reformulated or else additional measurements made. The latter may involve observing more or different states and/or judiciously perturbing the system with additional inputs. There is a vast body of knowledge pertinent to different disciplines as to the design of optimal input signals; for example, Sinha and Kuszta (1983) vis-a-vis control systems, and Godfrey (1983) for applications involving compartmental models in general (while Evans 1996 limits himself to their use for indoor air quality modeling).

The problem of *numerical identifiability* is related to the quality of the input-output data gathered, i.e., due to noise in the data and due to ill-conditioning of the correlation coefficient matrix (see Sect. 10.2.3). Godfrey (1983) discusses several pitfalls of compartmental modeling, one of which is the fact that difference in the sum of squares between different possible models tends to reduce as the noise in the signal increases. He also addresses the effects of limited range of data collected (neglecting slow transient or early termination of sampling or rapid transients; delayed start of sampling which can miss the initial spikes and limits the ability to extrapolate models back to the zero-time intercept values); effect of poorly spaced samples, and the effect of short samples. The three-room example will be used to illustrate some of these concepts in a somewhat ad hoc manner which the reader can emulate on his own, and enhance with different allied investigations.

(a) Effect of Noise in the Data The dynamic performance of the three rooms is given by Eq. 11.26. These models have been used to generate a sequence of 30 data points at one-minute intervals for each room. It is left to the reader to perform a reality check and verify that the correct values of the model parameters can be re-identified when no noise is introduced in the data. A better representation of reality is to introduce noise in the samples. Several amounts of noise and different types of distribution can be studied. Here, a sequence of normally distributed random noise with zero bias and standard deviation of 20, i.e., $\varepsilon(0, 20)$ have been generated to corrupt the simulated data sample. This is quite a small instrument noise considering that the maximum concentration to be read is 900 ppm. This data has been used to identify the parameters assuming that the system models are known to be the same tri-exponential equations. One would obtain slightly different values depending on the random noise introduced, and several runs would yield a more realistic evaluation of the uncertainty in the parameters estimated (this is the Monte Carlo simulation as applied to regression analysis). However, the results of only one trial are shown under column (a) in Table. 11.8. Note that though the model R^2 is excellent and the dynamic prediction of the models captures the "actual" behavior quite well (see Fig. 11.20), the parameters are quite different from the correct values, with the differences being room-specific. For example, the coefficients "a and c" for Room 3 are poorly reproduced probably because six parameters are being identified with only 30 data points. Further, this is a non-linear regression problem and good (i.e., reasonably close) starting values have to be provided to the statistical package. The selection of these starting values also affects the parameter estimation, and hence the need to perform numerous tests with different starting values.

Table 11.8 Results of parameter estimation for two cases of the three-room problem with the simulation data corrupted by normally distributed random noise $\varepsilon(0, 20)$

	Model parameters (Eq. 11.26)	Correct values (Eq. 11.26)	(a) With sampling at one minute intervals	(b) With high frequency sampling at 0.1 min intervals for first 10 min
Room 1	a	900	787.0	900.1
	b	−0.50	−0.439	−0.500
	Adj. R^2		96.6%	99.2%
Room 2	a	−1800	−1221.1	−1922.2
	b	−0.50	−0.707	−0.483
	c	1800	1118.8	1931.3
	d	−0.25	−0.207	−0.254
	Adj. R^2		98.0%	98.4%
Room 3	a	1500	3961.9	2706.1
	b	−0.50	−0.407	−0.436
	c	−9000	−11215.8	−8134.7
	d	−0.25	−0.280	−0.276
	e	7500	7239.0	5433.1
	f	−0.20	−0.208	−0.196
	Adj. R^2		97.1%	96.9%

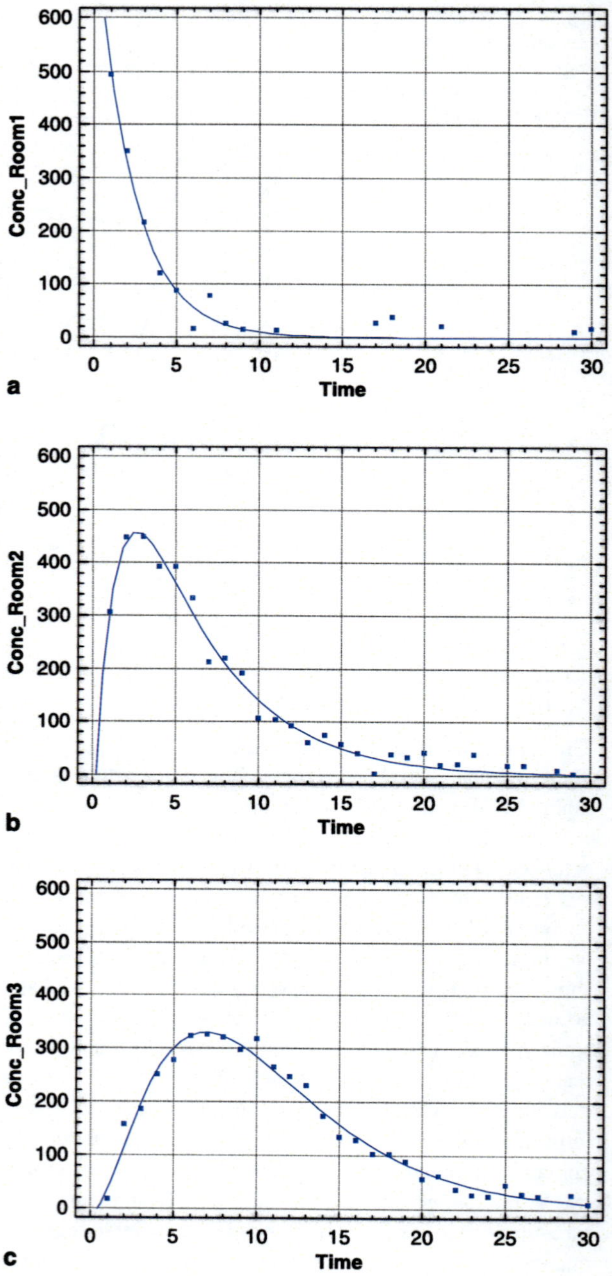

Fig. 11.20 Time series plots of measurements with random noise and identified models using Eq. 11.26 (case (a) of Table. 11.8). **a** First room. **b** Second room. **c** Third room

(b) Effect of Increasing Sampling Frequency Note that the concentration in Room 1 at time t=0 predicted by the identified model is 787.0 which is quite different than the correct value of 900 ppm. The differences in the two other

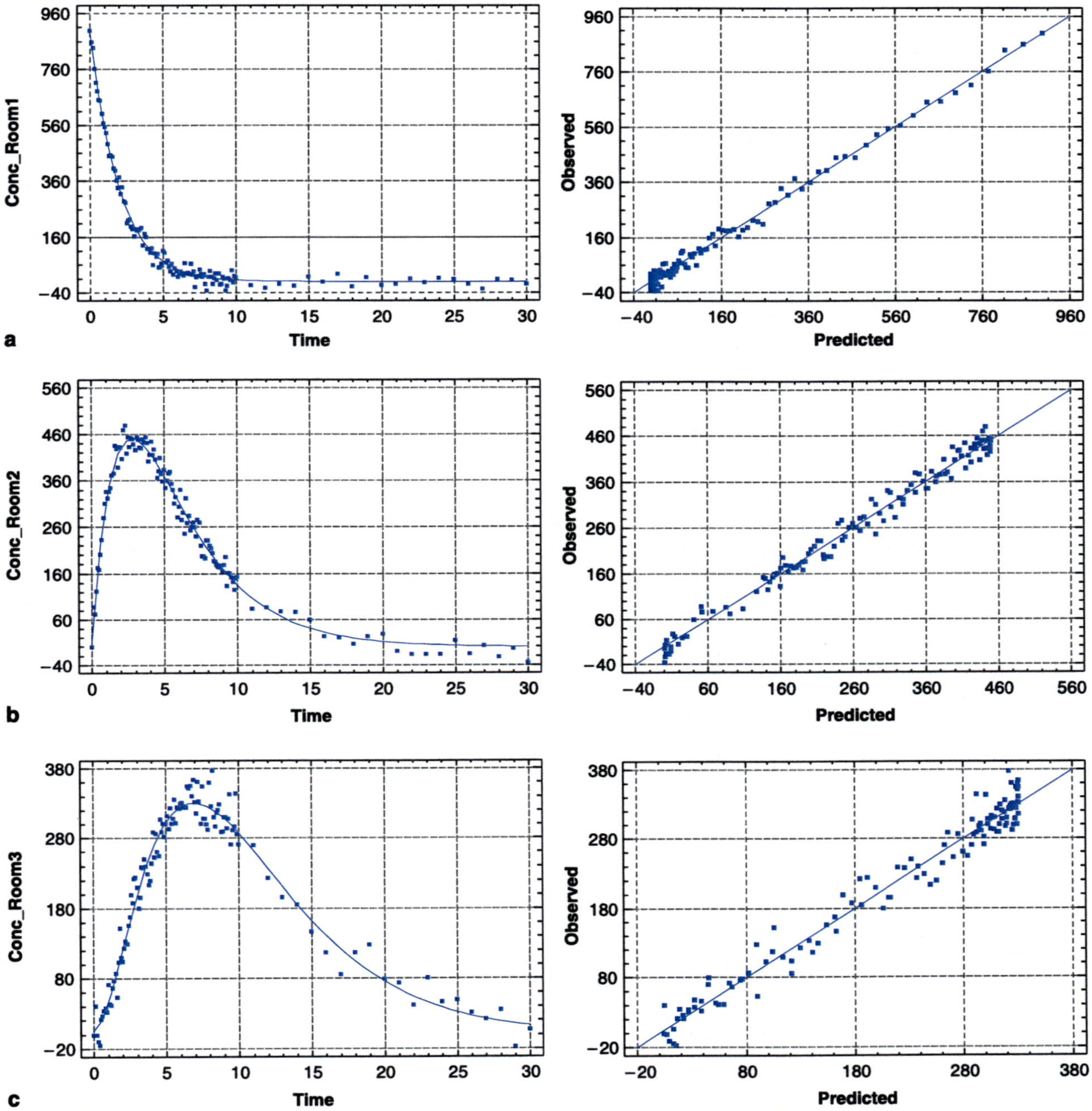

Fig. 11.21 Plot to illustrate how model parameter identification is improved if the sampling rate is increased to 0.1 min during the first 10 min when dynamic transients are pronounced. **a** First room. **b** Second room. **c** Third room

rooms are also quite large. This is a manifestation of poorly spaced sampling. It is intuitive to assume that parameter estimation would improve if more samples were collected during the period when the dynamic transients were large. This effect can be evaluated by assuming that samples were collected at 0.1 min for the first 10 min of the data collection period of 30 min. Note from Table. 11.8 that the parameter estimation for all rooms has improved greatly. The two parameters for Room 1 are almost perfect, while they are very good for Room 2. The exponent parameters for Room 3 (parameters b, d and f) are quite good while the amplitude parameters "a, c and e" for Room 2 are still biased. How well the models fit the data with almost no patterned residual behavior can be noted from Fig. 11.21.

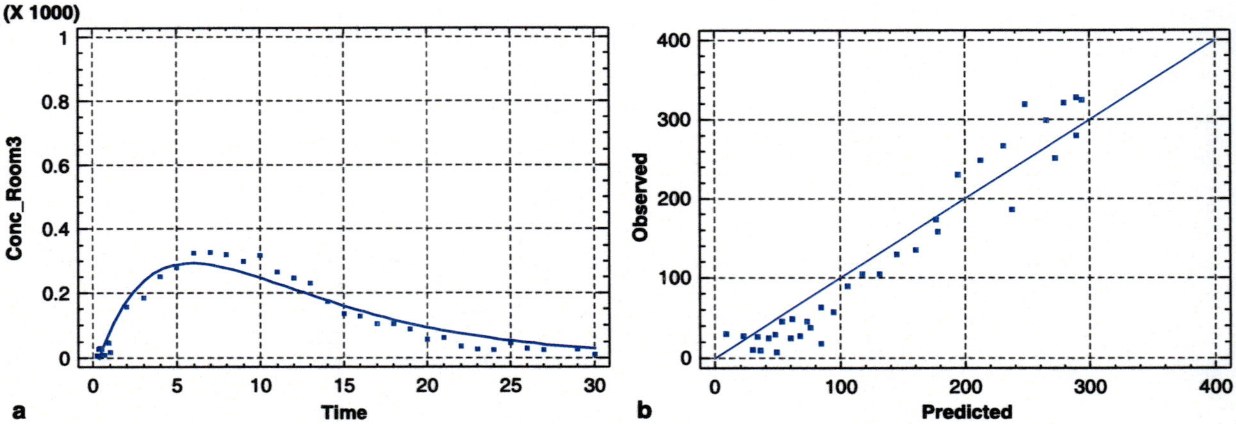

Fig. 11.22 Plot to illustrate patterned residual behavior when a lower order exponential model is fit to observed data. In this case, a two exponential model was fit to the data generated from the third compartment with random noise added. The adjusted R-square was 0.90. **a** Time series plot. **b** Observed versus predicted plot

(c) Effect of Fitting Lower Order Models How would selecting a lower order model affect the results? This aspect relates to system identification and not to parameter estimation. The simple case of regressing a bi-exponential model to the "observed" sample of Room 3 is illustrated in Fig. 11.22. In this case, there is a distinct pattern in the residual behavior which is unmistakable. In actual situations, this is a much more difficult issue. Godfrey (1983) suggests that, in most practical instances, one should not use more than three or four compartments. The sum of exponential models which the compartmental approach yields would require non-linear estimation which along with the dynamic response transients and sampling errors make robust system identification quite difficult, if not impossible.

11.4 Closure

11.4.1 Curve Fitting Versus Parameter Estimation

The terms "regression analysis" and "parameter estimation" can be viewed as synonymous, with statisticians favoring the former term, and scientists and engineers the latter. A common confusion is the difference between parameter estimation and "curve fitting". Curve fitting procedures are characterized by two degrees of arbitrariness (Bard 1974):

(a) the class of functions used is arbitrary and dictated more by convenience (such as using a linear model) than by the physical nature of the process generating the data;
(b) the best fit criterion is somewhat arbitrary and statistically unsophisticated (such as making inferences from least square regression results when it is strictly invalid to do so).

Thus, curve fitting implies a black-box approach useful for summarizing data and for interpolation. It should not be used for extrapolation. The equations and parameters determined from curve fitting provide little insight into the nature of the process. However, this approach is often adequate and suitable for many applications involving very complex phenomenon, and/or when the project budget allows for limited monitoring and analysis effort. Parameter estimation, on the other hand, is a more formalized approach which includes curve fitting at its simplest form. Grey box models, i.e., model structures derived from theoretical considerations are also an important sub-category of parameter estimation and can, moreover, include previous knowledge concerning the values of the parameters as well as the statistical nature of the measurement errors. Some professionals use the word *model fitting* in an attempt to distinguish it from curve fitting. Thus, in the framework of grey box inverse models, one attempts to identify the parameters in the model such that *their physical significance is retained*, thereby allowing better understanding of the basic physical processes involved. For example, parameters can be associated with fluid or mechanical properties. Some of the several practical advantages of using grey box models in the framework of the parameter estimation approach are that they provide finer control of the process, better prediction of system performance, and more robust on-line diagnostics and fault detection than one involving black-box models. The statistical process of determining *best* parameter values in the face of unavoidable measurement errors and imperfect experimental design is called parameter estimation.

11.4.2 Non-intrusive Data Collection

As stated in Sect. 6.1, one distinguishes between experimental design methods which can be performed in a laboratory-setting or under controlled conditions so as to identify models and parameters as robustly as possible, and observational or non-intrusive data collection. The latter is associated with systems under their normal operation and subject to whatever random and natural stimuli that perturb them. Three relevant examples of natural systems are the periodic behavior of predator-prey populations in a closed environment, the temporal changes in sunspot activity, and the seasonal changes in river water flow. One can distinguishes between two types of identification techniques relevant to non-intrusive data:

(i) *off-line* or batch identification where data collection and model identification are successive, i.e., data collection is done first, and the model identified later using the entire data stream. There are different ways of processing the data, from the simplest conventional ordinary least squares technique to more advanced ones (some of which are described in Chap. 10 and in this chapter);

(ii) *on-line or real-time* identification (also called adaptive estimation, or recursive identification) where the model parameters are identified in a continuous or semi-continuous manner *during* the operation of the system. Recursive identification algorithms are those where one normally processes the data several times so as to gradually improve the accuracy of the estimates. The basic distinction between on-line and off-line identification is that in the former case the new data is used to correct or update the existing estimates *without having to re-analyze* the data set from the beginning.

There are two disadvantages to on-line identification in contrast to off-line identification. One is that the model structure needs to be known before identification is initiated, while in the off-line situation different types of models can be tried out. The second disadvantage is that, with few exceptions, on-line identification, does not yield parameter estimates as accurate as do off-line methods (specially with relatively short data records). However, there are also advantages in using on-line identification. One can discard old data and only retain the model parameters. Another advantage is that corrective action, if necessary, can be taken in real time. On-line identification is, thus, of critical importance in certain fields involving fast reaction via adaptive control of critical systems, digital telecommunication and signal-processing. The interested reader can refer to numerous textbooks on this subject (for example, Sinha and Kuszta 1983).

11.4.3 Data Fusion and Functional Testing

The techniques and concepts presented in this chapter are several decades old, and though quite basic, are still relevant since they provide the basic foundation to more advanced and recent methods. Performance of engineered systems was traditionally measured by sampling, averaging and recording analog or digital data from sensors in the form of numeric streams of data. Nowadays, sensory data streams from multiple and disparate sources (such as video, radar, sonar, vibration,...) are easy and cheap to obtain. It is no surprise, then, that they are used to supplement data from traditional monitoring systems resulting in data that is more comprehensive, and which allow more robust decisions to be made. Such multi-sensor and mixed-mode data streams have led to a discipline called *data fusion,* defined as the use of techniques that combine data from multiple sources in order to achieve inferences, which will be more robust than if they were achieved by means of a single source. A good review of data fusion models and architectures is provided by Esteban et al. (2005). Application areas such as space, robotics, medicine, sensor networks have seen a plethora of allied data fusion methods aimed at detection, recognition, identification, tracking, change detection, decision making, etc. These techniques are generally studied under signal processing, sensor networks, data mining, and engineering decision making.

With engineered systems getting to be increasingly complex, several innovative means have been developed to check the sensing hardware itself. One such technique is *functional testing*; a term often attributed to software program testing with the purpose of ensuring that the program works the way it was intended while being in conformance with the relevant industry standards. It is also being used in the context of testing the proper functioning of engineered systems which are controlled by embedded sensors. For example, consider a piece of control hardware whose purpose is to change the operating conditions of a chemical digester (say, the temperature or the concentration of the mix). Whether the control hardware is operating properly or not, can be evaluated by intentionally sending test signals to it at some pre-determined frequency and duration, and then analyzing the corresponding digester output to determine satisfactory performance. If it is not, then the operator is alerted and corrective action to recalibrate the control hardware may be warranted. Such approaches are widely used in industrial systems, and have also been proposed and demonstrated for building energy system controls.

Problems

Pr. 11.1 Consider the data in Pr. 10.14 where the Wind Chill (WC) factor is tabulated for different values of ambient temperature and wind velocity. Use this data to evaluate different multi-layer perceptron (MLP) architectures assuming one hidden layer only. You will use 50% of the data for training, 25% as validation and 25% as testing. Compare model goodness of fit and residual behavior of the best identified MLP with the regression models identified in Pr. 10.14.

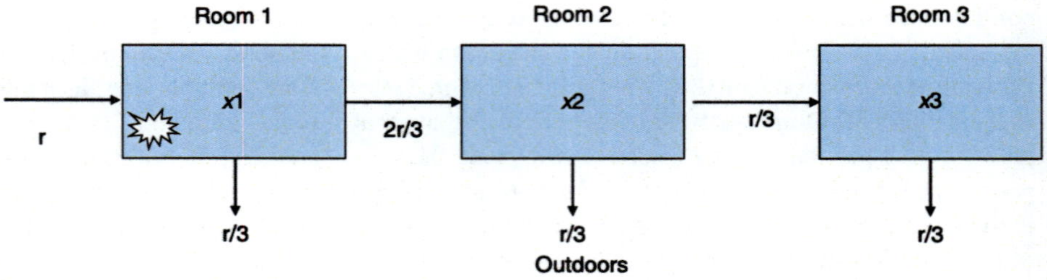

Fig. 11.23 A building with interconnected and leaky rooms

Pr. 11.2 Consider Fig. 1.5c showing the nth order model thermal network for heat conduction through a wall. The internal node temperatures T_s are the state variables whose values are internal to the system, and have to be expressed in terms of the outdoor temperature (variable u) and indoor air temperatures (variable y) and the individual resistors and capacitors of the system (parameters of the system). You will assume the following two networks, write the model equations for the temperature T_s at each internal node, reframe them in the state variable formulation (Eq. 11.17) and identify the elements of the A and B matrices in terms of the resistances and the capacitors:

(a) network with 2 capacitors and 3 resistors (3R2C)
(b) network with 3 capacitors and 4 resistors (4R3C).

Pr. 11.3 You will repeat the analysis of the three room compartmental problem presented in Sect. 11.3.5 for the configuration shown in Fig. 11.23.

(a) Follow the same procedure to derive the exponential equations for the dynamic response of each of the three rooms assuming the same initial conditions, namely: $x_1(0) = 900$, $x_2(0) = 0$, $x_3(0) = 0$
(b) You will now use these models to generate synthetic data with random noise and perform similar analysis as shown in Table. 11.8. Discuss your results.

References

ASHRAE 14-2002. Guideline 14-2002: Measurement of Energy and Demand Savings, American Society of Heating, Refrigerating and Air-Conditioning Engineers, Atlanta.

Bard, Y., 1974. *Nonlinear Parameter Estimation*, Academic Press, New York.

Claridge, D.E. and M. Liu, 2001. *HVAC System Commissioning*, Chap. 7.1 *Handbook of Heating, Ventilation and Air Conditioning*, J.F. Kreider (editor), CRC Press, Boca Raton, FL.

Clarke, J.A., 1993. Assessing building performance by simulation, *Building and Environment*, 28(4), pp.419–427.

Crump, K.S., 1984. An improved procedure for low-dose carcinogenic risk assessment from animal data, *Journal of Environmental Pathology, Toxicology and Oncology*, (5) 4/5:, pp 339–349.

Devore J., and N. Farnum, 2005. *Applied Statistics for Engineers and Scientists*, 2nd Ed., Thomson Brooks/Cole, Australia.

Edwards, C.H. and D.E. Penney, 1996. *Differential Equations and Boundary Value Problems*, Prentice Hall, Englewood Cliffs, NJ.

Esteban, J., A. Starr, R. Willetts, P. Hannah and P. Bryanston-Cross, 2005. A review of data fusion models and architectures: Towards engineering guidelines, *Neural Computing and Applications*, (14), pp. 273–281.

Evans, W. C., 1996. Linear Systems, Compartmental Modeling, and Estimability Issues in IAQ Studies, in Tichenor, B., *Characterizing Sources of Indoor Air Pollution and Related Sink Effects*, ASTM STP 1287, pp. 239–262.

Fausett, L., 1993. *Fundamentals of Neural Network: Architectures, Algorithms, and Applications*, Prentice Hall, Englewood Cliffs, NJ.

Franklin, G.F., J.D. Powell and A. Emami-Naeini,1994. *Feedback Control of Dynamic Systems*, 3rd Ed., Addison-Wesley, Reading, MA.

Godfrey, K., 1983. *Compartmental Models and Their Application*, Academic Press, New York.

Haykin, S., 1999. *Neural Networks*, 2nd ed., Prentice Hall, NJ.

Hammersley, J.M. and D.C. Handscomb, 1964. *Monte Carlo Methods*, Methuen and Co., London.

Helton, J.C. and F.J. Davis, 2003. "Latin hypercube sampling and the propagation of uncertainty of complex systems", *Reliability Engineering and System Safety*, vol. 81, pp. 23–69.

Hofer, E., 1999. "Sensitivity analysis in the context of uncertainty analysis for computationally intensive models", *Computer Physics Communication*, vol. 117, pp. 21–34.

Hornberger, G.M. and R.C. Spear, 1981. "An approach to the preliminary analysis of environmental systems", *Journal of Environmental Management*, vol.12, pp. 7–18.

Iman, R.L. and J.C. Helton, 1985. "A Comparison of Uncertainty and Sensitivity Analysis Techniques for Computer Models", Sandia National Laboratories report NUREG/CR-3904, SAND 84-1461.

IPCC, 1996. Climate Change 1995: The Science of Climate Change, Intergovernmental Panel on Climate Change, Cambridge University Press, Cambridge, UK.

Kammen, D.M. and Hassenzahl, D.M., 1999. *Should We Risk It*, Princeton University Press, Princeton, NJ

Kawashima, M., C.E. Dorgan and J.W. Mitchell, 1998, Hourly thermal load prediction for the next 24 hours by ARIMA, EWMA, LR, and an artificial neural network, *ASHRAE Trans. 98(2)*, Atlanta, GA.

Lam, J.C. and S.C.M. Hui, 1996. "Sensitivity analysis of energy performance of office buildings", *Building and Environment*, vol.31, no.1, pp 27–39.

Mayer, T., F. Sebold, A. Fields, R. Ramirez, B. Souza and M. Ciminelli, 2003. "DrCEUS: Energy and demand usage from commercial onsite survey data", *Proc. of the International Energy program Evaluation Conference*, Aug. 19–22, Seattle, WA.

Masters, G.M. and W.P. Ela, 2008. *Introduction to Environmental Engineering and Science*, 3rd Ed. Prentice Hall, Englewood Cliffs, NJ.

Miller, R.C. and J.E. Seem, 1991. Comparison of artificial neural networks with traditional methods of predicting return time from night or weekend setback, *ASHRAE Trans., 91(2)*, Atlanta, GA.,

Pindyck, R.S. and D.L. Rubinfeld, 1981. *Econometric Models and Economic Forecasts*, 2nd Edition, McGraw-Hill, New York, NY.

Reddy, T.A., 2006. Literature review on calibration of building energy simulation programs: uses, problems, procedures, uncertainty and tools, *ASHRAE Transactions*, vol.12, no.1, pp.177–196.

Reddy, T.A., I. Maor and C. Ponjapornpon, 2007a, "Calibrating detailed building energy simulation programs with measured data- Part I: General methodology", *HVAC&R Research Journal*, vol. 13(2), March.

Reddy, T.A., I. Maor and C. Ponjapornpon, 2007b, "Calibrating detailed building energy simulation programs with measured data- Part II: Application to three case study office buildings", *HVAC&R Research Journal*, vol. 13(2), March.

Saltelli, A., K. Chan and E.M. Scott (eds.) 2000. *Sensitivity Analysis*, John Wiley and Sons, Chichester.

Saltelli, A., 2002. "Sensitivity analysis for importance assessment", *Risk Analysis*, vol. 22, no.2, pp. 579–590.

Sinha, N.K. and B. Kuszta, 1983. *Modeling and Identification of Dynamic Systems*, Van Nostrand Reinhold Co., New York.

Sonderegger, R., J. Avina, J. Kennedy and P. Bailey, 2001.Deriving loadshapes from utility bills through scaled simulation, *ASHRAE seminar presentation*, Kansas City, MI.

Spears, R., T. Grieb and N. Shiang, 1994. "Parameter uncertainty and interaction in complex environmental models", *Water Resources Research*, vol. 30 (11), pp 3159–3169.

SPSS, 1997.Nural Connection- Applications Guide, SPSS Inc and Recognition Systems Inc., Chicago, IL.

Sun, J. and T.A. Reddy, 2006. Calibration of building energy simulation programs using the analytic optimization approach, *HVAC&R Research Journal*, vol.12, no.1, pp.177–196.

Wasserman, P.D., 1989. Neural Computing: Theory and Practice, Van Nostrand Reinhold, New York.

Winkelmann, F.C., B.E. Birdsall, W.F. Buhl, K.L. Ellington, A.E. Erdem, J.J. Hirsch, and S. Gates, 1993. DOE-2 Supplement, Version 2.1E, Lawrence Berkeley National Laboratory, November.

12 Risk Analysis and Decision-Making

As stated in Chap. 1, inverse modeling is not an end by itself but a precursor to model building needed for either better understanding of the process or for decision-making that results in some type of action. Decision theory is the study of methods for arriving at rational decisions under uncertainty. This chapter provides an overview of quantitative decision-making methods under uncertainty along with a description of the various phases involved. The decisions themselves may or may not prove to be correct in the long term, but the scientific community has reached a general consensus on the methodology to address such problems. Different sources of uncertainty are described, and an overall framework along with various elements of decision-making under uncertainty is presented. How Bayes' approach can play an important role in decision-making is also briefly presented. Decision problems are clouded by uncertainty, and/or by having to meet multiple objectives; this chapter also presents basic concepts of multi-attribute modeling. The role of risk analysis and its various aspects are covered along with applications from various fields. General principles are presented in somewhat idealized situations for better understanding, while several illustrative case studies are meant to provide exposure to real-world situations. The topics which relate to decision analysis are numerous, and at varying levels of maturity; only a basic overview of this vast body of knowledge has been provided in this chapter.

12.1 Background

12.1.1 Types of Decision-Making Problems and Applications

Decision-making is pervasive in real life; one makes numerous mundane decisions daily from what set of clothes to wear in the morning to what type of cuisine to eat for lunch. However, what one is interested here is the formal manner in which one goes about tackling more complex engineering and scientific problems involving several alternatives with bigger payoffs and penalties. It is assumed that the reader is familiar with basic notions taught in undergraduate classes on economic analysis of alternatives involving time value of money, simple payback analysis as well as present worth cash flow approaches. As stated in Sect. 1.6, decision theory is the study of methods for arriving at "rational" decisions under uncertainty. This theory, albeit in slightly different forms, applies to a broad range of applications covering engineering, health, manufacturing, business, finance, and government.

One can distinguish between two types of problems depending on the types of variables under study.

(i) The perhaps simpler application is the *continuous case* such as controlling an engineered existing system or operating it in a certain manner so as to minimize a *cost or utility* function subject to constraints. Such problems involve optimization of one (or several) continuous variables.
(ii) The *discrete case* such as evaluating different alternatives in order to select a course of action. This deals with a plethora of problems faced in numerous disciplines. In engineering, the explicit consideration of numerous choices in terms of design (selection of materials of components to how the components are assembled into systems) is a simple example. One could also consider decision analysis involving after-the-fact intervention where the analysis is enhanced by collecting data from the system while in operation, and subsequent analysis reveals that the design erred in some of its assumptions and, hence, discrete changes have to be made to the system or its operation. In environmental science such as high pollution in a certain area, one can visualize reaching decisions involving both engineering mitigation measures as well as policy decisions to remedy the situation and prevent future occurrences.

The decisions themselves may or may not prove to be correct in the long term, but the process provides a structure for the overall methodology. It is under situations when uncertainties dominate that this framework assumes crucial importance. Certain aspects of uncertainty have previously been discussed

T. Agami Reddy, *Applied Data Analysis and Modeling for Energy Engineers and Scientists*,
DOI 10.1007/978-1-4419-9613-8_12,

in this book. In Sect. 1.5, four different types of data uncertainty were described which relate to data types classified as qualitative, ordinal, metric and count (see Sect. 1.2.1). Propagation of measurement errors associated with metric data was addressed in Sects. 3.6 and 3.7. Uncertainties in regression model parameters when using least squares estimation and the resulting model prediction uncertainties were presented in Sects. 5.3.2, 5.3.4 and 5.5.2. Here, one deals with uncertainty in a larger system context, involving all the above aspects in some form or another, as well as other issues such as the various chance events and outcomes. A common categorization of uncertainty at the systems level involves:

(a) *Structural deficiency,* which, for the continuous variable problem, is due to an incomplete or improper formulation of the objective function model (such as use of improper surrogate variables, excluded influential variables, improper functional form) or of the constraints. For the discrete case, this could be due to overlooking some of the possible outcomes or chance nodes. This is akin to improper model or system identification.
(b) *Uncertainty in the parameters of the model,* which manifests itself by biased model predictions; this is akin to improper parameter estimation.
(c) *Stochasticity or inherent ambiguity* which no amount of additional measurements can reduce. This stochasticity can appear in the model parameters, or in the input forcing functions or chance events of the system, or in the behavior of the system outcomes, or in the vagueness associated with the risk attitudes of the user (one could adopt either fuzzy logic or follow the more traditional probabilistic approach). Thus, stochasticity can be viewed as a type of uncertainty (note that some professionals adopt the view that it is distinct from uncertainty and treat it as such).

The first two sources above can be grouped under *epistemic* or ignorance or lack of complete knowledge. This can be reduced as more metric data is acquired; the model structure is improved and the parameters identified more accurately. The third source is referred to as *aleotory uncertainty* and there are well accepted ways of analyzing such situations. The above considerations provide one way of classifying different decision-making problems (Fig. 12.1):

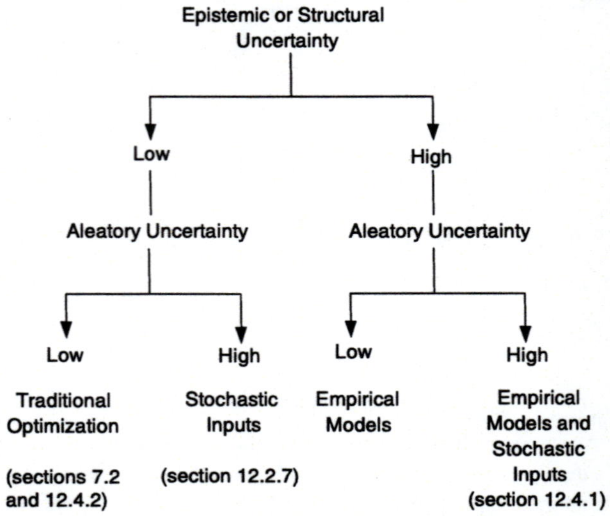

Fig. 12.1 Overview of different types of problems addressed by decision analysis and treated in this chapter along with section numbers. Traditional optimization corresponds to decision-making under low epistemic and aleatory uncertainties while the other cases can be viewed as decision-making under different forms of uncertainty

(a) *Problems involving no or low aleatory and epistemic uncertainty:* In such cases, the need for careful analysis stems from the fact that the problem to be solved is mathematically complex with various possible options, but the model structure itself is well-defined, i.e., robust, accurate and complete. The effect of uncertainty in the specification of the model parameters and/or inaccurate measurements in model inputs has a relatively minor impact on the model results. This is treated under *mathematical optimization* in traditional engineering and operations research disciplines, and most undergraduates are exposed to such techniques. Such problems fall under decision-making under certainty, and an overview has already been provided in Sect. 7.2. An illustrative example of how decision-making may play an important role even in such situations is given in Sect. 12.4.2.
(b) *Problems involving low epistemic but high aleatory uncertainty:* Queuing models studied in operations research fall in this category. Here, the arrival of customers who join queues for servicing is modeled as random variables while the time taken for servicing a customer may also be modeled as another random variable. Determining the number of service counters to keep open so that the mean waiting time for customers is less than a stipulated value is a typical example of such problems. This instance is discussed in Sect. 12.2.7.
(c) *Problems involving high epistemic but low aleatory uncertainty:* Cases falling in this category arise when the model structure is empirical or heuristic (such as logistic functions meant to model dose-response curves discussed in Sect. 10.4.4.), but the outcomes themselves are not complete and/or not well-defined. The occurrence of the events or inputs to the system may be probabilistic, but these are considered known in advance in statistical terms.
(d) *Problems involving both high epistemic and high aleatory uncertainties:* Such cases are the most difficult to deal with, and different analysts can reach widely different conclusions. Examples include societal costs due to air pollution from fossil-based power plants or societal cost of a major earthquake hitting California. The

treatment of such cases is very challenging and complex, and can be found in research papers (for example, Rabl 2005). A case study example is provided in Sect. 12.4.1.

(e) *Problems involving competitors with similar objectives:* Such cases fall under gaming problems or decision-making under situations where the multiple entities have potentially conflicting interests. There are two types of games: under conflict, i.e., the various competitors are trying to win the "game" at the expense of the others based on their selfish needs (called a non-cooperative situation to which the Nash equilibrium applies), or trying to reach a mutually-acceptable solution (called a cooperative situation) which maximizes the benefit to the whole group. There is a rich body of knowledge in this area and has great implications in international trade, business and warfare, but much less so in engineering and science applications; hence, these are not treated in this book.

All the above approaches (except the first) involve explicit consideration of the uncertainty during the process of reaching a solution, as against performing an uncertainty analysis after a solution or a decision has been made (as adopted during an engineering sensitivity analysis). These approaches require adopting formal risk analysis techniques whereby all uncertain and undesirable events are framed as risks, their probabilities of occurrence selected, the chain of events simplified and modeled, the gains and losses associated with different options computed, trade-offs between competing alternatives assessed, and the risk attitude of the decision-maker captured (Clemen and Reilly 2001).

Given its importance when faced with high epistemic uncertainty situations, risk analysis methods are discussed in Sect. 12.3. A simple way of differentiating between risk analysis and decision-making is to view the former as *event-focused* (what can go wrong? how likely is it? what are the consequences?), and mostly quantitative and empirical in nature, while decision-making is more comprehensive and includes, in addition, issues that are more qualitative and normative such as uncertain knowledge and uncertain future (Haimes 2004). Assuming different risk models can lead to different decisions. It is clear that risk analysis is a subset, albeit an important one, of the decision-making process. It is important to keep in mind that the real issue is not whether the uncertainty is large or not, but whether this may cause an improper or different decision to be made than the best one.

12.1.2 Engineering Decisions Involving Discrete Alternatives

This sub-section will present one example involving engineering decisions of selecting a piece of equipment among competing possibilities. This is a simple context-setting review example of optimization problems with low epistemic uncertainty.

Table 12.1 Problem specification for Example 12.1.1

	Pump A	Pump B
Purchase price ($)	4,000	6,000
Efficiency	80%	92%
Annual maintenance ($/year)	200	450
Life (years)	4	6

Example 12.1.1: *Simple example of an engineering decision involving equipment selection*

A designer is faced with the problem of selecting the better of two pumps of 2 kW each meant to operate 6,000 h/year with the attributes assembled in Table 12.1.

If unit price of electricity is $ 0.10/kWh, which pump is the better selection if time value of money is not considered? Note that no uncertainty as such is stated in the problem.

Annual operating expense

$$\begin{aligned}\text{for Pump A} &= (2\ \text{kW}/0.80) \times (6{,}000\ \text{h/year}) \\ &\quad \times (\$\ 0.1/\text{kWh}) + \$\ 200 = \$\ 1{,}700/\text{year} \\ \text{for Pump B} &= (2\ \text{kW}/0.92) \times (6{,}000\ \text{h/year}) \\ &\quad \times (\$\ 0.1/\text{kWh}) + \$\ 450 = \$1{,}750/\text{year}.\end{aligned}$$

Since both pumps have the same purchase price per year of life ($ 1,000/year), one would select the less efficient Pump A since it has a slightly lower annual operating cost. ■

The two options are sufficiently close that a sensitivity analysis is warranted. The traditional process adopted in engineering analysis is to identify the important parameters (in this problem, there are no "variables" as such) either by a formal sensitivity analysis or by physical intuition, assign a range of variability to the base values (say, the uncertainty expressed as a standard deviation), and look at the resulting changes in the final operating costs. For example, in the simple case treated above, the designer comes to know from another professional that the efficiency of Pump A is not 80% as claimed but closer to 75%. If this value was assumed, the annual operating expense for Pump A would turn out to be $ 1,800/year, and the designer would lean towards selecting Pump B instead. Obviously, other quantities assumed would also have some uncertainty, and evaluating the alternatives under the combined influence of all these uncertainties is a more realistic and complex problem than the simplistic one assumed above. If the uncertainties around the parameters are relatively small (say, in terms of the coefficient of variation CV, i.e., the uncertainty normalized by the mean value of around 5–10%), the problem would be treated as a low aleatory problem. Techniques relevant to this case are those studied under traditional optimization (discussed in Sect. 7.2). However, if the uncertainties are large (say, CV values of 20–50%) such as in many health and environmental issues, the problem would be treated as a high aleatory problem requiring a more formal probabilistic approach involving, say, Monte Carlo methods (this is treated in Sect. 12.2.7).

Formal analysis methods of the more complex classes (c) and (d) involving both aleatory and epistemic uncertainty are illustrated by way of examples in Sects. 12.2.2, 12.2.7 and 12.4.1. The simple pump selection example can be used to illustrate the type of considerations which would require such an approach. For example, the uncertainties of the parameters assumed in case (b) above are taken to be due to the randomness of the variables and not due to a lack of knowledge which additional equipment or system performance measurements can reduce. Thus, the issue of structural uncertainty did not arise. Say, the problem is framed as one where the amount of water to be pumped is not a fixed demand requiring the pumps to consume their rated 2 kW at all times, but a variable quantity of water which depends on say, the outdoor temperature (the hotter it is, more water is required, for example, to irrigate an agricultural field). One now has two models, one for predicting the water demand against temperature, and another for the power consumed by the pumps under variable flow (since the efficiency is not constant at part load performance). The regression models are unlikely to be perfect (i.e., have R^2 of 100%) due to overlooking other variables which also dictate water demand, and may be biased due to inadequate data point coverage of the entire performance map. If the models are excellent (say $R^2=95\%$), then the structural uncertainty is unlikely to be important, and whatever small structural uncertainties there may exist are clubbed with the random or aleatory uncertainties and treated together. However, if the models are poor, then the structural deficiencies would influence our analysis in a fundamentally different manner from the pure aleatory problem, and then a formal approach for case (c) would be warranted.

12.2 Decision-Making Under Uncertainty

12.2.1 General Framework

Unlike decision-making under certainty (discussed in Sect. 7.2 under traditional optimization), decision-making under uncertainty (or under risk) applies to situations when a course of action is to be selected when the results of each alternative course of action or outcomes of different possible chance events are not fully known deterministically. In such cases, the process of decision-making with discrete alternatives can be framed as a series of steps described below (Clemen and Reilly 2001):

(a) *Frame the **decision problem and pertinent objectives** in a qualitative manner*
This is often not as obvious as it seems. One may unintentionally be framing the "wrong problem" perhaps because the objectives were not clear at the onset. For example, the industrialist planning on building a photovoltaic (PV) module assembly factory at a particular location naturally wishes to maximize profit. However, he also considers a second attribute, namely creation of as many new jobs as possible in the factory. However, it could be that the real intent was to create prosperity in the region and not by direct hires. It is very likely that the subsequent decisions would be quite different. Hence, articulating and then framing the objective variable or function so that the real intent is captured is a critical first step. This phase, often referred to as "getting the context right" may require several iterations in cases when the decision-maker in unclear as to his inmost objectives or modifies them during the course of this step.

(b) *Identify decision alternatives or **actions***
This step would involve carefully identifying the various choices or alternatives or actions and characterizing them. For example, in the PV module factory instance, one can view the decision alternatives as technical and economic. The former would include the selection of the type of PV cell technology (such as say, single crystalline silicon or thin film), and the module size or power rating and voltage output,… which would maximize sales based on certain types of anticipated applications. Actions involving economic factors could involve selecting the capacity of the production line, and/or possible location (say, whether Malaysia or Vietnam or China).

(c) *Identify and quantify **chance events***
This step involves identifying unexpected factors which could affect the alternatives selected in step (b): economy turning sour, a breakthrough in a new type of PV cell, government withdrawing/reducing rebates to solar installations,..The chance events are probabilistic and have to be framed as either discrete or continuous. Continuous probability distributions (either objective or subjective) are often approximated by discrete ones for simpler treatment. This is discussed in Sect. 12.2.3. The chance events must be collectively exhaustive (i.e., include all possible situations), and be mutually exclusive (i.e., only one of the outcomes can happen). The term "states of nature" is also widely used synonymously to chance events.

(d) *Assemble the entire decision problem*
The interactions of the decomposed smaller (and, thus, more manageable) pieces of the problem are represented by influence diagrams and decision trees which provide a clear overall depiction of the structure of the entire problem. This aspect is discussed in Sect. 12.2.2.

(e) *Develop mathematical representations or models*
Here the **outcomes** (or consequences or payoffs) of each chance event and action are considered, and a structure to the problem is provided by framing the corresponding mathematical models along with constraints.

(f) *Identify sources and characterize magnitude of* ***uncertainties***
The issues relevant to this aspect can arise from steps (c)–(e) and have been previously described in Sect. 12.1.1.

(g) *Model user preferences and* ***risk attitudes***
Unlike traditional optimization problems, decisions are often not taken based simply on maximizing an objective function. Except for the simple types of problems, alternative or competing decisions have different amounts of risk. Different people have different risk attitudes, and thus are willing to accept different levels of risk. The same person may also have a different risk attitude under different situations. Thus, the utility function represents a way of mapping different units (for example, operating cost in dollars) into a "*utility value*" number more meaningful to the individual. How to model utility functions is discussed in Sect. 12.2.5. Further, as discussed earlier in step (a), there may be more than one objective on which the decision is to be made. How to consider multi-attributes and develop an objective function for the whole problem is an integral part of this step, and is also described in Sect. 12.2.6. Clearly, this step is an important one requiring a lot of skill, effort and commitment.

(h) ***Assemble*** *the objective function and* ***solve*** *the complete model along with uncertainty bounds*
The solution or the outcome of the entire model under low uncertainty would involve the types of continuous optimization techniques described in Sect. 7.2, as well as discrete ones illustrated through Example 12.1.1. However, when uncertainty plays a major role in the problem definition, it will have a dominant effect on the solution as well. One can no longer separate the solution of the model (objective function plus risk attitude) from its variability. Sensitivity or post-optimal analysis, so useful for low uncertainty problems, is to be supplemented by generating an uncertainty distribution about optimal solution. The latter can be achieved by Monte Carlo methods (described and illustrated in Sects. 12.2.7 and 12.4.2).

(i) *Perform* ***post-optimal analysis*** *and select action to implement*
The last step may involve certain types of post-optimal analysis before selecting on a course of action. Such analysis could involve reappraising certain factors such as chance events and their probability of occurrence, variants to risk attitudes, inclusion of more alternatives,… The concept of *indifference* is often used (illustrated in Sect. 12.4.2). Hence, this last step would involve reiterations till a preferred and satisfactory alternative or course of action is decided upon.

12.2.2 Modeling Problem Structure Using Influence Diagrams and Decision Trees

This section describes issues pertinent to step (d) above involving structuring the problem which is done once the objectives, decision alternatives and chance events have been identified. Two types of representations are used: influence diagrams and decision trees. An *influence diagram* is a simple graphical representation of a decision analysis problem which captures the decision-maker's current state of knowledge in a compact and effective manner. For example, the commercialization process of a new product launch involving essential elements such as decisions, uncertainties, and objectives as well as their influence on each other is captured by the influence diagram shown in Fig. 12.2. Different nodes of different shapes correspond to different types of variables; the convention is to represent decision nodes as squares, chance nodes as circles and payoff nodes as diamonds (or as triangles). Arrows (or arcs) define their relationship to each other. The convention when creating influence diagrams is to follow certain guidelines: (i) use only one payoff node, (ii) do not use cycles or loops, and (iii) avoid using barren nodes which do not lead to some other node (except when such a node makes the representation much more understandable).

Influence diagrams are especially useful for communicating a decision model to others and creating an overview of a complex decision problem. However, a drawback to influence diagrams is that their abstraction hides many details. It is difficult to see what possible outcomes are associated with an event or decision as many outcomes can be embedded in a single influence diagram decision or chance node. Often,

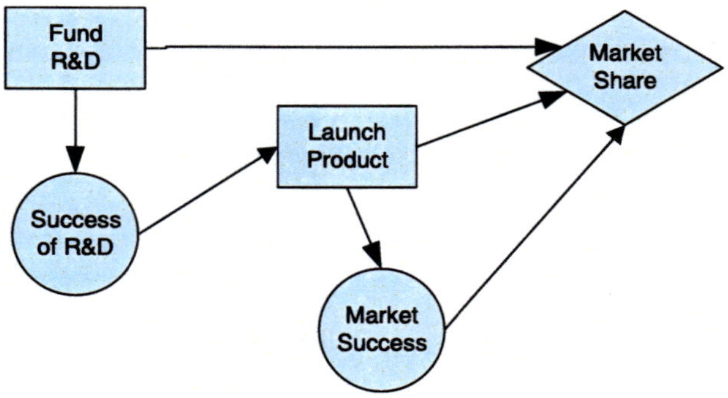

Fig. 12.2 Influence diagram for the process of commercialization of a product with the objective of achieving a sizeable market share. By convention, decision nodes are represented by *squares*, chance nodes by *circles* and payoff nodes by *diamonds* (or as *triangles*), while *arrows* (or *arcs*) define their relationship to each other

it is also not possible to infer the chronological sequence of events appearing during the decision-making process.

Decision trees (also called decision flow networks or decision diagrams), as opposed to influence diagrams, include all possible decision options and chance events with a branching structure. They proceed chronologically, left to right, showing events and decisions as they occur in time. All options, outcomes and payoffs, along with the values and probabilities associated with them, are depicted directly with little ambiguity as to the possible outcomes. Decision trees are popular because they provide a visualization of the problem which integrates both graphical and analytical aspects of the problem. In that sense, probability trees, already introduced and illustrated in Sect. 2.2, are closely related to decision trees. Decision trees do not have arcs. Instead, they use branches, which extend from each node. Branches are used as follows for the three main node types in a decision tree: (i) a decision node has a branch extending from it for every available option, (ii) a chance node has a branch for each possible outcome, and (iii) an end node has no branches succeeding it, and returns the payoff and probability for the associated path. Figure 12.3 is the corresponding decision tree for the product launch example whose influence diagram is shown in Fig. 12.2.

In summary, the influence diagram and decision tree show different kinds of information. Influence diagrams are excellent for displaying a decision's overall structure, but they hide many details. They show the dependencies among the variables more clearly than the decision tree. Influence diagrams offer an intuitive way to identify and display the essential elements, including decisions, uncertainties, and objectives, and how they influence each other. The diagram provides a high-level qualitative view under which the analyst builds a detailed quantitative model. This simplicity is useful since it shows the decisions and events in one's model using a small number of nodes. This makes the diagram very accessible, helping others to understand the key aspects of the decision problem without getting bogged down in details of every possible branch as shown in a decision tree. The decision tree, on the other hand, shows more details of all possible paths or scenarios as sequences of branches. It should be used at the latter stage of the analytical analysis. For simpler problems, the use of influence diagrams may be unnecessary, and one can resort to the use of decision trees directly for analysis.

Example 12.2.2: *Example involving two risky alternatives each with probabilistic outcomes*
This example illustrates the approach for a one-stage probabilistic problem with two risky alternatives. An oil company is concerned that the production from one of its oil wells has been declining and deems that the oil revenue can be increased if certain technological upgrades in the extraction process are implemented. There are two different technolo-

Fig. 12.3 Decision tree diagram corresponding to the influence diagram shown in Fig. 12.2

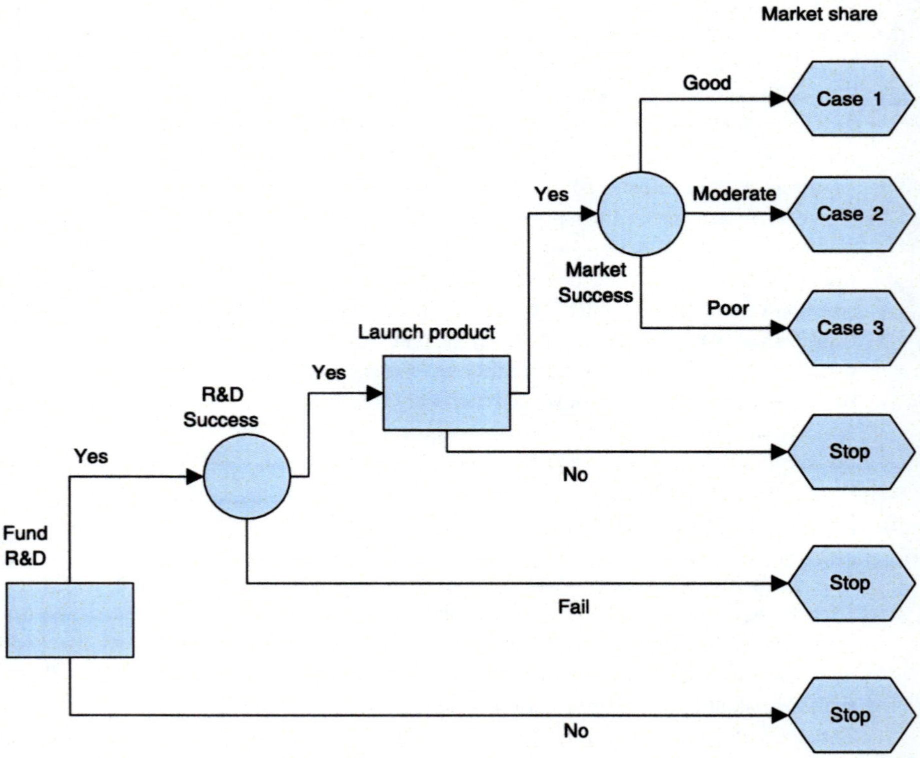

Table 12.2 Relevant data needed to analyze the two risky alternatives for the oil company (Example 12.2.2)

Alternative	Capital investment (millions of dollars)	State of economy	Probability of occurrence	Annual revenue over the base case of doing nothing (millions/year)
1	$ 320	Good	0.25	$ 120
		Average	0.60	$ 80
		Poor	0.15	$ 40
2	$ 100	Good	0.25	$ 50
		Average	0.60	$ 35
		Poor	0.15	$ 20

gies that can be implemented with different initial costs and different impacts on the oil revenue per year over a time horizon of 6 years. However, the oil revenues could change depending on whether the oil demand is good, average or poor (factors that depend on the overall economy and not on the oil company). Table 12.2 summarizes pertinent data for the three possibilities and the two competing alternatives. This example has only two factors: the technical upgrade alternatives, and the state of the economy (which in turn impacts oil demand). The influence diagram for this scenario is easy to generate and is depicted in Fig. 12.4.

Uncertainty regarding the state of the economy can be modeled by subjective probabilities, i.e., by polling expert economists and analysts. Probability in this case is to be viewed as a "likelihood" of occurrence rather than as the classical frequentist interpretation of long-run fraction. Economists of the company followed such a process and determined that the state of the economy over the next 6 years can be represented as chance events with probabilities of: good $p(G)=0.25$, average $p(A)=0.60$, and poor $p(P)=0.15$.

This is a *single stage decision* problem whose decision tree diagram is shown in Fig. 12.5. Such diagrams provide a convenient manner of visually depicting the problem, and also performing the associated probability propagation calculations under the classical frequentist view. If one neglects the time value of money, the total net savings under each of the three economic scenarios and for both alternatives are shown in the third column of Table 12.3. The expected value EV (some books use the term "expected worth") for each

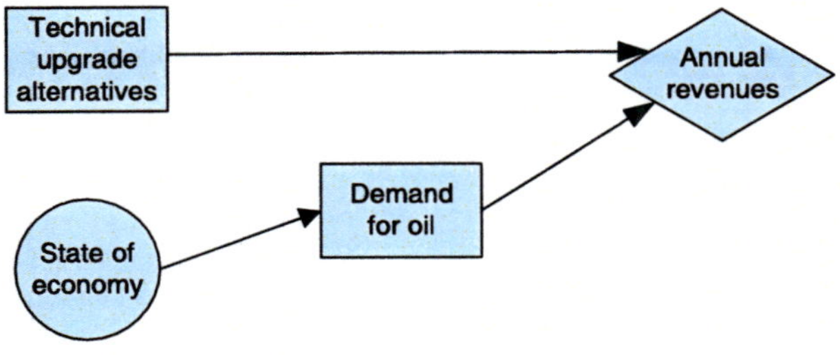

Fig. 12.4 Influence diagram for Example 12.2.2

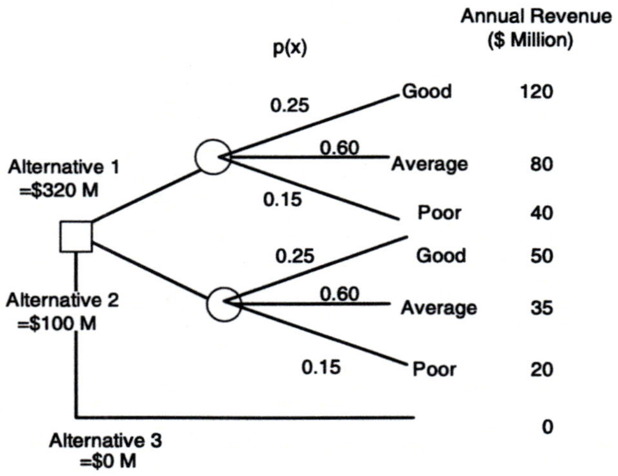

Fig. 12.5 Decision tree diagram for the oil company example. Note that this problem has three alternative action paths, three chance events (good, average, poor) and seven outcomes. The objective variable is the expected value EV shown in the last column of Table 12.3

Table 12.3 Calculation procedure of the expected value of both alternatives for the oil company

Alternative	Chance event-probability p(x)	Total net savings (TNS) over 6 years (millions $)	Expected value EV(x)=TNS * p(x)
1	0.25	−320+(6)(120)=400	100
	0.60	−320+(6)(80)=160	96
	0.15	−320+(6)(40)=−80	−12
		Total	184
2	0.25	−100+(6)(50)=200	50
	0.60	−100+(6)(35)=110	66
	0.15	−100+(6)(20)=20	3
		Total	119

outcome is shown in the last column by multiplying these values by the corresponding probability values. The total EV is simply the sum of these three values. Note that the annual savings are incremental values, i.e., over and above the current extraction system. Since the sums are positive, the analysis indicates that both alternatives are preferable to the current state of affairs, but that Alternative 1 is the better one to adopt since its EV is higher.

Though probabilities are involved, the analysis assumed them to be known without any inherent uncertainty. Hence, this is still a case of low epistemic and low aleotory uncertainties. A more realistic reformulation of this problem would include: (i) treating variability in the probability distributions (case "b" of our earlier categorization shown in Fig. 12.1), and (ii) including the inherent uncertainties in the models used to predict the annual revenues for each of the three economy states (case "d"). ■

12.2.3 Modeling Chance Events

(a) Discretizing probability distributions

Probability of occurrence of chance and uncertain events, and even subjective opinions, are often characterized by continuous probability distribution functions (PDF). Example 12.2.2 assumed such an approach to modeling the state of the economy with three distinct options. Replacing the continuous PDF by a small set of discretized events, especially for complex problems, simplifies the conceptualization of the decision, drawing the tree events as well as the calculation of the EVs. Two simple approximation methods are often used (Clemen and Reilly 2001):

(i) *Extended Pearson-Tukey method,* or the 3-point approximation, is said to be a good choice when the distribution is not well known. It works best when PDFs are symmetric; while some advocate its use even when it is not so. Rather than knowing the entire PDF, one needs to determine probabilities corresponding to only three points on the distribution, namely, the median value of the random variable, and the 0.05 and 0.95 fractiles. Even when the distribution is not well-known, probability values corresponding to these bounds (lower, upper and most likely) can be estimated heuristically. The attractiveness of this approach is that it has been found from experience (even though there is no obvious interpretation), that assigning probabilities of 0.63 for the median and 0.185 for the 0.05 and 0.95 fractiles gives surprisingly good results in many cases. Consider the CDF of hourly outdoor dry-bulb temperatures for Philadelphia, PA for a given year, shown in Fig. 2.6 and reproduced in Fig. 12.6. By this approach, the representative values of outdoor temperature would be 24.1°F, 55.0°F and 82.0°F corresponding to 0.05, 0.50 and 0.95 fractiles. This approach is illustrated in Fig. 12.6b where the fan representing the PDF is discretized into three branches denoting chance events to which probabilities of 0.185, 0.63 and 0.185 are assigned.

(ii) *Bracket median method* or the n-point approximation is a more complete and rigorous way of discretizing PDFs in cases where these are known with some certainty. It is simply found from the associated cumulative distribution function (CDF). The probability range (0–100%) is divided into n intervals which are equally likely to occur, i.e., each interval with (1/n) probability, and the median of each range is taken to be the discrete point representing that range. The higher the value of n, the better the approximation, but the more tedious the resulting analysis, though the discretization process itself is straightforward. Again considering Fig. 12.6, and for the case of $n = 5$, the representative values are easily read off from the y-axis scale from such a plot. The results are shown in Table 12.4 and plotted in Fig. 12.6b.

(b) Sequential decision-making

Situations often arise where decisions are to be made sequentially as against just once. For example, an electric utility would be faced with decisions as how best to meet anticipated load increases; such decisions may involve subsidizing energy conservation measures, planning for new generation types and capacity (whether gas turbines, clean coal, solar thermal, solar PV, wind,…), where to locate them,… As discussed in Sect. 2.2.4, conditional probabilities represent situations with more than one outcome (i.e., compound outcomes) that are sequential (successive). The chance result of the first stage determines the conditions of the next stage, and so on. The outcomes at each stage may be deterministic or

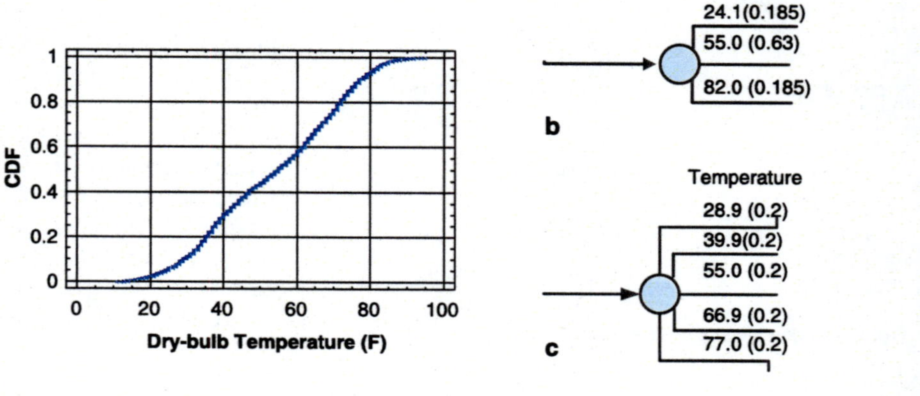

Fig. 12.6 Discretizing the continuous PDF for outdoor dry-bulb temperature (°F) of Philadelphia, PA (the CDF is copied from Fig. 2.6). **a** The continuous distribution, **b** the temperature values (and the associated probabilities for 0.05, 0.50 and 0.95 fractiles) following the three-branch extended Pearson-Tukey method, **c** those following the bracketed median method with five branches

Table 12.4 Discretizing the CDF of outdoor hourly dry-bulb temperature values for Philadelphia, PA shown in Fig. 12.6 into five discrete ranges of equal probability

Bracket median range	Median value	Representative value (°F)	Associated probability
0–0.2	0.1	28.9	0.2
0.2–0.4	0.3	39.9	0.2
0.4–0.6	0.5	55.0	0.2
0.6–0.8	0.7	66.9	0.2
0.8–1.0	0.9	77.0	0.2

may be probabilistic. Such a sequence of events can be conveniently broken down and represented by a series of smaller simpler problems by resorting to decision tree representation of the problem where the random alternatives can be assigned probabilities of occurrence. Such kinds of problems are called dynamic decision situations (note the similarity between such problems and dynamic programming optimization problems treated in Sect. 7.10). Uncertain events also impact such problems and they can do so at each stage; hence, it is necessary to dovetail them into the natural time sequence of the decisions. One can distinguish between two cases:

(i) when one has the luxury to wait till uncertain events resolve themselves before taking a decision (say, one waits for a few months to determine whether a bad economy is showing undeniable signs of recovery). This case is illustrated in Fig. 12.7.

(ii) when one has to take decisions in the face of uncertainty. Information known with certainty at a later stage may not have been so at an earlier stage when a course of action had to be determined; a non-optimal path may have been selected. This is similar to evaluating an earlier decision with 20/20 hindsight. Subsequent decisions have to be made which remedy this situation to a certain extent while considering future uncertainties as well. The following example illustrates this situation.

Example 12.2.3: *Two-stage situation for Example 12.2.2*
Consider the oil company problem treated in Example 12.2.2. This corresponds to a rather simple formulation of the decision-making process where a one-time decision is to be made. Assume that the first alternative (Alt#1) was selected involved the more expensive first cost upgrade (see Fig. 12.5). After a period of time during which the initial cost of $ 320 million was spent in technology upgrades and all measures relating to Alt#1 were implemented, the company economists revised their probability estimates of the state of the economy for the next 6 years to: p(G)=0.10, p(A)=0.20 and p(P)=0.70. This would result in a much lower EV easily determined as:

$$\text{Revised EV (Alt\#1)} = (400 \times 0.10) + (160 \times 0.2) + (-80 \times 0.70) = \$\ 16 \text{ million}$$

With hind sight, Alt#2 would have been the preferable choice since the corresponding revised EV(Alt#2)=56 million. However, what is done is done, and the company has to proceed forward and make the best of it. It starts with identifying alternatives (four are shown in Fig. 12.8) and evaluates them before making a second decision. Thus, the decision-making

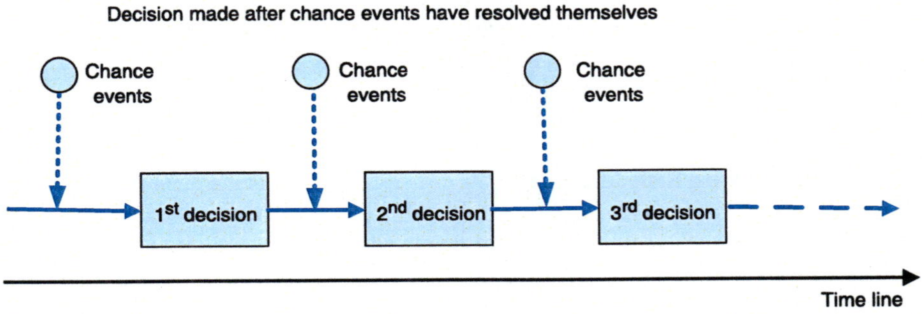

Fig. 12.7 Sequential decision-making process under the situation where one has the luxury to wait until uncertain or chance events resolve themselves before taking a decision

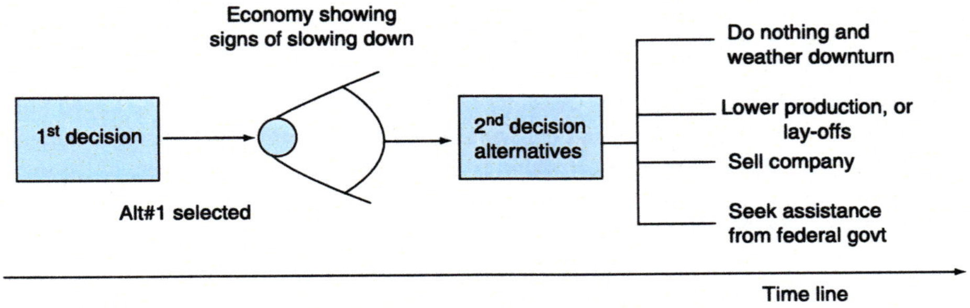

Fig. 12.8 Identification of second set of alternatives after Alt#1 was already selected

process is now being made before the uncertain event(s) resolves itself since, otherwise, the penalties may be more severe, and some of the alternative venues may have been closed. Businesses of all sorts (and even individuals) explicitly or implicitly follow such a sequential decision-making process over time since our very existence is dynamic and constantly subject to chance events that are unforeseen and unanticipated. ■

12.2.4 Modeling Outcomes

The outcomes of each of the discretized events can be quantified or modeled in a variety of ways. *Expected value* or worth or expected payoff criterion which represents the net profit or actual reward of the corresponding action is perhaps the most common, and has been illustrated in Example 12.2.2. The *maximum payoff* criterion is an obvious way of selecting the best course of action. This option will also result in the least *opportunity loss* (a term often used in decision-making analyses). However, a second criteria to consider is the risk associated with either alternative. A measure of risk characterizes the variability of the different courses of action, and this can be quantified by the standard deviation, or better still, by the normalized standard deviation, i.e., the coefficient of variation (CV). Thus for A#1, the standard deviation of the three outcomes [100,96,-12] is 63.5 while for A#2 the outcomes are [50,66,3] with 32.7 standard deviation. The associated CV values are 1.04 and 0.826 respectively. Hence, even though A#1 has a higher EV, it is much riskier. Hence, a risk-prone person may well decide to select A#2 even though its EV is lower. Thus, in summary, maximizing the EV and minimizing the exposure to risk are two usually conflicting objectives; how to trade-off between these objectives is basically a problem of decision analysis.

12.2.5 Modeling Risk Attitudes

In most cases, decisions are not taken based simply on maximizing an objective function or the expected value as discussed in Sect. 7.2. Except for the simple types of problems, alternative or competing decisions have different amounts of risk. Different people have different risk attitudes, and thus are willing to accept different levels of risk. Consider a simple instance when one is faced with a situation with two outcomes: (a) the possibility of winning $ 100 with 50% certainty or getting $ 0 with 50% certainty (like a gamble determined by flipping a coin), and (b) winning $ 30 with 100% certainty. The expected mean value is $ 50 for case (a), and if the decision is to be made based on economics alone, one would choose outcome (a). A risk-averse person

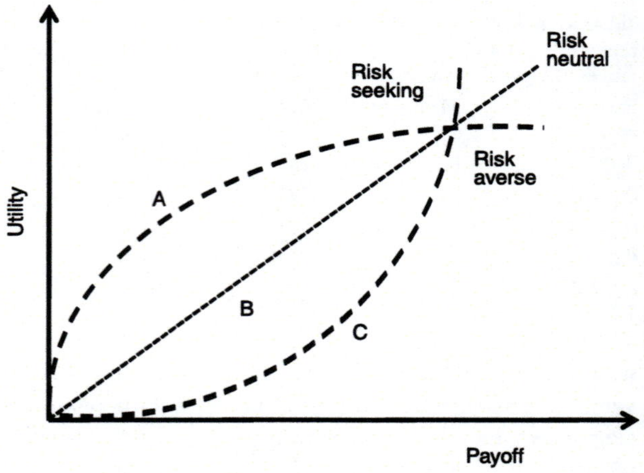

Fig. 12.9 Plot illustrating the shape of the utility functions that capture the three risk attitudes

would rather have a sure thing, and may opt for outcome (b). Thus, his utility (or the value he places to the payoff) compared to the outcome or payoff itself can be represented by the concave function shown in Fig. 12.9. This is a conservative attitude most people tend to have towards risk. However, there are individuals who are not afraid of taking risks, and the utility function of such risk-seeking individuals is illustrated by the convex function. A third intermediate category is the risk-neutral attitude where the utility is directly proportional to the payoff, and is represented by a linear function in Fig. 12.9. Another consideration is that the *amount* of payoff may also dictate the outlook towards risk. Say, instead of $ 1,000 at stake, one had $ 1 million, the same individual who would adopt an aggressive behavior towards risk when $ 100 was at stake, may become conservative minded in view of the large monetary amount. The above example is another manifestation of the law of diminishing returns in economic theory where one differentiates between the total returns and the marginal returns which is the incremental benefit or value perceived by the individual.

A utility is a numerical rating, specified as a graph, table or a mathematical expression assigned to every possible outcome a decision maker may be faced with. In a choice between several alternative prospects, the one with the highest utility is always preferred. The utility function represents a way to map different units (for example, operating cost in dollars) into a "utility" number to the individual which captures the user's attitude towards risk. Note that this utility number does not need to have a monetary interpretation. In any case, it assigns numerical values to the potential outcomes of different decisions made in accordance with the decision-maker's preference.

Table 12.5 Traditional present worth analysis

Type of AC	Initial cost	Annual operating cost	Traditional present worth analysis
High efficiency	$ 12,000	$ 3,000/year	12,000+7.72×3,000=$ 35,183
Normal	$ 8,000	$ 4,000/year	8,000+7.72×4,000=$ 38,880

Example 12.2.4: *Risk analysis for a homeowner buying an air-conditioner*
A homeowner is considering replacing his old air-conditioner (AC). He has the choice between a more efficient but more expensive unit, and one which is cheaper but less efficient. Thus, the high efficiency AC costs more initially but has lower operating cost. Assume that the discount rate is 5% and that the life of both types of AC units is 10 years.

Table 12.5 assembles the results of the traditional present worth analysis where the factor 7.72 is obtained from the standard cash flow analysis formula:

$$\left(\frac{P}{A}, i, n\right) = \left(\frac{P}{A}, 0.05, 10\right) = \frac{(1+i)^n - 1}{i(1+i)^n}$$
$$= \frac{(1.05)^{10} - 1}{0.05(1.05)^{10}} = 7.72$$

where P is the present worth, A is the annuity, i the discount rate and n the life (years).

Traditional economic theory would unambiguously suggest the high efficiency AC as the better option since its present worth is lower (in this case, one wants to minimize expenses). However, there are instances when the homeowner has to include other considerations. Let us introduce an element of risk by stating that there is a possibility that he will relocate and would have to sell his house. In such a case, a simple way is to use the probability of the homeowner relocating as the weight value. For the zero risk case, i.e., he will stay on in his house for the foreseeable future, he would weigh the up-front cost and the present value of future costs equally (i.e., utility weight of 0.5 for each since they have to add to 1.0 due to normalization consideration). If he is very sure that he will relocate, and given that he must replace his AC before putting his house up for sale on the market, he would minimize his cost by placing a utility weight of 1.0 to the initial cost (implying that this is the **only** criterion), and 0.0 for the annual operating costs. If, say, the probability of his moving is 30%, the utility weight factor would be 0.7 for the initial cost i.e., he is weighing the initial costs more heavily, while the utility weight for the annual operating costs would be (1 − 0.7) = 0.3. Thus, present worth for:

- High efficiency AC = 0.7 × 12,000 + 0.3 × (7.72 × 3,000) = $ 15,355
- Normal AC = 0.7 × 8,000 + 0.3 × (7.72 × 4,000) = $ 14,864

Thus, if a risk-averse attitude were to be adopted, the choice of the normal AC turns out to be the better option. Note, that there is no guarantee that this is the better choice in the long run. If the homeowner ends up not having to sell his house, then he has made a poor choice (in fact, he is very likely to end up paying more than that calculated above since the fact that electricity rates are bound to increase over time has been overlooked in the analysis). ■

The selection of the utility weights was somewhat obvious in the previous example since they were tied to the probability of the homeowner relocating to another location. The weights are to be chosen on some rational basis, say in terms on their relative importance to the analyst. A function commonly used for modeling utility is the exponential function which models a *constant risk attitude mind-set:*

$$U_{\exp(x)} = a - b \cdot \exp\left(-\frac{x - x_{\min}}{R}\right) \qquad (12.1)$$

where R is a parameter characterizing the person's risk tolerance, x_{min} is the minimum monetary value below which the individual would not entertain the project, and the coefficients "a" and "b" are normalization factors. Large values of R, corresponding to higher risk tolerance, would make the exponential function flatter, while smaller values would make it more concave which is reflective of more risk-averse behavior. The above function is a general form and can be simplified for certain cases; its usefulness lies in that it allows a versatile set of utility functions to be modeled. The following example illustrates these notions more clearly.

Example 12.2.5: *Generating utility functions*
Consider the case where a person wishes to invest in a venture. Assuming that his minimum cut-off threshold x_{min} = $ 500, the utility functions following Eq. 12.1 for two different risk tolerances: R = $ 5,000 and R = $ 10,000 are to be generated.

One needs to determine the numerical values of the coefficient a and b by selecting the limits of the function $U_{\exp(x)}$. A common (though not imperative) choice for the normalization range is 0–1 corresponding to the minimum and maximum positive payoffs. Then, for R = 5,000, with the lower limit set at $x = x_{min} = 500$ yields: $0 = a - b \cdot \exp(-\frac{500-500}{5{,}000})$ from where a = b. Thus, in this case, the utility function could have been expressed without the coefficient "b". For the upper limit, one takes $x = x_{max} = x_{min} + R$ yielding $1 = a - a \cdot \exp(-\frac{5{,}500-500}{5{,}000})$ from where a = 1.582 = b. Similarly for R = 10,000, one also gets a = 1.582 = b. The corresponding utility functions for a range of values for the variable x extending from −3,000 to 5,000 are plotted in Fig. 12.10a. Note that all payoff values below the minimum have negative utility values (i.e., a loss), while the utility function is steeper for the lower value of R. Thus, someone with a lower risk tolerance would assign a higher utility value for a given value of x, which is characteristic of a more risk-averse outlook.

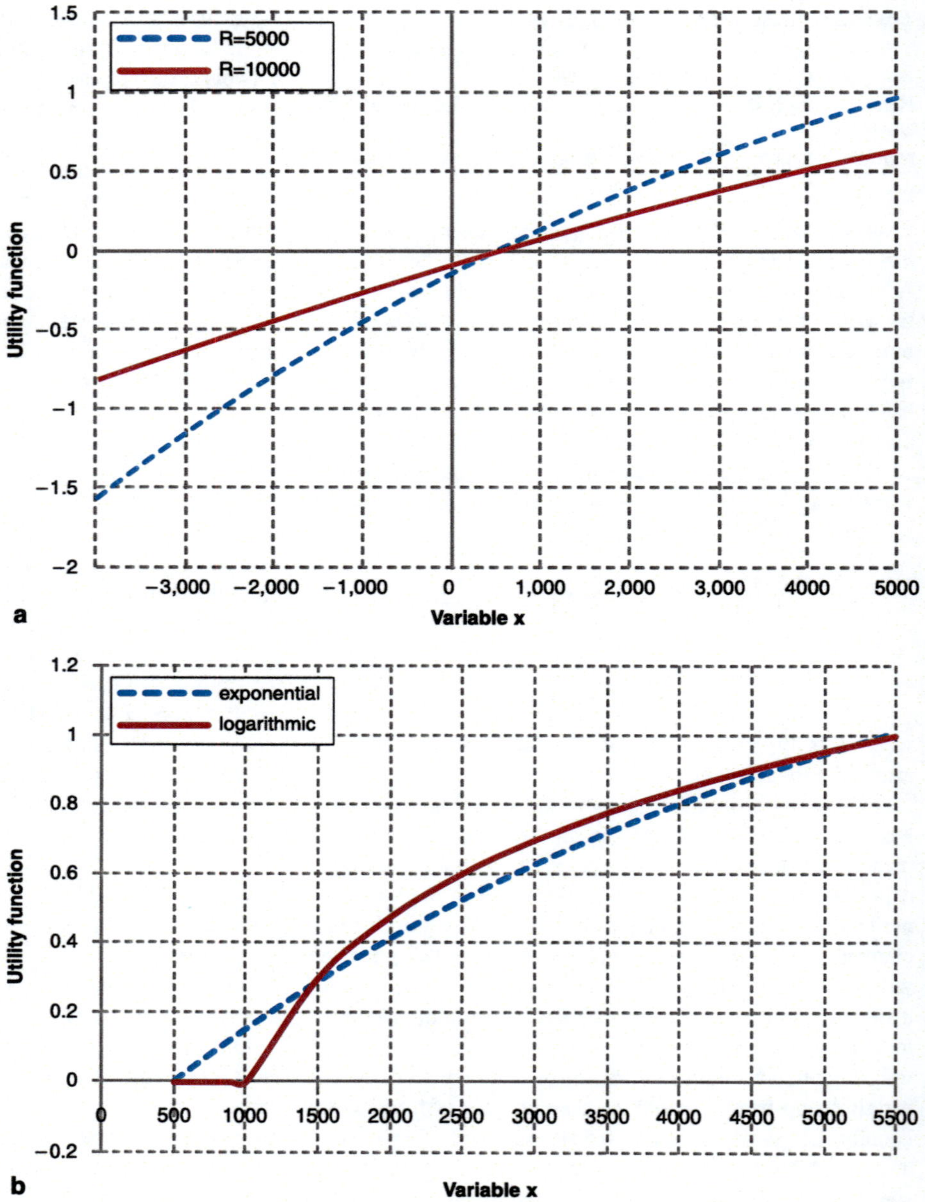

Fig. 12.10 **a** Utility functions of Example 12.2.5 ($x_{min} = 500$). **b** Comparison of the exponential and logarithmic utility functions using values from case (a)

The exponential utility function is used to characterize constant risk aversion, i.e., the risk tolerance is the same irrespective of the amount under risk. Another common attitude towards risk displayed by individuals is *the decreasing risk aversion* quality which reflects the attitude that people's risk tolerance decreases as the payoff increases. This is expressed as:

$$U_{\log(x)} = a + \log \frac{(x - x_{\min})}{R} \qquad (12.2)$$

Figure 12.10b provides a comparative evaluation of the two models on a common graph rescaled such that they have the same values at both extremities as shown, namely $x = x_{min} = 500$, and $x = (R - x_{min}) = 5{,}500$. One notes that as expected the logarithmic model has slightly higher utility values for higher values of the wealth variable indicating that is representative of a more risk-averse outlook. In any case, both models have the same general shape and are quite close to each other. ■

Example 12.2.6: Reanalyze the situation described in Example 12.2.2 using the exponential utility function given by Eq. 12.1. Normalize the function as illustrated in Example 12.2.5 such that $x_{min} = \$\ 2$ M, and consider two cases for the risk tolerance: $R = \$\ 100$ M, and $R = \$\ 200$ M.

Table 12.6 Expected values for the two cases being analyzed (Example 12.2.6)

Alternative	Chance event probability p(x)	Total net savings (TNS) over 6 years (millions $)	Case 1 X_{min}=$ 2 M, R=$ 100 M		Case 2 X_{min}=$ 2 M, R=$ 200 M	
			U(x)	U(x) * p(x)	U(x)	U(x) * p(x)
1	0.25	−320+(6)(120)=400	1.5524	0.3881	1.3657	0.3414
	0.60	−320+(6)(80)=160	1.2561	0.7537	0.8640	0.5184
	0.15	−320+(6)(40)=−80	−2.0099	−0.3015	−0.8018	−0.1203
		Total		0.8403		0.7396
2	0.25	−100+(6)(50)=200	1.3636	0.3409	0.9942	0.2486
	0.60	−100+(6)(35)=110	1.0448	0.6269	0.6601	0.3961
	0.15	−100+(6)(20)=20	0.2606	0.0391	0.1362	0.0204
		Total		1.0069		0.6650

Following Example 12.2.5, the coefficients in Eq. 12.1 are still found to be a=b=1.582. However, the utility values are not between 0 and 1. The results of this example are tabulated in Table 12.6.

Alt#2 is better under Case 1 which corresponds to the more risk-averse situation (i.e., under the lower value of R= $ 100 M) while Alt#1 is preferable in the other case. Though one may not have expected a reversal in the preferable alternative, the trend is clearly as expected. When one is more risk-averse, the possibility of losing money outweighs the possibility making more money.

This example is meant to illustrate that the selection of the risk tolerance value R is a critical factor during the decision-making process. One may well ask, given that it may be difficult to assign a representative value of R, whether there is an alternative manner of selecting R. A break-even approach could be adopted. Instead of selecting a value of R at the onset, the problem could have been framed as one of determining that value of R where both alternatives become equal. It is left to the reader to determine that this break-even value is R=$ 145 M. It is now easier for the decision-maker to gauge whether his risk aversion threshold is higher or lower than this break-even value, and thereby decide on the alternative to pursue. ■

12.2.6 Modeling Multi-attributes or Multiple Objectives

Optimization problems requiring the simultaneous consideration of several objectives or criteria or goals or attributes fall under *multi-attribute* or multiple objectives optimization. Many of these may be in conflict with each other, and so one cannot realize all of the goals exactly. One example involves meeting the twin goals of an investor who desires a stock with maximum return and with minimum risk; these are generally incompatible, and therefore unachievable. A second example of multiple conflicting objectives can be found in organizations that want to: maximize profits, increase sales, increase worker wages, upgrade product quality and reduce product cost, while paying larger dividends to stockholders and retaining earnings for growth. One cannot identify a solution where "all" objectives are met optimally, and so some sort of "compromise" has to be reached. Yet another example involves public policy decisions which minimize risks of economic downturn while replacing fossil fuel-based electric power production (currently, the worldwide fraction is around 85%) by renewable energy sources so that global warming and human health problems are mitigated. The analysis of situations involving multi-attribute optimization is a complex one, and is an area of active research. For the purpose of our introductory presentation, methods to evaluate multiple alternatives can be grouped into two general classes:

(a) ***Single dimensional methods***
The objective function is cast in terms of a common metric, such as the cost of mitigation or avoidance. To use this single criterion method, though, the utility of each attribute must be independent of all the others, while any type of function for the individual utility functions can be assumed. Suppose one has an outcome with a number of attributes, $(x_1, \ldots, x_m)$. Utility independence is present if the utility of one attribute, say $U_1(x_1)$, does not depend on the different levels of other attributes x_m. Say, one has individual utility functions $U_1(x_1), \ldots, U_m(x_m)$ for m different attributes x_1 through x_m. One simple approach is to use *normalized rank or non-dimensional scaling* wherein each attribute is weighted by a factor k_i determined from:

$$k_i = \frac{\text{outcome (i)} - \text{worst outcome}}{\text{best outcome} - \text{worst outcome}} \quad (12.3)$$

Thus, k_i values for each attribute U_i can be determined with values between 0 (the worst) and 1 (the best). Note that this results in the curve for $U_i(x_i)$ to be the same no matter how the other attributes change.

The *direct additive weighting method* is a popular subset of the single criteria weighing method because of its simplicity. It assigns compatible individual weights to each utility of different objectives or attributes, and then combines them as a weighted sum into a single objective function. The additive utility function is simply a weighted average of these different utility functions. For an outcome that has levels (x_1, ..., x_m) on the m objectives or attributes, one can define renormalized weights $k_i' = \frac{k_i}{m}$, and calculate the utility of this outcome as (Clemen and Reilly 2001):

$$U(x_1, ..., x_m) = k_1' U_1(x_1) + \cdots + k_m' U_m(x_m) = \sum_{i=1}^{m} k_i' U_i(x_i) \tag{12.4}$$

where

$$k_1' + k_2' + \cdots + k_m' = 1$$

Based on the above model, the utility of the outcome can be determined for each alternative, and the one with the greatest expected utility is the best choice. Note that the discounted cash flow or net present value analysis is actually a simple method of reconciling conflicting objectives where all future cash flows are "converted" to present value by way of discounting the time value of money. This is an example of an additive utility function with equal weights (see Example 12.2.4). Note that the above discussion applies to multi-attributes which do not interact which each other, i.e., are mutually independent. The interested reader can refer to texts such as Clemen and Reilly (2001) or to Haimes (2004) for modeling utility function with interacting attributes. An illustrative case study of the additive weighting approach is given in Sect. 12.4.2.

(b) ***Multi-dimensional methods***
Most practical situations involve considering multiple attributes, and decision-making under such situations is very widespread. However, in most cases a heuristic approach is adopted, often involving a quantitative matrix method. Even college freshmen in many disciplines are taught this method when basic concepts of evaluating different design choices are being introduced. The following example illustrates this approach.

Example 12.2.7: *Ascertaining risk for buildings*
Several semi-empirical methods to quantify the relative risk of building occupants to the hazard of a chemical-biological-radiological (CBR) attack have been proposed (see Sect. 12.3.4). A guidance document ASHRAE (2003) describes a multi-step risk analysis approach (see Sect. 12.3.1) which first involves defining an organization's (or a building's) exposure level. This approach assigns (with consultation with the building owner and occupants) levels to various attributes such as number of occupants, number of threats received towards the organization, critical nature of the building function, time to recover to a 80% functioning level after an incident occurs, monetary value of the building (plus product, equipment, personnel, information contained...), and the ease of public access to the building.

ASHRAE (2003) gives an example of an office building near a small rural town with 50 employees and typically five visitors at any time (refer to Table 12.7). The value of building is estimated to be $ 3 million, and the building function is low in criticality with only two threats received last year. Access to the building is restricted because card readers are required for entry. The expected recovery time is estimated to be 3 business days. The exposure level matrix (Table 12.7) shows the categories and attribute ranges levels estimated by the building management. Next, the levels specific to this circumstance are determined, and these are entered in the row corresponding to "Score". For example, the level for number of occupants is two, and so on. The third step involves selecting weighting factors for each of the attributes (which need to add up to unity or 100%), after which the weighted score for each factor is calculated. Finally, sum of calculated scores is deduced which provides a single metric of the building/organization's exposure level (=1.8 in this case). No action is probably warranted in this case. ■

It is obvious from the above example that the matrix method is highly subjective and empirical; in all fairness, it is

Table 12.7 Sample exposure level matrix (level 1–5) as suggested in ASHRAE (2003) for different hazards (Example 12.2.7)

Category	Number of people	Received threats	Critical nature of building	Recovery time	Dollar value of facility	Ease of public access
Level						
1	0–10	0–1	Low	<2 days	<$ 2 Million	Low
2	11–60	2–4	Low–Medium	2–14 days	$ 2–$ 10 M	Low–Medium
3	61–120	5–8	Medium	14–90 days	$ 10–$ 50 M	Medium
4	121–1,500	9–12	Medium–High	3–6 months	$ 50–$ 100 M	Medium–High
5	>1,500	>12	High	>6 months	>$ 100 M	High
Determined score[a]	2	2	1	2	2	3
Weighting factor	20%	20%	30%	20%	10%	10%
Calculated score[a]	0.4	0.4	0.3	0.4	0.2	0.3
Exposure level[a]	1.8	Sum of calculated scores				

[a] These values are specific to the building/organization being assessed in this example.

difficult, if not impossible, to quantify the various attributes in an analytical manner without resorting to very complex analytical methods which have their own intrinsic limitations. A more complete illustrative example of multi-attribute decision-making is presented in Sect. 12.4.2. The more rigorous manner of tacking multi-attribute problems is to analyze the attributes in terms of their original metrics. In such cases, decision analysis requires the determination of multi-attribute utility functions instead of a single utility function. These multi-attribute functions would evaluate the joint utility values of several measures of effectiveness toward fulfilling the various objectives. However, manipulation of these multi-attribute utility functions can be extremely complex, depending on the subjectivity of the metrics, the size of the problem and the degree of dependence among the various objectives. Hence, the suggested procedure is to use multi-dimensional analysis for initial screening of alternatives, reduce the dimension of the problem, and then resort to single-dimension methods for final decision-making.

Several techniques have been proposed for dealing with multi-dimensional problems. First, one needs to distinguish between objectives which are separate, and those that interact, i.e., are somewhat related (the latter is much harder to analyze). For the former instance, one modeling approach, called, *non-compensatory modeling,* is to judge the alternatives on a attribute-by-attribute basis. Sometimes, one alternative is better than all others in all attributes. This is a case of dominance, and there is no ambiguity as to which is the best option. Unfortunately, this occurs very rarely.

A second approach is to use the *method of feasible ranges,* where one establishes minimum and maximum acceptance values for each attribute, and then rejects alternatives whose attributes fall outside these limits. This approach is based on the *satisficing principle* which advocates that satisfactory rather than optimal performance or design is good enough for practical decision-making.

A third approach is to adopt a technique known as *goal programming* which is an extension of linear programming (Hillier and Lieberman 2001). The objectives are rank ordered in priority and optimized sequentially. First, specific numerical goals for each of objective are established, an objective function for each objective is formulated, and an optimal solution is sought that minimizes the weighted sum of the deviations of the objective functions from these respective target goals according to one of the following goals: a lower, one-sided goal which sets a lower limit that should be met; a upper, one-sided goal which sets an upper limit that should not be exceeded; two-sided goal where both level limits are set. Goal programming does not attempt to maximize or minimize a single objective function as does the linear programming model. Rather, it seeks to minimize the deviations among the desired goals and the actual results according to the priorities assigned. Note the similarity of this approach to the penalty function method (Sect. 7.3.4). The interested reader can refer to Haimes (2004) and other texts for an in-depth discussion of multi-dimensional methods.

12.2.7 Analysis of Low Epistemic but High Aleatory Problems

Let us now deal with situations where the structure of the problem is well defined, but aleatory uncertainty is present (see Fig. 12.1). For example, the random events impacting the system under study are characterized by a probability distribution and have inherent uncertainties large enough to need explicit consideration. Monte Carlo analysis (MCA) methods were first introduced for calculating uncertainty propagation in data (Sect. 3.7.3). In general, MCA is a method of analysis whereby the behavior of processes or systems subject to chance events (i.e., stochastic) can be better ascertained by artificially recreating the probabilistic occurrence of these chance events a large number of times. MCA methods are also widely used in decision-making problems, and serve to complement probabilistic analytical methods discussed earlier. Recall that, in essence, MCA is a "numerical process of repeatedly calculating a mathematical or empirical operator in which the variables within the operator are random or contain uncertainty with prescribed probability distributions" (Ang and Tang 2007). The following example will serve to illustrate the application of MCA to decision-making.

Example 12.2.8: *Monte Carlo analysis to evaluate alternatives for the oil company example*
Consider Example 12.2.2 dealing with the analysis of two risky alternatives faced by an oil company. The risk is due to the uncertainty in how the economy will perform in the future. Only three states were considered: good, average and poor, and discrete probabilities were assigned to them. The subsequent calculation of the expected value EV was straightforward, and did not have any uncertainty associated with it (hence, this is a low or no epistemic uncertainty situation). The probabilities associated with the future state of the economy are very likely to be uncertain, but the previous analysis did not explicitly recognize this variability. The analytic treatment of such variability in the PDF is what makes this problem a high aleatory one.

For this example, it was assumed that the probability distributions p(Good) and p(Average) are normally distributed with a Coefficient of Variation (CV) of 10% (i.e., the standard deviation is 10% of the mean), and consequently p(Poor) = 1 − p(Good) − p(Average). A software program was used to perform a MCA analysis with 1,000 runs. The results are plotted in Fig. 12.11 both as a histogram and as a cumulative distribution (which can be depicted as either descen-

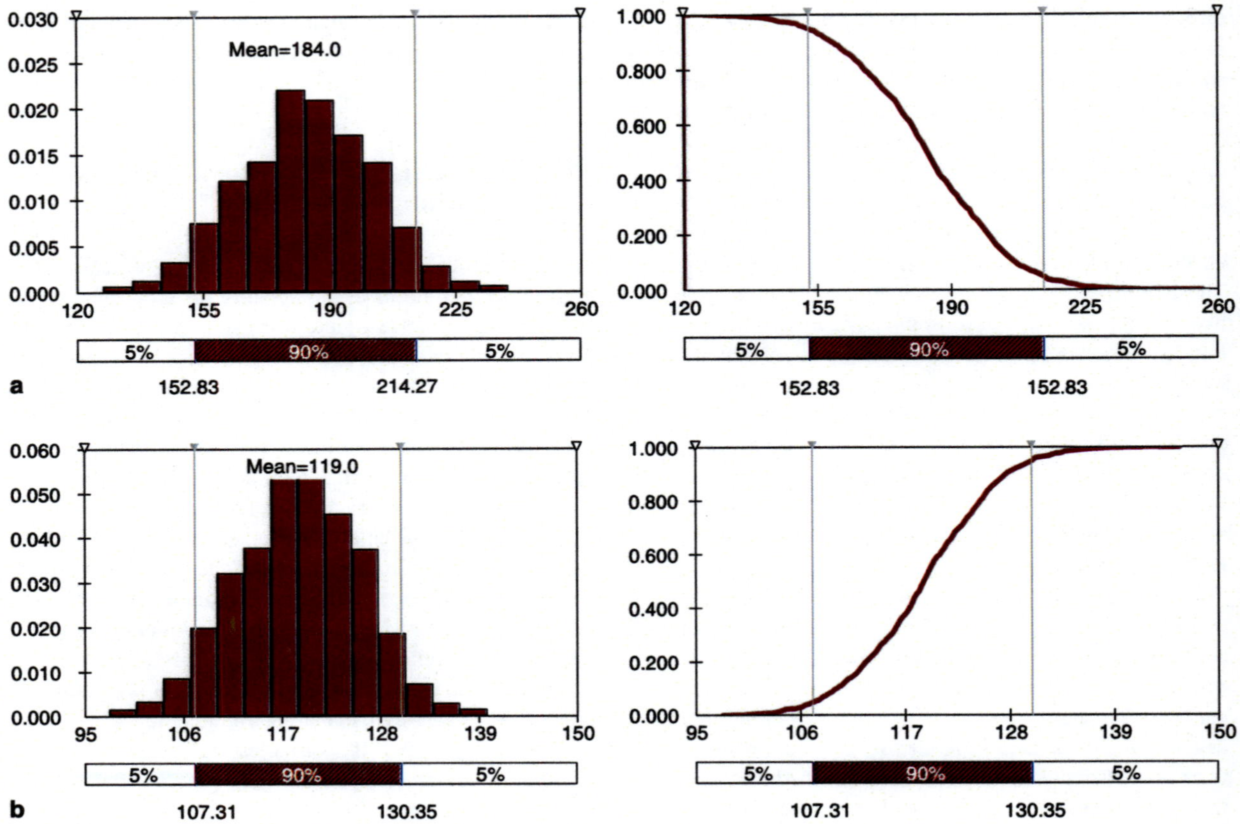

Fig. 12.11 Results of the Monte Carlo analysis with 1,000 runs (Example 12.2.8). **a** Histogram and descending cumulative distribution for the Expected Value (EV) for Alt#1. **b** Histogram and ascending cumulative distribution for the Expected Value (EV) for Alt#2

ding or ascending). Often, MCA results are plotted thus since these two figures provide useful complementary information regarding the shape of the distribution as well as values of the EV random variable at different percentiles. The 5 and 95% percentiles as well as the mean values are indicated. Thus, for Alt#1, the mean value of EV is 184.0 with the 90% uncertainty range being {152.8, 214.3}.One note that there is no overlap between the 5% value for Alt#1 and the 95% for Alt#2, and so in this case, Alt#1 is clearly the better choice despite the variability in the probability distributions of the state of the economy. With higher variability, there would obviously be greater overlap between the two distributions.■

12.2.8 Value of Perfect Information

The value of collecting additional information to reduce the risk, capturing heuristic knowledge or combining subjective preferences into the mathematical structure are intrinsic aspects of problems involving decision-making. As stated earlier, inverse models can be used to make predictions about system behavior. These have inherent uncertainties (which may be large or small depending on the specific situation), and adopting a certain inverse model over potential competing ones involves the consideration of risk analysis and decision-making tools. The issue of improving model structure identification by way of careful experimental design has been covered in Chap. 6, while the issue of heuristic information from experts merits more discussion. This section will present a method by which one can gauge the value of additional information, and thereby ascertain whether the cost of doing so is justified for the situation at hand.

An expert's information is said to be *perfect* if it is always correct (Clemen and Reilly 2001). Then, there is no doubt about how future events will unfold, and so, decision can be made without any uncertainty. The after-the-fact situation is of little value since the decision has to be made earlier. The real value of perfect information is associated with the benefit which it can provide before the fact. The following example will illustrate how the concept of *value of perfect information* can be used to determine an upper bound of the expected benefit, and thereby provide a manner of inferring the maximum additional monetary amount one can spend for obtaining this information.

Example 12.2.9: *Value of perfect information for the oil company*
Consider Example 12.2.2 where an oil company has to decide between two alternatives that are impacted depending on how the economy (and, hence, oil demand and sales) is likely to fare in the future. Perfect information would mean that that the future is known without uncertainty. One way of stating the value of such information compared to selecting Alt#1 is to say that the total net savings (TNS) would be \$ 400 M if the economy turns out to be good, \$ 160 M if it turns out to be average and a loss of \$ 8 M if bad (see Table 12.6). However, this line of enquiry is of little value since, though it provides measures of opportunity loss, it does not influence any phase of the current decision-making process. The better manner of quantifying the benefit of perfect information is to ascribe an expected value to it, called EVPI (expected value of perfect information) under the assumption that any of the three chance events (or states of nature) can occur.

The decision tree of Fig. 12.5 is simply redrawn as shown in Fig. 12.12 where the chance node with its three branches is the starting point, and each branch then subdivides into the two alternatives. Then, perfect information would have led to selecting Alt#1 during the good and average economy states (since their TNS values are higher), and Alt#2 during the poor economy state. The EVPI is given by:

$$\begin{aligned} \text{EPVI} &= 400 \times 0.25 + 160 \times 0.6 + 20 \times 0.15 \\ &= \$\ 199\ \text{M}. \end{aligned}$$

If the decision maker is leaning towards Alt#1 (since it has the higher EV as per Table 12.3), then the additional benefit of perfect information would be (\$ 199 M−\$ 184 M)= \$ 15 M. Hence, the option of consulting experts in order to reduce the uncertainty in predicting the future state of the economy should cost no more than \$ 15 M (and, preferably, much less than this amount). ■

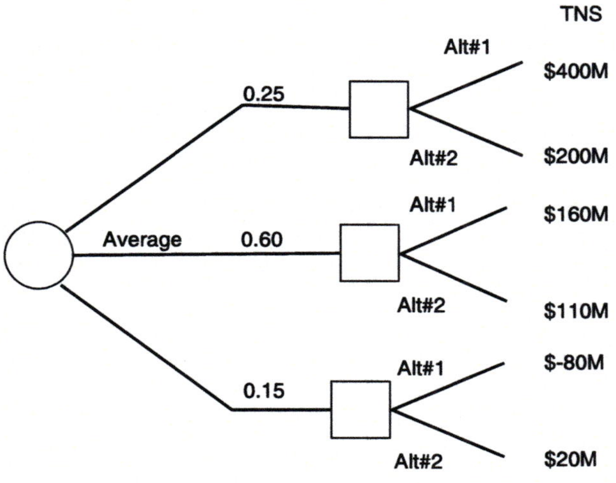

Fig. 12.12 Decision tree representation for determining expected value of perfect information (EVPI). TNS is the total net savings (Example 12.2.9)

12.2.9 Bayesian Updating Using Sample Information

The advantage of adopting a Bayesian approach whereby prior information about the population can be used to enhance the value of information from a collected sample has been discussed in Sect. 2.5 (from the point of view of sampling) and in Sect. 4.6 (for making statistical inferences). Sect. 2.5.1 presented relevant mathematical equations while Example 2.5.3 illustrated this process with an example where the prior PDF of defective pile foundations in a civil engineering project is modified after another pile is tested and found defective. If each defective proportion has a remediation cost associated with it, it is simple to correct the originally expected EV so as to be more reflective of the posterior situation. Thus, the Bayesian approach has obvious application in decision-making. Here, the prior information regarding originally assumed chance events (or states of nature) can be revised with information gathered from a sample. This may take the form of advice or opinion from an expert or even a sort of action such as polling or testing. A company wishing to launch a new product can perform a limited market release, gather information from a survey, and reassess their original sales estimates or even their entire marketing strategy based on this sample information. The following example illustrates the instance where prior probabilities of chance events are revised into posterior probabilities in light of sample information.

Example 12.2.10: *Benefit of Bayesian methods for evaluating real-time monitoring instruments in homes*
Altering homeowner behavior has been recognized as a key factor in reducing energy use in residences. This can be done by educating the homeowner on the electricity draw of various ubiquitous equipment (such as television sets, refrigerators, freezers, dishwashers and washers/dryers) implications of thermostat setup and setdown, switching off unneeded lights,... The benefits of this approach can be enhanced by a portable product consisting of a wireless monitoring cum display device called The Energy Detective (TED) which allows instantaneous energy use of various end-use devices to be displayed in real-time on a small screen along with an estimate of the electricity cost to-date into the billing period, as well as projections of what the next electric bill amount is likely to be. Such feed-back information is said to result in enhanced energy savings to the household as a result of increased homeowner awareness and intervention.

Table 12.8 Summary table of the Bayesian calculation procedure for Example 12.2.10

Chance events	(a) Energy reduction (f_i)	(b) Prior probability p_i	(c) Conditional probability L_i	(d) (b)×(c)	(e) Posterior probability $p'_i=(d_i)/\text{sum}(d)$
S_1	0–2% (1%)	0.10	0.1796	0.01796	0.088
S_2	2–4% (3%)	0.20	0.2182	0.04364	0.214
S_3	4–6% (5%)	0.30	0.1916	0.05748	0.282
S_4	6–8% (7%)	0.30	0.1916	0.05748	0.282
S_5	8–10% (9%)	0.10	0.2702	0.02702	0.133
Total		1.00		0.20358	1.000

A government agency is planning to implement this concept in the context of a major "greening" project involving several thousands of homeowners in a certain neighborhood of a city. Consultants to this project estimate the energy reduction potential f_i discretized into 5 bins (to be interpreted as chance events or states of nature) S_1 through S_5 along with the associated probability p_i of achieving them (shown under the first three columns of Table 12.8).

The prior expected average energy reduction throughout the population, i.e., the entire neighborhood=

$$\sum_{i=1}^{5} f_i \cdot p_i = (0.01)(0.1) + (0.03)(0.2) + (0.05)(0.3) + (0.07)(0.3) + (0.09)(0.1) = 0.051 \text{ or } 5.1\%.$$

The agency is concerned whether the overall project target of 5% average savings will be achieved, and decides to adopt a Bayesian approach before committing to the entire project. A sample of 20 households is selected at random, TED devices are installed with the homeowners educated adequately, and energy use before and after are monitored for say, 3 months each. The following sample information R_i was computed in terms of the energy savings for each of the five chance events:

- S_1-3 homes (i.e., in 3 homes, the energy savings were between 0–2%),
- S_2-4 homes, S_3-6 homes, S_4-6 homes, S_5-1 home.

Recall from Chap. 2 that the probability of event B with event A having occurred can be determined from its "flip" conditional probability:

$$p(B/A) = \frac{p(A/B) \cdot p(B)}{p(A/B) \cdot p(B) + p(A/\bar{B}) \cdot p(\bar{B})} \tag{12.5}$$

In the context of this example, the conditional probabilities or the likelihood function needs to be computed. Assuming the energy reduction percentage x to be a random *binomial* variable, the conditional probabilities or the likelihood function for this sample are calculated as shown below (listed under the fourth column in the table):

$$p(R_1/S_1) = p(x = 3/p = 0.1) = \binom{20}{3}(0.1)^3(0.9)^{17} = 0.1796$$

$$p(R_2/S_2) = p(x = 4/p = 0.2) = \binom{20}{4}(0.2)^4(0.8)^{16} = 0.2182$$

$$p(R_3/S_3) = p(x = 6/p = 0.3) = \binom{20}{6}(0.3)^6(0.7)^{14} = 0.1916$$

$$p(R_4/S_4) = p(x = 6/p = 0.3) = \binom{20}{6}(0.3)^6(0.7)^{14} = 0.1916$$

$$p(R_5/S_5) = p(x = 1/p = 0.1) = \binom{20}{1}(0.1)^1(0.9)^{19} = 0.2702$$

The posterior probabilities are easily determined following Eq. 12.5 with the computational procedure also stated in the table. As expected, the energy reduction amounts are modified in light of the Bayesian updating. For example, chance event S_3 which had a prior of 0.3 now has a posterior of 0.282, and so on. The revised average energy reduction throughout the population, i.e., the entire neighborhood= $\sum_{i=1}^{5} f_i \cdot p'_i = 0.0531$ or 5.31% which reinforces the fact that the target of 5% average energy reductions could be met. This is the confirmation brought about by the information contained in the sample of 20 households. It is obvious that the 20 households selected should indeed be random and a representative sample of the entire neighborhood (which is not as simple as it sounds—see Sect. 4.7 where the issue of random sampling is discussed). Also, the larger the sample, the more accurate the prediction results. However, there is a financial cost and a time delay associated with the sampling phase. The concept of value of perfect information, indicative of the maximum amount of funds that can be spent to gather such information (see Sect. 12.2.8), can also be extended to the problem at hand. Bayesian methods applied to decision-making problems embody a vast amount of literature, and the interested reader can refer to numerous advanced texts in this subject (for example, Gelman et al. 2004). ■

12.3 Risk Analysis

12.3.1 Formal Treatment of Risk Analysis

Without uncertainty, there is no risk. Risk analysis is a quantitative tool or process which allows for sounder decision-making by explicitly treating risks associated with issue at hand. Risk is a ubiquitous aspect of everyday life. It has different connotations in both every-day and scientific contexts, but all deal with the *potential* effects of a loss (financial, physical,…) caused by an undesired event or hazard. This section will describe the basic concepts of risk and risk analysis in general, while a case study example is presented in Sect. 12.4.1.

The analysis of risk can be viewed as a more formal and scientific approach to the well-known Murphy's Law. It consists of three sub-activities elaborated below. Risk analysis provides a framework for determining the relative urgency of problems and the cost-effective allocation of limited resources to reduce risk (Gerba 2006). Thus, preventive measures, remedial actions and control actions can be identified and targeted towards sources and situations which are most critical. The formal study of risk analysis arose as a discipline in the 1940s usually attributed to the rise of the nuclear industry. It has subsequently been expanded to cover numerous facets of our everyday life.

Though different sources categorize them a little differently, the formal treatment of risk analysis includes three specific and interlinked aspects (NRC 1983; Haimes 2004; USCG 2001):

(a) *risk assessment* which involves three sub-activities:
   (i) *hazard identification:* identifying the sources and nature of the hazards (either natural or man-made),
   (ii) *probability of occurrence:* estimating the likelihood or frequency of their occurrence expressed as a probability. Recall that in Sect. 2.6, three kinds of probabilities were discussed: objective or absolute, relative, and subjective probabilities.
   (iii) *evaluating the consequences* (monetary, human life,…) were they to occur. This can be done by one of three approaches: *Qualitative,* which is based on common sense or tacit knowledge of experienced professionals, and wherein guidance is specific to measures which can/ought to be implemented without explicitly computing risk. Generally, this type of heuristic approach is extensively used during the early stages of a new threat (such as that associated with recent extraordinary incidents); *Empirical,* which is based on some simple formulation of the risk function that involves combining heuristic weights to some broad measures characterizing the system; *Quantitative,* which is based on adopting scientific and statistical approaches that have the potential of providing greater accuracy in applications where the hazards are well-defined in their character, their probability of occurrence and their consequences.

Several books have been written on the issue of quantitative risk assessment, both in general terms (such as that by Haimes 2004) and for specific purposes (such as microbial by Haas et al. 1999). Quantitative risk assessment methods are tools based on accepted and standardized mathematical models that rely on real data as their inputs. This information may come from a random sample, previously available data, or expert opinion. The basis of quantitative risk assessment is that it can be characterized as the probability of occurrence of an adverse event or hazard multiplied by its consequence. Since both these terms are inherently such that they cannot be quantified exactly, a major issue in quantitative risk assessment is how to simulate, and thereby determine, confidence levels of the uncertainty in the risk estimates. Very sophisticated probability based statistical techniques have been proposed in the published literature involving traditional probability distributions in conjunction with Monte Carlo and bootstrap techniques as well as artificial intelligence methods such as fuzzy logic (Haas et al. 1999).

There has been a certain amount of skepticism by policy and decision makers towards quantitative risk assessment models even when applied to relatively well-understood systems. The reasons for this lack of model credibility have been listed by Haimes (2004) and include such causes as naïve or unrealistic models, uncalibrated models, poorly skilled users, lack of multi-objective criteria, overemphasis on computer models as against tacit knowledge provided by skilled and experienced practitioners. Nonetheless, the personal opinion of several experts is that these limitations should not be taken as a deterrent in developing such models, but be taken as issues to be diligently addressed and overcome in the future.

(b) *risk management* which is the process of controlling risks, weighing alternatives and selecting the most appropriate action based on engineering, economic, legal or political issues. Risk management deals with how best to control or minimize the specific identified risks through remedial planning and implementation. These include (i) enhanced technical innovations intended to minimize the consequences of a mishap, and (ii) increased training of concerned personnel in order to both reduce the likelihood and the consequences of a mishap (USCG 2001). Thus, good risk management and control cannot prevent bad things from happening altogether,

but they can minimize both the probability of occurrence as well as the consequences of a hazard. Risk management includes *risk resolution* which narrows the set of remedial options (or alternatives) to the most promising few by determining their *risk leverage* factor. This measure of their relative cost-to-benefit is computed as the difference in risk assessment estimates before and after the implementation of the specific risk action plan or measure divided by its implementation cost.
Risk management also includes putting in place response and recovery measures. A major natural disaster occurs in the U.S. on an average of 10 times/year with minor disasters being much more frequent (AIA 1999). Once such disasters occur, the community needs to respond immediately and provide relief to those affected. Hence, rapid-response relief efforts and longer-term rebuilding assistance processes have to be well-thought out and in place beforehand. Such disaster response efforts are typically coordinated by federal agencies such as the Federal Emergency Management Agency (FEMA) along with national and local volunteer organizations.

(c) *risk communication* which can be done both on a long-term or short-term basis, and involves informing the concerned people (managers, stakeholders, officials, public,…) as to the results of the two previous aspects. For example, at a government agency level, the announcement of a potential terrorist threat can lead to the implementation of certain immediate mitigation measures such as increased surveillance, while on an individual level it can result in people altering their daily habits by, say, becoming more vigilant and/or buying life safety equipment and storing food rations.

The above aspects, some have argued in recent years, are not separate events but are interlinked since measures from one aspect can affect the other two. For example, increased vigilance can deter potential terrorists and, thus, lower the probability of occurrence of such an event. As pointed out by Haimes (2004), risk analysis is viewed by some as a separate, independent and well-defined discipline as a whole. On the other hand, there are others who view this discipline as being a sub-set of *systems engineering* that involves (i) improving the decision-making process (involving planning, design, and operation), (ii) improving the understanding of how the system behaves and interacts with its environment, and (iii) incorporating risk analysis into the decision-making process. The narrower view of risk analysis can, nonetheless, provide useful and relevant insights to a variety of problems. Consequently, its widespread appeal has resulted in it becoming a basic operational tool across the physical, engineering, biological, social, environmental, business and human sciences areas, which in turn has led to an exponential demand for risk analysts in recent years (Kammen and Hassenzahl 1999).

12.3.2 Context of Statistical Hypothesis Testing

Statistical hypothesis testing involving Type I and Type II errors have been discussed in Sect. 4.2.2. Recall that such statistical decisions inherently contain probabilistic elements. In other words, statistical tests of hypothesis do not always yield conclusions with absolute certainty: they have in-built margins of error just like jury trials sometimes hand down wrong verdicts. Hence, there is the need to distinguish between two types of errors. Concluding that the null hypothesis is false when in fact it is true was called a Type I error, and represented the probability α (i.e., the pre-selected significance level) of erroneously rejecting the null hypothesis. This is also called the "false negative" or "false alarm" rate. The flip side, i.e. concluding that the null hypothesis is true when in fact it is false, was called a Type II error and represented the probability β of erroneously accepting the alternate hypothesis, also called the "false positive" rate.

The two types of error are inversely related. A decrease in probability of one type of error is likely to result in an increase in the probability of the other. Unfortunately, one cannot simultaneously reduce both by selecting a smaller value of α. The analyst usually selects the significance level depending on the tolerance, or seriousness of the consequences of either type of error specific to the circumstance. Hence, this selection process can be viewed as a part of risk analysis.

Example 12.3.1: *Evaluation of tools meant for detecting and diagnosing faults in chillers*

Refer to Example 2.5.2 in Sect. 2.5.1 where the case of fault detection of chillers from performance data is illustrated by a tree diagram from which Type I and II errors can be analyzed. Let us consider an application where one needs to evaluate the performance of different automated fault detection and diagnosis (FDD) methods or tools meant to continuously monitor large chillers and assess their performance. Reddy (2007) proposed a small set of quantitative criteria in order to perform an impartial evaluation of such tools being developed by different researchers and companies. The evaluation is based on formulating an objective function where one minimizes the sum of the costs associated with false positives and those of missed opportunities. Consider the FDD tool as one, which under operation, first sorts or flags incoming system performance data into either fault-free or faulty categories with further subdivisions as described below (Fig. 12.13):

(a) *False negative rate* denotes the probability of calling a faulty process good, i.e. missed opportunity loss (Type II error);
(b) *Correct fault-free detection rate* denotes the probability of calling a good process good;
(c) *False positive rate* denotes the probability of calling a good process faulty, i.e. false alarm (Type I error);

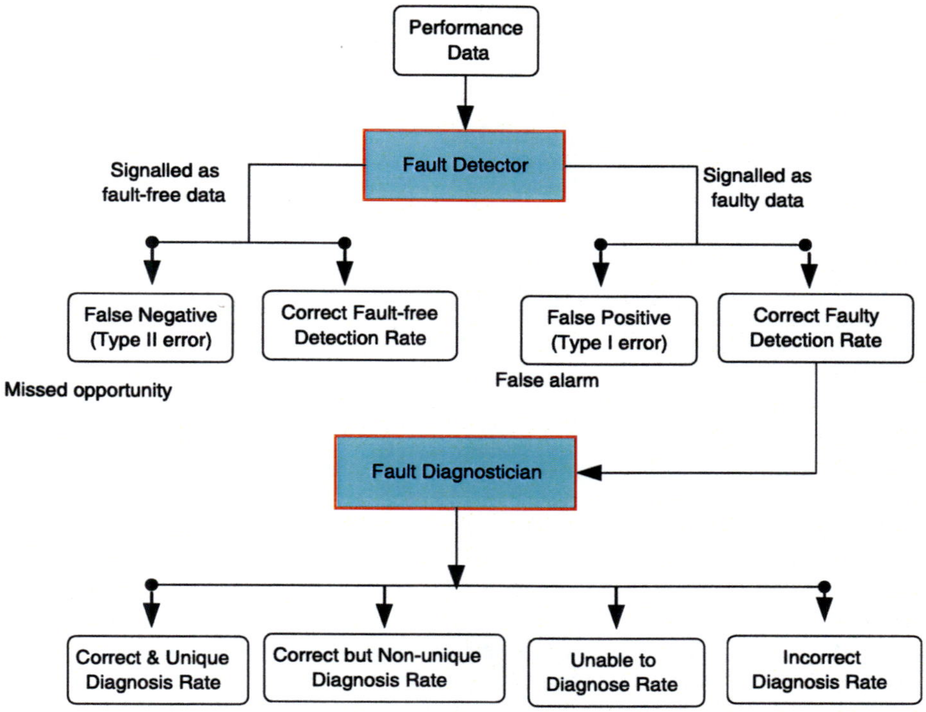

Fig. 12.13 Evaluation procedure for detecting and diagnosing faults while monitoring an engineering system. (From Reddy 2007)

(d) *Correct faulty detection rate* denotes the probability of calling a faulty process faulty.

Note that in case only one crisp fault detection threshold value is used, probabilities (a) and (d) should add to unity, and so should (b) and (c). Thus, if the correct fault-free detection rate is signaled by the FDD tool as 95%, then the corresponding false alarm rate is 5%. Once a fault has been detected correctly, there are four possibilities, each of which has a cost implication (in terms of technician's time to deal with it—see Fig. 12.13):

- *Correct and unique diagnosis,* where the fault is correctly and unambiguously identified;
- *Correct but non-unique,* where the diagnosis rules are not able to distinguish between more than one possible fault;
- *Unable to diagnose,* where the observed fault patterns do not correspond to any rule within the diagnosis rules;
- *Incorrect diagnosis,* where the fault diagnosis is done improperly.

The evaluation of different FDD methods consisted of two distinct aspects:

(a) *how well is fault detection done?* Given the rudimentary state of practical implementation of FDD methods in HVAC equipment, say chillers, it would suffice for many service companies merely to know whether a fault has occurred; they would then send a service technician to diagnose and fix the problem; and

(b) *how well does the FDD methodology perform overall?* Each of the four diagnosis outcomes will affect the time needed for the service technician to confirm the suggested diagnosis of the fault (or to diagnose it).

Varying the fault detection threshold affects the *sensitivity of detection,* i.e., the Correct Fault–Free Detection Rate and the False Positive Rate are affected in compensatory and opposite ways. Since these rates have different cost implications, one cannot simply optimize the total error rate. Instead, FDD evaluation can be stated as a minimization problem where one tries to minimize the sum of penalties associated with false positives (which require a service technician to needlessly respond to the service call) as against ignoring the alarm meant to signal the onset of some fault (which could result in an extra operating cost incurred due to excess energy use and/or shortened equipment life). The frequency of occurrence of different faults over a long enough period (say, 3 months of operation) need to be explicitly considered as well as their energy cost penalties. Such considerations led to the formulation of objective functions which were applicable to any HVAC&R system. Subsequently, these general expressions were then tailored to the evaluation of FDD methods and tools appropriate to large chillers, and specific numerical values of several of the quantities appearing in the normalized rating expression were proposed based on discussions with chiller manufacturers and service companies as well as analysis of fault-free and faulty chiller performance data gathered in a laboratory from a previous research study. This methodology was used to assess four chiller FDD methods using fault-free data and data from intentionally introduced faults of different types and severity collected from a laboratory chiller (Reddy 2007).

12.3.3 Context of Environmental Risk to Humans

One of the most important shifts in environmental energy policy was the acceptance in the late 1980s of the role of risk assessment and risk management in environmental decision-making (Masters and Ela 2008). One approach to quantifying such environmental risks (as against accidents where the causality is direct and obvious) is in terms of premature mortality or death. Everyone will eventually die, but being able to identify specific causes which are likely to decrease a person's normal life span is important. The government can then take decisions on whether the mitigation or cost of reducing that risk level is warranted in terms of its incremental benefit to society as a whole. Rather than focusing on one source of risk and regulating it to oblivion, the policy approach adopted is to regulate risks to a comparable level of consequence. This approach recognizes that there is a tradeoff between the utility gain due to increased life expectancy and the utility loss due to decreased consumption (Rabl 2005), and so a rational policy should seek to optimize this tradeoff, i.e., maximize the total life-time utility. The general scientific framework or methodology for evaluating risks in a quantitative manner is well established, with the issue of how best to apply/tailor it to the context or specific circumstance still being in varying stages of maturity. An important limitation is the lack of complete quantitative data needed to exercise the assessment models, along with the realization that such data may never be entirely forthcoming in many application areas.

The occurrence, and then avoidance, of environmental risks are not deterministic events, and have to be considered in the context of probability. In Sect. 2.6, the notion of relative probabilities was described, and Table 2.10 provided values of relative risks of various causes of deaths in the U.S. However, such statistics provide little insight in terms of shaping a policy or in defining a mandatory mitigation measure. For example, 24% die of cancer, but what specifically led to this cancer in the first place cannot be surmised from Table 2.10. For risk assessment to be useful practically, one has to consider the root causes of these cancers and, if possible, reduce their impact on just those individuals who are most at risk. Hence, environmental risks are stated as *incremental probabilities* of death to the exposed population (referred to as *statistical deaths*) which is much more meaningful. The USEPA and the Food and Drug Administration (FDA) restrict the amount of chemicals or toxins a person can be exposed to, based on some scientific studies reflective of the best current knowledge available while presuming an incremental probability range. USEPA has selected toxic exposure levels which pose incremental lifetime cancer risks to exposed members of the public in the range of 10^{-6}–10^{-4}, or one extra death due to cancer from a million to 10,000 exposed individuals. If one considers the lower risk level, and assuming the US population to be 300 million, one can expect 300 additional deaths, which spread over a typical 70 year life expectancy would translate to about four extra deaths per year. Table 12.9 assembles activities that increase mortality risk by one in a million, the level assumed for acceptable environmental risk. Risks from various activities such as smoking, living in major cities, flying are shown; these should be taken as estimates and may be contested or even modified over time. Note that this is a dated study (30 years or more), and many of the numbers in the table may no longer be realistic. Conditions change over time (for example, living in New York city or Boston has much less adverse air pollution risk these days), and periodic revision of these risks is warranted.

Table 12.9 Activities that increase risk by one in a million. (From Wilson 1979)

Activity	Type of risk
Smoking 1.4 cigarettes	Cancer, heart disease
Drinking ½ l of wine	Cirrhosis of the liver
Spending 1 h in a coal mine	Black lung disease
Living 2 days in New York or Boston	Air pollution
Travelling 300 miles by car	Accident
Flying 1,000 miles by jet	Accident
Traveling 10 miles by bicycle	Accident
Traveling 6 min in a canoe	Accident
Living 2 summer months in Denver (vs sea level)	Cancer by cosmic radiation
Living 2 months with a cigarette smoker	Cancer, heart disease
Eating 40 tablespoons of peanut butter	Liver cancer
Living 50 miles within 5 miles of a nuclear reactor	Accidental radiation release

Dose response models were introduced earlier in Sect. 1.3.3 and treated more mathematically in Sects. 10.4.4 and 11.3.4. The linear slope of the dose-response curve of a particular toxin is called the *potency factor* (PF), and is defined as (Masters and Ela 2008):

$$\text{Potency factor} = \frac{\text{Incremental lifetime cancer risk}}{\text{Chronic daily intake (mg/kg-day)}} \tag{12.6}$$

where chronic daily intake (CDI) is the dose averaged over an entire lifetime per unit kg of body weight. Rearranging the above equation:

$$\text{Incremental lifetime cancer risk} = \text{CDI} \times \text{PF} \tag{12.7}$$

The USEPA maintains a database on toxic chemicals called the Integrated Risk Information Systems (IRIS) where information on PF can be found. Table 12.10 assembles to-

Table 12.10 Toxicity data for selected carcinogens. (USEPA website, www.epa.gov/iris)

Chemical	Potency factors $(mg/day/kg)^{-1}$	
	Oral route	Inhalation route
Arsenic	1.75	50
Benzene	2.9×10^{-2}	2.9×10^{-2}
Chloroform	6.1×10^{-3}	8.1×10^{-2}
DDT	0.34	–
Methyl chloride	7.5×10^{-3}	1.4×10^{-2}
Vinyl chloride	2.3	0.295

xicity data of selected carcinogens. Note that the statistics in Table 2.10 are based on actuarial data which may be taken to be accurate. However, data in Table 12.10 is based on toxicological studies, usually with some assumptions on model structure and parameter values. In that sense, they are estimates which may contain large variability because of the uncertainties inherent in the extrapolations: (i) from lab tests on animals to human exposure, and (ii) from high dosage levels to the much lower dosage levels to which humans are usually exposed to.

Example 12.3.2:[1] R*isk assessment of chloroform in drinking water*

Drinking water is often disinfected with chlorine. Unfortunately, an undesirable byproduct is chloroform which may be cancerous. Suppose a person weighing 70 kg drinks 2 l of water with chloroform concentrations of 0.10 mg/l (the water standard) every day for 70 years.

(a) What is the risk of this individual?
From Table 12.10, the potency factor PI for oral route= 6.1×10^{-3} $(mg/day/kg)^{-1}$.
The daily chronic intake CDI=(0.10 mg/l) · (2 l/day)/ 70 kg= 0.00286 mg/kg-day
Thus, the incremental lifetime risk=CDI×PF= $0.00286\times 6.1\times10^{-3}=17.4\times10^{-6}$ i.e., the extra risk over a 70 year period is only 17 in one million.

(b) How many extra cancers can be expected per year in a town with a population of 100,000 if each individual is exposed to the above risk?
Expected number of extra cancers per year

$$= (100{,}000)\cdot\frac{17.4}{10^6}\cdot\frac{1}{70} = 0.024 \text{ cancers/year}$$

Clearly this number is much smaller than the total number of deaths by cancer of all sorts, and hence, is lost in the "background noise". This example serves to illustrate the statement made earlier that it is difficult, if not impossible, to attribute an individual death to a specific environmental effect. ■

Another problem with environmental risk assessment is that health and safety are *moral sentiments* (like freedom, peace and happiness), and are not absolutes which can be quantified impartially. They are measured intangibly by the absence of their undesirable consequences which are also difficult to quantify (Heinsohn and Cimbala 2003). This leads to much controversy even today from various stakeholders about environmental risk assessment in general-skepticism from scientists about the ambiguous and uncertain dose-response relationships used, to certain segments of the population expressing overblown concerns, and other segments feeling that risks are over-stated.

Yet, another issue is that risks have to be distinguished by whether they are *voluntary or involuntary*. Recreational activities such as skiing or riding a motorcycle, or lifestyle choices such as smoking, can lead to greatly increased risks to the individuals, but these are choices which the individual makes voluntarily. The government can take certain partial mitigation measures such as mandating helmets or increasing the tax on cigarettes. However, there are some voluntary risks (such as living in New York City, for example—see Table 12.9) which the government can do little about. Involuntary risks are those which are imposed on individuals because of circumstances beyond their control (such as health ailments due to air pollution), and people tend to view these in a more adversarial manner.

A final issue is *risk perception* which greatly influences how an individual views the threat, and this falls under subjective probability (discussed in Sect. 2.6). Figure 2.31 clearly demonstrates the wide disparity in how different professionals view the adverse impact of global warming on gross world product. At an individual level, the perception may be greatly influenced by either the lack of control on the event or the uncertainty surrounding the issue or by excessive media propaganda. For example, natural risks such as earthquakes or floods are accepted more readily (even though they cause much more harm) than man-made ones (such as plane crashes). Further, the uncertainty surrounding risks of radiation exposure to people living close to nuclear power plants or high-tension electric transmission lines has much to do with the current overblown concerns due to lack (or perceived lack) of scientific knowledge. When the bearers of risk do not share the cost of reducing the risk, extravagant remedies are apt to be demanded.

12.3.4 Other Areas of Application

This section will provide brief literature reviews on the application of risk analysis to different disciplines.

(a) ***Engineering***

DeGaspari (2002) describes past and ongoing activities by the American Society of Mechanical Engineers (ASME) on managing industrial risk, and quotes experts as stating that: (i) risk analysis with financial tools can benefit a company's

[1] From Masters and Ela (2008) by © permission of Pearson Education.

bottom line and contribute to safety, (ii) a full quantitative analysis can cost 10 times as much as a qualitative analysis, and (iii) fully quantitative risk analysis provides the best bet for optimizing plant performance and corporate values for the inspection/maintenance investment while addressing safety concerns. Lancaster (2000) investigates the major accidents in the history of engineering and gives reasons why they occurred. The book gives many statistics for different types of hazards and cost for each type of disaster. Additionally, chapters on human error are also included.

There is extensive literature on risk analysis as applied to nuclear power plants, nuclear waste management and transportation, as well as more mundane applications in mechanical engineering. A form of risk analysis that is commonly used in the engineering field is *reliability analysis*. This particular analysis approach is associated with the probability distribution of the time a component or machine will operate before failing (Vose 1996). Reliability has been extensively used in mechanical and power engineering in general, and in the field of machine design in particular. It is especially useful in modeling the likelihood of a single component of the machine failing, and then deducing the failure risk of several components placed in series or parallel. Reliability can be viewed as the risk analysis of a system due to mechanical or electrical failures, whereas traditional risk analysis in other areas deals with broader scenarios.

Risk analysis has also been adopted in several other fields, for example, during building construction (McDowell and Lemer 1991). In this application, risk analysis deals with cost and schedule:

(i) *Cost risk analysis* is modeled as a discrete possible event, where the cost of the building is compared to a payback period, (ii) *Schedule risk analysis* deals with the connection between tasks that influence the construction time. Often, penalties must be paid if a building is not completed within the stipulated time period.

(b) *Business*

Risk analysis has found extensive applications in the business realm, where several solutions may be posed, but only one is the best possible scenario since competing solutions always involve risk. In a marketing application, a sample can be taken from a random population, and through risk analysis and modeling, a marketing campaign can be designed. From a marketing standpoint, a company can identify the kind of campaign the public best responds to, and alter their marketing accordingly. Risk assessment techniques are also commonly employed in the business realm in order to help make important decisions such as whether to invest in a venture, or where to optimally site a factory or business. Such techniques are often rooted in financial modeling, where the risk is directly related to the monetary pay-off in the end. There are four major categories of decisions in the business world that utilize risk assessment (Evans and Olson 2000): (a) acceptance or rejection of a proposal based on either net present value or internal rate of return; (b) selection of the best choice among mutually exclusive alternatives; for example selecting a fuel source among wood, oil, or natural gas would dependent on several factors such as price, availability, and growth rate; (c) selection of the best choice among non-mutual alternatives; (d) decisions containing a degree of uncertainty which involve calculating the expected opportunity loss and return to risk ratios, by creating decision trees and using Monte Carlo simulation techniques.

(c) *Human Health, Epidemiology & Microbial Risk Assessment*

Risk assessment is also commonly used when human health concerns are a factor. Haas et al. (1999) outline the primary areas where risk assessment is applied in health situations, and give a process for performing a risk assessment. This area of study is concerned with the impact of exposure to defined hazards on human health. Epidemiology, which is a subset of human health, is the "study of the occurrence and distribution of disease and associated injury specified by person, place, and time" (Haas et al. 1999), while microbial assessment is concerned only with the disease and its opportunity to spread.

The risk assessment process in health situations consists of four steps: (i) *Hazard Identification* which describes the health effects that are the result of human exposure to any type of hazard; (ii) *Dose-Response Assessment* which correlates the amount of time of the exposure to the rate of incidence of infection or sickness; (iii) *Exposure Assessment* which determines the size and nature of the population that was exposed to the hazard, and also how the exposure occurred, the amount, and the total elapsed time of exposure; and (iv) *Risk Characterization* which integrates the information from the above steps to calculate the implications for the general public's health, and calculate the variability and uncertainty in the assessment. A book chapter by Gerba (2006) deals specifically with health-based and ecological risk assessment.

(d) *Extraordinary Events*

Several agencies such as EPA, OSHA, NIOSH, ACGIH, ANSI have developed quantitative metrics such as PEL, TLV, MAC, REL (see for example, Heinsohn and Cimbala 2003) which specify the threshold or permissible levels for short-term and long-term exposure. The bases of these metrics are similar in nature, but these metrics differ in their threshold values because they target slightly different populations. For example, NIOSH is more concerned with the workplace environment (where the work force is generally healthier), while EPA's concern is with the general public (that includes the young and elderly as well, and under longer exposure times).

NIOSH has developed TLV-C (threshold limit values-ceiling) guidelines for maximum chemical concentration levels that should never be exceeded in the workplace. EPA defines extreme events as once-in-a-life time, non-repetitive or rare exposure to airborne contaminants for not more than 8 h. Though only "chemicals" are mentioned, the definition could be extended to apply to biological and radiological events as well. Consequently, EPA formed a Federal Advisory Committee with the objective of developing scientifically valid guidelines called AEGLs (Acute Exposure Guideline Levels) to help national and local authorities as well as private companies deal with emergency planning, prevention and response programs involving chemical releases. The AEGL program further defines 3 levels (Level 1 thresholds for no adverse effects; Level 2 thresholds with some adverse effects; and Level 3 with some deaths) and also defines acute exposure levels at different exposure times: 10 min, 30 min, 60 min, 4 h and 8 h. Another parallel effort is aimed at developing a software program called CAMEO (Computer-Aided Management of Emergency Operations) to help in the planning and response to chemical emergencies. It is to be noted that exposure levels in CAMEO are different from AEGLs, and do not include exposure duration.

Recent world events (such as 9/11) have spurred leading American engineering societies (such as IEEE, ASCE, ASME, ASHRAE,...) as well as several federal and state agencies to form expert working groups with the mission to review all aspects of risk analysis as they apply to critical infrastructure systems under extreme man-made events. The journal of Risk Analysis devoted an entire special issue (vol. 22, no 4., 2002) with several scholarly articles on the role of risk analysis applied to this general problem, on government safety decisions, on how to use risk analysis as a tool for reducing risk of terrorism, and on the role of risk analysis and management. Studies of a more pragmatic nature have been developed by several organizations dealing with applying risk analysis and management methodologies and software specific to extreme event risk to critical infrastructure. Several private, state and federal entities and organizations have prepared numerous risk assessment guidance documents. ASTM and ASCE have developed guides specific to naturally occurring extreme events applicable to buildings and indoor occupants from a structural viewpoint. Sandia National Laboratory has numerous risk assessment methodology (RAM) programs for both terrorist and natural events; one well know example, is the RAMPART software tool (Hunter 2001) which performs property analysis and ranking for General Services Administration (GSA) buildings due to both natural and man-made risks. It allows assessing risks due to terrorism, natural disaster and crime in federal buildings nationwide by drawing on a database of historic risks for different disasters in different regions in the U.S. It can also be adapted to other types of critical facilities such as embassies, school systems and large municipalities.

There has also been a flood of documents on risk analysis and mitigation related to extreme events in terms of indoor air quality (IAQ) risks to building occupants (reviewed in a report by Bahnfleth et al. 2008). The current thinking among fire hazard assessment professionals (which is perhaps the most closely related field to extreme event IAQ) is that, since the data needs for the risk evaluation model are unlikely to be known with any confidence, it is better to adopt a **relative risk approach,** where the building is evaluated based on certain pre-selected set of extreme event scenarios, rather than an absolute one (Bukowski 2006). Several semi-empirical methods to quantify the relative risk of building occupants to the hazard of a chemical-biological-radiological (CBR) attack have been proposed. For example, Kowalski (2002) outlines a method which involves considering four separate issues (hazard level, number of occupants, building profile and vulnerability), assigning weights between 0–100% for each of them, and deducing an overall weighted measure of relative risk for the entire building. Appendix C of the ASHRAE guidance document (ASHRAE 2003) describes a multi-step process involving defining a building category (based on factors such as number of occupants, number of threats received, time to restore operation, monetary value of the building…), assigning relative weights of occupant exposure, assigning point values for severity, determining severity level in each exposure category, and finally calculating an overall score or rank from which different risk reduction measures can be investigated if a critical threshold were to be exceeded. Example 12.2.7 illustrated a specific facet of the overall methodology.

12.4 Case Study Examples

12.4.1 Risk Assessment of Existing Buildings

This section will present a methodology for assessing risk in existing large office buildings[2]. The conceptual quantitative model is consistent with current financial practices, and is meant to provide guidance on identifying the specific risks that need to be managed most critically in the building under consideration. This involves identifying and quantifying the various types and categories of hazards in typical buildings, and proposing means to deal with their associated uncertainties. The methodology also explicitly identifies the vulnerabilities or targets of these hazards (such as occupant safety, civil and operating costs, physical damage to a building and its contents, and failure of one or several of the major building systems), as well as considers the subtler fact that different stakeholders of the building may differ in their perception as to the importance of these vulnerabilities to their businesses. Finally, the consequences of the occurrence of these risks are quantified in terms of financial costs consistent with the current business practice of insuring a building and its oc-

[2] Based on a paper by Reddy and Fierko (2004).

cupants. The emphasis in this example is on presenting the conceptual methodology rather than accurate quantification of the risks as they pertain to an actual building.

Similar to a business portfolio manager advising a client on how to allocate funds for retirement savings based on the client's age and risk tolerance, the person analyzing risk in a building must first consider the type of stakeholder (say, the owner or the tenant in a leased building scenario). The owner may be more concerned with the risk to the civil construction and to the basic amenities which are his responsibilities, whereas the tenant may be more concerned with the cost of replacing the business equipment along with lost revenue should a deleterious event occur. Finally, both may be liable to be sued by the occupants if they were to be harmed. Hence, one needs to start with the stakeholder.

There are several stakeholders in a building. They include anyone who has an interest in the design, construction, financing, insurance, occupancy, or operation of a building. This list includes, but is not limited, to the following groups of people: (i) building owner/landlord, (ii) building occupants/tenants, (iii) architects/engineers, (iv) local code officials, (v) township/city representatives, (vi) neighbors, (vii) contractors/union officials, (viii) insurance companies, and (ix) banks/financing institutions. Let us consider only the first group: *building owner*. This example focuses on the risks associated with an occupied leased commercial building, and not with the design, code compliance, and construction of such a structure.

The building owner or landlord is defined as the person who finances and operates the building on a daily basis. This person is concerned more with the physical building and the financial impacts on the operation of such a structure. The concerns of the building owner lie in the areas of operating costs and the physical building. The building owner is assumed not be a regular occupant of the building, rather this role will be performed from a remote location. On the other hand, the occupants of the building are defined as the people who work in the building on a daily basis. A typical occupant is a company that leases space from the building owner. Although the occupants are concerned with the physical building to some degree, this group is much more concerned with the well-being of its employees and the company-owned contents inside of the building. This group is also less sensitive to financial impacts on the operation of the building[3], since it is assumed that their lease rate is not sensitively tied to fluctuations in building operations cost. Additionally, the tenant can be viewed as a part-time occupant who has the option of leaving the building after their lease has expired. The situation where the occupant is also the owner of the building is not be considered in this risk assessment study, although the proposed model can be altered to fit these conditions.

[3] The building occupants are not totally insensitive to operating costs since they would generally pay for the utilities.

Table 12.11 Description of different hazard categories which impact specific building elements

Building element	Hazard category	Description
Civil	Natural	Natural events that affect the civil construction of a building such as earthquakes, floods, and storms
	Intentional	Actions that are purposely committed and designed to harm the physical building such as bombings and arson
	Accidental	Actions that are not committed intentionally but have serious results such as unintentional fires and accidents
Direct Physical	Crime	Actions that only affect the occupants and not the physical building, such as robbery or assault
	Terrorist	An act of terrorism that is intended to affect the occupants only, such as hostage situations
	Bio & IAQ	Contamination of air in order to harm building occupants
Cybernetic	Intentional	Sabotage, hacking in IT networks
	Accidental	Not committed on purpose, but which result in harm, such as computer crashes
MEP Systems	Accidental	The failure of mechanical, electrical or plumbing systems as well as telecom, fire safety equipment
Operation Services	Unanticipated	Impact of the fluctuations of utility prices and operation and maintenance that are required to operate the building

The methodology consists of the following aspects:

(a) Hazard Categories and Affected Building Elements
Different building elements that can be damaged should different hazards occur need to be identified. The five building elements that are susceptible are shown in Table 12.11 along with their hazard categories.

(b) Vulnerable Targets in a Building
Targets are those which are affected by different building hazards, namely occupants, property replacement and revenue/loss. Table 12.12 lists these along with sub-targets.

(c) Applicability Matrix
The hazards identified have an impact on the stakeholders listed previously in different ways. One set of stakeholders may be more sensitive to a certain risk event than to others. The risk analysis methodology explicitly considers this linkage between targets and sub-targets to affected building elements and associated hazard categories. This mapping is done by defining an applicability matrix which depends on the type of stakeholder for whom the risk analysis is being performed. The applicability matrix is binary in nature (i.e., numerical values can be 0 or 1, with 0 implying "not applicable", and 1 implying "applicable"). Table 12.12 depicts such an applicability matrix (AM) from the perspective of the owner of a leased commercial building. For example, the building owner may not view revenue loss vulnerabilities

Table 12.12 Applicability Matrix (AM) from the perspective of the owner of a leased building

Vulnerable targets	Sub-targets	Building elements	Civil structure (building specific)			Direct physical (individual)			Cybernetic		MEP systems	Operation services
		Hazard categories	Natural	Intentio-nal	Acciden-tal	Crime	Terrorist	Bio & IAQ	Intentional	Accidental	Accidental	Unantici-pated
Occupants	Short-term		1	1	0	1	1	1	0	0	0	0
	Long-term		1	1	0	1	1	1	0	0	0	0
Property replace-ment	Physical building		1	1	1	0	0	1	0	0	0	0
	Contents		1	1	1	0	0	0	1	0	1	0
	Indoor environment		1	0	1	0	0	1	1	0	1	0
	Building systems		1	1	1	0	0	1	1	1	1	1
Revenue loss	Operating cost		0	1	1	1	1	0	0	0	1	0
	Cost of utilities		0	0	0	0	0	0	0	0	0	1
	Lost business		0	0	0	0	0	0	0	0	1	1

due to lost business or cost of utilities to be his responsibility. Hence, the corresponding cells have been assigned a value of zero in most cases).

(d) Importance Matrices

One approach to modeling the importance with which a particular stakeholder views a specific target type or sub-target is the continuous utility function approach described in Sect. 12.2.5. Another is to do so using fuzzy theory (see, for example, McNeill and Thro 1994). Fuzzy numbers provided by the stakeholder are used along with their uncertainty characterized by a symmetrical triangular membership function. The importance matrix IM1 for stakeholders versus target type on one hand, and that for target type and sub-target (called IM2) are shown in Fig. 12.14. From a building owner perspective, property replacement is more crucial than say, occupant hazards, and the fuzzy values of 0.6 and 0.1 for IM1 shown reflect this perception. The tenant is likely to view the relative importance of these two targets differently, which was the reason for our initial insistence that one should start first and foremost with the concerned stakeholder. Further, the building owner is likely to be more concerned with the short-term, as against their long term exposure (since occupants change and it is more difficult to prove the owner's culpability), which is translated into fuzzy values of 0.9 and 0.1 respectively in Fig. 12.14 corresponding to IM2. The numerical values have no factual or historic basis nor have they been deduced from surveys of a specific stakeholder class. They are merely illustrative numbers for the purposes of this example. Note that these fuzzy numbers are **conditional**, i.e., they add up to unity at each level. The membership function is characterized by only one number (representing a plus/minus range around the estimate) since a symmetric triangular function is assumed in order to keep the data gathering as simple as possible. The actual input fuzzy data needs to be refined over time and be flexible enough to reflect, first the actual perception of a class of stakeholder, and second, that of a specific individual stakeholder depending on his preferences, circumstances, and special concerns.

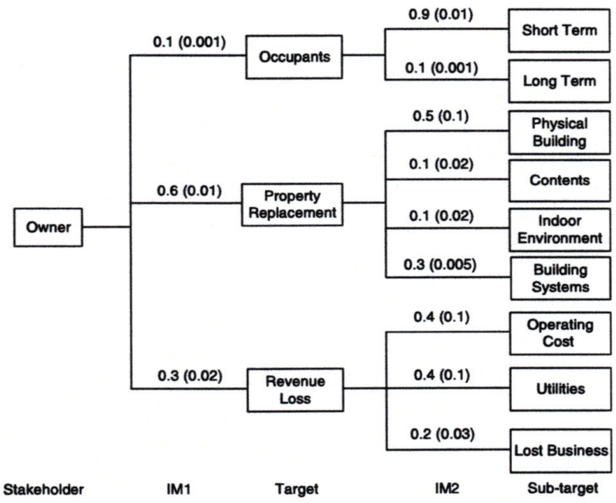

Fig. 12.14 Overall Risk Assessment Tree Diagram by *Target Category* with numerical values of the Importance Matrix (IM1) between stakeholder (in this case, the building owner) and target types, and of Importance Matrix (IM2) between targets and sub-target types. The numbers are fuzzy values (which sum to unity) with associated uncertainty in parenthesis (assuming symmetric triangular membership)

(e) Hazard Event Probabilities

Hazard categories proposed have been described above, and also summarized in Table 12.12. One needs finer granularity in the hazard categories by defining hazard events since each of them need different risk management and alleviation measures. The various event categories assumed are shown in Table 12.13. Note from Fig. 12.15 that natural civil hazards can result from five different events (hurricane, earthquake, tornado, flood and windstorm). The list of events is conside-

Table 12.13 Event probabilities $P_{i,j}$ (absolute probability of occurrence per year) for different hazard events and associated costs

Affected building element	Hazard category (i)	Hazard event (j)	Probability p_{ij}	Associated cost (C_h) ($/year)
Civil structure	Natural	Hurricane	0.005	I_d (building)
		Earthquake	0.0005	
		Tornado	0.001	
		Flood	0.01	
		Winter storm	0.0005	
	Intentional	Arson	0.005	
		Bombing	0.002	
		Terrorism	0.003	
	Accidental	Fire	0.005	
		Others	0.001	
Direct physical	Crime	Robbery	0.01	I_d(occupant)
		Assault	0.01	
		Homicide	0.005	
		Rape	0.005	
	Terrorist	Hostage	0.008	
		Hijacking	0.005	
		Murder	0.003	
	Bio & IAQ	Intentional	0.02	
		Accidental	0.01	
		Sick building	0.03	
Cybernetic	Intentional	Hacking/outside	0.01	C_{cyb}
		Hacking/inside	0.02	
		Industrial sabotage	0.02	
	Accidental	Crash	0.01	
		Power outage	0.08	
		Power surge	0.07	
MEP Systems	Accidental	HVAC/Plumbing	0.003	$C_{M\&E}$
		Electrical	0.002	
		Telecom	0.001	
		Security	0.002	
		Fire alarm	0.001	
		BMS	0.002	
Increase in Operation Services	Unanticipated	Fuel price	0.01	$C_{O\&M}$
		Elec. Price	0.008	
		Utility cost	0.005	
		Labor cost	0.007	

red to be exhaustive; a necessary condition for framing all chance events (as stated in Sect. 12.2.3). This categorization should be adequate to illustrate the risk assessment methodology proposed. It is relatively easy to add, or remove, specific hazard events, or even regroup some of them, which adds to the flexibility of the proposed methodology.

Each of the events will have an uncertainty associated with them (as shown in Table 12.13). These are best represented by a hazard event probabilities or absolute annual probability (contrary to conditional probabilities, these will not sum to unity) of occurrence of certain hazard events, and a distribution to characterize its uncertainty. The absolute probabilities assigned to specific hazard events, will depend on such considerations as climatic and geographic location of the city, location of building within the city, importance and type of building. These could be obtained through the research of historical records, as was done for the RAMPART database (Hunter 2001). In order to keep the assessment methodology simple, the variability associated with these event probabilities has been overlooked. Monte Carlo analysis could be used to treat such variability, as illustrated in Sect. 12.2.7.

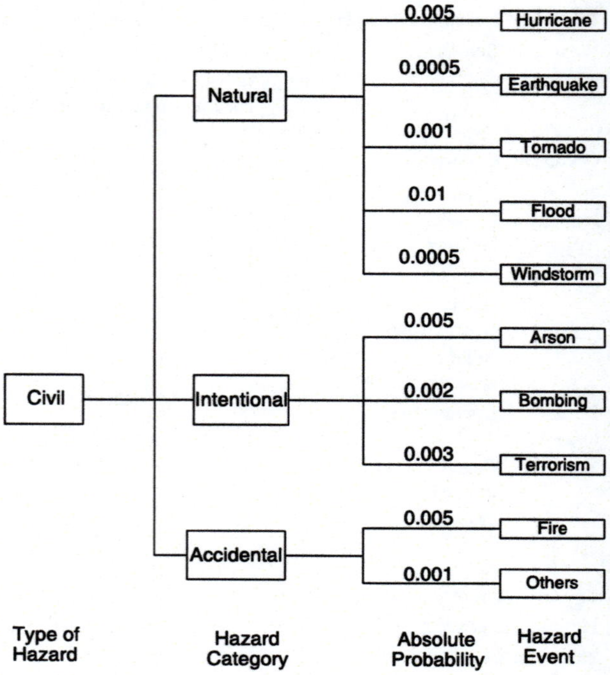

Fig. 12.15 Tree Diagram for various hazards affecting the civil element from the perspective of building owner (numerical values of absolute probabilities correspond to the case study)

(f) Hazard Event Costs

The consequences, or cost implications, of the occurrence of different hazard events from the perspective of the stakeholder (in this case, the building owner) need to be determined in order to complete the risk assessment study. The numerical values of these costs are summarized in the last column of Table 12.14, and are described below. Replacement costs for specific hazard events are difficult to determine, and more importantly, these costs are not reflective of the actual cost incurred by the building owner (unless he is self-insured). Most frequently, the building owner insures the building along with its contents and occupants with an insurance company to which he pays annual premiums (say, I_{BH} for civil

Table 12.14 Building specific financial data assumed in the case study from the perspective of the building owner

Description	Symbol	Assumed values	Calculated values
Building initial (or replacement) cost	C_I	\$ 15,000,000	
Net return on investment	ROI	15% per year	\$ 2,250,000/year
Number of occupants	N_{occup}	500	
Total amount of insurance coverage against occupant lawsuits	C_{Law}	\$ 10,000,000	
Occupant hazard insurance premium	I_{OH}	\$ 200/occupant/year	\$ 100,000/year
Building hazard insurance premium	I_{BH}	$(2\% * C_I)$ per year	\$ 300,000/year
Insurance deductible – for building – for occupants	I_d	 $(5\% * C_I)$ $(5\% * C_{Law})$	 \$ 750,000 (bldg) \$ 500,000 (occupants)
Annual building maintenance and utility cost	$C_{O\&M}$	$(5\% * C_I)$ per year	\$ 750,000/year
Replacement cost of MEP equipment	$C_{M\&E}$	\$ 3,000,000 per year	
Cost to recover from computer software failure	C_{cyb}	\$ 50,000	

Note that the current methodology overlooks variability or uncertainty in these data.

construction and I_{OH} for occupants) whether or not a hazard occurs (see Table 12.14). Hence, the additional cost faced by the building owner when civil and/or direct physical hazards do occur is actually the insurance deductible I_d. On the other hand, financial risks due to accidental MEP and cybernetic hazards ($C_{M\&E}$ and C_{cyb}) are considered to be direct expenses which the owner incurs whenever these occur. The monetary consequence of risk due to an unanticipated increase in operation services (maintenance, utility costs,...) is taken to be $C_{O\&M}$ which can be assumed to be a certain percentage of the total building cost that is spent yearly on operations, maintenance and utility costs. Most of the above costs need to be acquired from the concerned stakeholder.

(g) Computational Methodology

Recall that a decision tree is a graphical representation of all the choices and possible outcomes with probabilities assigned according to existing factual information or professional input. Figure 12.14 shows the stakeholder's importance towards various targets (characterized by the IM1 matrix) and sub-targets (characterized by IM2 matrix) shown in Table 12.12. Next these sub-targets are mapped onto the hazard categories using the binary information contained in the Applicability Matrix (AM). Finally, each of the affected building elements and the associated hazard categories are made to branch out into their corresponding hazard events with the associated absolute probability values. Once a probability is assigned to a specified risk event, a monetary value also needs to be associated with it which is representative of the cost of undoing or repairing the consequences of that event. By multiplying the monetary value and the probability for each hazard event, a characterization of the expected monetary value of risk becomes apparent. Addition of all these values will give an overall characterization of the building. Areas of high monetary risk can thus be easily identified and targeted for improvement. Mathematically, the above process can be summarized as follows:

Expected annual monetary risk specific to building element

$$h = C_h \sum_k \sum_m \sum_i \sum_j (IM1_k \cdot IM2_{k,m} \cdot AM_{m,i} \cdot p_{i,j}) \quad (12.8)$$

where i the hazard category, j the hazard event, k the target and m the sub-target.

Multiplication rules for fuzzy numbers are relatively simple and are described in several texts (for example, McNeill and Thro 1994). This is required for considering the propagation of uncertainty through the various sequential stages of the computation. Thus, in conjunction with computing the estimate of a risk, a range of numbers can also be computed reflective of the perceived importance to the stakeholder vis-à-vis specific targets and sub-targets.

(h) Illustrative example

A hypothetical solved example is presented to illustrate the entire methodology. A commercial building with 500 occupants has been assumed with the numerical values for AM, IM1, IM2, and P_j shown in Table 12.12, Fig. 12.14 and Table 12.13. Financial inputs and assumptions are shown in Table 12.14. The replacement cost of the building (C_I) is assumed to be \$ 15,000,000 with the ROI for the building owner to be 15% per year. The gross annual income is assumed to be \$ 3,400,000 per year, while annual expenditures include \$ 300,000 (or 2%.C_I) as building hazard insurance premium, \$ 100,000 per year (or \$ 200 per occupant) as occupant insurance premium and \$ 750,000 (or 5%. C_I) as building maintenance and utility costs. MEP replacement cost is estimated to be \$ 3,000,000, and cost to recover from a complete computer software failure is estimated to be \$ 50,000. The insurance deductibles for civil structure and occupants are 5% of C_I and C_{Law} (where C_{Law} is the total insurance coverage against occupant lawsuits), namely \$ 750,000 and \$ 500,000 respectively.

The results of the risk assessment are shown in Table 12.15 at the whole building level. The building owner spends $ 1,150,000 per year, with a net income of $ 2,250,000. The monetary mean value of the total risks is $ 68,285, i.e., about 3.0% of his net income. He may decide that this is within his tolerance threshold, and do nothing. On the other hand, he may calculate the risk based on the upper limit value of $ 102,415 shown in Table 12.15, and decide that 4.6% is excessive. In this case, he would want some direction as to how to manage the risk. This would require information about individual outcomes of various events. Figure 12.16 provides

Table 12.15 Monetary risks on the five building elements affected by various hazards considered ($/year)

Building Element	Mean	Lower limit	Upper limit
Civil	$ 17,895	$ 10,170	$ 25,620
Direct Physical	$ 24,260	$ 13,380	$ 35,140
Cybernetic	$ 2,190	$ 1,410	$ 2,970
M/E Failure	$ 15,840	$ 6,270	$ 25,410
Operations	$ 8,100	$ 2,925	$ 13,275
Total	$ 68,285	$ 34,155	$ 102,415

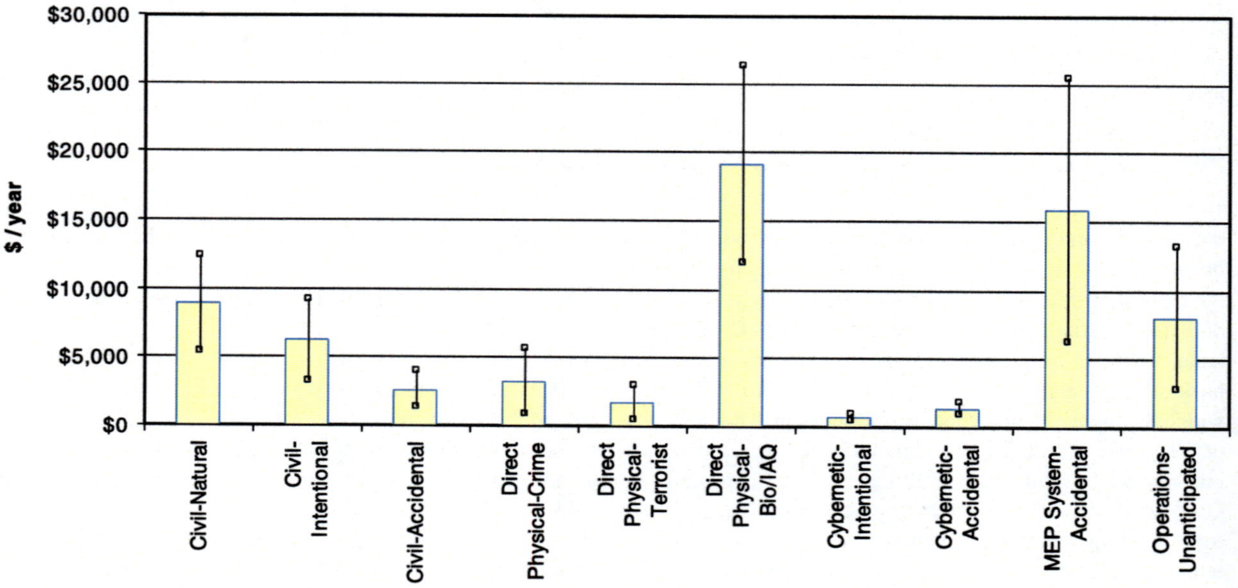

Fig. 12.16 Risk assessment results at the *Hazard Category* level along with uncertainty bands

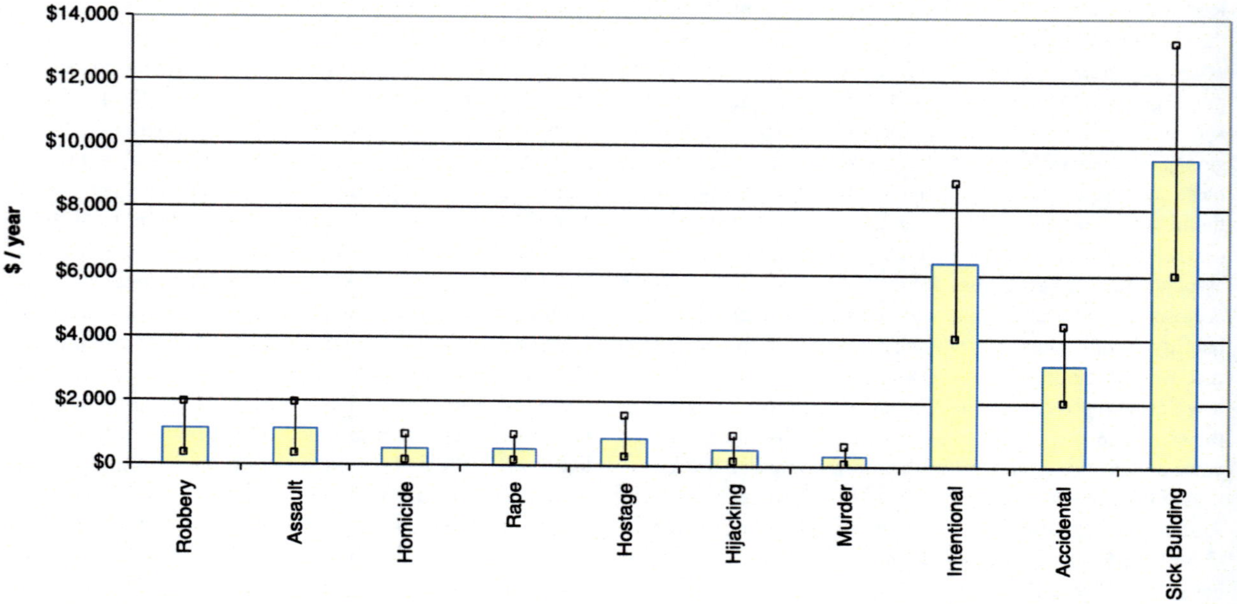

Fig. 12.17 Risk assessment results at the *Hazard Event* level specific to the Direct Physical Hazard category along with uncertainty bands

further specifics about individual hazard categories while Fig. 12.17 shows details of the largest hazard (namely "Direct Physical Hazard"). One notes that the "Sick Building" hazard is prone to the highest risk followed by "Intentional" attacks. Thus, the subsequent step involving determining the specific mitigation measures to adopt (such as paying for additional security personnel, improving IAQ by implementing proper equipment,...) would fall under the purview of decision-making.

12.4.2 Decision Making While Operating an Engineering System

Supervisory control of engineering systems involves optimization which relies on component models of individual equipment. This case study describes a decision-making situation involving an engineering problem where one needs to operate a hybrid cooling plant consisting of three chillers of equal capacity (two vapor compression chillers-VC and one absorption chiller-AC), along with auxiliary equipment such as cooling towers, fans, pumps,...[4] The cooling plant is meant to meet the loads of a building over a planning horizon covering a period of 12 h. This is a typical case of dynamic programming (addressed in Sect. 7.10). The cooling loads of the building are estimated at the start of the planning horizon, and whatever combination of equipment deemed to result in minimum operating costs is brought on-line or switched off. The optimization needs to consider chiller performance degradation under part load operation as well as the fact that starting chillers has an associated energy penalty (a certain start-up period is required to reach full operational status). However, there is a risk that the building loads are improperly estimated which could lead to a situation where there is inadequate cooling capacity, which also has an associated penalty. Minimizing the operating cost and minimizing the risk associated with loss of cooling (i.e. inadequate cooling capacity) are two separate issues altogether; how to trade-off between these objectives while considering the risk attitude of the operator is basically a decision analysis problem. The structure of the engineering problem is fairly well known with relatively low stochasticity and ambiguity; this is why this problem would fall into the low epistemic and low aleotory category.

(a) Modeling uncertainties

There are different uncertainties which play a role, and they do so in a nested manner. A suitable classification of different types of uncertainties is based on the nature of the source of uncertainty described at more length below.

(i) *Model-Inherent Uncertainty* which includes uncertainties of the various component models used for dynamic optimization that arise from inaccurate or incomplete data and /or lack of perfect regression to the response model. *Equipment model uncertainties* (assumed additive) can be dealt with by including an error term in the model. A general form of a multivariable ordinary least-squares (OLS) regression model with one response variable y and p regressor variables x can be expressed as:

$$y_{(n,1)} = x_{(n,p)}\beta_{(p,1)} + \varepsilon_{(n,1)}(0, \sigma^2) \tag{12.9}$$

where the subscripts indicate the number of rows and columns of the vectors or matrices. The random error term ε is assumed to have a normal distribution with variance σ^2 and no bias (mean is zero). Since the ε terms in the models are assumed to be independent variables, their variances can be assumed to be additive. The independent variable is the energy consumption estimates of electricity for the VC and associated components and gas for the AC. The model uncertainties expressed as CV-RMSE of the regression models for the total electricity and gas use (P_{elec} and P_{gas}) for each of the five different combinations or groups (G1-G5) of equipment operation are shown in Table 12.16.

(ii) *Process-Inherent Uncertainty,* due to control policies, are usually implemented by lower-level feedback controllers whose set point tracking response will generally be imperfect due to actuator constraints and un-modeled time-varying behavior, nonlinearities, and disturbances. The control implementation uncertainty on the chilled water temperature and the fan speed can be represented by upper and lower bounds. Based on the typical actuator constraints for these two variables, the accuracy of the temperature control is often in the range of 0.5~1.5°C, while accuracy of fan speed control would be 2~5%.

(iii) *External Prediction Uncertainty* which includes the errors in predicting the system driving functions such as the building thermal load profiles, and electricity rate profiles. *Load prediction uncertainty* can be treated either as unbiased, uncorrelated Gaussian noise or as unbiased correlated noise (only the results of the former are presented here). The distributed deviations around the true value are assumed such that their variance grows linearly with the look-ahead horizon. This

Table 12.16 Summary of CV-RMSE values for the component regression models for the hybrid cooling plant

Group	Description of operating equipment	CV-RMSE (%)	
		Electricity use (P_{ele})	Gas use (P_{gas})
G1	One VC	4.0	–
G2	One AC	3.0	5.0
G3	Two VC	5.7	–
G4	One VC and one AC	5.0	5.0
G5	All three chillers	5.7	5.0

VC vapor compression chiller, *AC* absorption chiller

[4] Adapted from Jiang and Reddy (2007) and Jiang et al. (2007).

corresponds to a quantity whose prediction performance deteriorates linearly with time. Mathematically, this noise model can be expressed as

$$y_{t,i} = x_{t+i}\left[1 + \varepsilon\left(0, \sigma_i^2\right)\right] \quad (12.10a)$$

$$\sigma_i = p\frac{i}{L} \quad (12.10b)$$

where $y_{t,i}$ is the output variable at hour t and look-ahead hourly interval i; x_t is the input at hour t; L is the planning horizon (in our case, L = 12 h) and p is a scalar characterizing the noise magnitude.

The dynamic optimization involved determining the optimal scheduling and control strategy which resulted in minimum energy cost for a specified hourly building cooling load profile over a 12 h period under a specified electric price signal. The process-inherent uncertainty was combined with the equipment uncertainties by using a simple multiplier. This strategy was called *optimal deterministic strategy* (ODS). Different noise levels were investigated p = {0, 0.05, 0.10, 0.15, 0.2} though in practice, one would expect p ≈ 0.05. For example, at the end of 12 h, the error bands represented by p = 0.2 would cover 20% of the mean value. Figure 12.18 illustrates the extent to which the load prediction error keeps increasing with the look-ahead hour.

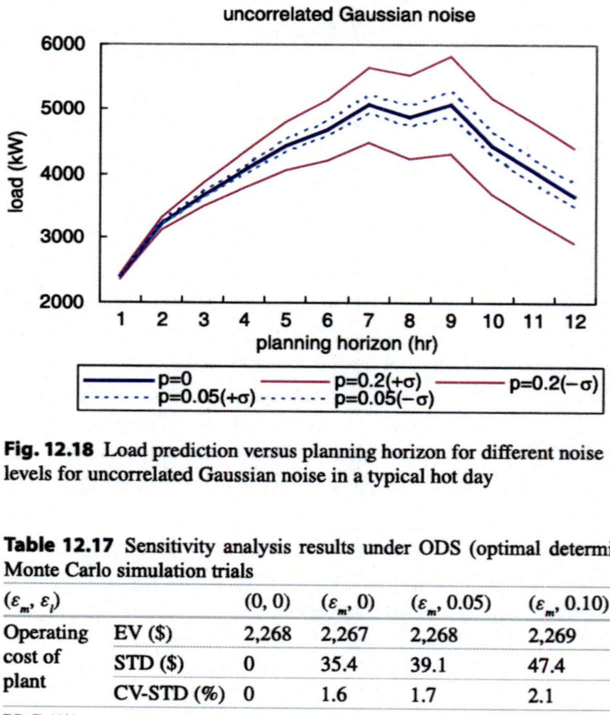

Fig. 12.18 Load prediction versus planning horizon for different noise levels for uncorrelated Gaussian noise in a typical hot day

Latin Hypercube Monte Carlo (LHMC) simulation with 10,000 trials was then performed to the ODS operating strategy under different diurnal conditions. The results are summarized in Table 12.17 from which the following observations can be made:

(i) As expected, the coefficient of variation of the standard deviation (CV-STD) values of operating cost over the 10,000 trials increase with increasing model error. When the model error is doubled, the CV-STD values are approximately doubled as well. Also, as expected, CV-STD values of the operating cost increase with increasing load prediction uncertainty.

(ii) The model-inherent uncertainty is relatively more important than the load prediction uncertainty. For example, if the uncertainty profile is changed from (ε_m, 0.05) to (ε_{m*2}, 0.05), the CV-STD of the minimal plant operating cost is increased from 1.7 to 3.2% (Table 12.17). On the other hand, if the load prediction error is doubled while the model uncertainty is kept constant, the CV-STD of operating cost is only increased from 1.7 to 2.1%. However, the model-uncertainty has less effect on probability of loss of cooling (PLC) capability than the load prediction error.

(iii) Under a practical uncertainty condition of (ε_m, 0.05), model-inherent uncertainty and load prediction uncertainty seem to have little effect on the overall operating cost of the hybrid cooling plant under optimal operating strategy with the CV-STD being around 2%.

(b) Decision analysis with multi-attribute utility framework

The above sensitivity analysis revealed that different uncertainty scenarios have different expected values (EV) and CV-STD values for the cost of operating the hybrid cooling plant. The risk due to loss of cooling capability is also linked with the uncertainty in the building cooling load prediction. The probability of loss of cooling capability is determined under each specified uncertainty scenario and diurnal condition, with LHMC simulation used to generate numerous trials of the building cooling load profile. Subsequently, for each trial, one can again generate numerous cases using

Table 12.17 Sensitivity analysis results under ODS (optimal deterministic strategy) assuming uncorrelated Gaussian noise and with 10,000 Monte Carlo simulation trials

(ε_m, ε_l)		(0, 0)	(ε_m, 0)	(ε_m, 0.05)	(ε_m, 0.10)	(ε_m, 0.15)	(ε_m, 0.2)	(ε_{m*2}, 0)	(ε_{m*2}, 0.05)	(ε_{m*2}, 0.2)
Operating cost of plant	EV ($)	2,268	2,267	2,268	2,269	2,271	2,274	2,267	2,268	2,273
	STD ($)	0	35.4	39.1	47.4	59.3	70.8	70.7	73.3	93.2
	CV-STD (%)	0	1.6	1.7	2.1	2.6	3.1	3.1	3.2	4.1
PLC (%)		0	0	0	0.17	1.9	4.5	0	0	4.6

Note: (ε_m, ε_l) describes the uncertainty profile, ε_m is the actual model uncertainty in the hybrid cooling plant, while ε_{m*2} implies that the uncertainty of the model and the process control combined is two times the actual model uncertainty; ε_l is the load prediction uncertainty with values being assumed to be 5, 10, 15 and 20%. EV is the expected value of operating cost, SD is the standard deviation of operating cost and PLC is the probability of loss of cooling

Table 12.18 Sensitivity analysis results under RAS (risk averse strategy) assuming uncorrelated Gaussian noise and with 10,000 Monte Carlo simulation trials

$(\varepsilon_m, \varepsilon_l)$		(0, 0)	$(\varepsilon_m, 0)$	$(\varepsilon_m, 0.05)$	$(\varepsilon_m, 0.10)$	$(\varepsilon_m, 0.15)$	$(\varepsilon_m, 0.2)$	$(\varepsilon_{m*2}, 0)$	$(\varepsilon_{m*2}, 0.05)$	$(\varepsilon_{m*2}, 0.2)$
Operating cost of plant	EV ($)	2,919	2,922	2,922	2,923	2,923	2,924	2,922	2,922	2,924
	STD ($)	0	37.6	39.6	45.3	53.4	62.9	75.2	76.2	90.6
	CV-STD (%)	0	1.3	1.4	1.6	1.8	2.2	2.6	2.6	3.1
PLC (%)		0	0	0	0	0	0	0	0	0

LHMC to simulate the effect of component model errors on the energy consumption of the chillers under ODS. This nested uncertainty computation thus captures the effect of both sources of uncertainty, namely that of load prediction and due to the component models plus the process control uncertainty. The total number of occurrences of loss of cooling capability is divided by the total number of simulation trials to yield the probability of the loss of cooling capability (PLC) under specified uncertainty scenario and optimal operating strategy. PLC values are also shown in Table 12.17.

(c) Risk-averse strategy

Certain circumstances warrant operating the cooling plant differently than that suggested by the optimal deterministic strategy (ODS). For example, the activities inside the building may be so critical that they require the cooling load be always met (such as in a pharmaceutical company, for example). Hence, one would like to have excess cooling capability at all times even if this results in extra operating cost. One could identify different types of risk averse strategies (RAS). For example, one could bring an extra chiller online only when the reserve capacity is less than a certain amount, say 10%. Alternatively, one could adopt a strategy of always having a reserve chiller online; this is the strategy selected here.

Sensitivity analysis results are summarized in Table 12.18. One notes that there is no loss of cooling under any of the uncertainty scenarios, but the downside is that operating cost is increased. In this decision analysis problem, three attributes are considered: EV, CV-STD of operating cost, and PLC. The CV-STD is a measure of uncertainty surrounding the EV. In a decision framework, higher uncertainty results in higher risk, and so this attribute needs to also figure in the utility function. Since these three attributes do not affect and interact with each other, it is reasonable to assume them to be additive independent, which allows for the use of an *additive multi-attribute utility function* of the forms (similar to Eq. 12.4):

$$U_{ODS}(EV, CV-STD, PLC) = k_{1,ODS}U_{1,ODS}(EV) + k_{2,ODS}U_{2,ODS}(CV-STD) + k_{3,ODS}U_{3,ODS}(PLC) \quad (12.11)$$

$$U_{RAS}(EV, CV-STD, PLC) = k_{1,RAS}U_{1,RAS}(EV) + k_{2,RAS}U_{2,RAS}(CV-STD) + k_{3,RAS}U_{3,RAS}(PLC) \quad (12.12)$$

Table 12.19 Utility values for EV, CV-STD and PLC under ODS and RAS strategies normalized such that they are between 0 and 1 representing the worst and best values respectively

		Case 1	Case 2	Case 3	Case 4	Case 5
$(\varepsilon_m, \varepsilon_l)$		$(\varepsilon_m, 0)$	$(\varepsilon_m, 0.05)$	$(\varepsilon_m, 0.10)$	$(\varepsilon_m, 0.15)$	$(\varepsilon_m, 0.2)$
ODS	$U_{1,ODS}$(EV)	1.00	1.00	1.00	1.00	1.00
	$U_{2,ODS}$(CV-STD)	0.88	0.84	0.68	0.48	0.28
	$U_{3,ODS}$(PLC)	1.00	1.00	0.96	0.58	0.00
RAS	$U_{1,RAS}$(EV)	0.00	0.00	0.00	0.00	0.00
	$U_{2,RAS}$(CV-STD)	1.00	0.88	0.64	0.32	0.00
	$U_{3,RAS}$(PLC)	1.00	1.00	1.00	1.00	1.00

where U is the utility value, EV is the expected value of total operating cost over the planning horizon, CV-STD is the coefficient of variation of the standard deviation of the total operating cost from all the Monte Carlo trials, PLC is the probability of loss of cooling capability, and k denotes the three weights normalized such that:

$$k_{1,ODS} + k_{2,ODS} + k_{3,ODS} = 1,$$

and

$$k_{1,RAS} + k_{2,RAS} + k_{3,RAS} = 1.$$

The benefit of using the additive utility function has already been described in Sect. 12.2.6. Further, risk neutrality, i.e., a linear utility function, has been assumed. The terms U_1, U_2 and U_3 under both ODS and RAS are normalized by dividing them by the differences of worst value and best value of the three attributes respectively so that these values are bounded by values of 0 and 1 (see Eq. 12.3). The normalized utilities of the three attributes at other levels ranging from worst to best can be determined by linear interpolation. Table 12.19 lists some of utility values of the three attributes under five different combinations or cases of model and load prediction uncertainties. Hence, knowing the weight *k*, one can calculate the overall utility under different uncertainty scenarios. However, it is difficult to assign specific values of the weights since they are application specific. If occasional loss of cooling can be tolerated, then k_3 can be either set to zero or given very low weight. A general heuristic guideline is that the weight for expected value (k_1) be taken to be 2–4 times

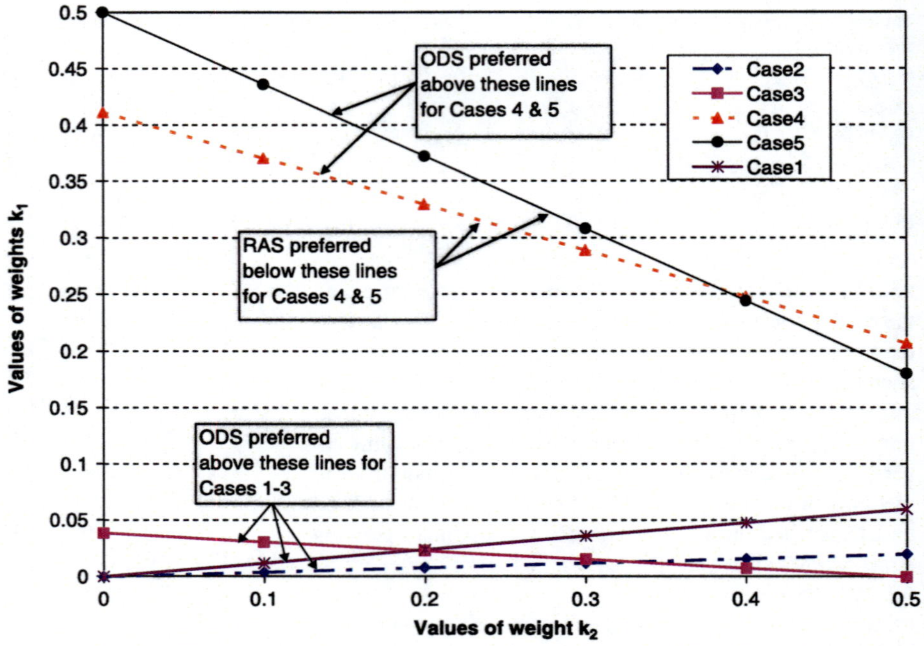

Fig. 12.19 Indifference plots for the five cases shown in Table 12.19 indicating the preferred operating strategies for different combinations of attribute weights k_1 and k_2. Thus, for case 5 and for a selected value of $k_2=0.2$, the ODS is superior if the operator deems U(EV) to have a weight $k_1>0.375$.

greater than that of variability of the predicted operating cost (k_2) (Clemen and Reilly 2001).

Indifference curves can be constructed to compare different choices. An indifference curve is the locus of all points (or alternatives) in the decision maker's evaluation space among which he is indifferent. Usually, an indifference curve is a graph showing combinations of two attributes to which the decision-maker makes has no preference for one combination over the other. The values for the multi-attribute utility functions $U_{ODS}(EV, CV-STD, PLC)$ and $U_{RAS}(EV, CV-STD, PLC)$ can be calculated based on the results of Tables 12.17 and 12.18. The model uncertainty is found to have little effect on the probability of loss of cooling capability, and is not considered in the following decision analysis; fixed realistic values of ε_m shown in Table 12.16 are assumed. All points on the indifference curves have the same utility, and hence, separate the regions of preference between ODS and RAS. These are shown in Fig. 12.19 for the five cases. These plots are easily generated by equating the right hand terms of Eqs. 12.11 and 12.12 and inserting the appropriate values of k_1 from Table 12.19. Thus, for example, if k_2 is taken to be 0.2 (a typical value), one can immediately conclude from the figure that ODS is preferable to RAS for case 5 provided $k_1>0.375$ (which in turn implies $k_3=1-0.2-0.375=0.425$). Typically the operator is likely to consider the attribute EV to have an importance level higher than this value. Thus, the threshold values provide a convenient means of determining whether one strategy is clearly preferred over the other, or whether precise estimates of the attribute weights are required to select an operating strategy.

Figure 12.19 indicates that for cases 1–3 (load prediction error of 0–10%), ODS is clearly preferable, and that it would, most likely, be so even for cases 4 and 5 (load prediction errors of 15 and 20%). From Fig. 12.20 which has been generated assuming $k_2=0$ (i.e. no weight being given to variability of the operating cost), it is clear that the utility curve is steeper as k_1 decreases. This means that the load prediction error has a more profound effect on the utility function when the expected value of operating cost has a lower weight; this is consistent with the earlier observation that the load prediction uncertainty affects the loss of cooling capability. One notes that RAS is only preferred under a limited set of conditions. While the exact location in parameter space of the switchover point between the two strategies may change from application to application, this

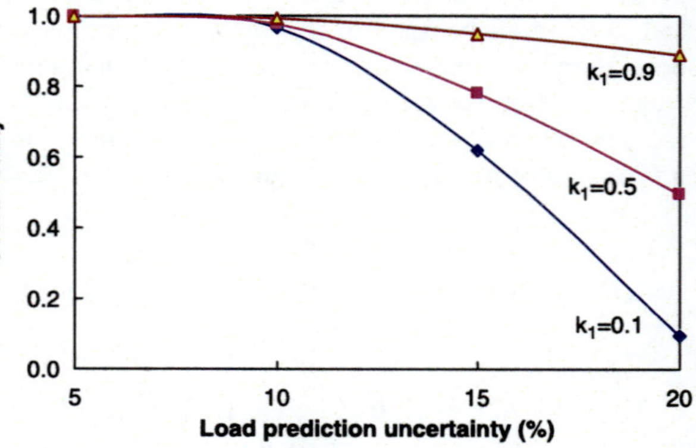

Fig. 12.20 Overall utility of ODS vs load prediction uncertainty assuming $k_2=0$

approach supports the idea that well characterized systems may be operated under ODS except when the load prediction uncertainty is higher than 15%. Of course, this result is specific to this illustrative case study and should not be taken to apply universally.

Problems

Pr.12.1 Identify and briefly discuss at least two situations (one from everyday life and one from an engineering or environmental field) which would qualify as problems falling under the following categories:

(a) Low epistemic and low uncertainty
(b) Low epistemic and high uncertainty
(c) High epistemic and low aleatory
(d) High epistemic and high aleatory

Pr. 12.2 The owner of a commercial building is considering improving the energy efficiency of his building systems using both energy management options (involving little or no cost) and retrofits to energy equipment. Prepare:

(a) an influence diagram for this situation
(b) a decision tree diagram (similar to Fig. 12.3)

Pr. 12.3 An electric utility is being pressured by the public utility commission to increase its renewable energy portfolio (in this particular case, energy conservation does not qualify). The decision-maker in the utility charges his staff to prepare the following (note that there are different possibilities):

(a) an influence diagram for this situation,
(b) a decision tree diagram.

Pr. 12.4 Compute the representative discretized values and the associated probabilities for the Gaussian distribution (given in Appendix A3) for:

(a) a 3-point approximation,
(b) a 5-point approximation.

Pr. 12.5 Consider Example 12.2.4.

(a) Rework the problem including the effect of increasing unit electricity cost (linear increase over 10 years of 3% per year). What is the corresponding probability level of moving at which both AC options break even, i.e., the indifference point?
(b) Rework the problem but with an added complexity. It was stated in the text (Sect. 12.2.6a) that discounted cash flow analysis where all future cash flows are converted to present costs is an example of an additive utility function. Risk of future payments is to be modeled as increasing linearly over time. The 1st year, the risk is low, say 2%, the 2nd year is 4% and so on. How would you modify the traditional cash flow analysis to include such a risk attitude?

Pr. 12.6 *Traditional versus green homes*
An environmentally friendly "green" house costs about 25% more to construct than a conventional home. Most green homes can save 50% per year on energy expenses to heat and cool the dwelling.

(a) Assume the following:
  - It is an all-electric home needing 8 MWh/year for heating and cooling, and 6 MWh/year for equipment in the house
  - The conventional home costs $ 250,000 and its life span is 30 years
  - The cost of electricity is $ 15 cents with a 3% annual increase
  - The "green" home has the same life and no additional value at the end or 30 years can be expected
  - The discount rate (adjusted for inflation) is 6% per year.
(b) Is the "green" home a worthwhile investment purely from an economic point of view?
(c) If the government imposes a carbon tax of $ 25/ton, how would you go about reevaluating the added benefit of the "green" home? Assume 170 lbs of carbon are released per MWh of electricity generation (corresponds to California electric mix).
(d) If the government wishes to incentivize such "green" construction, what amount of upfront rebate should be provided, if at all?

Pr. 12.7 *Risk analysis for building owner being sued for indoor air quality*
A building owner is being sued by his tenants for claimed health ailments due to improper biological filtration of the HVAC system. The suit is for $ 500,000. The owner can either settle for $ 100,000 or go to court. If he goes to court and loses, he has to pay the lawsuit amount of $ 450,000 plus the court fees of $ 50,000. If he wins, the plaintiffs will have to pay the court fees of $ 50,000.

(a) Draw a decision tree diagram for this problem
(b) Construct the payoff table
(c) Calculate the cutoff probability of the owner winning the lawsuit at which both options are equal to him.

Pr. 12.8[5] Plot the following utility functions and classify the decision-maker's attitude towards risk:

(a) $U(x) = \dfrac{100 + 0.5x}{250} \qquad -200 \le x \le 300$

(b) $U(x) = \dfrac{(x+100)^{1/2}}{20} \qquad -100 \le x \le 300$

(c) $U(x) = \dfrac{x^3 - 2x^2 - x + 10}{800} \qquad 0 \le x \le 10$

[5] From McClave and Benson (1988) by © permission of Pearson Education.

Table 12.20 Payoff table for Problem 12.9

Alternative	Chance events S_1 (p=0.1)	S_2 (p=0.3)	S_3 (p=0.4)	S_4 (p=0.2)
A#1	$ 2.5 k	$ 0.5 k	$ 0.9 k	$ 2.9 k
A#2	$ 3.0 k	$ 3.4 k	$ 0.1 k	$ 2.4 k
A#3	$ 2.2 k	$ 1.1 k	$ 1.8 k	$ 1.3 k
A#4	$ 0.2 k	$ 3.4 k	$ 3.7 k	$ 0.7 k

Table 12.21 Data table for Problem 12.11

Percentage level of design goal met (%)	Probability p(L)	Annual savings in operating cost ($/year)
90	0.25	3,470
70	0.40	2,920
50	0.25	2,310
30	0.10	1,560

(d) $U(x) = \dfrac{0.1x^2 + 10x}{110} \quad 0 \le x \le 10$

Pr. 12.9 Consider payoff Table 12.20 with four alternatives and four chance events (or states of nature).

(a) Calculate the alternative with the highest expected value (EV).
(b) Resolve the problem assuming the utility cost function is exponential with zero minimum monetary value and risk tolerance R=$ 2 k.

Pr. 12.10 *Monte Carlo analysis for evaluating risk for property retrofits*

The owner of a large office building is considering making major retrofits to his property with system upgrades. You will perform a Monte Carlo analysis with 5,000 trials to investigate the risk involved in this decision. Assume the following:

- The capital investment is $ 2 M taken to be normally distributed with 10% CV.
- The annual additional revenue depends on three chance outcomes:
  $ 0.5 M under good conditions, probability p(Good)=0.4
  $ 0.3 M under average conditions, p(Average)=0.3
  $ 0.2 M under poor conditions, p(Poor)=0.3
- The annual additional expense of operation and maintenance is normally distributed with $ 0.3 M and a standard deviation of $ 0.05.
- The useful life of the upgrade is 10 years with a normal distribution of 2 years standard deviation.
- The effective discount rate is 6%.

Pr. 12.11[6] *Risk analysis for pursuing a new design of commercial air-conditioner*

A certain company which manufactures compressors for commercial air-conditioning equipment has developed a new design which is likely to be more efficient than the existing one. The new design has a higher initial expense of $ 8,000 but the advantage of reduced operating costs. The life of both compressors is 6 years. The design team has not fully tested the new design but has a preliminary indication of the efficiency improvement based on certain limited tests have been determined. The preliminary indication has some uncertainty regarding the advantage of the new design, and this is quantified by a discrete probability distribution with four discrete levels of the efficiency improvement goal as shown in the first column of Table 12.21. Note that the annual savings are incremental values, i.e., the savings over those of the old compressor design.

(a) Draw the influence diagram and the decision tree diagram for this situation
(b) Calculate the present value and determine whether the new design is more economical
(c) Resolve the problem assuming the utility cost function is exponential with zero minimum monetary value and risk tolerance R=$ 2,500
(d) Compute the EVPI (expected value of perfect information)
(e) Perform a Monte Carlo analysis (with 1,000 trials) assuming random distributions with 10% CV for p(L = 90) and p(L = 30), 5% for p(L = 70). Note that p(L = 50) can be to be determined from these, and that negative probability values are inadmissible, and should be set to zero. Generate the PDF and the CDF and determine the 5% and the 95% percentiles values of EV
(f) Repeat step (e) with 5,000 trials and see if there is a difference in your results
(g) Repeat step (f) but now consider the fact that the life of both compressors is normally distributed with mean of 6 years and standard deviation of 0.5 years.

Pr. 12.12[7] *Monte Carlo analysis for deaths due to radiation exposure*

Human exposure to radiation is often measured in rems (roentgen-equivalent man) or millirem (mrem). The cancer risk caused by exposure to radiation is thought to be approximately one fatal cancer per 8,000 person-rems of exposure (e.g., one cancer death if 8,000 people are exposed to one rem each, or 10,000 exposed to 0.8 rems each,…). Natural radioactivity in the environment is thought to result in an exposure of 130 mrem/year.

(a) How many cancer deaths can be expected in the United States per year as a result (assume a population of 300 million).
(b) Reanalyze the situation while considering variability. The exposure risk of 8,000 person-rems is normally

[6] Adapted from Sullivan et al. (2009) by © permission of Pearson Education.

[7] From Masters and Ela (2008) by © permission of Pearson Education.

distributed with 10% CV, while the natural environment radioactivity has a value of 20% CV. Use 10,000 trials for the Monte Carlo simulation. Calculate the mean number of deaths and the two standard deviation range.
(c) The result of (b) could have been found without using Monte Carlo analysis. Using formulae presented in Sect. 2.3.3 related to functions of random variables, compute the number of deaths and the two standard deviation range and compare these with the results from (b).

Pr. 12.13 Radon is a radioactive gas resulting from radioactive decay of uranium to be found in the ground. In certain parts of the country, its levels are elevated enough that when it seeps into the basements of homes, it puts the homeowners at an elevated risk of cancer. EPA has set a limit of 4 pCi/L (equivalent to 400 mrem/year) above which mitigation measures have to be implemented. Assume the criterion of one fatal cancer death per 8,000 person-rems of exposure.
(a) How many cancer deaths per 100,000 people exposed to 4 pCi/L can be expected?
(b) Reanalyze the situation while considering variability. The exposure risk of 4pCi/L is a log-normal distribution with 10% CV, while the natural environment radioactivity is normally distributed with 20% CV. Use 10,000 trials for the Monte Carlo simulation. Calculate the mean number of deaths and the 95% confidence intervals.
(c) Can the result of (b) be found using formulae presented in Sect. 2.3.3 related to functions of random variables? Explain.

Pr. 12.14 *Monte Carlo analysis for buying an all-electric car*
A person is considering buying an all-electric car which he will use daily to commute to work. He has a system at home of recharging the batteries at night. The all-electric car has a range of 100 miles while the minimum round trip of a daily commute is 40 miles. However, due to traffic congestion, he sometimes takes a longer route which can be represented by a log-normal distribution of standard deviation 5 miles. Moreover, his work requires him to visit clients occasionally for which purpose he needs to use his personal car. Such extra trips have a lognormal distribution of mean of 20 miles and a standard deviation of 5 miles.
(a) Perform a Monte Carlo simulation (using 10,000 trials) and estimate the 99% probability of his car battery running dry before he gets home on days when he has to make extra trips and take the longer route.
(b) Can you verify your result using an analytical approach? If yes, do so and compare results.
(c) If the potential buyer approaches you for advice, what types of additional analyses will you perform prior to making your suggestion?

Pr. 12.15 *Evaluating feed-in tariffs versus upfront rebates for PV systems*
One of the promising ways to promote solar energy is to provide incentives to homeowners to install solar PV panels on their roofs. The PV panels generate electricity depending on the collector area, the efficiency of the PV panels and the solar radiation falling on the collector (which depends both on the location and on the tilt and azimuth angles of the collector). This production would displace the need to build more power plants and reduce adverse effects of both climate change and health ailments of inhaling polluted air. Unfortunately, the cost of such electricity generated in yet to reach grid-parity, i.e., the PV electricity costs more than traditional options, and hence the need for the government to provide incentives.

Two different types of financing mechanisms are common. One is the *feed-in tariff* where the electricity generated at-site is sold back to the electric grid at a rate higher than what the customer is charged by the utility (this is common in Europe). The other is the *upfront rebate* financial mechanism where the homeowner (or the financier/installer) gets a refund check (or a tax break) per Watt-peak (W_p) installed (this is the model adopted in the U.S.). You are asked to evaluate these two options using the techniques of decision-making, and present your findings and recommendations in the form of a report. Assume the following:
- The cost of PV installation is $ 7 per Watt-peak (W_p). This is the convention of rating and specifying PV module performance (under standard test conditions of 1 kW/m^2 and 25°C cell temperature)
- However, the PV module operates at much lower radiation conditions throughout the year, and the conventional manner of considering this effect is to state a capacity factor which depends on the type of system, the type of mounting and also on location. Assume a capacity factor of 20%, i.e., over the whole year (averaged over all 24 h/day and 365 days/year), the PV panel will generate 20% of the rated value
- Life of the PV system is 25 years with zero operation and maintenance costs
- The average cost of conventional electricity is $ 0.015/kWh with a 2% annual increase
- The discount rate for money borrowed by the homeowner towards the PV installation is 4%.

References

AIA, 1999. *Guidelines for Disaster Response and Recovery Programs*, American Institute of Architects, Washington D.C.
Ang, A.H.S. and W.H. Tang, 2007. *Probability Concepts in Engineering*, 2nd Ed., John Wiley and Sons, USA
ASHRAE, 2003. Risk *Management Guidance for Health and Safety under Extraordinary Incidents*, report prepared by the ASHRAE

Presidential Study Group on Health and Safety under Extraordinary Incidents, Atlanta, January.
Bahnfleth, W.H., J. Freihaut, J. Bem and T.A. Reddy, 2008. Development of Assessment Protocols for Security Measures - A Scoping Study, Subtask 06-11.1: Literature Review, report submitted to National Center for Energy Management and Building Technologies, prepared for U.S. Department of Energy, 62 pages, February.
Bukowski, R.W., 2006. An overview of fire hazard and fire assessment in regulation. *ASHRAE Transactions,* 112(1), January.
Clemen, R.T. and Reilly, T., 2001. *Making Hard Decisions with Decision Tools*, Brooks Cole, Duxbury, Pacific Grove, CA
DeGaspari, J., 2002."Risky Business", *ASME Mechanical Engineering Journal*, p.42-44, July.
Evans, J. and D. Olson. 2000. *Statistics, Data Analysis, and Decision Modeling*, Prentice Hall, Upper Saddle River, NJ.
Gelman, A., Carlin, J.B., Stern, H.S. and Rubin, D.B., 2004. *Bayesian Data Analysis*, 2nd Ed., Chapman and Hall/CRC, Boca Raton, FL.
Gerba, C. P. 2006. Chapter 14: Risk Assessment. Environmental and Pollution Science. I. L. Pepper, Gerba C.P. and Brusseau, M., *Academic Press*: 212-232.
Haas, C., J. Rose and C. Gerba, 1999. *Quantitative Microbial Risk Assessment*. John Wiley and Sons: New York, NY
Haimes, Y.Y., 2004. *Risk Modeling, Assessment, and Management*, 2nd Edition, Wiley Interscience, Hoboken, NJ
Heinsohn, R.J. and Cimbala, J.M., 2003, *Indoor Air Quality Engineering*, Marcel Dekker, New York, NY
Hillier, F.S. and G.J. Lieberman, 2001, *Introduction to Operations Research*, 7th Edition, McGraw-Hill, New York,
Hunter, R., 2001. "RAMPART Software", Sandia National Laboratories, http://www.sandia/gov/media/NewsRel/NR2001/rampart.htm
Jiang, W. and T.A. Reddy, 2007, "General methodology combining engineering optimization of primary HVAC&R plants with decision analysis methods - Part I: Deterministic analysis", *HVAC&R Research Journal*, vol. 13(1), pp. 93-118, January.
Jiang, W., T.A. Reddy and P. Gurian, 2007, "General methodology combining engineering optimization of primary HVAC&R plants with decision analysis methods - Part II: Uncertainty and decision analysis", *HVAC&R Research Journal,* 13 (1): 119-140, January.
Kammen, D.M. and Hassenzahl, D.M., 1999. *Should We Risk It*, Princeton University Press, Princeton, NJ
Kowalski, W.J., 2002. *Immune Building Systems Technology*, McGraw-Hill, New York
Lancaster, J., 2000. *Engineering Catastrophes: Causes and Effects of Major Accidents*, Abington Publishing, Cambridge, England.
McClave, J.T. and P.G. Benson, 1988. *Statistics for Business and Economics*, 4th Ed., Dellen and Macmillan, London
Masters, G.M. and W.P. Ela, 2008. *Introduction to Environmental Engineering and Science*, 3rd Ed. Prentice Hall, Englewood Cliffs, NJ
McDowell, B. and A. Lemer, 1991 *Uses of Risk Analysis to Achieve Balanced Safety In Building Design and Operations*, National Academy Press, Washington, DC.
McNeill, F. and E. Thro, 1994. *Fuzzy Logic, A Practical Approach*,. AP Professional, Boston, MA.
NRC 1983. Risk Assessment in the Federal Government: Managing the Process. Washington, DC, National Academy of Sciences, National Research Council.
Rabl, A. 2005. How much to spend for the protection of health and environment: A framework for the evaluation of choices, *Review paper* for Les Cahiers de l'Institut Veolia Environment, Ecole des Mines, Paris, France
Reddy, T.A. and Fierko, J. 2004. "Uncertainty-based quantitative model for assessing risks in existing buildings." *ASHRAE Transactions*. 110 (1), Atlanta, GA, January
Reddy, T.A., 2007. Formulation of a generic methodology for assessing FDD methods and its specific adoption to large chillers, *ASHRAE Transactions*, 114 (2), Atlanta, GA, June
Sullivan W.G., Wicks, E.M. and Koelling, C.P., 2009. *Engineering Economy*, 14th Edition, Prentice-Hall, Upper Saddle River, NJ
USCG, 2001. *Risk-Based Decision-Making*, United States Coast Guard, http://www.uscg.mil/hq/g-m/risk
Vose, D., 1996. *Quantitative Risk Analysis, A Guide to Monte Carlo Simulation Modeling*,. John Wiley and Sons: New York, NY.
Wilson, R., 1979. Analyzing the daily risks of life, *Technology Review*, 81(4): 41-46.

Appendix

A: Statistical Tables

Table A.1 Binomial probability sums $\sum_{x=0}^{r} b(x; n, p)$

		p									
n	*r*	0.10	0.20	0.25	0.30	0.40	0.50	0.60	0.70	0.80	0.90
15	0	0.2059	0.0352	0.0134	0.0047	0.0005	0.0000				
	1	0.5490	0.1671	0.0802	0.0353	0.0052	0.0005	0.0000			
	2	0.8159	0.3980	0.2361	0.1268	0.0271	0.0037	0.0003	0.0000		
	3	0.9444	0.6482	0.4613	0.2969	0.0905	0.0176	0.0019	0.0001		
	4	0.9873	0.8358	0.6865	0.5155	0.2173	0.0592	0.0094	0.0007	0.0000	
	5	0.9978	0.9389	0.8516	0.7216	0.4032	0.1509	0.0338	0.0037	0.0001	
	6	0.9997	0.9819	0.9434	0.8689	0.6098	0.3036	0.0951	0.0152	0.0008	
	7	1.0000	0.9958	0.9827	0.9500	0.7869	0.5000	0.2131	0.0500	0.0042	0.0000
	8		0.9992	0.9958	0.9848	0.9050	0.6964	0.3902	0.1311	0.0181	0.0003
	9		0.9999	0.9992	0.9963	0.9662	0.8491	0.5968	0.2784	0.0611	0.0023
	10		1.0000	0.9999	0.9993	0.9907	0.9408	0.7827	0.4845	0.1642	0.0127
	11			1.0000	0.9999	0.9981	0.9824	0.9095	0.7031	0.3518	0.0556
	12				1.0000	0.9997	0.9963	0.9729	0.8732	0.6020	0.1841
	13					1.0000	0.9995	0.9948	0.9647	0.8329	0.4510
	14						1.0000	0.9995	0.9953	0.9648	0.7941
	15							1.0000	1.0000	1.0000	1.0000
20	0	0.1216	0.0115	0.0032	0.0008	0.0000					
	1	0.3917	0.0692	0.0243	0.0076	0.0005	0.0000				
	2	0.6769	0.2061	0.0913	0.0355	0.0036	0.0002	0.0000			
	3	0.8670	0.4114	0.2252	0.1071	0.0160	0.0013	0.0001			
	4	0.9568	0.6296	0.4148	0.2375	0.0510	0.0059	0.0003			
	5	0.9887	0.8042	0.6172	0.4164	0.1256	0.0207	0.0016	0.0000		
	6	0.9976	0.9133	0.7858	0.6080	0.2500	0.0577	0.0065	0.0003		
	7	0.9996	0.9679	0.8982	0.7723	0.4159	0.1316	0.0210	0.0013	0.0000	
	8	0.9999	0.9900	0.9591	0.8867	0.5956	0.2517	0.0565	0.0051	0.0001	
	9	1.0000	0.9974	0.9861	0.9520	0.7553	0.4119	0.1275	0.0171	0.0006	
	10		0.9994	0.9961	0.9829	0.8725	0.5881	0.2447	0.0480	0.0026	0.0000
	11		0.9999	0.9991	0.9949	0.9435	0.7483	0.4044	0.1133	0.0100	0.0001
	12		1.0000	0.9998	0.9987	0.9790	0.8684	0.5841	0.2277	0.0321	0.0004
	13			1.0000	0.9997	0.9935	0.9423	0.7500	0.3920	0.0867	0.0024
	14				1.0000	0.9984	0.9793	0.8744	0.5836	0.1958	0.0113
	15					0.9997	0.9941	0.9490	0.7625	0.3704	0.0432
	16					1.0000	0.9987	0.9840	0.8929	0.5886	0.1330
	17						0.9998	0.9964	0.9645	0.7939	0.3231
	18						1.0000	0.9995	0.9924	0.9308	0.6083
	19							1.0000	0.9992	0.9885	0.8784
	20								1.0000	1.0000	1.0000

T. Agami Reddy, *Applied Data Analysis and Modeling for Energy Engineers and Scientists*,
DOI 10.1007/978-1-4419-9613-8, © Springer Science+Business Media, LLC 2011

Table A.2 Cumulative probability values for the Poisson distribution

$$P(x; \lambda, t) = \Pr[X \le x] = \sum_{k=0}^{x} p(k; \lambda, t)$$

	λt									
x	1.0	2.0	3.0	4.0	5.0	6.0	7.0	8.0	9.0	10.0
0	0.3679	0.1353	0.0498	0.0183	0.0067	0.0025	0.0009	0.0003	0.0001	0.0000
1	0.7358	0.4060	0.1991	0.0916	0.0404	0.0174	0.0073	0.0030	0.0012	0.0005
2	0.9197	0.6767	0.4232	0.2381	0.1247	0.0620	0.0296	0.0138	0.0062	0.0028
3	0.9810	0.8571	0.6472	0.4335	0.2650	0.1512	0.0818	0.0424	0.0212	0.0103
4	0.9963	0.9473	0.8153	0.6288	0.4405	0.2851	0.1730	0.0996	0.0550	0.0293
5	0.9994	0.9834	0.9161	0.7851	0.6160	0.4457	0.3007	0.1912	0.1157	0.0671
6	0.9999	0.9955	0.9665	0.8893	0.7622	0.6063	0.4497	0.3134	0.2068	0.1301
7	1.0000	0.9989	0.9881	0.9489	0.8666	0.7440	0.5987	0.4530	0.3239	0.2202
8		0.9998	0.9962	0.9786	0.9319	0.8472	0.7291	0.5926	0.4557	0.3328
9		1.0000	0.9989	0.9919	0.9682	0.9161	0.8305	0.7166	0.5874	0.4579
10			0.9997	0.9972	0.9863	0.9574	0.9015	0.8159	0.7060	0.5830
11			0.9999	0.9991	0.9945	0.9799	0.9466	0.8881	0.8030	0.6968
12			1.0000	0.9997	0.9980	0.9912	0.9730	0.9362	0.8758	0.7916
13				0.9999	0.9993	0.9964	0.9872	0.9658	0.9262	0.8645
14				1.0000	0.9998	0.9986	0.9943	0.9827	0.9585	0.9165
15					0.9999	0.9995	0.9976	0.9918	0.9780	0.9513
16					1.0000	0.9998	0.9990	0.9963	0.9889	0.9730
17						0.9999	0.9996	0.9984	0.9947	0.9857
18						1.0000	0.9999	0.9993	0.9976	0.9928
19							0.9999	0.9997	0.9989	0.9965
20							1.0000	0.9999	0.9996	0.9984
21								1.0000	0.9998	0.9993
22									0.9999	0.9997
23									1.0000	0.9999
24										0.9999
25										1.0000

Source: Lawrence L. Lapin, *Quantitative Methods for Business Decisions*, 4th ed. Copyright © 1987 Harcourt Brace Jovanovich, Inc., New York. Reproduced by permission of the publisher.

	λt									
x	11.0	12.0	13.0	14.0	15.0	16.0	17.0	18.0	19.0	20.0
0	0.0000	0.0000	0.0000	0.0000	0.0000	0.0000	0.0	0.0	0.0	0.0
1	0.0002	0.0001	0.0000	0.0000	0.0000	0.0000	0.0000	0.0000	0.0000	0.0
2	0.0012	0.0005	0.0002	0.0001	0.0000	0.0000	0.0000	0.0000	0.0000	0.0000
3	0.0049	0.0023	0.0011	0.0005	0.0002	0.0001	0.0000	0.0000	0.0000	0.0000
4	0.0151	0.0076	0.0037	0.0018	0.0009	0.0004	0.0002	0.0001	0.0000	0.0000
5	0.0375	0.0203	0.0107	0.0055	0.0028	0.0014	0.0007	0.0003	0.0002	0.0001
6	0.0786	0.0458	0.0259	0.0142	0.0076	0.0040	0.0021	0.0010	0.0005	0.0003
7	0.1432	0.0895	0.0540	0.0316	0.0180	0.0100	0.0054	0.0029	0.0015	0.0008
8	0.2320	0.1550	0.0998	0.0621	0.0374	0.0220	0.0126	0.0071	0.0039	0.0021
9	0.3405	0.2424	0.1658	0.1094	0.0699	0.0433	0.0261	0.0154	0.0089	0.0050
10	0.4599	0.3472	0.2517	0.1757	0.1185	0.0774	0.0491	0.0304	0.0183	0.0108
11	0.5793	0.4616	0.3532	0.2600	0.1847	0.1270	0.0847	0.0549	0.0347	0.0214
12	0.6887	0.5760	0.4631	0.3585	0.2676	0.1931	0.1350	0.0917	0.0606	0.0390
13	0.7813	0.6815	0.5730	0.4644	0.3632	0.2745	0.2009	0.1426	0.0984	0.0661
14	0.8540	0.7720	0.6751	0.5704	0.4656	0.3675	0.2808	0.2081	0.1497	0.1049
15	0.9074	0.8444	0.7636	0.6694	0.5681	0.4667	0.3714	0.2866	0.2148	0.1565
16	0.9441	0.8987	0.8355	0.7559	0.6641	0.5660	0.4677	0.3750	0.2920	0.2211
17	0.9678	0.9370	0.8905	0.8272	0.7489	0.6593	0.5640	0.4686	0.3784	0.2970
18	0.9823	0.9626	0.9302	0.8826	0.8195	0.7423	0.6549	0.5622	0.4695	0.3814
19	0.9907	0.9787	0.9573	0.9235	0.8752	0.8122	0.7363	0.6509	0.5606	0.4703
20	0.9953	0.9884	0.9750	0.9521	0.9170	0.8682	0.8055	0.7307	0.6472	0.5591
21	0.9977	0.9939	0.9859	0.9711	0.9469	0.9108	0.8615	0.7991	0.7255	0.6437
22	0.9989	0.9969	0.9924	0.9833	0.9672	0.9418	0.9047	0.8551	0.7931	0.7206
23	0.9995	0.9985	0.9960	0.9907	0.9805	0.9633	0.9367	0.8989	0.8490	0.7875
24	0.9998	0.9993	0.9980	0.9950	0.9888	0.9777	0.9593	0.9317	0.8933	0.8432
25	0.9999	0.9997	0.9990	0.9974	0.9938	0.9869	0.9747	0.9554	0.9269	0.8878
26	1.0000	0.9999	0.9995	0.9987	0.9967	0.9925	0.9848	0.9718	0.9514	0.9221
27		0.9999	0.9998	0.9994	0.9983	0.9959	0.9912	0.9827	0.9687	0.9475
28		1.0000	0.9999	0.9997	0.9991	0.9978	0.9950	0.9897	0.9805	0.9657
29			1.0000	0.9999	0.9996	0.9989	0.9973	0.9940	0.9881	0.9782
30				0.9999	0.9998	0.9994	0.9985	0.9967	0.9930	0.9865
31				1.0000	0.9999	0.9997	0.9992	0.9982	0.9960	0.9919
32					0.9999	0.9999	0.9996	0.9990	0.9978	0.9953
33					1.0000	0.9999	0.9998	0.9995	0.9988	0.9973
34						1.0000	0.9999	0.9997	0.9994	0.9985
35							0.9999	0.9999	0.9997	0.9992
36							1.0000	0.9999	0.9998	0.9996
37								1.0000	0.9999	0.9998
38									1.0000	0.9999
39										0.9999
40										1.0000

Table A.3 The standard normal distribution (cumulative z curve areas)

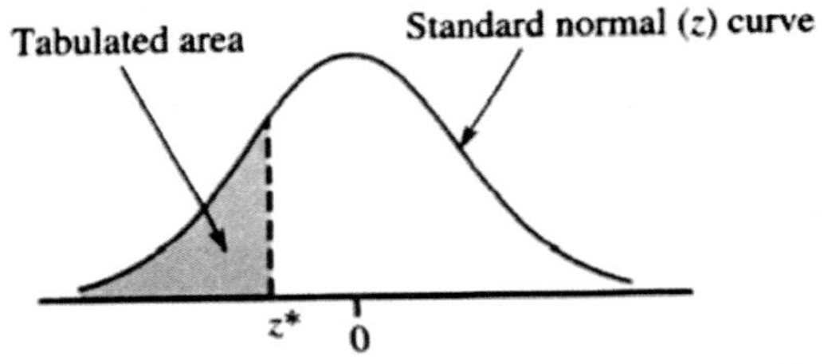

z^*	.00	.01	.02	.03	.04	.05	.06	.07	.08	.09
−3.8	.0001	.0001	.0001	.0001	.0001	.0001	.0001	.0001	.0001	.0000
−3.7	.0001	.0001	.0001	.0001	.0001	.0001	.0001	.0001	.0001	.0001
−3.6	.0002	.0002	.0001	.0001	.0001	.0001	.0001	.0001	.0001	.0001
−3.5	.0002	.0002	.0002	.0002	.0002	.0002	.0002	.0002	.0002	.0002
−3.4	.0003	.0003	.0003	.0003	.0003	.0003	.0003	.0003	.0003	.0002
−3.3	.0005	.0005	.0005	.0004	.0004	.0004	.0004	.0004	.0004	.0003
−3.2	.0007	.0007	.0006	.0006	.0006	.0006	.0006	.0005	.0005	.0005
−3.1	.0010	.0009	.0009	.0009	.0008	.0008	.0008	.0008	.0007	.0007
−3.0	.0013	.0013	.0013	.0012	.0012	.0011	.0011	.0011	.0010	.0010
−2.9	.0019	.0018	.0018	.0017	.0016	.0016	.0015	.0015	.0014	.0014
−2.8	.0026	.0025	.0024	.0023	.0023	.0022	.0021	.0021	.0020	.0019
−2.7	.0035	.0034	.0033	.0032	.0031	.0030	.0029	.0028	.0027	.0026
−2.6	.0047	.0045	.0044	.0043	.0041	.0040	.0039	.0038	.0037	.0036
−2.5	.0062	.0060	.0059	.0057	.0055	.0054	.0052	.0051	.0049	.0048
−2.4	.0082	.0080	.0078	.0075	.0073	.0071	.0069	.0068	.0066	.0064
−2.3	.0107	.0104	.0102	.0099	.0096	.0094	.0091	.0089	.0087	.0084
−2.2	.0139	.0136	.0132	.0129	.0125	.0122	.0119	.0116	.0113	.0110
−2.1	.0179	.0174	.0170	.0166	.0162	.0158	.0154	.0150	.0146	.0143
−2.0	.0228	.0222	.0217	.0212	.0207	.0202	.0197	.0192	.0188	.0183
−1.9	.0287	.0281	.0274	.0268	.0262	.0256	.0250	.0244	.0239	.0233
−1.8	.0359	.0351	.0344	.0336	.0329	.0322	.0314	.0307	.0301	.0294
−1.7	.0446	.0436	.0427	.0418	.0409	.0401	.0392	.0384	.0375	.0367
−1.6	.0548	.0537	.0526	.0516	.0505	.0495	.0485	.0475	.0465	.0455
−1.5	.0668	.0655	.0643	.0630	.0618	.0606	.0594	.0582	.0571	.0559
−1.4	.0808	.0793	.0778	.0764	.0749	.0735	.0721	.0708	.0694	.0681
−1.3	.0968	.0951	.0934	.0918	.0901	.0885	.0869	.0853	.0838	.0823
−1.2	.1151	.1131	.1112	.1093	.1075	.1056	.1038	.1020	.1003	.0985
−1.1	.1357	.1335	.1314	.1292	.1271	.1251	.1230	.1210	.1190	.1170
−1.0	.1587	.1562	.1539	.1515	.1492	.1469	.1446	.1423	.1401	.1379
−0.9	.1841	.1814	.1788	.1762	.1736	.1711	.1685	.1660	.1635	.1611
−0.8	.2119	.2090	.2061	.2033	.2005	.1977	.1949	.1922	.1894	.1867
−0.7	.2420	.2389	.2358	.2327	.2296	.2266	.2236	.2206	.2177	.2148
−0.6	.2743	.2709	.2676	.2643	.2611	.2578	.2546	.2514	.2483	.2451
−0.5	.3085	.3050	.3015	.2981	.2946	.2912	.2877	.2843	.2810	.2776
−0.4	.3446	.3409	.3372	.3336	.3300	.3264	.3228	.3192	.3156	.3121
−0.3	.3821	.3783	.3745	.3707	.3669	.3632	.3594	.3557	.3520	.3483
−0.2	.4207	.4168	.4129	.4090	.4052	.4013	.3974	.3936	.3897	.3859
−0.1	.4602	.4562	.4522	.4483	.4443	.4404	.4364	.4325	.4286	.4247
−0.0	.5000	.4960	.4920	.4880	.4840	.4801	.4761	.4721	.4681	.4641

Table A.4 Critical values of the t-distribution for confidence and prediction intervals

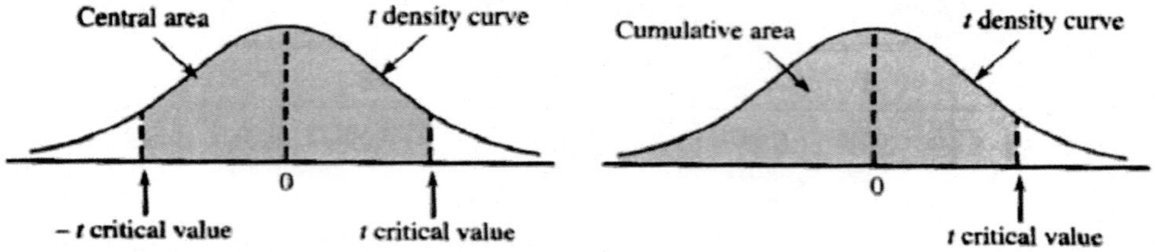

Central area = confidence/prediction level for two-sided interval:		80%	90%	95%	98%	99%	99.8%	99.9%
Cumulative area = confidence/prediction level for one-sided interval:		90%	95%	97.5%	99%	99.5%	99.9%	99.95%
Degrees of freedom	1	3.078	6.314	12.706	31.821	63.657	318.31	636.62
	2	1.886	2.920	4.303	6.965	9.925	22.326	31.598
	3	1.638	2.353	3.182	4.541	5.841	10.213	12.924
	4	1.533	2.132	2.776	3.747	4.604	7.173	8.610
	5	1.476	2.015	2.571	3.365	4.032	5.893	6.869
	6	1.440	1.943	2.447	3.143	3.707	5.208	5.959
	7	1.415	1.895	2.365	2.998	3.499	4.785	5.408
	8	1.397	1.860	2.306	2.896	3.355	4.501	5.041
	9	1.383	1.833	2.262	2.821	3.250	4.297	4.781
	10	1.372	1.812	2.228	2.764	3.169	4.144	4.587
	11	1.363	1.796	2.201	2.718	3.106	4.025	4.437
	12	1.356	1.782	2.179	2.681	3.055	3.930	4.318
	13	1.350	1.771	2.160	2.650	3.012	3.852	4.221
	14	1.345	1.761	2.145	2.624	2.977	3.787	4.140
	15	1.341	1.753	2.131	2.602	2.947	3.733	4.073
	16	1.337	1.746	2.120	2.583	2.921	3.686	4.015
	17	1.333	1.740	2.110	2.567	2.898	3.646	3.965
	18	1.330	1.734	2.101	2.552	2.878	3.610	3.922
	19	1.328	1.729	2.093	2.539	2.861	3.579	3.883
	20	1.325	1.725	2.086	2.528	2.845	3.552	3.850
	21	1.323	1.721	2.080	2.518	2.831	3.527	3.819
	22	1.321	1.717	2.074	2.508	2.819	3.505	3.792
	23	1.319	1.714	2.069	2.500	2.807	3.485	3.767
	24	1.318	1.711	2.064	2.492	2.797	3.467	3.745
	25	1.316	1.708	2.060	2.485	2.787	3.450	3.725
	26	1.315	1.706	2.056	2.479	2.779	3.435	3.707
	27	1.314	1.703	2.052	2.473	2.771	3.421	3.690
	28	1.313	1.701	2.048	2.467	2.763	3.408	3.674
	29	1.311	1.699	2.045	2.462	2.756	3.396	3.659
	30	1.310	1.697	2.042	2.457	2.750	3.385	3.646
	40	1.303	1.684	2.021	2.423	2.704	3.307	3.551
	60	1.296	1.671	2.000	2.390	2.660	3.232	3.460
	120	1.289	1.658	1.980	2.358	2.617	3.160	3.373
	∞	1.282	1.645	1.960	2.326	2.576	3.090	3.291

Table A.5 Critical values of the chi-square distribution

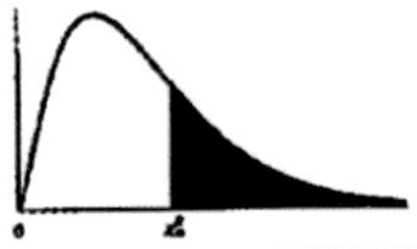

	α																			
v	0.995	0.99	0.98	0.975	0.95	0.90	0.80	0.75	0.70	0.50	0.30	0.25	0.20	0.10	0.05	0.025	0.02	0.01	0.005	0.001
1	0.0^4393	0.0^3157	0.0^3628	0.0^3982	0.00393	0.0158	0.0642	0.102	0.148	0.455	1.074	1.323	1.642	2.706	3.841	5.024	5.412	6.635	7.879	10.827
2	0.0100	0.0201	0.0404	0.0506	0.103	0.211	0.446	0.575	0.713	1.386	2.408	2.773	3.219	4.605	5.991	7.378	7.824	9.210	10.597	13.815
3	0.0717	0.115	0.185	0.216	0.352	0.584	1.005	1.213	1.424	2.366	3.665	4.108	4.642	6.251	7.815	9.348	9.837	11.345	12.838	16.266
4	0.207	0.297	0.429	0.484	0.711	1.064	1.649	1.923	2.195	3.357	4.878	5.385	5.989	7.779	9.488	11.143	11.668	13.277	14.860	18.466
5	0.412	0.554	0.752	0.831	1.145	1.610	2.343	2.675	3.000	4.351	6.064	6.626	7.289	9.236	11.070	12.832	13.388	15.086	16.750	20.515
6	0.676	0.872	1.134	1.237	1.635	2.204	3.070	3.455	3.828	5.348	7.231	7.841	8.558	10.645	12.592	14.449	15.033	16.812	18.548	22.457
7	0.989	1.239	1.564	1.690	2.167	2.833	3.822	4.255	4.671	6.346	8.383	9.037	9.803	12.017	14.067	16.013	16.622	18.475	20.278	24.321
8	1.344	1.647	2.032	2.180	2.733	3.490	4.594	5.071	5.527	7.344	9.524	10.219	11.030	13.362	15.507	17.535	18.168	20.090	21.955	26.124
9	1.735	2.088	2.532	2.700	3.325	4.168	5.380	5.899	6.393	8.343	10.656	11.389	12.242	14.684	16.919	19.023	19.679	21.666	23.589	27.877
10	2.156	2.558	3.059	3.247	3.940	4.865	6.179	6.737	7.267	9.342	11.781	12.549	13.442	15.987	18.307	20.483	21.161	23.209	25.188	29.588
11	2.603	3.053	3.609	3.816	4.575	5.578	6.989	7.584	8.148	10.341	12.899	13.701	14.631	17.275	19.675	21.920	22.618	24.725	26.757	31.264
12	3.074	3.571	4.178	4.404	5.226	6.304	7.807	8.438	9.034	11.340	14.011	14.845	15.812	18.549	21.026	23.337	24.054	26.217	28.300	32.909
13	3.565	4.107	4.765	5.009	5.892	7.041	8.634	9.299	9.926	12.340	15.119	15.984	16.985	19.812	22.362	24.736	25.471	27.688	29.819	34.527
14	4.075	4.660	5.368	5.629	6.571	7.790	9.467	10.165	10.821	13.339	16.222	17.117	18.151	21.064	23.685	26.119	26.873	29.141	31.319	36.124
15	4.601	5.229	5.985	6.262	7.261	8.547	10.307	11.037	11.721	14.339	17.322	18.245	19.311	22.307	24.996	27.488	28.259	30.578	32.801	37.698
16	5.142	5.812	6.614	6.908	7.962	9.312	11.152	11.912	12.624	15.338	18.418	19.369	20.465	23.542	26.296	28.845	29.633	32.000	34.267	39.252
17	5.697	6.408	7.255	7.564	8.672	10.085	12.002	12.792	13.531	16.338	19.511	20.489	21.615	24.769	27.587	30.191	30.995	33.409	35.718	40.791
18	6.265	7.015	7.906	8.231	9.390	10.865	12.857	13.675	14.440	17.338	20.601	21.605	22.760	25.989	28.869	31.526	32.346	34.805	37.156	42.312
19	6.844	7.633	8.567	8.907	10.117	11.651	13.716	14.562	15.352	18.338	21.689	22.718	23.900	27.204	30.144	32.852	33.687	36.191	38.582	43.819
20	7.434	8.260	9.237	9.591	10.851	12.443	14.578	15.452	16.266	19.337	22.775	23.828	25.038	28.412	31.410	34.170	35.020	37.566	39.997	45.314
21	8.034	8.897	9.915	10.283	11.591	13.240	15.445	16.344	17.182	20.337	23.858	24.935	26.171	29.615	32.671	35.479	36.343	38.932	41.401	46.796
22	8.643	9.542	10.600	10.982	12.338	14.041	16.314	17.240	18.101	21.337	24.939	26.039	27.301	30.813	33.924	36.781	37.659	40.289	42.796	48.268
23	9.260	10.196	11.293	11.689	13.091	14.848	17.187	18.137	19.021	22.337	26.018	27.141	28.429	32.007	35.172	38.076	38.968	41.638	44.181	49.728
24	9.886	10.856	11.992	12.401	13.848	15.659	18.062	19.037	19.943	23.337	27.096	28.241	29.553	33.196	36.415	39.364	40.270	42.980	45.558	51.179
25	10.520	11.524	12.697	13.120	14.611	16.473	18.940	19.939	20.867	24.337	28.172	29.339	30.675	34.382	37.652	40.646	41.566	44.314	46.928	52.619
26	11.160	12.198	13.409	13.844	15.379	17.292	19.820	20.843	21.792	25.336	29.246	30.435	31.795	35.563	38.885	41.923	42.856	45.642	48.290	54.051
27	11.808	12.878	14.125	14.573	16.151	18.114	20.703	21.749	22.719	26.336	30.319	31.528	32.912	36.741	40.113	43.195	44.140	46.963	49.645	55.475
28	12.461	13.565	14.847	15.308	16.928	18.939	21.588	22.657	23.647	27.336	31.391	32.620	34.027	37.916	41.337	44.461	45.419	48.278	50.994	56.892
29	13.121	14.256	15.574	16.047	17.708	19.768	22.475	23.567	24.577	28.336	32.461	33.711	35.139	39.087	42.557	45.722	46.693	49.588	52.335	58.301
30	13.787	14.953	16.306	16.791	18.493	20.599	23.364	24.478	25.508	29.336	33.530	34.800	36.250	40.256	43.773	46.979	47.962	50.892	53.672	59.702
40	20.707	22.164	23.838	24.433	26.509	29.051	32.345	33.66	34.872	39.335	44.165	45.616	47.269	51.805	55.758	59.342	60.436	63.691	66.766	73.403
50	27.991	29.707	31.664	32.357	34.764	37.689	41.449	42.942	44.313	49.335	54.723	56.334	58.164	63.167	67.505	71.420	72.613	76.154	79.490	86.660
60	35.534	37.485	39.699	40.482	43.188	46.459	50.641	52.294	53.809	59.335	65.226	66.981	68.972	74.397	79.082	83.298	84.58	88.379	91.952	99.608

Table A.6 Critical values of the F-distribution

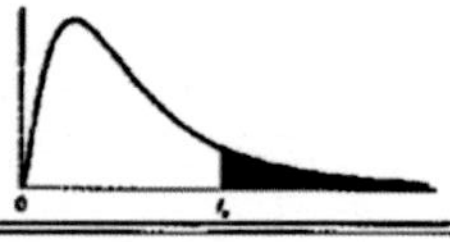

	$f_{0.05}(v_1, v_2)$								
v_2	1	2	3	4	5	6	7	8	9
1	161.45	199.50	215.71	224.58	230.16	233.99	236.77	238.88	240.54
2	18.51	19.00	19.16	19.25	19.30	19.33	19.35	19.37	19.38
3	10.13	9.55	9.28	9.12	9.01	8.94	8.89	8.85	8.81
4	7.71	6.94	6.59	6.39	6.26	6.16	6.09	6.04	6.00
5	6.61	5.79	5.41	5.19	5.05	4.95	4.88	4.82	4.77
6	5.99	5.14	4.76	4.53	4.39	4.28	4.21	4.15	4.10
7	5.59	4.74	4.35	4.12	3.97	3.87	3.79	3.73	3.68
8	5.32	4.46	4.07	3.84	3.69	3.58	3.50	3.44	3.39
9	5.12	4.26	3.86	3.63	3.48	3.37	3.29	3.23	3.18
10	4.96	4.10	3.71	3.48	3.33	3.22	3.14	3.07	3.02
11	4.84	3.98	3.59	3.36	3.20	3.09	3.01	2.95	2.90
12	4.75	3.89	3.49	3.26	3.11	3.00	2.91	2.85	2.80
13	4.67	3.81	3.41	3.18	3.03	2.92	2.83	2.77	2.71
14	4.60	3.74	3.34	3.11	2.96	2.85	2.76	2.70	2.65
15	4.54	3.68	3.29	3.06	2.90	2.79	2.71	2.64	2.59
16	4.49	3.63	3.24	3.01	2.85	2.74	2.66	2.59	2.54
17	4.45	3.59	3.20	2.96	2.81	2.70	2.61	2.55	2.49
18	4.41	3.55	3.16	2.93	2.77	2.66	2.58	2.51	2.46
19	4.38	3.52	3.13	2.90	2.74	2.63	2.54	2.48	2.42
20	4.35	3.49	3.10	2.87	2.71	2.60	2.51	2.45	2.39
21	4.32	3.47	3.07	2.84	2.68	2.57	2.49	2.42	2.37
22	4.30	3.44	3.05	2.82	2.66	2.55	2.46	2.40	2.34
23	4.28	3.42	3.03	2.80	2.64	2.53	2.44	2.37	2.32
24	4.26	3.40	3.01	2.78	2.62	2.51	2.42	2.36	2.30
25	4.24	3.39	2.99	2.76	2.60	2.49	2.40	2.34	2.28
26	4.23	3.37	2.98	2.74	2.59	2.47	2.39	2.32	2.27
27	4.21	3.35	2.96	2.73	2.57	2.46	2.37	2.31	2.25
28	4.20	3.34	2.95	2.71	2.56	2.45	2.36	2.29	2.24
29	4.18	3.33	2.93	2.70	2.55	2.43	2.35	2.28	2.22
30	4.17	3.32	2.92	2.69	2.53	2.42	2.33	2.27	2.21
40	4.08	3.23	2.84	2.61	2.45	2.34	2.25	2.18	2.12
60	4.00	3.15	2.76	2.53	2.37	2.25	2.17	2.10	2.04
120	3.92	3.07	2.68	2.45	2.29	2.18	2.09	2.02	1.96
∞	3.84	3.00	2.60	2.37	2.21	2.10	2.01	1.94	1.88

	$f_{0.05}(v_1, v_2)$									
v_2	10	12	15	20	24	30	40	60	120	∞
1	241.88	243.91	245.95	248.01	249.05	250.10	251.14	252.20	253.25	254.31
2	19.40	19.41	19.43	19.45	19.45	19.46	19.47	19.48	19.49	19.50
3	8.79	8.74	8.70	8.66	8.64	8.62	8.59	8.57	8.55	8.53
4	5.96	5.91	5.86	5.80	5.77	5.75	5.72	5.69	5.66	5.63
5	4.74	4.68	4.62	4.56	4.53	4.50	4.46	4.43	4.40	4.36
6	4.06	4.00	3.94	3.87	3.84	3.81	3.77	3.74	3.70	3.67
7	3.64	3.57	3.51	3.44	3.41	3.38	3.34	3.30	3.27	3.23
8	3.35	3.28	3.22	3.15	3.12	3.08	3.04	3.01	2.97	2.93
9	3.14	3.07	3.01	2.94	2.90	2.86	2.83	2.79	2.75	2.71
10	2.98	2.91	2.85	2.77	2.74	2.70	2.66	2.62	2.58	2.54
11	2.85	2.79	2.72	2.65	2.61	2.57	2.53	2.49	2.45	2.40
12	2.75	2.69	2.62	2.54	2.51	2.47	2.43	2.38	2.34	2.30
13	2.67	2.60	2.53	2.46	2.42	2.38	2.34	2.30	2.25	2.21
14	2.60	2.53	2.46	2.39	2.35	2.31	2.27	2.22	2.18	2.13
15	2.54	2.48	2.40	2.33	2.29	2.25	2.20	2.16	2.11	2.07
16	2.49	2.42	2.35	2.28	2.24	2.19	2.15	2.11	2.06	2.01
17	2.45	2.38	2.31	2.23	2.19	2.15	2.10	2.06	2.01	1.96
18	2.41	2.34	2.27	2.19	2.15	2.11	2.06	2.02	1.97	1.92
19	2.38	2.31	2.23	2.16	2.11	2.07	2.03	1.98	1.93	1.88
20	2.35	2.28	2.20	2.12	2.08	2.04	1.99	1.95	1.90	1.84
21	2.32	2.25	2.18	2.10	2.05	2.01	1.96	1.92	1.87	1.81
22	2.30	2.23	2.15	2.07	2.03	1.98	1.94	1.89	1.84	1.78
23	2.27	2.20	2.13	2.05	2.01	1.96	1.91	1.86	1.81	1.76
24	2.25	2.18	2.11	2.03	1.98	1.94	1.89	1.84	1.79	1.73
25	2.24	2.16	2.09	2.01	1.96	1.92	1.87	1.82	1.77	1.71
26	2.22	2.15	2.07	1.99	1.95	1.90	1.85	1.80	1.75	1.69
27	2.20	2.13	2.06	1.97	1.93	1.88	1.84	1.79	1.73	1.67
28	2.19	2.12	2.04	1.96	1.91	1.87	1.82	1.77	1.71	1.65
29	2.18	2.10	2.03	1.94	1.90	1.85	1.81	1.75	1.70	1.64
30	2.16	2.09	2.01	1.93	1.89	1.84	1.79	1.74	1.68	1.62
40	2.08	2.00	1.92	1.84	1.79	1.74	1.69	1.64	1.58	1.51
60	1.99	1.92	1.84	1.75	1.70	1.65	1.59	1.53	1.47	1.39
120	1.91	1.83	1.75	1.66	1.61	1.55	1.50	1.43	1.35	1.25
∞	1.83	1.75	1.67	1.57	1.52	1.46	1.39	1.32	1.22	1.00

*Reproduced from Table 18 of *Biometrika Tables for Statisticians*, Vol. I, by permission of E.S. Pearson and the Biometrika Trustees.

Table A.6 (continued)

v_2	$f_{0.01}(v_1, v_2)$								
	1	2	3	4	5	6	7	8	9
1	4052.18	4999.50	5403.35	5624.58	5763.65	5858.99	5928.36	5981.07	6022.47
2	98.50	99.00	99.17	99.25	99.30	99.33	99.36	99.37	99.39
3	34.12	30.82	29.46	28.71	28.24	27.91	27.67	27.49	27.35
4	21.20	18.00	16.69	15.98	15.52	15.21	14.98	14.80	14.66
5	16.26	13.27	12.06	11.39	10.97	10.67	10.46	10.29	10.16
6	13.75	10.92	9.78	9.15	8.75	8.47	8.26	8.10	7.98
7	12.25	9.55	8.45	7.85	7.46	7.19	6.99	6.84	6.72
8	11.26	8.65	7.59	7.01	6.63	6.37	6.18	6.03	5.91
9	10.56	8.02	6.99	6.42	6.06	5.80	5.61	5.47	5.35
10	10.04	7.56	6.55	5.99	5.64	5.39	5.20	5.06	4.94
11	9.65	7.21	6.22	5.67	5.32	5.07	4.89	4.74	4.63
12	9.33	6.93	5.95	5.41	5.06	4.82	4.64	4.50	4.39
13	9.07	6.70	5.74	5.21	4.86	4.62	4.44	4.30	4.19
14	8.86	6.51	5.56	5.04	4.69	4.46	4.28	4.14	4.03
15	8.68	6.36	5.42	4.89	4.56	4.32	4.14	4.00	3.89
16	8.53	6.23	5.29	4.77	4.44	4.20	4.03	3.89	3.78
17	8.40	6.11	5.18	4.67	4.34	4.10	3.93	3.79	3.68
18	8.29	6.01	5.09	4.58	4.25	4.01	3.84	3.71	3.60
19	8.18	5.93	5.01	4.50	4.17	3.94	3.77	3.63	3.52
20	8.10	5.85	4.94	4.43	4.10	3.87	3.70	3.56	3.46
21	8.02	5.78	4.87	4.37	4.04	3.81	3.64	3.51	3.40
22	7.95	5.72	4.82	4.31	3.99	3.76	3.59	3.45	3.35
23	7.88	5.66	4.76	4.26	3.94	3.71	3.54	3.41	3.30
24	7.82	5.61	4.72	4.22	3.90	3.67	3.50	3.36	3.26
25	7.77	5.57	4.68	4.18	3.85	3.63	3.46	3.32	3.22
26	7.72	5.53	4.64	4.14	3.82	3.59	3.42	3.29	3.18
27	7.68	5.49	4.60	4.11	3.78	3.56	3.39	3.26	3.15
28	7.64	5.45	4.57	4.07	3.75	3.53	3.36	3.23	3.12
29	7.60	5.42	4.54	4.04	3.73	3.50	3.33	3.20	3.09
30	7.56	5.39	4.51	4.02	3.70	3.47	3.30	3.17	3.07
40	7.31	5.18	4.31	3.83	3.51	3.29	3.12	2.99	2.89
60	7.08	4.98	4.13	3.65	3.34	3.12	2.95	2.82	2.72
120	6.85	4.79	3.95	3.48	3.17	2.96	2.79	2.66	2.56
∞	6.63	4.61	3.78	3.32	3.02	2.80	2.64	2.51	2.41

v_2	$f_{0.01}(v_1, v_2)$									
	10	12	15	20	24	30	40	60	120	∞
1	6055.85	6106.32	6157.28	6208.73	6234.63	6260.65	6286.78	6313.03	6339.39	6365.86
2	99.40	99.42	99.43	99.45	99.46	99.47	99.47	99.48	99.49	99.50
3	27.23	27.05	26.87	26.69	26.60	26.50	26.41	26.32	26.22	26.13
4	14.55	14.37	14.20	14.02	13.93	13.84	13.75	13.65	13.56	13.46
5	10.05	9.89	9.72	9.55	9.47	9.38	9.29	9.20	9.11	9.02
6	7.87	7.72	7.56	7.40	7.31	7.23	7.14	7.06	6.97	6.88
7	6.62	6.47	6.31	6.16	6.07	5.99	5.91	5.82	5.74	5.65
8	5.81	5.67	5.52	5.36	5.28	5.20	5.12	5.03	4.95	4.86
9	5.26	5.11	4.96	4.81	4.73	4.65	4.57	4.48	4.40	4.31
10	4.85	4.71	4.56	4.41	4.33	4.25	4.17	4.08	4.00	3.91
11	4.54	4.40	4.25	4.10	4.02	3.94	3.86	3.78	3.69	3.60
12	4.30	4.16	4.01	3.86	3.78	3.70	3.62	3.54	3.45	3.36
13	4.10	3.96	3.82	3.66	3.59	3.51	3.43	3.34	3.25	3.17
14	3.94	3.80	3.66	3.51	3.43	3.35	3.27	3.18	3.09	3.00
15	3.80	3.67	3.52	3.37	3.29	3.21	3.13	3.05	2.96	2.87
16	3.69	3.55	3.41	3.26	3.18	3.10	3.02	2.93	2.84	2.75
17	3.59	3.46	3.31	3.16	3.08	3.00	2.92	2.83	2.75	2.65
18	3.51	3.37	3.23	3.08	3.00	2.92	2.84	2.75	2.66	2.57
19	3.43	3.30	3.15	3.00	2.92	2.84	2.76	2.67	2.58	2.49
20	3.37	3.23	3.09	2.94	2.86	2.78	2.69	2.61	2.52	2.42
21	3.31	3.17	3.03	2.88	2.80	2.72	2.64	2.55	2.46	2.36
22	3.26	3.12	2.98	2.83	2.75	2.67	2.58	2.50	2.40	2.31
23	3.21	3.07	2.93	2.78	2.70	2.62	2.54	2.45	2.35	2.26
24	3.17	3.03	2.89	2.74	2.66	2.58	2.49	2.40	2.31	2.21
25	3.13	2.99	2.85	2.70	2.62	2.54	2.45	2.36	2.27	2.17
26	3.09	2.96	2.81	2.66	2.58	2.50	2.42	2.33	2.23	2.13
27	3.06	2.93	2.78	2.63	2.55	2.47	2.38	2.29	2.20	2.10
28	3.03	2.90	2.75	2.60	2.52	2.44	2.35	2.26	2.17	2.06
29	3.00	2.87	2.73	2.57	2.49	2.41	2.33	2.23	2.14	2.03
30	2.98	2.84	2.70	2.55	2.47	2.39	2.30	2.21	2.11	2.01
40	2.80	2.66	2.52	2.37	2.29	2.20	2.11	2.02	1.92	1.80
60	2.63	2.50	2.35	2.20	2.12	2.03	1.94	1.84	1.73	1.60
120	2.47	2.34	2.19	2.03	1.95	1.86	1.76	1.66	1.53	1.38
∞	2.32	2.18	2.04	1.88	1.79	1.70	1.59	1.47	1.32	1.00

Table A.7 Critical values of the correlation coefficient

The tabled entries represent the critical values of r, based on n pairs of observations, for testing the hypothesis $\rho=0$ at the .05 and .01 significance levels.

n	df	Two-tailed p		One-tailed p	
		.05	.01	.05	.01
4	2	.950	.990	.900	.980
5	3	.878	.959	.805	.934
6	4	.811	.917	.729	.882
7	5	.754	.875	.669	.833
8	6	.707	.834	.621	.789
9	7	.666	.798	.582	.750
10	8	.632	.765	.549	.715
11	9	.602	.735	.521	.685
12	10	.576	.708	.497	.658
13	11	.553	.684	.476	.634
14	12	.532	.661	.457	.612
15	13	.514	.641	.441	.592
16	14	.497	.623	.426	.574
17	15	.482	.606	.412	.558
18	16	.468	.590	.400	.542
19	17	.456	.575	.389	.529
20	18	.444	.561	.378	.515
25	23	.396	.505	.337	.462
30	28	.361	.463	.306	.423
40	38	.312	.402	.264	.367
60	58	.254	.330	.214	.300
120	118	.179	.234	.151	.212

For unlisted values of n, the critical value of r is given by

$$r=\frac{t_c}{\sqrt{t_c^2+n-2}},$$

where t_c is the critical t value associated with a given significance level and has $df=n-2$.

Table A.8 Critical values of the Studentized range ($\alpha=0.05$)

Error df	k = 2	3	4	5	6	7	8	9	10	11	12	13	14	15	16	17	18	19	20
1	18.0	27.0	32.8	37.1	40.4	43.1	45.4	47.4	49.1	50.6	52.0	53.2	54.3	55.4	56.3	57.2	58.0	58.8	59.6
2	6.08	8.33	9.80	10.9	11.7	12.4	13.0	13.5	14.0	14.4	14.7	15.1	15.4	15.7	15.9	16.1	16.4	16.6	16.8
3	4.50	5.91	6.82	7.50	8.04	8.48	8.85	9.18	9.46	9.72	9.95	10.2	10.3	10.5	10.7	10.8	11.0	11.1	11.2
4	3.93	5.04	5.76	6.29	6.71	7.05	7.35	7.60	7.83	8.03	8.21	8.37	8.52	8.66	8.79	8.91	9.03	9.13	9.23
5	3.64	4.60	5.22	5.67	6.03	6.33	6.58	6.80	6.99	7.17	7.32	7.47	7.60	7.72	7.83	7.93	8.03	8.12	8.21
6	3.46	4.34	4.90	5.30	5.63	5.90	6.12	6.32	6.49	6.65	6.79	6.92	7.03	7.14	7.24	7.34	7.43	7.51	7.59
7	3.34	4.16	4.68	5.06	5.36	5.61	5.82	6.00	6.16	6.30	6.43	6.55	6.66	6.76	6.85	6.94	7.02	7.10	7.17
8	3.26	4.04	4.53	4.89	5.17	5.40	5.60	5.77	5.92	6.05	6.18	6.29	6.39	6.48	6.57	6.65	6.73	6.80	6.87
9	3.20	3.95	4.41	4.76	5.02	5.24	5.43	5.59	5.74	5.87	5.98	6.09	6.19	6.28	6.36	6.44	6.51	6.58	6.64
10	3.15	3.88	4.33	4.65	4.91	5.12	5.30	5.46	5.60	5.72	5.83	5.93	6.03	6.11	6.19	6.27	6.34	6.40	6.47
11	3.11	3.82	4.26	4.57	4.82	5.03	5.20	5.35	5.49	5.61	5.71	5.81	5.90	5.98	6.06	6.13	6.20	6.27	6.33
12	3.08	3.77	4.20	4.51	4.75	4.95	5.12	5.27	5.39	5.51	5.61	5.71	5.80	5.88	5.95	6.02	6.09	6.15	6.21
13	3.06	3.73	4.15	4.45	4.69	4.88	5.05	5.19	5.32	5.43	5.53	5.63	5.71	5.79	5.86	5.93	5.99	6.05	6.11
14	3.03	3.70	4.11	4.41	4.64	4.83	4.99	5.13	5.25	5.36	5.46	5.55	5.64	5.71	5.79	5.85	5.91	5.97	6.03
15	3.01	3.67	4.08	4.37	4.59	4.78	4.94	5.08	5.20	5.31	5.40	5.49	5.57	5.65	5.72	5.78	5.85	5.90	5.96
16	3.00	3.65	4.05	4.33	4.56	4.74	4.90	5.03	5.15	5.26	5.35	5.44	5.52	5.59	5.66	5.73	5.79	5.84	5.90
17	2.98	3.63	4.02	4.30	4.52	4.70	4.86	4.99	5.11	5.21	5.31	5.39	5.47	5.54	5.61	5.67	5.73	5.79	5.84
18	2.97	3.61	4.00	4.28	4.49	4.67	4.82	4.96	5.07	5.17	5.27	5.35	5.43	5.50	5.57	5.63	5.69	5.74	5.79
19	2.96	3.59	3.98	4.25	4.47	4.65	4.79	4.92	5.04	5.14	5.23	5.31	5.39	5.46	5.53	5.59	5.65	5.70	5.75
20	2.95	3.58	3.96	4.23	4.45	4.62	4.77	4.90	5.01	5.11	5.20	5.28	5.36	5.43	5.49	5.55	5.61	5.66	5.71
24	2.92	3.53	3.90	4.17	4.37	4.54	4.68	4.81	4.92	5.01	5.10	5.18	5.25	5.32	5.38	5.44	5.49	5.55	5.59
30	2.89	3.49	3.85	4.10	4.30	4.46	4.60	4.72	4.82	4.92	5.00	5.08	5.15	5.21	5.27	5.33	5.38	5.43	5.47
40	2.86	3.44	3.79	4.04	4.23	4.39	4.52	4.63	4.73	4.82	4.90	4.98	5.04	5.11	5.16	5.22	5.27	5.31	5.36
60	2.83	3.40	3.74	3.98	4.16	4.31	4.44	4.55	4.65	4.73	4.81	4.88	4.94	5.00	5.06	5.11	5.15	5.20	5.24
120	2.80	3.36	3.68	3.92	4.10	4.24	4.36	4.47	4.56	4.64	4.71	4.78	4.84	4.90	4.95	5.00	5.04	5.09	5.13
∞	2.77	3.31	3.63	3.86	4.03	4.17	4.29	4.39	4.47	4.55	4.62	4.68	4.74	4.80	4.85	4.89	4.93	4.97	5.01

Table A.9 Critical values for Dunnett's method

Two-sided comparisons

	$k-1$ = number of treatment means (excluding control)								
$n-k$	1	2	3	4	5	6	7	8	9
5	2.57	3.03	3.29	3.48	3.62	3.73	3.82	3.90	3.97
6	2.45	2.86	3.10	3.26	3.39	3.49	3.57	3.64	3.71
7	2.36	2.75	2.97	3.12	3.24	3.33	3.41	3.47	3.53
8	2.31	2.67	2.88	3.02	3.13	3.22	3.29	3.35	3.41
9	2.26	2.61	2.81	2.95	3.05	3.14	3.20	3.26	3.32
10	2.23	2.57	2.76	2.89	2.99	3.07	3.14	3.19	3.24
11	2.20	2.53	2.72	2.84	2.94	3.02	3.08	3.14	3.19
12	2.18	2.50	2.68	2.81	2.90	2.98	3.04	3.09	3.14
13	2.16	2.48	2.65	2.78	2.87	2.94	3.00	3.06	3.10
14	2.14	2.46	2.63	2.75	2.84	2.91	2.97	3.02	3.07
15	2.13	2.44	2.61	2.73	2.82	2.89	2.95	3.00	3.04
16	2.12	2.42	2.59	2.71	2.80	2.87	2.92	2.97	3.02
17	2.11	2.41	2.58	2.69	2.78	2.85	2.90	2.95	3.00
18	2.10	2.40	2.56	2.68	2.76	2.83	2.89	2.94	2.98
19	2.09	2.39	2.55	2.66	2.75	2.81	2.87	2.92	2.96
20	2.09	2.38	2.54	2.65	2.73	2.80	2.86	2.90	2.95
24	2.06	2.35	2.51	2.61	2.70	2.76	2.81	2.86	2.90
30	2.04	2.32	2.47	2.58	2.66	2.72	2.77	2.82	2.86
40	2.02	2.29	2.44	2.54	2.62	2.68	2.73	2.77	2.81
60	2.00	2.27	2.41	2.51	2.58	2.64	2.69	2.73	2.77
120	1.98	2.24	2.38	2.47	2.55	2.60	2.65	2.69	2.73
∞	1.96	2.21	2.35	2.44	2.51	2.57	2.61	2.65	2.69

[a] Reproduced from C. W. Dunnett, "New Tables for Multiple Comparison with a Control," *Biometrics*, Vol. 20, No. 3, 1964.

Table A.10 Critical values of Spearman's rank correlation coefficient

The α values correspond to a one-tailed test of H_0: $\rho_S = 0$. The value should be doubled for two-tailed tests.

n	$\alpha = .05$	$\alpha = .025$	$\alpha = .01$	$\alpha = .005$
5	.900	—	—	—
6	.829	.886	.943	—
7	.714	.786	.893	—
8	.643	.738	.833	.881
9	.600	.683	.783	.833
10	.564	.648	.745	.794
11	.523	.623	.736	.818
12	.497	.591	.703	.780
13	.475	.566	.673	.745
14	.457	.545	.646	.716
15	.441	.525	.623	.689
16	.425	.507	.601	.666
17	.412	.490	.582	.645
18	.399	.476	.564	.625
19	.388	.462	.549	.608
20	.377	.450	.534	.591
21	.368	.438	.521	.576
22	.359	.428	.508	.562
23	.351	.418	.496	.549
24	.343	.409	.485	.537
25	.336	.400	.475	.526
26	.329	.392	.465	.515
27	.323	.385	.456	.505
28	.317	.377	.448	.496
29	.311	.370	.440	.487
30	.305	.364	.432	.478

Source: From E. G. Olds, "Distribution of Sums of Squares of Rank Differences for Small Samples," *Annals of Mathematical Statistics*, 1938, 9.

Table A.11 Critical values of T_L and T_U for the Wilcoxon rank sum for independent samples

Test statistic is the rank sum associated with the smaller sample (if equal sample sizes, either rank sum can be used).

a. $\alpha = .025$ one-tailed; $\alpha = .05$ two-tailed

n_2 \ n_1	3		4		5		6		7		8		9		10	
	T_L	T_U	T_L	T_U	T_L	T_U	T_L	T_U	T_L	T_U	T_L	T_U	T_L	T_U	T_L	T_U
3	5	16	6	18	6	21	7	23	7	26	8	28	8	31	9	33
4	6	18	11	25	12	28	12	32	13	35	14	38	15	41	16	44
5	6	21	12	28	18	37	19	41	20	45	21	49	22	53	24	56
6	7	23	12	32	19	41	26	52	28	56	29	61	31	65	32	70
7	7	26	13	35	20	45	28	56	37	68	39	73	41	78	43	83
8	8	28	14	38	21	49	29	61	39	73	49	87	51	93	54	98
9	8	31	15	41	22	53	31	65	41	78	51	93	63	108	66	114
10	9	33	16	44	24	56	32	70	43	83	54	98	66	114	79	131

b. $\alpha = .05$ one-tailed; $\alpha = .10$ two-tailed

n_2 \ n_1	3		4		5		6		7		8		9		10	
	T_L	T_U	T_L	T_U	T_L	T_U	T_L	T_U	T_L	T_U	T_L	T_U	T_L	T_U	T_L	T_U
3	6	15	7	17	7	20	8	22	9	24	9	27	10	29	11	31
4	7	17	12	24	13	27	14	30	15	33	16	36	17	39	18	42
5	7	20	13	27	19	36	20	40	22	43	24	46	25	50	26	54
6	8	22	14	30	20	40	28	50	30	54	32	58	33	63	35	67
7	9	24	15	33	22	43	30	54	39	66	41	71	43	76	46	80
8	9	27	16	36	24	46	32	58	41	71	52	84	54	90	57	95
9	10	29	17	39	25	50	33	63	43	76	54	90	66	105	69	111
10	11	31	18	42	26	54	35	67	46	80	57	95	69	111	83	127

Source: From F. Wilcoxon and R. A. Wilcox, "Some Rapid Approximate Statistical Procedures," 1964, 20–23.

Table A.12 Critical values of T_0 for the Wilcoxon paired difference signed rank test

ONE-TAILED	TWO-TAILED	$n = 5$	$n = 6$	$n = 7$	$n = 8$	$n = 9$	$n = 10$
$\alpha = .05$	$\alpha = .10$	1	2	4	6	8	11
$\alpha = .025$	$\alpha = .05$		1	2	4	6	8
$\alpha = .01$	$\alpha = .02$			0	2	3	5
$\alpha = .005$	$\alpha = .01$				0	2	3
		$n = 11$	$n = 12$	$n = 13$	$n = 14$	$n = 15$	$n = 16$
$\alpha = .05$	$\alpha = .10$	14	17	21	26	30	36
$\alpha = .025$	$\alpha = .05$	11	14	17	21	25	30
$\alpha = .01$	$\alpha = .02$	7	10	13	16	20	24
$\alpha = .005$	$\alpha = .01$	5	7	10	13	16	19
		$n = 17$	$n = 18$	$n = 19$	$n = 20$	$n = 21$	$n = 22$
$\alpha = .05$	$\alpha = .10$	41	47	54	60	68	75
$\alpha = .025$	$\alpha = .05$	35	40	46	52	59	66
$\alpha = .01$	$\alpha = .02$	28	33	38	43	49	56
$\alpha = .005$	$\alpha = .01$	23	28	32	37	43	49
		$n = 23$	$n = 24$	$n = 25$	$n = 26$	$n = 27$	$n = 28$
$\alpha = .05$	$\alpha = .10$	83	92	101	110	120	130
$\alpha = .025$	$\alpha = .05$	73	81	90	98	107	117
$\alpha = .01$	$\alpha = .02$	62	69	77	85	93	102
$\alpha = .005$	$\alpha = .01$	55	61	68	76	84	92
		$n = 29$	$n = 30$	$n = 31$	$n = 32$	$n = 33$	$n = 34$
$\alpha = .05$	$\alpha = .10$	141	152	163	175	188	201
$\alpha = .025$	$\alpha = .05$	127	137	148	159	171	183
$\alpha = .01$	$\alpha = .02$	111	120	130	141	151	162
$\alpha = .005$	$\alpha = .01$	100	109	118	128	138	149
		$n = 35$	$n = 36$	$n = 37$	$n = 38$	$n = 39$	
$\alpha = .05$	$\alpha = .10$	214	228	242	256	271	
$\alpha = .025$	$\alpha = .05$	195	208	222	235	250	
$\alpha = .01$	$\alpha = .02$	174	186	198	211	224	
$\alpha = .005$	$\alpha = .01$	160	171	183	195	208	
		$n = 40$	$n = 41$	$n = 42$	$n = 43$	$n = 44$	$n = 45$
$\alpha = .05$	$\alpha = .10$	287	303	319	336	353	371
$\alpha = .025$	$\alpha = .05$	264	279	295	311	327	344
$\alpha = .01$	$\alpha = .02$	238	252	267	281	297	313
$\alpha = .005$	$\alpha = .01$	221	234	248	262	277	292
		$n = 46$	$n = 47$	$n = 48$	$n = 49$	$n = 50$	
$\alpha = .05$	$\alpha = .10$	389	408	427	446	466	
$\alpha = .025$	$\alpha = .05$	361	379	397	415	434	
$\alpha = .01$	$\alpha = .02$	329	345	362	380	398	
$\alpha = .005$	$\alpha = .01$	307	323	339	356	373	

Source: From F. Wilcoxon and R. A. Wilcox, "Some Rapid Approximate Statistical Procedures," 1964, p. 28.

Table A.13 Critical values of the Durbin-Watson statistic for 5% significance level

	$k = 1$		$k = 2$		$k = 3$		$k = 4$		$k = 5$	
n	d_L	d_U	d_L	d_U	d_L	d_U	d_L	d_U	d_L	d_U
15	1.08	1.36	0.95	1.54	0.82	1.75	0.69	1.97	0.56	2.21
16	1.10	1.37	0.98	1.54	0.86	1.73	0.74	1.93	0.62	2.15
17	1.13	1.38	1.02	1.54	0.90	1.71	0.78	1.90	0.67	2.10
18	1.16	1.39	1.05	1.53	0.93	1.69	0.82	1.87	0.71	2.06
19	1.18	1.40	1.08	1.53	0.97	1.68	0.86	1.85	0.75	2.02
20	1.20	1.41	1.10	1.54	1.00	1.68	0.90	1.83	0.79	1.99
21	1.22	1.42	1.13	1.54	1.03	1.67	0.93	1.81	0.83	1.96
22	1.24	1.43	1.15	1.54	1.05	1.66	0.96	1.80	0.86	1.94
23	1.26	1.44	1.17	1.54	1.08	1.66	0.99	1.79	0.90	1.92
24	1.27	1.45	1.19	1.55	1.10	1.66	1.01	1.78	0.93	1.90
25	1.29	1.45	1.21	1.55	1.12	1.66	1.04	1.77	0.95	1.89
26	1.30	1.46	1.22	1.55	1.14	1.65	1.06	1.76	0.98	1.88
27	1.32	1.47	1.24	1.56	1.16	1.65	1.08	1.76	1.01	1.86
28	1.33	1.48	1.26	1.56	1.18	1.65	1.10	1.75	1.03	1.85
29	1.34	1.48	1.27	1.56	1.20	1.65	1.12	1.74	1.05	1.84
30	1.35	1.49	1.28	1.57	1.21	1.65	1.14	1.74	1.07	1.83
31	1.36	1.50	1.30	1.57	1.23	1.65	1.16	1.74	1.09	1.83
32	1.37	1.50	1.31	1.57	1.24	1.65	1.18	1.73	1.11	1.82
33	1.38	1.51	1.32	1.58	1.26	1.65	1.19	1.73	1.13	1.81
34	1.39	1.51	1.33	1.58	1.27	1.65	1.21	1.73	1.15	1.81
35	1.40	1.52	1.34	1.58	1.28	1.65	1.22	1.73	1.16	1.80
36	1.41	1.52	1.35	1.59	1.29	1.65	1.24	1.73	1.18	1.80
37	1.42	1.53	1.36	1.59	1.31	1.66	1.25	1.72	1.19	1.80
38	1.43	1.54	1.37	1.59	1.32	1.66	1.26	1.72	1.21	1.79
39	1.43	1.54	1.38	1.60	1.33	1.66	1.27	1.72	1.22	1.79
40	1.44	1.54	1.39	1.60	1.34	1.66	1.29	1.72	1.23	1.79
45	1.48	1.57	1.43	1.62	1.38	1.67	1.34	1.72	1.29	1.78
50	1.50	1.59	1.46	1.63	1.42	1.67	1.38	1.72	1.34	1.77
55	1.53	1.60	1.49	1.64	1.45	1.68	1.41	1.72	1.38	1.77
60	1.55	1.62	1.51	1.65	1.48	1.69	1.44	1.73	1.41	1.77
65	1.57	1.63	1.54	1.66	1.50	1.70	1.47	1.73	1.44	1.77
70	1.58	1.64	1.55	1.67	1.52	1.70	1.49	1.74	1.46	1.77
75	1.60	1.65	1.57	1.68	1.54	1.71	1.51	1.74	1.49	1.77
80	1.61	1.66	1.59	1.69	1.56	1.72	1.53	1.74	1.51	1.77
85	1.62	1.67	1.60	1.70	1.57	1.72	1.55	1.75	1.52	1.77
90	1.63	1.68	1.61	1.70	1.59	1.73	1.57	1.75	1.54	1.78
95	1.64	1.69	1.62	1.71	1.60	1.73	1.58	1.75	1.56	1.78
100	1.65	1.69	1.63	1.72	1.61	1.74	1.59	1.76	1.57	1.78

Source: J. Durbin and G. S. Watson, *Biometrika*, 38 (1951).

B: Large Data Sets

Problem 2.25 Generating cumulative distribution curves and utilizability curves from measured data

Table B 1 Measured global solar radiation values (All values are in MJ/m^2.) on a horizontal surface at Quezon City, Manila during October 1980. Taken from Reddy (1987)

Day	Solar time												
No.	06–07	07–08	08–09	09–10	10–11	11–12	12–13	13–14	14–15	15–16	16–17	17–18	Daily
1	0.209	0.837	1.256	1.591	1.214	1.759	1.005	1.172	0.963	0.461	0.126	0.042	10.63
2	0.126	0.335	0.712	1.172	1.465	1.130	0.796	0.712	0.544	0.544	0.293	0.084	7.91
3	0.084	0.419	0.754	0.921	1.005	0.963	0.461	0.502	0.419	0.293	0.209	0.084	6.11
4	0.126	0.754	2.010	1.968	2.303	2.177	2.094	1.800	1.926	0.251	0.461	0.209	16.08
5	0.628	1.214	1.884	2.428	2.387	2.010	2.010	1.298	2.177	1.759	0.921	0.335	19.05
6	0.209	0.754	2.387	2.596	2.931	3.057	2.722	2.805	1.759	1.214	0.712	0.042	21.19
7	0.293	0.544	0.837	1.800	2.596	3.475	2.847	2.135	1.800	0.921	0.837	0.209	18.30
8	0.419	1.382	2.219	2.010	2.010	2.219	2.094	2.177	1.800	0.837	0.335	0.126	17.63
9	0.461	1.340	2.303	2.094	2.387	2.763	2.596	2.722	0.963	0.377	0.293	0.126	18.42
10	0.335	0.502	1.800	3.057	2.847	1.633	2.135	0.377	1.675	1.298	0.712	0.126	16.50
11	0.377	1.130	1.842	2.010	2.094	2.219	2.470	2.303	0.502	0.167	0.084	0.042	15.24
12	0.209	0.670	0.754	0.754	0.963	1.172	2.973	2.805	1.465	0.419	0.126	0.000	12.31
13	0.167	0.796	0.963	2.010	1.800	1.884	0.544	1.130	1.884	1.172	0.754	0.084	13.19
14	0.335	0.419	1.675	1.591	2.094	2.219	2.345	2.177	1.424	0.335	0.084	0.000	14.70
15	0.126	1.382	1.884	2.094	2.596	2.973	2.596	2.094	1.968	0.419	0.251	0.042	18.42
16	0.419	0.754	1.549	2.847	2.470	2.428	0.251	0.126	0.209	0.419	0.419	0.126	12.02
17	0.502	1.256	2.010	2.052	2.763	2.805	2.428	2.554	1.968	0.754	0.712	0.084	19.89
18	0.126	0.377	1.633	2.345	2.303	1.256	1.130	0.628	0.586	0.544	0.419	0.126	11.47
19	0.293	1.214	1.717	2.010	2.805	2.428	0.754	0.544	0.879	1.298	1.298	0.419	15.66
20	0.126	0.419	0.377	0.293	0.209	0.293	0.335	0.837	1.256	0.754	0.335	0.084	5.32
21	0.126	0.335	0.335	0.335	0.335	0.419	0.502	0.293	0.335	0.293	0.167	0.084	3.56
22	0.084	0.084	0.126	0.461	0.419	1.172	2.387	1.130	0.628	0.126	0.042	0.000	6.66
23	0.251	1.382	2.177	2.931	2.638	2.219	2.680	0.377	0.167	0.126	0.000	0.000	14.95
24	0.126	0.335	1.172	2.345	2.303	1.800	2.554	1.424	1.675	0.251	0.209	0.126	14.32
25	0.209	0.670	0.712	1.591	2.554	2.805	3.350	2.722	2.010	1.884	0.754	0.126	19.39
26	0.377	1.130	2.094	2.219	2.596	2.596	2.303	2.052	1.968	1.172	0.544	0.126	19.18
27	0.419	1.089	1.968	2.219	2.428	2.638	2.261	1.968	1.424	0.963	0.335	0.126	17.84
28	0.293	0.544	0.837	2.177	0.879	0.419	0.502	0.544	0.754	0.335	0.084	0.042	7.41
29	0.000	0.042	0.084	0.377	0.628	1.465	0.963	0.712	0.754	0.586	0.167	0.042	5.82
30	0.335	1.130	1.591	1.759	2.596	2.428	0.712	2.010	1.884	1.633	0.796	0.126	17.00
31	0.084	0.209	0.461	0.754	1.591	2.177	2.470	2.847	1.842	1.424	0.502	0.126	14.49
Mean	0.254	0.756	1.359	1.768	1.942	1.968	1.783	1.515	1.278	0.743	0.419	0.107	13.89

† All values are in MJ/m^2.

Problem 4.15 Comparison of human comfort correlations between Caucasian and Chinese subjects

Table B 2 Experimental results of tests conducted with Chinese subjects (provided by Jiang, 2001). The values under the column P_v have been computed from psychrometric relations from T-db and RH values.

Females only					Combined				
T_db (° C)	RH	P_v (kPa)	PMV_meas	PPD_meas	T_db (° C)	RH	P_v (kPa)	PMV_meas	PPD_meas
20.3	0.50	1.194	-1.4	50	20.3	0.50	1.194	-1.35	40
21.5	0.30	0.776	-1.1	20	22.22	0.50	0.776	-0.8	18
21.3	0.50	1.277	-1.2	30	23	0.52	1.277	-0.8	15
22.22	0.50	1.354	-0.86	21.6	25	0.35	1.354	-0.22	5.6
21.8	0.70	1.846	-1	33.3	25	0.50	1.846	0.042	0
23	0.35	0.993	-0.85	21.4	25.7	0.54	0.993	0.22	0
23	0.52	1.475	-0.8	20	27	0.31	1.475	0.27	4.2
23	0.76	2.156	-0.25	0	26.4	0.73	2.156	0.6	18.2
25	0.35	1.109	-0.25	12.5	27.3	0.74	1.109	1.59	50
25	0.50	1.585	-0.083	0	21.5	0.30	1.585	-0.86	20
24.7	0.74	2.308	0.2	0	21.3	0.50	2.308	-0.91	22.8
25.7	0.54	1.793	0	0	21.8	0.70	1.793	-0.75	20.8
27	0.31	1.116	0.2	0	23	0.35	1.116	-0.75	16.7
27.1	0.33	1.195	0.33	8.4	23	0.76	1.195	-0.25	0
27.4	0.54	1.990	1.2	40	24.7	0.74	1.990	0.18	0
26.4	0.73	2.534	0.5	16.7	27.1	0.33	2.534	0.33	8.4
27.3	0.74	2.712	1.4	40	27.4	0.54	2.712	1.05	38.5
28.6	0.56	2.209	1.7	50	28.6	0.56	2.209	1.55	54.4

Males Only				
T_db (° C)	RH	P_v (kPa)	PMV_meas	PPD_meas
20.3	0.50	1.194	-1.2	30
21.5	0.30	0.776	-0.6	20
21.3	0.50	1.277	-0.67	16.7
22.22	0.50	1.354	-0.72	14.3
21.8	0.70	1.846	-0.75	8.3
23	0.35	0.993	-0.6	10
23	0.52	1.475	-0.8	10
23	0.76	2.156	-0.25	0
25	0.35	1.109	-0.2	0
25	0.50	1.585	0.167	0
24.7	0.74	2.308	0.17	0
25.7	0.54	1.793	0.3	0
27	0.31	1.116	0.33	8.4
27.1	0.33	1.195	0.33	8.4
27.4	0.54	1.990	1	37.5
26.4	0.73	2.534	0.7	20
27.3	0.74	2.712	1.73	57.1
28.6	0.56	2.209	1.42	58.3

Problem 5.13 Chiller performance data

Table B 3 Nomenclature similar to the figure for Example 5.4.3

	T_{cho} (˚F)	T_{cdi} (˚F)	Q_{ch} (tons)	P_{comp} (kW)	V (gmp)
1	43	85	378	247	1125
2	43	80	390	245	1125
3	43	75	399	241	1125
4	43	70	406	238	1125
5	45	85	397	257	1125
6	45	80	409	254	1125
7	45	75	417	250	1125
8	45	70	432	263	1125
9	50	85	430	267	1125
10	50	80	445	268	1125
11	50	75	457	269	1125
12	50	70	455	259	1125
13	43	85	340	247	500
14	43	80	360	249	500
15	43	75	375	248	500
16	43	70	387	246	500
17	45	85	360	258	500
18	45	80	379	259	500
19	45	75	394	258	500
20	45	70	405	255	500
21	50	85	390	268	500
22	50	80	407	267	500
23	50	75	421	266	500
24	50	70	437	267	500
25	43	85	368	248	813
26	43	80	382	247	813
27	43	75	393	244	813
28	43	70	401	240	813
29	45	85	388	258	813
30	45	80	401	256	813
31	45	75	411	253	813
32	45	70	420	249	813
33	50	85	420	269	813
34	50	80	435	268	813
35	50	75	447	267	813
36	50	70	459	269	813
37	45	85	397	257	1125
38	*45*	*82.5*	*358*	*225*	*1125*
39	*45*	*80*	*318*	*201*	*1125*
40	*45*	*77.5*	*278*	*178*	*1125*
41	*45*	*75*	*238*	*156*	*1125*
42	*45*	*72.5*	*199*	*135*	*1125*
43	*45*	*70*	*159*	*113*	*1125*
44	*45*	*67.5*	*119*	*91*	*1125*
45	*45*	*85*	*360*	*258*	*500*
46	*45*	*82.5*	*324*	*224*	*500*
47	*45*	*80*	*288*	*199*	*500*
48	*45*	*77.5*	*252*	*175*	*500*
49	*45*	*75*	*216*	*153*	*500*
50	*45*	*72.5*	*180*	*131*	*500*
51	*45*	*70*	*144*	*109*	*500*
52	*45*	*67.5*	*108*	*90*	*500*

Problem 5.14 Effect of tube cleaning in reducing chiller fouling

Table B 4 Nomenclature similar to the figure for Example 5.4.3

		Compressor	Chilled Water	Cooling Water	Refrigerant (R-11)		Cooling load
Date	Hour	P_{comp} (kW)	T_{cho} (°F)	T_{cdi} (°F)	Evap Temp. (°F)	Cond Temp. (°F)	Q_{ch} (Tons)
9/11/2000	0100	423.7	45.9	81.7	38.2	97.9	1290.7
	0200	417.7	45.8	82.0	38.3	97.9	1275.5
	0300	380.1	45.8	81.6	38.6	93.6	1114.9
	0400	366.0	45.6	81.6	38.6	96.1	1040.6
	0500	374.9	45.7	81.6	38.6	96.4	1081.5
	0600	395.0	45.7	81.7	38.3	94.1	1167.9
	0700	447.9	45.9	82.5	38.1	99.1	1353.8
	0800	503.8	46.0	82.5	37.3	100.1	1251.6
	0900	525.9	45.9	81.3	36.2	97.2	816.7
	1000	490.4	46.0	81.3	37.0	98.6	784.0
	1100	508.7	46.0	81.2	36.8	98.6	802.3
	1200	512.1	45.9	81.4	36.5	96.9	825.8
	1300	515.5	45.9	81.4	36.6	99.4	834.2
	1400	561.4	45.8	81.5	35.5	100.1	901.1
	1500	571.2	46.0	81.4	35.5	98.6	922.0
	1600	567.1	46.5	81.4	35.9	100.6	921.5
	1700	493.4	48.8	81.3	39.4	94.3	806.3
	1800	592.5	46.2	81.9	35.5	96.7	955.2
	1900	560.7	45.9	82.2	35.7	100.6	900.8
	2000	513.7	45.9	81.9	36.4	99.6	825.7
	2100	461.0	45.9	81.8	37.4	96.1	734.2
	2200	439.0	45.8	81.6	37.7	97.7	692.7
	2300	431.7	45.8	81.7	37.8	97.2	680.3
	2400	435.4	45.9	81.6	37.8	94.6	676.2
9/12/2000	0100	398.3	45.7	82.3	38.3	95.9	993.4
	0200	371.7	45.6	82.5	38.6	97.4	1121.6
	0300	369.3	45.5	82.5	38.5	96.7	1081.1
	0400	361.5	45.7	82.5	38.6	94.9	1056.9
	0500	374.0	45.8	82.6	38.7	97.4	1079.4
	0600	389.5	45.7	82.5	38.4	97.2	1116.0
	0700	394.4	45.8	82.6	38.4	95.9	1122.7
	0800	425.6	45.8	82.6	38.2	98.9	1237.0
	0900	475.1	45.9	83.1	37.9	99.6	1441.1
	1000	492.9	46.0	83.3	37.7	99.1	1522.0
	1100	531.5	46.0	83.4	37.0	102.2	1627.2
	1200	576.0	46.2	82.9	36.3	101.8	1181.6
	1300	603.6	46.5	82.6	35.7	101.3	955.9
	1400	514.9	48.3	83.3	39.0	95.1	1437.8
	1500	556.7	45.9	83.6	36.1	99.1	1696.9
	1600	548.2	45.9	83.6	36.1	99.1	1664.2
	1700	540.6	45.9	83.6	36.2	101.3	1651.6
	1800	556.8	45.9	83.6	36.2	101.1	1682.2
	1900	589.1	46.1	83.6	35.9	101.3	1736.3
	2000	584.3	46.1	83.5	35.7	102.9	1697.4
	2100	586.8	46.1	83.6	35.8	103.6	1657.4
	2200	572.4	45.9	83.5	35.6	101.8	1592.8
	2300	528.3	46.0	83.3	36.8	101.5	1464.5

Table B 4 (continued)

		Compressor	Chilled Water	Cooling Water	Refrigerant (R-11)		Cooling load
Date	Hour	P_{comp} (kW)	T_{cho} (°F)	T_{cdi} (°F)	Evap Temp. (°F)	Cond Temp. (°F)	Q_{ch} (Tons)
	2400	499.9	46.0	83.4	37.4	101.3	1416.0
9/13/2000	0100	471.8	46.0	83.4	37.8	98.6	1373.2
	0200	463.9	46.0	83.4	37.9	99.6	1365.8
	0300	469.7	46.0	83.3	37.8	100.6	1390.1
	0400	471.7	45.8	83.3	37.7	98.4	1390.7
	0500	464.2	45.9	83.3	37.9	99.9	1386.9
	0600	463.9	45.9	83.3	37.9	100.3	1389.5
	0700	482.8	46.0	83.3	37.7	98.9	1470.8
	0800	493.8	46.0	83.3	37.5	99.1	1513.6
	0900	511.3	46.0	83.4	37.3	101.8	1554.0
	1000	557.1	46.0	83.5	36.5	102.7	1658.2
	1100	573.2	46.0	83.5	36.1	101.3	1694.9
	1200	573.3	46.0	83.4	36.0	103.4	1681.8
	1300	577.2	46.1	83.4	36.1	103.2	1691.7
	1400	577.3	46.2	83.3	35.9	101.3	1702.1
	1500	580.0	46.4	83.3	36.0	103.4	1718.0
	1600	588.0	46.3	83.4	36.0	103.4	1739.4
	1700	588.9	46.3	83.3	35.7	101.8	1730.7
	1800	574.2	46.0	83.2	35.5	103.2	1684.7
	1900	555.6	46.0	83.0	36.3	102.0	1629.6
	2000	535.6	46.0	82.8	36.5	99.9	1603.6
	2100	497.8	46.0	82.8	37.3	100.8	1499.6
	2200	469.4	45.9	82.7	37.6	99.4	1429.5
	2300	446.1	45.9	82.4	37.9	96.9	1370.2
	2400	426.5	45.8	82.2	38.0	98.6	1306.7
Cleaning							
1/17/2001	0100	408.8	44.5	87.0	39.0	94.9	1125.1
	0200	400.6	44.5	86.9	39.2	95.1	1097.9
	0300	397.4	44.4	86.9	39.1	94.9	1070.8
	0400	402.0	44.5	86.8	39.1	95.4	1104.7
	0500	394.1	44.4	86.8	39.1	95.1	1062.5
	0600	398.8	44.5	87.0	39.2	95.6	1093.1
	0700	403.9	44.5	86.8	39.1	95.6	1102.9
	0800	418.2	44.5	86.9	39.0	96.1	1167.6
	0900	430.1	44.6	86.9	39.1	96.7	1201.5
	1000	457.4	44.6	87.0	38.9	97.7	1285.7
	1100	478.6	44.8	87.2	38.9	98.2	1355.1
	1200	499.2	44.8	87.3	38.7	98.9	1411.0
	1300	513.1	44.8	86.9	38.4	98.9	1460.5
	1400	530.1	44.8	87.0	38.1	99.1	1517.8
	1500	525.9	44.8	87.0	38.0	99.1	1492.4
	1600	513.4	44.8	87.0	38.1	98.9	1465.7
	1700	501.2	44.8	87.0	38.1	98.4	1437.0
	1800	493.4	44.8	87.2	38.3	98.4	1405.0
	1900	470.8	45.0	87.2	38.6	97.7	1352.2
	2000	462.6	45.0	87.1	38.7	97.7	1316.7
	2100	460.4	44.9	87.3	38.5	97.7	1301.9
	2200	443.3	44.7	87.2	38.5	97.2	1239.1
	2300	428.3	44.5	87.2	38.4	96.9	1194.8

Table B 4 (continued)

		Compressor	Chilled Water	Cooling Water	Refrigerant (R-11)		Cooling load
Date	Hour	P_{comp} (kW)	T_{cho} (°F)	T_{cdi} (°F)	Evap Temp. (°F)	Cond Temp. (°F)	Q_{ch} (Tons)
	2400	414.5	44.5	87.1	38.5	96.7	1140.3
1/18/2001	0100	406.4	44.4	87.1	38.5	96.4	1124.1
	0200	398.6	44.5	87.1	38.7	96.7	1089.6
	0300	402.9	44.4	87.0	38.6	96.4	1100.2
	0400	392.3	44.3	86.8	38.6	96.4	1057.7
	0500	402.3	44.3	87.0	38.5	96.9	1098.4
	0600	391.9	44.3	87.0	38.6	96.7	1053.8
	0700	411.8	44.4	87.0	38.5	97.4	1128.3
	0800	424.9	44.4	86.9	38.4	97.7	1177.8
	0900	442.4	44.5	86.6	38.2	97.9	1249.7
	1000	454.4	44.7	86.5	38.2	98.2	1296.4
	1100	473.0	44.7	86.7	38.1	98.9	1347.5
	1200	477.5	44.8	86.8	38.0	99.1	1354.7
	1300	484.3	44.7	87.0	38.0	99.4	1376.5
	1400	486.4	44.7	86.9	37.8	99.4	1387.4
	1500	487.4	44.7	86.8	37.8	99.4	1381.0
	1600	508.8	44.8	86.9	37.5	100.1	1442.1
	1700	506.7	44.8	86.8	37.4	100.1	1436.5
	1800	509.0	44.8	86.9	37.3	100.1	1443.1
	1900	503.5	44.8	86.9	37.3	99.9	1427.3
	2000	492.3	44.8	87.0	37.5	99.9	1407.8
	2100	477.9	44.7	87.0	37.5	99.6	1376.9
	2200	456.5	44.7	86.8	37.7	98.9	1302.3
	2300	445.7	44.6	86.8	37.8	98.9	1262.6
	2400	432.0	44.5	86.1	37.7	98.4	1240.9
1/19/2001	0100	426.9	44.5	85.7	37.7	98.9	1208.7
	0200	416.4	44.5	85.6	37.8	98.9	1189.5
	0300	416.4	44.5	85.6	37.9	99.1	1200.7
	0400	414.2	44.6	85.6	37.9	99.1	1200.5
	0500	411.6	44.5	85.8	37.9	99.4	1182.7
	0600	421.4	44.5	85.7	37.8	99.6	1209.2
	0700	429.2	44.6	85.8	37.8	100.1	1219.9
	0800	447.3	44.7	86.0	37.7	100.6	1280.4
	0900	481.0	44.7	86.1	37.4	101.3	1371.2
	1000	532.0	44.8	86.5	36.7	102.9	1498.4
	1100	535.1	44.8	86.8	35.5	103.2	1503.0
	1200	529.8	44.9	86.5	36.5	102.9	1491.9
	1300	503.3	44.9	86.5	36.7	102.5	1433.9
	1400	449.4	44.7	85.9	37.4	100.6	1287.2
	1500	517.0	44.9	86.3	36.7	102.5	1468.9
	1600	501.2	44.8	86.4	36.6	102.2	1450.2
	1700	443.6	44.8	86.0	38.1	100.8	1287.0
	1800	368.8	44.6	86.0	39.1	98.6	1045.5
	1900	376.4	44.6	85.9	39.2	98.9	1051.5
	2000	383.7	44.6	85.8	39.1	98.9	1084.7
	2100	390.2	44.6	85.9	39.1	99.6	1110.9
	2200	382.9	44.5	85.8	39.1	99.4	1089.8
	2300	364.5	44.5	85.7	39.2	98.9	999.2
	2400	354.3	44.4	86.1	39.2	99.1	964.0

Problem 9.4 Time series analysis of sun spot frequency per year from 1770–1869

Table B5 Time series analysis of sun spot frequency per year from 1770–1869. (From Montgomery and Johnson 1976 by permission of McGraw-Hill)

Year	n	Year	n	Year	n	Year	n
1770	101	1795	21	1820	16	1845	40
1771	82	1796	16	1821	7	1846	62
1772	66	1797	6	1822	4	1847	98
1773	35	1798	4	1823	2	1848	124
1774	31	1799	7	1824	8	1849	96
1775	7	1800	14	1825	17	1850	66
1776	20	1801	34	1826	36	1851	64
1777	92	1802	45	1827	50	1852	54
1778	154	1803	43	1828	62	1853	39
1779	125	1804	48	1829	67	1854	21
1780	85	1805	42	1830	71	1855	7
1781	68	1806	28	1831	48	1856	4
1782	38	1807	10	1832	28	1857	23
1783	23	1808	8	1833	8	1858	55
1784	10	1809	2	1834	13	1859	94
1785	24	1810	0	1835	57	1860	96
1786	83	1811	1	1836	122	1861	77
1787	132	1812	5	1837	138	1862	59
1788	131	1813	12	1838	103	1863	44
1789	118	1814	14	1839	86	1864	47
1790	90	1815	35	1840	63	1865	30
1791	67	1816	46	1841	37	1866	16
1792	60	1817	41	1842	24	1867	7
1793	47	1818	30	1843	11	1868	37
1794	41	1819	24	1844	15	1869	74

Problem 9.5 Time series of yearly atmospheric CO_2 concentrations from 1979–2005

Table B6 Time series of yearly atmospheric CO_2 concentrations from 1979–2005. (From Andrews and Jelley 2007 by permission of Oxford University Press)

Year	CO_2 conc	Temp. diff	Year	CO_2 conc	Temp. diff	Year	CO_2 conc	Temp. diff
1979	336.53	0.06	1988	350.68	0.16	1997	362.98	0.36
1980	338.34	0.1	1989	352.84	0.1	1998	364.9	0.52
1981	339.96	0.13	1990	354.22	0.25	1999	367.87	0.27
1982	341.09	0.12	1991	355.51	0.2	2000	369.22	0.24
1983	342.07	0.19	1992	356.39	0.06	2001	370.44	0.4
1984	344.04	–0.01	1993	356.98	0.11	2002	372.31	0.45
1985	345.1	–0.02	1994	358.19	0.17	2003	374.75	0.45
1986	346.85	0.02	1995	359.82	0.27	2004	376.95	0.44
1987	347.75	0.17	1996	361.82	0.13	2005	378.55	0.47

Problem 9.6 Time series of monthly atmospheric CO_2 concentrations from 2002–2006

Table B7 Time series of monthly atmospheric CO_2 concentrations from 2002–2006

	2002	2003	2004	2005	2006
Month	CO_2 concent.	CO_2 concent.	CO_2 concent.	CO_2 concent.	CO_2 concent.
1	372.4	374.7	377.0	378.5	381.2
2	372.8	375.4	377.5	379.0	381.8
3	373.3	375.8	378.0	379.6	382.1
4	373.6	376.2	378.4	380.6	382.5
5	373.6	376.4	378.3	380.7	382.5
6	372.8	375.5	377.4	379.4	381.5
7	371.3	374.0	375.9	377.8	379.9
8	370.2	372.7	374.4	376.5	378.3
9	370.5	373.0	374.3	376.6	378.6
10	371.8	374.3	375.6	377.9	380.6
11	373.2	375.5	377.0	379.4	–
12	374.1	376.4	378.0	380.4	–

Problem 9.8 Transfer function analysis using simulated hourly loads in a commercial building

Table B8 Transfer function analysis using simulated hourly loads in a commercial building

Month	Day	Hour	Tdb (°F)	Twb (°F)	Qint (kW)	Total Power (kW)	Cooling (Btu/h)	Heating (Btu/h)
8	1	1	63	60	594.8	844.6	3,194,081	427,362
8	1	2	63	61	543.4	791.8	3,152,713	445,862
8	1	3	63	61	543.4	789.8	3,085,678	446,532
8	1	4	64	61	543.4	791.7	3,145,654	452,646
8	1	5	64	61	655.8	906.4	3,217,818	434,740
8	1	6	65	63	879.9	1160.3	4,119,484	835,019
8	1	7	68	65	1107.6	1447.1	5,651,144	734,486
8	1	8	72	66	1086.4	1470.7	6,644,373	474,489
8	1	9	75	66	1086.3	1487.3	6,972,290	397,013
8	1	10	78	68	933.2	1392.6	8,161,122	335,897
8	1	11	80	69	934.6	1423.8	8,669,268	306,703
8	1	12	81	69	922.8	1412.4	8,642,472	328,010
8	1	13	81	69	934.8	1426.8	8,688,363	297,011
8	1	14	82	69	934.8	1429.2	8,730,540	287,861
8	1	15	80	67	933.7	1405.4	8,300,878	276,095
8	1	16	82	68	934.7	1434.6	8,877,782	256,967
8	1	17	81	67	787.1	1278.3	8,683,202	345,135
8	1	18	79	66	1153.3	1639.0	8,600,599	402,828
8	1	19	79	66	1358.7	1828.2	8,307,491	446,782
8	1	20	74	65	1448.6	1873.0	7,416,578	519,596
8	1	21	73	65	1345.4	1746.3	6,958,935	551,707
8	1	22	66	62	1104.7	1425.3	5,100,758	643,950
8	1	23	66	63	790.3	1096.2	4,781,506	748,713
8	1	24	65	63	646.3	932.0	4,227,319	365,618
8	2	1	64	61	594.7	856.9	3,551,712	394,635
8	2	2	63	61	543.3	794.8	3,239,986	430,030
8	2	3	64	62	543.5	799.9	3,400,810	443,896
8	2	4	63	60	543.2	785.0	2,924,189	434,391
8	2	5	65	62	656.0	916.0	3,522,530	447,170

8	2	6	65	62	879.6	1169.9	4,388,385	771,258
8	2	7	71	66	1108.6	1503.2	6,925,686	668,155
8	2	8	75	68	1087.7	1533.7	7,897,258	387,154
8	2	9	80	70	1089.8	1603.5	9,130,235	318,321
8	2	10	81	69	935.1	1447.1	9,021,254	287,550
8	2	11	84	70	935.9	1464.8	9,347,346	278,832
8	2	12	84	70	924.0	1450.0	9,264,431	307,150
8	2	13	85	71	936.6	1476.4	9,511,320	275,150
8	2	14	85	71	936.9	1491.1	9,731,617	251,269
8	2	15	85	71	937.4	1505.1	9,956,236	247,798
8	2	16	85	72	938.4	1529.7	10,342,235	247,176
8	2	17	84	71	789.8	1360.3	9,955,993	244,046
8	2	18	81	71	1156.8	1730.2	10,041,017	379,881
8	2	19	79	70	1361.4	1898.3	9,412,865	425,252
8	2	20	76	70	1451.7	1954.6	8,844,201	495,725
8	2	21	74	69	1347.8	1811.4	8,125,985	524,132
8	2	22	74	69	1108.4	1536.9	7,465,724	571,150
8	2	23	73	69	793.4	1192.4	6,868,408	648,661
8	2	24	72	68	648.9	1014.8	6,152,558	304,174
8	3	1	71	68	597.1	944.8	5,755,310	320,848
8	3	2	71	68	545.5	883.7	5,538,507	333,868
8	3	3	70	67	544.9	867.3	5,150,570	339,970
8	3	4	71	66	544.4	858.2	4,947,774	346,538
8	3	5	71	68	657.9	990.4	5,434,792	344,485
8	3	6	73	69	882.7	1262.4	6,514,805	687,770
8	3	7	77	71	1111.3	1560.6	7,936,272	608,746
8	3	8	80	74	1092.5	1644.8	9,760,724	377,956
8	3	9	83	75	1094.1	1705.1	10,594,630	301,946
8	3	10	85	76	941.2	1590.0	11,124,807	270,331
8	3	11	87	77	942.6	1627.3	11,613,918	246,362
8	3	12	87	76	929.6	1591.3	11,207,712	274,067
8	3	13	87	76	941.7	1608.5	11,338,916	237,845
8	3	14	89	77	943.2	1644.6	11,852,497	237,519
8	3	15	89	77	943.3	1651.3	11,887,520	237,746
8	3	16	89	76	942.4	1635.4	11,664,144	236,992
8	3	17	87	76	794.3	1468.9	11,436,989	238,088
8	3	18	87	77	1162.6	1867.1	11,894,025	363,009
8	3	19	85	75	1365.8	2014.9	11,001,879	407,210
8	3	20	85	75	1456.7	2088.4	10,839,194	454,740
8	3	21	79	74	1352.0	1929.0	9,960,708	487,662
8	3	22	78	76	1114.2	1667.6	9,627,610	528,198
8	3	23	78	75	798.1	1306.9	8,801,340	593,055
8	3	24	79	75	654.5	1144.3	8,512,977	280,531

C: Solved Examples and Problems with Practical Relevance

Example 1.3.1: Simulation of a chiller
Problem 1.7 Two pumps in parallel problem viewed from the forward and the inverse perspectives
Problem 1.8 Lake contamination problem viewed from the forward and the inverse perspectives

Example 2.2.6: Generating a probability tree for a residential air-conditioning (AC) system
Example 2.4.3: Using geometric PDF for 50 year design wind problems
Example 2.4.8: Using Poisson PDF for assessing storm frequency
Example 2.4.9: Graphical interpretation of probability using the standard normal table
Example 2.4.11: Using lognormal distributions for pollutant concentrations
Example 2.4.13: Modeling wind distributions using the Weibull distribution
Example 2.5.2: Forward and reverse probability trees for fault detection of equipment
Example 2.5.3: Using the Bayesian approach to enhance value of concrete piles testing
Example 2.5.7: Enhancing historical records of wind velocity using the Bayesian approach
Problem 2.23 Probability models for global horizontal solar radiation.
Problem 2.24 Cumulative distribution and utilizability functions for horizontal solar radiation
Problem 2.25 Generating cumulative distribution curves and utilizability curves with measured data.

Example 3.3.1: Example of interpolation.
Example 3.4.1: Exploratory data analysis of utility bill data
Example 3.6.1: Estimating confidence intervals
Example 3.7.1: Uncertainty in overall heat transfer coefficient
Example 3.7.2: Relative error in Reynolds number of flow in a pipe
Example 3.7.3: Selecting instrumentation during the experimental design phase
Example 3.7.4: Uncertainty in exponential growth models
Example 3.7.5: Temporal Propagation of Uncertainty in ice storage inventory
Example 3.7.6: Using Monte Carlo to determine uncertainty in exponential growth models
Problem 3.7 Determining cooling coil degradation based on effectiveness
Problem 3.10 Uncertainty in savings from energy conservation retrofits
Problem 3.11 Uncertainty in estimating outdoor air fraction in HVAC systems
Problem 3.12 Sensor placement in HVAC ducts with consideration of flow non-uniformity
Problem 3.13 Uncertainty in estimated proportion of exposed subjects using Monte Carlo method
Problem 3.14 Uncertainty in the estimation of biological dose over time for an individual
Problem 3.15 Propagation of optical and tracking errors in solar concentrators

Example 4.2.1: Evaluating manufacturer quoted lifetime of light bulbs from sample data
Example 4.2.2: Evaluating whether a new lamp bulb has longer burning life than traditional ones
Example 4.2.3: Verifying savings from energy conservation measures in homes
Example 4.2.4: Comparing energy use of two similar buildings based on utility bills- the wrong way
Example 4.2.5: Comparing energy use of two similar buildings based on utility bills- the right way
Example 4.2.8. Hypothesis testing of increased incidence of lung ailments due to radon in homes
Example 4.2.10: Comparing variability in daily productivity of two workers
Example 4.2.11: Ascertaining whether non-code compliance infringements in residences is random or not
Example 4.2.12: Evaluating whether injuries in males and females is independent of circumstance
Example 4.3.1: Comparing mean life of five motor bearings
Example 4.4.1: Comparing mean values of two samples by pairwise and by Hotteling T^2 procedures
Example 4.5.1: Non-parametric testing of correlation between the sizes of faculty research grants and teaching evaluations
Example 4.5.2: Ascertaining whether oil company researchers and academics differ in their predictions of future atmospheric carbon dioxide levels
Example 4.5.3: Evaluating predictive accuracy of two climate change models from expert elicitation
Example 4.5.4: Evaluating probability distributions of number of employees in three different occupations using a non-parametric test
Example 4.6.1: Comparison of classical and Bayesian confidence intervals
Example 4.6.2: Traditional and Bayesian approaches to determining confidence levels
Example 4.7.1: Determination of random sample size needed to verify peak reduction in residences at preset confidence levels

Example 4.7.2: Example of stratified sampling for variance reduction
Example 4.8.1: Using the bootstrap method for deducing confidence intervals
Example 4.8.2: Using the bootstrap method with a nonparametric test to ascertain correlation of two variables
Problem 4.4 Using classical and Bayesian approaches to verify claimed benefit of gasoline additive
Problem 4.5 Analyzing distribution of radon concentration in homes
Problem 4.7 Using survey sample to determine proportion of population in favor of off-shore wind farms
Problem 4.8 Using ANOVA to evaluate pollutant concentration levels at different times of day
Problem 4.9 Using non-parametric tests to identify the better of two fan models
Problem 4.10 Parametric test to evaluate relative performance of two PV systems from sample data
Problem 4.11 Comparing two instruments using parametric, nonparametric and bootstrap methods
Problem 4.15 Comparison of human comfort correlations between Caucasian and Chinese subjects

Example 5.3.1: Water pollution model between solids reduction and chemical oxygen demand
Example 5.4.1: Part load performance of fans (and pumps)
Example 5.4.3: Beta coefficients for ascertaining importance of driving variables for chiller thermal performance
Example 5.6.1: Example highlighting different characteristic of outliers or residuals versus influence points.
Example 5.6.2: Example of variable transformation to remedy improper residual behavior
Example 5.6.3: Example of weighted regression for replicate measurements
Example 5.6.4: Using the Cochrane-Orcutt procedure to remove first-order autocorrelation
Example 5.6.5: Example to illustrate how inclusion of additional regressors can remedy improper model residual behavior
Example 5.7.1: Change point models for building utility bill analysis
Example 5.7.2: Combined modeling of energy use in regular and energy efficient buildings
Example 5.7.3: Proper model identification with multivariate regression models
Section 5.8 Case study example effect of refrigerant additive on chiller performance
Problem 5.4 Cost of electric power generation versus load factor and cost of coal
Problem 5.5 Modeling of cooling tower performance
Problem 5.6 Steady-state performance testing of solar thermal flat plate collector
Problem 5.7 Dimensionless model for fans or pumps
Problem 5.9 Spline models for solar radiation
Problem 5.10 Modeling variable base degree-days with balance point temperature at a specific location
Problem 5.11 Change point models of utility bills in variable occupancy buildings
Problem 5.12 Determining energy savings from monitoring and verification (M&V) projects
Problem 5.13 Grey-box and black-box models of centrifugal chiller using field data
Problem 5.14 Effect of tube cleaning in reducing chiller fouling

Example 6.2.2: Evaluating performance of four machines while blocking effect of operator dexterity
Example 6.2.3: Evaluating impact of three factors (school, air filter type and season) on breathing complaints
Example 6.3.1: Deducing a prediction model for a 2^3 factorial design
Example 6.4.1: Optimizing the deposition rate for a tungsten film on silicon wafer
Problem 6.2 Full-factorial design for evaluating three different missile systems
Problem 6.3 Random effects model for worker productivity
Problem 6.6 2^3 factorial analysis for strength of concrete mix
Problem 6.8 Predictive model inferred from 2^3 factorial design on a large laboratory chiller

Example 7.3.3: Optimizing a solar water heater system using the Lagrange method
Section 7.8 Illustrative example: combined heat and power system
Example 7.10.1: Example of determining optimal equipment maintenance policy
Example 7.10.2 Strategy of operating a building to minimize cooling costs
Problem 7.4 Solar thermal power system optimization
Problem 7.5 Minimizing pressure drop in ducts
Problem 7.6 Replacement of filters in HVAC ducts
Problem 7.7 Relative loading of two chillers
Problem 7.8 Maintenance scheduling for plant
Problem 7.9 Comparing different thermostat control strategies

Example 8.2.1: Using distance measures for evaluating canine samples
Example 8.2.3: Using discriminant analysis to model fault-free and faulty behavior of chillers
Example 8.3.3: Using k nearest neighborhood to calculate uncertainty in building energy savings
Example 8.4.1: Using treed regression to model atmospheric ozone variation with climatic variables
Example 8.5.2: Clustering residences based on their air-conditioner electricity use during peak summer days

Example 9.1.1: Modeling peak demand of an electric utility
Example 9.4.1: Modeling peak electric demand by a linear trend model
Example 9.4.2: Modeling peak electric demand by a linear trend plus seasonal model
Example 9.5.2: AR model for peak electric demand
Example 9.5.3: Comparison of the external prediction error of different models for peak electric demand
Example 9.6.1: Transfer function model to represent unsteady state heat transfer through a wall
Example 9.7.1: Illustration of the mean and range charts
Example 9.7.2: Illustration of the p-chart method
Example 9.7.3: Illustration of Cusum plots
Problem 9.4 Time series analysis of sun spot frequency per year from 1770–1869
Problem 9.5 Time series of yearly atmospheric CO_2 concentrations from 1979–2005
Problem 9.6 Time series of monthly atmospheric CO_2 concentrations from 2002–2006
Problem 9.7 Transfer function analysis of unsteady state heat transfer through a wall
Problem 9.8 Transfer function analysis using simulated hourly loads in a commercial building

Example 10.3.3: Reduction in dimensionality using PCA for actual chiller data
Section 10.3.4 Chiller case study analysis involving collinear regressors
Section 10.3.6 Case study of stagewise regression involving building energy loads
Example 10.4.1: .MLE for exponential distribution
Example 10.4.2: MLE for Weibull distribution
Example 10.4.3: Use of logistic models for predicting growth of worldwide energy use
Example 10.4.4: Fitting a logistic model to the kill rate of the fruit fly
Example 10.4.5: Dose response modeling for sarin gas
Problem 10.3 Chiller data analysis using PCA
Problem 10.7 Non-linear model fitting to thermodynamic properties of steam
Problem 10.8 Logistic functions to study residential equipment penetration
Problem 10.9 Fitting logistic models for growth of population and energy use
Problem 10.10 Dose response model fitting for VX gas
Problem 10.14 Model identification for wind chill factor
Problem 10.16 Parameter estimation for an air infiltration model in homes

Section 11.2.2 Example of calibrated model development: global temperature model
Section 11.2.4 Case study: Calibrating detailed building energy simulation programs to utility bills
Section 11.3.4 Grey-box models and policy issues concerning dose-response behavior

Example 12.2.2: Example involving two risky alternatives each with probabilistic outcomes
Example 12.2.4: Risk analysis for a homeowner buying an air-conditioner
Example 12.2.5: Generating utility functions
Example 12.2.7: Ascertaining risk for buildings
Example 12.2.8: Monte Carlo analysis to evaluate alternatives for the oil company example
Example 12.2.9: Value of perfect information for the oil company
Example 12.2.10: Benefit of Bayesian methods for evaluating real-time monitoring instruments in homes
Example 12.3.1: Evaluation of tools meant for detecting and diagnosing faults in chillers
Example 12.3.2: Risk assessment of chloroform in drinking water
Section 12.4.1 Case study: Risk assessment of existing buildings
Section 12.4.2 Case study: Decision making while operating an engineering system
Problem 12.6 Traditional versus green homes
Problem 12.7 Risk analysis for building owner being sued for indoor air quality
Problem 12.10 Monte Carlo analysis for evaluating risk for property retrofits
Problem 12.11 Risk analysis for pursuing a new design of commercial air-conditioner
Problem 12.12 Monte Carlo analysis for deaths due to radiation exposure
Problem 12.14 Monte Carlo analysis for buying an all-electric car
Problem 12.15 Evaluating feed-in tariffs versus upfront rebates for PV systems

Index

T. Agami Reddy, *Applied Data Analysis and Modeling for Energy Engineers and Scientists*,
DOI 10.1007/978-1-4419-9613-8, © Springer Science+Business Media, LLC 2011